PARALLEL PROCESSING IN NEURAL SYSTEMS AND COMPUTERS

PARALLEL PROCESSING IN NEURAL SYSTEMS AND COMPUTERS

Edited by

Rolf ECKMILLER

Division of Biocybernetics
Department of Biophysics
Heinrich Heine University Düsseldorf
F.R.G.

Georg HARTMANN

Electrical Engineering Department
University of Paderborn
F.R.G.

Gert HAUSKE

Communication Science Department
Technical University München
F.R.G.

NORTH-HOLLAND
AMSTERDAM · NEW YORK · OXFORD · TOKYO

ELSEVIER SCIENCE PUBLISHERS B.V.
Sara Burgerhartstraat 25
P.O. Box 211, 1000 AE Amsterdam, The Netherlands

Distributors for the United States and Canada:

ELSEVIER SCIENCE PUBLISHING COMPANY INC.
655 Avenue of the Americas
New York, N.Y. 10010, U.S.A.

First published in 1990
Second impression, 1991

```
        Library of Congress Cataloging-in-Publication Data

Parallel processing in neural systems and computers / edited by Rolf
  Eckmiller, Georg Hartmann, Gert Hauske.
        p.    cm.
    "A selection of ... papers of the International Conference on
Parallel Processing in Neural Systems and Computers (ICNC) held in
Düsseldorf (FRG) from 19 to 21 March 1990"--Pref.
    Includes bibliographical references.
    ISBN 0-444-88390-8 (U.S.)
    1. Parallel processing (Electronic computers)--Congresses.
2. Neural computers--Congresses.   I. Eckmiller, Rolf, 1942-
II. Hartmann, G. (Georg), 1937-    .  III. Hauske, Gert.
IV. International Conference on Parallel Processing in Neural
Systems and Computers (1990 : Düsseldorf, Germany)
QA76.5.P314854   1990
304'.35--dc20                                             90-6719
                                                           CIP
```

ISBN: 0 444 88390 8

Printed in The Netherlands

PREFACE

This book compiles a selection of thoroughly reviewed papers of the International Conference on Parallel Processing in Neural Systems and Computers (ICNC) held in Düsseldorf (FRG) from 19 to 21 March, 1990.

ICNC (the 10th Cybernetics Congress of the DGK) assembled scientists from more than 15 nations representing the research fields of: Neuroscience, Artificial Intelligence, Neuroinformatics, and Parallel Computing, as well as Optical and Molecular Computing.

The 119 contributions in this book, which include 10 invited papers, are arranged in 13 chapters covering a wide range of topics including: Parallel Computers, Parallel Processing in Biological Neural Systems, Simulators for Artificial Neural Networks, Neural Networks for Visual and Auditory Pattern Recognition as well as for Motor Control, Artificial Intelligence, and Examples of Optical and Molecular Computing.

The readibility of this book was enhanced not only by detailed Author and Subject Indices but also by a separate alphabetical list for all references quoted in the individual contributions.

It is hoped that this book as a timely 'State of the Art Report' and at the same time an 'Interdisciplinary Reference Source for Parallel Processing' will prove to be useful as a catalyzer for more international cooperation and interdisciplinary collaboration among neuroscientists, physicists, engineers, and computer scientists in the challenge to: *Decipher Parallel Information Processes in Biology, Physics, and Chemistry,* and to: *Design Conceptually Similar Technical Parallel Information Processors.*

The thorough and efficient managerial and technical assistance of Claudia Berge, Miriam Buck, and Sandra Winter, which was crucial to meet the deadline for preparation of the final book manuscript, is gratefully acknowledged. Their work together with the expert work of the publisher made it possible to deliver this book in high quality within only four months after submission of the manuscripts.

Düsseldorf, March 1990   The Editors

ACKNOWLEDGEMENT OF SPONSORSHIP

The generous sponsorship of the following institutions, which made this International Conference on Parallel Processing in Neural Systems and Computers (ICNC) possible, is gratefully acknowledged:

Robert Bosch GmbH / Stuttgart (FRG)

Industrie- und Handelskammer zu Düsseldorf / Düsseldorf (FRG)

Deutsche Gesellschaft für Kybernetik (DGK) / München (FRG)

IBM Deutschland GmbH / Stuttgart (FRG)

Kontron Elektronik GmbH / Eching (FRG)

Minister für Wissenschaft und Forschung des Landes Nordrhein-Westfahlen (MWF in NRW) / Düsseldorf (FRG)

Nixdorf Computer AG / Paderborn (FRG)

PARACOM GmbH / Aachen (FRG)

PARSYTEC GmbH / Aachen (FRG)

Philips GmbH / Aachen (FRG)

Siemens AG / München (FRG)

Sun Microsystems GmbH / Ratingen (FRG)

Carl Zeiss / Oberkochen (FRG)

TABLE OF CONTENTS

Section 1: General Introduction

Neuroinformatics and Cybernetics (*Invited Paper*)
G. HAUSKE (München, FRG) — 3

Concerning the Emerging Role of Geometry in Neuroinformatics
R. ECKMILLER (Düsseldorf, FRG) — 5

Philosophical Concepts of Computational Neuroscience
K. MAINZER (Augsburg, FRG) — 9

Section 2: Development and Application of Parallel Computers

Parallel Processing Using Memory-Based Connectionist Networks
C.-C. CHEN (Brussels, Belgium) — 15

Accelerating Convergence of Symmetric Neural Networks on a Local
  Memory Multiprocessor
J.P. COUGHLIN, R.H. BARAN (Towson, USA) — 19

Why Do Neural Network Researchers Ignore Stochastic Computers?
R. MASSEN (Konstanz, FRG) — 23

A Programmable Highly Parallel Architecture: Functional Definition and
  Performance Evaluation
G. MAZARE, E. PAYAN (Grenoble, France) — 27

A Genetic Algorithm for Massively Parallel Computers
P. SPIESSENS, B. MANDERICK (Brussels, Belgium) — 31

A Transputer System for the Recognition of Human Faces by Labeled
  Graph Matching
R.P. WÜRTZ, J.C. VORBRÜGGEN (Frankfurt, FRG),
  C. v.d. MALSBURG (Los Angeles, USA) — 37

CORDIC Processor Arrays for Adaptive Least Squares Algorithms
B. YANG, J.F. BÖHME (Bochum, FRG) — 43

Section 3: New Concepts of Information Processing in the Brain

Computer Simulation of Neural Networks with Chemical Markers
A. ADAMOPOULOS, P. ANNINOS (Alexandroupolis, Greece) — 49

Dynamics of Activity in Neuronal Networks Give Rise to Fast
    Modulations of Functional Connectivity
K.-H. BOVEN, A. AERTSEN (Tübingen, FRG)   53

Using Proprioceptive Information to Control Movement to a Target
J. DEAN (Bielefeld, FRG)   57

A New Dynamic Model of Receptive-Field Organization
H.R. DINSE, J. BEST, K. KRÜGER (Bochum, FRG)   61

Direct Observation of Neural Assemblies During Neocortical
    Representational Reorganization
H.R. DINSE (Bochum, FRG), G.H. RECANZONE, M.M. MERZENICH
    (San Francisco, USA)   65

Large-Scale Simulation of a Self-Organizing Neural Network:
    Formation of a Somatotopic Map
K. OBERMAYER, H. RITTER, K. SCHULTEN (Urbana-Champaign, USA)   71

Synchronized Activity in the CA1-Region of the Rat Hippocampus Reveals
    Spatio-Temporal CSD-Patterns of Action Potentials
D. PLENZ, A. AERTSEN (Tübingen, FRG)   75

Emulation of Biology-Oriented Neural Networks
S.J. PRANGE (Berlin, FRG)   79

Are Fractal Dimensions a Good Measure for Neuronal Activity?
H. PREISSEL A. AERTSEN (Tübingen, FRG), G. PALM
    (Düsseldorf, FRG)   83

A Dynamic Theory of Coordination of Discrete Movement
G. SCHÖNER (Bochum, FRG)   87

Spike Arrival Times: A Highly Efficient Coding Scheme for Neural Networks
S.J. THORPE (Paris, France)   91

Section 4: Parallel Processing in the Visual System

Correlated Rhythmic Activity of Neurons in Monkey Visual Cortex
F. AIPLE, J. KRÜGER (Freiburg, FRG)   97

Feature Linking Across Cortical Maps via Synchronization
R. ECKHORN, H.J. REITBOECK, P. DICKE, M. ARNDT,
    W. KRUSE (Marburg, FRG)   101

Synchronization of Oscillatory Responses: A Mechanism for
    Stimulus-Dependent Assembly Formation in Cat Visual Cortex
A.K. ENGEL, P. KÖNIG, C.M. GRAY, W. SINGER (Frankfurt, FRG)   105

Parallel Distributed Processing of Configural Features of Moving Objects
in the Toad's Visual System
J.-P. EWERT, W.W. SCHWIPPERT, T.W. BENEKE (Kassel, FRG) 109

The Thalamo-Cortical Feedback-System: A Nonlinear Network-Model for
Binocular Interaction
D. HEINKE, F. GIANNAKOPOULOS, J. BEST (Bochum, FRG) 113

Segregation of Oscillatory Responses by Conflicting Stimuli –
Desynchronizing Connections in Neural Oscillator Layers
P. KÖNIG, T.B. SCHILLEN (Frankfurt, FRG) 117

Segregation of "Meaning" and "Importance" of Neuronal Messages
J. KRÜGER, J.D. BECKER (Freiburg, FRG) 121

Retinal Sampling Grids and Space-Variant Image Compression
H.A. MALLOT, G.-J. GIEFING (Bochum, FRG) 125

Towards a Network Theory of Cortical Areas
H.A. MALLOT, W. VON SEELEN (Bochum, FRG) 129

Processing of Figure and Background Motion in the Visual System
of the Fly (*Invited Paper*)
W. REICHARDT (Tübingen, FRG) 133

Simulation of Delayed Oscillators with the *MENS* General Purpose
Modeling Environment for Network Systems
T.B. SCHILLEN (Frankfurt, FRG) 135

Coherency Detection by Coupled Oscillatory Responses – Synchronizing
Connections in Neural Oscillator Layers
T.B. SCHILLEN, P. KÖNIG (Frankfurt) 139

A Model for Neuronal Oscillations in the Visual Cortex
H.G. SCHUSTER, P. WAGNER (Kiel, FRG) 143

Temporal Dynamics in Neuronal Microcircuitry
F. WÖRGÖTTER, D.M. KAMMEN, B. BRANDT (Pasadena, USA) 147

Section 5: Self-Organization and Learning in Neural Networks

Properties of an Adaptive Perceptron Algorithm
J.K. ANLAUF, M. BIEHL (Giessen, FRG) 153

Learning Symbol Processing with Recurrent Networks
R. GOEBEL (Marburg, FRG) 157

Training Strategies for Probabilistic RAMs
D. GORSE, J.G. TAYLOR (London, UK) 161

Learning in Neural Nets by Genetic Algorithms
J. HEISTERMANN (München, FRG)   165

A Neural Network which Adapts its Structure to a Given Set of Patterns
S. JOCKUSCH (Göttingen, FRG)   169

Efficient Learning Algorithms for Single-Layered Neural Networks
N.B. KARAYIANNIS, A.N. VENETSANOPOULOS (Toronto, Candada)   173

Internal Representations and Associative Memory (*Invited Paper*)
T. KOHONEN (Espoo, Finland)   177

Hebbian Learning of Principal Components
A. KROGH, J.A. HERTZ (Copenhagen, Denmark)   183

The Anti-Hebb Rule Derived from Information Theory
H. KÜHNEL, P. TAVAN (München, FRG)   187

Dynamical Learning in Networks with Sparse Connectivity
K.E. KÜRTEN (Bochum, FRG)   191

Self-Organization versus Programming in Massively Parallel Systems:
    A Case Study
B. MANDERICK, F. MOYSON (Brussels, Belgium)   195

Top-Down Learning in Modular Feed-Forward Networks
A. OSSEN (Berlin, FRG)   201

Genetic Generation of Backpropagation Trained Neural Networks
W. SCHIFFMANN, K. MECKLENBURG (Koblenz, FRG)   205

Temporal-Difference-Driven Learning in Recurrent Networks
J. SCHMIDHUBER (München, FRG)   209

Associative Memory versus Linear Filter: Distinct Operational Modes
    of Stable Artificial Neural Systems
B. SCHÜRMANN, I. LEUTHÄUSSER, J. HOLLATZ (München, FRG)   213

Competitive Sequence Learning
C. WINDHEUSER, J. KINDERMANN (St. Augustin, FRG)   217

Section 6: Evaluation of Artificial Neural Networks

Ideal Neurons for Neural Computers
I. ALEKSANDER (London, UK)   225

On the Biologically Motivated Derivation of Kohonen's Self-Organizing
    Feature Maps
R. ACKER, A. KURZ (Darmstadt, FRG)   229

All-or-None Connections of Variable Weights?
A. DE CALLATAY (La Hulpe, Belgium)   233

Learning to Predict the Consequences of One's Own Actions
F. CECCONI, D. PARISI (Rome, Italy)   237

Various Roles for Inhibitory Interneurons in the Formation of
     Associative Memory
I.E. DAMMASCH, J.R. WOLFF (Göttingen, FRG)   241

Soluble Low-Activity Networks with Nonlinear Synapses
H. ENGLISCH (Leipzig, GDR), J.L. VAN HEMMEN (München, FRG)   245

Implementation of Fuzzy Production Systems with Neural Networks
W. EPPLER (Karlsruhe, FRG)   249

A Neural Network for Job Sequencing
L. FANG, W.H. WILSON (Bedford Park, Australia), T. LI (Clayton, Australia)   253

Increasing the Storage Capacity and Shaping the Basins of Attraction of
     Neural Nets
S. GEVA, J. SITTE (Brisbane, Australia)   257

Drift and Diffusion in Backpropagation Networks
G. RADONS, H.G. SCHUSTER, D. WERNER (Kiel, FRG)   261

Solving Constraint Problems Using Feedback Neural Networks
H.N. SCHALLER (München, FRG)   265

The Exact Evaluation of Memory Capacity for New Stochastic Versions
     of the Hopfield Net
P. WHITTLE (Cambridge, UK)   269

Sparse Coding: The Link between Low Level Vision and Associative Memory
C. ZETZSCHE (München, FRG)   273

The Convergence of Parallel Boltzmann Machines
P.J. ZWIETERING, E.H.L. AARTS (Eindhoven, The Netherlands)   277

Section 7: Hardware and Software Simulators for Neural Networks

A Transputer Based General Simulator for Connectionist Models
H.P. ERNST, B. MOKRY, Z. SCHRETER (Zürich, Switzerland)   283

Neural Pascal: A Language for Neural Network Programming
H.P. GUMM (New Paltz, USA), F.B. HERGERT (München, FRG)   287

Modeling of Neuronal Systems on Transputer Networks
M. MIGLIORE, G.F. AYALA, S.L. FORNILI (Palermo, Italy)   291

A Development Tool for Neural Networks Simulations on Transputers
M. MIKSA (Marburg, FRG)   295

A Software Environment for Flexible and Rapid Prototyping of
    Neural Network Applications
J.A.G. NIJHUIS, P.E. DE HAAN, L. SPAANENBURG, F. WARKOWSKI
    (Stuttgart, FRG)   299

Systolic Simulation of Multilayer, Feedforward Neural Networks
N. PETKOV (Erlangen, FRG)   303

Artificial Neural Network for Real-Time Task Allocation in Fault-Tolerant,
    Distributed Processing System
P.W. PROTZEL (Hampton, USA)   307

Distributed Processing Hardware for Realization of Artificial Neural Networks
P. RICHERT, G. HESS, B. HOSTICKA, M. KEPSER, M. SCHWARZ
    (Duisburg, FRG)   311

Multiprocessor Simulation of a Self-Organizing Neural Network on NERV
R. STOTZKA, R. HAUSER, R. MÄNNER (Heidelberg, FRG)   315

IC^3: A Neural ASIC for Real-Time Prototyping
F. WARKOWSKI, L. SPAANENBURG, J.A.G. NIJHUIS (Stuttgart, FRG)   319

New Concepts for Information Processing in Connectionistic Systems
T. WASCHULZIK, H. GEIGER, M. ARNOLDI, D. BÖLLER,
    A. NISCHWITZ, W. BRAUER (München, FRG)   323

Section 8: Neural Networks for Visual Pattern Recognition

Recogniton of New Spatio-Temporal Patterns by Adaptive Junction
Y. AJIOKA, Y. ANZAI, H. AISO (Yokohama, Japan)   331

Hierarchical Spin Model for Stereo Interpretation Using Phase Sensitive
    Detectors
R. DIVKO, K. SCHULTEN (München, FRG)   335

A Shift Invariant Network Utilizing the GRT
M. FANG, G. HÄUSLER (Erlangen, FRG)   339

Applying the ART1 Architecture to a Pattern Recognition Task
E.C.D.B.C. FILHO, D.L. BISSET (Kent, UK)   343

Neural Network Models for Visual Pattern Recognition (*Invited Paper*)
K. FUKUSHIMA (Osaka, Japan)   351

$\Sigma\Pi$-Networks for Motion and Invariant Form Analyses
H. GLÜNDER (München, FRG)   357

Self Organization of a Network Linking Features by Synchronization
G. HARTMANN, S. DRÜE (Paderborn, FRG) 361

A Self-Organizing Network for Complete Feature Extraction
J. RUBNER, K. SCHULTEN, P. TAVAN (München, FRG) 365

A Self-Organising Neural Net for Depth Movement Analysis
F. SEYTTER (Edinburgh, UK) 369

Section 9: Neural Networks for Auditory Pattern Recognition

Sequence Analysis in Feedback Multilayer Perceptrons
H.-U. BAUER, T. GEISEL (Frankfurt, FRG) 375

A Neural Net for Recognition and Storing of Spoken Words
H. BEHME (Göttingen, FRG) 379

Adaptive Resonance Theory: Neural Network Architectures for
   Self-Organizing Pattern Recognition (*Invited Paper*)
G.A. CARPENTER, S. GROSSBERG (Boston, USA) 383

Word Recognition with a Recurrent Neural Network
F. KOWALEWSKI, H.W. STRUBE (Göttingen, FRG) 391

A Feedback Network for Temporal Pattern Recognition
T.B. LUDERMIR (London, UK) 395

Spectral Processing of Harmonic Complex Tones and Pitch by PDP Networks
R.W.W. TOMLINSON, W. TREURNIET (Darmstadt, FRG and Ottawa, Canada) 399

Section 10: Neural Networks for Motor Control

Resistive Network Approach for Obstacle Avoidance in Trajectory Planning
J. BECKMANN (Düsseldorf, FRG) 405

A Simple Network Controlling the Movement of a Three Joint
   Planar Manipulator
H. CRUSE, M. BRÜWER (Bielefeld, FRG) 409

Optimal Solution of Underdetermined Linear Matrix Equations in
   Neural Networks
W.J. DAUNICHT, H. WERNTGES (Düsseldorf, FRG) 413

Inverse Kinematics with Obstacle Avoidance Implemented as a DEFAnet
W.J. DAUNICHT, M. LADES, H. WERNTGES, R. ECKMILLER
   (Düsseldorf, FRG) 417

Parallel Neural-Net Path-Planner in Hypercubes and Transputers
A.W. HO, G.C. FOX (Pasadena, USA) 421

xiv   *Table of Contents*

A Control Concept for a Multifingered Robot Gripper Based on
Neuron-Like Associative Memories
M. HORMEL (Darmstadt, FRG)   427

Learning of Visuomotor-Coordination of a Robot Arm with Redundant
Degrees of Freedom
T. MARTINETZ, H. RITTER, K. SCHULTEN (Urbana-Champaign, USA)   431

Delta Rule-Based Neural Networks May Be Applicable to a Control Task
that Does Not Provide a "Teacher"
H.W. WERNTGES (Düsseldorf, FRG)   435

Section 11: Selected Applications for Neural Networks

Parallel Algorithms for Channel Assignment in Cellular Mobile Radio Systems:
The Neural Network Approach
M. DUQUE ANTÓN, D. KUNZ (Hamburg, FRG)   441

Storing Cycles in Analog Neural Networks
T. GENCIC, M. LAPPE, G. DANGELMAYR, W. GÜTTINGER
(Tübingen, FRG)   445

Recognition of Patterns and Movement Patterns by a Synergetic Computer
(*Invited Paper*)
H. HAKEN (Stuttgart, FRG)   451

Pattern Recognition in Damaged Hopfield Networks
E. KOSCIELNY-BUNDE (Hamburg, FRG)   459

Section 12: Parallel Processing in Artificial Intelligence

Parallel Process Interfaces to Knowledge Sytems
(*Invited Paper: Cremers*)
K.H. BECKS, W. BURGARD, A.B. CREMERS, A. HEMKER, A. ULTSCH
(Dortmund and Wuppertal, FRG)   465

Motion Detection by Correlation and Voting
S. BOHRER, H.H. BÜLTHOFF, H.A. MALLOT (Bochum, FRG
and Providence, USA)   471

The Development of "Symbolic Behaviour" in Natural and Artificial
Neural Networks
G.D.A. BROWN, M. OAKSFORD (Bangor, UK)   475

Apparent Motion and Other Mysteries (*Invited Paper*)
J.A. FELDMAN (Berkeley, USA)   479

Learning to Produce Discriminative Object Descriptions: On the Representation
of Rules in a PDP-Net
R. MANGOLD-ALLWINN (Mannheim, FRG)   487

Toward a Computational Architecture for Monocular Preattentive
    Segmentation
H. NEUMANN, H.S. STIEHL (Hamburg, FRG)   491

Modelling Attention in a Connectionist Speech Production Model
U. SCHADE, H.-J. EIKMEYER (Bielefeld, FRG)   495

Learning of Control Knowledge for Symbolic Proofs with
    Backpropagation Networks
A. ULTSCH, R. HANNUSCHKA, U. HARTMANN, V. WEBER
    (Dortmund, FRG)   499

Intelligent Dimensional Data-Reduction by a Topological Map
L. VERCAUTEREN, R.A. VINGERHOEDS, L. BOULLART
    (Ghent, Belgium)   503

Parallel Associative Processes in Information Retrieval
M. WETTLER, R. RAPP (Paderborn, FRG)   509

Distributed Parallel Cooperative Problem-Solving with Voting and
    Election System of Neural Learning Networks
B.-T. ZHANG, G. VEENKER (Bonn, FRG)   513

Section 13: Optical and Molecular Computing

Novel Logic and Architectures for Molecular Computing (*Invited Paper*)
J.R. BARKER (Glasgow, UK)   519

New Laser Techniques for Quasi-Molecular Storage (*Invited Paper*)
D. HAARER (Bayreuth, FRG)   525

Chaos, Cooperation, and Associative Memory in Nonlinear Pictorial
    Feedback Systems
G. HÄUSLER (Erlangen, FRG)   533

Optically Controlled Multistable Quantum Systems:
    Possible Realization of a Molecular Computer?
G. MAHLER, W.G. TEICH (Stuttgart, FRG)   539

Learning in Optical Neural Networks (*Invited Paper*)
D. PSALTIS, D. BRADY, K. HSU (Pasadena, USA)   543

References from All Contributions   549

List of Contributors   599

Author Index   611

Subject Index   623

Section 1
General Introduction

Parallel Processing in Neural Systems and Computers
R. Eckmiller, G. Hartmann and G. Hauske (Editors)
© Elsevier Science Publishers B.V. (North-Holland), 1990

Neuroinformatics and Cybernetics

Gert HAUSKE *
Lehrstuhl für Nachrichtentechnik
Technische Universität München
Arcisstr. 21, D–8000 München 2
F R G

The International Conference on "Parallel Processing in Neural Systems and Computers" is also the 10th Congress under the auspices of the German Society of Cybernetics (Deutsche Gesellschaft für Kybernetik/DGK). The relation between the recently established and rapidly growing discipline Neuroinformatics and the slightly more traditional discipline Cybernetics is by no means accidental, but is characterized by close familiar bonds. The precise degree, however, of the relationship between Neuroinformatics and Cybernetics (legitimate or adopted child ?) is open for speculations.

First, a brief definition of the term Cybernetics shall be given. It was N. Wiener, the great American mathematician, who 1949 coined this term in a book entitled "Cybernetics – or Control and Communication in the Animal and the Machine". The basic idea behind the cybernetical approach was to investigate the capabilities of systems (here described as control and communication) independent of their origin and nature. Cybernetics therefore represents an essentially interdisciplinary approach, however, it is not assumed to be a scientific discipline of its own. The German Society of Cybernetics adopted this definition of Cybernetics and regards it as principal task to promote the exchange of scientific ideas within the context mentioned.

A brief historical review reveals that the use of analogies between biological and technical systems has been frequently used as scientific method since the 17th century (typical example: Descartes). But only since the middle of our century the rapid development of mathematical and technical concepts including the possibility to use digital computers enabled us to describe complex capabilities of systems similar to human performance. A quite recently establisehed branch of this development is given by the investigation of self–organizing (parallel) systems which operate without control of a guiding program. Such system are met in a widespread variety of disciplines like mathematics (cellular automata), physics (spin glasses), chemistry (crystalization), and biology (neural structures) to give only a few typical examples. These systems posess interesting and important properties which are not immediately derived from the local structural relations between their elements, but are a result of globally operating interactions. Systems of this kind can not be decomposed into functional subunits because their whole represents more than the mere sum of their components (Aristotle).

As an application of self organizing systems relatively complex performances can now be realized in the sensory domain (processing, recognition, and interpretation of patterns), in the motor domain (planning, processing, and execution of movements), in the

---

(*) President of the German Society of Cybernetics

sensory–motor domain (execution of movements under sensory control), and in the cognitive domain (problem solving, expert systems). A critical question is whether the cybernetical approach pursued by Neuroinformatics is really more successful and reasonable than the classical technical proceeding. At first the fact should be mentioned that biological systems and their capabilities serve as ideal prototypes for the design of the respective technical systems. The existence of biological systems with highly developed capabilities on the other hand guarantees the solution of the task considered. It should also be kept in mind that biological systems were forced to develop their performance over a long time of experience and under a wider variety of boundary conditions than usually encountered in technical systems. Therefore it is obvious and reasonable to imitate the principles of biological solutions. A basic requirement for this procedure, however, is an interdisciplinary cybernetical approach balanced between biological and mathematical/technical sciences. An understanding of the functioning of complex biological systems implies a mastership in the treatment of formal mathematical concepts and only on the basis of this understanding a further development of biologically orientated technology is possible.

For an application of artificial neural networks as computing units a number of properties typical for the biological nervous system are essential. These include parallel processing with high data rate, associative and distributed memory with high robustness against errors, and the ability to learn. Whether recently developed concepts fulfill all necessary demands on the capabilities of biological systems (e.g. capabilities already shown by the relative low level system of a free flying insect) is questionable. In relation to brains artificial neural networks represent small sections of well understood local connectivity. Not very much is known about the precise functioning of long ranging fibres which represent quite a fraction of all nervous tissue in the brain. Only recently reliable information about globally interacting brain regions during the execution of certain tasks became available. So with some interest we might speculate about the capabilities of a larger type of system consisting of interconnected neural nets. Of some importance will also be the fact that signals in real nervous systems usually have to pass certain preprocessing stages before they are fed into the actual neural net area. An adequate adjustment of signal–related preprocessing leading to an appropriate internal representation in order to make central processing more effective may be of considerable importance.

Finally I wish to refer to a more general aspect in connection with technological development. It may be assumed that traditional technical development occurs more or less independent from biological models and biological environment, but, on the other hand, an increasing interaction with biological systems (particularly humans) exists. These interactions in our case predominantly take place in the informational sector. Since a realistic goal can only be a symbiosis between technical and biological systems informational technology primarily is addressed to reconsider its role. It is simply unreasonable to restrict the development of information processing machines designed to support or substitute human performance exclusively to technicians. Despite obvious improvements the majority of such machines still operate in a clumsy and difficult to handle manner. The text processing software I am using during the preparation of these lines may serve as an example in this direction. A basic requirement for the design of future information processing machines must include knowledge about the structure of infomation in the sense of human utility. These questions in my opinion can only be solved by a widespread interdisciplinary cooperation of natural and technical sciences on one side, biological sciences on the other side, and perhaps an inclusion of humanities as additional partner. The interdisciplinary cybernetical approach of Neuroinformatics will have to occupy a key position in this direction.

Parallel Processing in Neural Systems and Computers
R. Eckmiller, G. Hartmann and G. Hauske (Editors)
© Elsevier Science Publishers B.V. (North-Holland), 1990

God ever geometrizes - PLATO

CONCERNING THE EMERGING ROLE OF GEOMETRY IN NEUROINFORMATICS

Rolf ECKMILLER *

Department of Biophysics
Heinrich-Heine-Universität Düsseldorf
Düsseldorf (FRG)

1. HISTORY

As a result of interactions between certain biological neural networks, namely human brains and the physical environment, these brains began to register (quote by A. Einstein, 1920: "How can it be that mathematics, being after all a product of human thought independent of experience, is so admirably adapted to the objects of reality?") certain regularities and rules (Bell, 1937). These rules which eventually became known as mathematical theorems have always been a mixed result of spatio-temporal events in the physical environment and sensori-motor as well as associative events within human brains.

Mathematics (with strong ties to philosophy and only few connections to natural science) evolved historically mainly as two sets of rules (or theory spaces), namely Algebra and Geometry (Fig. 1) (Courant and Robbins, 1941; Davis and Hersh, 1986; v.d. Waerden, 1983). Individual scientists have a preference (possibly due to a dominance of the left or right brain hemisphere; see: Bell,1937; Briggs, 1988) for either algebraic-analytical thinking (e.g. Boole, Lagrange) or geometric-topological thinking (e.g. Lobatchewsky, Riemann). It is generally accepted that a given physical problem P can be represented algebraic-analytically as P_A or geometric-topologically as P_G. Algebra and geometry are considered to be fundamentally different by some scientists (comparable to the difference between particle theory and wave theory in physics), whereas the more popular view holds that both representations are always interchangeable and that algebraic-analytical representations are even superior.

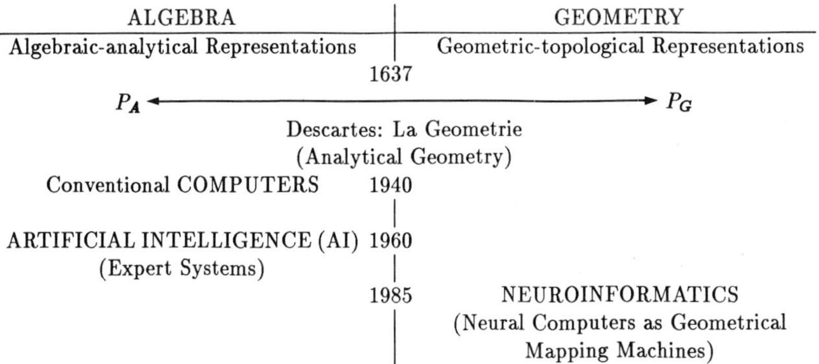

Figure 1: History of mathematics and corresponding information technologies.

*) Supported by Ministry for Research and Technology (BMFT, under grant No. ITR 8800F6)

The desire to find equivalent algebraic representations of geometrically represented problems P_G reached a culmination point in the book chapter "La Geometrie" by R. Descartes (1637).

Descartes invented (presumably incorporating some ideas by P. Fermat) the coordinate system and founded the new mathematical field of "Analytical Geometry" with the main topic to find equivalent representations for a given P_A or P_G in the respectively opposite theory space (Fig. 1). In subsequent centuries (Bell,1937; Yaglom, 1988) the fundamental dichotomy between these two theory spaces became more and more blurred as the number of examples with seemingly equivalent representations increased. However, physicists, mathematicians, and engineers even today are dealing with many problems represented only either as P_A or as P_G. Approximation methods (e.g.: method of finite elements; Dwoyer et al., 1988) for mappings of P_G onto nearly equivalent P_A are of limited value, especially if nonlinearities, higher dimensionalities, and temporal events have to be considered.

It is important to note that geometry comprises a vast number of rules without algebraic-analytical equivalences,which evidently form the basis for many physical events including biological information processing. A few examples are listed below:

A) Ratio U/D of circular circumference U and diameter D always yields a constant, namely the irrational and even transcendental number π. (L. Euler: "...transcend the power of algebraic methods.")

B) The length of the diagonal of a square is an irrational number, even though its side length may be a rational number.

C) Only triangles can be used to envelop a volume by a minimal number (4) of polygons.

Around 1940 began the development of software-driven computers. Computers are listed under algebra (left side in Fig. 1), since any mathematical problem to be solved by such a computer has to be represented algebraic-analytically in order to be entered as software. This basic fact has the dramatic and generally overlooked consequence that the use of software-driven computers typically limits the user to algebraic-analytical representations of a given problem. The far reaching impact of this consequence may be elucidated by the bizarre assumption that physicists would be forced (by using certain instruments) to restrict their research to the realm of the wave theory, thus ignoring other (and often conflicting) results from the realm of the particle theory.

Around 1960 research efforts were initiated to generate so-called 'intelligent' functions (a special domain of biological neural networks), such as: pattern recognition, associative memory, and movement control of redundant robots or autonomous vehicles. Given the alternative of the two approaches: special software (e.g. new symbolic languages) for conventional computers or special hardware (e.g. perceptron-like neural net hardware) most researchers concentrated on novel software for intelligent functions, thus starting the new field of "Artificial Intelligence". Conventional AI-research is also listed under algebra (Fig. 1), since software approaches require algebraic-analytical representations. About 25 years later (around 1985), it became more and more evident among AI experts that the special software approach would find its limits at the level of 'expert systems'. One of the most plausible reasons for the limits of conventional (software based) AI-research was the notion that algebraic-analytical representations require mathematical theories of the underlying 'intelligent' functions. However, such theories for pattern recognition or redundant robot control in real time in a natural environment are presently not available.

This sobering evaluation in the early 80s was the decisive factor for a strong revival of the special hardware approach in AI-research under various headings such as: neural networks for computing, connectionism, or neuroinformatics. The most important merit of this approach is the fact that numerous examples of biological existence proof exist. Any fly (presumably born with a hard-wired net of well under 10^6 neurons which doesn't change during its lifetime) is capable of mapping visual trajectories in real time onto obstacle avoiding flight trajectories (before gracefully landing at the ceiling). Many frogs (with considerably larger neural nets and the ability to change their net topology by means of selforganization during learning) can motionlessly monitor a fly's flight trajectory and with reasonable success generate a prey-catching movement.

Such biological neural networks are information processing systems, which presumably:
1. do not represent the sensori-motor mapping operations (or behavior) algebraic-analytically as in a software-driven computer, but rather 2. embed the mapping operations in the highly parallel dynamical neural net topology, and 3. acquire these geometric-topological representations partly during evolution and partly during their lifetime by means of selforganization and learning in interaction with the physical environment.
Accordingly, both biological and artificial neural networks (abbreviated as NNs) are listed under geometry (right side in Fig. 1). In other words, it is postulated here that software-driven conventional computers and selforganization-driven neural networks (neural computers) correspond to two separate mathematical theory spaces with only limited overlap.

2. CURRENT STATE

At present most research in neuroinformatics, which can be considered as a novel approach in AI is still based on software simulations of artificial NNs (with the same synchronicity- and real-time problems as in simulations of highly parallel analog electronic circuits) with certain algebraically derived learning rules. The radical change from algebraic-analytical to geometric-topological representations is just beginning. This development will be boosted by the commercial availability of fully parallel and asynchronously operating neural net hardware with user-specifiable initial net topology and dynamics, as well as learning rules being embedded in the net topology (rather than separate from the net as in most current software simulations). Several companies and research institutes are currently designing neural net hardware, which will operate as a new kind of co-processor in conventional workstations or as stand alone system (Eckmiller/Universität Düsseldorf; Hammerstrom/Adaptive Systems, Beaverton; Hecht Nielsen/HNC, San Diego; Hosticka/Fraunhofer Institut, Duisburg; Mead/Caltech, Pasadena). The available neural net software simulators and the emerging hardware simulators offer information processing far beyond current mathematical theories. In other words, neuroinformatics is at present largely an experimental research field like experimental physics or engineering. Intuition and concepts borrowed from other areas (especially neuroscience) spur this field; one never knows in advance whether the combination of a certain initial net topology, a certain learning rule, and a certain set of learning opportunities will yield a satisfying selforganization with convergence toward a final net topology and - most importantly - mapping functions with desired precision and generalization.
Most of the mathematical theories and methods being presently used for simulation of artificial NNs belong to the algebraic-analytical theory space. They include:
A) Linear systems theory; B) Adaptive filter theory; C) Dynamical systems theory; D) Variation calculus; E) Fuzzy set theory.

3. FUTURE DEVELOPMENTS

The biological existence proof guarantees that NNs are (in principle) capable of generating various mapping operations, which were acquired by selforganization of the dynamical net topology during a learning phase. Accordingly, biological and artificial NNs can be described as Geometrical Mapping Machines. However, no mathematical theory is available, which could predict the relation between dynamical neural net topology, learning rule, set of learning opportunities, and the desired mapping operation. Given the fact that more than half of our present knowledge in mathematics was discovered in the last 50 years (Peterson, 1988), there is hope that future research efforts can be successfully directed towards novel mathematics of information processing in neural networks. The expected emphasis on geometry and dynamical net topology will certainly be also valuable for theories of other parallel information processing systems (e.g.: optical computers and molecular computers).

A number of mathematical theories and methods with origins in the geometric-topological theory space could form a base for the new theory to be developed. They include:

F) Various geometry theories (Abelson and diSessa, 1980; Edmondson, 1987; Kapur and Mundy, 1989; Morgan, 1988; Thompson and Stewart, 1986),

G) Various topology theories (Brown, 1988; Grünbaum and Shepard, 1987),

H) Graph theory,

I) Cellular automata theory (Toffoli and Margolus, 1987),

J) Theory of fractal geometry (Mandelbrot, 1982), and

K) Knot theory (Kauffman, 1987).

As in most other previous technological developments (e.g.: feedback control systems, mechanical clocks and automata, conventional computers) the experimental, engineering design of selforganizing NN-hardware (neural computers) for various special purposes (sensory, associative learning, or motor) precedes the development of the corresponding theory of geometrical mapping machines. Geometry and topology play a more crucial role in NN-design than in any other information technology. This demands a new breed of experts ("network architects"), who will be as important as software engineers were for conventional computers.

In summary, neuroinformatics is heading toward computer technologies, which are capable of geometric-topological representations of specific mapping functions. Progress is under way to merge the technologies of neural computers (Eckmiller and v.d. Malsburg, 1988) with those of conventional computers, thus creating a genuinely novel hybrid computer generation.

REFERENCES

[1] Abelson, H.; diSessa, A.: Turtle geometry. MIT Press - Cambridge, 1980, paperback 1986

[2] Bell, E.T.: Men of mathematics - The lives and achievements of the great mathematicians from Zeno to Poincaré. Simon & Schuster - New York, 1937, 1965, and 1986

[3] Briggs, J.: Fire in the crucible. St. Martin's Press - New York, 1988

[4] Brown, R.: Topology - A geometric account of general topology, homotopy types and the fundamental groupoid. Horwood-Chichester, 1988

[5] Courant, R.; Robbins, H.: What is mathematics? Oxford University, Press - New York, 1941 and 1969

[6] Davis, P.J.; Hersh, R.: Descartes dream - The world according to mathematics. Harcourt Brace Javanovich, 1986, and Penguin Books - London, 1988

[7] Descartes, R.: Discours DE LA METHODE pour bien conduire sa raison & chercher, La verité dans les sciences, plus LA DIOPTRIQUE. LES METEORES. Et LA GEOMETRIE. Qui sont des essais de cete METHODE. Ian Maire - Leyden, 1637

[8] Dwoyer, D.L.; Hussaini, M.Y.; Voigt, R.G. (eds.): Finite elements - Theory and application. Springer - New York, 1988

[9] Eckmiller, R.; v.d.Malsburg, C.: Neural computers. Springer-Heidelberg, 1988, reprinted 1989

[10] Edmondson, A.C.: A Fuller explanation - The synergetic geometry of R. Buckminster Fuller. Birkhäuser - Boston, 1987

[11] Grünbaum, B.; Shepard, G.C.: Tilings and patterns. Freeman - New York, 1987

[12] Kapur, D.; Mundy, J.L. (eds.): Geometric reasoning. MIT Press - Cambridge, 1989

[13] Kauffman, L.H.: On knots. Princeton University, Press - Princeton, 1987

[14] Mandelbrot, B.: The fractal geometry of nature. Freeman - New York, 1982

[15] Morgan, F.: Geometric measure theory. Academic Press - Boston, 1988

[16] Peterson, I.: The mathematical tourist - Snapshots of modern mathematics. Freeman - New York, 1988

[17] Thompson, J.M.T.; Stewart, H.B.: Nonlinear dynamics and chaos - Geometrical methods for engineers and scientists. Wiley - New York, 1986, reprinted 1988

[18] Toffoli, T.; Margolus, N.: Cellular automata machines. MIT Press - Cambridge, 1987

[19] v.d.Waerden, B.L.: Geometry and algebra in ancient civilizations. Springer-Heidelberg, 1983

[20] Yaglom, I.M.: Felix Klein and Sophus Lie - Evolution of the idea of symmetry in the nineteenth century. Birkhäuser - Basel, 1988

Parallel Processing in Neural Systems and Computers
R. Eckmiller, G. Hartmann and G. Hauske (Editors)
© Elsevier Science Publishers B.V. (North-Holland), 1990

PHILOSOPHICAL CONCEPTS OF COMPUTATIONAL NEUROSCIENCE

Klaus Mainzer

Lehrstuhl für Philosophie mit Schwerpunkt
Analytische Philosophie/Wissenschaftstheorie
Universität Augsburg
D – 8900 Augsburg, Fed. Rep. of Germany

Computational Neuroscience tries to transfer concepts of brain function to
artificial nets for the design of neural computers. The brain as product of natural evolution provides the general principles of a new computer technology.
Philosophically, this approach favors a *naturalized epistemology* which is
founded by an *interdisciplinary research program* of mathematics, physics,
neurobiology, and computer science. Several concepts of research such as parallelism, non–linearity, self–organization, complex system theory, cellular automata etc. are merged in the new paradigm of connectionism which gives new
insight in the philosophy of human nature and mind.

1. THE NEUROCOMPUTER REVOLUTION IN THE PHILOSOPHY OF SCIENCE

In the *history of science and philosophy* brain was illustrated by *technical models* of the
most advanced machinery. Thus, during the century of mechanization, the brain functions
were thought of as hydraulic pressures which are conducted along the nerves to operate on
the muscles. With the beginning of electrotechnics brain was compared with telegraphs or
telephone switchboards. Since the development of computers brain was identified with the
most advanced hardware generations (Mainzer [7]). But, there were only *models* and
analogies between human brain and technology. A *theory* of the 'cerebral computer' as
product of natural evolution still failed.

In 1958 J. von Neumann emphasized that the *mathematical language* of digital and
program–controlled computers was well established, while that of brain was unknown. His
concept of *cellular automata* gave first hints to a mathematical theory of living organisms
which are conceived as self–reproducing networks of cells. *Self–organization* and the
evolution of *complex open systems* are necessary conditions of life that needed an
explanation before the principles of brain could be derived. Modern *thermodynamics* and
statistical mechanics deal with the spontaneous formation of macroscopic patterns or
structures in complex systems far from the thermal equilibrium via self–organization.
Physical, chemical, and biological examples of synergetics are well–known. Even the
dynamics of neural networks are going to be explained in this framework (Haken [5]). The
coming together of physics, molecular neurobiology, and the growth in computational
ability has brought about the final switch in the study of neural networks. From the
viewpoint of *philosophy of science* we get a *reductionistic research program* that should
allow to explain neurocomputational self–organization as a natural consequence of
physical, chemical, and neurobiological evolution by common principles.

2. MATHEMATICAL CONCEPTS OF NEUROSCIENCE AND EPISTEMIC CATEGORIES

It is the task of epistemology to make the *philosophical categories* precise that stand behind the mathematical concepts of neuroscience. In the 18th century, I.Kant tried to introduce the philosophical categories which found the axioms of Newtonian mechanics. It is the main feature of his epistemology that recognition does not arise by passive impressions of the external world on the 'tabula rasa' of our brain. Recognition in the Kantian sense is an active process producing models of the world by a priori categories. The *spatial* and *temporal order* of physical events is reduced to our a priori Euclidean forms of intuition. *Perception* in the Kantian sense is an active information processing regulated by a priori anticipations. The causal connection of events is philosophically made possible by an a priori category of *causality.*

In Newtonian mechanics, causality is mathematically described by differential equations which determine the motion of a physical body completely. As causal *interactions* were reduced to two bodies (for instance the gravitational interaction of sun and earth in Kepler's law), classical physicists of the 18th century described physical processes by *linear dynamics.* Thus, the whole universe seemed to be a predictable and calculable clock (at least for a Laplacian spirit) if the initial conditions are well known.

One goal of computational neuroscience is to provide a mathematical explanation for informational self−organization similar to that given by thermodynamics to dissipative self−organization. According to the *category of causality,* organizing systems are structured by principles which differ fundamentally from that of algorithmically controlled computers. Whereas in a sequential computer the *global state* is controlled by programming instructions, organizing systems are dominated by *locally acting laws and rules.* Global order has to arise from interaction between local entities such as neurons in the case of neural networks. Thus, self−organization of complex networks is a property of the ensemble, not predictable a priori from the observation of an isolated unit: The whole is, philosophically spoken, greater than the sum of the parts. The *non−linear dynamics* of neural networks is thought to exhibit non−periodic behavior approaching the attractors of *noise* and *chaos.* Thus, a mechanization of the brain in the sense of classical determinism must be excluded.

Our brain is not as good in arithmetic operations as a digital computer. But when it comes to *epistemic actions* such as association, categorization, generalization, and learning, the advantages of massively interconnected networks are obvious. New neural network structures with *learning algorithms* (Hopfield systems, Boltzmann machines etc.) that can be reduced to the thermodynamic framework of our research program have been introduced recently. The *Boltzmann machine,* for instance, is a stochastic network architecture with nondeterministic processor elements and a distributed knowledge representation which is described mathematically by an energy function. Historically, a first reference to the relationship between massively parallel systems and the Boltzmann distribution dates from J. von Neumann. He presumed that theories of brain and computation were much less combinatorial, and much more analytical, and closer to thermodynamics (in the form it was received from Boltzmann) which in some of its aspects comes nearest to manipulating and measuring information. In general it is the aim of a learning algorithm to diminish the information−theoretic measure of the discrepancy between the brain's internal model of the world and the real environment via self−organization.

Internal models of the external world need *geometric representation.* In the history of philosophy there was a controversial dispute as to which geometry is the right one (Kant, Helmholtz, Einstein etc.). From the viewpoint of neuroscience, every subsystem of the cerebral network has its own internal geometric language with intrinsic generalized coordinates of the corresponding multidimensional and non−orthogonal frames. The information transfer between the different subsystems (for instance from sensory frame to motor apparatus) is represented by tensor transformations. Thus, the *intrinsic natural space structure of brain* is mathematically represented by *tensor geometry*, and not by the very special Euclidean geometry.

Concerning the *time structure*, cellar automata are assumed to work synchronously in the sense that a global clock sets the overall timing for each cell of the network. But, we must be aware that our natural 'cerebral computer' (Baron [1]) is an open system of agents which operate with no global control, incomplete information and a high degree of communication. *Synchronization* would cost a too high price for components which are separated by long distances. Furthermore, the behavior of the environment in which a complex open system is embedded is not predictable by the system itself. Thus, the system has to operate *asynchronously* with information of the outside world while engaging in different internal tasks. The cerebral networks have, philosophically spoken, their own *intrinsic time*.

A further epistemic category concerns *perception*. In the sense of empiristic and sensualistic epistemology (Locke, Hume), perceptions are only sensory stimuli *passively* impressed on the 'tabula rasa' of the brain. Modern neuroscience supports an *active* processing of sensory stimuli, be they auditory, visual, or tactile, arriving at specific sense organs or receptor cells and generating trajectories or *space–time functions* of neural activity (Eckmiller [3]). The available information can be decoded as a pattern of stimulation intensity at sensory location as a function of time on a neural network. Such an intensity–modulated *trajectory* can be stored for purposes of recognition, association, or generation of corresponding motor trajectories. Intelligent robots are currently being designed that consist of various neural net modules with those special purpose functions. It is assumed that each sensory module is connected with an internal representation module via specific sensory coordinate transformations. These internal representations of the external space and of movement trajectories form a neural space net which is the *intrinsic apparatus of perceptions*. It is obvious that modern neuroscience supports a kind of epistemology which assumes intrinsic transformational structures of recognition.

3. OUTLOOK ON A NEUROCOMPUTATIONAL EPISTEMOLOGY

In the traditional epistemology only few philosophers (Descartes, Helmholtz, James etc.) pay attention to the cerebral conditions of human feelings and recognition. Today, computational neuroscience favors a *naturalized epistemology* which is based on an interdisciplinary research program. We call it *'neurocomputational epistemology'*. As it depends on a developing research program, it is *not an a priori doctrine* in the traditional sense of philosophical epistemology. Nevertheless, neurocomputational epistemology assumes *intrinsic transformational structures*. But these epistemic categories are involved in the research process itself, and therefore they are *fallible* and *hypothetical*.

Now we can conclude some structural categories of the brain which are reducible to the principles of natural evolution in a mathematically precise manner. I remind of *non–linear causality, holism, self–organization*, and *intrinsic space–time structure* of the brain. In modern theory of evolution the emergence of physical, chemical, and biological macroscopic forms is explained by the same general principles. Thus, we get a general framework of complex dynamical systems to describe nature on several hierarchical levels from its atomic and molecular structure to the self–organization of brain. In this sense modern epistemology must be embedded in a philosophy of nature founding the principles of (non–linear) complex dynamical systems. I call it the *'naturalistic turn'* from *epistemology* to *philosophy of nature* which is obviously the more fundamental discipline of philosophy.

It is a consequence of this framework that the cerebral apparatus cannot be reduced to the simple *materialistic view* of mechanism. Even the *dualistic distinction* of 'mind' and 'matter' does not seem to be adequate. In the sense of Aristotle we may say that 'mind' and 'matter' or 'form' and 'substance' are no separated entities, but complementary functions of a *complex structural nature* with its own *intrinsic principles* (Mainzer [5]). It is the goal of computational neuroscience to explain these principles mathematically.

This is, of course, a *reductionistic* research program which may anguish some people to become "nothing more than a complex machine". Indeed the *history of science* seems to support the permanent loss of man's dominant position in nature step by step. With the Copernican astronomy mankind lost its central position in the universe. With Darwin's biology humans were reduced to the evolution of animals. In Freud's psychoanalysis man was no longer a superior being which is dominated by reason, but driven by animal instincts. With the theory of 'cerebral computer' man seems to loose his last superior position in nature. His intelligence is reduced to a complex cerebral mechanism which may be overcome by "intelligent computers"." There is a world–wide *crisis of technological acceptance* caused by sometimes irrational feelings and frights which we must take in earnest. But the reductionism of scientific hypotheses and theories does not mean an *ethical reductionism*. The worth of man does not depend on his position in natural evolution, on the explanation of his cerebral functions or his simulation by machines, but on ethical values such as they are written down in the declaration of human rights. The protection of these rights is a main topic in the *ethics of modern technology*. In order to pursue this ethical goal it is necessary to understand the basic nature of human reasoning, feeling, and acting better and better. In this sense neurocomputational epistemology gives also new insights in *anthropology*, and overcomes traditional doctrines. It is part of a *fertile interdisciplinary research program* which pushes forward a revolutionary computer technology.

REFERENCES

[1]   Baron, R.J., The Cerebral Computer. An Introduction to the Computational Structure of the Human Brain (Lawrence Erlbaum Ass., Hillsdale, New Jersey, 1987)

[2]   Churchland, P., Neurophilosophy: Toward a Unified Science of the Mind–Brain (MIT Press, Cambridge Mass., 1986)

[3]   Eckmiller, R., v.d.Malsburg, C. (Eds.), Neural Computers (Springer Verlag, Berlin 1989)

[4]   Guckenheimer, J., Holmes, P., Nonlinear Oscillations, Dynamical Systems, and Bifurcations of Vector Fields (Springer Verlag, Berlin 1983)

[5]   Haken, H. (Ed.), Neural and Synergetic Computers (Springer Verlag, Berlin 1988)

[6]   Mainzer, K., Symmetrien der Natur. Ein Handbuch zur Natur– und Wissenschafts-philosophie (De Gruyter Verlag, Berlin 1988)

[7]   Mainzer, K., Die Evolution intelligenter Systeme, Z.f.Semiotik XII/1 (1990): Zeichen im Gehirn? Semiotik und Künstliche Intelligenz

Section 2
Development and Application
of Parallel Computers

Parallel Processing in Neural Systems and Computers
R. Eckmiller, G. Hartmann and G. Hauske (Editors)
© Elsevier Science Publishers B.V. (North-Holland), 1990

PARALLEL PROCESSING USING MEMORY-BASED CONNECTIONIST NETWORKS

Chung-Chih CHEN

Artificial Intelligence Laboratory
Free University of Brussels
Pleinlaan 2
1050 Brussels, Belgium

Memory-based connectionist networks are connectionist models using value-coding as the representational paradigm. Since value-coding is a table look-up strategy, such networks inherently demand more memory to perform a task. The advantages of memory-based networks are fast learning and algorithmic simplicity. Two applications of such networks are shown in this paper: robot path planning and speech recognition. We argue that memory-based connectionist networks are suitable for the parallel implementations on fine-grained SIMD machines and optical computers.

1. INTRODUCTION

There are many different learning schemes proposed for connectionist networks [1]. On one extreme, some networks use compact representations. They learn by repeatedly adjusting a set of *global* parameters to find the optimal values. For example, back-propagation adjusts the weights. On the other extreme, some networks use non-compact representations. They learn by just adding new *local* parameters for table look-up. For example, sigma-pi learning builds new connections for new input/output pairs [2]. While the former class needs fewer parameters to represent the desired input/output mapping, it is slow. In other words, it saves space but wastes time. On the contrary, the latter class is fast (one-shot learning), but needs more memory. We refer to the latter class as "memory-based connectionist networks".

Most researches in connectionist networks have focused on compact representations, while relatively few are memory-based. It is true that networks using back-propagation have better generalization properties because they optimize the global parameters. Memory-based networks still should not be neglected, because they have other advantages which may be more useful than generalization. For example, the algorithms used by memory-based networks are simpler. One main reason for investigating connectionist networks is that they are massively parallel, so they are suitable for the massively parallel machines like the Connection Machine or the DAP machine which are more and more popular. Since such fine-grained SIMD machines have a lot of simple one-bit processors, they are more suitable for the situations where a lot of data and simple computations are used [3]. Besides, the memory-based architectures are also suitable for optical computers which can provide a large amount of connections [4]. From this viewpoint we feel that memory-based networks will be better than those using compact representations.

We have simulated two memory-based connectionist networks for robot path planning [5] and speech recognition [6] respectively. Both networks need a lot of data (units and connections) and use only very simple computations.

2. ROBOT PATH PLANNING

To plan a path for a manipulator in an environment with obstacles, the simplest way is to do it in its configuration space. The problem is that the configuration space approach takes a lot of time to transform obstacles in real space into obstacles in configuration space, since such transformation uses inverse kinematics which is an ill-posed one-to-many mapping. We have built a memory-based network, which is called GGT (Grid-to-Grid Transformation), to solve this difficult transformation problem. GGT uses grid models (value-coding) to represent both real space and configuration space. So GGT is a mapping from the real space grid (RSG) to the configuration space grid (CSG). GGT learns this mapping by exploring CSG and then building the connections from RSG to CSG. During the performance mode, GGT automatically transforms the obstacles and the target in RSG into their counterparts in CSG through the connections. This transformation is just like a simple spreading-activation process. To plan a path in CSG, GGT uses the bi-directional propagation search which spends only half of the time needed by the standard Lee grid method (which is an one-directional search).

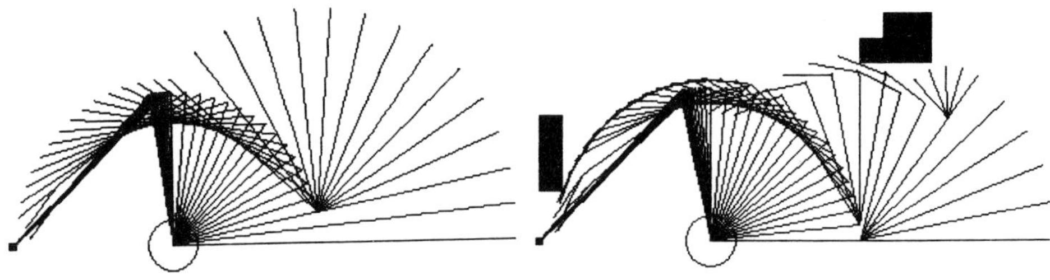

FIGURE 1	FIGURE 2
Reaching the target without obstacle avoidance.	Reaching the target with obstacle avoidance.

Figure 1 and 2 show the simulation of a 3-joint robot arm. The circle in the left bottom of each picture is the main axis (the first joint) of the arm. The large polygon is the obstacle. The small rectangle is the target. Initially the arm lies horizontally along the x-coordinate. The robot automatically transforms the target in RSG into many possible targets in CSG, and reaches a CSG target with the shortest path. Although real space is represented by a grid, the target can be anywhere in a grid cell. An interpolation strategy is adopted in order to move within a cell without being constrained by the cell granularity. When the arm is close enough to the target, its gripper will aim at the target, and gradually reach it using the interpolation. The distance error can be made as small as possible in general cases.

3. SPEECH RECOGNITION

Speech recognition is a difficult problem because of its temporal nature and the possibility of the presence of noise. Our task here is to recognize 4 different phonemes (B, D, E and V) [7]. Each phoneme pattern consists of 12 time frames, and each time frame consists of 16 frequency bands. Hence, each phoneme is represented by 192 energy levels (each is a real value from 0 to 1). Our memory-based network has three layers: the input layer, the hidden layer and the output layer. The input layer and the hidden layer consist of 192 units each. The output layer consists of 4 units (one for each phoneme). Each hidden unit $H(i,j)$ has connections from its corresponding input unit $I(i,i)$ and from the neighbors $I(i+a,j+b)$ of $I(i,j)$. $H(i,j)$ is also connected to each output unit $O(k)$. (We also use I, H and O to represent the activation of the unit respectively.) The index i is constrained by the number of time frames in each speech pattern; j by the number of frequency bands; k by the number of possible outputs; a by the number of neighboring time frames; b by the number of neighboring frequency bands. Each $H(i,j)$ has a look-up table $T(a,b,d(I(i,j)),d(I(i+a,j+b)),k)$ (each entry of the table is just one bit) to record the relations between $I(i,j)$ and $I(i+a,j+b)$ during learning from the training set. The index d is a discretization function constrained by the number of possible energy discretizations.

The network learns by setting appropriate bits of the look-up tables. When a training pattern is presented to the input layer, each hidden unit sets the bits of its look-up table according to the following rule:

$$T(a,b,d(I(i,j)),d(I(i+a,j+b)),k) = 1$$
for all the legal values of a,b,i,j;
and k is the output which the pattern belongs to.

After learning, when a speech pattern is presented to the input layer for recognition, each hidden unit will cast some votes to each output unit of the output layer depending on how well the pattern matches the previously learned patterns. The output unit which receives the most votes is the correct output. The voting rule is:

$$O(k) = \Sigma T(a,b,d(I(i,j)),d(I(i+a,j+b)),k)$$
where $k=0,1,2,3$;
the summation is applied to all legal values of a,b,i,j.

TABLE 1

Recognition Performance (%) (Total No. of Patterns = 768)	Number of Training Patterns				
	101	201	372	523	668
on Training Set	100	100	100	100	100
on Test Set	68	72	72	71	83

Table 1 shows the performances of our network with different sizes of the training set. We see that our network always remembers all training cases 100% correctly. When the training set is small, the generalization to the test set remains at about 72% correctly. But when the training set is large enough to exceed some critical point, the generalization improves dramatically. The network can generalize to the test set 83% correctly. Several factors contribute to the good performance of our simple network. One is that we choose the neighborhood relations between input units which capture the temporal nature of speech. Since in our network the neighbors of one input unit extend to three time frames before the unit and after the unit respectively, they provide the necessary temporal information for speech processing. Another reason is that the look-up table in each hidden unit is able to capture the nonlinearity of the input/output mapping. The combinatorial relations (or value-coding) between one input unit and any of its neighbors are recorded in the look-up table. And in fact, such combinatorial relations can be regarded as a very simplified approximation of a discrete 2-dimensional Gauss distribution.

4. CONCLUSIONS

The main advantages of our memory-based connectionist networks are that they are very fast and very simple. They are very fast because they use only very simple bit-setting and voting (or vetoing) computations and they use one-shot learning. There are not any complex floating-point number computations or mathematical equations. Their simple architectures make them suitable for the parallel implementations on the fine-grained SIMD machines or optical computers. For example, GGT needs a lot of connections between RSG and CSG which can be implemented optically. Another advantage of our networks is that they remember all training patterns without any errors. This feature may be quite useful in some situations where learning specific instances is more important than generalization. In contrast to typical connectionist networks which have the interference problem, our networks also have the advantage of interference-free learning. Learning new patterns causes no interference with the previously trained patterns.

ACKNOWLEDGEMENTS

I want to thank Jo Decuyper for valuable discussions and comments on this paper. Ludo Cuypers helped me a lot in using the machines and text formatting.

REFERENCES

[1] Hinton, G.E., Connectionist Learning Procedures, Artificial Intelligence 40 (1989), pp. 185-234.

[2] Mel, B.W., MURPHY: A Neurally-Inspired Connectionist Approach to Learning and Performance in Vision-Based Robot Motion Planning, Center for Complex Systems Research, Beckman Institute, University of Illinois (1989).

[3] Hillis, W.D. and Steele, G.L.Jr., Data Parallel Algorithms, Comm. of the ACM, Vol. 29 (1986), No. 12, pp. 1170-1183.

[4] Abu-Mostafa, Y.S. and Psaltis, D., Optical Neural Computers, Scientific American, March 1987.

[5] Chen, C.C., GGT: A Connectionist Network for Path Planning and Obstacle Avoidance, AI-Lab., Free University of Brussels (1989).

[6] Chen, C.C., A Memory-Based Connectionist Network for Speech Recognition, AI-Lab., Free University of Brussels (1989).

[7] Lang, K.J. and Hinton, G.E., The Development of the Time-Delay Neural Network Architecture for Speech Recognition, Tech Report CMU-CS-88-152 (1988), Department of Computer Science, Carnegie Mellon University.

Parallel Processing in Neural Systems and Computers
R. Eckmiller, G. Hartmann and G. Hauske (Editors)
© Elsevier Science Publishers B.V. (North-Holland), 1990

ACCELERATING CONVERGENCE OF SYMMETRIC NEURAL NETWORKS ON A LOCAL MEMORY MULTIPROCESSOR

J. P. COUGHLIN and R. H. BARAN

Department of Mathematics
Towson State University
Towson, Maryland, USA

and Naval Surface Warfare Center

Symmetrically interconnected (Hopfield) neural networks are partitioned into subnets which are then assigned to concurrent processors and treated using the asynchronous (Glauber) dynamics. Numerical tests using Inmos Transputers support the hypothesis that acceleration is proportional to the number of processors (subject to mild restrictions.)

1. INTRODUCTION

Hopfield [1] portrayed a class of neural networks in terms of an electronic circuit in which high gain amplifiers feed back to each other through a symmetric array of coupling conductances. The circuit exhibits global asymptotic stability, meaning that the electronic variables always settle to fixed points at local minima of the computational energy, which is the Lyapunov function of the system of differential equations describing the ideal circuit [2].

When we try to simulate the future applications of these analog neural networks using digital computers, two considerations lead us away from the differential equations. First, the real circuit, characterized by component variablity, parasitics, noise and jitter, may not conform to the model. Second, the differential equations are harder to handle numerically than discrete dynamics. The Glauber dynamics is a logical departure, since it guarantees downhill motion in the energy landscape at zero temperature [3]. In addition, the statistical properties of such physical systems at nonzero temperature are governed by Glauber's "master equation" [4]. Recall too that Hopfield's original stochastic model [5] was implicitly based on these assumptions. We study the feasibility of accelerating the Glauber dynamics on local memory multiprocessor systems in the expectation of finding that speedup in proportion to the number of processors can be approached in large neural network calculations almost irrespective of the particular problem.

2. GLAUBER DYNAMICS

Let there be N McCulloch-Pitts neurons in a fully connected network described by the symmetric NxN weight matrix $\mathbf{W}$. The initial state (at time t=0) is $\mathbf{x}(0)$, where the components of the network state vector $\mathbf{x}(t) = (x_1(t),\ldots,x_N(t))$ are binary 0-1 variables.

The fixed, externally-applied inputs to the network are given by
$\mathbf{y} = (y_1,\ldots,y_N)$. At zero temperature, the dynamical equation is

$$x_i(t+1) = \begin{cases} \underline{1}\left[\sum_{j=1}^{N} W_{ij}x_j(t) + y_i\right] & \text{if } i = U(t) \\[2em] x_i(t) & \text{otherwise,} \end{cases} \tag{1}$$

where $\underline{1}[y] = [1+\text{sign}(y)]/2$ is the unit step and $\{U(t),\ t=0,1,2,\ldots\}$ is a sequence of independent random variables that are uniformly distributed among the first N integers.

2.1. Coordinated Isolation

Partition the fully interconnected network of N neurons into M subnets in order to distribute the computational burden among M processors. Let $n = N/M$ be the number of neurons in each subnet; and let S(m) be a set which contains the indices of all neurons in the mth subnet. Since every neuron belongs to exactly one subnet, use the notations

  m(i) = the subnet to which the ith neuron belongs and

  S[i] = the set of indices of all neurons sharing the same
         subnet as the ith neuron = S(m(i)).

The coordination phase. With input vector $\mathbf{y}$ and initial state $\mathbf{x}(0)$ as before, compute the net input received by the ith neuron from sources outside its own subnet:

$$\tilde{y}_i(1) = \sum_{j \notin S[i]} W_{ij}x_j(0) + y_i \tag{2}$$

This computation is carried out for every i in S(m) to initialize the mth subnet. Computation of $\tilde{\mathbf{y}}(1) = [\tilde{y}_1(1),\ldots,\tilde{y}_N(1)]$ completes the coordination phase. Since (2) involves N-n inner product steps (ips), initialization of S(m) takes (N-n)n ips. Because the processors work concurrently, the time required is proportional to $D_1 = (N-n)n$ ips.

The Isolated Phase. Each processor operates independently, randomly selecting and updating the neurons in its subnet, treating the outside units as if they were clamped to $\mathbf{x}(0)$. The duration of this isolated phase is λn steps, which is λ sweeps of the subnet. Call λ the coordination period and assume $\lambda \geq 1/n$. Note that λ is measured in sweeps per cycle. Its reciprocal is the coordination rate. For every i in S(m), the new dynamical equation is like (1) except that the summation is over all $j \notin S(m)$ and the sequence $\{U(t)\}$ is replaced by $\{U_m(t)\}$, which is uniform on S(m). This routine is followed by every processor, resulting in a new state $\mathbf{x}(\lambda n)$. The duration of the isolated phase is $D_2 = n^2\lambda$ ips, since each iteration of (3) requires n interconnects.

The Communication Phase. The amount of time spent in the communication phase will depend on the details of the multiprocessor. In a rectangular network of processors, such as we made with Inmos Transputers, the duration is dominated by the time needed to transmit information to the processors which occupy the corners of the network. Let r be the ratio of the time to transmit the state of a neuron to the time required by one interconnect. Then $D_3 = (M-1)nr/2 \approx Nr/2$ ips is the duration of the communication phase.

The communication phase of the first cycle replaces $\mathbf{x}(0)$ with $\mathbf{x}(\lambda n)$ in the local memory of each processor. Now the second cycle begins with a coordination phase in which $\tilde{\mathbf{y}}(2)$ is computed by substituting the components of $\mathbf{x}(\lambda n)$ for those of $\mathbf{x}(0)$ in the right side of equation (2), and so on. The duration of a cycle is proportional to

$$D = D_1 + D_2 + D_3 = n^2(M + \lambda - 1) + Nr/2 \quad \text{ips}.$$

2.2. Convergence Time and Acceleration

Every cycle witnesses the interrogation of $(\lambda n)M = \lambda N$ neurons. Defining a sweep as N interrogation steps, this is λ sweeps of the network. The mean time to convergence is $\mu(M,\lambda)$ sweeps, the dependence on other parameters suppressed. Since attainment of a stable state will go untested in the isolated phase, $\mu(M,\lambda)$ will be measured as $\lambda\bar{k}$, where $\bar{k}$ is the observed mean <u>cycles</u> to convergence: $\mu(M,\lambda) = \lambda\bar{k}(M,\lambda)$. Note that $\mu(1,1/N) \equiv \beta$ is the mean convergence time for the pure dynamics.

The multiprocessing scheme allows state transitions which go uphill in the energy landscape. Therefore we measure its <u>inefficiency</u> as $\zeta(M,\lambda) = \mu(M,\lambda)/\beta$.

Let $T(M,\lambda)$ be the mean convergence time (in ips). Then $T(M,\lambda) = D\bar{k}$. For the single processor, $T(1,1/N) = N^2\beta$. The multiprocessor accelerates convergence by a factor of

$$T(1,1/N)/T(M,\lambda) = \beta N^2/(D\bar{k})$$

$$= (\lambda/\zeta)M[1 + (\lambda-1)/M + Mr/2N]^{-1} .$$

For $N \gg Mr$ and $M \gg \lambda$, the acceleration increases roughly as λ/ζ times the number of processors.

3. EXPERIMENTS

We used a 4x4 rectangular network of Inmos T414 Transputers, each with 130 KBytes of local memory, hosted by an IBM AT compatible. Weights and states were represented with integer precision (4 bytes). Using 240 neurons with from 2 to 15 processors, the outer product rule was used to create $\mathbf{W}$'s from pattern sets of up to 10 random vectors. Each <u>trial</u> consisted in (i) generating a new pattern set, (ii) computing the components of $\mathbf{W}$, (iii) generating an initial state vector with the same statistics as the patterns, and (iv) cycling until convergence, which was tested between cycles. 128 trials were averaged to obtain the mean convergence time for parameter combinations of interest. Figure 1 shows the inefficiency as a function of coordination period with 12 and 15 processors, using the average of all the results for $M \leq 8$ for the baseline mean convergence time. The inefficiency is essentially independent of M. The fitted curve is

$$\zeta(\lambda) = 1 + \lambda^2/(\lambda + 3).$$

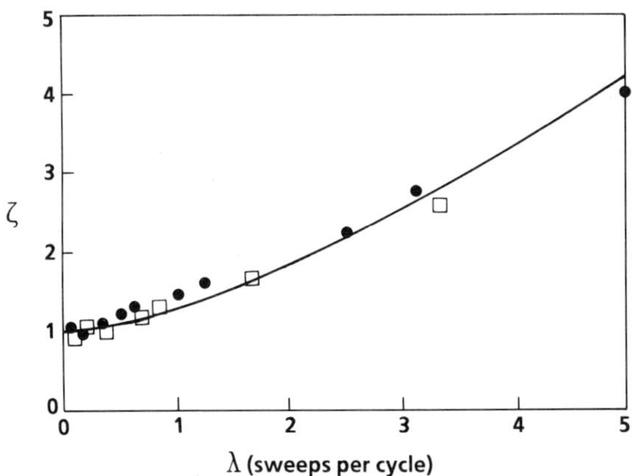

Figure 1. Inefficiency (ζ) versus coordination period (λ) using the transputer-based multiprocessor with M = 15 (solid circles) and M = 12 (open squares). Each plotted point represents the mean of approx. 128 trials.

4. CONCLUSIONS

The network state converges in several sweeps irrespective of the number of processors (so long as the processors are much fewer than the neurons). The acceleration gained by extending the isolated phase--to reduce the "run time" cost of coordinating the subnets--is not appreciable above three of four sweeps per cycle.

ACKNOWLEDGEMENT

This work was supported by the U.S. Office of Naval Research through the Naval Surface Warfare Center's Independent Research Program.

REFERENCES

[1] Hopfield, J.J., Neurons with graded responses have collective computational properties like those of two-state neurons, **Proc. Nat. Acad. Science USA** 81 (1984) 3088-3092.
[2] Cohen, M.A., and Grossberg, S., Absolute stability of global pattern formation and memory storage in neural networks, **IEEE Trans. Systems Man and Cybernetics** 13 (1983) 815-825.
[3] Crisanti, A., and Sompolinsky, H., Dynamics of spin systems with randomly asymmetric bonds: Ising spins and Glauber dynamics, **Physical Review A**, 37 (1988) 4865-4874.
[4] Glauber, R.J., Time-dependent statistics of the Ising model, **Journal of Mathematical Physics** 4 (1963) 294-307.
[5] Hopfield, J.J., Neural networks and physical systems with emergent collective computational abilities, **Proc. Nat. Acad. Science USA** 79 (1982) 2554-2558.

Parallel Processing in Neural Systems and Computers
R. Eckmiller, G. Hartmann and G. Hauske (Editors)
© Elsevier Science Publishers B.V. (North-Holland), 1990

WHY DO NEURAL NETWORK RESEARCHERS IGNORE STOCHASTIC COMPUTERS ?

Robert Massen

Transfer Centre Constance for Image Processing
Reichenaustr. 81c D-7750 Constance FRG

Most N.N. paradigms simulate the way by which biological neurons are supposed to *process* information. The inherent randomness of biological information *coding* as random spike discharges is mostly overlooked. Stochastic Computers, known since 1969, code information into the probability of occurence of a pulse in a *random pulse train*. This leads not only to a very noise-immune data encoding but allows as well to build extremely simple bit-serial arithmetic processing units with just a few logic gates . We remember the basics of stochastic computers and show up the application to the design of digital Stochastic Neural Network chips.

1. NON - BIOLOGICAL INFORMATION ENCODING IN ACTUAL ARTIFICIAL NEURAL NETWORKS

Most known N.N. simulate at their best what is assumed to be the principles of *information processing* in biological neuron systems: summation of weighted input signals and non-linear transfer function to produce an output which is spread to a large number of other neurons. No matter what technology is used for implementing N.N., be it with analog or digital electrical or optical signals, all these N.N. computers consists of a large number of multipliers, summers and non-linear transfer function elements.

The *coding of information* is however still done in complete opposition to the natural model. This one uses the short-time statistics of random bursts of pulse spikes, i.e. a *probabilistic mapping* of the input signal strength into a statistical parameter of a random carrier.The actual artificial N.N. however use a one-to-one *deterministic mapping* of the input signal strength into the amplitude of an analog or into the bit pattern of a digital electrical or optical signal.

Randomness appears in actual N.N. mostly as unpleasant noise or as a way to prevent a learning system to get trapped in a local optimum: weights are initialized to random values, noise is injected to keep adaptation going on. Stochastic Computers however, the roots of which go back as far as to von Neuman[1] and which had a burst of active research between 1964 and 1980 [2], [3], [4], [5] ,[6] rely on an inherent *probabilistic coding* of signals which comes very close to the biological model. Stochastic coding leads to extremely simple circuits for the basic arithmetic computing blocks, an attractive feature in the design of dense N .N. chips.This papers remembers a few basic principles of stochastic computers and sketches their possible use in integrated N.N. chips.

2. STOCHASTIC ENCODING OF INFORMATION

A stochastic encoder is basically a tunable random pulse generator. The probability of occurence for a pulse, i.e. the mean pulse rate, is controlled by the analog or digital input signal to be encoded in such a way, that

$$p(x) \; = \; U_{in}/U_{max} = D_{in}/D_{max} \qquad\qquad /1/$$

$p(x)$ is the probability that the binary random pulse train $x(t)$ $\{o,1\}$ assumes a value of 1 at time t; Uin is the amplitude of an analog voltage signal, Din the value of a digital code-word to be encoded. The probability of a pulse in the stochastic pulse train is thus proportional to the normalized input signal. A stochastic computer works on probabilities ,i.e. in the range of $\{o....1\}$. The random pulse train is in the simplest case of coding unipolar input signals. We will for case of simplicity confine to this simplest single-line encoding scheme in this paper. Fig. 1 shows the basic circuit for encoding an analog voltage signal and a digital multi-bit signal into a random pulse train with approbriate probability. Both the *Analog-to-Stochastic Converter* (ASC) as well as the *Digital-to-Stochastic Converter* (DSC) use a high-frequency noise source and a comparator.The random pulse output may be clocked or asynchronous. The noise generator should deliver successive pulses which are statistically independent. This stochastic encoding shows some unusual features:

1. in the same way as with regular pulse rate systems, it is an analog signal mapping (the continuous mean pulse rate) using a discrete binary carrier.
2. using non-weighted bits in a code of infinite word length, it is an extremely noise-proof system with graceful degradation in case of bit errors.
3 unlike classical digital systems, it is a system with adaptive accuracy. As information is recovered through a pulse counting process, we can at any moment decide for a fast but unprecise or for a slow but accurate response.
4. due to the random nature of the pulse train, basic arithmetic processors can be realized with just a few gates, allowing for the integration of a very large number of such processors on a single chip.

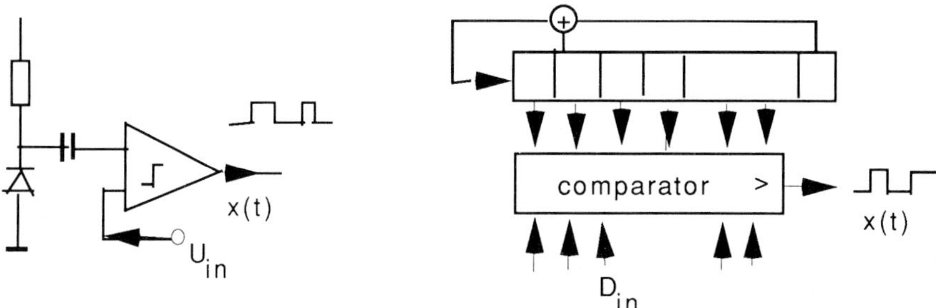

Analog-to-Stochastic Converter   FIGURE 1   *Digital-to-Stochastic Converter*

The analog or digital input signal is encoded into the probability of a random pulse train x(t) by comparing the input signal to an equiprobable analog (semi-conductor) or a digital (Pseudo-Random Binary Number) noise source

3. STOCHASTIC ARITHMETIC COMPUTING ELEMENTS

Probability is the parameter an input signal is mapped to. We may thus use the basic axioms on the probability of random joint and disjoint events as design rules for stochastic arithmetic circuits.
If we AND two statistically independent binary random pulses with probabilities p(x1) resp. p(x2), we obtain for the probability of the output of the AND gate

$$p(x3) = \mathbf{P}\,[x1\ \text{AND}\ x2] = p(x1) * p(x2) \qquad /2/$$

A multiplier in a stochastic computer is just a single AND-gate ! (Fig. 2 left)

Using the theorem on the union of disjoint events, we can design a processor for scaled summation of two stochastically encoded signals (Fig. 2 right)

$$p(x5) = \mathbf{P}\,[x3\ \text{OR}\ x4] = \mathbf{P}\,[(\,x1\ \text{AND}\ o.5)\ \text{OR}\ (x2\ \text{AND}\ o.5)]$$

$$= 0.5 * \{\,p(x1) + p(x2)\} \qquad /3/$$

A stochastic summer requires just 3 gates and a p= 0.5 noise source. Two AND multipliers act as multiplexers for achieving mutual exclusive pulse trains p(x3) and p(x4) which can summed by simple OR-ing.

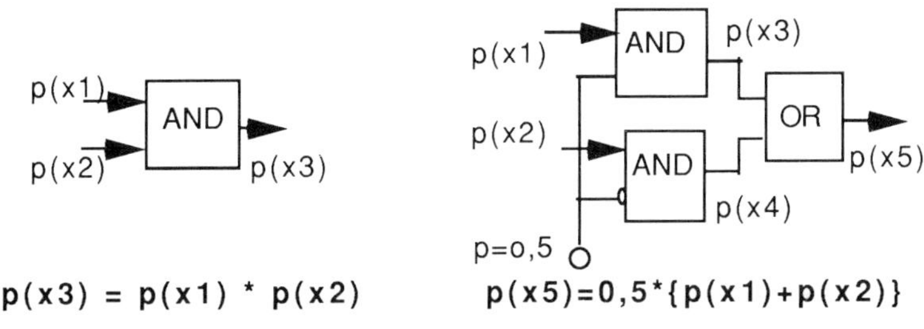

p(x3) = p(x1) * p(x2)   **p(x5)=0,5*{p(x1)+p(x2)}**

FIGURE 2

Stochastic multiplier (left)   and   stochastic summer (right)

Controllable non-linear transfer functions can be achieved with simple sequential circuits (Fig. 3) : an RS-flipflop with LL-inhibit generates a set of useful non-linear functions (see reference[2]):

$$p(Q) = \frac{p(S)*\{1-p(R)\}}{p(S)+P(R)-2*p(S)*p(R)}$$

FIGURE 3

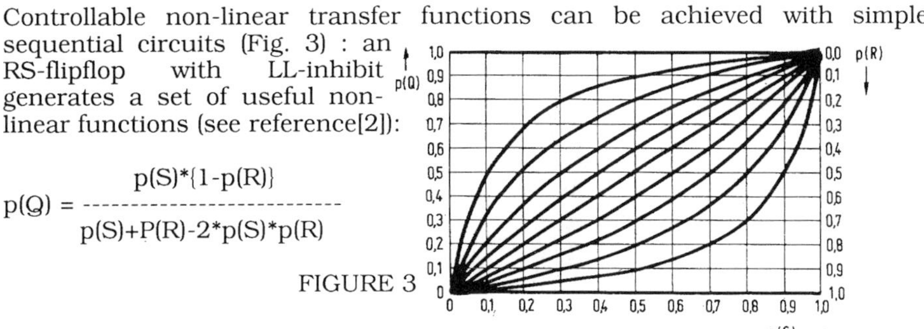

4. STOCHASTIC NEURAL NETWORKS (SNN)

The extremely simple hardware of stochastic computing elements and the inherent randomness of the carrier make them an interesting candidate for N.N. chips.Fig. 4 gives as an example a stochastic implementation of a single perceptron element for a 4-dimensional input feature vector **X** .Thanks to the bit-serial coding, the stochastic neurons are simple and large numbers can be integrated on a single chip.The conversion from classical analog or digital inputs to the stochastic coding will of course just take place once in a complete network and thus add not more complexity to a stochastic N.N. then analog-to-digital conversion does to a classical deterministic N.N.

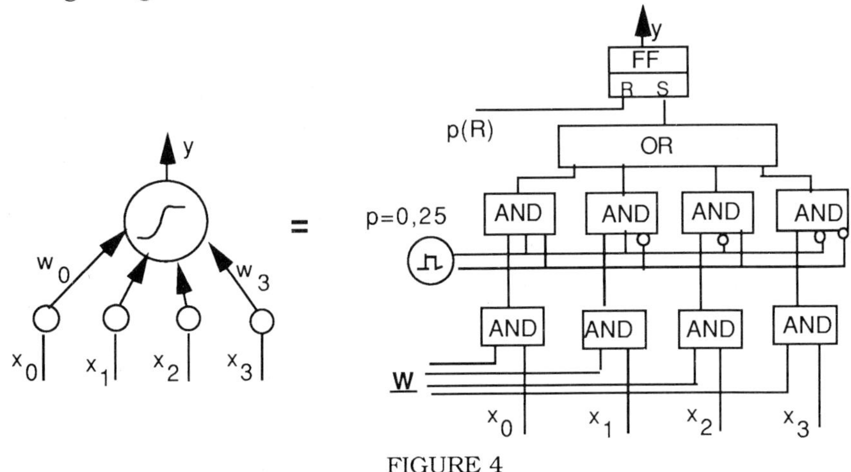

FIGURE 4

A stochastic perceptron neuron . All inputs are random pulses

The price for this simplicity is a rather low response time. The accuracy of the estimated pulse probability is proportional to $1/\sqrt{n}$; %- accuracy is typical for SNN . Ultimately, single photon gating would be the ideal random carrier for a stochastic signal encoding. These ideas have been dismantled 15 years ago by the well known argument, that stochastic computers are hardwired and not programmable. Their advantage in view of a connectionist computer architecture has not been recognized. Time for a revision has come.

REFERENCES

[1] Neumann,von   Probabilistic logics. Automata Studies, ed. C.E. Shannon, Princeton University Press 1956

[2] Massen, R.   Stochastische Rechentechnik. Hanser Verlag 1977

[3] Gaines, B.R.   Stochastic Computing Systems. in: Advances in Information Systems Sc. J.Tou, Plenum Press 1969

[4] Mars,P.   Stochastic and deterministic averaging processors.
Poppelbaum, W.J.   IEE Digital Electronics and Computing Series. Peregrinus Ltd. 1981

[5] Castanié, F,   Estimation de moments par quantification à référence stochastique. PhD Institut Polytec Toulouse 1977

[6] N.N   Proceedings of 1rst Int. Symposium on Stochastic Computing.org. Inst. Polytechnique de Toulouse 1978

Parallel Processing in Neural Systems and Computers
R. Eckmiller, G. Hartmann and G. Hauske (Editors)
Elsevier Science Publishers B.V. (North-Holland), 1990

A PROGRAMABLE HIGHLY PARALLEL ARCHITECTURE : FUNCTIONAL DEFINITION AND PERFORMANCE EVALUATION

Guy MAZARE, Eric PAYAN

Laboratoire de Genie Informatique
INPG
46 Avenue Félix Viallet
38031 Grenoble Cedex
France

With the recent development of massively parallel architectures (from large Transputer Network to the XILINX) there is a growing need for new programing methods. We propose here a new programing method which can be used to solve many problems as : neural network programing, digital signal processing[1], systolic array,...

This job is part of a larger project : a programmable highly parallel architecture which will also be presented.

1. INTRODUCTION

The realization of highly parallel architectures (more than 1000 processing elements) is now possible thanks to the progress of VLSI technology.

Many topologies and architectural designs for processing arrays (including globally synchronous systolic arrays [2] and globally asynchronous wavefront array [3]) have recently been proposed. Most existing algorithms for such arrays were developed for problems with inherent regularity (vector and matrix operations, FFT...). Many algorithms do not have underlying regularity and, therefore are not suitable for these arrays.

Classical low level programing inconvenients still exist on these architectures : poor instruction set, high debugging complexity, but also parallel processing add new problems : process synchronization, interlock... All these problems make the manual programing of a large network very difficult.

On a data-flow computer, data processing is represented as a graph, where each node is an operator which starts to compute when all his operands are available.

We suggest that mapping a Data Flow graph extracted from a program (written in the data-flow langage LUSTRE[4]) on a cellular network with global communications [5] could be an easy and efficient way to program highly parallel architectures.

2. THE HARDWARE CONCEPT

2.1 The network

The hardware architecture is a regular network of asynchronous cells. Each cell performs a simple local function and is surrounded by eight one-way-driving buffers, one in each way of the four directions. A flip-flop based mechanism performs the mutual exclusion of two cells sharing a buffer.

2.2 Message transmission mechanism

Each cell includes a routing mechanism which allows the transmission of the message to any

cell through the network. A message is made of two parts :
- an information field carrying the value intended to the application implemented on the network.
- a routing field carrying information required by the message transmission mechanism. The routing algorithm needs the relative displacement dx,dy to the addressed cell and a value for input register choice since non commutative operations can be implemented .

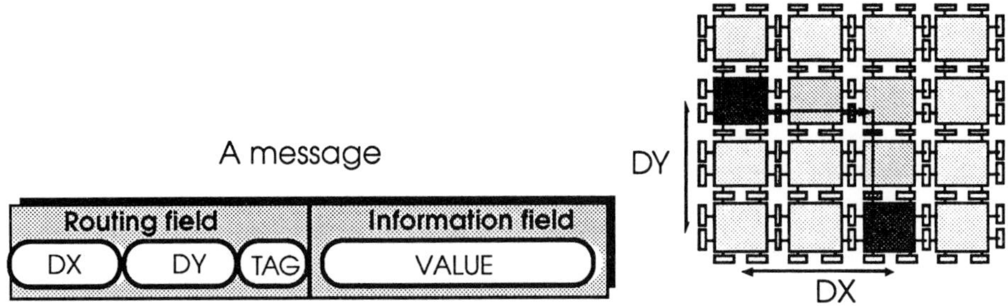

2.3 The cell

Each cell performs simultaneously two concurrent tasks :
- the routing of messages from input to output buffers
- the processing of its dedicated algorithm.
These two tasks communicate through buffers, which are managed by the routing part in the same way as the others.

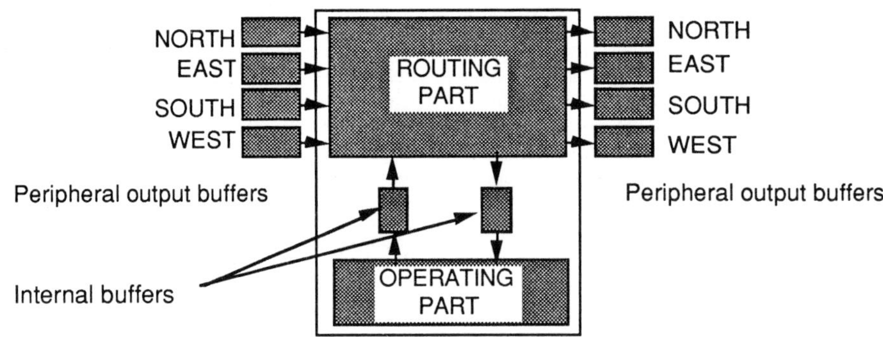

Computing a data flow program onto our array of processors consists of four phases:
1 : DF - program translation
2 : DFG mapping onto array
3 : Array initialization
4 : Program execution

The asynchronous cell operations and the variable length of the message paths introduce significant speed differences in the message routing . Some cells may receive their inputs and send their outputs faster than others. As we want to be sure that each message sent can be received, and because there is no FIFO file between cells, we have to synchronize them.
The use of acknowledge messages seems to be the best compromise between hardware complexity and maximum speed.
The algorithm implemented in the processing part of the cell is :

Send acknowledges

```
WHILE true DO
    Wait for operands
    Wait for acknowledges
    Send acknowledges
    Compute function
    Send result
END-WHILE
```
Some other applications using this network have already been studied by our team:
- logical simulation : a chip including 4 cells has been realized [6] .
- Image reconstruction : a chip has been realized[7]
- neural network with a delta learning rule [8]

3. PERFORMANCE STUDIES

3.1 Functional definition

We first used the parallel language OCCAM [9] running on the INMOS Transputer to simulate the network. The instruction set of our cell was defined and tested.This instruction set is very close to the LUSTRE language, the only differences are:
-Data in the network have no type (a boolean is only a special integer value)
-There is no special value for NIL so the implanted operators can not be strict with respect to NIL.
These problems can easily be solved: the strong semantic of LUSTRE allows a static verification for data types and NIL value.

3.2 Performance evaluation

For larger (already more than 1000 cells) and more precise network simulation we have programed a complete simulation tool, including : Data flow graph extraction, mapping (by a simulated annealing method) and network simulation.

As the network is asynchronous, each operator has a different computing delay (depending of its function). If we use the 1cell=1node; mapping faster cell will have to wait for the slower one. By mapping several "slow" nodes on one cell we can increase the network activity.

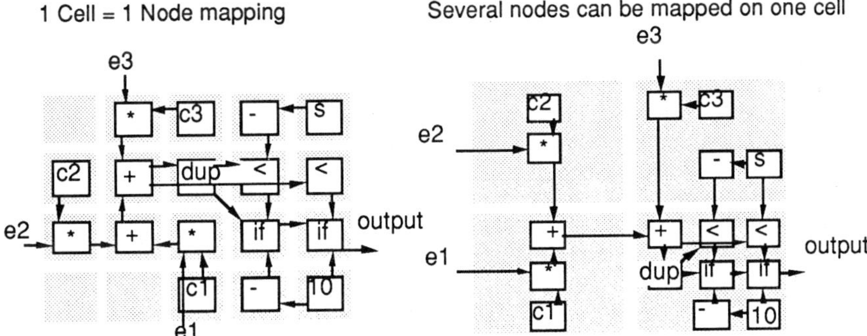

One of the main results of our simulation is the answer to the question : how many data-flow nodes shall we map on one cell to obtain the best result? we simulated a fast Fourier transform with several mappings (from 304 nodes on 4 cells to 304 nodes on 100 cells) and with several communication delays (proc/rout=4 means that the average delay needed by an operator to compute his function is 4 time slower than the delay needed to route a message from one side of the cell to the other one). We obtain the following result:

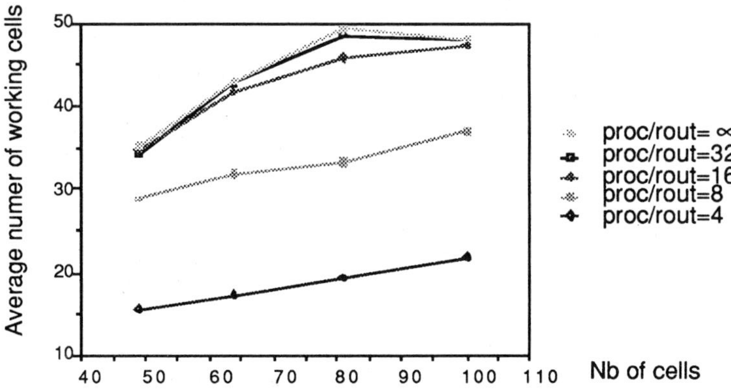

The best result appears to be obtained with a few nodes on each cell.

4. CONCLUSION

We believe that highly parallel architecture programing can be a real problem without specifi compilation technique (especially on non regular problems). With the complete compilation and simulation tool we had developed, we can already achieve the following goal:
- Independence between program and architecture (the same program can be mapped on various network size)
- The total mapping automatically done is totally automatized so that the programer does not have to worry about network topology, parallelism extraction...
- Many obtimization can be done (critical path detection ..) but the network utilization is already good.

REFERENCES

[1]. G. Mazaré E. Payan "A programmable Highly Parallel Architecture for Digital Signal Processing" 1989 IEEE Internatinal symposium on circuits and systems.
[2]. H.T. Kung "Why systolic architectures?," IEEE Computer, vol.15 January 1982
[3]. S.Y. Kung, "On Supercomputing with Systolique/Wavefront Array Processors" Proceeding of the IEEE, Vol 72 No7 july 1984.
[4]. J.A. Plaice, N. Halbwatchs "LUSTRE V2: User's Guide and Reference Manual" Laboratoire de Genie Informatique Octobre 1987.
[5]. P. Objois "Reseau de cellules integre: mecanisme de communication inter-cellulaire et application a la simulation logique." Thesis, Institut National Polytechnique de Grenoble septembre 1988.
[6]. R. Cornu-Emieux "Reseau de cellules integre : Etude d'architectures pour des applications de CAO de VLSI" Thesis, Institut National Polytechnique de Grenoble septembre 1988.
[7]. D. Lattard G. Mazaré "Image Reconstruction using an Original Asynchronous Cellular Array" 1989 IEEE Internatinal symposium on circuits and systems.
[8]. B. Faure, G. Mazaré "A VLSI Asynchronous Cellular Architecture Dedicated to Multilayered Neural Networks" proceedings of nEuro '88.
[9]. INMOS Limited "OCCAM programming manual" Prentice Hall, 1984

Parallel Processing in Neural Systems and Computers
R. Eckmiller, G. Hartmann and G. Hauske (Editors)
© Elsevier Science Publishers B.V. (North-Holland), 1990

A GENETIC ALGORITHM FOR MASSIVELY PARALLEL COMPUTERS

Piet SPIESSENS and Bernard MANDERICK

Artificial Intelligence Laboratory
Free University of Brussels
email: piet@arti.vub.ac.be
       bernard@arti.vub.ac.be

This paper describes a genetic algorithm for massively parallel computers. In this algorithm, individuals of the population are placed on a planar grid and selection and crossover are restricted to small neighborhoods on that grid. The presented algorithm which we call FG is compared with the standard genetic algorithm. We also discuss the performance of the algorithm as a function of the population size and as a function of the size of the neighborhoods on the grid.

1. INTRODUCTION

In Artificial Intelligence, the process known as *adaptation* is becoming increasingly more important. In Robotics, for instance, it is becoming clear that it is not possible to program a robot so that it can efficiently cope with every situation it will encounter. Therefore, a robot has to be able to adapt itself to its (changing) environment. More generally, many problems in Machine Learning can be viewed as adaptation problems.

While it is clear that these problems are very hard for AI researchers, the adaptation problem in the biological context has been successfully handled by natural evolution. Ever since the beginning of AI, this observation has inspired many researchers, but the first to successfully adopt the evolutionary paradigm was Holland [1] with his class of algorithms known as Genetic Algorithms.

Genetic Algorithms (GAs) are essentially optimization techniques based on concepts from population genetics. A GA is an iterative procedure which maintains a fixed-size population of candidate solutions. During each iteration step, called a generation, the structures in the current population are evaluated, giving each structure a so-called fitness value. Then, in a process akin to natural selection, each structure is reproduced in proportion to its fitness. In order to introduce new structures in the population, the population is then subjected to genetic operators. The most important of these operators is called crossover. Under the crossover operator, two structures in the new population exchange portions. Another genetic operator is the mutation operator, which randomly changes elementary bits of a structure.

Empirical studies have demonstrated the capabilities of GAs in many diverse areas including function optimization, scheduling problems and dynamic system control. In Machine Learning, GAs have been incorporated as the discovery component in Classifier Systems. More information on GAs and their applications can be found in the monograph by Goldberg [2].

2. PARALLEL GENETIC ALGORITHMS

The classical GA as introduced by Holland in the 70s is a sequential algorithm (from now on, we will call these sequential GAs R-algorithms, as in [1]). For some applications this poses some problems since an R-algorithm may require many generations and a large number of individuals in the population. Time limitations make it infeasible then to run the algorithm. To alleviate this problem this research investigates a parallel version of the algorithm.

The characteristics of GAs that make them difficult to parallelize are the selection process and the crossover operator. Selection is based on the distribution of the fitness over the *whole* population (i.e. selection depends on global information) and crossover is applied to *randomly* selected individuals.

Recently, results on parallel genetic algorithms have been reported, e.g. [3,4]. However, what essentially happens in the cited approaches is that a number of R-algorithms run in parallel and interact from time to time by exchanging individuals. In other words, these are coarse-grained parallel algorithms.

In contrast, we propose a fine-grained parallel algorithm (FG). Instead of coupling a number of R-algorithms, we parallelize the algorithm itself. The basic idea is to put the individuals of the current population on a planar grid and to restrict the selection and crossover operators to small neighborhoods on that grid. This way, selection in FG depends only on local information: to select individuals only the fitness-distribution in their neighborhood has to be known.

The advantage of a fine-grained parallel algorithm with only local communication is clear: it can easily be implemented on fine-grained massively parallel computers such as, for instance, the Connection Machine [5] or the DAP [6].

3. OUTLINE OF THE ALGORITHM

In the FG-algorithm, the individuals of the current population are put on a planar grid. Each individual is assigned a fixed-size neighborhood. The size of the neighborhood is determined by a parameter *range*. For instance, if range equals one then the neighborhood of a given individual i consists of nine individuals (this includes the i).

Since local selection and crossover are used, each step in the algorithm can be done in parallel for all individuals (see Figure 1). This makes the run time of the algorithm on a parallel computer independent of the population size. In contrast, the run time of an R-algorithm grows linearly with the size of the population. In the best case, the run time of an FG-algorithm is *population size* times faster than the run time of an R-algorithm.

```
PARFOR each grid element el in the grid DO

    Initialization : Randomly generate an individual to occupy el.
    Evaluation : Compute the fitness of individual(el)

    REPEAT

        Selection : Calculate the cumulative fitness-distribution in neighborhood(el, range).
                    Select according to this distribution an individual from
                    neighborhood(el, range) to become the new individual(el).
        Crossover : Randomly select an individual from neighborhood(el, range).
                    Crossover this individual and individual(el) with probability crossover rate.
                    Choose an individual from the offspring as the new individual(el).
        Mutation : Mutate individual(el) with probability mutation rate.
        Evaluation : Compute the fitness of individual(el).

    UNTIL <some end-criterion>
```

Figure 1 : Outline of the FG-algorithm.

4. COMPARISON BETWEEN R-ALGORITHMS AND FG-ALGORITHMS

In this section we compare our parallel algorithm with the standard sequential algorithm. The syntax we will use for both algorithms is FG(*population-size, crossover-rate, mutation-rate, range*) and R(*population-size, crossover-rate, mutation-rate*), respectively.

We compare both algorithms using the online, offline and best-so-far performance measures [7]. However, since each generation is evaluated in parallel in FG, we cannot use the standard performance measures for GAs. These standard online and offline performance are defined as the average of the function evaluations so far and the average of the best function evaluations so far, respectively. Instead, we average over the mean fitness and best fitness per population. The online and offline performance at generation T are now defined as

$$\text{Online(T)} = \sum_{t=0}^{T} \frac{\text{mean(t)}}{T}$$

where mean(t) is the mean fitness of the population at generation t, and,

$$\text{Offline(T)} = \sum_{t=0}^{T} \frac{\text{best(t)}}{T}$$

where best(t) is the fitness of the fittest individual of the population at generation t.

4.1. A First Comparison

There is a lot of experimental evidence [7,8] concerning the optimal parameter settings of R-algorithms for both the online and offline performance. For instance, a crossover rate of 0.6, a mutation rate of 0.001 and a population size of 50 are a good first choice. In a first experiment we compared FG(49, 0.6, 0.001, 1) and R(49, 0.6, 0.001). We have used a population size of 49 instead of 50 because we work with square grids in FG. Both algorithms are tested on function minimization tasks. The test functions are four of the five standard test functions used in [7].

The performance evaluation proceeded as follows: each FG and R was tested on the 4 test functions over 100 generations for a number of runs. For each run, both algorithms were given the same initial population. This way, we reduced the chance that one algorithm was favored by an unusual initial population. To deal with the variance in the performances of the individual runs, we averaged the performances over 20 runs. The performance measures at generation 100 are shown in Figure 2.

Function	Performance	FG	R
f1	Online	2.833296	2.709526
	Offline	0.019216	0.019112
	Best-so-far	0.000046	0.000056
f2	Online	47.7812	46.6265
	Offline	0.3032	0.282
	Best-so-far	0.2573	0.2004
f3	Online	22.186295	21.620747
	Offline	11.0415	10.8205
	Best-so-far	10.85	10.15
f5	Online	168.94079	173.65917
	Offline	14.42135	16.19737
	Best-so-far	9.28335	9.62393

Figure 2. The performance of FG(49, 0.6, 0.001, 1) and R(49, 0.6, 0.001) on f1, f2 , f3 and f5.

For the first three functions, the performances of FG are worse (except for the best-so-far performance of f1). For f5, the performance of FG is clearly better. For this function with its small peaks, R gets easily stuck in one of the suboptimal peaks. FG, on the other hand, with its distributed population, explores a number of peaks in parallel.

Here, we have an indication of an additional advantage of our algorithm. FG seems to be able to explore more thoroughly the search space. For complicated functions, like f5 and most of the functions encountered in the real world, this is certainly necessary. For simpler, unimodal functions, like the first two ones, this extensive exploration is not necessary.

4.2. Comparison on a Genetically-hard Function

The first section has shown that FG-algorithms can compete with R-algorithms on DeJong's testbed of functions. But what about more complicated functions? To find out, we have taken a Walsh function W also used by Tanese [4]. W is a genetically-hard function [9], and it has a huge search space (the domain of W is 2^{64}).

We have tried out a number of FG-algorithms and FG(400, 0.65, 0.001, 1) showed a good best-so-far performance. The best-so-far performances of this algorithm and of R(400, 0.6, 0.0078125) and R(400, 0.6, 0.001) are displayed in Figure 3. The first of the two R-algorithms was chosen because it is used by Tanese and the second R-algorithm has the same mutation rate as the FG-algorithm. Recall that we use pure selection while Tanese uses the elitist strategy, i.e. the best individual is preserved in the next generation. This explains why our results for R(400, 0.6, 0.0078125) differ from Tanese's results. The displayed data are the average of 15 runs and each run took 1000 generations.

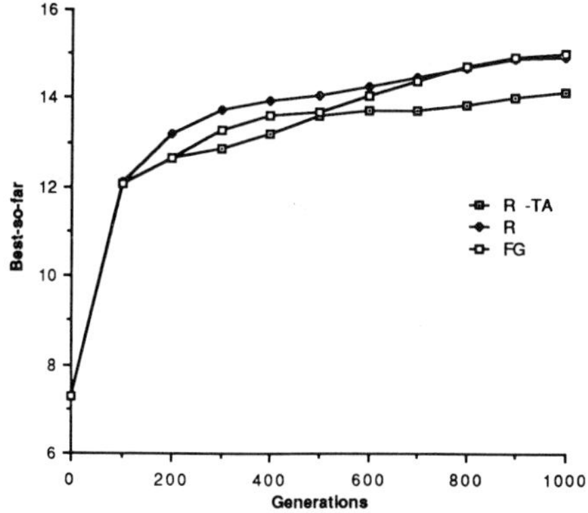

Figure 3. The best-so-far performances of FG(400, 0.65, 0.001, 1) (FG in the figure), R(400, 0.6, 0.0078125) (R-TA in the figure), and R(400, 0.6, 0.001) (R in the figure) on the Walsh-function W.

4.3. Discussion

Finding a proper balance between exploration and exploitation is the determining factor in the performance of genetic algorithms. Too much exploitation may result in premature convergence. Especially for the offline performance it is important that the search space is explored sufficiently, certainly if the function is multimodal.

Since FG-algorithms had a better offline performance on the multimodal functions f5 and W than comparable R-algorithms, we hypothesize that FG-algorithms explore the search space more fully than R-algorithms. We believe this because the selection pressure in the FG-algorithm is less than in the R-algorithm. And this is a consequence of local selection: good individuals at one place can only be propagated to neighboring places on the grid. In contrast, in the R-algorithm, good individuals can take over a large part of the population in one step.

5.　FG AND THE SIZE OF THE POPULATION

Goldberg [10] has developed a theory of optimal population size for binary-coded genetic algorithms. His theory shows that the optimal size is an exponentially growing function of the length of the individuals, e.g. the optimal size for f1 is 116, for f2 is 51, for f3 is 2240, and for f5 is 262. Besides that, the theory seems to imply that it is not useful to work with larger population sizes. Or even, that the performance could get worse when the size increases. The latter has been confirmed by Pettey et al. [3]. With a coarse-grained parallel genetic algorithm, they got optimal performances for f1, f2, f3 and f5 with population sizes near the predicted optimal sizes and no significant improvements were obtained for larger populations.

In a number of experiments, we have investigated the dependence of the performance of the FG-algorithm on the population size. We have found that there probably is no such thing as an optimal population size for FG. The offline and best-so-far performances improved clearly for each test function as the population size increased. Figure 4a shows the offline performance for f3 for population sizes of 49, 100, 196, 400, 784 and 1024 individuals. In our experiments, the online performance showed little variance with an increasing population size. Consequently, the population size should be chosen as big as possible in FG. Robertson [11] obtained similar results with his parallel genetic algorithm. Recall also that on a parallel machine, the run times for all these population sizes are the same.

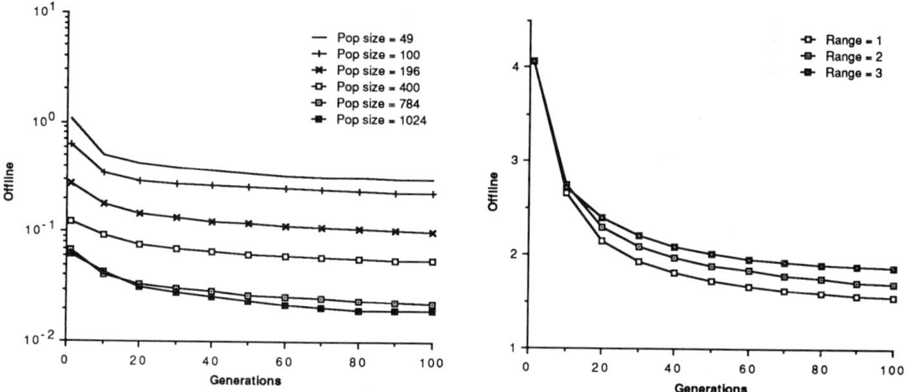

Figure 4. (a) Offline performance for f3 for population sizes 49, 100, 196, 400, 784 and 1024. (b) Offline performance for f5 with population size 1024. The range of the neighborhood varies from 1 to 3.

6.　FG AND THE RANGE OF THE NEIGHBORHOODS

The run time of a parallel implementation of the FG-algorithm is a function of the range parameter: the larger the range, the longer the run time, and for a very large range, the FG-algorithm reduces to the R-algorithm. Therefore, it is of interest to study the effect of the range on the performance of the FG-algorithm.

We tried out 3 different range values: 1, 2 and 3 (with neighborhood sizes of 9, 25 and 49 individuals, respectively) on populations with sizes of 100, 196, 400, 784 and 1024 individuals. Test function f1 was excluded from the experiments since FG always found the maximum value within a few generations.

Since f2 and f3 do not seem to need the extra exploration FG offers, we expected performance measures for these functions to increase with the range. And indeed they did, but not significantly. For a population size of 784 individuals we even obtained best offline and best-so-far performances with range 2. In general however, these measures certainly got better as the neighborhood range increased and the online performance again seemed unaffected.

For f5, we expected the contrary: better performances with smaller neighborhoods. The experiment confirmed this assumption since we got the best offline and best-so-far performance with range 1 (again, online performance showed little variance). Figure 4b shows the offline performance for f5 with a population size of 1024 individuals while the range of the neighborhood varies from 1 to 3.

From the neighborhood experiments we can conclude that FG is certainly able to obtain satisfactory results with moderately sized neighborhoods.

7. CONCLUSION

Genetic Algorithms are a powerful abstraction of the adaptation process as found in natural evolution. In practice however, their applicability is limited by their sequential nature. Therefore, we introduced a fine-grained parallel genetic algorithm called the FG-algorithm. Its most important characteristics are local selection and local crossover. As a consequence of local selection, there is less selection pressure and a tendency towards more exploration of the search space. We tested the FG-algorithm in a number of environments and compared it with the R-algorithm. We also studied the relationship between performance and range of the neighborhoods and between performance and the size of the population. The most attractive feature of the algorithm is that it can easily be implemented on massively parallel computers. Since the run time will be independent of the population size in such an implementation, this opens interesting perspectives for experimenting with very large populations.

ACKNOWLEDGEMENTS

Thanks to professor Luc Steels, the director of the AI Lab, for initiating us in the domain of complex dynamics and selectionism. We also wish to thank Tony Bell for his valuable comments. Piet Spiessens is sponsored by IMPULS and Bernard Manderick by IWONL.

REFERENCES

[1] Holland, J.H., Adaptation in Natural and Artificial Systems (University of Michigan Press, Ann Arbor, 1975)

[2] Goldberg, D.E., Genetic Algorithms in Search, Optimization, and Machine Learning (Addison-Wesley, Reading, 1989)

[3] Pettey, C.P., Leuze, M.R. and Grefenstette, J.J., A Parallel Genetic Algorithm, in: Grefenstette, J.J. (ed.), Proceedings of the Second International Conference on Genetic Algorithms (Lawrence Erlbaum, Hillsdale, 1987)

[4] Tanese, R., A Parallel Genetic Algorithm for a Hypercube, in: Grefenstette, J.J. (ed.), Proceedings of the Second International Conference on Genetic Algorithms (Lawrence Erlbaum, Hillsdale, 1987)

[5] Hillis, D., The Connection Machine (MIT Press, Cambridge, 1986)

[6] DAP Series Technical Overview (Active Memory Technology Ltd, Reading, 1988)

[7] DeJong, K.A., An Analysis of the Behavior of a Class of Genetic Adaptive Systems, Ph.D. Dissertation, University of Michigan, Ann Arbor (1975)

[8] Grefenstette, J.J., Optimization of Control Parameters for Genetic Algorithms, IEEE Transactions on Systems, Man, and Cybernetics, Vol. SMC-16, No. 1 (1986)

[9] Bethke, A.D., Genetic Algorithms as Function Optimizers, Ph.D. Dissertation, University of Michigan, Ann Arbor (1981)

[10] Goldberg, D.E., Optimal Initial Population Size for Binary-coded Genetic Algorithms, Technical Report TCGA No. 85001, University of Alabama, Tuscaloosa (1985)

[11] Robertson, G., Population Size in Classifier Systems, in: Proceedings of the Fifth International Conference on Machine Learning (Morgan Kaufmann, San Mateo, 1988)

Parallel Processing in Neural Systems and Computers
R. Eckmiller, G. Hartmann and G. Hauske (Editors)
© Elsevier Science Publishers B.V. (North-Holland), 1990

A Transputer System for the Recognition of Human Faces by Labeled Graph Matching[*]

Rolf P. Würtz [†] *Jan C. Vorbrüggen*[†] *Christoph von der Malsburg*[‡]

Abstract

The method of labeled graph matching has been introduced as an extension to both associative memory and classical neural networks. The concept's ability to tackle cognitive tasks is demonstrated here with a programme that recognizes images of human faces. Its parallel implementation on a network of transputers in OCCAM achieves recognition times below one minute.

1   General method

We are going to give only a very brief overview of the idea of labeled graph matching here. For an in-depth discussion, see [1, 2, 3]. Abstract objects are represented by labeled graphs. Once edge and vertex labels have been chosen in a way that they describe the essential features of an object while discarding unimportant ones, the problem of recognition is reduced to graph matching. Unlike with conventional neural networks, relations between objects can again be coded as labeled graphs, without any need for new cells. This is an answer to the serious problem of combinatorial explosion.

Our system for face recognition is meant to demonstrate the capabilities of this concept, which is in no way limited to vision. Currently, it should not be regarded as a mature mechanism for object recognition because it still lacks many ingredients, most prominently figure-ground separation and invariance to lighting conditions.

In contrast to the style of AI, our system needs no specialization to the application domain. It has been tested with equal success on the recognition of faces and of vehicles.

2   Image representation

A crucial part of the method is the right choice of the vertex labels. We transform the image $I(\vec{x})$ by convolution with so-called Morlet wavelets [4]. Their definition, continuous Fourier transform and the related linear operator are as follows ($\vec{x}, \vec{\omega}, \vec{x_0}, \vec{k} \in \mathbf{R}^2$):

$$\psi_{\vec{k}}(\vec{x}) := n_{k,\sigma} \, \exp\left(-\frac{\vec{k}^2 \vec{x}^2}{2\sigma^2}\right) \exp\left(i\vec{k}\vec{x}\right) \tag{1}$$

$$(\mathcal{F}\psi_{\vec{k}})(\vec{\omega}) = m_{k,\sigma} \, \exp\left(-\frac{\sigma^2(\vec{\omega} - \vec{k})^2}{2k^2}\right) \tag{2}$$

$$(\mathcal{W}I)(\vec{k}, \vec{x_0}) := \int \psi_{\vec{k}}(\vec{x_0} - \vec{x})\, I(\vec{x})\, d^2x \ . \tag{3}$$

[*]This work was made possible by a grant from the German Ministry of Science and Technology (ITR-8800-H1). Contributions were made from grant 88-0274 by AFOSR.

[†]Max-Planck-Institut für Hirnforschung, Deutschordenstr. 46, D–6000 Frankfurt 71, F. R. G.
[‡]Dept. of Computer Science, University of Southern California, Los Angeles, CA 90089–0782, USA

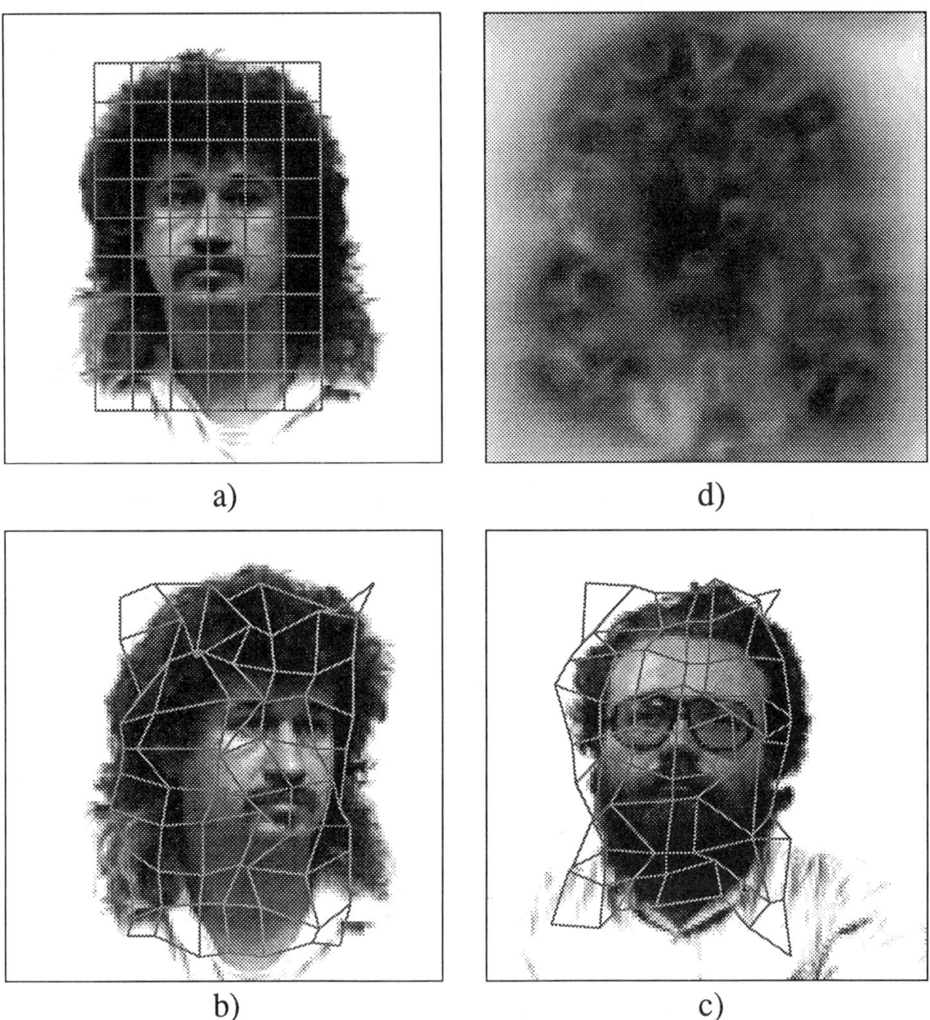

a)   d)

b)   c)

Figure 1: Objects, images, and graphs. **a)** Sample object: Picture of a face, overlayed with the graph of stored jets. (For visualization, only nearest neighbour connections are shown; the actual graph is a complete one.) **b)** Picture of the same person, but looking in a different direction. The overlayed graph is the final result of the matching process with the object in a). **c)** Picture of a different person and the graph matched to a). In our system, this process is repeated for a number of stored objects, and the one giving the best value of the minimized cost function is selected. Final values of C_{total} for a), b), c) are -70.0, -57.7 and -55.8, respectively. **d)** A jet from the tip of the nose in a) was selected and compared with all jets of the same picture. The result is displayed as a grey-level image, black coding for the best match. This gives an impression of the "cost surface" which drives the minimization process.

This family of functions is well adapted to our problem for the following three reasons:

- They are optimally localized in both space and frequency domain [5].
- The shape of the wavelets and the process of convolution are justified by neurophysiological results (see, e.g., [6, 7]). Burr *et al.* build a vision model based on the squared sum of the responses of pairs of simple cells with even and odd symmetry (absolute value of the complex convolution) and provide psychophysical evidence for the applicability of this model to human vision [8].
- The self similarity inherent in the definition of wavelets [4] makes the transform commute with translations, rotations and scaling of the image domain, although this is, of course, limited by the sampling density.

In order to improve matching behaviour, we take the modulus of $\mathcal{W}I$ and normalize within frequency levels (ϕ is the direction angle of $\vec{k}$):

$$\mathcal{J}I(\vec{k}, \vec{x}_0) = \frac{|\mathcal{W}I|}{\int |\mathcal{W}I|\, d^2x\, d\phi}\,. \tag{4}$$

The restriction of $\mathcal{J}I$ to some fixed $\vec{x}_0$ is a local descriptor of the image near this point and will be referred to as a *jet*. These jets provide the vertex labels for our graphs, the vertices themselves being points in the image. The vertex set will be denoted by V, the edge set by E.

3 Object comparison

Similarity between jets is measured by means of a function $\mathcal{S}(\mathcal{J}^O, \mathcal{J}^I)$. Currently we are using one of the simplest choices, the cosine of the angle between two jets, as derived from the scalar product. Similarities are added up over all pairs of corresponding points in the two graphs. The total cost function contains in addition a topology term:

$$C_{total} := \lambda C_{top} + C_{jet} = \lambda \sum_{(i,j)\in E} (\vec{\Delta}^I_{(i,j)} - \vec{\Delta}^O_{(i,j)})^2 - \sum_{i\in V} \mathcal{S}(\mathcal{J}^O, \mathcal{J}^I)\,. \tag{5}$$

$\vec{\Delta}^I_{(i,j)}$ is the vector connecting vertices i and j in the image domain, $\vec{\Delta}^O_{(i,j)}$ in the object domain. The factor λ controls the relative weight of the topology constraint. C_{total} is minimized by using the Metropolis algorithm at temperature zero. We terminated minimization once a certain number of consecutive moves failed to improve it. For examples of resulting graphs, see figure 1.

4 General Purpose Farm Software

We implemented the programme on a network of transputers, using the OCCAM 2 language. Grey-level pictures are acquired with a transputer-controlled frame grabber. The network consisted of up to 16 T800s.

The three main components — two-dimensional FFT, Morlet transform, and matching the object database to an image — all consist of a number of independent tasks which operate on a subset of the data. This makes them well suited to parallelization using the farm concept, where a controller hands out tasks to any number of identical, anonymous workers.

The typical chain of events when using a farm is as follows:

- The controller asks the user what to do next.
- It broadcasts any global (read-only) data required to all workers. This will include information on what action was requested.
- A "run" is started, which consists of a set of tasks. Each worker is handed a task; when it has returned the results, it is given a new one, until there are no more tasks available. At this point, the controller waits for any remaining results to come in.

This sequence is repeated until the programme is terminated. Note that one user-requested action may require a number of runs by the farm.

To make life easier for the developer, we designed a set of routines which implement a general purpose farm. The user just provides two modules (one for the controller, one for the workers) which use a simple and safe interface via OCCAM channels. On the controller, the farm software takes care of distributing tasks to the workers and collecting the results, routing messages as necessary. It also gives the user control of acquisition and display of performance data (see section 6). On the worker processors, the farm software provides an interface to the routing system and handles the read-out of monitoring data. It also allows the reception and buffering of the next command from the router while the previous one is being worked on; thus, an optimal overlap of computation and communication is achieved.

5   Network Topology and Routing Software

Suited to the structure of a farm is some sort of tree topology, with the controller at its root. In order to achieve maximal flexibility, the router system supports any tree-like structure, the only limitation being the number of transputer links. A network reconfiguration can be accomplished without changes to the user software, only the network description has to be modified.

After initialization is complete, each router runs two parallel processes. The **up** process receives messages coming in from all active "down" links and from the local worker process. It then sends these messages on its "up" link. The **up** process on the controller hands all incoming messages to the farm software, which passes them on to the user. The **down** process receives message from the "up" link (or from the farm software, in case of the controller processor). It examines the address field of every message and then passes it on to either the local worker or, after examining its routing table, to the appropriate "down" link. If the address has the reserved value 0, it is a broadcast packet and is sent down all active links and to the local worker.

6   Performance Figures

The Morlet transform (3) was based on a set of kernels with 6 values of $|\vec{k}|$, spaced by half octaves, each frequency level containing 8 orientations differing by 22.5°, for a total of 48 convolutions. They were computed using the relation

$$\psi_{\vec{k}} * I = \mathcal{F}^{-1}(\mathcal{F}\psi_{\vec{k}} \cdot \mathcal{F}I). \tag{6}$$

Table 1 gives some performance indicators for our programme. The two-dimensional FFT was parallelized as two runs, one performing all one-dimensional row FFTs, followed by all column FFTs. In the Morlet transform, it is used to compute $\mathcal{F}I$, and every worker calculates $\mathcal{F}\psi_{\vec{k}}$ from the analytical formula (2). Every task then consists of multiplication with the appropriate kernel, the inverse FFT, and normalization (see (4)) of the result. Elapsed time was measured using the transputer clock. To determine load balancing, a statistical programme counter sampling technique was used. The load on the controller, measured in this way, is reported in table 1.

Timing for the graph matching phase depends strongly on the objects and their number. As a reference, a single Monte Carlo step takes $\sim 2.5\,\mathrm{ms}$. One match requires between 2500 and 5000 steps, of which some 10% improve C_{total}. Matching a database of 30 objects on 12 transputers thus takes about 20 seconds.

Table 1: Performance of parallel FFT and Morlet transform implementation for different numbers N of worker processors. A sampling interval of 3 ms was used; this has an overhead of less than one percent. **FFT**: Times given are for both forward and inverse FFT of an image of size 128×128 pixels containing a vertical stripe pattern. The serial part is calculated as 0.08 s. Note the impact of increased controller utilization on the relative speedup, especially for 12 processors. It limits the useful network size to about 16 processors. **Morlet transform**: Times given include the time to perform the forward FFT of the data. Measurement conditions are as above. The serial part is calculated as 1.37 s. In this case, the bottleneck is the communication of the results from the workers to the controller.

	128×128 FFT			Morlet Transform		
N	Elapsed Time (s)	Controller Load (%)	Relative Speedup	Elapsed Time (s)	Controller Load (%)	Relative Speedup
1	4.975	8.2	1.00	153.05	2.5	1.00
2	2.489	16.4	1.00	76.53	4.9	1.00
4	1.285	32.2	0.97	38.95	9.8	0.98
6	0.884	44.8	0.93	26.66	14.6	0.95
8	0.689	56.7	0.89	20.70	18.8	0.92
12	0.558	76.9	0.65	15.47	26.0	0.79

Acknowledgments: We would like to thank Wolf Singer for providing excellent working conditions in Frankfurt. Peter König, Jörg Lange, and Joachim Buhmann contributed to this work through innumerable hints and discussions.

References

[1] J. Buhmann, J. Lange, C. v. d. Malsburg, J. C. Vorbrüggen, and R. P. Würtz. Object recognition in the dynamic link architecture — parallel implementation on a transputer network. In B. Kosko, editor, *Neural Networks: A Dynamical Systems Approach to Machine Intelligence*. Prentice Hall, New York, 1990. In print.

[2] Christoph von der Malsburg. Pattern recognition by labeled graph matching. *Neural Networks*, 1:141–148, 1988.

[3] Christoph von der Malsburg. The correlation theory of brain function. Technical report, Max-Planck-Institute for Biophysical Chemistry, Postfach 2841, Göttingen, FRG, 1981.

[4] J.M. Combes, A. Grossmann, and Ph. Tchamitchian, editors. *Wavelets, Time-Frequency Methods and Phase Space*. Springer, Berlin, Heidelberg, New York, 1989.

[5] D. Gabor. Theory of communication. *J. Inst. Elec. Eng. (London)*, 93:429–457, 1946.

[6] S. Marčelja. Mathematical description of the responses of simple cortical cells. *Journal of the Optical Society of America*, A 70(11):1297–1300, 1980.

[7] J.P. Jones and L.A. Palmer. An evaluation of the two-dimensional gabor filter model of simple receptive fields in cat striate cortex. *Journal of Neurophysiology*, pages 1233–1258, 1987.

[8] D.C. Burr, M.C. Morrone, and D. Spinelli. Evidence for edge and bar detectors in human vision. *Vision Research*, 29(4):419–431, 1989.

Parallel Processing in Neural Systems and Computers
R. Eckmiller, G. Hartmann and G. Hauske (Editors)
© Elsevier Science Publishers B.V. (North-Holland), 1990

CORDIC Processor Arrays for Adaptive Least Squares Algorithms

B. Yang and J.F. Böhme

Department of Electrical Engineering
Ruhr University Bochum, 4630 Bochum, F.R.G.

This paper is concerned with highly concurrent and numerically stable adaptive least squares (LS) algorithms and parallel implementations. In particular, a multichannel least squares lattice (MLSL) algorithm that solves a family of adaptive LS problems is developed. Array architectures based on CORDIC processors for parallel implementations are presented.

1. INTRODUCTION

Adaptive LS algorithms have received much attention over the past decade for applications in system identification, spectrum estimation, speech processing, echo cancellation, channel equalization, beamforming and direction finding [1]. They require a large amount of computations. For multichannel (m channels) linear prediction and FIR filtering (p taps), for example, $O(pm^2)$ operations every data sample have to be computed. On the other side, many applications in real time signal processing systems demand a high computation speed. Solutions to this problem can be found by using massively parallel/pipelined computing architectures, i.e. high performance vector computers, transputer systems, systolic arrays etc. In this paper, we present a CORDIC processor array to implement a MLSL algorithm for solving the multichannel prediction and filtering problem mentioned above.

2. ADAPTIVE LS ALGORITHMS

There are different adaptive LS problems. Using the conventional linear combiner, we wish to estimate $x(t)$ by a linear combination of $y_1(t), y_2(t), \cdots, y_m(t)$ in the LS sense. In single channel FIR filtering, $x(t)$ should be estimated by a linear combination of $y(t), y(t-1), \cdots, y(t-p+1)$. Both problems are special cases of the multichannel FIR filtering problem. In this paper, we only consider the general case.

We assume that the samples of m reference signals $\underline{y}(t) = [y_1(t), y_2(t), \cdots, y_m(t)]'$ at discrete times $t \geq 1$ being available. The purpose is to design a m-channel p-tap FIR filter to estimate a desired signal $x(t)$ in the LS sense. To be more specific, the weight vector $\underline{w}_p(t)$ of mp filter coefficients is to be chosen to minimize the sum of residual squares $\sum_{i=1}^{t} \lambda^{t-i}[x(i) - \underline{w}_p'(t)\underline{y}_p(i)]^2$, where $\underline{y}_p(t) = [\underline{y}(t)', \underline{y}(t-1)', \cdots, \underline{y}(t-p+1)']'$ is the data vector involved and $0 < \lambda \leq 1$ is an exponentially weighting factor. This problem is known to be closely coupled with linear predictions in which we estimate $\underline{y}(t+1)$ (forward prediction) and $\underline{y}(t-p)$ (backward prediction) by linear combinations of the same data. Let $\underline{f}_p(t+1), \underline{b}_p(t)$ and $e_p(t)$ denote the forward prediction, backward prediction and joint process estimation errors, i.e. the differences between the estimated and actual values of $\underline{y}(t+1), \underline{y}(t-p)$ and $x(t)$, respectively. The following set of time and order lattice recursions for recursively computing $\underline{f}_p(t+1), \underline{b}_p(t)$ and $e_p(t)$ have been derived [2,3],

Prediction

$$\text{Forward prediction error}: \quad \underline{f}_p(t) = \underline{f}_{p-1}(t) - \Delta_{p-1}(t)\mathbf{C}_{p-1}^{-b}(t-1)\underline{b}_{p-1}(t-1), \quad (1)$$

$$\text{Backward prediction error}: \quad \underline{b}_p(t) = \underline{b}_{p-1}(t-1) - \Delta_{p-1}'(t)\mathbf{C}_{p-1}^{-f}(t)\underline{f}_{p-1}(t), \quad (2)$$

$$\text{Forward error covariance}: \quad \mathbf{C}_p^f(t) = \lambda\mathbf{C}_p^f(t-1) + \underline{f}_p(t)\underline{f}_p'(t)/\gamma_p(t-1), \quad (3)$$

$$\text{Backward error covariance}: \quad \mathbf{C}_p^b(t) = \lambda\mathbf{C}_p^b(t-1) + \underline{b}_p(t)\underline{b}_p'(t)/\gamma_p(t), \quad (4)$$

Cross error covariance : $\quad \boldsymbol{\Delta}_p(t) \;=\; \lambda\boldsymbol{\Delta}_p(t-1) + \underline{f}_p(t)\underline{b}'_p(t-1)/\gamma_p(t-1)\,,$ $\qquad(5)$

Likelihood variable : $\quad \gamma_p(t) \;=\; \gamma_{p-1}(t) - \underline{b}'_{p-1}(t)\mathbf{C}^{-b}_{p-1}(t)\underline{b}_{p-1}(t)\,,$ $\qquad(6)$

$\underline{\text{Joint process estimation}}$

Joint estimation error : $\quad e_p(t) \;=\; e_{p-1}(t) - \underline{\Delta}'^{x}_{p-1}(t)\mathbf{C}^{-b}_{p-1}(t)\underline{b}_{p-1}(t)\,,$ $\qquad(7)$

Cross error covariance : $\quad \underline{\Delta}^{x}_p(t) \;=\; \lambda\underline{\Delta}^{x}_p(t-1) + \underline{b}_p(t)e_p(t)/\gamma_p(t)\,.$ $\qquad(8)$

Now, the central question of interest is how to reorganize and to compute these matrix operations suitably. We investigate a method that is efficient, numerically stable and highly concurrent [4]. We assume the prediction error covariance matrices $\mathbf{C}^f_p(t)$ and $\mathbf{C}^b_p(t)$ being positively definite. Let $\mathbf{C}^f_{p-1}(t) = \mathbf{R}'^f_p(t)\mathbf{R}^f_p(t)$ and $\mathbf{C}^b_{p-1}(t-1) = \mathbf{R}'^b_p(t)\mathbf{R}^b_p(t)$ be the Cholesky factorizations, where $\mathbf{R}^f_p(t)$ and $\mathbf{R}^b_p(t)$ are upper triangular with positive diagonal elements. For ease of descriptions, we define the following variables,

$$
\begin{aligned}
\tilde{\gamma}_p(t) &= \sqrt{\gamma_p(t)}\,, & \tilde{e}_p(t) &= e_p(t)/\sqrt{\gamma_p(t)}\,, \\
\underline{\tilde{f}}_p(t) &= \underline{f}_p(t)/\sqrt{\gamma_p(t-1)}\,, & \underline{\tilde{b}}_p(t) &= \underline{b}_p(t)/\sqrt{\gamma_p(t)}\,, \\
\underline{\beta}^f_p(t) &= \mathbf{R}'^{-f}_p(t)\underline{f}_{p-1}(t)\,, & \underline{\beta}^b_p(t) &= \mathbf{R}'^{-b}_p(t)\underline{b}_{p-1}(t-1)\,, & (9) \\
\mathbf{X}^f_p(t) &= \mathbf{R}'^{-f}_p(t)\boldsymbol{\Delta}_{p-1}(t)\,, & \mathbf{X}^b_p(t) &= \mathbf{R}'^{-b}_p(t)\boldsymbol{\Delta}'_{p-1}(t)\,, \\
\underline{X}^x_p(t) &= \mathbf{R}'^{-b}_p(t)\underline{\Delta}^x_{p-1}(t-1)\,.
\end{aligned}
$$

After considerable algebraic manipulations, it can be shown that the complete set of lattice recursions (1–8) is equivalent to two orthogonal transformations,

$$
\mathbf{Q}^f_p(t)\begin{bmatrix} \sqrt{\lambda}\mathbf{R}^f_p(t-1) & \underline{0} & \sqrt{\lambda}\mathbf{X}^f_p(t-1) \\ \underline{\tilde{f}}'_{p-1}(t) & \tilde{\gamma}_{p-1}(t-1) & \underline{\tilde{b}}'_{p-1}(t-1) \end{bmatrix}
$$

$$
= \begin{bmatrix} \mathbf{R}^f_p(t) & \underline{\beta}^f_p(t) & \mathbf{X}^f_p(t) \\ \underline{0}' & \tilde{\gamma}_p(t) & \underline{\tilde{b}}'_p(t) \end{bmatrix}\,, \qquad (10)
$$

$$
\mathbf{Q}^b_p(t)\begin{bmatrix} \sqrt{\lambda}\mathbf{R}^b_p(t-1) & \underline{0} & \sqrt{\lambda}\mathbf{X}^b_p(t-1) & \sqrt{\lambda}\underline{X}^x_p(t-1) \\ \underline{\tilde{b}}'_{p-1}(t-1) & \tilde{\gamma}_{p-1}(t-1) & \underline{\tilde{f}}'_{p-1}(t) & \tilde{e}_{p-1}(t-1) \end{bmatrix}
$$

$$
= \begin{bmatrix} \mathbf{R}^b_p(t) & \underline{\beta}^b_p(t) & \mathbf{X}^b_p(t) & \underline{X}^x_p(t) \\ \underline{0}' & \tilde{\gamma}_p(t-1) & \underline{\tilde{f}}'_p(t) & \tilde{e}_p(t-1) \end{bmatrix}\,. \qquad (11)
$$

The two orthogonal matrices $\mathbf{Q}^f_p(t)$ and $\mathbf{Q}^b_p(t)$ with $\mathbf{Q}'^f_p(t)\mathbf{Q}^f_p(t) = \mathbf{Q}'^b_p(t)\mathbf{Q}^b_p(t) = I$ are designed to eliminate the lower left row vectors $\underline{\tilde{f}}'_{p-1}(t)$ and $\underline{\tilde{b}}'_{p-1}(t-1)$ while preserving the upper triangular structure of $\mathbf{R}^f_p(t-1)$ and $\mathbf{R}^b_p(t-1)$. They can be easily implemented by m Givens rotations each. In this way, the scaled versions $\underline{\tilde{f}}_p(t), \underline{\tilde{b}}_p(t), \tilde{e}_p(t-1)$ of the desired prediction and FIR filtering errors are recursively computed by applying a sequence of plane rotations.

3.   CORDIC ALGORITHM AND CORDIC PROCESSOR

The basic operation of the MLSL algorithm is a plane rotation in two variants, Cartesian to polar coordinates conversion for computing the rotation angle and the rotation of a two-component vector by a given angle. Although these operations can be performed by conventional multiplier-adder based processors supported by software, there are more dedicated hardware solutions, especially the CORDIC (COordinate Rotation DIgital Computer) processor.

The CORDIC algorithm, in the original version designed by Volder [5], consists of iterative operations on a three-component vector

$$
\begin{cases} x_{i+1} &= x_i - \sigma_i 2^{-S_i}y_i \\ y_{i+1} &= y_i + \sigma_i 2^{-S_i}x_i \qquad (i = 0, 1, \cdots, n)\,, \\ z_{i+1} &= z_i - \sigma_i \arctan(2^{-S_i}) \end{cases} \qquad (12)
$$

where $\sigma_i = \pm 1$ and $\{S_i\}$ forms a non-decreasing integer sequence. For a given set $\{S_i\}$, Eq. (12) requires only shift-add operations. Now, if $\{S_i\}$ and the input data $\{x_0, y_0, z_0\}$ fulfil some convergence conditions, $\{\sigma_i\}$ can be suitably chosen to force y_n or z_n to zero. The resulting data of the other iterations in (12) are given by

$$y_n \to 0 \quad \left\{ \begin{array}{rcl} x_n & = & K\sqrt{x_0^2 + y_0^2} \\ z_n & = & z_0 + \arctan(y_0/x_0) \end{array} \right. , \quad z_n \to 0 \quad \left\{ \begin{array}{rcl} x_n & = & K(x_0 \cos z_0 - y_0 \sin z_0) \\ y_n & = & K(x_0 \sin z_0 + y_0 \cos z_0) \end{array} \right. \quad (13)$$

with $K = \prod_{i=0}^{n-1} \sqrt{1 + 2^{-2S_i}}$. This means, the CORDIC algorithm enables to compute a Cartesian to polar coordinates conversion or a vector rotation by only a sequence of shift-add operations, except for the scaling factor K. Methods to compensate K are known in the literature. They require an additional number (approximately $\lceil n/4 \rceil$) of shift-add operations. We refer to [6] for a brief overview.

Clearly, CORDIC processors are regular and modular in structure. They have a high data throughput in a pipeline realization. Moreover, they require the same computation time for different operations and are easily programmable. They provide powerful arithmetic units in highly parallel/pipelined computing architectures [7], especially if rotation operations have to be computed as in the MLSL algorithm.

4. PARALLEL IMPLEMENTATION

Fig. 1 shows a parallel implementation of the presented MLSL algorithm. It consists of p identical filter sections seperated by delays D in the backward prediction path. Each filter section implements the two orthogonal transformations (10,11). These are matrix operations having a very regular structure so that they can be themselves implemented by array architectures. Depending on the performance requirements, there exist many space-time tradeoffs for the complexity distribution. A structure allowing maximum parallelism is indicated in Fig. 1, where Eq. (10,11) are implemented on two trapezoidal arrays of processing elements (PEs). For space reason, we only discuss the lower trapezoidal array (Fig. 2) for implementing Eq. (10).

In Fig. 2, the array architecture used for implementation is a straightforward mapping of the matrix structure involved in (10). The quantities $\mathbf{R}_p^f(t-1)$ and $\mathbf{X}_p^f(t-1)$ are stored in the PEs of the array, each matrix element in one PE. They are updated at each time step. The input data $\tilde{f}_{p-1}(t), \tilde{\gamma}_{p-1}(t-1)$ and $\tilde{b}_{p-1}(t-1)$ are put from left into the array and propagate from left to right. The m rotation angles are calculated in the circle diagonal PEs. They are CORDIC processors operating in the mode $y_n \to 0$. The angles propagate from top to bottom. The internal square PEs operating in the mode $z_n \to 0$ perform plane rotations with angles received. In this way, the resulting output data at the right side of the array are just the desired quantities $\tilde{\gamma}_p(t)$ and $\tilde{b}_p(t)$.

5. CONCLUSION

We have presented an adaptive MLSL algorithm for solving the multichannel linear prediction and FIR filtering problem. We have proposed a CORDIC processor array for parallel implementation as a natural solution. Certainly, because the algorithm exhibits a high degree of inherent concurrence, it can also be implemented by parallel architectures as vector computers and transputer systems.

References

[1] S. Haykin, *Adaptive Filter Theorey*. Englewood Cliffs: Prentice-Hall, 1986.

[2] B. Friedlander, "Lattice filters for adaptive filtering," *Proc. IEEE*, vol. 70, no. 8, pp. 829–867, 1982.

[3] P. S. Lewis, "Multichannel adaptive least squares - relating the "Kalman" recursive least squares (RLS) and least squares lattice (LSL) adaptive algorithms," in *Proc. IEEE ICASSP*, pp. 1926–1929, April 1988.

[4] B. Yang and J. F. Böhme, "On a parallel implementation of the multichannel adaptive least-squares lattice filter," in *Proc. U.R.S.I. ISSSE*, Sep. 1989.

[5] J. E. Volder, "The CORDIC trigonometric computing technique," *IRE Trans. on Electronic Computers*, vol. 8, pp. 330–334, 1959.

[6] J. F. Böhme and B. Yang, "Highly parallel and pipelined computing using CORDIC processors and applications to computer graphics, signal processing and linear algebra," in *Proc. German-Chinese Electronics Week*, VDE Verlag, Düsseldorf, Mar. 1990.

[7] H. M. Ahmed, J. M. Delosme, and M. Morf, "Highly concurrent computing structures for matrix arithmetic and signal processing," *IEEE Computer*, vol. 15, no. 1, pp. 65–82, 1982.

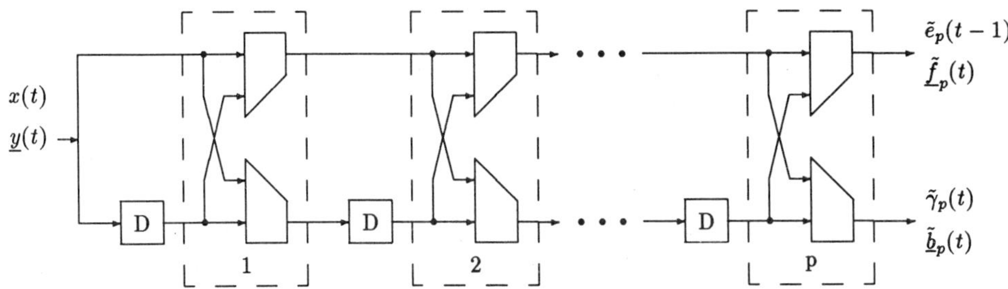

Figure 1: Parallel implementation of the MLSL algorithm

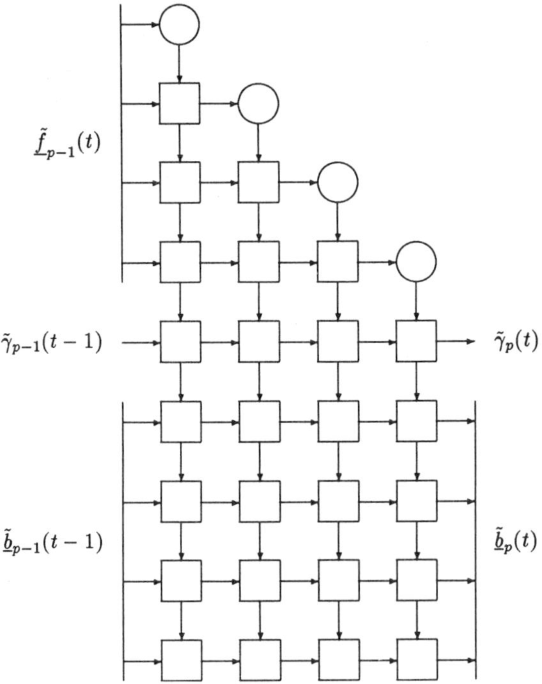

Figure 2: Details of the lower trapezoidal array

Section 3
New Concepts
of Information Processing
in the Brain

Parallel Processing in Neural Systems and Computers
R. Eckmiller, G. Hartmann and G. Hauske (Editors)
© Elsevier Science Publishers B.V. (North-Holland), 1990

COMPUTER SIMULATION OF NEURAL NETWORKS WITH CHEMICAL MARKERS

Adam ADAMOPOULOS and Photios ANNINOS

Department of medicine
University of Thraki
Alexandroupolis, Greece

The dynamics of neural networks consisting of discrete populations of
formal neurons have been studied through computer simulation. Among
the basic assumptions of the model is that neurons are divided in subpopulations depending of the number of the chemical markers involved.
Interconnections exist only among neurons carrying the same or similar type of chemical markers. These networks have been shown to produce multistability phenomena supporting self-maintaining activity in
the form of cycling modes of fixed periods. The level of activation
as well as the period of the observed cyclic modes were found to depend on the level of initial activity. These results suggest that
neural networks may respond in specific manner to specific stimuli.

1. INTRODUCTION

The development and study of mathematical and computer simulation models are
two of the most valuable alternative methods to investigate the structure and
function of the central nervous system. In the early future,it is expected that
these methods are to become more and more attractive to the investigators as the
capicity and availability of the digital computers used will increase. This will
result drastically to the gradually improvement of the realism of the computeri-
sed methods, as it is happened in the recept past (1-3). In this paper we are
presenting results obtained by examining the dynamics of neural networks with
chemical markers using a theoretical and a computer simulation approach, which
had been developed in our previous work (4,5). Both models use physiologically
realistic parameters and have been proved to be very fruitfull.Among others the
theoretical model has been shown to produce evoked potentials(6), background
EEG (7,8) and epileptifom-like activity (9,10). The study is primarily concerned
with the network ability to appear multiple stable states and oscillatory activity (cyclic modes).

2. FORMULATION AND ASSUMPTIONS OF THE MODELS

2.1. General specifications

For both models, the neural networks are assumed to be constructed of formal neurons. Neurons are bistable elements as it was postulated by McCulloch & Pitts
(11) and operate synchronously in discrete time. Depending on the action potential they are characterized either excitatory or inhibitory. A neuron is fired
at one instant if the total arriving PSP´s exceed a specified threshold, and
produce the appropriate PSP after a fixed time interval, the synaptic delay τ.
After firing, neurons are insensitive to further stimulation for a period of time which was called refractory period r. Refractory period may in general take
any integer value; for our purposes r was given the values r=1 (if refractori-
ness is assumed) and r=0 otherwise. A basic assumption of the model is that neural connections are set up by means of chemical markers (5). Thus, the network
is constructed by a set of subpopulaitons characterized by a specific chemical

marker carried by the individual cells. The information of synaptic connections effected by these markers is restricted exclusively between neurons carrying the same type of marker.

2.2. Theoretical model

As it have been postulated in previous papers (4,5) the only significant dynamical variable which we are considering in the theoretical model is the level of activity α_n, i.e. the fractional number of neurons firing at specific time $t=n\tau$. The activity is a scalar quantity and does not specify which neurons are firing in the network. Given the initial activity α_n, the expectation value of the activity at the next time step $t=(n+1)$ have been calculated in detail in (5). Assuming a network consisting of A neurons which are characterized by N markers, the expectation value of the activity at the next time step will be given by:

$$<\alpha_{n+1}> = (1-\alpha_n) \sum_{J=I}^{N} m_j \sum_{M=0}^{M_{max,j}} \sum_{I=0}^{I_{max,j}} R_{M,j} Q_{I,j} \left(1 - \sum_{L=0}^{n_j-1} P_{L,j}\right) \qquad (4)$$

where $R_{M,j}$, $Q_{I,j}$, $P_{L,j}$ are the probabilities that a neuron carrying the jth marker will receive M external, I inhibitory and L excitatory inputs respectively. The upper limits of the sums $M_{max,j}$, $I_{max,j}$ and n_j are the total numbers of the external inputs of inhibitory inputs and the minimum number of excitatory necessary to trigger a neuron inputs respectively. A typical $<\alpha_{n+1}>$ v's α_n curve is shown in Fig.1, whereas in Fig.2 is shown the time course of the activity for the same network.

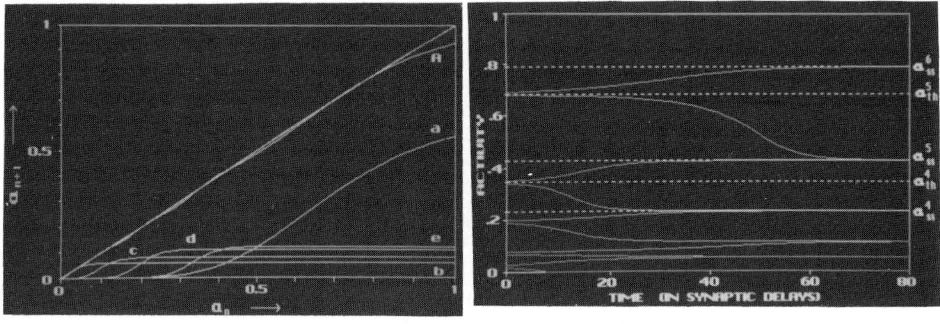

FIGURE 1	FIGURE 2

$<\alpha_{n+1}>$ v's α_n plot for a network with five markers. A is the total activity whereas a through e are the contributions of each marker.

Time course of neural activity for the network of Fig. 1. α_{ss} are stable states whereas α_{th} are unstable states.

2.3. Computer simulation model

What follows is a brief description of the used neuronal simulation approach. In the spirit of previous research (4) given a network consisting of A neurons, the first step is to establish the neuronal connectivity matrix (K_{lm}) which defines the connections between individual neurons. The subscripts l and m indicate the post-and presynaptic neurons respectively, whereas the coupling coefficient k_{lm} denotes the synaptic strength of the connection. k may take either positive or negative values depending on the type of the synaptic neuron (excitatory or inhibitory respectively). Macroscopic parameters of the model such as the synaptic strength (k), the number of excitatory or inhibitory connections per neuron (μ^+, μ^-), and the threshold (θ) are followed to vary between a maximum and a minimum value in order to produce more realistic behavior. After establishing all desired

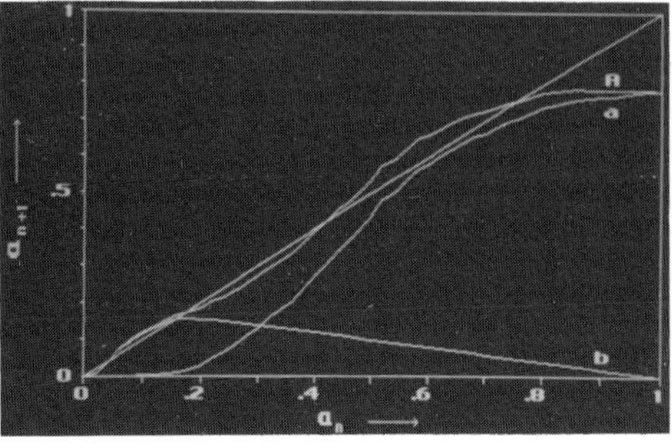

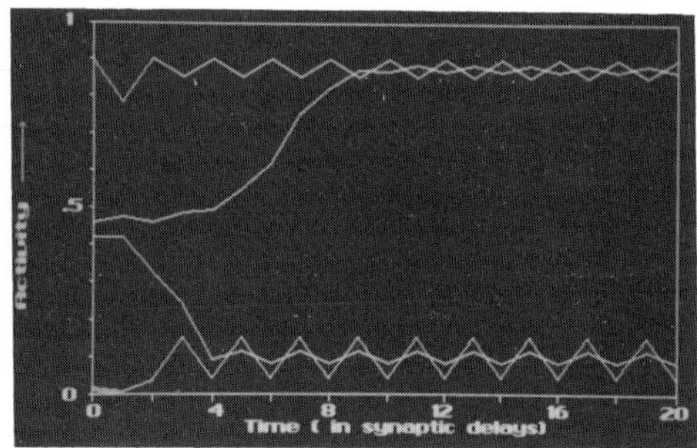

FIGURE 3

$\langle\alpha_{n+1}\rangle$ v´s α_n for a simulated network with two markers a and b.

FIGURE 4

Time course of neural activity for the network of Fig. 3.

interconnections, the network is activated by specifying the set of neurons which will fire at time t=0. One synaptic delay later all neurons linked to those initially active will receive the appropriate inputs. All these inputs arriving at a neuron are summed instantly and if the sum exceeds the neuron threshold then the neuron will fire. In such a way all active neurons at the next time step are specified. The firing neurons for t=nτ define the state vector $\vec{a}_n$. Subsequently, the condition of appearance of cyclic mode in the network is that for which in the two intants t=nτ and t=(n+k)τ, the equation $\vec{a}_n=\vec{a}_{n+k}$ is filled. Results obtained using this model are shown in Figs.3-6.

3. RESULTS

In Fig.1 is plotted $\langle\alpha_{n+1}\rangle$ v´s α_n for a network with five markers. The intersects for the curve with the diagonal line fill the condition $\langle\alpha_{n+1}\rangle=\alpha_n$ indicating the steady states of the network (stable and unstable ones). In Fig. 2 is shown the time course of the activity for several initial values of activity. It is clear that the stable state which will be reached after a finite period of time depends on the given initial activity. This can be explained assuming that between two consecutive stable states always exists an unstable one. If the initial activity is greater that the unstable state the activity tends to go to the upper stable state, otherwise tends to go to the lower one. Computer simulation was used for similar analysis. In Fig.3 and Fig.4 is shown plots of $\langle\alpha_{n+1}\rangle$ v´s α_n and the time course of the activity for a network with two markers. Furthermore it was of interest to explore the time it will take the network to reach a stable state as a function of the initial activity (Fig.5). It is clear that the closer the initial activity to a stable state the shorter it will take the network to reach this state. With such plots an alternative representation of steady states arises. Stable states are represented as "wells", whereas unstable ones are represented as "peaks". Network activity moves from one well to another if the initial activity exceeds the intermediate "peak" value. Such a behavior is in a great similarity in classical mechanics with a classical particle moving in a multi-well potential.These plots are in excellent qualitative agreement with our theoretical findings. Finally, in Fig. 6 is illustrated the time course of the activity for another network with two markers. Marker a exhibits a cyclic mode of period 6 after 22 time steps,whereas marker b exhibits a cyclic mode of period 60 after 28 time steps. As a result the total activity exhibits cyclic mode after 28 time steps with a period of 60 time steps,i.e. the Least Common Product of the cycling periods of the two markers.

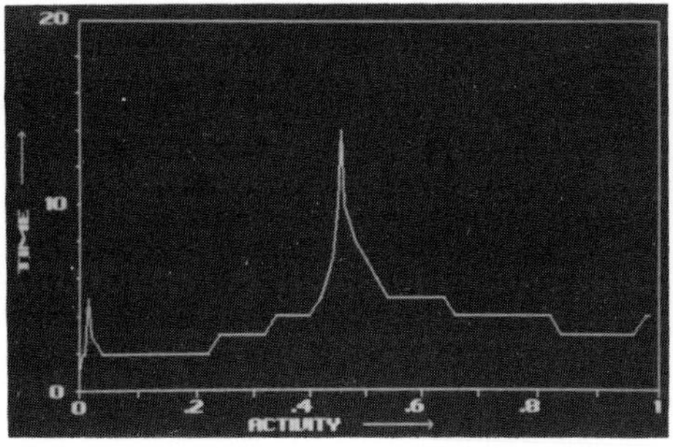 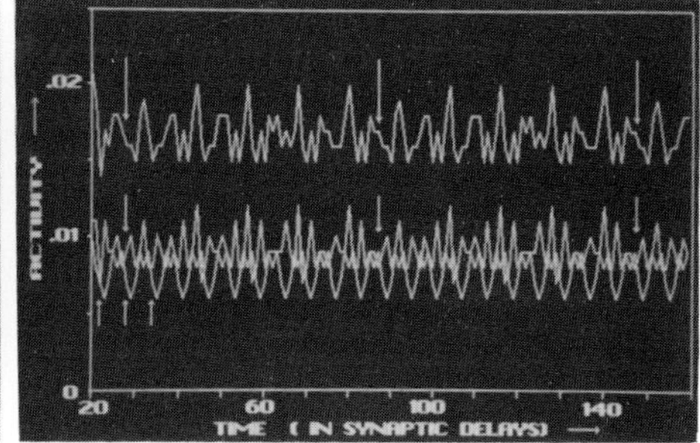

FIGURE 5   FIGURE 6
Time to reach a stable state against i- Time course of the activity for a net-
nitial activity for the network of Fig.3 work with two markers. Arrows indicate
                              periods of cyclic activity.

4. DISCUSSION

Computer simulations of randomly interconnected neural networks have shown to
produce oscillatory (cyclic) activity which may represent a memory of a particular stimuli (12). These networks are shown to have multiple stability phenomena
and the final state of the network was found to depend on the given initial activity. These findings are in agreement with the ones of our theoretical model
(5,13). Different initial activities may represent different kinds of stimuli,
or different intensities of the same stimulus. According to the initial conditions the neural activity finally may reach to unique state and cyclic mode, differentiating and memorizing in such a way different stimuli.

5. REFERENCES

(1) Kohonen,T., Neuroscience, 2 (1977) 1065.
(2) MacGregor,R.J. & Lewis,E.R., Neural modeling. electrical signal processing
    in the nervous system (Plenum,New York,1977).
(3) MacGregor,R.J., Neural and Brain Modeling (Academic Press,New York,(1987).
(4) Anninos,P.A., Beek,B., Csermely,T.J., Harth,E.M. & Pertile,G., Journal of
    Theoretical Biology **26** (1970) 121.
(5) Anninos,P. & Kokkinidis,M., Journal of Theoretical Biology **109** (1984) 95.
(6) Anninos,P.A., Kybernetik, **13** (1973) 24.
(7) Anninos,P.A., Kybernetik, **11** (1972) 5.
(8) Anninos,P.A., Zenone,S. & Elul,R., Journal of Theoretical Biology **103** (1983)
    339.
(9) Anninos,P.A., & Cyrulnic,R., Journal of Theoretical Biology **66** (1977) 695.
(10) Anninos,P.A., & Anogianakis,G., Computer simulation studies to deduce the
    structure and function of the human brain, in: Cotterill, R.M.J., (ed.),
    Computer simulation in brain science (Cambridge University Press, 1988).
(11) McCulloch,W.S. & Pitts,W., Bulletin of Mathematical Biophysics 5 (1943) 115.
(12) Anderson,P., Gillow,M. & Rudjord,T., Journal of Physiology **185** (1966) 418.
(13) Anninos,P., A neural model with multiple memory domains, in: Eckmiller,R. &
    Malsburg,C.v.d. (eds), Neural Computers (NATO ASI Series F vol. 41.,Springer
    -Verlag, Berlin, Heidelberg, 1988).

Parallel Processing in Neural Systems and Computers
R. Eckmiller, G. Hartmann and G. Hauske (Editors)
© Elsevier Science Publishers B.V. (North-Holland), 1990

DYNAMICS OF ACTIVITY IN NEURONAL NETWORKS GIVE RISE TO FAST MODULATIONS OF FUNCTIONAL CONNECTIVITY

Karl-Heinz BOVEN and Ad AERTSEN

Max-Planck-Institut für biologische Kybernetik, D-7400 Tübingen, FRG

Multi-unit spike trains recorded from different brain areas provide evidence of fast stimulus-locked modulations of neuronal interaction. This indicates that the functional connectivity among neurons may be highly dynamic. Possible mechanisms underlying such dynamic network organization have been investigated. In particular, it appears that the experimentally observed, context-sensitive fast modulation of functional connectivity can be explained by the dynamics of network activity, without having to assume fast synaptic plasticity.

Introduction

Connectivity among neurons is usually measured by cross-correlating their spike trains, recorded under some appropriate stimulus condition(s). The ordinary cross-correlogram presents a time-averaged count of near-coincident spikes; in general it is composed of contributions from quite different origins, such as stimulus-induced modulations of firing rates and various types of neuronal interactions. Appropriate normalization is therefore essential when comparing results across experiments; only then may we interpret residual correlation as indicative of (possibly stimulus-dependent) *effective connectivity* among the neurons. Recently such a normalization was developed, the *Joint-PSTH* [1], based on a stimulus-locked temporal decomposition of the ordinary cross-correlogram. This normalization allows one to separate stimulus-induced correlation from interneuronal correlation; moreover it enables a study of the *dynamic* properties of neuronal connectivity.

Application to physiological multi-unit spike trains from several different preparations for the first time produced evidence for fast stimulus-locked modulations of effective connectivity [1,2]. These findings point at dynamic cooperativity as an emergent property of neuronal assembly organization in the brain. They suggest that the usual concept of neurons with static interconnections of fixed or only slowly changing efficacy (during learning), is no longer appropriate. Instead one should distinguish between *structural* (or anatomical) connectivity on the one hand and *functional* (or effective) connectivity on the other. Whereas the former can presumably be described as (quasi) stationary, the latter may be highly dynamic, with time constants of modulation in the range of tens to hundreds of milliseconds.

Two types of explanation for this phenomenon can be envisaged: a local one and a global one. On the one hand the synapses might actually modulate their efficacy on such a short timescale [3], on the other hand the modulation of 'effective connectivity' might be a reflection of the dynamics of activity in the entire network [4]. In order to investigate the latter possible explanation, i.e. the influence of network activity on the efficacy of intrinsically constant connections, we have studied the behaviour of artificial neuronal networks under different activity conditions, both in simulation and with analytic methods.

2. Effective Connectivity as a Function of Network Activity

A scheme of the network we analysed is shown in the inset in Fig. 1a. A spontaneously active neuron (1) drives a second neuron (2), which gets additional input from a pool of N independent, spontaneously firing neurons. All synaptic connections α and β have fixed and moderately weak strength, i.e. spikes arriving at any particular junction give rise to epsp's with constant and clearly sub-threshold magnitude. We studied the behavior of the effective connectivity between neurons (1) and (2) as a function of the activity arriving at (2) from the pool, for constant connectivity α and different values of pool coupling β.

2.1 Simulation

Each of the neurons is represented as a nonlinear system described by four state variables, the properties of which are determined by specifying conductances, thresholds and time constants [5]. This representation allows us to monitor both the neuron's membrane potential and its spike firing behaviour.

The mean firing rate λ of all spontaneous neurons was set to about 10 spikes/sec. The overall activity from the pool could be varied by changing either the number of neurons N or the firing rate λ of each of the member neurons. The latter manipulation mimics the effect of driving the pool with a stimulus S.

The effective connectivity α' between neurons (1) and (2) was measured by crosscorrelating the spike activity from the two: α' equals the number of correlated events N_c (i.e. the net area of the peak in the correlogram) divided by the number of presynaptic events N_1 [6]. Curves of the behavior of α' as a function of pool activity $N \cdot \lambda$ for various choices of β are shown in Fig. 1a.

Two observations can be made:
1. The efficacy α' varies strongly with pool activity $N \cdot \lambda$, even though the synapse itself is kept at a fixed strength α (= 0.2) throughout all simulations. With increasing pool activity the efficacy of the connection initially increases strongly to reach a maximum, after which it slowly decays again. Dynamic changes of pool activity would thus show up as equally fast changes in effective connectivity.
2. The dynamic range of the efficacy α' increases with decreasing pool coupling β. Moreover, for smaller β the maximum of α' is actually higher, but is only reached at larger values of the pool activity. These findings are especially relevant for large, weakly coupled nets, of which the mammalian cortex is a typical example.

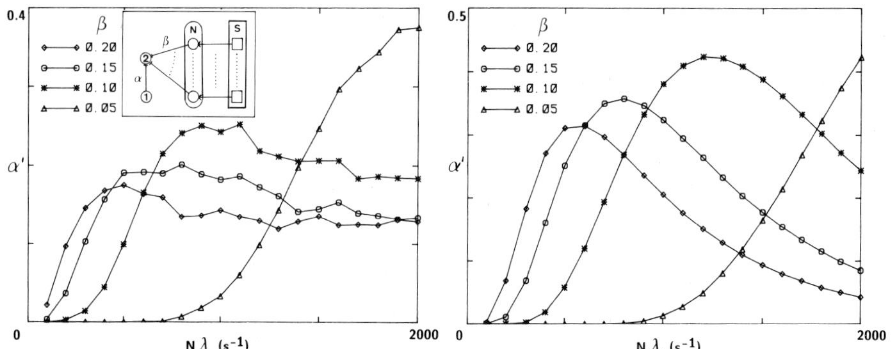

Figure 1. Effective connectivity α' as a function of network activity $N \cdot \lambda$, for different values of pool coupling β. Fig. 1a: simulation results, Fig. 1b: analytic results.

2.2 Analytical Calculation

In order to obtain an analytic expression for the effective connectivity we assume (1) that the membrane potential at the site of spike initiation is the result of linear, spatio-temporal integration of incoming spike trains, and (2) that a spike is generated whenever this membrane potential crosses a fixed threshold Θ.

The contribution to the membrane potential of neuron 2, due to the pool of N independent Poisson processes, each with constant rate λ, connected to neuron 2 with strength β can be described as 'shot-noise' (i.e. a linearly filtered (time constant τ) Poisson process).

For reasonably large N, this can be approximated by a Gaussian distribution with a mean value $\mu = N\cdot\lambda\cdot\beta$ and variance $\sigma^2 = N\cdot\lambda\cdot\beta^2/2\tau$.

Following [7] the firing rate λ_2 of neuron 2 can then be calculated as

$$\lambda_2 = 1/\sqrt{2\pi} \int_{(\Theta-\mu)/\sigma}^{\infty} \exp(-y^2/2)\ dy$$

The effective connectivity α' from neuron 1 to neuron 2 equals the increase in firing rate λ_2 upon the arrival of a spike from neuron 1, i.e. upon a shift of the membrane potential over an amount α

$$\alpha' = 1/\sqrt{2\pi} \int_{(\Theta-\mu-\alpha)/\sigma}^{(\Theta-\mu)/\sigma} \exp(-y^2/2)\ dy$$

The curves corresponding to this analytic expression for the effective connectivity α' are shown in Fig. 1b. One observes that they exhibit essentially the same behavior as the simulation results in Fig. 1a : an initial fast increase until a maximum is reached, followed by a slower decay.

3. Fast Modulation of Synaptic Efficacy

The dynamic effect of variations of the network activity on the functional connectivity α' was investigated by rapidly varying the pool activity as a function of time. To this end the pool neurons were driven by a periodic stimulus S (T = 100 ms), causing them to fire at a mean rate of 16.9 spikes/s for 30 ms, followed by an increase to 19.3 spikes/s for 40 ms and again 16.9 spikes/s for the remaining 30 ms. Stimulation parameters were selected such, that the spike train of each of the pool neurons can be described by a rate-modulated Poisson process and that the Poisson processes of different neurons remained independent. The pool size was chosen to be N=100, all synaptic couplings α and β were set to the weak value of 0.05.

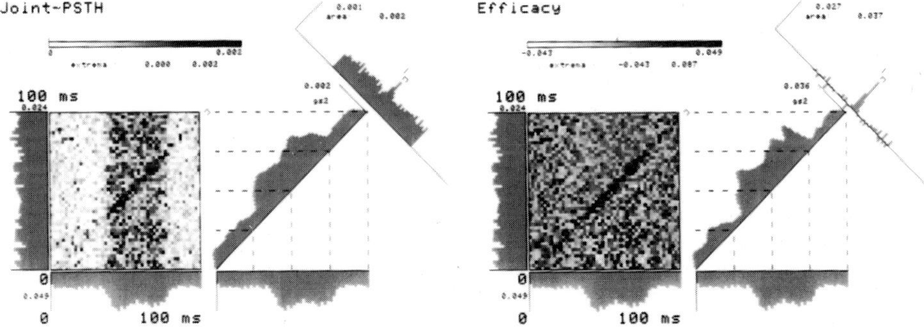

Figure 2. Fast modulation of synaptic efficacy by variation of network activity. Fig. 2a: 'raw' dynamic correlation, Fig. 2b: dynamic synaptic efficacy. Numbers of spikes: 4813 (1), 6923 (2); number of stimuli: 5000.

The dynamic correlation between the spike trains from neurons 1 and 2 was analysed using the Joint-PSTH [1]. Fig. 2a shows the 'raw' correlation. The time-dependent firing probability of neuron 1 is shown along the vertical axis, that of neuron 2 along the horizontal axis. One clearly observes how the time variation of the pool activity is reflected in a similar modulation of the firing rate of neuron 2. The matrix represents the probability of coincidences as a function of ongoing time and of time delay. The diagonal of the matrix corresponds to zero delay (i.e. exact coincidence); time delay increases with growing distance to the diagonal. The right-hand side of

Fig. 2a shows the temporal distribution of coincidences with a delay corresponding to the connection between neurons 1 and 2 (the histogram from bottom left to top right), and the time-averaged distribution of coincidences as a function of time delay (i.e. the usual crosscorrelation; top left to bottom right).

Appropriate normalization of the dynamic correlation for variations in firing rate of both neurons 1 and 2 leads to the dynamic synaptic efficacy (Fig. 2b). The variations in the time course along the diagonal clearly show that - even though the synaptic coupling between neurons 1 and 2 was kept constant throughout the entire simulation - the modulation of the pool activity results in a strong modulation of the synaptic efficacy of the (1-2) - pathway, with a time constant of only a few milliseconds. Also the magnitude of the efficacy along its time course closely follows the numeric values predicted by the statically determined curves in Fig. 1a.

4. Discussion

The demonstrated effect of fast, activity-related changes of synaptic efficacy provides a possible explanation for the experimentally observed fast modulations of firing correlation in multi-neuron activity.

This activity-related modulation of effective connectivity might serve as a fast 'gating' mechanism for synaptic pathways, i.e. modulating the effectivity of the (1-2) - pathway through variation of the activity carried by a different, diffuse projection onto distal dendrites of neuron 2. Furthermore it might provide a mechanism for the successive ignition of selected neural assemblies ('phase sequences', [8]) within an interconnected network by dynamic modulation of the input activity to that network. The latter aspect bears a close relation to the concept of the 'skeleton filter' [9].

The proposed mechanism does not have to invoke intrinsic rapid modulation of synaptic efficacy [3]; obviously it does not exclude it as a possible additional mechanism either. Further investigation, both theoretically and experimentally is required to clear this issue.

Acknowledgements

We thank Michael Erb, George Gerstein, Günther Palm and Hubert Preißl for helpful discussions, and Volker Staiger for skillful assistance in preparing the Figures.

References

[1] Aertsen et al., J.Neurophysiol. 61:900-917 (1989); Palm et al., Biol.Cybern. 59:1-11 (1988)
[2] Gerstein et al., IEEE Trans. Biomed. Engin. 36:4-14 (1989)
[3] Von der Malsburg, Ber. Bunsenges. Phys. Chem. 89: 703-710 (1985)
[4] Boven and Aertsen, In: Elsner and Singer (eds.) Dynamics and plasticity in neuronal systems. Stuttgart, New York: Thieme (1989); Erb et al., ibid.
[5] Simulator modified from MacGregor, Neural and Brain Modeling. London: Academic Press (1987).
[6] Aertsen et al., Brain Research 340: 341-354 (1985)
[7] Abeles, Local Cortical Circuits - Studies of Brain Function, vol.6. New York: Springer (1982)
[8] Hebb, The organization of behavior. New York: Wiley (1949)
[9] Sejnowski, in Hinton,Andersen: Parallel Models of Associative Memory (1981)

Parallel Processing in Neural Systems and Computers
R. Eckmiller, G. Hartmann and G. Hauske (Editors)
© Elsevier Science Publishers B.V. (North-Holland), 1990

USING PROPRIOCEPTIVE INFORMATION TO CONTROL MOVEMENT TO A TARGET

Jeffrey Dean

Abteilung für Biokybernetik und Theoretische Biologie
Universität Bielefeld, Postfach 8640
D-4800 Bielefeld 1, FRG

A simple neural network was used to test alternative means of coding
proprioceptive information for the simple task of placing one limb
close to another. This task requires the control network to associate
an appropriate configuration of the moving limb with each position of
the target limb. Tarsus positions and spatial relationships are most
easily coded in terms of Cartesian coordinates, but limb configurations are more naturally coded in terms of joint angles. The transformation between these alternative coordinate systems would appear to
present a difficult calculation for a nervous system. Therefore the
performance of a simple neural network was studied when the required
association was formulated in Cartesian coordinates, joint angles, or
combinations of both. The simulation was based on the target movement
shown by the stick insect when walking on a flat surface. The results
show that a simple 3x3 network can perform adequately simply by asso-
ciating joint configurations with joint configurations. Hence, the
simplest hypothesis is that the nervous system uses joint coordinates
in controlling the target movement and avoids the complex coordinate
transformation.

1. INTRODUCTION

Proprioceptive information provides a sense of body position which can be used
to guide movement. A simple example is the target movement shown by the stick
insect during walking (1,2). When the rear leg swings forward, the tarsus moves
to a position slightly behind the tarsus of the ipsilateral middle leg (1). When
the substrate is discontinuous, this spatial coordination increases the chance
that the rear leg will find a foothold at the end of its swing (1). A similar
coordination occurs between middle and front legs. The behavior depends upon
information from proprioceptors on the target leg (2,3). These proprio- ceptors
provide information on the angles at the three major leg joints.

For an observer, the behavior is most easily described in terms of Cartesian
coordinates: the moving leg is placed a constant distance behind the target leg
or a given change in the position of the target tarsus elicits a parallel shift
in the end-point of the moving tarsus. However, the configuration of a leg is
most naturally described in terms of the angles at the joints--joint coordinates. This is the form in which proprioceptive information on the position of
the target leg is delivered to the central nervous system and the form in which
motor commands are sent to the muscles which move the leg to the desired position. Cartesian and joint coordinates provide equivalent information, but the
transformation from one coordinate system to the other is neither trivial nor
linear.

The focus of the present study is the question of whether the simple nervous
system of the stick insect needs to perform the complex coordinate transformation that is required by an analytic algorithm for controlling the target

movement. Alternatively, the control problem can be described as an association
task in which the appropriate configuration of the rear leg must be associated
with each position of the target leg. The ability of a simple neural network to
learn this association was used to evaluate control strategies based upon
representations of leg position in Cartesian or joint coordinates. (In the
present simulations, learning was used as a convenient means to find appropriate connections for the network; the use of this method is not meant to imply
that synaptic connections in the stick insect arise in the same way.)

2. METHODS

In the simulations described here a simple, 3x3 network was trained to associate final configurations of the moving leg with configurations of the target
leg. Activity in the three input units corresponded to a target leg position
coded in either Cartesian or joint coordinates. Activity in the three output
units was interpreted as the final position of the moving leg, again coded in
either Cartesian or joint coordinates. Each unit represented one coordinate of
the tarsus position (Cartesian coordinates) or one of the three leg joints
(joint coordinates). The results reported here concern movements in two dimensions: this corresponds to the case of an insect walking with a constant body
height over a flat surface.

The first step was to create a set of 28 patterns. Each pattern contained a
middle leg position and the corresponding rear leg position. For simplicity the
test patterns were calculated so as to place the tarsus of the rear leg at the
same position as the tarsus of the middle leg rather than a constant distance to
the rear. The lengths of leg segments (coxa 1.5 mm, trochanter+femur 12.3 mm,
tibia 11.8 mm) and the separation between the insertions of the legs on the
thorax (11 mm) were based on measurements from the stick insect (4), but middle
and rear legs were given the same dimensions. The 28 patterns corresponded to
tarsus positions in a grid within the ipsilateral area which can be reached by
both the rear and the middle legs. The grid interval was 5 mm.

Four sets of patterns were constructed—one for each of the four possible
combinations of Cartesian and joint coordinates for coding the position of the
target leg and that of the moving leg. In four separate trials the network was
trained on each set of patterns using the back-propagation algorithm (5). The
same initial weights were used for all four trials. Learning was followed for
5000 epochs in which all 28 patterns were tested; the sum of the squared error
over all 28 patterns was recorded at regular intervals. The squared error on
each pattern was recorded for the final state of the network. This sequence was
repeated for 10 different initial weights.

3. RESULTS

The simple network was able to learn the association in each of the four
formulations (Figure 1). Both the rate of learning and the final performance
were best when the positions of both legs were formulated in terms of Cartesian
coordinates of the tarsus position. For the three other trials, the final performance of the network was not significantly different. Of these three, the
association of leg positions in terms of Cartesian coordinates with target leg
positions given in terms of joint angles was learned the slowest; the association of leg positions in terms of joint angles with target leg positions given
in terms of Cartesian coordinates was learned the fastest.

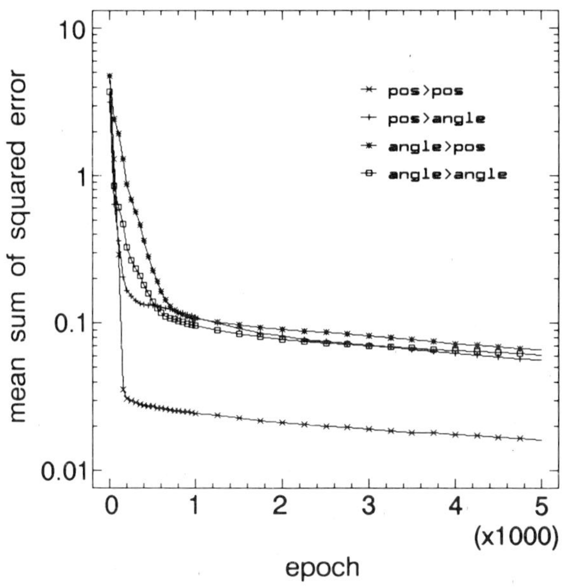

FIGURE 1

Learning curves for the four different formulations of the association. The form
of the association is indicated by the symbols: e.g. pos^3angle is the
association of a rear leg position in terms of joint angles (output) with a
target leg position given in terms of Cartesian coordinates of the tarsus
(input). Each curve is the mean of ten simulations. The error is the difference
between the correct output pattern and the actual output. This error corresponds to the error on the association task; for outputs formulated as joint
angles, it is not equivalent to the error in tarsus position.

The mean errors on the individual patterns showed the same differences. For the
association of Cartesian coordinates with Cartesian coordinates the mean error
did not vary much from one pattern to another: all were low. For the other three
formulations, mean errors were pronounced for patterns on the margin of the
grid, particularly for those along the medial border. Within the ten trials for
a given formulation, the same patterns consistently led to large errors.

In order to illustrate what the errors determined for the association problem
mean for the accuracy of the target movement, the tarsus positions were calculated from the joint angles produced by the final state of the network in one
trial after training to associate joint angles with joint angles. Figure 2
compares the accuracy of the control for two formulations. It shows that the
network made small, relatively unsystematic errors when the task was to associate Cartesian coordinates with Cartesian coordinates. When the task was
learned in terms of joint angles, then the network made larger errors for
positions on the margin of the area which can be reached by both legs.

4. CONCLUSIONS

The results show that a simple network is sufficient to control the target
movement described for the stick insect and at the same time indicate how
proprioceptive information may be coded for this control. That the network

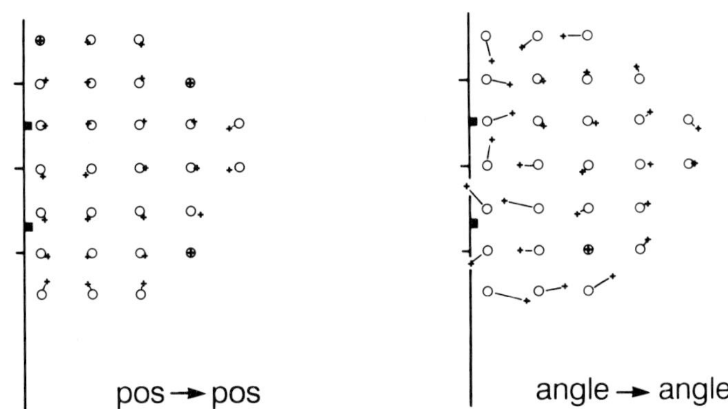

FIGURE 2
Comparison of the network's control of tarsus placement when trained to associate Cartesian coordinates with Cartesian coordinates (left) or joint angles
with joint angles (right). The body of the insect is illustrated schematically
by the line at the left; the two black squares indicate the insertions of the
middle and rear legs. The circles show the grid of middle leg positions tested;
the associated crosses indicate the positions of the rear leg tarsus produced by
the network.

performs best when the positions of both target leg and moving leg are specified in Cartesian coordinates is not surprising. In this formulation, the task
simply requires that a constant given by the separation between the insertions
of the legs be added to the position of the middle leg tarsus in order to obtain
the required position of the rear leg tarsus. The additive nature of the network
processing is well suited to perform this linear operation. However, the insect
does not have direct access to a Cartesian representation of the tarsus
position: it must begin with proprioceptive information on joint angles.

The surprise is that the simple network relying on additive mechanisms performs
so well with the other three formulations although an analytical solution
requires a non-linear coordinate transformation. In particular, the results
suggest that the task is especially difficult when the middle leg position is
given in joint angles and the Cartesian coordinates for the tarsus of the rear
leg must be produced. More significantly, the association of a particular rear
leg configuration in terms of joint angles with a middle leg configuration given
in terms of joint angles leads to a performance that is not noticeably worse
than that found in behavioral measurements on stick insects. This algorithm is
appealling because the nervous system is spared the need to perform a non-linear
coordinate transformation.

REFERENCES

(1)   Cruse, H., Physiological Entomology 4 (1979) 121.
(2)   Dean, J. and Wendler, G., J. exp. Biol. 103 (1983) 75.
(3)   Cruse, H., Dean, J. and Suilmann, M., J. Comp. Physiol. A 154 (1984) 695.
(4)   Cruse, H., J. Comp. Physiol. 112 (1976) 235.
(5)   McClelland, J.L. and Rummelhart, D.E., Explorations in parallel distributed
      processing. A Handbook of models, programs, and exercises (MIT Press,
      Cambridge, Mass, 1988)

Parallel Processing in Neural Systems and Computers
R. Eckmiller, G. Hartmann and G. Hauske (Editors)
Elsevier Science Publishers B.V. (North-Holland), 1990

A NEW DYNAMIC MODEL OF RECEPTIVE-FIELD ORGANIZATION

Hubert R. O. DINSE, Johannes BEST, Katharina KRÜGER

Institut für Neuroinformatik
Abteilung für Theoretische Biologie
Ruhr-Universität of Bochum
4630 Bochum, West Germany

Temporal oscillations in the firing pattern of neuronal responses are a characteristic feature of cortical neurons. Their receptive field and their tuning characteristics are built up over time. The spatial and temporal behaviour of cortical neurons is not separable.

1. INTRODUCTION

The concept of receptive fields (RFs) as one of the basic elements for information processing has been the guiding principle in the investigation of sensory systems since the work of Hartline in 1938 [1]. The most accepted RF-concept, proposed by Hubel & Wiesel in 1962 [2] for visual area 17, was based on static response properties and can be interpreted as a spatial filter or feature detector. According to their concept, the selectivity of cortical neurons for the geometrical dimensions of a stimulus, is generated by the spatial arrangement of the excitatory and inhibitory afferent inputs onto a neuron. One major problem of this interpretation is the lack of a genuine anisotropic dendritic field organization that matches the electrophysiological tuning characteristics of single cells [3]. In contrast, there is a remarkable uniformity in the architecture of the neocortex. It can be described as 2-dimensionally extended sheets of excitatory and inhibitory neurons that are vertically segregated into laminae constituting a type of multi-input/multi-output system, highly interconnected via intracortical feedback loops.

These anatomical constrains were incorporated into a computer model of cortical architecture [4]. The main outcome of the model was: 1. processing is strongly layer specific; 2. temporal oscillations occur; 3. positive feedback loops are responsible for the observed temporal response dynamics; 4. functional neuronal selectivity is derived through successive iterations within the feedback system; and 5. spatio-temporal behaviour is no longer separable into two independent terms.

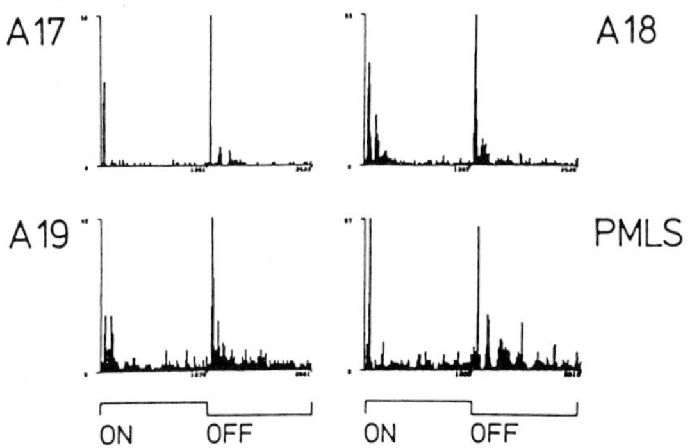

Fig. 1: Examples of PSTHs for neurons recorded in areas 17, 18, 19 and PMLS, respectively. Whole field on-off stimulation, which covered the entire receptive field. Repetition rate 2.5 sec (1.25 sec on, 1.25 sec off). Abcsissa: 2.5 sec; ordinate: spikes/bin. Bin width 10 ms. Time course of stimulation is indicated on the bottom.

In the following we report on experiments performed in the cortical visual areas 17, 18, 19 and PMLS of the cat. The outcome of physiological investigations support the extension of the RF-concept to the term spatio-temporal receptive field.

2. OSCILLATIONS IN NEURONAL RESPONSES

Oscillatory behaviour in the neuronal discharge is a fundamental attribute of cortical neurons (cat visual cortex [5]; monkey striate cortex, J. Krüger, pers. communication; rat somatosensory cortex, Dinse & Merzenich, unpublished; cat auditory cortex [6]). Our investigations in the visual cortex of the cat were performed with so-called whole-field stimulation, in which the size of the stimuli was adjusted to cover the entire excitatory RF. Most of the cortical neurons show a clear oscillatory type of response (Fig. 1). At least two thirds of all cortical neurons of our sample have an oscillatory type of response with a mean frequency of 7 to 8 Hz. This frequency decreases with the duration of oscillation and the first peak amplitude is mostly the largest. This response pattern is best described as a *damped aperiodic oscillation* with a length of up to 600 msec. The number of peaks (7 to 8) observed is different and the latencies to the first peak are shortest for area 18 neurons and longest for area 19 neurons. Area 17 and PMLS are intermediate [7].

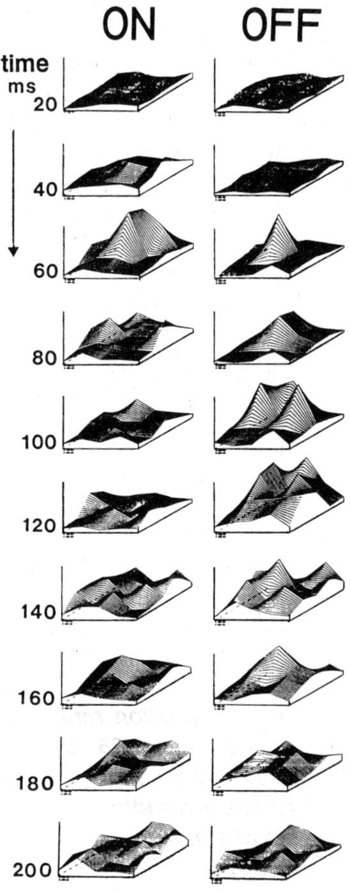

3. DYNAMICS OF RF-ORGANISATION

In order to evaluate the distribution of excitation within the receptive field as a function of both space *and* time, we subdivided receptive fields into 5 by 5 to 8 by 8 subfields and stimulated them in random order with spots of light, which were of the same size as the diameter of the small subfields. We obtained 25 to 64 single PSTHs, which were used to compile temporal sequences of response planes [8]. As these phenomena are basically dynamic, a more adequate illustration can be achieved by

Fig. 2: Sequences of response planes for a neuron recorded in area 17. Time steps 20 ms. On-off stimulation within a grid of 5x5 subfields was used. Size of the stimulation matrix was 2.5 deg (each subfield 0.5 deg). Ordinate is spikes/time step (in this example spikes/20 msec). This sequence illustrates the 2-dimensional RF organization for each moment (i.e. each 20 ms time step) over the first 200 ms after photic stimulation. The RF organization changes dramatically over time. As each spatial location exhibits a temporal characteristic of its own, space and time are not-separable in cortical cells that show such or similar behaviour.

viewing this sequence on a video in slow motion. The response planes obtained in this way were different for the on- and off-response of the same cell in approximately 80 % of all neurons. A distinct temporal response structure may be restricted to either on- or off- responses. Note that the on- and off-responses exhibit a quite different response pattern and the dimensions of the excitatory part of the RF in each time step clearly differ over time. In the sequences of response planes, the excitation appears as a 'wave' spreading over the entire receptive field. Each subfield within the receptive field exhibits its own temporal characteristic, i.e the oscillatory pattern elicited at different locations within the receptive field may be different, which is a clear indication that space and time are not separable.

4. TIME DEPENDENCE OF RF PROPERTIES

Orientation selectivity and hypercomplexity (end-stopping property) can be regarded as the most prominent RF properties of cortical visual neurons. They describe specific neuronal responses to orientations or lengths of an elongated stimulus.

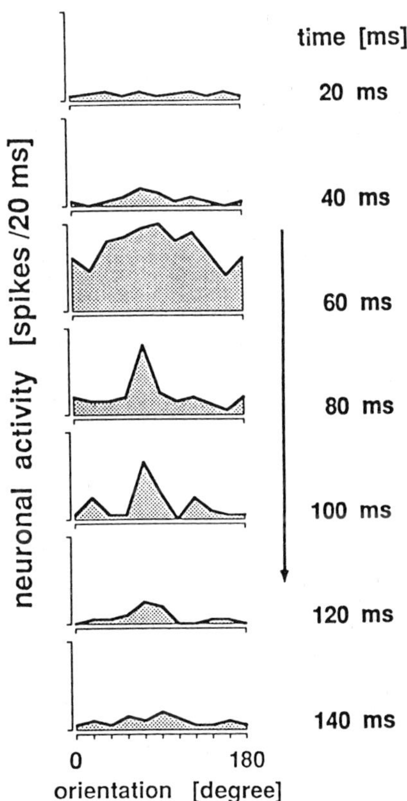

Accordingly, we used stationary flickered bars of lights (0.2 to 0.4 Hz) of different orientations and different lengths positioned in the RF centre. Similarly to the time course of RF-size, we investigated the time course of orientation selectivity and hypercomplexity. This was done by compiling sequences of tuning curves for time steps of 20 msec. In this case, each tuning curve describes the actual selectivity of a cell at that moment. In this way we avoid the temporal average that is usually put up with compiling conventional tuning curves based on measures of maximal spike rate. Following on-off stimulation with flashed bars of light presented at different orientations the neurons exhibit a first initial state of brisk excitation that does not exhibit a selectivity for any orientations.

Fig. 3: Sequence of orientation tuning curves found in area 17. Time step 20 ms. Stimulus was a bar of light, flickered at 0.4 Hz and set at different orientations. Abscissa: stimulus orientation in deg; ordinate: number of spikes elicited during each 20 ms time step.

After this initial stage, orientation selectivity evolves, indicated by a high level of selectivity for a given orientation. The time dependence of RF properties was also found for hypercomplexity. It is noteworthy that in all cases selectivity evolves from an early state of non-selectivity. Selectivity during the initial period of excitation and development of activity into an unspecific state was never observed.

5. FUNCTIONAL CONSEQUENCES

This spatio-temporal response structure enables neurons to analyze environmental stimuli in terms of elementary events rather than spatial features. These elementary events are the 'primitives' of early natural vision such as movements, accretion and deletion of points, and looming. However, the described dynamics were obtained under conditions, which allowed the neurons to reach steady state, and no contextual clues were available. Including clues such as optical flow information, can be assumed to have a strong impact on the observed dynamics. When moving stimuli were used, speed and direction of stimulus motion became crucial variables that determined the overall response characteristics, thus rendering RFs time- and stimulus-variant.

Both the considerable length of the temporal modulated response and the observation that high level of stimulus selectivity is reached delayed and independent of maximal spike counts, argues in favour of an essential role of a time code. Finally, the comparison of the time course of the temporal modulated response in different cortical areas, provides little evidence for a serial way of processing. In contrast, information seems to be processed in parallel, but temporally distributed. Interareal time dependent interaction could be a functional substrate for positive feedback acting beyond areal borders throughout the entire visual system. It could be the substrate that enables the formation of assemblies or functional groups across long distances.

The occurence of fully non-separable spatio-temporal RFs and the occurence of time dependence of RF properties in the visual cortex is evidence for the relevance of intracortical feedback. As shown theoretically by [4] and [9], spatially distributed feedback systems generally result in this type of spatio-temporal-interactions. Accordingly, cortices provide the framework of structure that leads to abundant positive feedback, which in turn favours the generation of oscillatory response pattern, as has been found in 4 cortical areas so far investigated. A manifestation of multiple positive feedback systems are non-separability of space and time, which makes RF organization time and stimulus dependent.

REFERENCES

[1] H.K. Hartline: The response of single optic nerve fibres of the vertebrate eye to illumination of the retina. Am. J. Physiol. 121, 400-415 (1938)

[2] D.H. Hubel, T.N. Wiesel: Receptive fields, binocular interaction and functional architecture in the cat's visual cortex. J. Physiol. Lond. 160, 106-154 (1962)

[3] C.D. Gilbert, T.N. Wiesel: Morphology and intracortical projections of functionally characterized neurons in the cat visual cortex. Nature 280, 120-125 (1979)

[4] G. Krone, H.A. Mallot, G. Palm, A. Schuz: Spatio-temporal receptive fields: a dynamical model derived from cortical architectonics. Proc. R. Soc. Lond. B 226, 421-444 (1986)

[5] H.R.O. Dinse: Informationsverarbeitung im visuellen System der Katze. Neuronale Grundlagen. Thieme Verlag, Stuttgart (1989a)

[6] C.E. Schreiner, P.X. Jaris: Intrinsic oscillations in the primary auditory cortex of cats. IUPS Satellite Symposium on Hearing. UC, San Francisco, pp. 81 (1986)

[7] J. Best, S. Reuss, H.R.O. Dinse: Lamina-specific differences of visual latencies following photic stimulation in the cat striate cortex. Brain Res. 385, 356-360 (1986)

[8] J.K. Stevens, G.L. Gerstein: Spatiotemporal organization of cat lateral geniculate receptive fields. J. Neurophysiol. 39, 213-238 (1976)

[9] W. von Seelen, H.A. Mallot, F. Giannakopoulos: Characteristics of neuronal systems in the visual cortex. Biol. Cybern. 56, 37-49 (1987)

Parallel Processing in Neural Systems and Computers
R. Eckmiller, G. Hartmann and G. Hauske (Editors)
© Elsevier Science Publishers B.V. (North-Holland), 1990

DIRECT OBSERVATION OF NEURAL ASSEMBLIES DURING NEOCORTICAL REPRESENTATIONAL REORGANIZATION

Hubert R. Dinse [1], *Gregg H. Recanzone* [2] & *Michael M. Merzenich* [2]

[1] Institut für Neuroinformatik, Theoretische Biologie,
Ruhr - Universität Bochum, D-4630 Bochum, West Germany
[2] Dept. Otolaryng., Coleman Laboratories, University of California,
San Francisco, CA 94143, USA

By studying correlated neuronal activity within neural assemblies before, during and after its rapid modification by intracortical microstimulation (ICMS), we were able to quantitatively demonstrate that this special case of cortical representational plasticity involves changes in the memberships and extents of nerve cell assemblies. Coupled neuronal groups enlarge progressively during ICMS. Equivalently short latencies to cutaneous stimulation are recorded all across these expanded groups, which are many times larger in area than are coupled neural assemblies recorded prior to stimulation. The observed changes in correlated activity are directly related to quantitatively assessible changes of cortical receptive field organization and thus cortical topographic maps.

These experiments confirm that cell assemblies or neuron groups in adults can be dynamically modulated in their functional connectivity, spatial extents and memberships. It is concluded that these dynamics play a central role in adult representational plasticity.

1. INTRODUCTION

One characteristic feature of mammalian cortex is the topographic representation of the physical world within the brain. The resulting two-dimensional maps in different sensory systems have long been regarded as fixed, hard-wired entities, modifiable during development but rather rigid in adults.

However, recent experiments have clearly demonstrated that there is a considerable capacity for functional reorganization in adult primates (and other mammals) following different experimental manipulations [for overview see 1-3]. Most interestingly, hand surface representations in the neocortex are reorganized in normal monkeys by tactile behaviors in which restricted skin surfaces are differentially stimulated in a digit contact, pressure-regulation task [4], or in a temporal discrimination task [5].

On the basis of the details of receptive field and topography changes recorded in these and other related experiments, it has been hypothesized that representational dynamics involves mechanisms of *input-selection* [6] which occur by cooperation within and between coupled groups of neurons [1,2,6-8]. Based on neuroanatomical observations of cortical feedback architecture, the idea of cell assemblies as a fundamental cortical processing level has been put forward [9]. In addition, recent studies of temporal aspects of receptive field organization in cat visual system revealed the existence of intrinsic dynamics that alter and modify RF properties [10].

To test these hypotheses, we have used intracortical microstimulation (ICMS), which produces rapid changes in cortical topographic representations that are comparable in dimension to those produced by the above-described experiments [11-15]. In this series of experiments, we simultaneously recorded from pairs of units at separated cortical loci to study the spatial extent and modifiability of correlated activity. We also determined how receptive field organization, response magnitudes, response latencies and cortical topographies were changed as neuron assembly coupling was altered by ICMS.

2. METHODS

Changes in skin representation were produced by ICMS in acute experiments performed in adult barbiturate-anesthetized rats (*Sprague-Darley*) and in adult monkeys (*Aotus trivirgatus, Saimiri sciureus*). Parameters of ICMS: 5 µA charge-balanced pulses, 200 µsec/phase delivered in 300 pps bursts of 40 ms duration, applied at 1/sec. To obtain PSTHs of neuron responses, a feedback-controlled (Chubbuck) skin stimulator was used to apply rectangularly ramped and sinusoidal stimuli of different intensity. Four different protocols were followed to investigate and relate alterations of neuronal discharge to representational changes.

PROTOCOL 1: Assessment of representational changes before and after ICMS by 30 - to 50 - penetration maps.

procedure: Control maps of parts of the hand representation in area 3b of new world monkeys were established by means of multiple penetrations separated by 150 to 200 microns. After 4 to 8 h of ICMS, topographic maps were again derived over the same representational sectors.

PROTOCOL 2: Quantitative assessment of ICMS-induced alterations of receptive field (RF) organization determined by measuring RF response planes, RF profiles, sensitivity profiles and sensitivity equivalences.

procedure: a) *RF response planes*. Responses to quantitatively controlled stimulation at different sites within the RF were sampled in random order, and used to compile off-line 2-dimensional landscapes of excitation throughout RFs. b) *RF profiles*. Same as response planes, but along one dimension. c) *Sensitivity profile*. Neuronal response strengths were plotted as a function of stimulus amplitude. Neuronal activity measures used were spike frequency and onset latency. d) *Sensitivity equivalence*. As in *c*, but for pairs of neurons that were simultaneously recorded and stimulated. In this case, the response of one neuron was used as a standard against which the response strength of the second neuron was defined.

PROTOCOL 3: Recording pairs of neurons simultaneously at separated cortical loci.

procedure: Spontaneous, on-going neuronal activity was recorded on two independent glass microelectrodes, or on solid state multichannel microelectrodes that had 4 recording sites separated at 75 microns which allowed a reliable assessment of distance between neuron pairs. Correlated activity was measured by computing cross correlograms for a tau of +/- 50 ms. Correlation strength (CORR) was expressed as the weighted sum of the ratio of the number of correlated events within windows of 1 to 20 ms around tau = 0 and the total number of correlated events during +/- 50 msec around tau = 0. Strength of correlated activity was measured a) as a function of cortical distance before and after ICMS, and b) as a function of ICMS-induced reorganization of cortical maps.

PROTOCOL 4: Time course of ICMS-induced representational changes.

procedure: The time course was investigated by simultaneously recording from pairs of neurons while ICMS was applied through one microelectrode and the effects of ICMS were monitored at the other, or at a control time, and for different times after initiation of the ICMS. The time course was assessed for a) correlation strength (CORR); b) RF size; c) RF position; d) RF overlap; and e) sensitivity equivalence.

3. RESULTS

Experiments were performed in 24 rats and 3 new world monkeys. In the following the main experimental results are described according to the 4 protocols described in Methods.

PROTOCOL 1: Intracortical microstimulation alters cortical skin representation maps in SI of adult rats and in area 3b of adult owl and squirrel monkeys. Changes are limited to an area of about 1 mm in diameter around the stimulation site. These changes were usually asymmetric, commonly revealing an elongated zone of reorganization. These changes can be related to alterations of the underlying RF organization. Close to the ICMS site (within 100 to 400 microns) there was a shift of RF position within the skin field [13]. For recording sites further away there is an enlargement of RF sizes, with RFs extending towards the stimulation-site RF. Both effects contributed to an overall several-fold enlargement of the cortical zone over which RFs overlap with the stimulation-site RF.

PROTOCOL 2: The above experiment related cortical map alterations to alterations of RF organization. They led us to quantitatively investigate how RFs are changed during ICMS. Measurements of *RF response planes* and *RF profiles* confirmed the above described results. We found that the internal RF structure becomes distorted by ICMS in a very systematic way. Under control conditions, cutaneous RFs can best be described in terms of probability distributions of excitation. However, these Gaussian-like RF profiles tend not to be symmetric, but show considerable anisotropic distributions of excitation. Thresholding those distributions resulted in RF contours that were in general agreement with those obtained by handplotting. After ICMS, *RF profiles* showed a distortion of their inner distributions of weights. Those RFs close to the stimulation site shifted their maximal response locations towards the centres of the stimulation site RF without changes in the extents of the distributions. In the end, these RFs could correspond closely with the stimulation site RF. In RFs located further away from the stimulation site, the location of maximal excitation was uneffected, while a skin region that was not driven under control conditions and that extended towards the stimulation site became excitable, leading to an enlargement of the excitatory field distributions specifically in the direction of the stimulation site RF.

Sensitivity equivalence was established by comparing sensitivity profiles for simultaneously recorded pairs of neurons. Under control conditions, when the two neurons had clearly non-overlapping RFs, the sensitivity profiles showed sharp differences in all parameters of response sensitivities, i.e. in threshold, in steepness and in saturation. After ICMS, for skin stimulation in the center of the original cortical stimulation sites RF, the recording site responded so that their thresholds, steepness of their sensitivity functions and response saturation levels commonly came to match the sensitivity and characteristic transfer properties at the stimulation site hundreds of microns across the cortex.

Systematic examination of measures of neural activity revealed that generally response latency was considerable more sensitive in detecting changes of neuronal excitability than response magnitude measures such as spike frequency. This finding suggests that during representational plasticity neuronal discharge is less effected in terms of spike counts, but can more appropriately described in terms of changes of the temporal pattern of the responses.

PROTOCOL 3: Under the assumptions that map changes reflect cooperative processing within a large number of single but interconnected elements we investigated the temporal interactions of pairs of neurons during this ICMS-induced plasticity by means of cross correlogram analysis to describe quantitatively changes in neuronal cooperativity. To avoid problems associated with the separation of stimulus-driven activity from on-going activity, only spontaneous on-going activity was used to measure correlated activity [16].

As a first step, correlated activity was measured before and after ICMS for pairs of neurons separated by different distances. These measurements revealed that correlated activity dropped to chance level at separations of 200 to 250 microns. After various periods of ICMS, correlation strength invariably increased for neuron pairs in this central zone. Most notable, however, was an emergence of correlated activity for neuron pairs separated by 300 to 800 or more microns. This emergent functional coupling was similar in strength to that observed for the closely spaced neuron pairs in the central zone reported under control conditions.

Changes in correlated activity were related to changes of the underlying map by combining cross correlogram analysis with the above-described mapping techniques. These results revealed that changes of correlated activity were restricted to those regions of cortex that underwent reorganization of their skin representation. Increases of correlated activity were highest close to the stimulation site, but were seen for neuron pairs more than 800 microns away from the stimulation site. On the other hand, when pairs were separated by only 300 to 400 microns, but one recording site was clearly outside the reorganized region, flat correlograms were obtained.

It is important to note that we never encountered exclusively narrowly peaked correlograms indicating monosynaptic interaction due to direct excitation or common input. By contrast, we observed broad peaks with a half-width of 10 to 20 ms that were always centered around tau = 0. For closely spaced neuron pairs, some cross correlograms for neuron pairs could be interpreted as representing superposition of a narrow peak with an additional broader one. We conclude that the broadly peaked correlograms can best be explained not in terms of direct connections, but in terms of positively coupled neurons that share some common input sources. According to this view, ICMS changes the state of small neuron assemblies by changing the intrinsic temporal discharge pattern. The observed widths of the peaked correlograms in the range of 20 ms indicate the crucial time scale during which those shifts of temporal synchronized discharge patterns and thus modulation of neuron memberships and extents of cell assemblies might occur.

PROTOCOL 4: The time course of ICMS-induced plasticity was studied by monitoring the effects through one microelectrode while applying ICMS throught the other one. This study was limited by the fact that the data acquisition for one point took about 5 to 8 min, which made it impossible to sample with a finer temporal resolution. However, clear effects were obtained after an initial stimulation period of ICMS of about 15 min. These effects increased progressively during the next several hours and seemed to reach a steady state after about 3 to 4 hours. When stimulation was terminated after 5 to 8 hours, enduring effects were recorded for several hours, i.e. for as long as we maintained these preparations. The observed time courses of ICMS-induced changes in correlation strength, RF size and sensitivity equivalence closely paralleled one another.

4. CONCLUSIONS

These results reveal that the effective functional conectivity and spatial extents of coupled neurons in cortical *groups* or *assemblies* and the extents of a cortical zone over which RFs overlap can be dramatically altered by very low current level intracortical microstimulation, effectively exciting cortical neurons and processes over a cortical zone not greater than 5 to 10 microns across [11]. This experimental finding supports earlier arguments that cortical groups or assemblies are dynamically determined and provides direct evidence that discharge coincidence plays a crucial role in the formation and modification of functional neuronal groups [1,2,6-9,16-18]. That is consistent with earlier arguments that receptive field organization and representational topography must be time-based [1,2,6,7,10]. Measured correlograms indicate that the time constant for coincidence of discharge pattern is of the order of 10 to 20 ms.

Changes in cortical group coupling have been recorded over long distances within just a few minutes of initiation of ICMS, with stimulation schedules and response magnitudes that plausibly reflect those generated by natural stimulation. This rapid and powerful modulation of neuronal assemblies is consistent with the hypothesis that temporally based intracortical self-organizing processes operate continually to establish neuronal group constituencies. This would result in highly idiosyncratic and continually changing cortical networks.

Supported by NIH grant NS-10414, the Coleman Fund, the Max Kade Foundation and the Deutsche Forschungsgemeinschaft. The authors thank Dr. David Anderson and his collegues at the University of Michigan, who provided us with the solid state multichannel electrodes used in some of these experiments.

REFERENCES

[1] Merzenich, M.M., Jenkins, W.M. and Middlebrooks, J.C., in: Edelman, G.M., Cowan, W.M. and Gall, W. (eds.), Dynamic Aspects of Neocortical Function (Wiley, New York, 1984) pp. 397.

[2] Merzenich, M.M., Recanzone, G.H., Jenkins, W.M., Allard, T. and Nudo, R.J., in: Rakic, P., Singer, W. (eds.), Neurobiology of Neocortex. Dahlem Konferenzen 1988, (Wiley, New York, 1988) pp. 41.

[3]   Rasmusson, W.J., J. Comp. Neurol. 205 (1982) 313.

[4]   Jenkins, W.M., Merzenich, M.M., Ochs, M.T., Allard, T. and Guic-Robles, E., J. Neurophysiol. 63 (1990) in print.

[5]   Recanzone, G.H., Jenkins, W.M., Hradek, G.T., Schreiner, C.E., Grajski, K.A. and Merzenich, M.M., Neurosci. Abstr. 15 (1989) 1223.

[6]   Edelman, G.M. and Finkel, L., in: Edelman, G.M., Cowan, W.M. and Gall, W. (eds.), Dynamic Aspects of Neocortical Function (Wiley, New York, 1984) pp. 653.

[7]   Merzenich, M.M., Nelson, R.J., Kaas, J.H., Stryker, M.P., Jenkins, W.M., Zook, J.M., Cynader, M.S. and Schoppmann, A., Comp. Neurol. 258 (1987) 281.

[8]   von der Malsburg, C., in: Changeux, J.P., Konishi, M. (eds.), The Neural and Molecular Basis of Learning. Dahlem Konferenzen 1987, (Wiley, New York, 1987) pp. 411.

[9]   Braitenberg, V., in: Palm, G., Aertsen, A.M.H. (eds.), Brain Theory (Springer, New York, 1986) pp. 81-96.

[10] Dinse, H.R.O., Krüger, K. and Best, J., in: Krüger, J. (ed.), Neuronal Cooperativity - Models and Experiments (Springer, New York) in print.

[11] Asanuma, H., Fernandez, J., Scheibel, M.E. and Scheibel, A.B., Exp. Brain Res. 20 (1974) 315.

[12] Nudo, R. and Merzenich, M.M., Neurosci. Abstr. 13 (1987) 1596.

[13] Recanzone, G.H. and Merzenich, M.M., Neurosci. Abstr. 14 (1988) 223.

[14] Dinse, H.R.O. and Merzenich, M.M., Neurosci. Abstr. 15 (1989) 1223.

[15] Dinse, H.R.O. and Merzenich, M.M., in: N. Elsner, W. Singer (eds.), Dynamic and Plasticity in Neuronal Systems (Thieme Verlag, 1989) pp. 12.

[16] Aertsen, A.M., Gerstein, G.L., Habib, M.K. and Palm, G., J. Neurophysiol. 61 (1989) 900.

[17] Edelman, G.M. and Mountcastle, V.B., The Mindful Brain (1978) Cambridge, MIT Press.

[18] Gerstein, G.L., Bedenbaugh, P. and Aertsen, A.M., IEEE 36 (1989) 4.

[19] Shaw, G.L., Silverman, J.C. and Pearson, J.C., in: Palm, G., Aertsen, A.M.H. (eds.), Brain Theory (Springer, New York, 1986) pp. 177.

Parallel Processing in Neural Systems and Computers
R. Eckmiller, G. Hartmann and G. Hauske (Editors)
© Elsevier Science Publishers B.V. (North-Holland), 1990

Large-Scale Simulation of a Self-organizing Neural Network: Formation of a Somatotopic Map

K. Obermayer, H. Ritter and K. Schulten

Beckman-Institute and Department of Physics
University of Illinois at Urbana/Champaign, Urbana, IL 61801

Abstract:

The "somatotopic map" of the body surface of animals and humans reflects an ordered, neighborhood preserving connectivity between tactile skin receptors and cortical neurons. The border between adjacent areas representing different body parts changes even in adult animals, which shows, that the map is dynamically maintained. This makes this system an interesting model for the investigation of cortical plasticity.

In this paper we present a large-scale simulation study of a neural network model for the formation and readaptation of a somatotopic map of the inner hand surface. The network, which is based on an algorithm developed by Kohonen [3,4], contains 16,384 neurons and 800 tactile receptors, giving a total of 13,107,200 adaptive connections.

The simulation assumes random initial connections between receptors and neurons, and the formation of the map proceeds during randomly applied, local stimuli. The number of necessary adaptive steps was surprisingly small, which indicates, that a high dimensional input space facilitates the ordering process. Neurons with double and multiple receptive fields can emerge during simulation. The network readapts upon partial deprivation of sensory input much in the same way as is found in experiments [8].

1. Introduction

Topographic representations of sensory surfaces within the cortex are a widespread architectural feature of the brain of higher animals. They can be found in nearly all sensory and motor areas within the brain - in the visual, auditive and somatosensory fields as well as in the motor-cortex [1,2,12,13,14] - and it is an intriguing notion that maps of more abstract features might even play a role on higher processing levels.

In the somatosensory system the complete body surface is mapped onto a certain part of the cortex called somatosensory cortex. In a series of experiments on the cortical representation of the hand surface in owl monkeys Merzenich et al. [7,8] characterized this topographic representation and showed, that the somatotopic map is dynamically maintained even in adult animals. After depriving a certain area of the cortex from its sensory inputs (e.g. by nerve section or by amputation of a digit) part of this area is "silent" upon stimulation of the receptor surface. But within the following months the representations of the adjacent digital areas expand and approach each other in the reorganizing map.

Several models have been designed to explain the formation and plasticity of topographic maps within the cortex. Malsburg and Willshaw [5] introduced a model to explain the development of a topographical projection by fibers growing out of a neuronal sheet into another layer. Pearson et al. [9] proposed a medium scale neural network model of locally connected excitatory and inhibitory cells receiving projections from two receptor sheets corresponding to the glabrous and dorsal surfaces of the hand. A certain topographic order was established at the beginning, but the receptive fields were still large and the projection fuzzy. Following repeated stimulation of the receptor sheet the "model cortex" breaks into a pattern of "neuronal groups" of specific cells with similar receptive fields forming a topographic map. Contrary to the Malsburg-Willshaw model part of the formation process was due to an input-selection mechanism based on synaptic plasticity. A third type of network is based on an algorithm proposed by Kohonen [3,4]. In this algorithm each neuron can access all inputs at the beginning, which corresponds to anatomical connections between all cells and all receptors on the receptor surface. Formation of a map proceeds by input-selection alone. Although this algorithm in its original version does not include much biological detail it embodies a (minimal) set of clearcut principles, which allow to mimic the experimentally observed behaviour.

In previous work Kohonen's algorithm has been successfully applied to modelling the formation and readaptation of maps in the auditive [6] and somatosensory [10] cortex as well as the formation of maps of more abstract "semantic" features pertaining to language [11]. The simulations were restricted to relatively small networks with maximally a few hundred neurons, which received their input from a low dimensional input space. Although the results showed good qualitative agreement with the experimental findings it remained unclear, how the features of the model system will change, when the system is scaled up to a biologically more realistic size.

In this paper we report a simulation of a fairly large scale neural network comprising 16,384 neurons, which are connected to 800 receptors providing an 800-dimensional input vector. The simulation was carried out on a Connection Machine CM-2 using one processor for each neuron and its connections.

2. Model and Algorithm:

Our model consists of a hand-shaped sensory surface containing 800 randomly distributed tactile receptors and a two-dimensional grid of 16,384 formal "neurons" (128×128) modelling the corresponding somatosensory area within the cortex. Each neuron is connected to each receptor leading to a total of 13 million adaptive connections, whose initial strengths are choosen at random and which are modified during the simulation. The tactile stimuli are modelled by localized Gaussian excitation profiles, which describe the output r_i of receptor i at position $\vec{x}_i$ as a function of its distance from the center $\vec{x}_s$ of the stimulus by:

$$r_i = A \exp\left[-(\vec{x}_i - \vec{x}_s)^2/\sigma_r^2\right] \tag{1}$$

The width σ_r and the "intensity" A of the stimuli were held constant throughout the simulation. For each adaptation step the center $\vec{x}_s$ of the stimulus was choosen at random. Each neuron (k, l) at position $\vec{y}_{kl}$ of the neuronal sheet computes a weighted sum:

$$o_{kl} = \sum_i w_{kli} r_i \tag{2}$$

over all receptor outputs, where w_{kli} denotes the connection strength from receptor i to neuron (k, l). The input for each neuron is therefore described by an 800-dimensional vector $\vec{r} = (r_1, r_2...r_{800})^T$. Following the algorithm of Kohonen the neuron (r, s) with the maximal sum o_{rs} is selected and the output o_{kl} of the neuron (k, l) is replaced by a Gaussian output function $h_{rs;kl}$ centered at the position $\vec{y}_{rs}$ of the selected neuron:

$$h_{rs;kl}(t) = \exp\left[-(\vec{y}_{rs} - \vec{y}_{kl})^2/\sigma_h^2(t)\right] \tag{3}$$

The width $\sigma_h(t)$ decreases during the simulation from an initial value σ_i to a final value σ_f to allow the neurons to specialize for a certain part of the input space. The introduction of the output-function $h_{rs;kl}$ is an algorithmic "shortcut" to account for the effect of lateral connections between the neurons. The connection strengths are then changed according to a Hebb-type learning rule:

$$w_{kli}(t + 1) = (w_{kli}(t) + \epsilon(t)h_{rs;kl}(t) \cdot r_i)/\Sigma_i w_{kli}(t) \tag{4}$$

The learning step width $\epsilon(t)$ decreases exponentially from an initial value ϵ_i to a final value ϵ_f.

3. Results of the Simulations:

Fig. 4.1 shows the initial state of the network. On the left we see the receptor surface. Each receptor location is indicated by a black dot. In the centermost picture, different brightness values for these receptor dots have been used to indicate the initial random connection strengths from the receptors to a typical neuron. The rightmost picture shows the view on the model cortex. Each pixel represents a neuron (k, l), and its gray value coincides with the gray value of the location $\vec{s}_{kl}$ (defined by: $\vec{s}_{kl} = \sum_i w_{kli}\vec{x}_i$) of its receptive field center in Fig. 4.1 (left). The almost uniform gray results

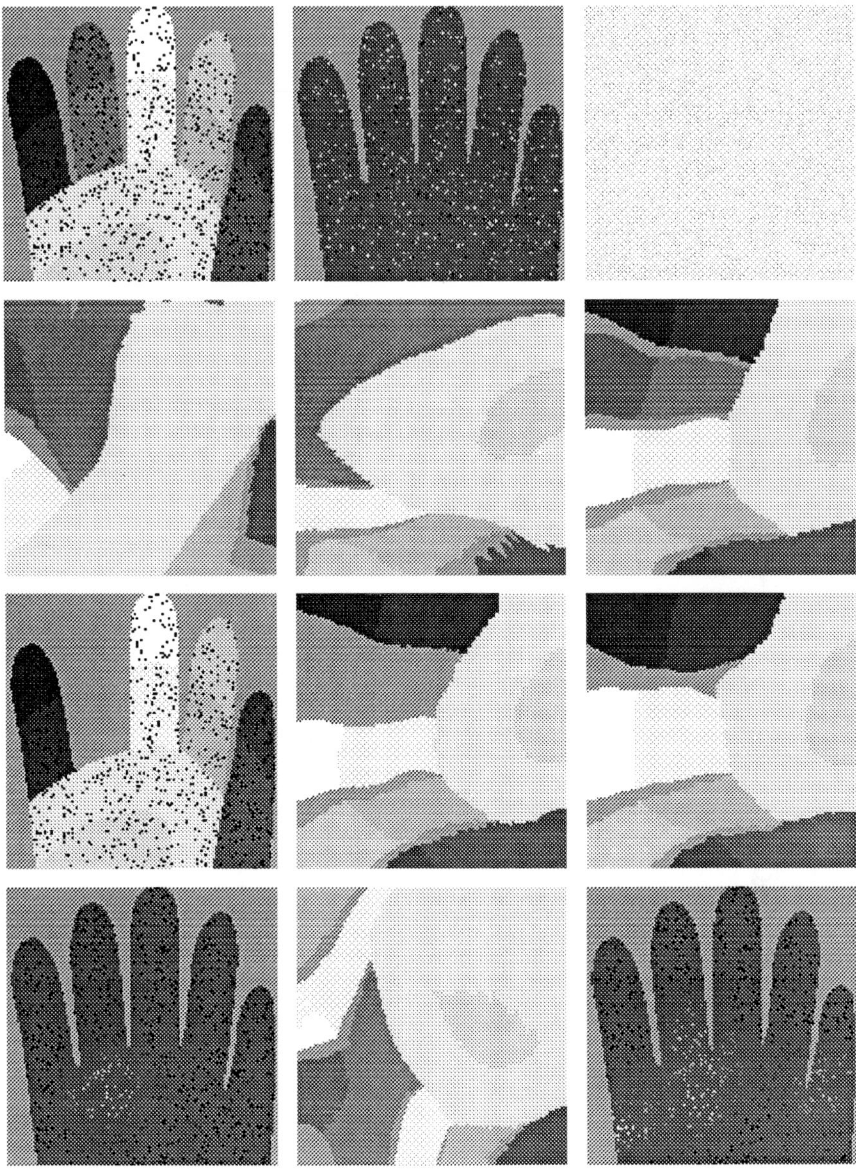

Fig.4.1 (top row): Initial state. Left: receptor surface, center: typical receptive field, right: neuronal sheet. **Fig.4.2** (row 2): somatotopic map after 200, 1200 and 10000 steps. **Fig.4.3** (row 3): "Amputated hand", and map immediately and 10000 steps after "amputation". **Fig.4.4** (bottom row): left: typical receptive field of adapted map, center: imperfectly ordered map, right: example of multiple receptive field.

from the initial connectivity: each cell's receptive field is diffusely extended over the entire receptor surface. Consequently all positions $\vec{s}_{kl}$ lie in the center of the sensory surface.

After about 200 stimuli the neurons begin to specialize (Fig. 4.2 left). The receptive fields shrink as the cells get more and more specific, but cells with double and multiple receptive fields may develop, especially during a stage, where the map is fragmented and several representations of the sensory surface compete (see below). After 1200 stimuli a crude map of the hand-surface has emerged with four fingers already present in the correct topographic order (Fig. 4.2 center), but it took about 10,000 stimuli to complete and stabilize the map (Fig. 4.2 right). Parameter values of this simulation were $\sigma_i=40$, $\sigma_f=20$ (units of the 128×128 grid), $\sigma_r=0.12$, $\epsilon_i=0.2$, $\epsilon_f=0.1$.

The number of adaptive steps necessary to form the map is less than the number of cells in the network. Previous simulations of Kohonen networks using low-dimensional input vectors required a much larger number of steps (20,000 steps for a network with 900 cells and a two-dimensional input-space [10]) although the network was much smaller. Note that σ_h is only 30% of the diameter of the receptor surface. These results indicate, that a high-dimensional input space strongly facilitates the topographic ordering process.

Fig. 4.3 shows the simulation of the "amputation" experiment performed by Merzenich et al. [8] ($\sigma_i=20$, $\sigma_f=10$, $\sigma_r=0.12$, $\epsilon_i=0.1$, $\epsilon_f=0.05$). The fourth digit of our model hand was removed (Fig. 4.1 left), depriving the neurons of the corresponding part of the neuronal sheet from their sensory input (Fig. 4.3 center). During subsequent stimulation the somatotopic map gradually readapts and the expanding representations of the adjacent areas approach each other. After 10,000 further stimuli the new map has stabilized (Fig. 4.3 right). It is interesting to note, that the "cortical" reorganization almost exclusively affects the neighboring areas of the deprived region, although each cell is in principle connected to each receptor: The size of the "digit 1" and "digit 2" representations changes little, while the size and location of the "digit 3" and "digit 5" areas changes dramatically as it is found in the corresponding experiments.

Fig. 4.4 (left) shows a typical receptive field for a neuron in the topographic map of Fig. 4.2 right. It is well localized, with width roughly of the diameter of the applied stimulus. Note that although most of the synaptic strengths decay to zero except for those whose receptors form the receptive field, each cell keeps its connections to all receptors so that synapses can be "revived" if the stimulus distribution changes.

Fig. 4.4 (center) shows the result of another simulation leading to an incompletely ordered map. The second, third and fourth digit are mapped twice to the neuronal sheet. Configurations like this are generally accompanied by patches of cells with double or multiple receptive fields (Fig. 4.4 right). Neurons of this type allow to map distant locations on the receptor surface onto neighboring locations of the cortical sheet without introducing a discontinuity in the response properties of the cells. These neurons have been found within the somatosensory cortex, but they are seldom, perhaps because of their correlation with defects in the topographic maps.

4. References:

[1] Lemon R. (1988), Trends in Neur. Sci. (Vol. 11) 11:501

[2] Kaas J.H. et al. (1983), Annu. Rev. Neurosci. 6:325

[3] Kohonen T. (1982a), Biol. Cybern. 43:59

[4] Kohonen T. (1982b), Biol. Cybern. 44:135

[5] von der Malsburg C., Willshaw D.J. (1977), PNAS USA 74:5176

[6] Martinetz T. et al. (1988), SGAICO-Proceedings "Connectionism in Perspective", Zurich 1988

[7] Merzenich M.M. et al.(1983), Neurosci. 10, 3:639

[8] Merzenich M.M. et al.(1984), J. Comp. Neu. 224:591

[9] Pearson J.C. et al. (1987), J. Neurosci. 12:4209

[10] Ritter H., Schulten K. (1986), Biol. Cybern. 54:99

[11] Ritter H., Kohonen T. (1989), Biol. Cybern. 61:241

[12] Sparks D.L., Nelson J.S. (1987), TINS 10:312

[13] Suga N., O'Neill W.E. (1979), Science 206:351

[14] Tunturi A.R. (1952), Am. J. Physiol. 162:489

Parallel Processing in Neural Systems and Computers
R. Eckmiller, G. Hartmann and G. Hauske (Editors)
© Elsevier Science Publishers B.V. (North-Holland), 1990

SYNCHRONIZED ACTIVITY IN THE CA1-REGION OF THE RAT HIPPOCAMPUS REVEALS SPATIO-TEMPORAL CSD-PATTERNS OF ACTION POTENTIALS

Dietmar PLENZ and Ad AERTSEN

Max-Planck-Institut für biologische Kybernetik, D-7400 Tübingen, FRG

Using the one-dimensional current source density (CSD) analysis in the CA1 region of the rat hippocampal slice, we demonstrate stereotyped 'finger prints' of epsp's and action potentials in this part of the brain. On the basis of their spatio-temporal patterns, the measured CSD-profiles can be divided into two classes. In the simpler class we only find two elementary patterns: one corresponding to the epsp, the other to an action potential, originating in the cell body region. The second, more complex class in addition shows the pattern of another, slightly earlier action potential, which is shifted with respect to the second action potential in the direction of the proximal dendritic region.

1. Introduction

In order to investigate the spatio-temporal spread of activity in brain tissue, several recording techniques can be applied. In the context of an investigation into the space-time dynamics of neuronal activity in brain slices we are currently combining the optical recording technology, using voltage-dependent fluorescent dyes [1] and current source density (CSD) analysis [2].
Most experimental findings using the CSD-technique are based on recordings in the cerebellum and the neo-cortex, in particular the visual cortex (references can be found in [3]). In this paper we present results of application of the CSD-technique to slices from the adult rat hippocampus. Special emphasis will be on the description and interpretation of the spatio-temporal patterns revealed in the current distribution.

2. Methods

For the preparation of hippocampus slices, adult rats (Lewis) at the age of 4 to 6 weeks were anaesthetized with ketamine (150 mg ketamine/kg ip) and decapitated. The brains were quickly removed and cooled down in a 4 °C ACSF-solution (124 mMol NaCl, 1.25 mMol KH_2PO_4, 5 mMol KCl, 3.6 mMol $CaCl_2$, 1.3 mMol $MgSO_4$, 26 mMol $NaHCO_3$, 9.4 mMol D-Glucose). For the first part of the preparation 25 mg ketamine/100 ml ACSF-solution was added (ACSFKet). Transverse slices (400 μm thick) of the hippocampus were cut with an egg slicer. Slices were stored in ACSFKet, saturated with Carbogen ($95\%O_2/5\%CO_2$) at 33.0 ±0.1 °C on thin pieces of micropur filter. The flow rate in the storage chamber was set at 1 ml/min. After half an hour the ACSFKet solution was exchanged for pure ACSF.

For recording, the slices were transferred to the recording chamber, which was continuously perfused with ACSF at a flowrate of 0.5 ml/min at a highly constant temperature (33.0±0.1 °C). The slices were lying submerged on a thin glass coverslip and were held down by a nylon net. For the electrical stimulation of the Schaffer collaterals we used monopolar tungsten electrodes. Pulses with a duration of 50 μs and a constant amplitude in the range of 150-500 μA were applied at a frequency of 0.2 Hz. We recorded extracellular field potentials with a glass micropipette, filled with 3M NaCl (6-12M). Recordings were made at a constant depth in the range between 70 and 100 μm below the surface of the slice. Successive recording positions were separated by 30 μm and were situated along a linear electrode-path, orthogonal to the CA1 pyramidal cell layer (Fig. 1c: rec., electrode-path). At each recording site 10 consecutive responses were recorded and used for averaging later on. We started the experiment once the extracellular field potential became stable for at least 30 min. During the experiment the stability of the recording was checked by a second electrode measuring the field potential at one location in CA1 throughout the entire experiment (Fig. 1c: cont.).

Signals were recorded with a conventional electrophysiological setup, bandpass filtered between 1 Hz and 1 kHz, digitized at a rate of 3.3 kHz and stored on a PDP 11/73. From there the data were transferred to a Vax 750 for further processing. From the sets of 30 to 35 field potentials per electrode-path (each potential being the result of averaging over 10 trials) we computed CSD-profiles according to the method described by Mitzdorf and Singer [4]. We made the assumptions that (1) the extracellular space has the properties of an Ohmic conductor with nearly constant conductivity in the direction of the electrode-path and (2) that the electric field is quasi-static. In that case the extracellular CSD-profile is proportional to the second order spatial derivative of the field potential distribution. Furthermore, because of the only slight changes of the field potentials along the CA1 layer as compared to the profound changes along the axis perpendicular to the CA1 layer [5], we approximated the true CSD-distribution by applying the one-dimensional CSD-method [4] according to the formula:

$$\frac{d^2\Phi(z)}{dz^2} \simeq \frac{\Phi(z+n\cdot\Delta z) - 2\cdot\Phi(z) + \Phi(z-n\cdot\Delta z)}{(n\cdot\Delta z)^2}$$

where $\Phi(z)$ is the field potential at point z, Δz is the sampling interval between adjacent locations, and $n\cdot\Delta z$ is the differentiaton grid size. In all CSD-profiles shown here the sampling interval was 30 μm and n was set equal to 2.

3. Results

We calculated 14 CSD-profiles from 14 field potential distributions in as many slice experiments. Based on the spatio-temporal characteristics, these CSD-profiles could be subdivided into two qualitatively different classes. From each of these classes we present a typical example.

Figure 1a shows a typical field potential distribution (27 traces) from the first and simpler class; the corresponding CSD-distribution is given in Fig. 1b. The lower traces originate from the outer region of CA1, while the upper traces refer to the distal dendritic region (cf. the electrode-path Fig. 1c). The horizontal axis denotes time, with the stimulus onset on the left. The location of the stimulating and control electrodes are indicated in Fig. 1c. Following the convention for displaying CSD-profiles, sinks are drawn upwards and sources downwards.

In the upper half of Fig. 1b (stratum radiatum) the CSD-profile shows a sink and an adjacent source below it. The lower half of Fig. 1b (strat. pyramidale & strat. orientale) is considerably more complex, showing first (i.e. on the left) a central sink with corresponding sources above and below it, immediately followed by a nearly inversed sink/source distribution to the right. Both sink/source profiles are schematically drawn in Fig. 1d. Note that the first two sources in the lower part are very narrow in time and have an increasing delay with growing spatial distance to the central sink.

We propose that this CSD-profile is generated by synaptic input from the Schaffer collaterals, which causes the early sink/source distribution in the stratum radiatum, whereas the CSD-profile in the lower half of the picture corresponds to the highly synchronized action potentials generated in the stratum pyramidale. In this view the sink in the upper part is caused by sodium influx compensated by passive sources in the distal dendritic region. The primary sink in the lower part is caused by sodium influx compensated by passive sources in the axonal and the proximal dendritic regions of CA1 cells, whereas the following source is generated by potassium efflux and active or passive compensations from adjacent locations. A more detailed explanation of the spatio-temporal characteristics of the various features in the CSD-profile obviously should incorporate the effect of the propagation of action potentials.

Keeping these elementary patterns for epsp's and action potentials in mind, it is easy now to interpret the results in Fig. 2. This picture presents a typical field potential distribution and the corresponding CSD-analysis from the second, more complex class of CSD-profiles. Although a visual comparison of the field potential distributions in Figs. 1a and 2a does not reveal any significant difference, this is clearly not the case for the corresponding CSD-patterns. In Fig. 2b the CSD-profile is composed of three elementary patterns, schematically drawn in Fig. 2c.

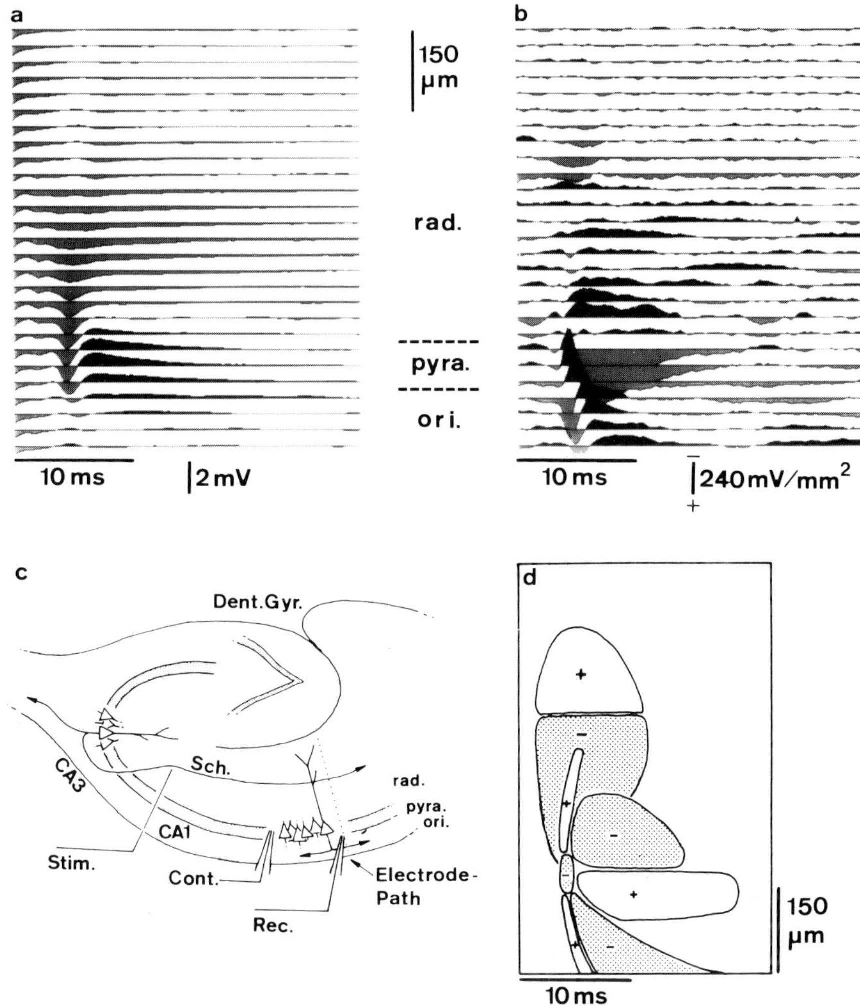

Figure 1. Typical example from the simple class of CSD-profiles in the CA1 region of the rat hippocampus slice. 1a: Field potential distribution; 1b: CSD-profile; 1c: sketch of the recording situation; 1d: schematic diagram of profile in Fig. 1b. Sinks are drawn upwards, sources downwards.

The upper pattern again originates from synaptic input in the distal dendritic region. In the lower part we clearly distinguish *two* action potential patterns, shifted in space and time. An early action potential can be seen in the zone of the inner cellbody layer and/or proximal dendritic region; a second, slightly later action potential is revealed in the outer cellbody layer.

4. Discussion

In the CSD-analysis we assumed that the extracellular space has the properties of an Ohmic conductor with nearly constant conductivity in the direction of the electrode-path and that the electric field is quasi-static. At first sight, the CA1 anatomy does not seem to support these assumptions. Nevertheless, the highly stereotyped nature of the elementary patterns in our results together with their plausible explanation in terms of currents generated by epsp's and action potentials provide a posteriori support of these simplifying assumptions.

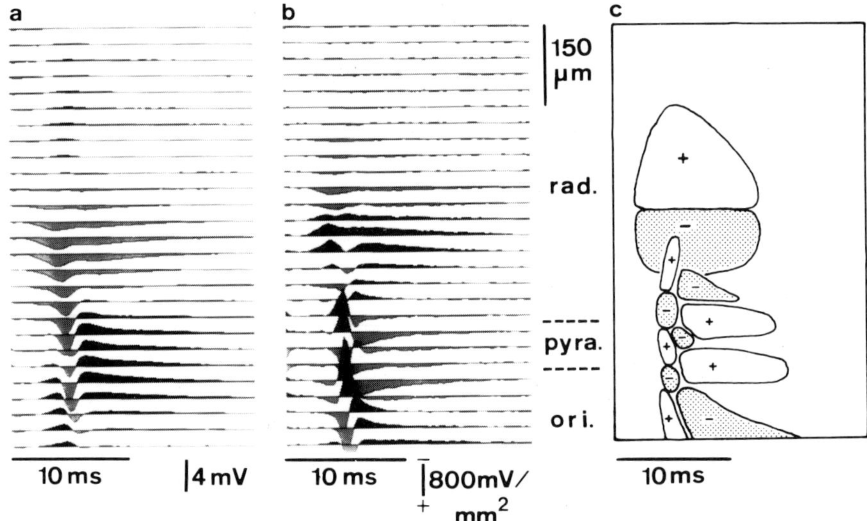

Figure 2. Typical example from the complex class of CSD-profiles in the CA1 region of the rat hippocampus slice. 2a: Field potential distribution; 2b: CSD-profile; 2c: schematic diagram of profile in Fig. 2b.

The typical CSD-pattern we were able to show in the hippocampus CA1 region can be explained by currents which are generated by synaptic input from the Schaffer collaterals and by the time course of currents generated by action potentials near or in the cellbody layer. This means that the earlier idea that action potentials cannot be revealed in CSD-analysis, due to scattering in space and time [2] does not hold for the hippocampus (see also [6]).
It appears to be possible to show a 'fingerprint' of an action potential in CSD-plots, presumably due to the very orderly neuroanatomical organization of the CA1 region and the highly synchronized firing of CA1 pyramidal cells.

Furthermore, as shown in Fig. 2 the CSD-analysis proves to be a very useful tool to separate action potentials which are generated in the hippocampus at different locations in space and different moments in time. Based on the anatomical location, we speculate that the lower of the two action potential patterns arises in the cell body layer, whereas the upper of the two corresponds to a dendritic action potential, arising, or at least revealed in the proximal dendritic region. More detailed research into the underlying mechanisms is currently in progress.

Acknowledgements

We thank Jürgen Bolz for his support in establishing the slice technique and Volker Staiger for skillful assistance in preparing the Figures.

References

[1] Bonhoeffer, T. et al., Proc. Natl. Acad. Sci. USA 86: 8113-8117 (1989)
[2] Nicholson, C., Freeman, J.A., J. Neurophysiol. 38: 356-368 (1975)
[3] Mitzdorf, U., Physiol. Rev. 65: 37-100 (1985)
[4] Mitzdorf, U., Singer, W., Exp. Brain Res. 33: 371-394 (1978)
[5] Novak, J.L., Wheeler, B.C., Brain Res, 497: 223-230 (1989)
[6] Miyakawa, H., Kato, H., Brain Res, 399:303-309 (1986)

Parallel Processing in Neural Systems and Computers
R. Eckmiller, G. Hartmann and G. Hauske (Editors)
© Elsevier Science Publishers B.V. (North-Holland), 1990

EMULATION OF BIOLOGY-ORIENTED NEURAL NETWORKS

Stefan J. Prange

Institute for Microelectronics
Technical University of Berlin
Berlin 12, Germany

The simulation of asynchronous, dynamic neural networks on conventional digital computers based on the "von-Neumann"-machine needs a lot of computing time. Realtime processing can only be achieved by the connection of several computers. Microelectronics offers the possibility to design specific hardware for the emulation of biology-oriented neural networks [1,2].

Starting with the electrical phenomena at the membrane of a nerve cell, a simplified model was developed, a simulation program was implemented, an electronic circuit was realized as an integrated circuit and a hardware emulator consisting of these integrated circuits was constructed. Using the emulator the synaptic transmission parameters can be adjusted, pulses can be put into the network and output pulses can be read out via personal computer.

1. THE NEURON IN BIOLOGY

The smallest data processing unit in the central nervous system is the neuron, the nerve cell. The human brain contains about 15 billion neurons. Each neuron possesses 1000 to 10000 inputs and one output, the axon, that further branches out to 1000 up to 10000 inputs. The connections between neurons are called synapses. A synapse consists of the sending cell's presynaptic terminal, the synaptic cleft and the postsynaptic membrane [3].

Inside a nerve cell a resting-potential is maintained and restored after a disturbance. An incoming nerve pulse causes the synapse to pour out a transmitting substance into the synaptic cleft. Depending on the kind of the transmitting substance the potential of the postsynaptic membrane is changing into a positive or negative direction (Fig. 1). The duration of this potential change and the quantitative effect on the sumpotential of the cell body is depending on the size of the synapse, on the amount and kind of transmitting substance and on the distance between the synapse and the axon hillock. The sum of all these potential changes is converted into a series of nerve pulses at the axon hillock. The amplitude of these pulses is constant. The frequency of pulses is determined by a threshold- and saturation-characteristic (Fig. 3) [4,5,6].

2. THE EMULATION OF NEURAL NETWORKS

In an emulation the computer is only used to adjust parameters, to provide input data and to read out output data. The emulation is running on an own model hardware, and not algorithmically on the computer as in a simulation. Additional external input sources and output devices besides the computer can easily be applied to the model hardware.

At the Institute for Microelectronics of the Technical University of Berlin an emulator controlling a network with 16 electronic model neurons was developed (Fig. 2). Via personal computer the transmission parameters of the synapses can be adjusted, the input pulses can be specified and the

Stefan J. Prange, Technische Universität Berlin, Institut für Mikroelektronik,
Jebensstr. 1, Sekr. J13, D-1000 Berlin 12, Tel.: (30) 314-26705, FAX: (30) 314-24597

output pulses can be read out. The processing inside the network is done in realtime on the special neural network hardware. The frequency range the net is working with is defined by the capacitors used. A change of the frequency range can easily be achieved by a change of these capacitors.

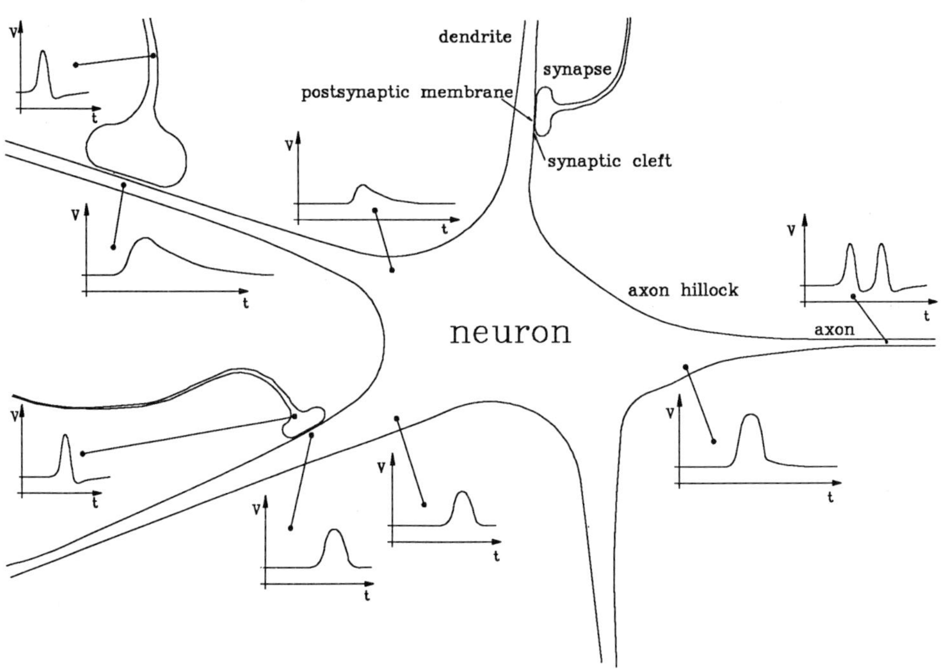

FIGURE 1
Schematic diagram of a nerve cell with all potentials during
the arrival of two nerve pulses via excitatory synapses

3. THE ELECTRONIC NEURON-MODEL

Based on the idea that the application of biology-oriented neural networks offers advantages for some kinds of dynamic signal processing applications a neuron-model orientated on biology was evolved and an analog integrated bipolar circuit on the B1000-transistor-array of AEG was developed, fabricated and tested. The chip-area of the integrated circuit is 4.5mm·5mm. The array contains about 200 npn- and 100 pnp-transistors. About 200 of them are used. The circuit was fabricated with one metal layer with a line width of 12μm [7].

The circuit contains one neuron with eight synapses each. It has the following features: A synapse building block widens an incoming rectangular pulse into a triangular one (Fig. 3). The slope of the trailing edge can be regulated by an external voltage. Besides the (positive or negative) weighting is adjusted by a second external voltage. The signals preprocessed as described are summed up and put into a voltage-to-frequency-converter with a linearized threshold- and saturation-characteristic. The frequency range depends on external capacitors [8].

4. THE SIMULATION OF NEURAL NETWORKS

A simulation program was additionally developed to predict the behaviour of our model and its suitability for different tasks. Furthermore the simulation is used to test the ability of optimization strategies to adjust the different parameters of the network, so that desired transmitting characteristics can be achieved [9].

In the simulation the nerve pulses are assumed to be infinitesimally short, that means that they are only represented by a discrete point in time. The postsynaptic potential is assumed to jump instantaneously and descend linearly to zero (Fig. 3). So the transmitting characteristics of a synapse are exactly defined by the (positive or negative) height of the jump and the slope of the descent. The voltage-to-frequency-conversion of the axon hillock can be realized by deciding individually for each point of time, whether to fire a nerve pulse or not. The probable next pulse has to be calculated with respect to the time-dependent potential. Our simulation program uses a linearized conversion-characteristic and the potential's mean value over time to calculate the next pulse.

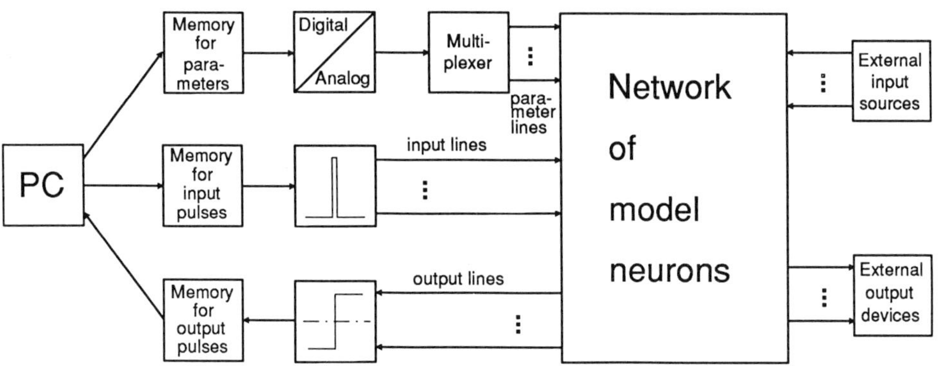

FIGURE 2
Block diagram of the emulator

5. PERSPECTIVE

The simulation of asynchronous, dynamic neural nets in realtime is impossible for actual computers based on the "von-Neumann"-machine. The realistic emulation of biology-oriented neural nets with several hundred neurons will only be possible with an application specific hardware. In contrast to many existing neuron-models, the presented one imitates the potentials at the membrane of a biological nerve cell much more exactly. An emulator using such a model can in first respect be applied by neurophysiologists as a tool for imitating biological processes [10]. However, this biology-oriented neural network with dynamic synaptic transmission characteristics and two parameters per synapse is a starting point for fault-tolerant dynamic signal processing, e.g. auditory signal processing, instead of using static neural network models with extensive preprocessing [11]. The major aim of the presented work is the integration of an emulator for biology-oriented neural networks in modern CMOS-technology.

                    S.J. Prange

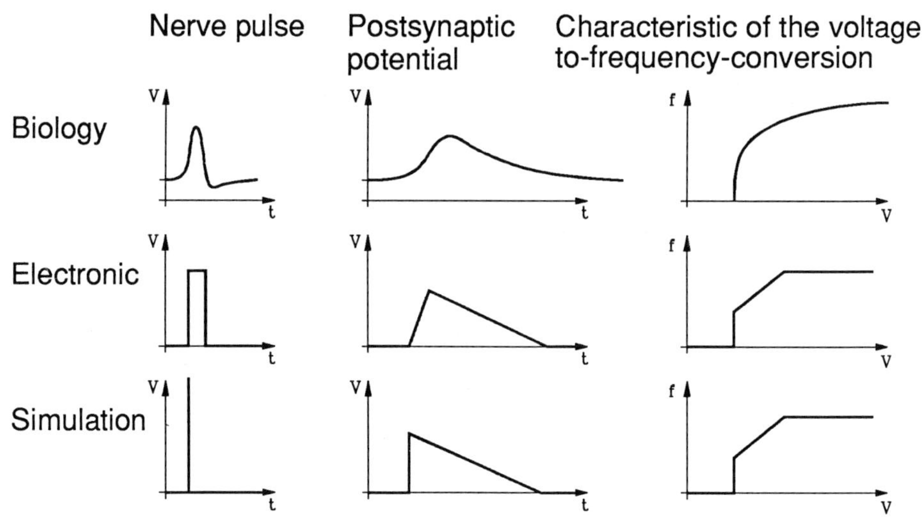

FIGURE 3
Comparison of the characteristics

REFERENCES

[1] Klar, H. and Ramacher, U., Microelectronics for Artificial
     Neural Networks, VDI-Verlag GmbH, Düsseldorf 1989
[2] Mead, C., Analog VLSI and Neural Systems, Addison
     Wesley, 1989
[3] 'Gehirn und Nervensystem', Spektrum der Wissenschaft:
     Verständliche Forschung, Heidelberg 1987
[4] Fohlmeister, J., Electrical Processes Involved in the
     Encoding of Nerve Impulses, Biol. Cybernetics 36,
     103-108 (1980)
[5] Stein, R. B., Leung, K. V., Mangeron, D., Oguztöreli,
     M. N., Improved Neuronal Models for Studying Neural
     Networks, Kybernetik 15, 1-9 (1974)
[6] Nelson, P. P., General Input-Output Relations for the
     Electrical Behaviour of Neurons, Kybernetes 1980, Vol. 9,
     123-131
[7] Prange, S., Aufbau eines Neuronenmodells mit Hilfe einer
     analogen, kundenspezifischen Schaltung, Diplomarbeit am
     Institut für Mikroelektronik der TU Berlin, Nov. 1988
[8] Prange, S., Neuronenmodell als analoge integrierte
     Schaltung, 4. Treffpunkt Medizintechnik, 17.2.1989, 42-43
[9] Rechenberg, I., Evolutionsstrategie, Frommann-Holzboog,
     Stuttgart 1973
[10] Reitboeck, H. J., Eckhorn, R., Arndt, M. and Dicke, P., A
     Model of Feature Linking via Correlated Neural Activity,
     in: Haken, H., Synergetics of Cognition, Springer, 1989
[11] Kohonen, T., The 'Neural' Phonetic Typewriter, IEEE
     Computer, March 1988, 11-22

Parallel Processing in Neural Systems and Computers
R. Eckmiller, G. Hartmann and G. Hauske (Editors)
© Elsevier Science Publishers B.V. (North-Holland), 1990

ARE FRACTAL DIMENSIONS A GOOD MEASURE FOR NEURONAL ACTIVITY?

Hubert PREIßL[1], Ad AERTSEN[1] and Günther PALM[2]

[1]Max-Planck-Institut für biologische Kybernetik, D-7400 Tübingen, FRG
[2]Vogt-Institut für Hirnforschung, Universität Düsseldorf, D-4000 Düsseldorf, FRG

Over the last years it has become increasingly popular to apply methods from non-linear dynamical system theory to biological systems [1]. In the context of brain research it was claimed that both pulse train activity of single neurons [2] and continuous activity as measured by the EEG [3] should be described as the output of a deterministic chaotic system.
We show that a fundamental discrepancy exists between the fractal dimension of a continuous process and of pulse trains generated from that process, which implies that fractal dimension analysis is not an appropriate tool for the characterization of pulse train measurements. This result has major implications not only for the interpretation of neuronal data, but also for various other fields in biology and physics.

1. Introduction

In recent years there has been a growing interest in the description of dynamics in complex systems, both at the theoretical and experimental level. It was shown that one can distinguish between stochastic and deterministic, non-periodic ('chaotic') processes; various measures for the characterization of chaotic systems were developed [4]. This approach was applied in such different fields as ecology, climatology, fluid dynamics, quantum physics and life sciences [5]. In the field of brain research, some researchers attempted to adapt these methods to the analysis of single neuron spike trains [2], others to the description of more global electrophysiological measurements such as the EEG [3]. In this approach, both types of signals are analysed as the output of a deterministic chaotic system, as opposed to the usual analysis which relies on methods derived from the theory of stochastic processes. A further reason for an increased interest in the investigation of dynamics in the nervous system may be found in recent reports on oscillatory behavior of single neurons and groups of neurons in the visual system [6], and on dynamic properties of connectivity in physiological neuronal nets [7].

The dynamics of a single neuron can in principle be determined from two different experimental observations: the membrane potential, which is a *continuous* signal, and the train of action potentials ('spikes'), described as a *point process*. The first can be measured with intracellular electrodes or with optical methods (using voltage-sensitive dyes), the second also by the more common (and less tedious) technique of extracellular recording. Most of single neuron electrophysiology is in fact based on such extracellular spike train recordings. Since it was not clear whether the choice between the continuous signal and the pulse train might influence the outcome of the analysis (see also [8]), we decided to investigate this problem in more detail. To this end we analysed several well-known dynamical systems (Lorenz, Rössler, Henon [9]), each one with a strange attractor and chaotic dynamics, focusing on the issue of continuous versus discrete time series observations.

2. Chaotic Dynamics and Fractal Dimensions

Since the work of Lorenz [10] it is known that non-linear, dissipative dynamical systems can exhibit chaotic behavior. The attractor of such a system may be characterized by a non-integer or *'fractal'* dimension, in which case it is called a *'strange attractor'*. The most popular among the various current definitions for fractal dimensions is the *'correlation dimension'*. This can be determined using an algorithm developed by Grassberger and Procaccia (G-P) [11], based on considerations by Renyi [12]. It involves calculating the correlation integral $C(\varepsilon)$, which determines the number of state vectors which fall within a region with size ε. For small ε, $C(\varepsilon)$ behaves like a power of ε: $C(\varepsilon) \sim \varepsilon^{\sigma}$. The exponent σ is called the 'correlation dimension'; it can be determined as the slope of the $\ln C(\varepsilon)$ vs. $\ln \varepsilon$ curve in the so-called 'scaling region' (sr), i.e. where this curve is linear.

In most experimental situations the differential equations for the system under investigation and, consequently, the state variables, are unknown. The experimenter usually only has access to consecutive measurements of a single scalar observable y(t). A considerable step forward was made when it was shown that the attractor of the underlying system can be reconstructed from such a time series by a procedure known as *'embedding'* [13]. A vector in an n-dimensional *'pseudo state space'* is defined from the measurement by taking an appropriately spaced 'comb' of values $(y(t),y(t+\tau)...,y(t+(n-1)\tau))$; different values of t correspond to different vectors in that space. The *'embedding theorem'* of Takens [13] states that, if y is a smooth function mapping the original attractor to R, and if $n \geq 2m+1$ (m is the dimension of the attractor), the reconstructed attractor in pseudo state space is diffeomorphic to the original attractor: both have the same dimension.

The embedding theorem and the G-P algorithm provide the means of analysing the dynamics of a system under experimental observation. One records a time series, embeds it in an n-dimensional space (usually starting with n=1) and determines a 'dimension' d with the G-P algorithm. This procedure is repeated with increasing n. Initially the 'dimension' d will increase with n. However, if the underlying system has a low-dimensional attractor with a correlation dimension σ, the 'dimension' d will saturate at this value for increasing n. In the scaling region the $\ln C(\varepsilon)$ vs. $\ln \varepsilon$ curves for different n will consequently become parallel.

3. Characterization of Continuous Signals and Point Processes

This standard approach was applied in our analysis to the attractors of some well-known dynamical systems. Our goal was to investigate the results from the observation of continuous signals as opposed to point processes derived from the same system. In order to obtain the two types of data we simulated the Lorenz-system [14] and recorded the time course of one of the system variables, the z-component (continuous signal) as well as the series of time intervals between successive, positive-going level-crossings of a fixed amplitude threshold Z (pulse train; cf. Fig. 1). In the latter case these time intervals were regarded as the dynamic system variable as in [15].

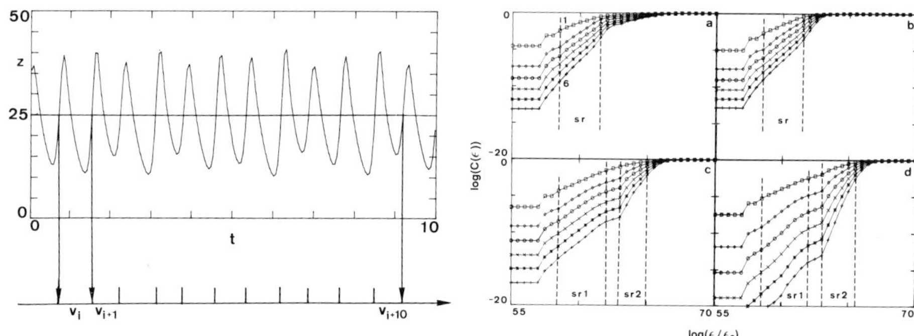

Fig. 1. Generation of pulse train observations from the z-component of the Lorenz system: the dynamic variable for the reconstruction procedure is the sequence of time intervals v_i between successive, positive-going level crossings at a fixed threshold value Z=25. Note that in this case the vectors are defined by taking sequences of adjacent v_i values.

Fig. 2. Results of dimension analysis for pulse trains obtained with different thresholds Z on the z-component of the Lorenz system: Z=35 (a), Z=25 (b), Z=15 (c), and Z=12 (d). In each case the reconstruction was performed with 5000 intervals; the embedding dimension ranges from 1 to 6. For details of the algorithm see [21].

The attractor of the system has a correlation dimension of 2.06 [11]. According to the embedding theorem, a proper reconstruction of this attractor can be obtained in a pseudo state space with a dimension of at least 6. Using the time course of the z-component, it was possible to obtain the correct correlation dimension of 2.06. Rather to our surprise, however, the pulse train measurements gave quite different results, as illustrated in Figure 2 for 4 different values of the threshold Z. Only for Z=35 (Fig. 2a) and Z=25 (Fig. 2b) were we able to determine a single correlation dimension: for Z=35 it was 1.92, whereas for Z=25 it amounted to 1.79. Clearly these two values differ from each other as well as from the correlation dimension of the continuous system. With a threshold at Z=15 (Fig. 2c) one obtains two scaling regions (sr1, sr2); this is even more obvious with a threshold at Z=12 (Fig. 2d). For Z=15 the correlation dimension equals 1.52 in sr1, whereas in sr2 it cannot be determined at all: the 'dimension' d continues to grow with the embedding dimension; for Z=12 neither of the two scaling regions allows to determine a correlation dimension. Incidentally, we note that similar pictures have been obtained for measurements from a system composed of two non-interacting subsystems with different amplitudes [4].

We made similar observations for the other (x and y) components of the Lorenz-system, as well as for the Rössler- and the Henon-system. In addition we considered other types of pulse train measurements, e.g. the sequence of intervals between consecutive points of entry into an arbitrarily selected box in pseudo state space, or the sequence of intervals between maxima in a selected component. In *none* of these cases was it possible to obtain the correct correlation dimension of the attractor from analysis of pulse train measurements. Finally we analysed the membrane potential and the simultaneously recorded spike train from a neuronal network simulator which employs a pseudo-random number generator [16]. Again the two different types of measurements gave different results for the correlation dimension.

4. Conclusions

We conclude that pulse train measurements (like spike trains from single neurons, or peaks in electrocardiograms) do *not* allow the characterization of an attractor, even if the underlying system is a deterministic dynamical system: the attractor reconstructed from the pulse trains is not diffeomorphic to the underlying attractor. The reason is that a pulse train is not the result of a smooth mapping of the underlying continuous process, such as would be required in the embedding theorem [13]. This discrepancy between continuous measurements and pulse train measurements has major implications for the interpretation of neuronal data, and also in various other fields in biology and physics with similar types of measurements [8,17]. The relationship between the dimensions for the two types of signals, as well as the observed threshold-dependence of pulse train measurements is the subject of current investigation.

In addition to this discrepancy, the application of these methods in fields like neurophysiology faces further, as yet unresolved questions. For example, spatially extended systems with many interacting components, of which the brain is surely an example, may exhibit dominating, long transients [18], which make it impossible to observe the attractor. Furthermore, such systems may have multiple, coexisting attractors [19]. This argues that the spatio-temporal dynamics of spatially extended systems cannot be characterized by means of an attractor reconstructed from the measurement of a single variable, irrespective of whether it is a single-neuron pulse train or a continuous signal such as a membrane potential or some form of spatio-temporal summation like the EEG. Consequently, attractor characteristics such as fractal dimensions and Lyapunov exponents would be of little value here [20].

Methods from dynamical system theory may potentially contribute new insights to various fields, including neuroscience. However, one should be aware of the fact that fractal dimension analysis is not an appropriate tool for the characterization of pulse train (or interval) measurements, nor does it seem to be an adequate descriptor for the dynamics of large systems such as the brain.

Acknowledgements

We thank Professors G.L. Gerstein, P. Grassberger and O.E. Rössler for helpful discussions, Volker Staiger for skilful assistance in producing the figures, and Shirley Würth for improving the English.

REFERENCES

[1]   R. Pool, *Science* **243**, 604 (1989).

[2]   P.E. Rapp, I.D. Zimmermann, A.M. Albano, G.C. de Guzman, N.N. Greenbaun, *Phys.Lett.* **110A**, 335 (1985); A.M. Albano et al., in *Dimensions and entropies in chaotic systems*, G. Mayer-Kress, Ed. (Springer, Berlin, 1986), pp. 231-240.

[3]   A. Babloyantz, in *Dimensions and entropies in chaotic systems*, G. Mayer-Kress, Ed. (Springer, Berlin, 1986), pp. 241-245; A. Babloyantz and A. Destexhe, *Proc.Natl.Acad.Sci USA* **83**, 3513 (1986); W.J. Freeman, *Biol.Cybern.* **56**, 139 (1987); A. Skarda and W.J. Freeman, *Behavioral and Brain Sciences* **10**, 161 (1987).
      However, see also S.P. Layne, G. Mayer-Kress, J. Holzfuss, in *Dimensions and entropies in chaotic systems*, G. Mayer-Kress, Ed. (Springer, Berlin, 1986), pp. 246-256, and I. Dvorak, J. Siska, *Phys.Lett.* **118A**, 63 (1986).

[4]   for a review see: J.P. Eckmann and D. Ruelle, *Rev.Mod.Phys.* **57**, 617 (1985).

[5]   R. Pool, *Science* **243**, 25; 310; 604; 893; 1290 (1989); J.P. Crutchfield, J.D. Farmer, N.H. Packard, R.S. Shaw, *Sci.Am.* **225(6)**, 38 (1986).

[6]   C.M. Gray and W. Singer, *Proc.Natl.Acad.Sci USA* **86**, 1698 (1989); C.M. Gray, P. König, A.K. Engel, W. Singer, *Nature* **338**, 334 (1989); R. Eckhorn et al., *Biol.Cybern.* **60**, 121 (1988)

[7]   A. Aertsen et al., *J.Neurophys.* **61**, 900 (1989); G.L. Gerstein, P. Bedenbaugh, A. Aertsen, *IEEE Trans.Biomed.Engin.* **36**, 4 (1989).

[8]   A. Babloyantz and A. Destexhe, *Biol.Cybern.* **58**, 203 (1988).

[9]   J.M.T. Thompson and H.B. Stewart, *Nonlinear dynamics and chaos* (John Wiley and Sons, Chichester, 1986).

[10] E.N. Lorenz, *J. Atmospheric Sciences* **20**, 130 (1963).

[11] P. Grassberger and I. Procaccia, *Phys.Rev.Lett.* **50**, 346 (1983); P. Grassberger and I. Procaccia, *Physica* **9D**, 189 (1983).

[12] A. Renyi, *Wahrscheinlichkeitsrechnung* (VEB Verlag, Berlin, 1962).

[13] F. Takens, in *Dynamical systems and turbulence*, Rand and Young, Eds. (Springer, Berlin, 1981), pp. 366-381; N.H. Packard, J.P. Crutchfield, J.D. Farmer, R.S. Shaw, *Phys.Rev.Lett.* **45**, 712 (1980).

[14] For the simulation of the Lorenz system we used the parameters given in J.G. Caputo and P. Atten, *Phys.Rev.A* **35**, 1311 (1987).

[15] R.S. Shaw, *The dripping faucet as a model chaotic system* (Ariel Press, Santa Cruz, 1984).

[16] K.-H. Boven and A. Aertsen, in *Dynamics and plasticity in neuronal systems*, N. Elsner, W. Singer, Eds. (Thieme, Stuttgart-New York, 1989), p. 444.

[17] e.g. K. Aoki, O. Ikezawa, N. Migubayashi, K. Yamamoto, *Physica* **134B**, 288 (1985); M. Warden, W. Floeder, F. Waldner, in *Proceedings of 23rd Congress Ampere on Magnetic Resonance*, B. Maravigilia, F. De Luca, R. Campanella, Eds. (Rome, 1986), pp 272-273.

[18] J.P. Crutchfield and K. Kaneko, *Phys.Rev.Lett.* **26**, 2715 (1988).

[19] G. Mayer-Kress and K. Kaneko, *J.Stat.Phys.* **54**, 1489 (1989).

[20] H. Chate and P. Manneville, *Phys.Rev.Lett.* **58**, 112 (1987).

[21] T.S. Parker and L.O. Chua, *Proc. IEEE*, **75**, 982 (1987).

Parallel Processing in Neural Systems and Computers
R. Eckmiller, G. Hartmann and G. Hauske (Editors)
© Elsevier Science Publishers B.V. (North-Holland), 1990

A dynamic theory of coordination of discrete movement*

Gregor Schöner

Institut für Neuroinformatik, Ruhr-Universität, D-4630 Bochum, FRG

Dynamic theories of the coordination of movement are generalized to provide a unified treatment of discrete and rhythmic movement coordination. Tendencies to synchronize discrete movement, but also to perform such movements sequentially are treated. Dynamic stability of coordination patterns as an important concept of this theory gives rise to predictions on remote compensation of perturbations.

1. Introduction

Recently, dynamic theories have been proposed to understand the patterns of coordination of rhythmic movements [10]. The basic concepts, which are motivated by physical theories of pattern formation [4], are the description of such coordination patterns by collective variables and the identification of pattern dynamics as equations of motion of these collective variables. In such a theory, the stability is an important property of a pattern and the various measures of stability have led to prediction, in particular near phase transitions involving loss of stability of a coordination pattern [6]. In a second step, the concept of behavioral information was introduced [11] to express the adaptation of coordination patterns to various behavioral requirements (environmental, intentional, memorized) in terms of contributions to the pattern dynamics that attract the coordination system to the required pattern.

In the present contribution we provide a unified description of posture, discrete, and rhythmic movement in terms of pattern dynamics. The stress is on temporal order in such movements, as captured by stable and reproducible timing [3]. By restricting the theory to one spatial dimension per component we avoid the problem of spatial trajectory formation in a first step. Coordination of discrete movement is then captured in terms of the relative timing of the movements of two components. This is motivated by experiments [7] that found a tendency to synchronize discrete movements of two limbs as a characteristic signature of coordination.

The chosen level of description is that of the end-effector kinematics, that is, end-effector position, x, and velocity, v. These variables are collective with respect to the underlying neuro-muscular degrees of freedom and must be interpreted with sufficient abstraction as expressing the coordination activity of the central nervous system in a given function. The measured mechanical end-effector position and velocity are uniquely determined by these variables in an unperturbed stable movement pattern. Perturbations applied to the mechanical degrees of freedom affect the modelled level of coordination only when they are sufficiently strong [2]. The key step in achieving a dynamic description of movement is to identify intrinsic dynamics in terms of initial and target postural states. These intrinsic dynamics determine the timing properties of discrete and rhythmic movement. The intention to move is captured as behavioral information that stabilizes a position-velocity curve (a part of a limit cycle), but does not completely specify timing. Coordination arises from coupling of the dynamics of different end-effectors [5].

---

  * *Work performed in part at the Center for Complex Systems, FAU, Boca Raton, Florida with support through NIMH (MH 42900-01) and ONR (N00014-89-J-1191). Present support through a grant from BMFT, W. Germany (ITR8800K4).*

2. Dynamic model for one component

The intrinsic dynamics of the collective variables, (x, v), must be chosen such that the phase diagram contains three types of phase portraits: (a) one globally stable fixed point (posture); (b) two stable fixed points (initial and target posture in discrete movement); (c) one globally stable limit cycle (rhythmic movement). Such dynamics can be modelled, for example, by expanding the vectorfield of a second order dynamics in powers of x and v and truncating the expansion when the phase diagram becomes sufficiently rich [9]. A particularly simple parametrization is possible if we refer to an exactly solvable nonlinear oscillator model [1]. This system reads:

$$\dot{x} = v$$
$$\dot{v} = -f_{\text{intr}}(x, v) - (a^2 + \omega^2)x + 2av - 4bx^2v + 2abx^3 - b^2x^5 \tag{1}$$

and has only three model parameters a, ω^2, and $b(> 0$ for bounded solutions; note that ω^2 may be negative). The phase diagram of Eq. (1), shown in the (ω^2, a) plane in Fig. 1, is exactly of the required nature.

Discrete movement is modelled as follows: By task definition two posture states, the initial and the target position at $A = \pm\sqrt{(a + |\omega|)/b}$ are intrinsically stable (region (c) or (d)). The intention to move acts as behavioral information stabilizing a limit cycle for a period of time that approximates a half-cycle time. After that time the intrinsic dynamics again dominate so that the system relaxes to the target fixed point. The dynamic system with behavioral information is:

$$\dot{x} = v$$
$$\dot{v} = -f_{\text{intr}}(x, v) + \chi_{[t_0, t_0 + \Delta t]}(t) \; c_{\text{int}} \; f_{\text{int}}(x, v) \tag{2}$$

where

$$\chi_{[t_0, t_0 + \Delta t]}(t) = \begin{cases} 1 & \text{for } t \in [t_0, t_0 + \Delta t] \\ 0 & \text{elsewhere} \end{cases}$$

Here the temporal evolution of an intention to move has been modelled very roughly by an abrupt onset at time t_0 and an abrupt offset after the time interval Δt. A simple functional form of the contribution of behavioral information is

$$f_{\text{int}} = -x \tag{3}$$

which moves the system into region (e), and thus stabilizes a limit cycle, for sufficient strength, c_{int} of the intention to move.

Fig. 2 shows a simulation of this dynamical system: Two postural states are stable initially, with the system prepared in one of them. An intention to move is present in the time interval of length Δt, The time, T_0, when the half-distance is covered, can be used as a measure of the movement time. It can be shown, that the timing characteristics of the resulting discrete movement depend only weakly on the parameter Δt (and not on the parameter t_0) [9]. Therefore, the movement time is determined mostly by the intrinsic dynamics and relationships between amplitude and movement time as in Fitts' law [3] can be captured.

3. Coordination of discrete movement

The coordination of two end-effectors is understood as the result of coupling two single-effector dynamics such as Eq. (2). The coupling function can be adapted from the well-studied rhythmic case [5], where in-phase and anti-phase relative timings have to be accounted for. A mathematical form reads:

$$\dot{x}_1 = v_1$$
$$\dot{v}_1 = -f_{1,\text{tot}}(x_1, v_1) - k(v_1/v_{1,p} - v_2/v_{2,p}) - l(v_1/v_{1,p} - v_2/v_{2,p})(x_1/A_1 - x_2/A_2)^2$$
$$\dot{x}_2 = v_2$$
$$\dot{v}_2 = -f_{2,\text{tot}}(x_2, v_2) - k(v_2/v_{2,p} - v_1/v_{1,p}) - l(v_2/v_{2,p} - v_1/v_{1,p})(x_2/A_2 - x_1/A_1)^2$$
$$(4)$$

Here $f_{i,\text{tot}}$ is the sum of intrinsic vectorfield and behavioral information in Eq. (2), where we admit different parameter values for each component. The velocities are normalized by dividing by peak velocity $v_{i,p} = |\omega_i| A_i$, where A_i is the amplitude of the movement.

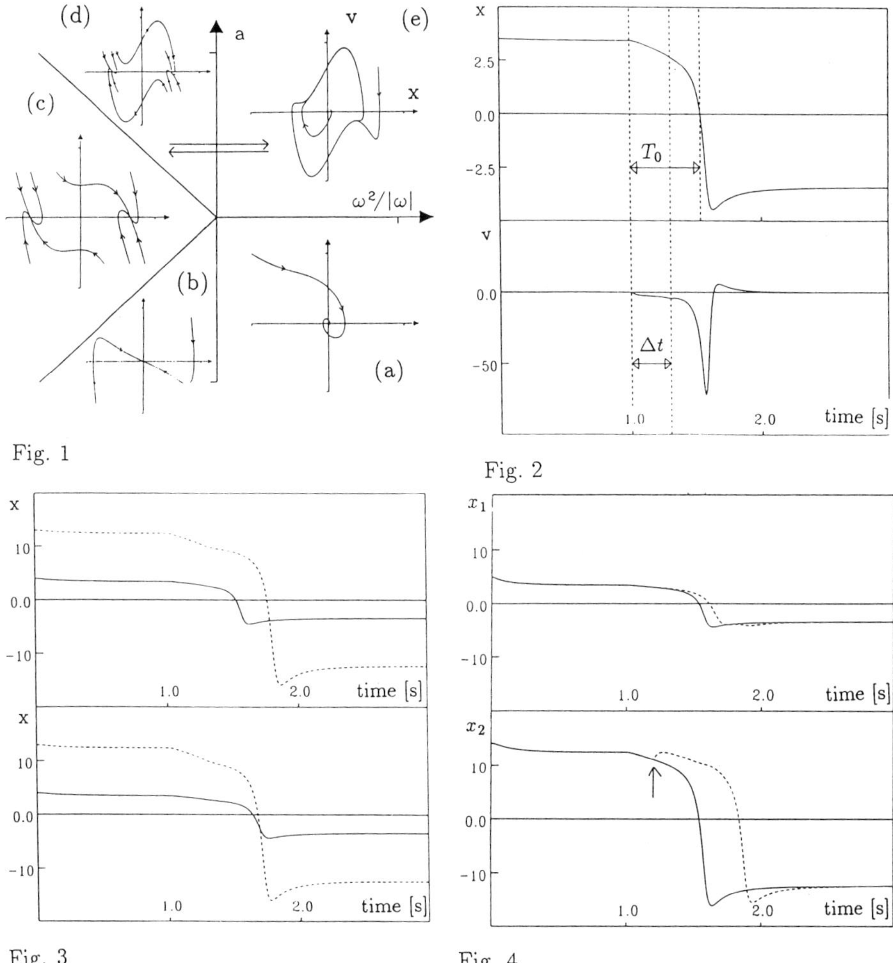

Fig. 1

Fig. 2

Fig. 3　　　　　　　　　　　　Fig. 4

Such coupling accounts for the tendency to synchronize discrete movement [7]. In Fig. 3 we show trajectories of two movements that are at different ends of the Fitts' law relation: one movement is far and slow (dashed line, top panel), the other short and fast (solid line, top panel). When the two components are coupled (bottom panel), the movement times are almost identical.

In such a theory the synchronization is due to the formation of a stable pattern. Therefore, the system compensates perturbations to restore the stable relative timing of the two components. This is illustrated in the simulation of Fig. 4 where an unperturbed trial (solid) is shown together with a perturbed trial (dashed). The perturbation in one component (arrow) induces a compensatory response in the other limb which attempts to restore synchronzition. This prediction could be observed in experiments in which one limb in a two limb task is delayed or accelerated. Hints at such remote compensatory effects are contained in experiments on articulatory coordination during speech [8].

What is the discrete analogon of anti-phase locking? By introducing a delay between the onsets of intention for the two limbs we can identify the corresponding tendency in discrete movement: For small delay we find synchronization. But when the delay is sufficiently large, the nonlinear coupling further delays the late starter und the two movements are performed sequentially. This is illustrated in Fig. 5, where the difference between the uncoupled (solid) and coupled (dashed) cases shows the tendency to synchronize for small delay (left bottom) and the tendency to sequential movement for large delay (right bottom). The onset times of the intention to move are marked in the figure as $t_{0,1}$, $t_{0,2}$, and $t'_{0,2}$, resp. For further details see [9].

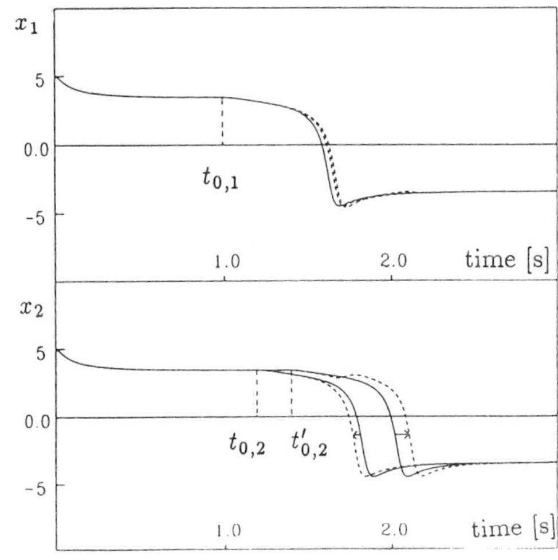

Fig. 5

References

[1] Gonzalez DL, Piro O. *Phys. Rev.* **A36**, 4402 (1987).

[2] Feldman AG. *Biofizika* **11**, 667 (1967).

[3] Fitts PM. *J. Exp. Psychol.* **47**, 381 (1954).

[4] Haken H. *Synergetics*, Springer, Berlin, 1983 (3rd. ed.)

[5] Haken H, Kelso JAS, Bunz H. *Biol. Cybern.* **51**, 347 (1985).

[6] Kelso JAS, Schöner G, Scholz JP, Haken H. *Physica Scripta* **35**, 79 (1987)

[7] Kelso JAS, Southard DL, Goodman D. *Science* **203**, 1029 (1979)

[8] Kelso JAS, Tuller B, Vatikiotis-Bateson E, Fowler CA. *J. Exp. Psychol.: Hum. Perc. Perf.* **10**, 812 (1984)

[9] Schöner G. *Biol. Cybern.* (submitted)

[10] Schöner G, Kelso JAS. *Science* **239**, 1513 (1988)

[11] Schöner G, Kelso JAS. *J. Theor. Biol.* **135**, 501 (1988)

Parallel Processing in Neural Systems and Computers
R. Eckmiller, G. Hartmann and G. Hauske (Editors)
© Elsevier Science Publishers B.V. (North-Holland), 1990

SPIKE ARRIVAL TIMES : A HIGHLY EFFICIENT CODING SCHEME FOR NEURAL NETWORKS

Simon J. THORPE

Institut des Neurosciences

Département des Neurosciences de la Vision

Université Pierre et Marie Curie

Quai St. Bernard,

Paris. France

1. Introduction - Temporal constraints on neural coding.

The speed with which visual processing occurs in the human brain puts heavy constraints on the way in which information is processed. Indeed, we have recently argued that processing is so fast, that a great deal of visual processing must be achieved using a single feed-forward pass through the visual system (Thorpe, 1988; Thorpe & Imbert, 1989). Briefly, the argument is as follows. (1) Neurophysiological data have shown that neurons in the temporal cortex have highly selective responses (to faces for example), which start with latencies that are often in the range 100 to 140 ms (Perrett, Rolls & Caan, 1982). (2) Anatomically, it would appear that there are at least 10 synaptic stages between the photoreceptors of the retina and these visually responsive neurons in the temporal lobe (i.e., 2 in the retina, 1 in the lateral geniculate nucleus, 2 each in V1, V2, and V3, and one more in inferotemporal cortex) (see DeYoe & Van Essen, 1988; Zeki & Shipp, 1988). (3) Since the 10 synaptic layers must be crossed in 100 msec, this implies that each layer has only an average of 10 msec processing time. (4) Given that the firing rates of cortical neurones are typically in the range 0 to 100 spikes per second, it follows that a neuron in any given layer will only generate 1 or perhaps 2 spikes before neurons in the next layer have to respond.

Clearly, this puts major constraints on the way in which information is coded. For example, it is unreasonable to propose that firing rate per se can be used as an accurate code for analogue values. One possibility is that with two spikes, one can use the interval between them as a code - the closer the spikes, the stronger the activation. However, the evidence indicates that a large amount of information can be transferred using only 1 spike per neuron.

In this paper, I would like to suggest what I believe is a novel way of coding information in neural networks. It will be proposed that analog information can be encoded by the relative arrival times of spikes. It will also be shown how such a scheme allows very rapid resolution of the classic "Winner-takes-all" problem.

---

* Funded by ESPRIT Grant 3149 - Multisensory control of movement.

2. The spike arrival time hypothesis.

It is a well known neurophysiological fact that the time taken for a neuron to reach its threshold for generating a spike depends on the strength of the stimulus - the stronger the stimulus, the faster the neuron depolarizes, and the sooner it generates a spike. Imagine, therefore, what happens when an image is flashed onto a simple retinal array of sensory cells. If the cells in the retina are sensitive to local intensity, a wavefront of spikes will be generated in the optic nerve, with the leading spikes corresponding to the points on the retina where the intensity is highest. As a result, even if each retinal cell generated one, and only one, action potential, the complete intensity distribution of the image could be coded by the relative timing of spikes in the different fibres of the optic nerve. Obviously, the example given here is not biologically realistic - retinal ganglion cells do not simply code the local intensity level. Rather, their activity is more clearly related to local contrast. Nevertheless, the argument is just the same - whatever the parameter coded by the input array, the relative arrival time of spikes can be used to code analog values with considerable precision, and this, even if only one spike per fibre is available.

Of course, in order for relative spike arrival times to be a useful code, it is necessary for the next processing layer to be able to make use of such information. Is such a process physiologically plausible? The answer would appear to be a definite yes. Poggio and Koch (1986) have recently proposed that synaptic veto mechanisms could exist in the retina with the property that if a spike from one input arrives before another, it could block the transmission of the second spike. Furthermore, it has been known for some time that neurons in the auditory system are sensitive to the relative arrival times of auditory stimuli in the two ears - such information is used in determining the spatial position of auditory stimuli (Knudsen et al, 1987). Another example concerns the processing of sensory information in electric fish. Heiligenberg and his collaborators have recently reported the some neurons are sensitive to electrical pulse arrival time differences in the microsecond range (Kawasaki, Rose & Heiligenberg, 1988).

In all the above examples, the differences in spike arrival time are due to timing differences in the arrival times of actual physical stimuli, be they points of light, auditory stimuli to the left and right ears, or electrical discharges to the left and right sides of a fishes body. What is being proposed here is simply that the nervous system could also make use of differences in spike arrival times resulting not from temporal differences in the stimulus itself but rather the delays introduced by the coding.

3. The problem of continuous input.

While it is relatively easy to see how an image flashed on the retina could lead to a information-rich wavefront of spikes in the optic nerve, it may be less obvious how the visual system could cope with continuously changing information. However, one important factor that may well reduce this problem are eye-movements. It has recently been reported that transmission in the visual pathways in inhibited during eye movements, and that this inhibition is relaxed as soon as the eye movement has terminated (Lal & Friedlander, 1989). It could be that this could provide the synchronization necessary to allow relative spike arrival times to be used later in the system.

Another possibility for providing the synchronization may be found in the oscillatory activity found in many parts of the nervous system. In particular, it is possible that the 50 Hz oscillations recently reported in the visual cortex by Gray and Singer (1989) and Eckhorn and coworkers (1988) could be used to this effect. They have reported that neurons tend to fire during the negative going part of the local field potential oscillations. More strongly activated neurons would tend to fire earlier in the cycle, and this information could be used during processing by subsequent stages.

4. An application : The "Winner-Takes-All" problem.

One standard problem in connectionist models is the so-called "Winner-takes-all" problem. Consider an array of units, with varying levels of activity. We want to select only the most active one. One commonly used method is to introduce inhibition between all the units. Since the most active unit inhibits the others the most strongly, after a number of iterations the activity in all units except the most active will decrease until in the end, only one unit remains active. The problem is that the number of iterations can be quite large. Although it is possible to optimize such parameters as the amount of lateral inhibition, it is clear that at least 3 iterations are required (Feldman & Ballard 1982; personal communication).

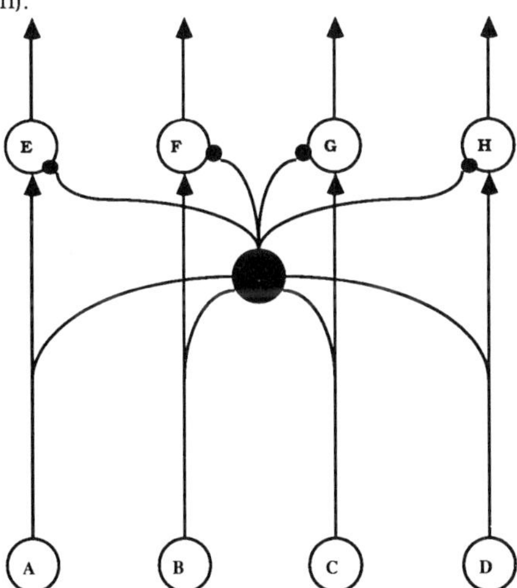

Figure 1. A network using relative spike arrival time to solve the Winner-Takes-All problem.

The figure below illustrates how, if spike arrival times are used as an analog code, solving the Winner-takes-all problem is almost trivial. The four input units (A-D) have direct connections to four output units (E-H), but they also have collateral connections to an inhibitory interneuron (shown in black). Imagine that all four input units generate spikes, but that the spike in unit B is initiated a few milliseconds before the others. Unit F will respond to its input as usual, but because B's spike also activates the inhibitory interneuron, the other three units will either fail to respond, or respond with longer latencies   . As a result, the output pattern will correspond to a Winner-Take-All function of the input, but unlike conventional neural algorithms, the time taken can be as short as a few milliseconds.

5. Perspectives.

The coding scheme proposed here would appear to provide a powerful way of increasing the processing speed of neural networks, and may provide an important insight into how the visual system can process highly complex visual inputs with only 100-150 msec of processing time. It may also be the case that such a coding scheme could be used in artificial neural network systems. An analog retina in which the individual pixel elements generate impulses with a delay that depended on local image intensity would be relatively easy to construct. Subsequently, circuits of the type described in the preceding section could be used to "pick off" the most active inputs as well as   performing   other more complicated operations.

6. Bibliography.

DeYoe E.A. & Van Essen D.C. (1988) *Concurrent processing streams in monkey visual cortex,* Trends in Neurosciences, **11**, 219.

Feldman J.A. & Ballard D.H. (1982) *Connectionist models and their properties.* Cognitive Science, **6**, 205-254.

Gray C.M. & Singer W. (1989) Stimulus-specific oscillations in orientation columns of cat visual cortex. *Proc. National Academy of Sciences,* **86**, 1698-1702.

Kawasaki M., Rose G. & Heiligenberg W. (1988) *Temporal hyperacuity in single neurons of electric fish.* Nature, **336**, 173.

Knudsen E.I., du Lac S. & Esterley S.D. (1987) *Computational maps in the brain.* Annual Review of Neuroscience, **10**, 41-66.

Lal R. & Friedlander M.J. (1989) *Gating of retinal transmission by afferent position and movement signals.* Science, **243**, 93.

Perrett D.I., Rolls E.T. & Caan W.C. (1982) *Visual neurons responsive to faces in the monkey temporal cortex.* Experimental Brain Research, **47**, 329-342.

Poggio T. & Koch C. (1986) *Synapses that compute motion.* Scientific American, **256**, 46-71.

Thorpe S.J. (1988) *Traitement d'images chez l'homme.* Techniques et Sciences Informatiques, **7**, 517-525.

Thorpe S.J. & Imbert M. (1989) *Biological constraints on connectionist models.* In Pfeiffer R. et al (ed.) "Connectionism in Perspective". Amsterdam : North-Holland. p 63-92.

Zeki S. & Shipp S. (1988) *The functional role of cortical connections.* Nature, **335**, 311-316.

Section 4
Parallel Processing in the Visual System

Parallel Processing in Neural Systems and Computers
R. Eckmiller, G. Hartmann and G. Hauske (Editors)
© Elsevier Science Publishers B.V. (North-Holland), 1990

CORRELATED RHYTHMIC ACTIVITY OF NEURONS IN MONKEY VISUAL CORTEX*

Franz AIPLE and Jürgen KRÜGER

Neurol. Univ. Klinik
Hansastr. 9, D-7800 Freiburg, F.R.G.

In addition to central peaks about 50 ms wide many auto- and cross-correlograms of spike trains obtained in a small region of monkey visual cortex during visual stimulation showed several lateral peaks. They reflect coupled rhythmic activities of different neurons with a frequency of about 10 Hz. This periodic activity is probably induced by rhythmic common input, but it could also constitute an emergent property of the network of cortical neurons.

1 INTRODUCTION

The functioning of the central nervous system is supposed to involve functional groups of neurons that cooperate in a coherent manner. Coordination of activity must occur in a dynamic manner in time and could perhaps be linked to rhythmic activity patterns. Periodic correlated firing patterns of cortical neurons were reported in different animals and on different time scales in a number of laboratories [1,2]. Some models of neural networks are based explicitly on evolution in time and consider cyclic spatio-temporal patterns of activity as quasistable states [3]. In the present paper we show correlated rhythmic firing of neurons in the primary visual cortex of monkeys and discuss its possible origin.

2 METHODS

Using a 30-fold multi microelectrode neuronal impulse activities were recorded simultaneously in a small region of the primary visual cortex of anesthetized monkeys [4]. Coloured large fields and other stimuli were presented repeatedly during the recording. Cross correlation histograms ("raw correlograms") and shift predictors that account for stimulation induced correlation [5] were calculated for all pairs of spike trains using a bin width of 11 ms. Only the **"net correlograms"** obtained by subtraction of the shift predictor from the corresponding raw correlogram were analyzed that show neuronal correlation without stimulus induced correlation. The positions of broad peaks were defined by the center of gravity of the peaks. The relative phase of the periodic firing patterns was defined by the average position of the two lateral peaks to the left and the right of the central peak.

3 RESULTS

Cross correlation histograms, briefly correlograms, of neuronal spike trains show the average temporal relationship between impulses of different neurons. Broad peaks about 30 to 60 ms wide were clearly visible in most correlograms and were centered near time lag 0. In two out of four animals most correlograms showed secondary broad

---

*This work was supported by the "Synergetik" program of the *Volkswagen Foundation*.

peaks on both sides of the central one. In the other two animals secondary peaks were rare, but occassionally present. A typical example is shown in Fig.1.

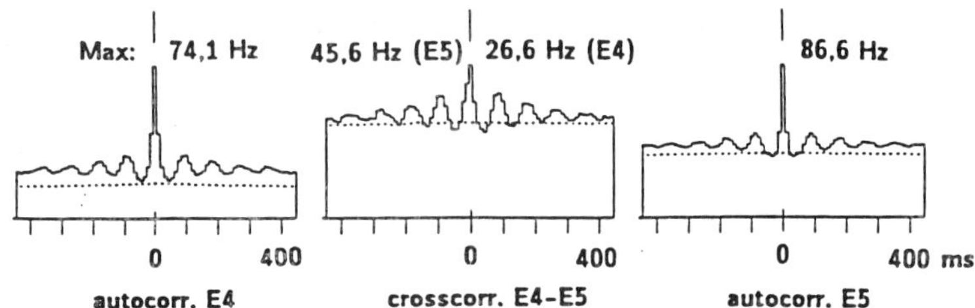

Fig.1 Correlograms of two neurons in layer VI. Binwidth 11 ms. 4-5 periods of rhythmic correlated activity are visible.

Generally the secondary peaks can be seen in auto correlograms as well as cross correlograms. The peaks were positioned at time lags of about 100 ms indicating rhythmic firing with a frequency of approximately 10 Hz.

The rhythmic correlated bursts can even be seen in the unaveraged spike trains that are displayed in Fig.2.

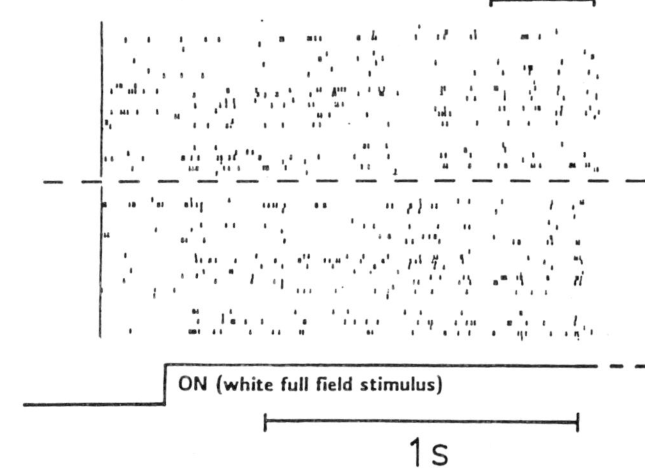

Fig.2 Unaveraged activities of all electrodes (individual traces within every block) for two stimulus repetitions (the two blocks). Every tick stands for a spike.

In general the rhythmic correlated firing was not phase locked to specific stimulus conditions. However, coloured large field stimuli that were turned on abruptly and presented binocularly were found to synchronize the rhythmic activity. In Fig.2 rhythmic groups of spikes occurred at the **same** time instants **shortly** after stimulus onset in the two different stimulus sweeps shown. One second later strong rhythmic activity reappeared in the time interval marked by a black bar on top of the plot. However the rhythms in the two sweeps were shifted against each other, i.e. they were **no longer synchronous** to the stimulus onset. The synchronizing effect of the stimulus can better be seen in Fig.3 which shows peri-stimulus-time-histograms (PSTH) of the summed activity of all neurons in two cortical layers.

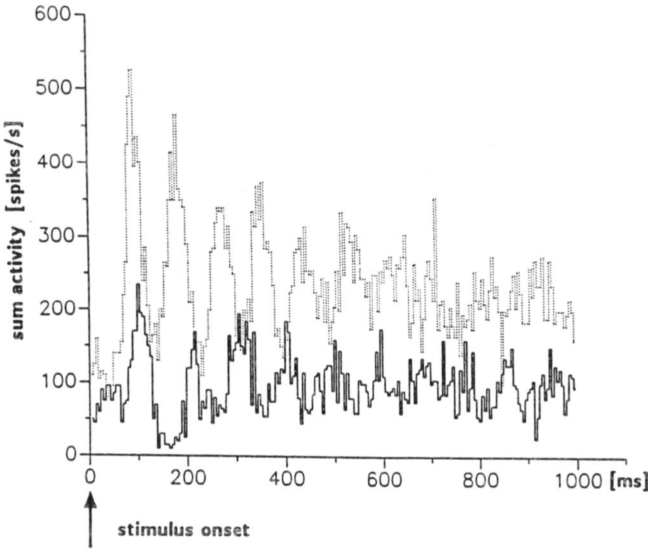

Fig.3 Sum response of all 30 neurons upon onset of white large field stimulus for cortical layer IVc (dotted line) and layer III (solid line).

The distance of the response peaks corresponds to the 100-ms-period found in correlograms. In layer III the second and later response peaks (solid line) occur more than 25 ms later than the corresponding responses in the input layer IVc (dotted line).

As published earlier [6], the central peaks seen in cross correlograms were not always centered on time lag 0. The positions of central peaks in layer IVc showed a systematic shift between the cases where one eye was stimulated as compared to the other eye: if the correlation was observed between a left-eye dominated neuron and a right-eye dominated one, the peak shift always indicated that preferentially spikes in the well-responding neurons preceded spikes in the poorly responding neurons. The lag was up to 10 ms. This was also true for the rhythmic correlation represented by the secondary peaks, i.e. the central and the secondary peaks were shifted to the same side.

4 DISCUSSION

The frequency of the periodic activities of 10 Hz corresponds to the frequency of alpha waves in the EEG. Since a long time rhythmic synchronous firing of large populations of neurons has been supposed to be the origin of EEG waves [7]. This view was supported by experiments using single cell and EEG recordings.

The present study raises the question whether the reported rhythmic firing is due to modulation by extra-cortical inputs or whether it is a pure intracortical phenomenon.
   1. In the thalamus rhythmic activities of a frequency of about 10 Hz are known to exist [8]. Thalamic input to either cortex or lateral geniculate nucleus could modulate the activities of the observed neurons. In analogy to central peak positions [6] the phase shift of the rhythmic activities when changing the stimulated eye could be explained if the primary modulation due to the periodic input activity were restricted to the activity of neurons dominated by the stimulated eye. The rhythmic activity would reach these

neurons earlier than the neurons dominated by the other eye and thus lead to the observed phase shift in correlograms.

2. Cortical layers are multiply interconnected within each area and between areas. This forms the anatomical basis of a model by Krone et al. [9] in which the activity reaches first the input layer IVc and cycles several times through different cortical layers. This in turn can lead to periodic changes of activity of single neurons. The observed coupling of the rhythmic activity to stimuli supports this point of view. The lag of the second and later response peaks in layer III with respect to layer IVc could be related to cycling of activity, but a period of about 100 ms seems quite long for an intracortical loop. The explanation of the phase shift between neurons of weak and strong activity seems to be more difficult with this hypothesis than with the first one.

we suggest that the observed rhythmic activity reflects rather a periodic modulation of the input to the cortical network than an emergent property of the network itself.

5 CONCLUSION

Periodic spike activities observed on different time scales in different laboratories raise the question of the function they subserve. Up to now only few network models incorporate time in an essential way. To include temporal evolution beyond "standard dynamics" seems necessary for the understanding of cortical function.

REFERENCES

[1] Eckhorn, R., Bauer, R., Jordan, W., Brosch, Kruse, W., Munk, M., Reitböck, H.J., Coherent Oscillations: A Mechanism of Feature Linking in the Visual Cortex? Multiple Electrode and Correlation Analysis in the Cat. Biol.Cybern. 60, 121-130, 1988.

[2] Gray, C.M., König, P., Engel, A.K., Singer, W., Oscillatory Responses in Cat Visual Cortex Exhibit Inter-Columnar Synchronization Which Reflects Global Stimulus Properties. Nature 338, 334-337, 1989.

[3] Silverman, D.J., Shaw, G.L., Pearson, J.C., Associative Recall Properties of the Trion Model of Cortical Organization. Biol.Cybern. 53, 259-271, 1986.

[4] Krüger, J., Aiple, F., Multi-microelectrode investigation of monkey striate cortex: spike train correlations in the infragranular layers. J. Neurophysiol. 60, 798-828, 1988.

[5] Perkel, D.H., Gerstein, G.L., and Moore, G.P., Neuronal spike trains and stochastic point processes. II. Simultaneous spike trains. Biophys. J. 7: 419-440, 1967.

[6] Aiple, F., Krüger, J., Neuronal synchrony in monkey striate cortex: interocular signal flow and dependency on spike rates. Exp. Brain Res. 72, 141-149, 1988.

[7] Adrian, E.D., Matthews, R., The Action of Light on the Eye. Part I. The Discharge of Impulses in the Optic Nerve and its Relation to the Electric Changes in the Retina. J.Physiol. 63, 378-414, 1927.

[8] Steriade, M. and Deschenes, M., The Thalamus As a Neuronal Oscillator. Brain Res.Rev. 8, 1-63, 1984.

[9] Krone, G., Mallot, H., Palm, G., Schüz, A., Spatio-Temporal Receptive Fields: A Dynamical Model Derived from Cortical Architectonics. Proc.Roy.Soc.Lond.B 226, 421-444, 1986.

Parallel Processing in Neural Systems and Computers
R. Eckmiller, G. Hartmann and G. Hauske (Editors)
© Elsevier Science Publishers B.V. (North-Holland), 1990

FEATURE LINKING ACROSS CORTICAL MAPS VIA SYNCHRONIZATION

R. Eckhorn, H.J. Reitboeck, P. Dicke, M. Arndt and W. Kruse

Department of Biophysics
Philipps University
Marburg, Federal Republic of Germany

It is still unknown how information from different cortical maps is integrated into coherent percepts. We recently discovered a process of dynamic interactions between cell assemblies in the visual cortex that could function as a global integration mechanism in sensory and motor systems: stimulus-specific synchronizations of 35-80 Hz were found in remote cell assemblies within and between primary visual cortex areas of the cat. Based on these neurophysiological findings we modelled temporally coded feature linking in a network consisting of two neural layers ("cortex maps") with modulatory lateral and feedback connections. The model's performance is demonstrated in examples of region linking, where rhythmic responses to spatially separate moving stimulus regions become phase-locked via feedback linking connections from a "higher cortex map". This behaviour can be interpreted as serving the definition of spatial and temporal continuity of a global feature's neural representation during the synchronization period. Our results support the hypothesis that stimulus-induced synchronization might be a general principle for the coding of relations within and between sensory and motor systems.

1. INTRODUCTION

Stimulus related oscillations of neural activities were recently discovered in the primary visual cortex of monkey [4] and cat [2,3,5]. These findings stimulated the proposal that synchronization between stimulus-specific oscillatory activities might serve to transitory link local visual features into coherent global percepts: our (and other groups) results show that *signal oscillations* are induced in local assemblies that can be synchronized among geographically remote cortex positions by appropriate global stimulus features [2,3,6].

2. EXPERIMENTAL RESULTS FROM CAT VISUAL CORTEX

Stimulus-induced oscillations (35-80 Hz) were found in all three types of neural signals that were recorded with our multiple microelectrode technique [7]. They were most pronounced and most often found in local field potentials (LFPs), they were less often visible in multiple unit activities (MUA) and were less pronounced and not so often observed in single-cell spike trains. In most recording positions the oscillatory components had a correlated tuning with the classical single cell RF-properties (e.g., orientation preference) [2,3,5]. *Oscillation spindles* occured with durations of about 80-250 ms, separated by intervals of stochastic activity, and their *response latencies* were found to be significantly longer than those of visually evoked cortical potentials (VECPs),(Fig.1). VECPs indicate *stimulus-forced* ("single-shot") synchronizations (see conclusions).

Stimulus-specific synchronizations were found among all layers of vertical cortex columns, and between spatially remote assemblies of the same and of different cortex areas (A17 and A18) if the assemblies code common visual features (Fig. 2). Synchronizations generally have zero phase difference, even if the assemblies are located in different visual cortex areas [2,3].

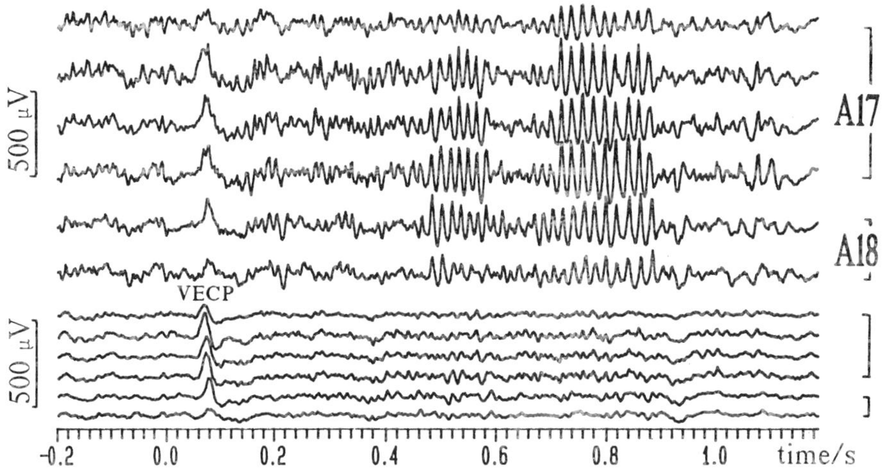

FIGURE 1

Stimulus-induced synchronization between two cortical areas (A17 and A18) in response to a drifting grating (0.7 cycles/deg swept at 10 deg/s; movement starts at t=0 and continues for 900 ms; binocular stimulation; RFs of A18-neurons overlapped those of A17). Upper panel: single-sweep LFP responses. Lower panel: averages of 10 LFP-responses to identical stimuli. Note that the oscillatory components are averaged out while the VECP at 60-110 ms remains visible.

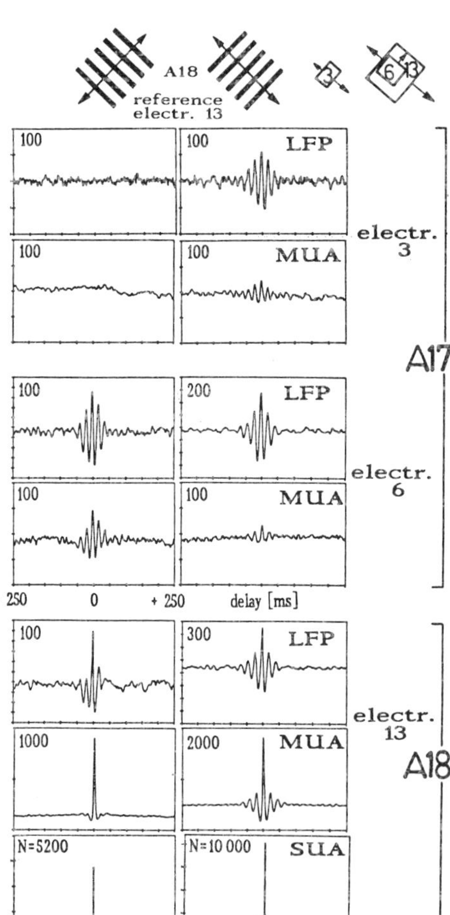

FIGURE 2

Synchronization between two visual cortex areas (A17 and A18), dependent on stimulus movement direction (same stimulus parameters as for Fig.1; stimulus directions and relative RF-positions indicated above). Each panel shows cross correlations between A18-spikes from electrode 13 and LFPs or MUA from an electrode as indicated. SUA panels show auto-coincidence functions of spikes on reference el.13. (Numbers in upper left corners: ordinate scalings in μV). Note the high stimulus specificities of synchronization between single cell and mass activities in the two cortex areas.

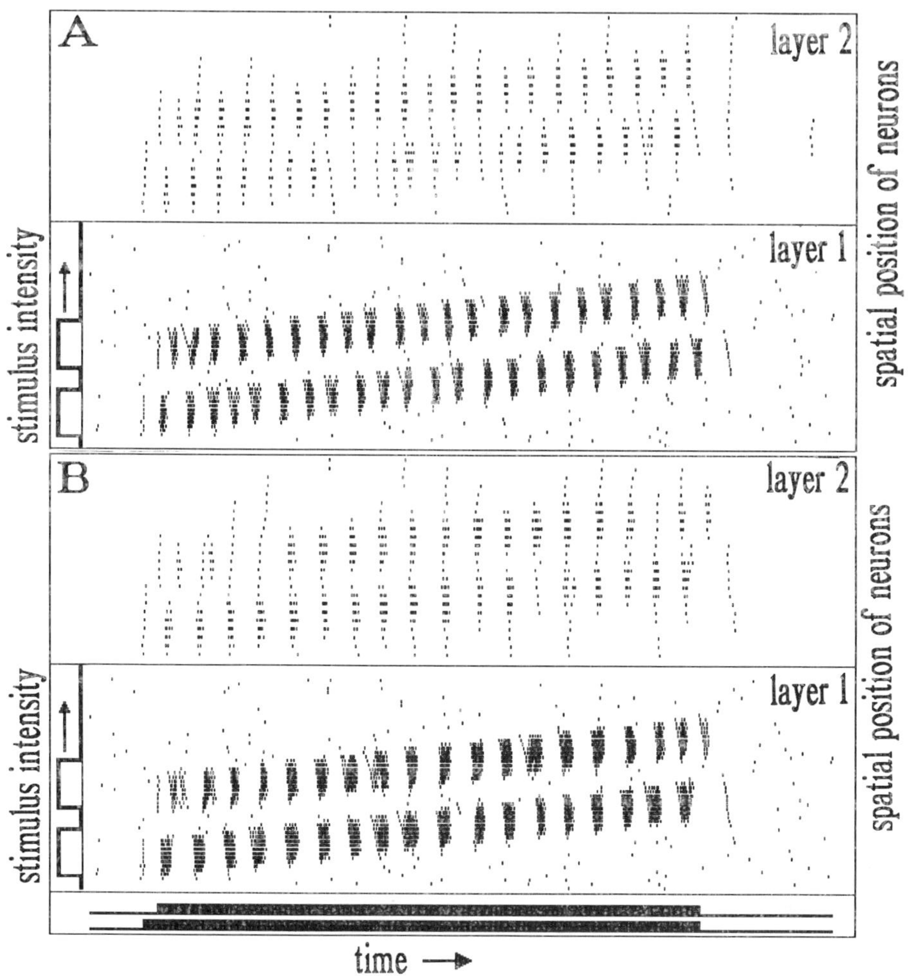

FIGURE 3

Simulation results: "Spike trains" (dots) at the outputs of the two layers. Stochastic maintained input activity was applied at all layer-1 feeding inputs. Black bars indicate stimulus time courses. (For more details see text).

3. RESULTS FROM NEURAL NETWORK MODEL

In the simulations we used a *model neuron* that has leaky integrators at its synapses and a dynamic threshold with negative feedback. Our model neuron has an important feature that assures linking between neurons without major degradation of the "receptive field" properties: Each dendrite has *feeding* and *linking inputs*. Feeding inputs drive the neurons mainly by afferent visual signals via classical (PSP-) synapses, while linking inputs act modulatory on the postsynaptic "feeding signals" and are driven by lateral and feedback connections [3].

The *neural network* used in the present simulations consists of 2 one dimensional layers with 50 neurons in the first and 23 in the second layer. In both layers each neuron is connected to 4 neighbors at each side with positive laterally declining strength at the linking synapses [3]. Neurons in layer-1 are directly driven via feeding connections by "spatial visual stimuli" while each layer-2 neuron receives feeding connections from 4 layer-1 neurons. In terms of receptive fields (RFs) this means that layer-2 RFs completely overlap 4 layer-1 RFs. Modulatory linking between layer-2 and layer-1 is made by (positive) feedback via linking connections (each layer-2 neuron sends 4 connections back to layer-1). The choice of the model parameters (such as time constants and coupling strengths and the numbers of convergent and divergent connections) does not crucially influence the simulation results.

Fig.3 shows *results from simulations* in which two spatially separate "bright stimulus regions" are switched on, one after another, and then move with constant velocity until they are switched off simultaneously. In the simulation Fig.3B the network is "complete" (as specified above), while for Fig.3A the feedback connections are left out. In the latter cases synchronization occurs within each subfield immediately after "stimulus onset" due to the lateral linking connections in layer-1. However, the horizontal connections do not bridge the gap between the two regions, i.e. their rhythmic activities are not correlated. With the addition of feedback from layer-2 to layer-1 (Fig.3B), phase-locked rhythmic burst discharges do occur within "stimulated regions" in both layers. Synchronization between the stimulated subfields in layer-1 due to influences from the "higher visual map" transiently links the two regions. We want to point out that linking connections, even when they converge from neurons with remote or much broader RFs, do not deteriorate single cell RF-properties, because their modulatory influence mainly causes phase shifts in the output spike patterns [2,3].

4. CONCLUSIONS

We extend our previous concept of temporal feature linking [2,3] by the proposal that *two types of synchronization* can support temporal linking: *stimulus-forced* (single-shot) and *stimulus induced* (repetitive) *synchronization*. Stimulus-forced synchronization can be accomplished by near-simultaneous transients at feeding inputs that mutually enhance one another via a linking network. In the case of stimulus-induced synchronizations the network itself generates repetitive synchronized discharges necessary for feature linking (sustained signals at the feeding inputs). Both types of synchronization occur successively in natural vision. Stimulus-forced synchronization is assumed to establish fast but "crude sketches" of association codes [1] in the visual cortex, whereas repetitive internally generated synchronization allows the formation of complex and sophisticated association codes in iteration steps. We showed that feedback from higher cortical maps can bridge spatial gaps in lower maps by synchronizing the activities in the subfields.

ACKNOWLEDGEMENTS

We thank Dr. R. Bauer, W. Jordan, M. Munk, M. Brosch and T. Schanze for help in experiments and data reductions, and U. Thomas, J.H. Wagner and W. Lenz for their expert technical support. Grants came from the Deutsche Forschungsgemeinschaft (DFG): Ba 636/4-1,2, Re 547/2-1, Ec 53/4-1 and from Stiftung Volkswagen: I/35695, I/64605.

REFERENCES

[1] Damasio AR, Multiregional Retroactivation, Cognition (in press).
[2] Eckhorn R, Bauer R, Jordan W, Brosch M, Kruse W, Munk M, Reitboeck HJ, Biological Cybernetics 60 (1988) 121-130.
[3] Eckhorn R, Reitboeck HJ, Arndt M, Dicke P, A Neural Network for Feature Linking via Synchronous Activity, in: Cotterill RMJ (ed), Models of Brain Function (Cambridge Univ Press, 1989) pp. 1-18.
[4] Freeman WJ, van Dijk BW, Brain Res 422 (1987) 267-276.
[5] Gray CM, Singer W, Neurosci Suppl (1987) 1301P.
[6] Gray CM, König P, Engel AK, Singer W, Nature 338 (1989) 334-337.
[7] Reitboeck HJ, IEEE-SMC 13 (1983) 676-682.

Parallel Processing in Neural Systems and Computers
R. Eckmiller, G. Hartmann and G. Hauske (Editors)
© Elsevier Science Publishers B.V. (North-Holland), 1990

SYNCHRONIZATION OF OSCILLATORY RESPONSES:
A MECHANISM FOR STIMULUS-DEPENDENT ASSEMBLY
FORMATION IN CAT VISUAL CORTEX

Andreas K. Engel, Peter König, Charles M. Gray and Wolf Singer

Max-Planck-Institut für Hirnforschung
Deutschordenstr. 46, 6000 Frankfurt 71, F.R.Germany

1. INTRODUCTION

Current concepts in the neurobiology of vision assume that object perception requires the segmentation of visual scenes and the segregation of figures from ground [2,11,12,16,17]. It is hypothesized that a primary step involves the local extraction of specific features of an object [14]. Subsequently, to permit the identification of the object, these local features have to be evaluated for coherencies which can be used as a segregation criterion [12]. For this purpose, a variety of cues can be employed such as the continuity or co-linearity of contours, the common direction of movement, similiarities of the spatial frequency content, interocular disparity and colour [2,11,12,17]. On the cortical level, this requires the interaction of spatially distributed neuronal feature detectors to yield a cell assembly which encodes the object presented in the visual field. In the model suggested originally by Hebb [10], assemblies are defined by the concurrent elevation of the average firing frequency of the participating cells. However, it has been recognized that this mechanism of assembly formation remains necessarily ambigous, because in natural scenes several objects are usually present on a complex background all of which activate many interspersed feature detectors [1,3,16]. Therefore it was proposed that the precise temporal coincidence between individual spikes rather than just similar average rates of firing are important to distinguish the cells of an assembly [1,3]. More recently, it was suggested that synchronization of responses with a regular *periodic* temporal structure could be particularly useful to permit the coexistence of several distinct functional assemblies in the same region of the cortical network [16].

It is now well established that neuronal responses with an oscillatory modulation of the firing frequency occur in a variety of cortical areas. In particular, the origin and functional role of oscillatory responses have been investigated in the olfactory system [5,6]. In our laboratory, stimulus-dependent oscillations in a frequency range of 40-60Hz have been discovered in the visual cortex of the cat [7,8]. This finding, which was subsequently confirmed by [4], prompted us to investigate whether oscillatory responses can synchronize between spatially separate cell groups. The results of these experiments demonstrate that such an intercolumnar synchronization does indeed occur in cat striate cortex, depending on the orientation preferences and the spatial separation of the recorded cells [9]. In this paper we will discuss evidence for the hypothesis that the synchronization of oscillatory responses is also markedly influenced by features of the visual stimulus.

2. METHODS

We recorded multiunit activity from area 17 of anesthetized and paralyzed cats. In most experiments, we used arrays of multiple electrodes with interelectrode distances ranging from 0.4 to 12mm. Peri-stimulus-time histograms, auto-correlation functions (ACF) and cross-correlation functions (CCF) were computed to determine the response amplitudes, the degree of rhythmicity of the responses and the interaction between cells at different recording sites, respectively. The frequency and amplitude modulation of the correlograms were quantified by fitting a damped sine wave (Gabor) function to the data (Fig. 2). Since the modulation of the correlograms usually had a certain offset, we normalized the Gabor function amplitude with respect to this offset. This measure permits the quantification of the oscillatory modulation independently of variations in the absolute strength of the response. Here, we term this measure "normalized modulation amplitude" (NMA). To examine the feature dependence of oscillatory responses, we varied systematically the

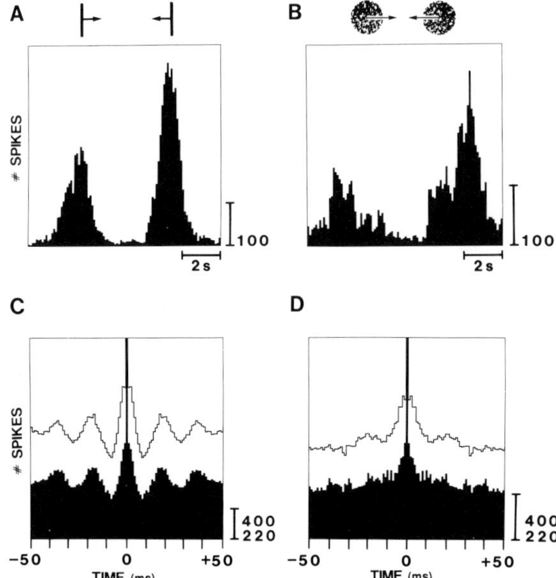

Fig.1. Rhythmicity of the response is eliminated by stimulation with random dot patterns. In this example, the cells responded almost equally well to a light bar (**A**) and a random dot texture (**B**) moved in the same direction. The texture always covered the entire receptive field. The auto-correlograms indicate a significant oscillatory modulation of the light bar response (**C**), whereas the response to the texture stimulus shows a random temporal structure (**D**). ACFs for the first stimulus direction are shown with filled bars. The scale bars indicate numbers of spikes.

orientation, velocity and length of moving light-bar stimuli. In addition, we tested the effects of monocular stimulus presentation (routinely, stimuli were applied binocularly), the influence of conflicting stimuli, of random dot patterns and of stationary stimuli on the oscillatory modulation of the responses. Moreover, we studied the effect of suboptimal stimulus orientations, monocular stimulation, stationary flashed light bars and conflicting stimuli on the interaction between cells at different recording sites.

3. RESULTS

We observed that all stimulus parameters tested, except orientation, influenced the temporal structure of the responses at individual recording sites: (1) The response frequency increases with stimulus velocity (10 out of 11 cases). The NMA of the ACF shows only minor variations with changing stimulus velocity. Above a cut-off velocity of 5-7deg/sec, the oscillatory behaviour disappears, although the cells still respond strongly. The range of stimulus velocities which elicited oscillatory responses was consistently lower than that which evoked maximum firing rates. (2) The NMA of the ACFs increases as a function of stimulus length within the boundaries of the receptive field (11 out of 11). In 5 out of these 11 cases, stimuli extending beyond the receptive field produced further enhancement of the NMA. (3) Binocular cells frequently exhibit an enhanced NMA of the ACF when stimuli are applied through both eyes, in particular if the cells show binocular summation (9 out of 16). (4) The combined presentation of an optimal and an orthogonal conflicting stimulus reduces the NMA as compared to the optimal stimulus alone (16 out of 21 cases). (5) Responses evoked by moving random dot patterns show little or no rhythmicity, as illustrated in Figure 1 (17 out 19 cases). (6) If moving stimuli are replaced by light bars of the same orientation which are flashed ON and OFF within the receptive field, the oscillatory modulation disappears, although the cells respond well to the stationary stimulus (9 out of 13).

Our analysis of the feature dependence of the cross-columnar synchronization of oscillatory responses yielded the following results: (1) Response synchronization is facilitated by stimulation of the cells with their preferred orientation but does not necessarily depend on it. If the cells at the two recording sites have overlapping receptive fields, a synchronization (as

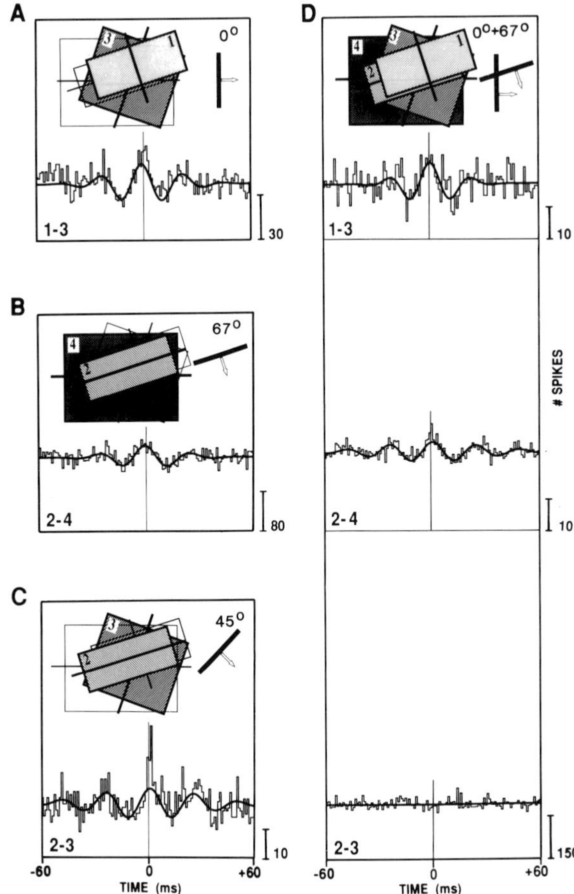

Fig.2. Effect of conflicting stimuli on the cross-columnar interaction. We recorded simultaneously from 4 different cell groups which were narrowly tuned to orientation preferences of 157 (site 1), 67 (2), 22 (3) and 90 deg (4), as indicated in **D**. The electrode separation was 400μm. The figure compares responses to stimulation with single light bars of 0 (**A**), 67 (**B**) and 45 (**C**) deg orientation with responses to combined presentation of 0 and 67 deg light bars (**D**). For each stimulus condition, the shading of the receptive fields indicates the sites at which responses were elicited. The CCF computed from the respective responses is shown with unfilled histogram bars. The thick continous line represents the Gabor function which was fitted to the correlogram. Note that the CCF of (2-3) in **D** does not show an oscillatory modulation, i.e. the Gabor function amplitude was not significantly different from 0. Scale bars indicate the number of spikes.

indicated by a CCF with a periodic modulation) is obtained for all stimuli which evoke oscillatory responses at both recording sites (19 out of 19 cases). Thus, a synchronization can also be observed between columns with orthogonal orientation preferences [9]. (2) Monocular stimulation attenuates the cross-columnar correlation. The periodic modulation of the CCF disappears (mostly due to the reduction of the ACF modulation), and only a center peak remains in the CCF which is slightly reduced in amplitude (9 out of 15). (3) Stationary stimuli reduce the degree of correlation even to a larger extent. In 3 out of 8 cases tested, a single peak with low amplitude remained in the CCF. In 4 out of 8 cases, the correlation is completely absent. (4) Preliminary evidence indicates that conflicting stimuli also affect the intercolumnar synchronization (2 out of 2). An example is illustrated in Figure 2, where we recorded from cells in 4 different columns with overlapping receptive fields but alternating orientation preferences. Stimulation with a single light bar yielded a synchronization between those sites responding to the respective orientation, i.e. between site 1 and 3 with a vertical light bar (Fig. 2A), between 2 and 4 with a 67deg stimulus (Fig. 2B), and between 2 and 3 with an intermediate orientation (Fig. 2C). The simultaneous presentation of a 0 and 67deg light bar strongly activated all four sites (Fig. 2D). As expected, with this stimulus a response synchronization was observed between sites 1-3 and 2-4. However, no correlation occurred *between* these two pairs, e.g. between sites 2 and 3, although at both sites strong responses were evoked with sufficient temporal overlap. Thus, the conflicting stimulus segregates the recorded cells in two different groups according to their orientation preferences, although with a *single* stimulus they correlate in all possible combinations.

4. CONCLUSIONS

As outlined above, our working hypothesis implies first, that the visual cortex can be envisioned as a system of coupled oscillators and second, that assembly formation is mediated by transient synchronization of such oscillators depending on coherences in the feature domain which are evaluated to permit segmentation of the visual scene [16,17]. Several testable postulates can be deduced from this hypothesis, which are consistent with the experimental data: (1) *Interactions should not occur between similar feature detectors only.* Our experiments with suboptimal stimulus orientations clearly demonstrate the possibility of synchronization of feature detectors with widely differing orientation specificity (see also [9]). This may be required to form assemblies coding for edges or curvature. (2) *Neuronal responses should synchronize in a stimulus-dependent manner. In particular, the synchronization should be influenced by features indicating the coherence of the visual stimulus.* As shown in this study, a number of stimulus features clearly influence the cooperativity within a local group of cells (as reflected in the ACF of multi-unit activity) as well as the synchronization between spatially separate cortical columns. These results suggest that no direct inferences about the underlying anatomical circuitry can be made from the correlograms [13,15]. Rather, the synchronization of oscillatory responses represents a functional coupling which may show rapid changes. This functional connectivity is particularly sensitive to features reflecting the coherence of the visual stimulus. The importance of spatial continuity is demonstrated by our experiments with conflicting stimuli and random dot texture (Fig. 1). In addition, the global similiarity of orientation and coherence of motion also influence the intercolumnar synchronization. Evidence for this was obtained in cases where we recorded from two cell groups with similar orientation preferences and non-overlapping receptive fields which were aligned co-linearly [9]. In these experiments, the strongest synchronization was observed with a single continous contour stimulating both fields simultaneously. If a gap was inserted into this stimulus, synchronization became weaker. The interaction disappeared completely, if the two stimulus fragments were moved in opposite directions over the receptive fields [9]. Finally, binocular stimulation in appropriate retinotopic correspondence seems to be required for the local and cross-columnar synchronization of neuronal responses. (3) *Evidence should be obtained for a stimulus-dependent segregation of synchronized cells into two different assemblies.* According to our working hypothesis, the cells constituting an assembly should oscillate in synchrony, whereas no consistent phase-relationship should be observed between different assemblies. The mechanism proposed would permit a rapid "switching" of cells from one assembly to another by subtle changes of frequencies and relative phases, according to variations of the stimulus. Our experiments with conflicting stimuli provide a direct test of this prediction (Fig. 2). The superposition of two light bars with different orientations leads to a segregation of the responding columns into two assemblies which are distinguished by the phases of their oscillations and coexist in the same region of cortex. However, using a single light bar as a stimulus, formerly uncorrelated cells can now synchronize and recombine to another assembly. In conclusion, these results demonstrate that transient coupling of neuronal oscillators may provide a suitable mechanism of assembly formation in cat visual cortex.

REFERENCES

[1] Abeles, M. (1982) Local Cortical Circuits. An Electrophysiological Study. Springer, Berlin.
[2] Ballard, D.H., Hinton, G.E. and Sejnowski, T.J. (1983) Nature 306: 21-26. [3] Crick, F. (1984) Proc.natn.Acad.Sci.USA 81: 4586-4590. [4] Eckhorn, R., Bauer, R., Jordan, W., Brosch, M., Kruse, W., Munk, M. and Reitboeck, H.J. (1988) Biol.Cybern. 60: 121-130. [5] Freeman, W.J. (1975) Mass Action in the Nervous System. Academic Press, New York. [6] Freeman, W.J. (1988) In: Basar, E. Dynamics of Sensory and Cognitive Processing by the Brain pp. 19-29. Springer, Berlin. [7] Gray, C.M. and Singer, W. (1987) Soc.Neurosci.Abstr. 13: 404.3. [8] Gray, C.M. and Singer, W. (1989) Proc.natn.Acad.Sci.USA 86: 1698-1702. [9] Gray, C.M., König, P., Engel, A.K. and Singer, W. (1989) Nature 338: 334-337. [10] Hebb, D.O. (1949) The organization of behaviour. Wiley, New York. [11] Julesz, B. (1981) Nature 290: 91-97. [12] Marr, D. (1982) Vision. Freeman, San Francisco. [13] Toyama, K. (1988) In: Rakic, P. and Singer, W. Neurobiology of Neocortex pp. 203-217. Wiley, New York. [14] Treisman, A.M. and Gelade, G. (1980) Cogn.Psychol. 12: 97-136. [15] Ts'o, D.Y., Gilbert, C.D. and Wiesel, T.N. (1986) J.Neurosci. 6: 1160-1170. [16] v.d.Malsburg, C. and Schneider, W. (1986) Biol.Cybern. 54: 29-40. [17] v.d.Malsburg, C. and Singer, W. (1988) In: Rakic, P. and Singer, W. Neurobiology of Neocortex pp. 69-99. Wiley, New York.

Parallel Processing in Neural Systems and Computers
R. Eckmiller, G. Hartmann and G. Hauske (Editors)
© Elsevier Science Publishers B.V. (North-Holland), 1990

PARALLEL DISTRIBUTED PROCESSING OF CONFIGURAL FEATURES OF MOVING OBJECTS IN THE TOAD'S VISUAL SYSTEM

J.-P. Ewert, W.W. Schwippert, and T.W. Beneke

Abteilung Neurobiologie
Universität Kassel
D-3500 Kassel, Bundesrepublik Deutschland

Abstract. Mammals and amphibians have developed different principles to discriminate object orientations. We show how a toad distinguishes the orientations of elongated objects relative to the direction of their movements.

1. INTRODUCTION

Detecting the orientations of elongated work pieces on assembly lines is one problem for perceptual robotics. Suppose an engineer is asked to construct an opto-electronic device that selects elongated objects traveling in the orientation of their longer axes from the same objects whose longer axes are oriented perpendicular to the direction of movement. The first that comes to mind is *explicit computing* of each possible object orientation and movement direction with the aid of asymmetric processors, which will require an enormous amount of processing steps. Neural circuitries in vertebrates solve comparable problems in different ways:

The highly complex structural and functional differention of the "orientation columns" in the visual cortex of mammals allows a great number of neurons with differently oriented asymmetric visual receptive fields to *explicitly* detect any possible object orientation [1].

Amphibians, lacking such a telencephalic cortical organization, apply a different principle [2] called *implicit computing* [3]. The central visual system of frogs and toads involves relatively few types of neurons with radially symmetrical excitatory receptive fields. Unlike mammalian "orientation detectors", here, the asymmetry required for discriminating the orientation of stripes relative to the direction of movement lies in the time domain of the stimulus parameters to be processed: changing the stimulus area a_p *perpendicular to the* direction of object movement influences spatial properties; changing the stimulus area a_i *in the* direction of movement influences both spatial and temporal properties. Importantly, these assignments are invariant under the object's direction of movement [4,5]. There is evidence to show here (i) that the figural features a_i and a_p are processed in parallel in different neural networks, (ii) that "con"-figuration is achieved through subtractive network interaction, and (iii) that "configuration detectors" expresssing this kind of network interaction send their axons directly to the bulbar reticular formation, an area known to be involved in motor pattern generation.

2. BEHAVIORAL RESULTS

The Fig. 1A shows the prey-catching activity of toads in response to objects whose features a_p and a_i are varied at different proportions. In this system of feature relationships, stripes oriented in the direction of their movement

($a_i > a_p$, signifying optimal "prey") and those oriented perpendicular to the direction of movement ($a_p > a_i$, signifying "non-prey" or "threat") are at "opposite sites" of a physiologically relevant stimulus continuum (Fig. 1A and B) [4,5]. Searching for visual neurons involved in this pattern discrimination, therefore, we have looked for cells sensitive to a_i/a_p relationships that determine moving objects' *dynamic configurations*.

3. PARALLEL PROCESSING OF a_i AND a_p IN PRETECTAL AND TECTAL NETWORKS

Visual information, prefiltered in the retina, is fed in parallel to the mesencephalic optic tectum and to various diencephalic nuclei such as the pretectal thalamus; these structures are interconnected (Fig. 2A) (for review see [5]). In the tectum and the pretectum neurons have been recorded which

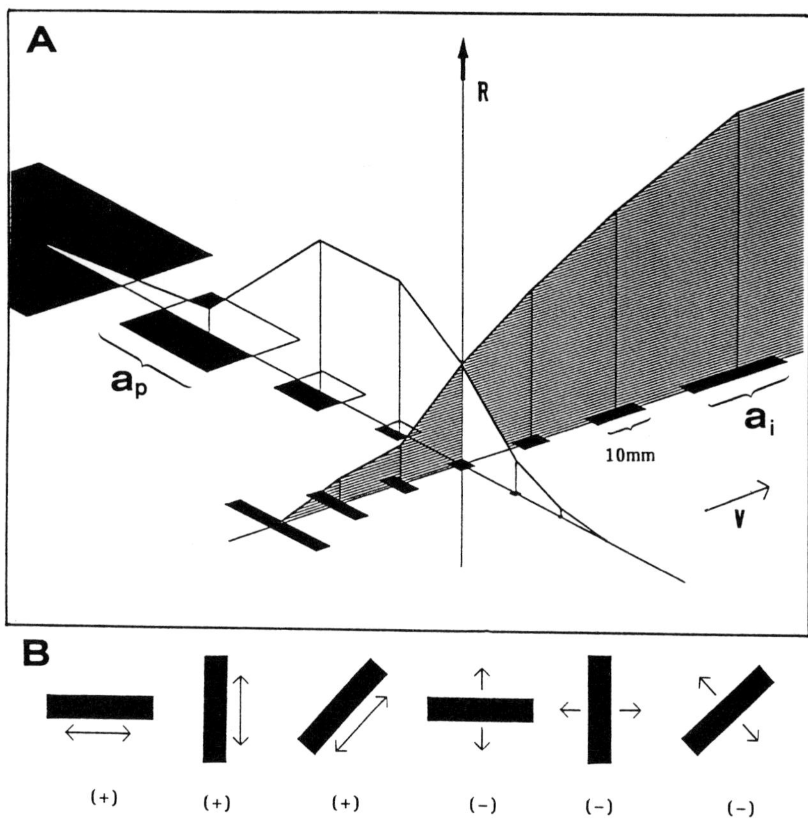

FIGURE 1

A) Prey-catching activity (R) of a toad in response to changes of configural features a_i/a_p of rectangular black objects moving at constant velocity (v) against a white background; movement direction is indicated by the arrow. (Note that for stripes oriented in the direction of movement: $R \rightarrow 0$ if $a_i \rightarrow \infty$). After Ewert [4]. B) Different orientation of a rectangular 3mm x 20mm black stripe relativ to the directions of its movement (indicated by double arrows); (+) refers to prey and (−) to non-prey.

distinguish moving configural objects that are smaller than the diameters of
the radially symmetrical excitatory receptive fields [5,6]. While the thalamic
pretectal class TH3 neurons largely resemble and emphasize the *spatial
summation* properties of converging retinal (class R3 and R4) ganglion cells,
the *spatiotemporal* sensitivities of class T5.1 and T5.2 tectal neurons
(collecting inputs of R2 and R3 cells) probably take advantage of tectal
intrinsic lateral excitation.

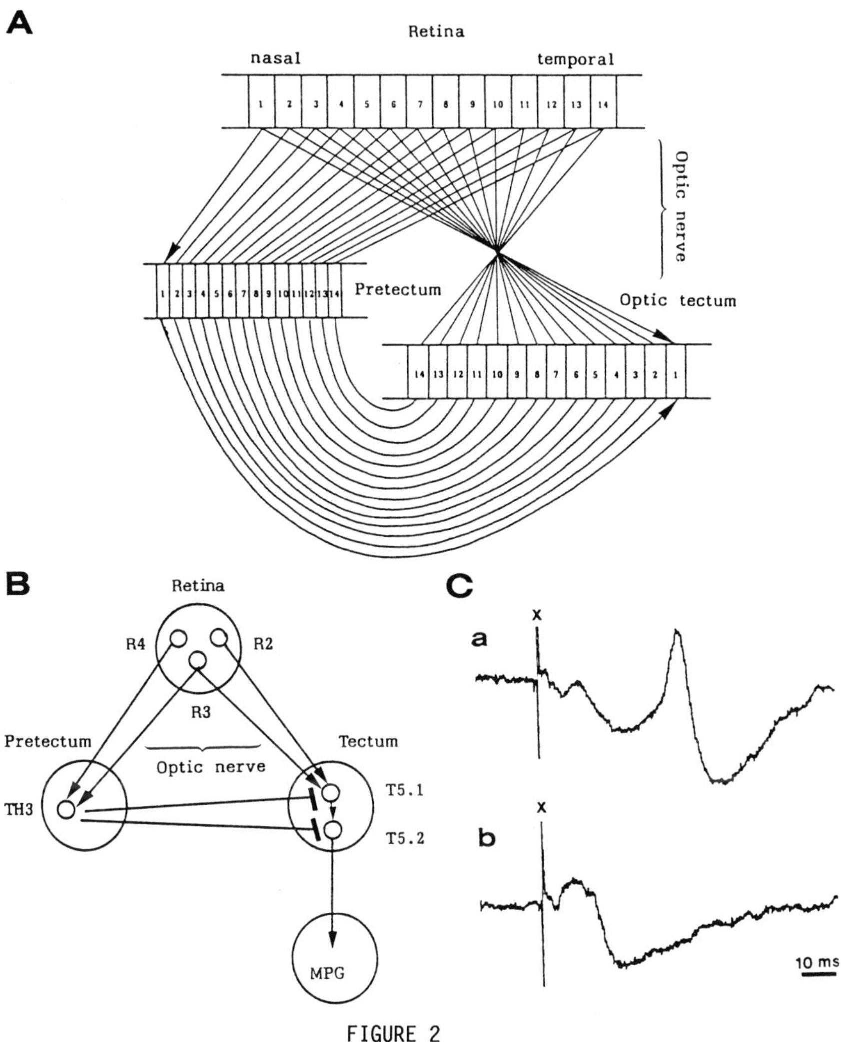

FIGURE 2

A) Schematic illustration showing parallel innervation of the toad's pretectum
and optic tectum via components of the optic nerve, topographic
representations of the retina, and pretecto-tectal connectivity. B) Parallel
processing retinal inputs to thalamic pretectal and tectal networks; arrows
denote excitatory connections and lines with bars inhibitory influences; MPG,
motor pattern generating structures in the medulla oblongata. C) Intracellular
activity of a tectal cell recorded in response to electrical stimulation (x)
of the entire optic nerve (a), and response of the same cell to electrical
stimulation only in the pretectum (b).

4. OBJECT DISCRIMINATION THROUGH SUBTRACTIVE NETWORK INTERACTION

The activities of T5.2 neurons in response to moving visual objects express an interaction of a_i and a_p, in principle similar to the motor output shown in Fig. 1. The selectivity of these "configuration detectors" can be explained by pretecto-tectal inhibitory inputs. The hypothesis on subtractive network interaction (Fig. 2B) [2,4,7] is supported by new experimental results:

Data from extracellular recordings: (a) Studies of a T5.2 neuron in paralyzed toads show that this neuron's configural selectivity is lost shortly after an ipsilateral pretectal lesion: each moving object, irrespective of size and configuration, excessively activates the neuron, probably due to dominating tectal intrinsic mutual excitatory processes. (b) Comparable recordings from freely moving toads [8] demonstrate that the change in configural selectivity of T5.2 responses, pre- and post-lesion, is resembled by a comparable change in the prey-selecting behavior.

Data from intracellular recordings: (a) A T5.2 cell displays mainly excitatory postsynaptic activity (ERSPs>IPSPs) in response to a stripe oriented in the direction of movement, but inhibitory activity (IPSPs>EPSPs) if the same stripe is oriented perpendicular to the direction of movement [9]. (b) Electrical stimulation of the entire optic nerve elicits both EPSP and IPSP activities in a tectal cell (Fig. 2Ca); however, electrical stimulation in the pretectum leads to increased inhibitory postsynaptic activity in the same tectal cell (Fig. 2Cb). (c) Iontophoretic labeling [9] shows that T5.2 neurons project their axons toward the medial reticular formation of the medulla oblongata (see also [10]), an area in which also cells with T5.2 properties could be recorded [11].

5. CONCLUSION

The principle of *implicit computing* the orientations of moving elongated objects - taking advantage of parallel information processing and subtractive network interaction - is economic and fast in its execution. Our current data suggest that the circuitry shown in Fig. 2B can also be influenced by modulatory telencephalic inputs to pretectal cells, which, e.g., may explain modification of feature discrimination by associative learning.

REFERENCES

[1] Hubel, D.H. and Wiesel, T.N., J. Physiol. 160 (1962) 106-154.
[2] Ewert, J.-P. and v. Seelen, W., Kybernetik 14 (1974) 167-183.
[3] Stevens, K.H., Behav. Brain Sci. 10 (1987) 387-388.
[4] Ewert, J.-P., Pflügers Arch. 308 (1969) 225-243.
[5] Ewert, J.-P., Behav. Brain Sci. 10 (1987) 337-405.
[6] Ewert, J.-P. and v. Wietersheim, A., J. Comp. Physiol. 92 (1974) 131-148.
[7] Arbib, M.A., The Metaphorical Brain 2 (Wiley, New York, 1989).
[8] Schürg-Pfeiffer, E., Behavior-Correlated Properties of Tectal Neurons in Freely Moving Toads, in: Ewert, J.-P. and Arbib, M.A. (eds.), Visuomotor Coordination: Amphibians, Comparisons, Models, and Robots (Plenum, New York, 1989) pp. 451-480.
[9] Matsumoto, N., Schwippert, W.W. and Ewert, J.-P., J. Comp. Physiol. 159 (1986) 721-739.
[10] Satou, M. and Ewert, J.-P., J. Comp. Physiol. 157 (1985) 739-748.
[11] Schwippert, W.W., Beneke, T.W. and Ewert, J.-P., J. Comp. Physiol. (1990), in print.

Parallel Processing in Neural Systems and Computers
R. Eckmiller, G. Hartmann and G. Hauske (Editors)
© Elsevier Science Publishers B.V. (North-Holland), 1990

THE THALAMO-CORTICAL FEEDBACK-SYSTEM: A NONLINEAR NETWORK-MODEL FOR BINOCULAR INTERACTION*

Dietmar HEINKE[†], Fotios GIANNAKOPOULOS, Johannes BEST[‡]

Lehrstuhl für theoretische Biologie
Institut für Neuroinformatik
Ruhr-Universität Bochum
Universitätsstr. 150
D-4630 Bochum, Federal Republic of Germany

We present a nonlinear mathematical network model based on principles of cortical and subcortical architecture to describe binocular interaction in the thalamo-cortical feedback-system. The results of the computer aided simulations are discussed in the context of psychophysical experiments concerning spatial hysteresis, Panum's fusion area, lateral inhibition and binocular rivalry.

1 Introduction

A key position in the afferent pathway of mammalian sensory systems is held by gating stations, the thalamic nuclei. Here, the information from the periphery is not only relayed to other processing levels but also to a considerable extend controlled by cortifugal and ascending reticular pathways.

In the retinocortical pathway of the visual system the lateral geniculate nucleus (LGN) is the state where the information from the sense organs, the two eyes, is transmitted toward the neocortex (layer 4 and 6) and controlled by centrifugal fibers originating from layer 6 of the cortex.

The simplifyed structure of the present study constitutes of an LGN with two principle layers, one for each eye. Here the ascending stream of activity is transmitted via relay cells to another control station, the perigeniculate nucleus (PGN) and to the main target zone, the primary visual cortex, taken as a system of one input and one output layer. All these parts are highly interconnected via positive and negative feedback loops. An illustration of this intrinsic circuity is seen in Fig.1. The physiological base of the present model is the mammalian visual system [1],[5],[6],[7].

2 Model

To derive a mathematical description of the model the following assumption are made: The neurons are arranged in 2-dimensional layers that are homogenous and continuous. The

---

*supported by the Stiftung Volkswagenwerk, Grant I/62979
†present address: Institut für Regelungstechnik, Technische Hochschule Darmstadt
‡supported by a grant of the Max Plank Gesellschaft

neurons perform a summation over space which can be modelled by spatial convolutions with appropriate axonal and dendritic weight distributions. The synapses are represented by a low-pass with time-delay and have positive (excitatory) or negative (inhibitory) effect on the soma. The activity of the neuron is a nonlinear function of the total membran potential. The nonlinearity is due to the threshold and the refractory period of the neuron [4],[8].
The Figure 1 shows the connectivity scheme of the model (for more detail see [4]). It shows intrinsic coupling in the cortex including strong positive and weaker negative feedback between different layers. Within the thalamus, negative feedback is most important. Excitatory ascendent and descendent connections between cortex and thalamus are also included. While feedforward connections from the two eyes into the LGN are separated in layers, combined binocular influences are mediated by corticofugal feedback projections [1],[5],[6],[7].

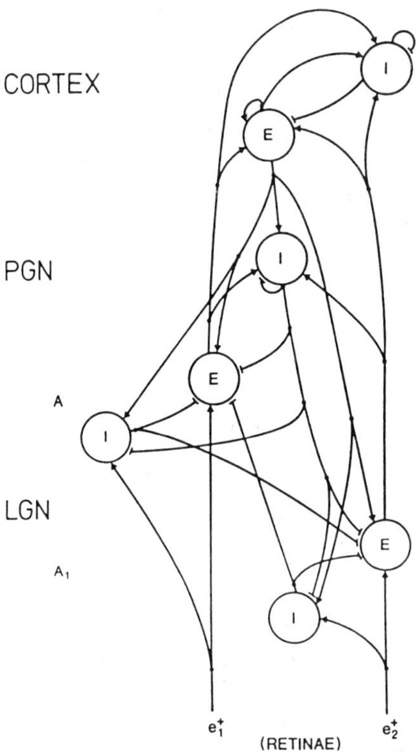

Figure 1: Connectvity scheme of the model. Each circle represents a 2-dimensional layer (cell population). The output of each layer can be either excitatory (E) or inhibitory (I). In addition → symbolizes a positiv influence on a layer and ⊣ a negative one.

3 Results

Based on this model, several simulations with monocular and binocular stimuli were made:
The eigen-behavior shows an oscillating neural activity in the cortex and in the talamus.
The envelope of the oscillation decreases exponentially. Thus, the spatio-temporal pattern of
activity can be described as a travelling wave spreading from the site of simulation [4]. Figure
2 shows the spatial distribution of activity at $t = 45\,ms$ after applying a spatial and temporal
δ-stimulus.

Upon monocular presentation of a bar stimulus, the model shows a dynamic type of lateral
inhibition. The inhibitory zones at adjacent sides of the bar oscillate in antiphase.

Binocular stimuli fuse to a coherent pattern of activity, if disparities are small (cf. Panum's
fusional area in psychophysics [2]). For disparities increasing in time (i.e. stimulus motion in
depth) fusion persists up to relatively large disparities; conversely, during decrease, fusion is
restored only at about the disparity limit for stationary stimuli (spatial hysteresis [3],[8]).

If binocular rivalry is introduced by presenting a horizontal bar to one eye and a vertical bar
to the other eye, inhibition-areas occur in the vicinity of the "cyclopean" intersection point.
This effect can also be observed in psychophysical experiments (Fig. 3).

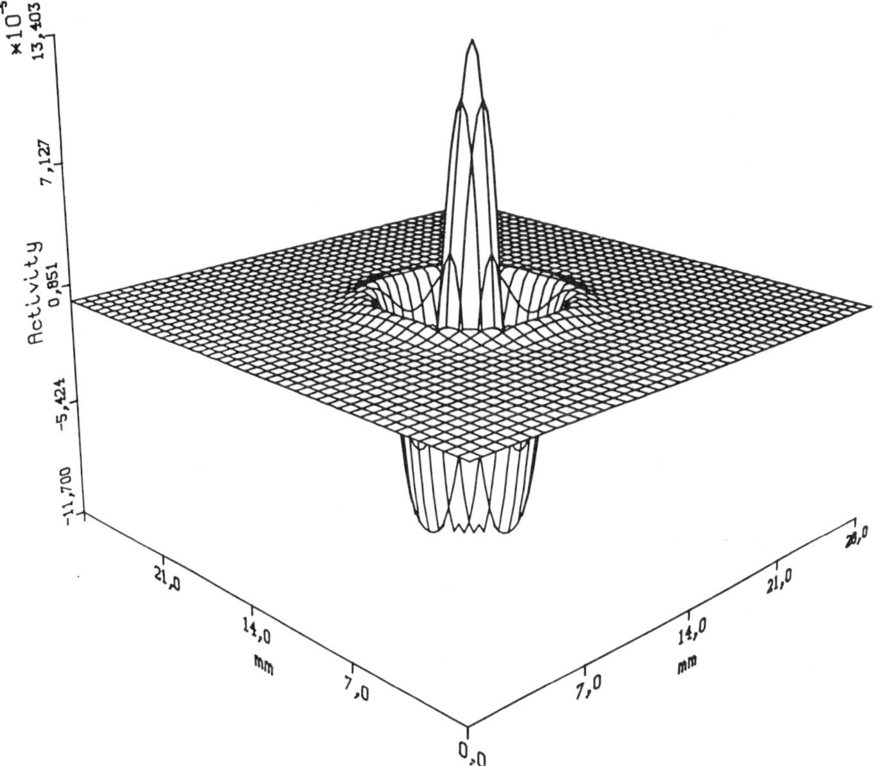

Figure 2: The Figure shows the spatial distribution of neural activity within
the cortex part of the model at $t = 45\,ms$ after applying a spatial and temporal
δ-stimulus.

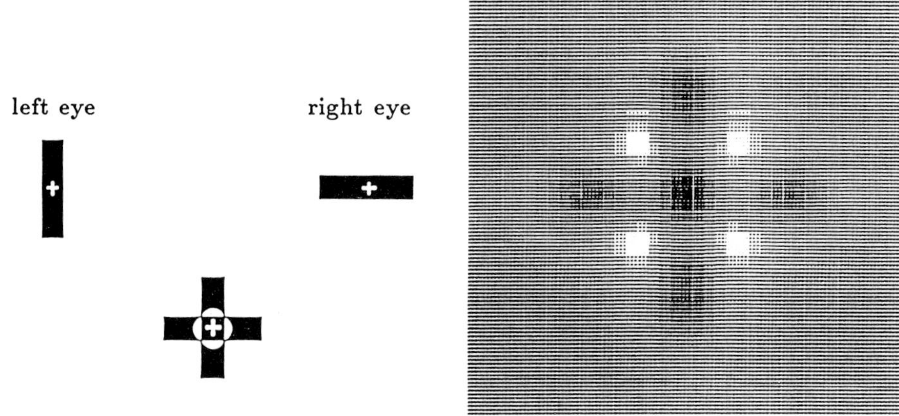

Figure 3: The left part of the Figure shows a psychophysical experiment [2]. A horizontal bar is presented to one eye and a vertical bar to the other eye. The result shows a inhibition area in the vicinity of the "cyclopean" intersection point. The right part shows the spatial distribution of activity within the cortex of the model as result of the binocular rivalry. For better visualization of the result a grey–value–image is used. For this image the neural activity is normalized to 64 grey–values, so that more active regions appear darker than less active ones.

Acknowledgements

We like to thank H. A. Mallot who constructively criticized the previous version. We are grateful to Ms. K. Rehbinder who carefully prepared the Figures 1 and 2.

References

[1] Ahlsen, G., Lindström, S. and Fu-Sun, Lo, Exp Brain Res. 58 (1985) 134.

[2] Campenhausen, C., von, Die Sinne des Menschen (Georg Thieme, Stuttgart New York, 1981).

[3] Fender, D. and Julez, B., J. opt. Soc. Am. 87, No. 6 (1967) 819.

[4] Giannakopoulos, F., Nichtlineare Systeme zur Beschreibung geschichteter neuronaler Strukturen (Dissertation, Universität Mainz, 1989) .

[5] Krone, G., Mallot, H. A., Palm, G. and Schüz, A., Proc. R. Soc.: Lond. 226 (1986) 421.

[6] Sherman, S. M. and Koch, C., Exp. Brain Res. 63 (1986) 1.

[7] Singer, W., Physiol.Rev. 57 (1977) 386.

[8] Wilson, H. R. and Cowan, J. D., Kybernetik 13 (1973) 55.

Parallel Processing in Neural Systems and Computers
R. Eckmiller, G. Hartmann and G. Hauske (Editors)
© Elsevier Science Publishers B.V. (North-Holland), 1990

SEGREGATION OF OSCILLATORY RESPONSES BY CONFLICTING STIMULI – DESYNCHRONIZING CONNECTIONS IN NEURAL OSCILLATOR LAYERS

Peter König and Thomas B. Schillen

Max-Planck-Institut für Hirnforschung
Deutschordenstr. 46, 6000 Frankfurt 71, F.R.Germany

Temporal information processing in the brain necessarily requires uncorrelated oscillation of distinct neuronal assemblies if it is to make use of the available phase space. This paper introduces delayed excitatory to excitatory unit coupling as a means of desynchronizing bulk oscillations within coupled oscillatory layers. This type of coupling shows stimulus-dependent formation of assemblies very similar to physiological evidence.

1 Introduction

Physiological evidence suggests the presence of temporal information processing in the brain [1, 2, 3, 4, 5, 6]. Theoretical considerations emphasize the importance of temporal coding as a solution to the binding problem [7, 8, 9]. Recent experimental data show the stimulus-dependent synchronization of oscillatory firing activities [5, 6] and coupling to different neural assemblies [6].

Information coding by coupled oscillations obviously has to avoid synchronized bulk oscillations within whole cortical areas. Usage of the available phase space necessarily requires uncorrelated oscillation of distinct neuronal assemblies.

Assemblies coding two partially overlapping but different objects in a visual scene ought to segregate by engaging in independent oscillatory patterns. As shown in the previous paper [10] synchronizing connections tend to bind the two assemblies together if they connect representations of a common feature, e.g. shared location, collinear orientation. In order to allow differentiating features, e.g. velocity, texture, etc., to segregate the assemblies representing the two objects a desynchronizing mechanism must be present.

In this paper we therefore address the question of desynchronizing connections within an oscillatory layer. We further show the stimulus-related segregation of oscillators that are coupled by both synchronizing and desynchronizing connections.

2 Desynchronizing an oscillator layer by delayed excitatory to excitatory unit connections

In this section we want to show the effect of delayed excitatory to excitatory unit coupling on the synchronicity of oscillatory layers.

In a 15×15 2-dimensional layer of isolated delayed Fermi oscillators [11] all oscillators are initialized to the same limit cycle starting condition. Through the entire simulation temperature is set equivalent to an amplitude variance of 0.1 of average peak amplitude. Fig 1 shows for $t < 0$ the layer's synchronicity – within temperature limits – for 8 arbitrarily chosen oscillators.

At $t = 0$ delayed excitatory connections are enabled that couple every oscillator's excitatory unit u_e to the excitatory units u_e of all next nearest neighbour oscillators. Coupling weights w_{ee}^{NNN} are isotropic with cyclic boundary conditions. Delay τ_{ee}^{NNN} is of the order

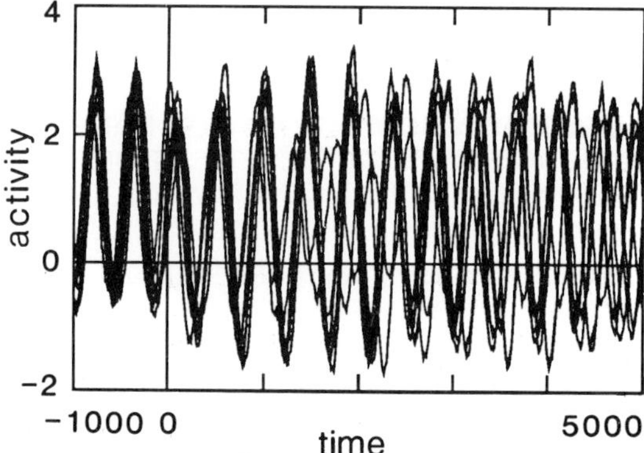

Figure 1: Desynchronizing an oscillator layer by delayed excitatory connections. Activity traces of 8 arbitrary units from a layer of delayed Fermi oscillators [11]. $t < 0$: Synchronous oscillations of isolated layer elements. $t > 0$: Introduction of delayed excitatory next nearest neighbour coupling w_{ee}^{NNN} (see text) rapidly desynchronizes the entire layer within a few oscillation cycles.

of the oscillator's intrinsic delay τ_{ei}. This is again well compatible with physiological delay times (cf. [10]). (All symbols as introduced in [11]).

For $t > 0$ Fig 1 demonstrates the layer's desynchronization within a few oscillation cycles. With the specified coupling every oscillator will excite all its nearest neighbour excitatory units simultaneously to its own inhibitory unit. This coupling, therefore, supports a nonzero phase relation between oscillators. Within the context of a 2-dimensional layer the local solutions can not all be reconciled with each other at the same time. This leads to a frustrated system with degenerate minimum energy, which in the presence of some noise will quickly drift through its pertaining set of solutions.

Above simulation shows that suitably coupled excitatory to excitatory unit connections may well be able to bring about desynchronization between different "neuronal" oscillators. This is also interesting with respect to the organization of GABAergic connections in area 17 of cat visual cortex [12].

Furthermore, the simultaneous introduction of short range synchronizing [10] and long range desynchronizing connections will lead to a finite correlation length in the resulting system.

3 Segregation of orientation-selective oscillatory responses by conflicting stimuli

We now proceed to investigate the stimulus-dependent assignment of oscillators to different "neuronal" assemblies, as suggested by experimental evidence [6].

For this purpose we use a 1-dimensional chain of 8 delayed Fermi oscillators incorporating now both synchronizing [10] and desynchronizing neighbour coupling. The length constant of the desynchronizing connections (next nearest neighbour) is taken to be larger than that of the synchronizing ones (nearest neighbour). In this example we use identical weights $(w_{ei}^{NN} = w_{ee}^{NNN})$ and delay times $(\tau_{ei}^{NN} = \tau_{ee}^{NNN} = \tau_{ei})$ for both coupling types. Each of the oscillators represents a receptive field (RF) of a different preferred orientation but at identical "retinal" location. For this simulation a continuous sequence of 8 orientations at 22.5 deg increments is being used (Fig 2A). All RFs are assumed to exhibit a Gaussian orientation tuning.

In correspondence to the experiment [6] we "record" from two oscillators of 45 deg different preferred orientations (112, 157 deg) (Fig 2A, hatched boxes) and present two distinct stimulus paradigms: (1) a single stimulus bar of intermediate orientation (135 deg) (Fig 2B-

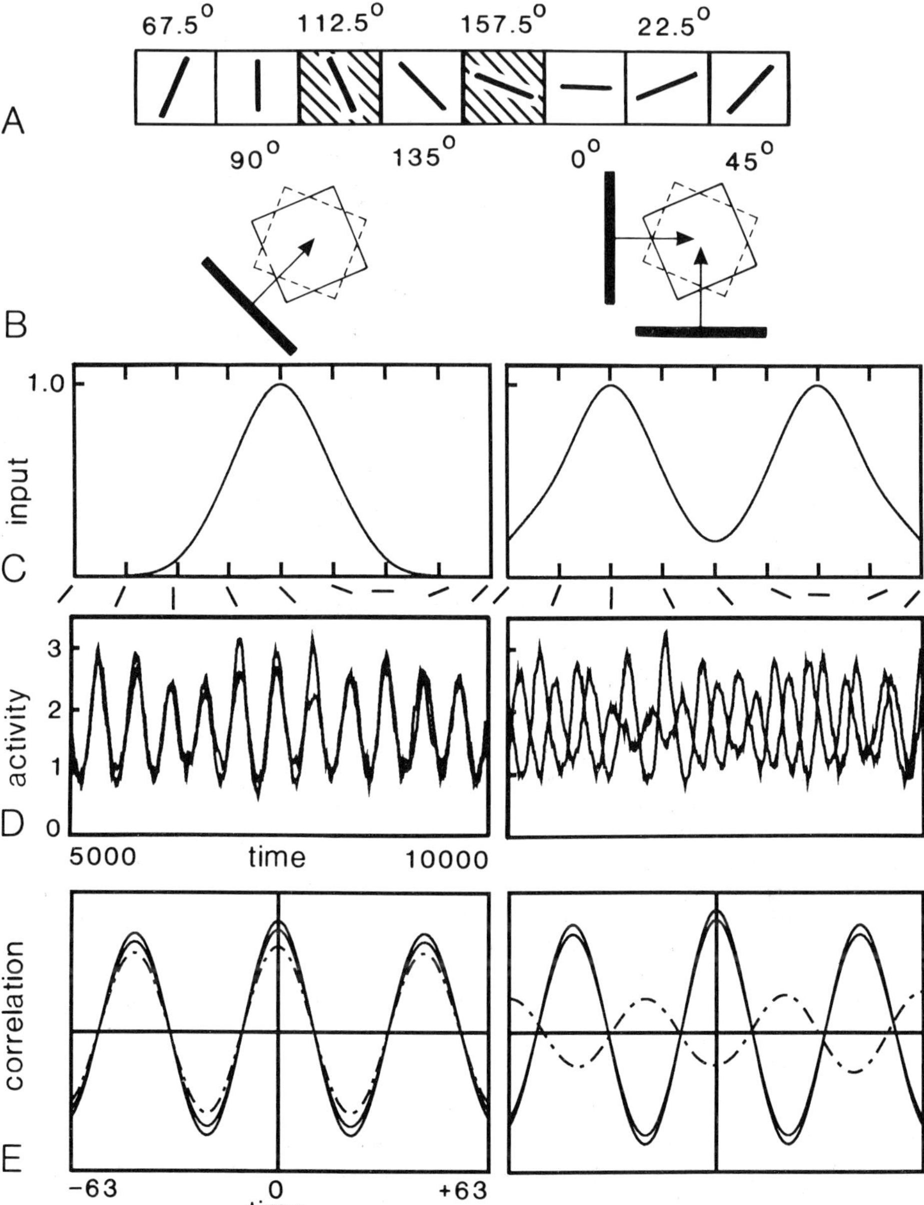

Figure 2: Simulation of an orientation-selective oscillator chain: Effect of a single stimulus and conflicting stimuli. (Explanation see text).

E, left column) and (2) two conflicting stimuli oriented at 0 and 90 deg (Fig 2B-E, right column). Panels (C) depict the corresponding external inputs to the oscillator chain in accordance with the assumed orientation tuning. Input to the monitored oscillators is set to be identical for both stimulus conditions. Panels (D) show the resulting activity traces. In the case of the single stimulus both oscillators are well synchronized, while for the conflicting stimuli they couple to the two assemblies representing the presented stimulus orientations and thus are out of phase to each other. This situation is also apparent in the oscillators' cross correlograms (Fig 2E).

The simulation shows the stimulus-dependent coupling of the two oscillators to different "neuronal" assemblies as seen in the experiment [6]. The contribution of the synchronizing connections provides the coupling of an oscillator also to assemblies representing suboptimal orientation preferences. This is again consistent with experimental evidence [5]. The inclusion of desynchronizing connections with a length constant larger than the synchronizing one introduces a finite correlation length. This is the origin of the decoupling in the case of the two conflicting stimuli. Without desynchronizing connections every sufficiently overlapping input configuration would readily synchronize completely.

Above results extend canonically to other stimulus modalities. E.g. the oscillator chain could equally well be interpreted as a sequence of RFs having different velocity preferences but the same preferred orientation (⟵ ⟵ · ⟶ ⟶).

This particular interpretation provides a natural extension of the results presented in [10] to the direction-selective coherency detection seen in the experiment [5]: Every nondirection-selective oscillator of [10] is mapped onto an oscillator chain representing a sequence of velocity preferences. Within each layer of a distinct velocity preference the synchronizing connections of [10] are maintained. Within each velocity chain at a distinct spatial location we implement the synchronizing / desynchronizing connections outlined in this paper. Extended in this way the model evaluates stimulus velocity as an additional coherency criterion besides stimulus continuity. This model then desynchronizes oscillations of two oppositely moving stimulus bars while synchronizing coherently moving stimuli that are sufficiently close to each other.

4 Conclusions

The results presented in this paper demonstrate that excitatory connections may well be suited to provide desynchronization between different oscillator assemblies.

The addition of long range desynchronizing connections to short range synchronizing connections leads to a finite correlation length. This allows the segregation of oscillators responding to partially overlapping stimuli of different objects. In this case the pertaining oscillators couple to different oscillating ensembles. In presence of a single intermediate stimulus, however, the same oscillators may well couple to each other again.

Extending above concepts to several feature modalities one predicts (1) synchronizing connections for the formation of object representations by coherent features and (2) desynchronizing connections to segregate representations of objects with differing features.

References

[1] Freeman, W.J., Mass Action in the Nervous System (Academic Press, New York, 1975). [2] Freeman, W.J., in: Basar, E., Dynamics of Sensory and Cognitive Processing by the Brain (Springer, Berlin, 1988) pp. 19-29. [3] Gray, C.M. and Singer, W., Soc.Neurosci.Abstr. 13 (1987) 404.3. [4] Gray, C.M. and Singer, W., Proc.natn.Acad.Sci.USA 86 (1989) 1698-1702. [5] Gray, C.M., König, P., Engel, A.K., and Singer, W., Nature 338 (1989) 334-337. [6] Engel, A.K., König, P., Gray, C.M. and Singer, W., Synchronization of Oscillatory Responses: A Mechanism for Stimulus-dependent Assembly Formation, this volume. [7] Crick, F., Proc.natn.Acad.Sci.USA 81 (1984) 4586-4590. [8] v.d.Malsburg, C. and Schneider, W., Biol.Cybern. 54 (1986) 29-40. [9] Damasio, A.R., Neural Computation 1 (1989) 123-132. [10] Schillen, T.B. and König, P., Coherency Detection by Coupled Oscillatory Responses ..., this volume. [11] Schillen, T.B.,Simulation of Delayed Oscillators with the *MENS* general purpose Modelling Environment for Network Systems, this volume. [12] Bueno-Lopez, J.L., Beaulieu, C., and Somogy, P., Europ.Neurosci.Abstr. 12 (1989) 34.8.

Parallel Processing in Neural Systems and Computers
R. Eckmiller, G. Hartmann and G. Hauske (Editors)
© Elsevier Science Publishers B.V. (North-Holland), 1990

SEGREGATION OF "MEANING" AND "IMPORTANCE" OF NEURONAL MESSAGES.

Jürgen KRÜGER and Jan Dirk BECKER

Neurologische Universitätsklinik, Hansastr. 9
7800 Freiburg, FRG

We propose that the temporal structures of spike trains carry the meaning, and the spike frequencies signal how important a particular contribution to that meaning is.

Identical repeated stimuli are known to evoke responses with highly variable spike frequencies. Only the responses averaged over many similarly reacting neurones are thought to carry a reliable information. However, already some years ago (recording with 30 microelectrodes from the visual cortex of anaesthetized monkeys) we have noted that groups of cortical neurones tended to covary [1]. We conclude that ensemble averaging cannot easily lead to reproducible responses. From other experiments we conclude that they should not: otherwise systematic efferent response modulations (e. g., by attention, [2]) would be annihilated.

How, then, can the identity of two stimuli be signalled? We have examined this question by analyzing similar data from the monkey striate cortex as those mentioned above. Our goal was to find the best way of recognizing identical stimuli out of several others on the basis of the response patterns recorded with 30 micro-electrodes. Although other types of stimuli were examined as well, we shall describe here the results for bars moving in 16 directions. Each stimulus, characterized by its movement direction, was applied about 10 times, and a spatiotemporal response pattern, averaged over these repetitions ("template") was associated to it. Then each stimulus was applied again, and its non-averaged evoked pattern was compared to all 16 templates. The most similar template was sought, hoping that it was the correct one. The aim was to select procedures and to vary parameters such that the score of correct recognitions became maximal.

Out of several similarity measures a correlation procedure was optimal. The spike trains were subjected to various normaliz-ations, and another option was to subtract the ongoing activity. To our surprise, the raw spike trains yielded the best results. The relevant response intervals (kept fixed, ranging between 300 and 700 ms) were subdivided into bins whose number was varied. Thus, each spatiotemporal response pattern consisted of an array of bin counts whose number of elements was 30 times the number of bins.

Using only one long bin spanning the entire response interval amounted to use the information as it is contained in the usual orientation polar plots, (which, however, normally are obtained by averaging) but to ignore all temporal structures, and to base the recognition on spike frequency alone. We were surprised at the extremely low performance even when most neurones were strongly orientation and direction selective (Fig. 1, bottom).

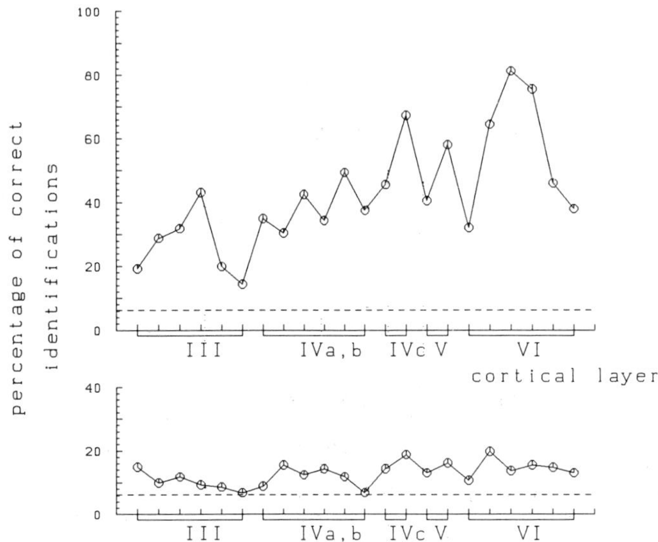

Fig. 1. Score of correctly recognized stimuli on the basis of non-averaged spike patterns. Dashed line: chance level.

Top: Each response of each neurone is characterized by a succession of spike counts in 50 ms-bins into which the relevant response interval (lasting 300 to 700 ms) is subdivided.

Bottom: Each response of each neurone is characterized by one spike count within the relevant response interval. This corresponds to the assumption of a spike frequency code. Although in particular in layer VI many neurones have very clearcut orientation and direction selectivities, the systematic stimulus-dependent variance is not sufficient for a reliable recognition in the presence of neuronal variability.

The score of correct recognitions rose considerably in all monkeys and cortical layers when the response interval was subdivided into several bins (Fig. 1, top). A broad optimum was found for bin widths between 30 and 100 ms. We intentionally degraded the spike rate information by replacing all bin counts > 1 by 1 but kept a 50 ms time resolution: the recognition score fell to a half but was still clearly superior to the above "spike frequency case".

Thus the meaning of the neural message is essentially carried by the temporal structure. The optimal resolution of about 50 ms reminds the width of prominent correlogram peaks observed in monkey striate cortex [3].

What is the purpose of different spike frequencies? In our example the so-called orientation-selective neurones were unable to signal orientations by their spike rates because of the large frequency variability. Variability can consist of the emission of "spontaneous" bursts, or changes of firing rates of a factor of 10 within a few seconds. The temporal response structure is much less variable unless response components disappear altogether during instants of powerful suppressive variability. Clearly, the strongest response bursts are most likely to survive, and therefore they are the most important elements to whose temporal information the meaning should be attached. Thus, we believe, as others before, that a high frequency burst of an oriented neurone contains the most essential information about the tilt angle. The new aspect is that the relevant information is in the <u>temporal</u> response structure. This coding principle offers the opportunity to modulate the importance of selected components of spatiotemporal excitation patterns by messages from higher brain centres, without altering the meaning carried by the patterns. It may be part of a more general principle according to which the meaning, or contents, is kept separate from control signals.

ACKNOWLEDGEMENTS

This work was supported by the Volkswagen Foundation (Schwerpunkt "Synergetik").

REFERENCES

1. Bach, M. & Krüger, J. <u>Exp. Brain Res</u>. <u>61</u>, 451-456 (1986).

2. Boch, R. <u>Exp. Brain Res</u>. <u>64</u>, 610-614 (1986).

3. Krüger, J. & Aiple, F. <u>J. Neurophysiol</u>. <u>60</u>, 798-828 (1988).

Parallel Processing in Neural Systems and Computers
R. Eckmiller, G. Hartmann and G. Hauske (Editors)
Elsevier Science Publishers B.V. (North-Holland), 1990

Retinal Sampling Grids and Space–Variant Image Compression*

Hanspeter A. Mallot and Gerd-Jürgen Giefing

Institut für Neuroinformatik, Ruhr–Universität, D–4630 Bochum, West Germany

We present a theory for the construction of mapping functions and sampling grids satisfying predefined distributions of areal magnification or retinal ganglion cell density. The results are discussed in terms of retinal receptive fields, technical image compression, and foveation of relevant image parts via eye-movements in an active–vision system.

1 INTRODUCTION

One striking feature of neural image processing is the space–variant resolution and sampling of the retinal image by the ganglion cells. It has been noted early on that a close relation exists between the ganglion cell density on the one hand and retinotopic mapping on the other [2]: if the same volume of cortical tissue is allocated to each ganglion cell (i.e., if cellular magnification [6] is constant), the density should be proportional to the areal magnification factor of the associated retinotopic mapping. The receptive field profiles of retinal ganglion cells must then be chosen such as to prevent aliasing in the space–variant sampling grid.

In this paper we review the relation between retinal ganglion cell density and receptive fields, areal magnification, and retinotopic mapping. We present a mathematical theory and numerical results on a "retinal input filter".

2 GANGLION CELL DENSITY AND AREAL MAGNIFICATION

Consider two subsets of the Euclidian plane, $\mathbf{S}, \mathbf{T} \subseteq \mathbf{R}^2$ with the interpretation that $\mathbf{S}$ is the source area of a topographic mapping (i.e., the retina) and $\mathbf{T}$ the target (e.g., the Lateral Geniculate). The retinotopic mapping is modelled as a coordinate transform $\mathcal{R} : \mathbf{S} \mapsto \mathbf{T}$. Denoting the polar coordinate transform by $\mathcal{P}$, we can summarize the notations:

$$\underbrace{(r,\varphi) \overset{\mathcal{P}}{\mapsto} (x_1, x_2)}_{\mathbf{S}} \overset{\mathcal{R}}{\longrightarrow} \underbrace{(y_1, y_2) \overset{\mathcal{P}^{-1}}{\mapsto} (s, \vartheta)}_{\mathbf{T}} \tag{1}$$

We denote the retinal ganglion cell density (or, equivalently, the desired space–variant resolution) by $\varrho(x_1, x_2)$ or by $\varrho^*(r, \varphi)$, respectively, and state the basic equation

$$|\det \mathrm{J}_{\mathcal{R}}(x_1, x_2)| \;\overset{!}{=}\; \varrho(x_1, x_2). \tag{2}$$

---

*Supported by the German Federal Department of Research and Technology (BMFT), Grant No. ITR8800K4

$|\det J_{\mathcal{R}}|$ is of course the areal magnification of $\mathcal{R}$ (for a more comprehensive discussion, see [5]). In polar coordinates, we have:

$$s|s_r \vartheta_\varphi - s_\varphi \vartheta_r| = r\varrho^*(r, \varphi). \tag{3}$$

The problem, then, is to solve the partial differential equation (2) for $\mathcal{R}$. Before discussing some special cases, a few remarks are in place:

1. The distribution of ganglion cell density ϱ does not completely determine the mapping function $\mathcal{R}$. That is, solutions of (2) are generally not unique in the following sense: Consider a mapping $\mathcal{M} : \mathbf{T} \mapsto \mathbf{T}$ with $|\det J_{\mathcal{M}}(y_1, y_2)| \equiv 1$, i.e. an equal-area transformation. Then, for any solution $\mathcal{R}_0$, $\mathcal{M} \circ \mathcal{R}_0$ is also a solution. For example, the areal magnification factors of the mapping functions for the visual areas 17, 18, and 19 in the cat are basically proportional to each other [3].

2. Fischer [2] uses the relation

$$\frac{\Delta A}{\pi R^2} = \frac{\Delta N}{k}$$

   where A is retinal area, πR^2 receptive field size, N representational area and k the constant overlap factor of ganglion cell receptive fields. This equation is similar to (2) if $\Delta N / \Delta A$ is taken to be the areal magnification factor and $k/\pi R^2$ corresponds to the ganglion cell density ϱ.

3. Schwartz [7] treats the sets $\mathbf{S}, \mathbf{T}$ as subsets of the complex plane. $\mathcal{R}$ then becomes a 1D coordinate transform and (2) reduces to an ordinary differential equation. Of course, this description can only yield conformal mappings as solutions for $\mathcal{R}$.

We consider three special cases for the solution of (2):

Case 1: *Radial separability.* We assume that the decrease of ganglion cell density with eccentricity is similar along all radii. That is, for each angle φ, there is a scaling factor $\kappa_1(\varphi)$ such that $\varrho^*(r, \varphi) = \tilde{\varrho}(r \cdot \kappa_1(\varphi))$ for some suitable $\tilde{\varrho}$. We choose $\tilde{\varrho}(r) := \varrho^*(r, 0)$, i.e., $\kappa_1(0) = 1$. Similar assumptions for $\mathcal{R}$ are slightly more complicated, since it is not clear, where the compression factor should be applied, to the domain or the range. We try the combined *Ansatz* $s(r, \varphi) = \kappa_3(\varphi)\sigma(r \cdot \kappa_2(\varphi))$, $\vartheta(r, \varphi) \equiv \varphi$ and obtain from (3):

$$\sigma_r(r\kappa_2(\varphi))\sigma(r\kappa_2(\varphi))\kappa_3^2(\varphi)\kappa_2(\varphi) = r\tilde{\varrho}(r\kappa_1(\varphi)), \tag{4}$$

which, using the substitution $u(r, \varphi) := \frac{1}{2}\sigma^2(r\kappa_2(\varphi))$, yields the equation:

$$\kappa_3^2(\varphi)\sigma^2(r, \kappa_2(\varphi)) = \frac{2}{\kappa_1(\varphi)}\int_0^{r\kappa_1(\varphi)} r'\tilde{\varrho}(r')dr' \tag{5}$$

We find a solution by setting $\kappa_1 = \kappa_2 = \frac{1}{\kappa_3} =: \kappa$:

$$s(r, \varphi) = \frac{\sigma(r\kappa(\varphi))}{\kappa(\varphi)} \quad \text{where} \quad \sigma(r) := \left(2\int_0^r r'\tilde{\varrho}(r')dr'\right)^{\frac{1}{2}} \tag{6}$$

Eq. 6 can be used as a model of visual streaks. For instance, if $\kappa(\varphi)$ is chosen to be $1 + e\cos\varphi$, the ganglion cell iso-density lines are ellipses of eccentricity e.

Case 2: *Rotationally symmetric ganglion cell distribution and mapping.* This is a specialization of case 1, where $\kappa(\varphi) \equiv 1$. In this case, (6) represents a "radial compression" with a

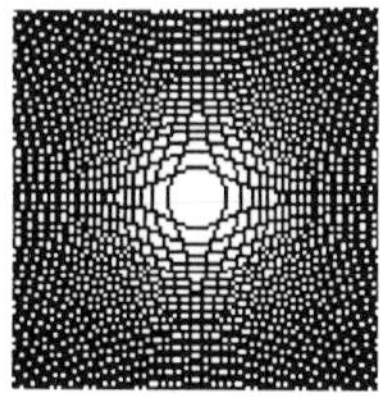

Figure 1: Central section of the space–variant sampling grid in **S** derived from a square grid in **T**. The grid is superimposed to a 512 × 512 image; sampling points are located at the white spots while pixels coinciding with black spots are ignored. The density follows an inverse power law.

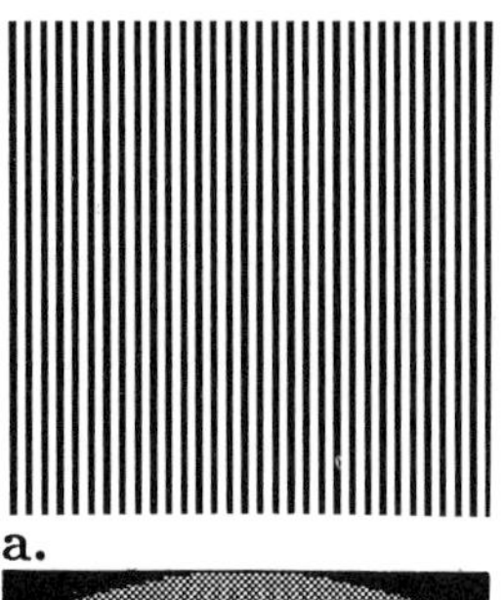

a.

b.

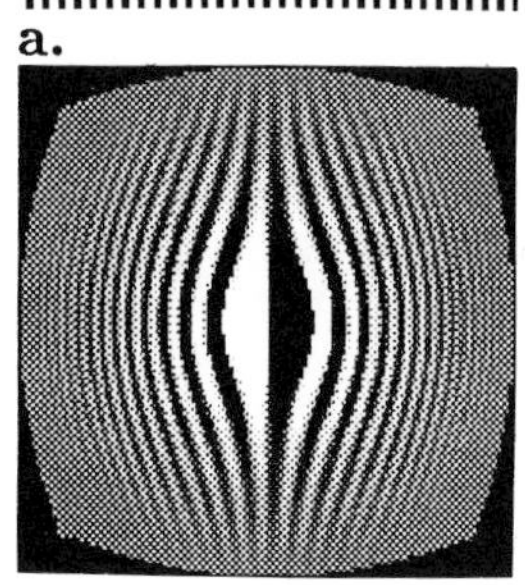

c.

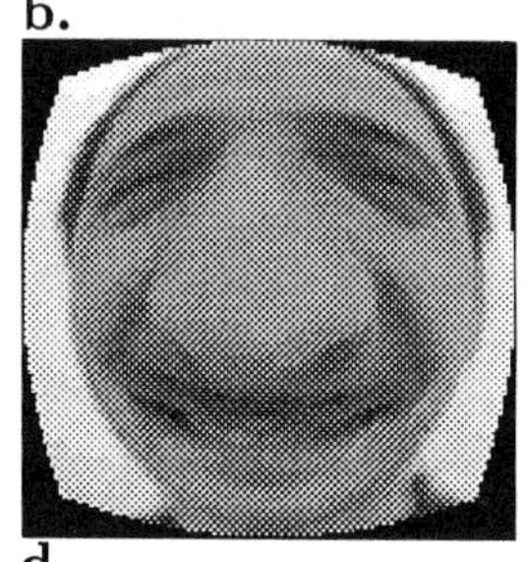

d.

Figure 2: **a.** and **b.** are the original images of 512 × 512 pixels shown at reduced size. **c.** and **d.** ("nose man") are the space–variant filtered and downsampled images (128 × 128 pixels). The central parts of the image are highly weighted by the sampling.

predefined areal magnification. For a special case ($\varrho^*(r) = r^{2(p-1)}$), an equivalent solution is given in [3, Eqs. 6,7].

Case 3: *Cartesian separability.* This case, where $\varrho(x_1, x_2) = f(x_1)g(x_2)$, has been studied in [5, Sect. 5.3]. It models simple ocular dominance stripes.

Case 4: *Conformal mappings.* If $\mathcal{R}$ is required to be conformal, the Cauchy–Riemann Equations apply and (2) takes the form

$$\left(\frac{\partial y_1}{\partial x_1}\right)^2 + \left(\frac{\partial y_1}{\partial x_2}\right)^2 = \varrho(x_1, x_2). \tag{7}$$

Clearly, in this situation, the complex notation [7] is more appropriate.

3 NUMERICAL SOLUTION

We are interested in constructing retinotopic mapping functions with arbitrary distributions of areal magnification. (Of course, there are obvious constraints for $\varrho(x_1, x_2)$, namely $\varrho \geq 0$ and $\int \int_S \varrho(x_1, x_2)dx_1 dx_2 = \|\mathbf{T}\|$.) We developed a simple algorithm that computes a retinal sampling grid from a density ϱ and a desired "cortical grid". The mapping procedure includes two steps:

Input filtering: The images Fig. 2a,b are filtered by a space–variant lowpass (i.e., Gaussians of the form $G((\mathbf{x} - \mathbf{x_0})/\|\mathbf{x_0}\|)$ simulating the space–variant resolution of the human eye. This is an approximation of optimal filter structures discussed in [5].

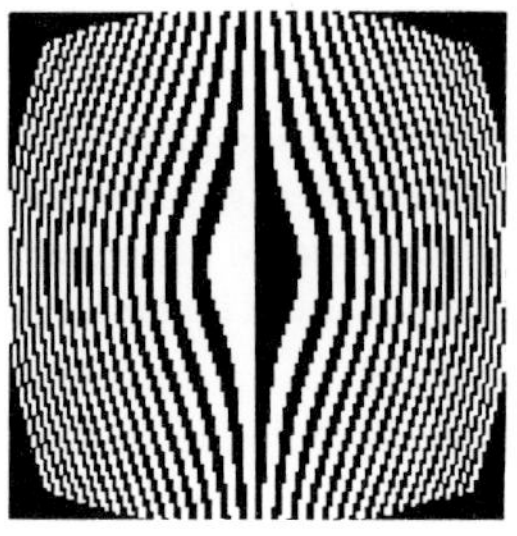

Figure 3: Aliasing effect in the peripheral image areas. Moiré-pattern are visible if the filter step is omitted.

Space-variant sampling: Compression of the image size is done by space–variant sampling. Using a square grid in **T** with constant density, the mapping is constructed by solving (6) for case 2. The distribution in the foveal center is seen in Fig. **2**. Each white dot represents a sampling point. Two examples for the effect of the algorithm are given in Fig. **2**.

The filtering step suppresses aliasing effects and provides interpolation functions for image reconstruction. Fig. **2** shows aliasing in the visual periphery if the filter step is omitted.

Peripheral loss of information must be compensated for by eye-movements driven by an attention control to deal with a natural environment. In these peripheral areas, fast α–like cells report temporal changes of the image. First tests show that even such vital tasks as the guidance of eye-movements can be performed on images compressed by the above algorithm.

4   OUTLOOK

- We plan to use the principle of space–variant sampling for image preprocessing in a mobile stereo camera system.
- A foveating system of this type has to rely on saccadic eye-movements in order to construct a representation of its environment [1].
- An additional step is to optimize the sampling with respect to the contents of the image. Highly textured or interesting areas could be mapped with greater resolution than others.
- The inherent information processing by the mapping step can be used to simplify further processing steps [4].

References

[1] D. H. Ballard. Technical Report TR218, Computer Science Department, University of Rochester, 1987.

[2] B. Fischer. *Vision Research*, 13:2113 – 2120, 1973.

[3] H. A. Mallot. *Biol. Cybern.*, 52:45 – 51, 1985.

[4] H. A. Mallot, E. Schulze, and K. Storjohann. In G. Dreyfus and L. Personnaz, editors, *Neural Networks from Models to Applications*, pages 560 – 569, Paris, 1989. I.D.S.E.T.

[5] H. A. Mallot, W. von Seelen, and F. Giannakopoulos. *Neural Networks*, 1990. in press.

[6] J. Myerson, P. B. Manis, F. M. Mieyin, and J. M. Allman. *Science*, 198:855 – 857, 1977.

[7] E. L. Schwartz. *Biol. Cybern.*, 25:181 – 195, 1977.

Parallel Processing in Neural Systems and Computers
R. Eckmiller, G. Hartmann and G. Hauske (Editors)
© Elsevier Science Publishers B.V. (North-Holland), 1990

Towards a Network Theory of Cortical Areas*

Hanspeter A. Mallot and Werner von Seelen

Institut für Neuroinformatik, Ruhr–Universität, D–4630 Bochum, FRG

In the connectivity of cortical neurons, one can distinguish two basic patterns: Intra–areal connectivity is local with roughly topological organization while the inter–area connectivity forms a complicated graph structure, the *Cortical Area Network* (CAN). Nodes of the CAN are entire brain areas with their intrinsic processing capabilities; the edges are the parallel fiber projections between different areas (neural mappings). In this papers, we present the network equations for the CAN and discuss three special cases: mapped filters, reciprocal feedback, and patchy connectivity. Applications to technical image processing and visual receptive field organization are discussed elsewhere [7, 4, 6, 5].

1   INTRODUCTION

The topolgy of cortical networks is characterized by two connectivity patterns called "A–and B–system" by Braitenberg [1]. We present a linear model for "Cortical Area Networks" (CAN), i.e. a graph built of two basic parts:

1. The **nodes** are formed by cortical areas with their intrinsic connectivity ("B–system") and the according computational capabilities. Here, we model intrinsic processing by convolutions [3].

2. The **edges** are formed by the mappings between the various cortex areas ("A–system"). With respect to the spatial organization, we distinguish *topographic maps* or coordinate transforms, *patchy maps* that occur when multiple input converges to a common target area, and *parametric maps* or 2D–histograms that encode stimulus properties into spatial position.

2   THE NETWORK EQUATIONS

Consider a network built out of n areas $\mathbf{A}_i \subseteq \mathbf{R}^2$, $i = 1, ..., n$. Within each area, processing is described by a convolution kernel $k_i : \mathbf{R}^2 \mapsto \mathbf{R}$. The distribution of excitation in area $\mathbf{A}_i$ is denoted by $e_i : \mathbf{A}_i \mapsto \mathbf{R}$. (In this linear theory, we allow for negative values of e and k to model inhibition.) The projection from one area to another is modelled by two parts: a mapping function $\mathcal{R}_{ij} : \mathbf{A}_i \mapsto \mathbf{A}_j$ and a connection strength $c_{ij} : \mathbf{A}_j \mapsto \mathbf{R}_0^+$. The connection strength is needed to allow for patchy connectivity where some regions of $\mathbf{A}_j$ receive more or

---

*Supported by the German Federal Department of Research and Technology (BMFT), Grant No. ITR8800K4

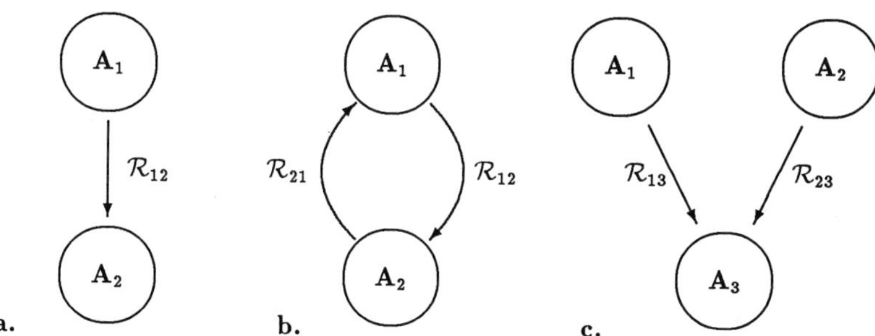

Figure 1: Simple CAN topologies. **a.** Cascade of mapping $\mathcal{R}_{12}$ plus intrinsic operation in $\mathbf{A}_2$. **b.** Reciprocal feedback between two areas $\mathbf{A}_1$ and $\mathbf{A}_2$. **c.** Convergence of two (patchy) maps $\mathcal{R}_{13}, \mathcal{R}_{23}$ into one common target area.

denser input from $\mathbf{A}_i$ than others. We can now state the linear equations for the CAN:

$$
e_j(\mathbf{x}_j) \;=\; \int_{\mathbf{A}_j} \underbrace{\sum_{i=1}^{n} c_{ij}(\mathbf{x}'_j) e_i(\mathcal{R}_{ij}^{-1}(\mathbf{x}'_j))}_{\text{total input}} \quad \underbrace{k_j(\mathbf{x}_j - \mathbf{x}'_j)}_{\text{intrinsic operation}} \quad dx'_j \tag{1}
$$

$$
\;=\; \sum_{i=1}^{n} \int_{\mathbf{A}_i} e_i(\mathbf{x}_i) \cdot K_{i,j}(\mathbf{x}_i; \mathbf{x}_j) dx_i \quad \text{where} \tag{2}
$$

$$
K_{i,j}(\mathbf{x}_i; \mathbf{x}_j) \;:=\; c_{ij}(\mathcal{R}_{ij}(\mathbf{x}_i))|\det \mathrm{J}_{\mathcal{R}_{ij}}(\mathbf{x}_i)| k_j(\mathbf{x}_j - \mathcal{R}_{ij}(\mathbf{x}_i)). \tag{3}
$$

All integrals are taken over 2D domains. Often, the following constraints hold:

$$
\sum_{i=1}^{n} \int_{\mathbf{A}_j} c_{ij}(\mathbf{x}_j) dx_j \;=\; \int_{\mathbf{A}_j} 1 dx_j = \|A_j\| \text{ for all } j \tag{4}
$$

$$
\mathcal{R}_{ij} \circ \mathcal{R}_{jk} \;=\; \mathcal{R}_{ik}. \tag{5}
$$

From the second condition, it follows $\mathcal{R}_{ii} \equiv id$, the identity; and $\mathcal{R}_{ij} = \mathcal{R}_{ji}^{-1}$.

3   MAPPED FILTERS

Consider the case $n = 2$, $c_{12} \equiv 1$, and $c_{21} \equiv c_{22} \equiv 0$ (Fig. 1a). This is a simple cascade where an input distribution e_1 is first mapped and then filtered to yield the output distribution e_2. The resulting "mapped filters" are a class of linear, space–variant operators generalizing the space–invariant operators developed in [10]. For a detailed discussion, see [7]. Here, we state one important result concerning the relation of retinotopic mapping and receptive field location.

In analogy to the neurophysiological procedure for measuring retinotopic maps, we define a *"mean topographic map"* for a mapped filter with mapping function $\mathcal{R}_{12}$ and intrinsic kernel k_2 by assigning to each point $\mathbf{x}_2 \in \mathbf{A}_2$ the centroid of the corresponding receptive field:

$$
\mathcal{T} : \mathbf{A}_2 \mapsto \mathbf{R}^2 \;\; ; \;\; \mathcal{T}(\mathbf{x}_2) := \frac{\int_{\mathbf{A}_1} \mathbf{x}_1 K_{12}(\mathbf{x}_1; \mathbf{x}_2) dx_1}{\int_{\mathbf{A}_1} K_{12}(\mathbf{x}_1; \mathbf{x}_2) dx_1}, \tag{6}
$$

where $K_{12}(\mathbf{x}_1; \mathbf{x}_2)$ is taken from (3). Note that $\mathcal{T}$ goes from the target area "backwards" to the source area (Fig. 2).

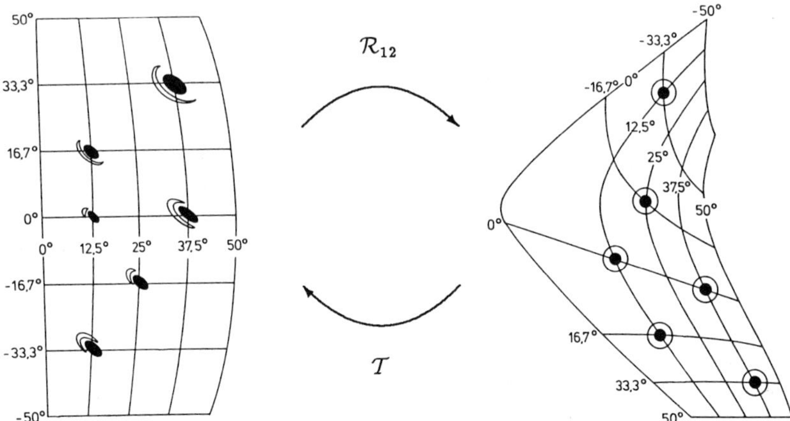

Figure 2: Mean topographic mapping theorem: Retinotopic mapping is modelled by the coordinate transform $\mathcal{R}_{12}$. In electrophysiology, the mapping $\mathcal{T}$, linking a recording site with the receptive field center can be measured. The theorem states the identity of $\mathcal{R}_{12}$ and $\mathcal{T}$ under certain conditions. The figures show contour lines of intrinsic kernels k_2 and "backprojected" receptive fields modelled by the kernel of Eq. (3). For details, see [4, 7].

Theorem: Consider a mapped filter with an intrinsic kernel k_2 and a topographic mapping $\mathcal{R}_{12}$. Let k_2 be rotationally symmetric, i.e $k_2(\mathbf{x_2}) = \hat{k}_2(\|\mathbf{x}_2\|)$ and the mapping $\mathcal{R}_{12}$ be conformal. Then the mean topographic mapping $\mathcal{T}$ is the inverse of $\mathcal{R}_{12}$ except for boundary effects.

The proof is given in [7]. This Theorem confirms the intuitive expectation that, in mapped filters, the center of gravity of the receptive field of a cortical cell (which is, of course, a location in the visual field) is mapped to the cortical position of that cell by the underlying mapping function (cf. Fig. 2).

4 RECIPROCAL FEEDBACK

The next case can be derived from (1) by setting $n = 2$, $c_{12} \equiv c_{21} \equiv 1$, $c_{11} \equiv c_{22} \equiv 0$, $\mathcal{R}_{12} = \mathcal{R}_{21}^{-1}$, and $k_1 \equiv k_2$ (Fig. 1b). As was pointed out by van Essen [2], most cortical connections are reciprocal in the sense that if area i projects to area j, the reverse projection exists too. The stronger relation defined here can, e.g., be found for the areas 17 and 18.

In this case, (1) can be simplified by the following consideration. First, we define a joint mapping $\mathcal{Q}$ by:

$$\mathcal{Q} : \mathbf{A}_1 \cup \mathbf{A}_2 \mapsto \mathbf{A}_1 \cup \mathbf{A}_2 \quad ; \quad \mathcal{Q}(\mathbf{x}) := \begin{cases} \mathcal{R}_{12}(\mathbf{x}) \text{ for } \mathbf{x} \in \mathbf{A}_1 \\ \mathcal{R}_{21}(\mathbf{x}) \text{ for } \mathbf{x} \in \mathbf{A}_2 \end{cases} \tag{7}$$

From $\mathcal{R}_{12} = \mathcal{R}_{21}^{-1}$, we have $\mathcal{Q} = \mathcal{Q}^{-1}$. For the joint distribution of excitation $e : \mathbf{A}_1 \cup \mathbf{A}_2 \mapsto \mathbf{R}$, we obtain from (1):

$$e(\mathbf{x}) = \int_{\mathbf{A}_1 \cup \mathbf{A}_2} e(\mathcal{Q}(\mathbf{x}'))k(\mathbf{x} - \mathbf{x}')d\mathbf{x}'. \tag{8}$$

The solutions of this feedback equation are the eigenfunctions of the mapped filter operation defined above. They depend on both, the geometry of the mappings $\mathcal{R}_{ij}$ and the intrinsic kernel k.

5   OCULAR DOMINANCE STRIPES

Consider two areas $\mathbf{A}_1$ and $\mathbf{A}_2$ projecting to one common target area $\mathbf{A}_3$ (Fig. 1c). The segregation of the two inputs, which is often found in this situation, can be modelled by suitable choices of the connection strength c_{i3}. In the case of ocularity stripes, we assume that $c_{13}(\mathbf{x}_3) = 1 - c_{23}(\mathbf{x}_3)$ for all $\mathbf{x}_3 \in \mathbf{A}_3$. For a simple model, $c_{ij}(x_3, y_3)$ might take the form $1 + \frac{1}{2}\sin(x_3)$ where x_3, y_3 are the components of $\mathbf{x}_3$. We distinguish two cases:

Transmission Grating Model: Suppose that $\mathcal{R}_{13} \equiv \mathcal{R}_{23}$. An example where the mapping is the identity is discussed by Schwartz [9]. Note that in the transmission grating model, those parts of the inputs that happen to be mapped to a $\mathbf{x}_3 \in \mathbf{A}_3$ with $c_{i3}(\mathbf{x}_3) \neq 1$ are partially deleted. In Schwartz' model, this affects one half of each image. The advantage of the transmission grating model is that it can be interpreted as a spatial amplitude modulation of the input stimuli.

Constant Cellular Magnification Model: This case avoids the problem of partial deletion. By "cellular magnification", we denote the number of cells in the target area that are concerned with processing of the information provided by one cell in the source area [8]. We require the cellular magnification to be constant (that is, no partial deletions occur), and obtain:

$$\left| det J_{\mathcal{R}_{i3}^{-1}}(\mathbf{x}_3) \right| \;=\; c_{i3}(\mathbf{x}_3) \text{ for } i \in \{1, 2\} \tag{9}$$

With (9) we obtain:

$$e_3(\mathbf{x}_3) \;=\; \sum_{i=1}^{2} \int_{\mathbf{A}_i} e_i(\mathbf{x}_i) \cdot k_3\left(\mathbf{x}_3 - \mathcal{R}_{i3}(\mathbf{x}_i)\right) \, d\mathbf{x}_i \tag{10}$$

The problem of reconstrucing a mapping function that satisfies (9) for a given pattern of input segregation has been studied in more detail in [7].

References

[1] V. Braitenberg. In R. Hein and G. Palm, editors, *Theoretical Approaches to Complex Systems, Lecture Notes in Biomathematics 21*, pages 171 – 188. Springer Verlag, Berlin etc., 1978.

[2] D. C. van Essen. In Alan Peters and Edward G. Jones, editors, *Cerebral Cortex, Volume 3: Visual Cortex*, pages 259 – 329. Plenum Press, New York and London, 1985.

[3] G. Krone, H. A. Mallot, G. Palm, and A. Schüz. *Proc. Roy. Soc. London B*, 226:421 – 444, 1986.

[4] H. A. Mallot. *Trends in Neurosciences*, 10:310 – 311, 1987.

[5] H. A. Mallot and R. Brittinger. In R. Cotterill, editor, *Models of Brain Function*, Copenhagen, 1989 (in press).

[6] H. A. Mallot, E. Schulze, and K. Storjohann. In G. Dreyfus and L. Personnaz, editors, *Neural Networks from Models to Applications*, pages 560 – 569, Paris, 1989. I.D.S.E.T.

[7] H. A. Mallot, W. von Seelen, and F. Giannakopoulos. *Neural Networks*, 1990. in press.

[8] J. Myerson, P. B. Manis, F. M. Mieyin, and J. M. Allman. *Science*, 198:855 – 857, 1977.

[9] E. L. Schwartz. *Biol. Cybern.*, 42:157 – 168, 1982.

[10] W. von Seelen. *Kybernetik (= Biol. Cybern.)*, 5:133 – 148, 1968.

Parallel Processing in Neural Systems and Computers
R. Eckmiller, G. Hartmann and G. Hauske (Editors)
© Elsevier Science Publishers B.V. (North-Holland), 1990

Processing of Figure and Background Motion

in the Visual System of the Fly

Werner Reichardt

Max-Planck-Institut für biologische Kybernetik
Spemannstrasse 38, D 7400 Tübingen, FRG

The visual system of the fly is able to extract different types of global retinal motion patterns as may be induced on the eyes during different flight maneuvers and to use this information to control visual orientation. The mechanisms underlying these tasks were analyzed by a combination of quantitative behavioral experiments on tethered flying flies (*Musca domestica*) and model simulations using different conditions of oscillatory large-field motion and relative motion of different segments of the stimulus pattern. Only torque responses about the vertical axis of the animal were determined.

The stimulus patterns consisted of random dot textures ('Julesz patterns') which could be moved either horizontally or vertically. Horizontal rotatory large-field motion leads to compensatory optomotor turning responses, which under natural conditions, would tend to stabilize the retinal image. The response amplitude depends on the oscillation frequency: It is much larger at low oscillation frequencies than at high ones. When an object and its background move relative to each other, the object may, in principle, be discriminated and then induce turning responses of the fly towards the object.

However, whether the object is distinguished by the fly depends not only on the phase relationship between object and background motion but also on the oscillation frequency. At all phase relations tested, the object is detected only at high oscillation frequencies. For the patterns used here, the turning responses are only affected by motion along the horizontal axis of the eye. No influences caused by vertical motion could be detected.

The experimental data can be explained best by assuming two parallel control systems with different temporal and spatial integration properties: The *LF- system* which is most sensitive to coherent rotatory large-field motion and mediates compensatory optomotor responses mainly at low oscillation frequencies. In contrast, the *SF-system* is tuned to small-field and relative motion and thus specialized to discriminate a moving object from its background; it mediates turning responses towards objects mainly at high oscillation frequencies.

The principal organization of the neural networks underlying these control systems could be derived from the characteristic features of the responses to the different stimulus conditions. The input to the model circuits responsible for the characterisitic sensitivity of the SF-system to small-field and relative motion is provided by retinotopic arrays of local movement detectors. The movement detectors are integrated by a large-field element, the output cell of the network. The synapses between the detectors and the output cells have nonlinear transmission characteristics. Another type of large-field elements ('pool cells') which respond to motion in front of both eyes and have characteristic direction selectivities are assumed to interact with the local movement detector channels by inhibitory synapses of the shunting type, before the movement detectors are integrated by the output cells.

The properties of the LF-system can be accounted for by similar model circuits which, however, differ with respect to the transmission characteristic of the synapses between the movement detectors and the output cell; moreover, their pool cells are only monocular. This type of network, however, is not necessary to account for the functional properties of the LF-system. Instead, intrinsic properties of single neurons may be sufficient. Computer simulations of the postulated mechanisms of the SF- and LF-system reveal that these can account for the specific features of the behavioral responses under quite different conditions of coherent large-field motion and relative motion of different pattern segments.

Parallel Processing in Neural Systems and Computers
R. Eckmiller, G. Hartmann and G. Hauske (Editors)
© Elsevier Science Publishers B.V. (North-Holland), 1990

SIMULATION OF DELAYED OSCILLATORS WITH THE *MENS* GENERAL PURPOSE MODELLING ENVIRONMENT FOR NETWORK SYSTEMS

Thomas B. Schillen

Max-Planck-Institut für Hirnforschung
Deutschordenstr. 46, 6000 Frankfurt 71, F.R.Germany

Recent theoretical and experimental work leads to the question of temporal information processing in the brain. This paper presents (1) a general purpose network simulator designed in particular for investigating this class of problems and (2) the simulation of nonlinear "neural" oscillators employing delayed coupling between excitatory and inhibitory units. The inclusion of coupling delays results in fundamentally changed and physiologically reasonable characteristics of the functional behaviour of conventional oscillators.

1 Introduction

Current theories of visual processing assume the parallel "low level" processing of object features like colour, texture, orientation, velocity, etc. [1, 2, 3, 4, 5, 6]. This processing is considered to occur through corresponding feature detectors involving spatially separated neuronal populations. With respect to the processing of natural scenes these concepts lead to the problem of binding distributed feature responses to unique representations of the pertaining objects [4, 6]. The conjunction of all possible feature patterns to dedicated neurons coding for particular objects is prohibited by the ensuing combinatorial explosion. As a consequence, temporal synchronization of the distributed responses to features of the same object has been suggested as a solution to the binding problem [4, 6, 7].

Stimulus-driven oscillations of neuronal firing activities have by now been found in various cortical areas [8, 9, 10, 11, 13]. Furthermore, a stimulus-dependent synchronization of these oscillations could be demonstrated for the cat visual cortex [12, 16]. Moreover, the stimulus-dependent coupling of a particular neuronal oscillator to different distinct oscillating ensembles has also been shown [16].

Accompanying our experimental work network simulations are being employed to investigate the temporal synchronization of neural oscillators [14, 15]. This paper presents (1) a general purpose network simulator designed in particular for the analysis of temporal information processing with respect to delayed signal propagation and (2) results from the simulation of delayed Fermi oscillators. The question of coupling this class of oscillators within an entire oscillatory layer will be addressed in the following two papers [14, 15].

2 The *MENS* general purpose network simulator

In order to investigate the implications of temporal coding in more detail a neural network simulator (Fig 1A) has been developed, the data structures (Fig 1B) and modularity of which make it a general purpose tool for many types of network applications. With respect to the investigation of oscillatory responses a data structure has been incorporated that allows individually delayed unit outputs [17] to represent the temporal relationship of neuronal signals. Furthermore, the data structure provides a way of deleting "deteriorated" synapses [17] after a period of learning as a means for saving computation time in initially highly connected networks. The design of stimulus input and display output additionally contributes

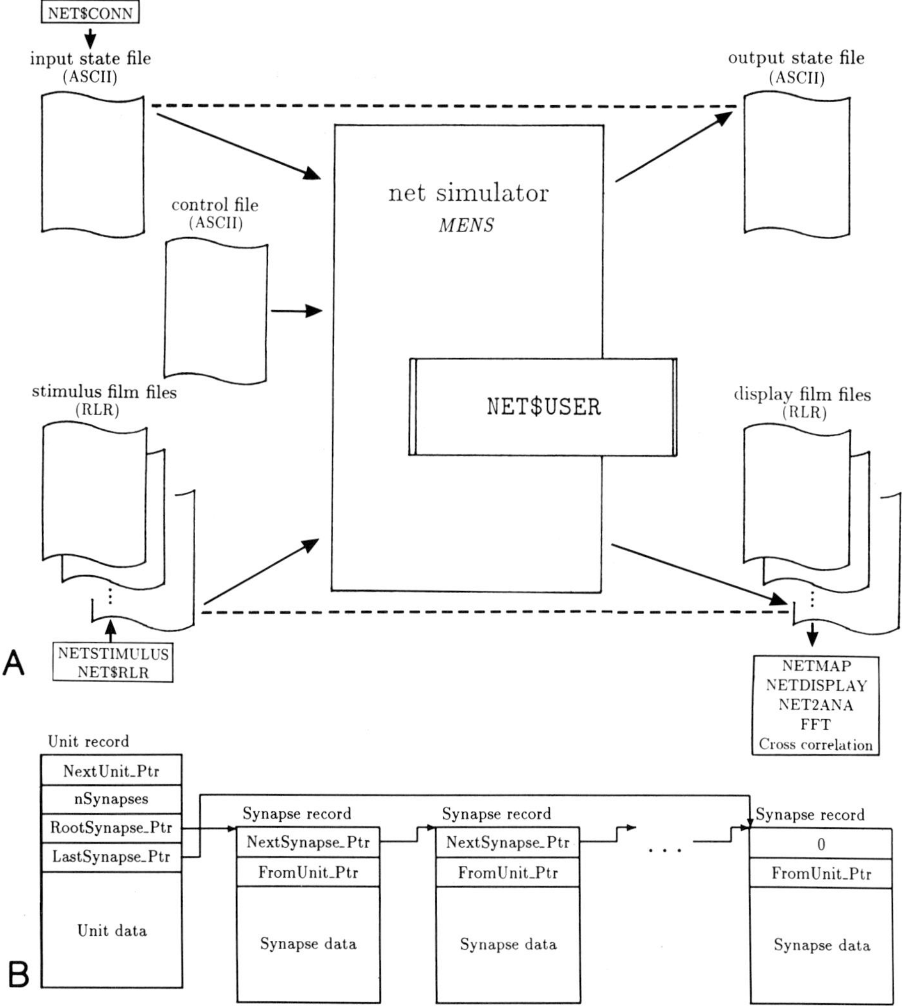

Figure 1: (A) Overview of the *MENS* simulator layout. (B) Example of data structure.

to computational economy by facilitating preprocessing of stimulus films [17] through the fixed input stages of multilayered networks. In order to save memory space the simulator software dynamically allocates memory only to such units, synapses, or delay queues that are actually in use in the current network.

The simulator package provides those tasks that are standard for most network simulations: specification of multilayered networks with an arbitrary number of layers, units per layer, and synaptic topology through a configuration file; unit and synapse parameters at the user's disposal; input to all unit and synapse parameters in every layer by stimulus films as e.g. stimulus, teacher, or gating input; output of all unit and synapse parameters in every layer into display films for monitoring the development of the network; modularity of user-specific unit transfer, learning, and integration functions; a simple command language to control build up, connection verification, execution, and dump of a network; a programming environment that supports the generation of stimulus and configuration files and the analysis

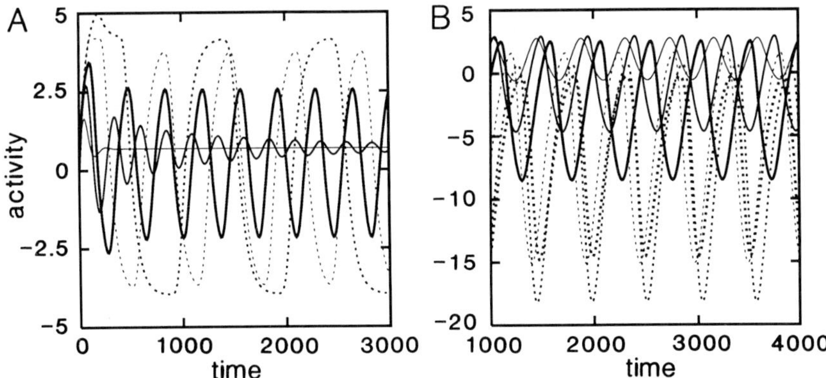

Figure 2: Delayed Fermi oscillators: (A) Effect of varying delay time $\tau \equiv \tau_{ei} = \tau_{ie}$; with increasing line thickness: solid: $\tau = 0, 20, 40$ steps, dashed: $\tau = 100, 200$ steps. $w_{ei} = 1.4$, $w_{ie} = -w_{ei}$. (B) Effect of varying coupling strength $w \equiv w_{ei} = -w_{ie}$; with increasing line thickness: solid: $w = 1.0, 1.5, 2.0$, dashed: $w = 3.0, 4.0, 5.0$. $\tau_{ei} = \tau_{ie} = 40$ steps.

of monitored display data.

The user has to supply only those functions that are specific for the intended application. This requires mainly filling in the desired unit transfer and learning functions. The simulator's data structure is designed to allow a layer-specific implementation of these functions, which makes it possible to adapt different sections of a larger network to the representation of different physiological functions. Additionally, the user can switch learning on or off for entire layers at run time as is suitable for the self-organization of different parts of the network. To give the user sufficient flexibility the simulator software supports different unit update and integration schemes. In particular, with respect to network simulations employing differential equations a Runge-Kutta algorithm is provided as an integral part of the simulator. Also, the location of the central integration procedure in the user module makes it easily adaptable to specific needs.

The problem-specific integration of transfer, learning, and integration functions makes the network simulator a general purpose tool for many types of network applications. So far it has been used to run network memories, back-propagation networks, and networks related to physiology.

3　Simulation of delayed nonlinear oscillators

Experimental evidence [8, 9, 11, 12] leads to the question of temporal information coding by neuronal oscillators. In order to address this question a nonlinear oscillator with inhibitory feedback and delayed coupling has been implemented. An excitatory unit u_e couples with delay τ_{ei} to an inhibitory unit u_i, which in turn projects back to unit u_e with delay τ_{ie}:

$$\dot{x}_e(t) = -\alpha_e\, x_e(t) - w_{ie}\, F(x_i(t - \tau_{ie})) + i_e(t)$$
$$\dot{x}_i(t) = -\alpha_i\, x_i(t) - w_{ei}\, F(x_e(t - \tau_{ei}))$$

with

$$F(x(t)) = \frac{1}{e^{\Theta - x(t)} + 1}$$

a Fermi function output nonlinearity and t: time, $x(t)$: unit activity, α: damping constant, w: coupling constant, τ: delay time, $i(t)$: external input, Θ: threshold.

The inclusion of coupling delays is motivated by two reasons: (1) Delays are naturally present in biological networks through synaptic transmission and finite conduction velocity and (2) coupling delays result in a fundamentally modified behaviour of the described oscillators.

Fig 2A shows the effect of increasing delay time τ on the oscillatory behaviour of the two coupled units. With no or too little delay the system relaxes to a stable fixpoint determined by coupling parameters and external input. Increasing delay time sufficiently transfers the system to a stable limit cycle. The oscillators time scale is given by parameters α^{-1}, w^{-1}, and τ. Note that the minimum delay time necessary to facilitate oscillation is in the order of 0.1 of the oscillation period length. This is very well compatible with physiological data [11].

A similar dependence of oscillatory behaviour as for delay also holds for input amplitude $i_e(t)$ (Data not shown). If input activity is too low the system is located at a stable fixpoint. Only a sufficiently high level of input will induce oscillations. (Note that the input activity is stationary and, thus, is not driving the oscillator by itself). This input dependence, therefore, allows a stimulus-dependent transition of coupled layer units between a nonoscillatory and an oscillatory state.

Fig 2B demonstrates the effect of varying coupling strength w. As can be seen, the oscillation frequency depends relatively little on the exact value of w. Similar frequency stabilities can also be found for variations of other parameters like output function threshold, slope, etc. This is again physiologically reasonable if oscillator synchronization is to be employed for information processing. E.g. the brain's synaptic efficacy will vary by biological variance and modifications through learning. If oscillation frequency were sensitively dependent on synaptic coupling strength learning might easily destroy the very synchronization that was the cause of the learned synaptic modification. The restriction to a limited frequency band thus renders synchronization less sensitive to exact biological parameters.

4   Conclusions

The inclusion of delays into a simple inhibitory feedback oscillator fundamentally changes the oscillator's behaviour. In particular, delay and frequency characteristics turn out to be physiologically reasonable. These effects do not depend critically on specific parameters or the chosen form of nonlinearity.

The effects of coupling these oscillators through synchronizing and desynchronizing delay connections within an oscillatory layer will be addressed in [14, 15].

References

[1] Ballard, D.H., Hinton, G.E., and Sejnowski, T.J., Nature 306 (1983) 21-26. [2] Julesz, B., Nature 290 (1981) 91-97. [3] Marr, D., Vision (Freeman, San Francisco, 1982). [4] v.d.Malsburg, C. and Schneider, W., Biol.Cybern. 54 (1986) 29-40. [5] v.d.Malsburg, C. and Singer, W., in: Rakic, P. and Singer, W., Neurobiology of Neocortex (Wiley, New York, 1988) pp. 69-99. [6] Damasio, A.R., Neural Computation 1 (1989) 123-132. [7] Crick, F., Proc.natn.Acad.Sci.USA 81 (1984) 4586-4590. [8] Freeman, W.J., Mass Action in the Nervous System (Academic Press, New York, 1975). [9] Freeman, W.J., in: Basar, E., Dynamics of Sensory and Cognitive Processing by the Brain (Springer, Berlin, 1988) pp. 19-29. [10] Gray, C.M. and Singer, W., Soc.Neurosci.Abstr. 13 (1987) 404.3. [11] Gray, C.M. and Singer, W., Proc.natn.Acad.Sci.USA 86 (1989) 1698-1702. [12] Gray, C.M., König, P., Engel, A.K., and Singer, W., Nature 338 (1989) 334-337. [13] Eckhorn, R., Bauer, R., Jordan, W., Brosch, M., Kruse, W., Munk, M., and Reitboeck, H.J., Biol.Cybern. 60 (1988) 121-130. [14] Schillen, T.B. and König, P., Coherency Detection by Coupled Oscillatory Responses ..., this volume. [15] König, P. and Schillen, T.B., Segregation of Oscillatory Responses by Conflicting Stimuli ..., this volume. [16] Engel, A.K., König, P., Gray, C.M. and Singer, W., Synchronization of Oscillatory Responses: A Mechanism for Stimulus-dependent Assembly Formation, this volume. [17] Schillen, T.B., Introduction to designing a neural network simulator — The *MENS* modelling environment for network systems, Comp.Appl.Biosci., in print

Parallel Processing in Neural Systems and Computers
R. Eckmiller, G. Hartmann and G. Hauske (Editors)
© Elsevier Science Publishers B.V. (North-Holland), 1990

COHERENCY DETECTION BY COUPLED OSCILLATORY RESPONSES – SYNCHRONIZING CONNECTIONS IN NEURAL OSCILLATOR LAYERS

Thomas B. Schillen and Peter König

Max-Planck-Institut für Hirnforschung
Deutschordenstr. 46, 6000 Frankfurt 71, F.R.Germany

Recent experimental data suggest temporal information processing as a potential mechanism for solving the binding problem in the brain. In particular, there is evidence of stimulus coherency leading to synchronized oscillations of firing activity in cat visual cortex. This paper demonstrates (1) zero phase lag synchronization of oscillatory layers by delayed excitatory to inhibitory unit coupling and (2) the ability of this type of coupling to exhibit stimulus coherency-dependent cross correlations of different "neuronal" assemblies as seen in the experiment.

1 Introduction

Stimulus-related oscillatory firing activity of neural responses has been found in the olfactory bulb [1, 2] as well as in cat visual cortex [3, 4, 5, 6, 7]. Furthermore, recent experimental data from area 17 of cat visual cortex demonstrate coding of stimulus-coherency by synchronization of neural oscillations [5]: Two cells with nonoverlapping rececptive fields (RF) are stimulated by two collinear light bar segments. Coherently moving bar segments of small or no gap distance lead to a pronounced cross correlation amplitude of the cells' responses. Incoherent stimulus movement or too large a separation of RFs abolishes this correlation. This information could be found neither in the post stimulus time histogram nor in the cells' autocorrelations.

As has been suggested previously [8, 9, 10] temporal coding may provide a means for solving the binding problem in the brain. Stimulus-dependent coupled oscillations could represent a potential mechanism for an object representation by assembly formation that avoids the ambiguity problems of rate coded assemblies. In this case an assembly coding for an object could be defined and identified by synchronization of the oscillatory responses of the pertaining spatially distributed feature detectors.

2 Synchronizing an oscillator layer by delayed excitatory to inhibitory unit connections

Before investigating the response of an oscillatory layer to stimulus bar segments we first want to demonstrate the effect of delayed excitatory to inhibitory unit coupling as such.

We start with a 15 × 15 2-dimensional layer of isolated oscillators. The basic oscillatory elements in this layer are delayed Fermi oscillators as described in [11]. Initially, all oscillators are randomly distributed through all limit cycle phases by a process of simulated annealing. Random distribution has been verified by a phase scatter plot of all oscillators' states. Annealing is stopped at a temperature level equivalent to an amplitude variance of $\frac{1}{3}$ of average peak amplitude. This temperature is maintained through the entire simulation. The situation of isolated random phase oscillations is visible at negative time values in the left part of Fig 1, which shows the activity traces of 8 units arbitrarily chosen from the oscillator layer.

At $t = 0$ delayed excitatory nearest neighbour connections are introduced. These additional connections couple an oscillator's excitatory unit u_e to the inhibitory units u_i of all

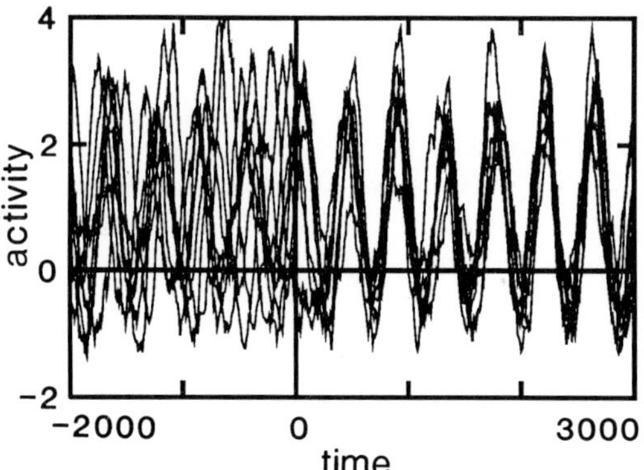

Figure 1: Synchronizing an oscillator layer by delayed excitatory connections. Activity traces of 8 arbitrary units from a layer of delayed Fermi oscillators [11]. $t < 0$: Random phase oscillations of isolated layer elements. $t > 0$: Introduction of delayed excitatory nearest neighbour coupling w_{ei}^{NN} (see text) rapidly synchronizes the entire layer without phase lag within very few oscillation cycles.

nearest neighbour oscillators. Coupling weights w_{ei}^{NN} are isotropic and cyclic boundary conditions are imposed. Coupling delays τ_{ei}^{NN} are chosen to be of the order of the oscillator's intrinsic delays, in this case $\tau_{ei}^{NN} = \tau_{ei} = \tau_{ie}$. (All symbols as introduced in [11]).

As apparent from Fig 1 for $t > 0$ the introduction of w_{ei}^{NN}-connections is able to synchronize the entire oscillator layer within very few oscillation cycles. We want to emphasize that the described form of neighbour coupling results in a synchronization of zero phase lag, as is required by experimental evidence [5]. With the specified type of connectivity and range of delays every oscillator will excite all its neighbouring inhibitory units simultaneously to its own. By this timing every oscillator is promoting synchronous oscillations of its neighbouring oscillators without phase lag relative to itself.

Note that with this type of coupling the observed correlation length of synchronicity within the layer is much larger than the implemented coupling length. Note also that the coupling delay time τ_{ei}^{NN} is small compared to the oscillation period length. Furthermore, the delay time is approximately 0.1 period length. Considering 1 ms synaptic delay, 1 $mm \cdot ms^{-1}$ signal propagation velocity and an oscillation period of 20 ms [4] this agrees well with the order of physiological data.

This simulation demonstrates that delayed nearest neighbour coupling can provide rapid synchronization of stimulus-dependent oscillations within very few oscillation cycles without phase lag across large correlation lengths.

3   Coherency detection by coupled oscillations (nondirection-selective case)

We now turn to the investigation of oscillatory layer responses to stimulus bar segments.

A 10×20 2-dimensional layer of delayed Fermi oscillators with cyclic boundary conditions is configured with nearest, next nearest, and double next nearest neighbour coupling. As in the previous section the coupling is of the delayed excitatory to inhibitory unit type (w_{ei}^{NN}, w_{ei}^{NNN}, w_{ei}^{NNNN}). Coupling weights are again isotropic and represent a Gaussian distribution of synaptic connectivity. To allow symmetry breaking a temperature level is maintained as described above.

Each oscillator in the layer is interpreted to represent an entire retinal receptive field (RF). In this example we restrict ourselves to the case of nondirection-selective cells, all of a single identical orientation preference. With this interpretation an oscillator's external input $i_e(t)$ [11] reflects the presence of a moving light bar within the pertaining RF. Correspondingly,

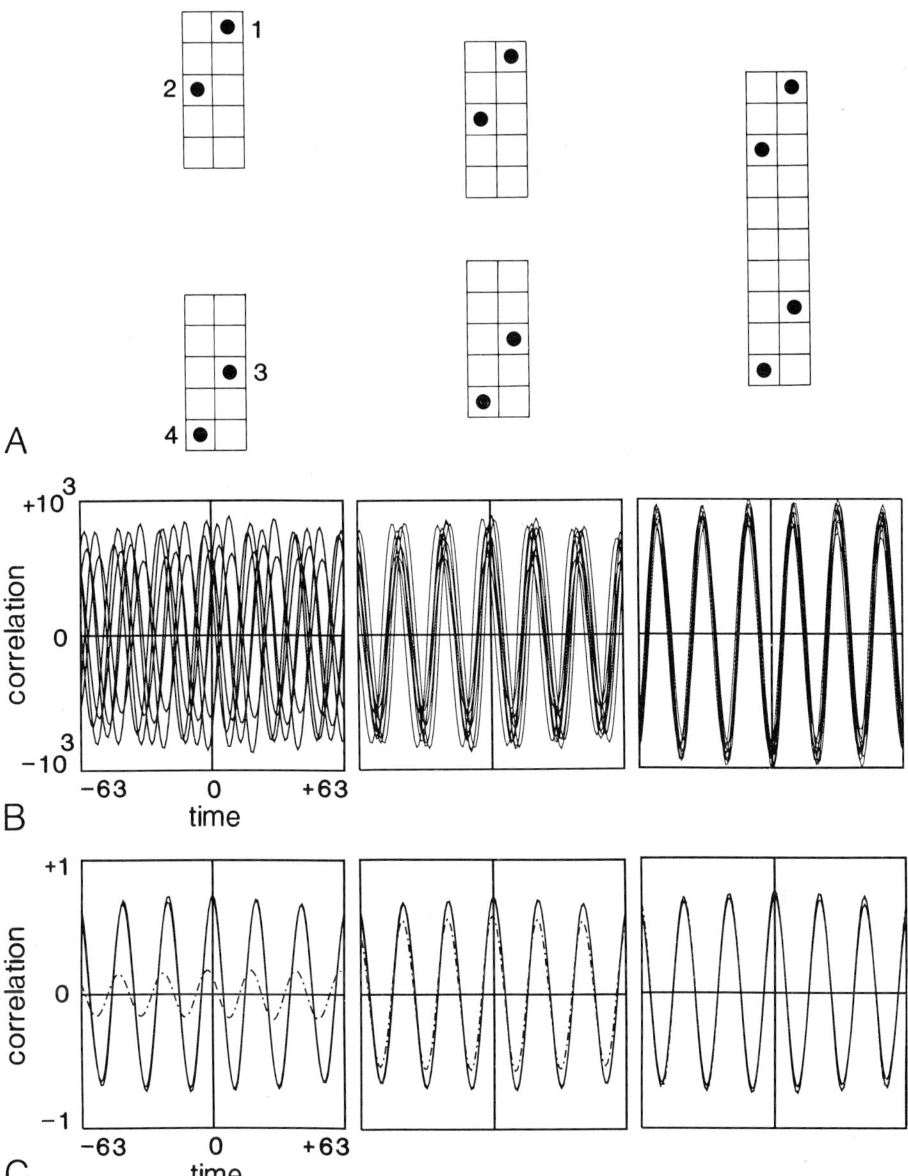

Figure 2: Effect of stimulus continuity on cross correlations in an oscillator layer. (A) Input configurations for gap distances 4, 2, and 0; analysed oscillators marked by numbered dots. (B) Cross correlations (2–3) for 10 epochs of 50 cycles. (C) Cross correlations (1–2), (3–4) (solid) compared to cross correlations (2–3) (dashed); normalized (see text).

movement of a light bar on the retina will provide a stimulus to the covering map of RFs and their pertaining oscillators.

Fig 2 depicts the simulated reponse of the oscillator layer to two moving light bar segments separated by varying gap distances. The area activated by a single bar segment was set to 2×5 oscillators. The gap distances shown here span 4, 2, and 0 oscillator positions. The data for each gap distance are presented within a separate column of the figure.

The top row in Fig 2 shows the distributions of external input to the oscillator layer. The numbered dots in each bar segment mark the locations of those oscillators which have been chosen for the cross correlation analysis presented in the middle and bottom row of the figure. The middle row depicts across-gap cross correlations (2–3) for 10 epochs of approximately 50 cycles each. The bottom row shows cross correlations (1–2) and (3–4) compared to (2–4) normalized by the mean of auto correlation center values.

As seen in the previous section all oscillators within every single bar segment's area are tightly coupled and cross correlations show zero phase lag. This observation applies to all three gap distances.

In the case of no gap distance (Fig 2, right column) the two bar segments form a contiguous long bar which is completely coupled without phase lag across its entire area. Cross correlation (2–3) virtually coincides with correlations (1–2) and (2–4).

In the other extreme with gap distance exceeding the range of synchronizing connections (Fig 2, left column) coupling is restricted to within each bar segment's area. Between segments activity oscillations shift through all phases resulting in a minimum cross correlation.

With an intermediate gap distance (middle column) coupling between the bar segments' responses is still established but is less stringent. Phase differences between segments vary somewhat around zero leading to a reduced amplitude in the cross correlogram. Note, however, that as in the case of the contiguous long bar there is no phase lag between the oscillations induced by the two segments as required by experimental observation [5].

4   Conclusions

The results presented in this paper demonstrate that neighbour coupling of the described delayed excitatory to inhibitory unit type is well suited to establish zero phase lag synchronization within oscillatory layers. This synchronization exhibits correlation lengths that are large compared to coupling lengths. Furthermore, an oscillatory layer with this type of coupling qualitatively shows the coherence relation in response to "moving light bar" segments required by physiological data [5]. What is lacking at this stage without direction-selective cells is the loss of synchronicity in response to stimulus bars moving in opposite direction [5].

As a consequence of long correlation lengths the problem arises of how to avoid bulk synchronization within entire oscillatory layers. This question will be the topic of the following paper [12]. A canonical extension of the basic principles presented here to oppositely moving stimuli will also be outlined in brief in [12]. This will involve direction-selective RFs with the inclusion of desynchronizing delayed connections.

References

[1] Freeman, W.J., Mass Action in the Nervous System (Academic Press, New York, 1975). [2] Freeman, W.J., in: Basar, E., Dynamics of Sensory and Cognitive Processing by the Brain (Springer, Berlin, 1988) pp. 19-29. [3] Gray, C.M. and Singer, W., Soc.Neurosci.Abstr. 13 (1987) 404.3. [4] Gray, C.M. and Singer, W., Proc.natn.Acad.Sci.USA 86 (1989) 1698-1702. [5] Gray, C.M., König, P., Engel, A.K., and Singer, W., Nature 338 (1989) 334-337. [6] Engel, A.K., König, P., Gray, C.M. and Singer, W., Synchronization of Oscillatory Responses: A Mechanism for Stimulus-dependent Assembly Formation, this volume. [7] Eckhorn, R., Bauer, R., Jordan, W., Brosch, M., Kruse, W., Munk, M., and Reitboeck, H.J., Biol.Cybern. 60 (1988) 121-130. [8] v.d.Malsburg, C. and Schneider, W., Biol.Cybern. 54 (1986) 29-40. [9] Damasio, A.R., Neural Computation 1 (1989) 123-132. [10] Crick, F., Proc.natn.Acad.Sci.USA 81 (1984) 4586-4590. [11] Schillen, T.B.,Simulation of Delayed Oscillators with the *MENS* general purpose Modelling Environment for Network Systems, this volume. [12] König, P. and Schillen, T.B., Segregation of Oscillatory Responses by Conflicting Stimuli ..., this volume.

Parallel Processing in Neural Systems and Computers
R. Eckmiller, G. Hartmann and G. Hauske (Editors)
© Elsevier Science Publishers B.V. (North-Holland), 1990

A Model for Neuronal Oscillations in the Visual Cortex

H.G. Schuster and P. Wagner
Institut für Theoretische Physik, Universität Kiel, D-2300 Kiel, FRG

It is shown that the dynamical behaviour of columns in the visual cortex can be described by switchable limit cycle oscillators if one starts from a model of two interacting excitatory and inhibitory neuron populations. A sparse, long but finite ranged coupling between different columns on the neuronal level induces a dynamical interaction between the oscillators in the limit cycle description. This leads to a stimulus dependent synchronization which reflects global features of the stimulus, as observed experimentally.

It has recently been demonstrated experimentally that neurons in spatially separate columns in cortical area 17 of adult cats can synchronize their oscillatory responses [1]. The synchronization has on the average no phase difference, depends on the spatial separation and the orientation preferences of the cells and is influenced by global stimulus properties. The last property has been demonstrated beautifully in two cases in which pairs of recording sites were $7mm$ apart. Receptive fields of the neurons at the two sites had a common orientation specifity and were aligned so that they could be stimulated by a single long bar of light (see Fig. 3b of [1]). Synchronization occurred only when the two columns were stimulated with a single long bar. When they were simultaneously activated by two shorter bars the correlation disappeared. This means that synchronization between different columns depends on relations between features in different parts of the visual field - whether this figure is connected or not.

In this article we would like to address the following questions:

- What is the origin of the observed oscillations within each column i.e. how could they be explained on a neuronal level?

- What architecture is necessary such that different columns are coupled dynamically in such a way that their responses reflects global features of the stimulus i.e. that there is synchronization without phaseshift for stimulation with a long bar and no synchronization for two separated small bars?

We start from the following model. Each orientation specific column is described by two subpopulations of neurons whose axons end with excitatory or inhibitory synapses, respectively [2]. If we make the biologically reasonable assumption that within a column each neuron is coupled to a macroscopic number ($\approx 10^3$) of other cells we find that the dynamical behaviour of a single column is well described by a limit cycle oscillator [3] which can be switched on and off by an external stimulus via a Hopf bifurcation. This means that the oscillations result from a competition between excitatory and inhibitory stimuli exchanged between two macroscopically coupled subpopulations within a column. Next we couple subpopulations of different columns via sparse interactions which connect a column with about 10 others whose distances d have been chosen randomly according to an exponential probability distribution $P(d) \propto \exp(-d/\lambda)$, $\lambda \approx 7$, see [4]. After linearization in the interaction strengths and elimination of fast variables [5] it is shown that the dynamics of coupled columns can be described by phases of dynamically coupled limit cycle oscillators whose frequencies and interaction strengths depend on the external stimulus, i.e. only activated oscillators with frequencies different from zero interact and there is no interaction between an active and a silent oscillator. Fig. 1 shows the dependence of the frequency ω and Fig. 2 the interaction strength K between two oscillators as a function of the external stimulus P.

This picture of switchable limit cycle oscillators also explains the experiment of Gray and Singer [1] as shown in Figs 3 and 4.

We would like to stress the novel features of our model:

- The dynamic coupling between columns leads in contrast to a constant coupling to a sharp foreground background separation without introducing synchronization of background activities.

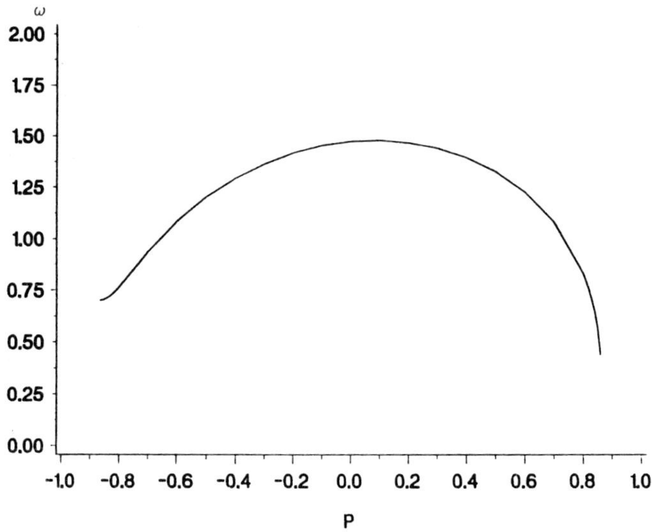

Figure 1: Frequency ω of a limit cycle as a function of the strength of the external stimulus P.

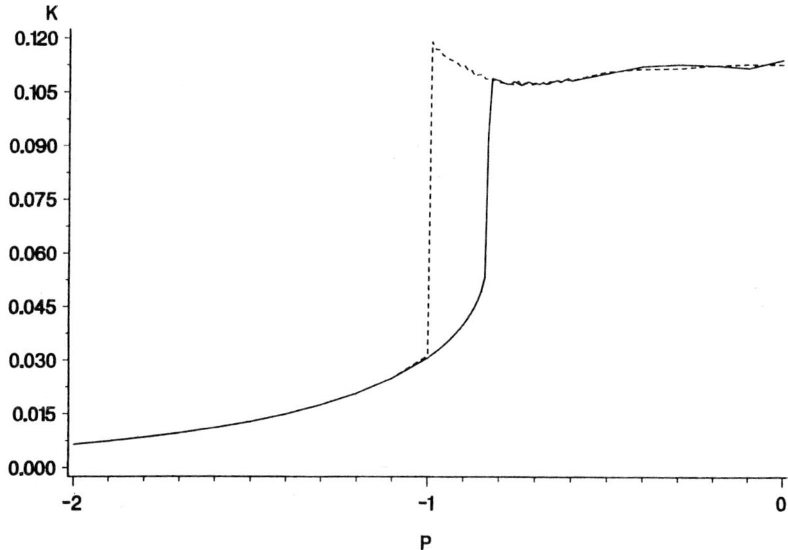

Figure 2: Time averaged dynamical coupling K as a function of the strength of the external stimulus P.

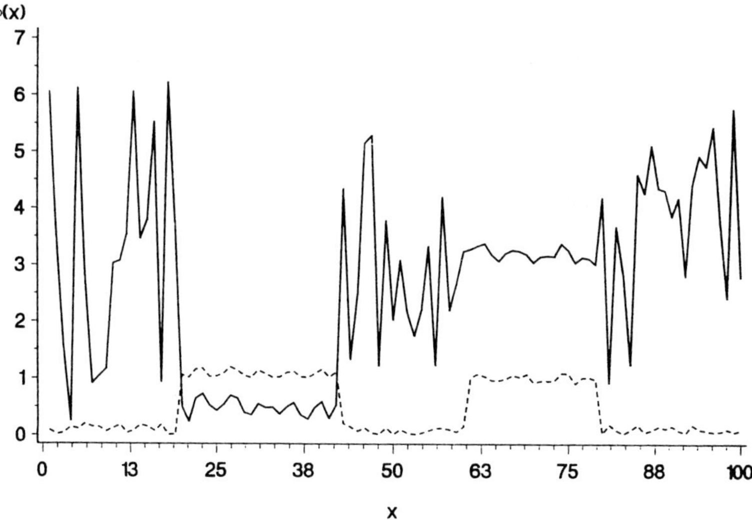

Figure 3: Input profile (dashed line) and phases of limit cycle oscillators as function of the inter-columnar spacing x. Input corresponds to two active spots generated by two small bars. Within the spots the oscillations are synchronized but the relative phases between the spots are not.

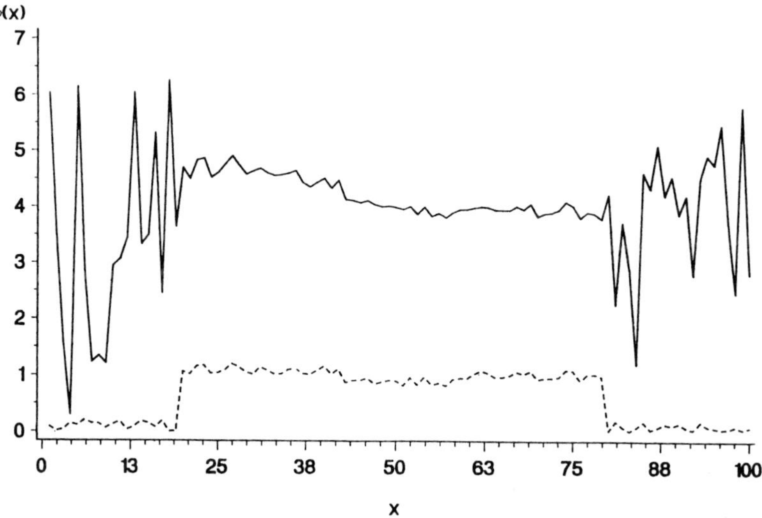

Figure 4: Input profile (dashed line) and phases of limit cycle oscillators as function of the inter-columnar spacing x, stimulation by a single long bar. The synchronized region covers the activated region but is decoupled from the background.

- The long ranged sparse coupling leads a) to a proper synchronization between columns in the experiment with a single long bar (fig. 4) which cannot be achieved with nearest neighbor interactions [6] and b) it avoids synchronization if columns are stimulated with two small bars which are disconnected (fig. 3); this cannot be achieved with the so called comparator model of Kammen et al. [6] which yields - despite a possible, but difficult to achieve foreground background separation - always to a synchronization between columns.

Our model makes the following prediction. If there are indeed couplings of one column to many others as assumed above then one should find experimentally synchronization if one column is stimulated by a small bar and many nearby other columns are stimulated by many small bars which are however disconnected from the first one i.e. one could simulate the result obtained with a long bar by using instead a single small bar and many small bars in parallel.

One of us (P. W.) acknowledges support from the SFB 185 "Nichtlineare Dynamik".

References

[1] C.M. Gray, P. Koenig, A.K. Engel and W. Singer, Nature **338** (1989), 334.

[2] H.R. Wilson and J.D. Cowan, Biophys. J. **12** (1972), 1.

[3] G.B. Ermentrout and J.D. Cowan, J. Math. Biol. **7** (1979), 265.

[4] B. Baird, Physica **22D** (1986), 150.

[5] Y. Kuramoto, "Chemical Oscillations, Waves, and Turbulence", Springer-Verlag Berlin New York Tokyo, 1984.

[6] D.M. Kammen, P.J. Holmes and C. Koch: "Cortical Architecture and Oscillations in Neuronal Networks: Feedback Versus Local Coupling" in "Models of Brain Function", R.M.J. Cotterill (Ed.).

Parallel Processing in Neural Systems and Computers
R. Eckmiller, G. Hartmann and G. Hauske (Editors)
© Elsevier Science Publishers B.V. (North-Holland), 1990

TEMPORAL DYNAMICS IN NEURONAL MICROCIRCUITRY

Florentin WÖRGÖTTER[1] Daniel M. KAMMEN[2]   and Brian BRANDT

Computation and Neural Systems Program,
California Institute of Technology
Pasadena, CA 91125, USA

Temporal coordination is apparent in the activity of populations of cortical cells.
We have explored the response characteristics of small circuits – or modules –
of biophysically realistic neurons and discuss the role of temporal modulation in
information processing.

1. Introduction

Recent electrophysiological experiments have shown stimulus specific temporal variability in
the response properties of individual neurons in the striate cortex. Changes in receptive field
profiles [1] and evidence for information multiplexing in single neuron firing [2] have been
supplemented by equally intriguing observations of stimulus specific oscillatory activity in
the firing rates of cells separated by as much as 7 mm on the cortical surface [3,4].
A number of theoretical models utilize the temporal firing pattern of cells or cell populations
to extract signal information. In the "feature linking" hypothesis [5], for example, images
are segregated by the temporal coherence in the firing of the cell populations that subserve
distinct perceptual figures in the scene. Such theories emphasize the importance of temporal correlations in neuronal activity. The precise mechanisms used in temporal processing,
however, are not understood.
The *temporal signature* of neuronal modules is of paramount importance in determining the
significance of these experimental results and must be understood prior to an analysis of the
mechanisms underlying cortical information processing [6].
We have explored the temporal behavior of many realistic small circuits of biophysically
accurate neurons and in this short paper we will show the results from the analysis of one
example circuit.

2. Methods

We have used a detailed biophysical neural network simulation package [7] to model the
neurons as multi-channel units obeying Hodgkin-Huxley dynamics. The circuits investigated
range from simple nearest neighbor excitatory or inhibitory pairings to circuits consisting of
small networks of reciprocally connected excitatory and inhibitory cells. An auto- and

[1]Present address: Institut für Neurophysiologie, Ruhr Universität Bochum, 4630 Bochum, FRG
[2]To whom correspondence should be addressed.
F.W. was supported by the *Deutsche Forschungsgemeinschaft* grant WO 388-1; D.M.K. is the recipient
of a Weizman Postdoctoral Fellowship. The authors gratefully acknowledge the support from the Air
Force Office of Scientific Research and of a NFS Presidential Young Investigator Award and from the
James S. McDonnell Foundation to C. Koch

cross- correlation analysis of the output activity of the cells was performed for a range of synaptic connection weights and axonal and synaptic delay times. Power spectra were computed from the correlation functions to determine oscillatory behavior. Fig.1 outlines the analysis scheme that we applied to each basic circuit for each set of parameter values.

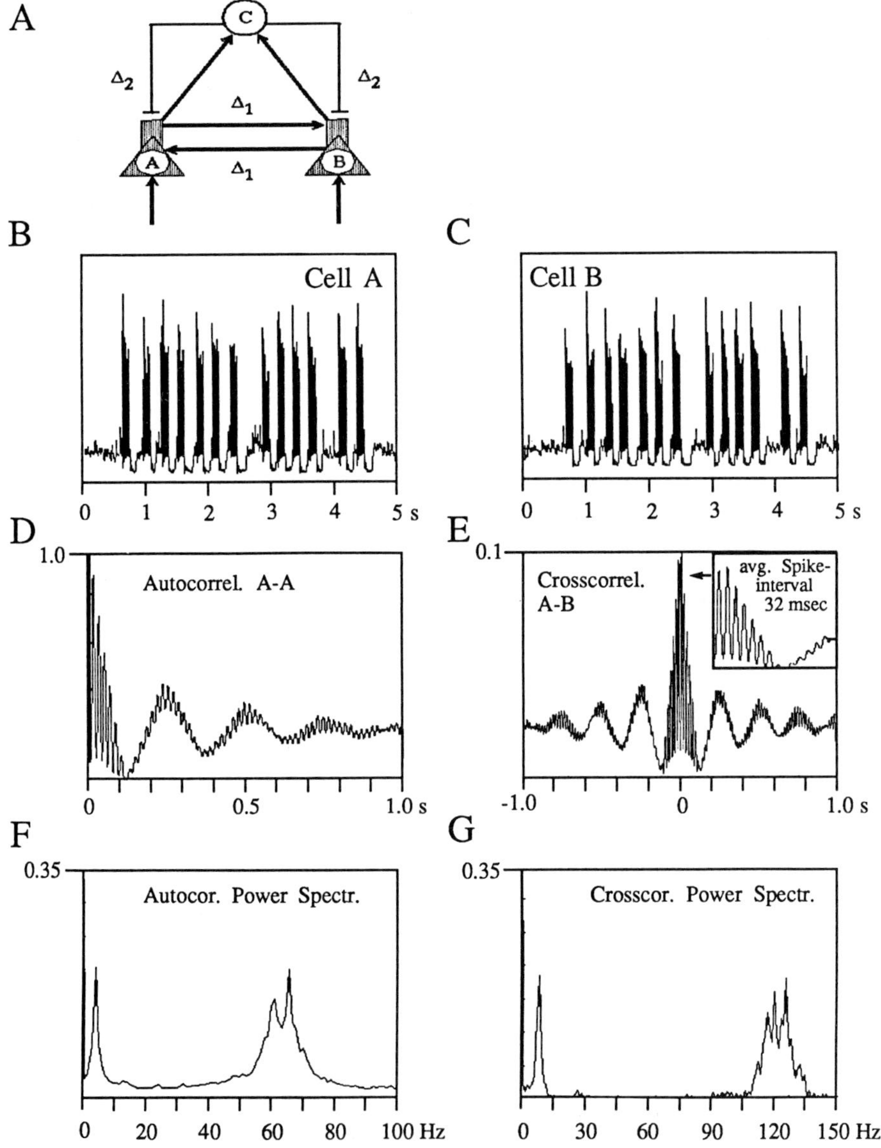

Fig.1) Circuit diagram (A) and analysis of coherent firing pattern (B-G) for a given set of axonal delays (Δ_1 = 2 msec, Δ_2 = 50 msec). For further explanation see text.

3. Results

In Fig.1A we sketch the diagram of the circuit discussed below. In this scheme two coupled excitatory cells (A and B) receive parallel uncorrelated input and inhibitory feedback from a single interneuron (C). Figs.1B and 1C show the response of cells A and B to 4 seconds of sustained stimulation. The spike train actually consists of *bursts* of spikes that are interrupted by quiescent intervals, the duration of which is related to the onset of inhibition as determined by Δ_2, in the circuit. In Fig.1D the auto-correlogram of cell A is presented and in Fig.1E the cross correlogram of cell A and cell B is plotted along with an inset demonstrating the sizable 32 *msec* phase delay that corresponds to the mean interspike interval between cell A and cell B. The power spectra (Fig.1F,G) show two peaks, one at low frequency corresponding to the time between bursts, the other at higher frequency representing the interspike spacing within each burst. The peak at 120 Hz in Fig.1G occurs at twice the frequency of the corresponding peak in Fig.1F. This reflects the ratio between a half (A to B) and the full excitatory loop (A to B to A).

The two delays, Δ_1 and Δ_2, determine the onset time of the positive and negative feedback loops, respectively. We found that the time delays are far more significant in determining the circuit response than the synapse strengths or precise channel dynamics for parameter values commensurate with known pyramidal cell biophysics.

When the excitatory and inhibitory pathways both act rapidly no regular firing pattern can be detected; in fact the cells are simply stimulus driven (Fig.2 bottom left corner, the size of the symbols indicates the strength of coherency). As depicted in Fig.2, increasing the time delay in the inhibitory pathway results in an increasingly strong oscillatory firing (bottom right corner of Fig.2) and the cells exhibit *bursting* behavior. The reason for this is clear: as the ratio between the time for excitatory to inhibitory coupling Δ_1/Δ_2 is decreased the

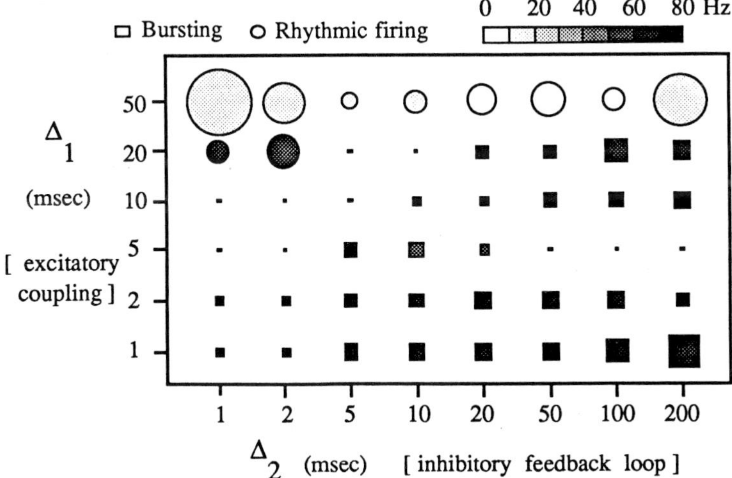

Fig.2) Characteristic of the temporal behavior of cell A from the circuit shown in Fig.1A. The strength of the oscillations was determined from the power spectra. Bursting occurs at about 60 Hz, rhythmic firing at about 8-15 Hz. For further explanation see text.

direct excitatory coupling causes one cell to fire whenever the other cell receives input sufficient to elicit a spike. The cells thus enter in a positive feedback loop that continues until the onset of inhibition occurs and the spiking is terminated. When the inhibition decreases and the cells depolarize the cycle can begin again. The burst frequency is, of course, determined by the rapidity of the positive feedback time scale (Δ_1 loop).

The other generic type of coherent activity that the cells can perform is *rhythmic*, or steady, firing which occurs when Δ_1/Δ_2 is large. Firing of either cell A or cell B quickly and recurrently inhibits the activity of both cells and bursting behavior is prevented. Direct excitation then occurs delayed, not until the rapid inhibition has ceased, and elicits a spike. Slow, rhythmic, firing of cells A and B results with a period of $8-15Hz$ (circles in Fig.2).

4. Discussion

This study presents the basic protocol on analysis of circuits of highly realistic neurons. The multi-dimensional nature of the solution-space of even small circuits necessitates a simulation based approach to understand the dynamics of the system. The basic circuit shown here exhibits the two simplest types of correlated neuronal activity: bursting and rhythmic firing. Phase shifts in the coupling between cells A and B are large and predictable. However, already the inclusion of a fourth cell results in a dramatically changed behavior. Interrupting the direct excitatory interaction between the cells and including another excitatory cell in the circuit that couples cells A and B indirectly while also driving the inhibitory interneuron will eliminate the phase shift between cells A and B. This result is consistent with the findings of Gray *et al.* [3] who reported phase shifts close to zero between cortical neurons that are driven with similar stimuli. The strong changes in temporal behavior already resulting from small alterations in the circuitry emphasizes the importance of comparative studies quantifying the *temporal signature* of module like circuit types.

References

[1] Dinse, H. R. O., Krüger, K., and Best, J. Temporal aspects of cortical information processing. Cortical architecture, oscillations and non-separability of spatio-temporal receptive field organization. Krüger, J. (ed.), Neural Cooperativity: Models and Experiments. (Freiburg, Springer Verlag 1989).

[2] Richmond, B. J., Optican, L. M., Podell, M. and Spitzer, H., Temporal encoding of two-dimensional patterns by single units in primate inferior temporal cortex. I. Response characteristics. *J. Neurophysiol.*, **57** (1987) 132 - 146.

[3] Gray, C. M., König, P., Engel, A. K. and Singer, W., Oscillatory responses in cat visual cortex exhibit inter-columnar synchronization which reflects global stimulus properties. *Nature*, **338** (1989) 334 - 337.

[4] Eckhorn, R., Bauer, R., Jordan, W., Kruse, W., Munk, M. and Reitboeck, H. J., Coherent oscillations: a mechanism for feature linking in the visual cortex? *Biol. Cybern.* **60** (1988) 121 - 130.

[5] von der Malsburg, C. and Schneider, W., A neural cocktail-party processor. *Biol. Cybern.*, **54** (1986) 29 - 40.

[6] Kammen, D. M., Holmes, P. J. and Koch, C. Cortical architecture and oscillations in neuronal networks: feedback versus local coupling. Cotterill, R. M. J. (ed.) Models of Brain Function (Cambridge University Press: Cambridge 1989).

[7] Wilson, M. A. and Bower, J. M. The simulation of large-scale neural networks, in: Koch, C. and Segev, I. (eds.), Methods in Neuronal Modeling. (MIT Press, Cambridge, 1989) 291 - 334.

Section 5
Self-Organization
and Learning
in Neural Networks

Parallel Processing in Neural Systems and Computers
R. Eckmiller, G. Hartmann and G. Hauske (Editors)
© Elsevier Science Publishers B.V. (North-Holland), 1990

PROPERTIES OF AN ADAPTIVE PERCEPTRON ALGORITHM

J.K. Anlauf and M. Biehl

Institut für Theoretische Physik
Justus–Liebig–Universität
D–6300 Giessen, FRG

A new learning algorithm for neural networks of spin–glass type is proposed. The patterns can be presented either sequentially or in parallel. Both versions relax exponentially towards the Perceptron of optimal stability if such a matrix exists. If such a solution does not exist the parallel version of the algorithm minimizes a semi–quadratic cost function. The properties of the resulting coupling matrix are studied with the help of the replica trick for random patterns and a surprising jump of the cost function as a function of the capacity α at $\alpha=2$ is found.

In the following we present a new learning algorithm for neural networks of spin glass type [1–3] which either yields the so called Perceptron matrix of optimal stability [4,5] or, in case no such solution exists, still finds an "optimal solution" by minimizing a certain cost function defined below. The method is called the AdaTron–algorithm because it applies the concept of adaptive learning known from other rules (e.g. Adaline [6,7]) to the Perceptron problem. In the following some analytic results on the solution of the optimization problem are presented.

A system of N Ising spins $S_i=\pm1$, $i=1,...,N$, coupled through synaptic efficacies J_{ij} can serve as a simple model of a neural network. The dynamics of such a system may be defined by a zero–temperature Monte Carlo process:

$$S_i(t+1) = \text{sign}\left[\sum_j J_{ij}S_j(t)\right].$$

Given a set of p patterns $\vec{\xi}^{\nu}=\left[\xi^{\nu}_1,...,\xi^{\nu}_N\right]^T$, $\xi^{\nu}_i=\pm1$, $\nu=1,...,p$, the aim of Perceptron–learning is to adjust the couplings J_{ij} such that these p patterns become fixed points of the dynamics. This is true, if the so called field E^{ν}_i of pattern ν fulfills

$$E^{\nu}_i = \xi^{\nu}_i \sum_j J_{ij} \xi^{\nu}_j \ge c > 0 \text{ for } i=1,...,N, \ \nu=1,...,p.$$

The stability Δ_i in row i, defined as

$$\Delta_i = \min_{\nu} \left\{\Delta^{\nu}_i\right\}, \text{ where } \Delta^{\nu}_i = E^{\nu}_i \ \bigg/ \ \sqrt{\sum_j J^2_{ij}}$$

____

*This work was supported by the Stiftung Volkswagenwerk and the Höchstleistungsrechenzentrum Jülich.

does not depend on the actual value of c (c=1 will be assumed in the following). For high values of Δ_i large basins of attraction for the pattern retrieval are expected [5]. Thus, the aim is to calculate the matrix with maximal Δ_i in every row i. This can be expressed as a quadratic optimization problem [4]: minimize $\sum_j J_{ij}^2$ subject to constraints $E_i^\nu \geq 1$. Note, that the intensively studied projector matrix [8,9] gives optimal stability subject to $E_i^\nu=1$. So called local learning rules are of importance from a biological point of view [1]. Here the coupling J_{ij} is modified by adding $\delta J_{ij}=a_i \xi_i^\nu \xi_j^\nu$ when pattern ν is presented to the network and the resulting matrix can be written as

$$J_{ij} = \frac{1}{N} \sum_\nu x_i^\nu \xi_i^\nu \xi_j^\nu.$$

The x_i^ν are called embedding strengths. Self–couplings will be excluded in the following: $J_{ii}=0$. Since different rows of the matrix can be treated independently we consider one row only and omit the index i. In the method of Diederich and Opper [7], similar to the Adaline scheme [6], the patterns are presented sequentially several times and x^μ is changed by $\delta x^\mu=1-E^\mu$ when pattern μ is presented. The algorithm was proven to yield the projector matrix for a given set of patterns [7]. Opper [3,10] calculated the distribution of the embedding strengths $x^{*\nu}$ for the optimal Perceptron and pointed out that two classes of patterns have to be distinguished which fulfill either ($x^{*\nu}>0$ and $E^{*\nu}=1$) or ($x^{*\nu}=0$ and $E^{*\nu}\geq 1$). Thus, one subset of patterns does not have to be embedded at all and for the other one the optimal Perceptron is in fact the projector matrix (in row i). Simple Adaline learning could be applied to the latter, giving the same matrix with $x^{*\nu}\geq 0$ only [10]. Unfortunately the subsets are not known in advance. The new AdaTron–algorithm overcomes this difficulty by applying Adaline to <u>all</u> patterns but simply restricting the x^ν to non–negative values:

$$\delta x^\mu = \max \left\{ -x^\mu,\ \gamma\,(1-E^\mu) \right\}.$$

The patterns are presented repeatedly, either in parallel or sequentially. The algorithm can be shown to converge to the optimal Perceptron for $0<\gamma<2$ in the sequential version if a solution exists [11]. In the parallel AdaTron γ has to be sufficiently small in order to ensure convergence. Numerical studies show an exponential relaxation towards the solution [11]. Hence, the AdaTron is a much more efficient method to calculate the optimal Perceptron than the minover–algorithm introduced recently [5] which shows an algebraic decay of error. Furthermore it can be shown for $\alpha=p/N>2$ where no Perceptron exists [4], that the parallel AdaTron minimizes the cost function

$$g = \frac{1}{N} \sum_\nu (1-E^\nu)^2 \Theta(1-E^\nu).$$

That is, the difference between the fields and their desired value contributes quadratically to the total cost, but only for fields lower than 1. Note that Adaline minimizes $\frac{1}{N}\Sigma_\nu(1-E^\nu)^2$ including also the stable patterns with $E^\nu>1$ [9]. For random patterns we were able to calculate the minimal value of g with the help of the replica trick, assuming replica–symmetry [4,12]. We found

$$g(\alpha) = Q(\alpha\Phi(1/\sqrt{Q})-1)^2 \quad \text{where} \quad \Phi(z) = \int_{-\infty}^{z} \frac{dx}{\sqrt{2\pi}} \exp\left[-\frac{x^2}{2}\right]$$

and $Q=\frac{1}{N}\Sigma_j J_j^2$ is the solution of the saddle point equation

$$G(z) = \int_{-\infty}^{z} \frac{dx}{\sqrt{2\pi}} (z-x)^2\exp\left[-\frac{x^2}{2}\right] = \alpha\Phi^2(z) \qquad (z = 1/\sqrt{Q}).$$

For $\alpha<2$ Gardner has shown [4] that the optimal Perceptron has a norm Q that is the solution to $G(1/\sqrt{Q})=1/\alpha$. Since it is always possible to find a coupling matrix with $E^\nu\geq1$ in this case, $g(\alpha)=0$ is valid for $\alpha<2$.

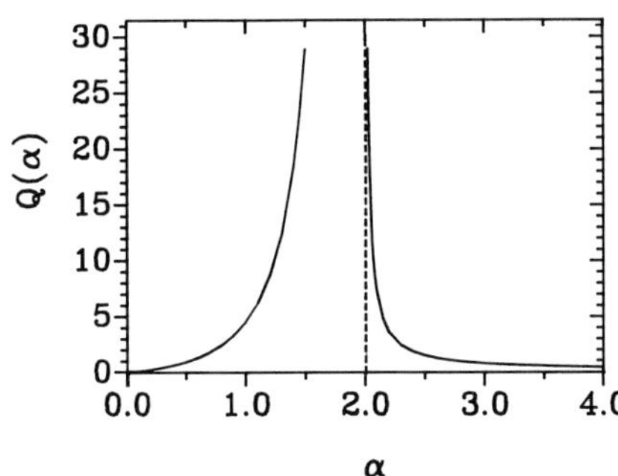

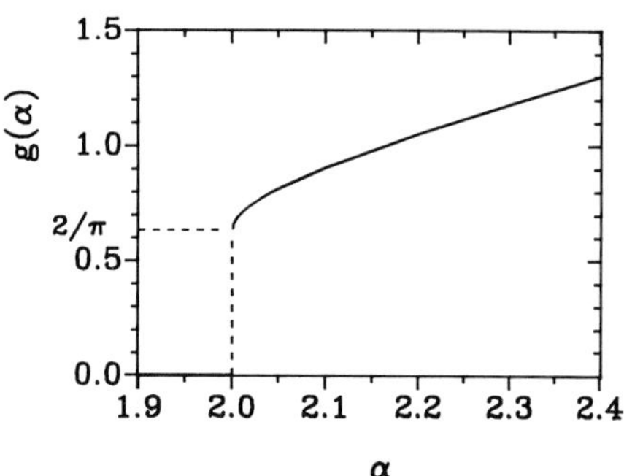

Fig.1: $Q(\alpha)=1/N \Sigma_j J_{ij}^2$ as found   Fig.2:   minimized cost

by the AdaTron   function versus α

Q is plotted against α in Fig.1. It diverges as α approaches 2, reflecting the difficulty to distinguish between solvable and unsolvable problems in this range. In Fig.2 the function $g(\alpha)$ is shown. It is zero up to $\alpha=2$ where it jumps to the finite value $2/\pi$. We calculated the field distribution $P(E)$ for $\alpha>2$ and found

$$P(E) = \frac{1}{\sqrt{2\pi}\sigma} \exp\left[-\frac{1}{2\sigma^2}\left[E - \frac{1}{\alpha\Phi(1/\sqrt{Q})}\right]^2\right] \qquad \text{for} \quad E<1$$

$$P(E) = \frac{1}{\sqrt{2\pi Q}} \exp\left[-\frac{E^2}{2Q}\right] \qquad \text{for} \quad E>1$$

with

$$\sigma = \sqrt{Q}\left[1 - \frac{1}{\alpha\Phi(1/\sqrt{Q})}\right].$$

In the limit $\alpha \rightarrow 2^+$ we have a finite width $\sigma = \sqrt{2/\pi}$ and $\alpha \Phi(1/\sqrt{Q})$ tends to 1. This shows that half of the patterns have fields $E^\nu < 1$ for $\alpha \rightarrow 2^+$, whereas in the limit $\alpha \rightarrow 2^-$ half of the patterns have fields $E^\nu = 1$ [13]. The sudden change of the field distribution leads to the surprising jump of the cost function $g(\alpha)$. This unexpected behaviour was also confirmed by preliminary computer simulations. For the fraction of patterns that cannot be stabilized at all we find $P(E<0) \rightarrow \Phi(-\sqrt{\pi/2}) = 0.105$ in the limit $\alpha \rightarrow 2^+$. For $\alpha \rightarrow \infty$ this fraction increases to 1/2 as expected. This quantity is important for learning in networks with hidden units, where single Perceptrons are needed that stabilize as many patterns as possible [14–16]. Because of the drastic change in the cost function at the point where a Perceptron cannot be found any more, we believe that the AdaTron is a very efficient method to distinguish between solvable and unsolvable problems. This task often occurs in the construction of feed–forward multi–layer networks [15–16]

Acknowledgement
We would like to thank W. Kinzel, M. Opper, and P. Kuhlmann for many stimulating discussions.

References
[1]   J.J. Hopfield, Proc. Natl. Acad. Sci. USA 79, 2554 (1982)
[2]   D.J. Amit, H. Gutfreund, and H. Sompolinsky, Phys. Rev. Lett. 55, 1530 (1985)
[3]   W. Kinzel and M. Opper, in Physics of Neural Networks, ed. by J.L. van Hemmen, E. Domany, and K. Schulten, (Springer, Berlin, Heidelberg, in press)
[4]   E. Gardner, J. Phys. A 21, 257 (1988)
[5]   W. Krauth and M. Mezard, J. Phys. A 20, L 745 (1987)
[6]   B. Widrow and M.E. Hoff, WESCON Convention, Report IV, 96 (1960)
[7]   S. Diederich and M. Opper, Phys. Rev. Lett. 58, 949 (1987)
[8]   L. Personnaz, I. Guyon, and G. Dreyfus, Phys. Rev. A 34, 4217 (1986)
[9]   M. Opper, Europhys. Lett. 8, 389 (1989)
[10]   M. Opper, Phys. Rev. A 38, 3824 (1988)
[11]   J.K. Anlauf and M. Biehl, to be published in Europhysics Letters 1990
[12]   E. Gardner and B. Derrida, J. Phys. A. 21, 271 (1988)
[13]   L.F. Abbott and T.B. Kepler, Preprint Brandeis University 1989
[14]   P. Rujan in INNS: International Joint Conference on Neural Networks 1989, (IEEE TAB Neural Network Committee, 1989)
[15]   M. Mezard and J.P. Nadal, J. Phys. A 22, 2191 (1989)
[16]   T. Grossman, Complex Systems 2, 555 (1989)

Parallel Processing in Neural Systems and Computers
R. Eckmiller, G. Hartmann and G. Hauske (Editors)
© Elsevier Science Publishers B.V. (North-Holland), 1990

Learning Symbol Processing with Recurrent Networks

Rainer Goebel

Department of Psychology

Gutenbergstr. 18

3550 Marburg, West-Germany

1 Introduction

Distributed connectionist models possess nice properties: learning from examples, generalization to novel inputs, content addressable memories, fault tolerance etc. On the other hand, they have difficulties with symbolic computations since they don't provide simple ways to handle complex data structures, to establish variable bindings, and to perform logical reasoning.

Many researchers try to build connectionist architectures with more "symbolic power". Some work in this direction has been done on local (structured) connectionist models (e.g. Shastri [1]). However good results are also obtained using distributed representations by adding complex control mechanisms (e.g. Touretzky and Hinton [2]). Unfortunately, both approaches suffer from the following problem: almost all connections in such networks has to be "hand-wired" which requires a large amount of human effort. These models are *implementations* of symbolic schemes (as production systems or semantic nets), thus losing many advantages of simple distributed connectionist models.

In the following, I suggest a different perspective: using rather simple distributed connectionist networks and making them *learn* symbolic computations. The learning algorithm itself should construct appropriate symbolic representations (see also Pollack [3]) and operations.

My central assumption is: distributed connectionist models can learn symbolic computations if they succeed in representing time "intelligently". Modeling high-level cognitive functions – like natural language understanding, planning and problem solving – requires appropriate representations of the time dimension. Therefore, connectionist networks are required which can process *sequences* of input elements, *storing* information as long as needed, and *combining* information of different time events to produce desired output values. *Recurrent networks* are able to do just this kind of processing.

2 Applying back-propagation to recurrent networks

2.1 Back-Propagation through time

Every recurrent network operating in (discrete) time events can be *"unfolded"* to a feedforward network with identical behavior to which the normal back-propagation algorithm can be applied. In two simulations of recurrent networks, Rumelhart et. al. [4] used *synchronous* update: the activation of unit i at time step t_n is computed using only activation values of time step t_{n-1} (see Figure 1). Therefore the units can be updated in an arbitrary order. At the end of a sequence, error signals are propagated back through the unfolded network. To make appropriate weight changes every unit must remember its activity at each point in time.

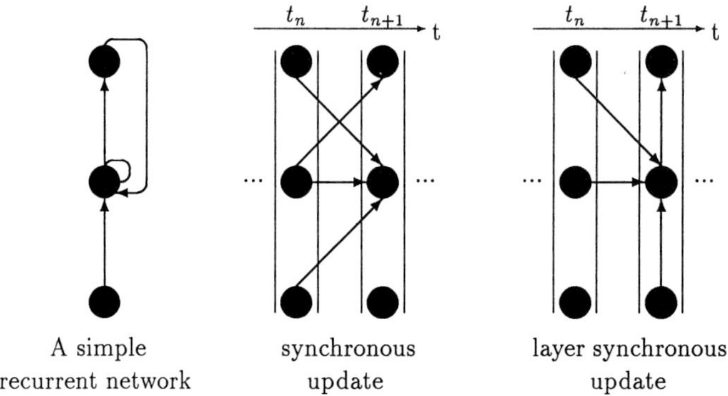

Figure 1: back-propagation through time

2.2   Sequential Recurrent Back-Propagation (SRBP)

The following modifications enhance the performance of the described "unfolding" procedure leading to the Sequential Recurrent Back-Propagation algorithm (SRBP).

First, instead of using synchronous update, "layer synchronous" update is used: at each time step, information flows from bottom to top as in feedforward nets. Therefore units must be updated in this order. In order to update unit i at time step t_n the activation value of unit j at the same time step is used if unit j is below unit i. (see Figure 1).

The second modification is to use both sigmoid and linear units within the network. Linear units improve the performance considerably because they generate better back-propagation of error signals, greater "dynamical range" and better "memory" properties.

The third modification is to incorporate a modified version of Fahlmann's [5] "Quick Prop" learning algorithm which yields in at least ten times less learning epochs than the normal weight change procedure.

The last modification is to introduce the number of back- propagation steps through time as a new parameter. With this modification SRBP can run continuously. This means it is not necessary to wait until the end of a training sequence is reached before starting back-propagation through time. This is required in cases where the sequence length is very long or not known a priori. (Williams and Zipser [6] have derived the Real Time Recurrent Learning algorithm (RTRL) which makes appropriate weight changes at each time step but requires a very large amount of storage and computational resources.)

The recurrent networks of Jordan [7] and Elman [8] can be seen as special cases of the SRBP algorithm using layer synchronous update and one back step through time. (Jordan and Elman actually copy activation values to lower layers)

3　Working memory: restricting generalization

From a psychological point of view, the activation of (some) units inside the network can be regarded as the "working memory". If these units are explicitly trained, one can introduce "hints" in the form of desired output values for intermediate stages of processing. This may lead to much better generalization behavior. Also, complex cognitive functions can be trained incrementally by way of simpler subtasks.

4　Simulation experiments

In the following sections, two examples show first steps in learning symbolic computations.

4.1　Evaluating simple Lisp expressions

The first example shows how a recurrent network develops fixed length representations of variable length lists. It also shows how symbolic operations like car, cdr, cadr, cddr and quote can be learned.

A network with 20 input units, 40 hidden units (fully interconnected) and 10 output units was constructed (see Figure 2). The input to the network was a Lisp-list – one element at each time step – followed by one of the following Lisp-primitives: car, cdr, cadr, cddr or quote. As an example, consider the expression "(cdr '(A B))". In the first processing cycle the opening paranthesis '(' is presented, then 'A', 'B' and the closing paranthesis. Until now, the network is not forced to produce any specific output value. In the next time step the Lisp function 'cdr' is presented and now the network must produce the desired output elements, first '(', then 'B' and finally the closing paranthesis. The network was trained with different numbers of training patterns (20, 50 or 100). It learns the whole training set and generalizes well to many - but not all - new lists.

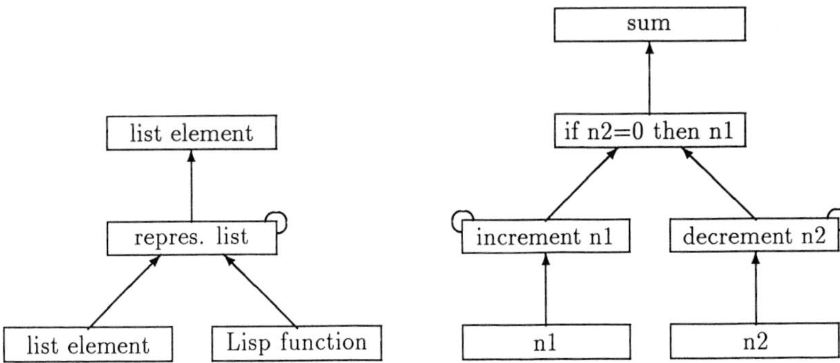

Figure 2: Evaluating Lisp expressions　　　Figure 3: Learning to add

4.2　Learning adding through counting

This example shows how a psychological theory about successive mental states can be incorporated in these networks. A recurrent network was trained to learn adding two numbers with values between 0 and 5. The network consisted of 10 input units (representing the two numbers), 15 units in a first hidden layer, 10 units in the second hidden layer and 10 output units representing the sum of the two numbers (see Figure 3). The first hidden layer was partitioned: 10 fully connected units to process the first number and five fully connected units to process the second number. These 15 units represent the "working memory". After presenting the two numbers they must be copied to the first hidden layer. At each successive time step, the first number is to be incremented by one and the second number decremented by one. The output units are to produce zero values until the second number is zero. Then the output units are to produce the incremented first number. If, for example, the numbers 4 and 3 are presented, the network is trained to have the following intermediate states in the working memory: 4,3; 5,2; 6,1; 7,0.

The designed network learns the training set consisting of eleven sequences in only 60 learning epochs and generalizes perfectly to all other (25) possible tasks.

5   Discussion

Despite the simple simulation experiments (compared to symbolic architectures), I think that the proposed modeling scheme points in a promising direction: simple recurrent networks can be trained to follow "sequential programs" like traditional symbol processing systems while preserving the advantages of distributed connectionist networks. In the future I explore whether more difficult problems are also learnable with such recurrent networks. I think that problem solving processes – like learning to program Lisp functions – are especially promising because problem solving requires both the power of distributed connectionist models and of symbolic processing.

References

[1]   Shastri, L. (1988) Semantic Nets: An Evidential Formalization and Its Connectionist Realization. Los Altos, CA: Morgan Kaufmann.

[2]   Touretzky, D.S. and Hinton, G.E. (1988) A distributed connectionist production system. Cognitive Science 12(3), pp. 423-466.

[3]   Pollack, J.B. (1988) Recursive Auto-Associative Memory: Devising Compositional Distributed Representations. Proceedings of the Tenth Annual Conference of the Cognitive Science Society, pp. 33-39

[4]   Rumelhart, D.E., Hinton, G.E. and Williams, R.J. Learning internal representations by error propagation. In Rumelhart, D.E. and McClelland, J.L., (eds.), Parallel Distributed Processing. Volume I, MIT Press, Cambridge, MA, 1986.

[5]   Fahlman, S.E. (1988) Faster Learning Variations on Back-Propagation: An Empirical Study. In Touretzky, D., Hinton, G. and Sejnowski, T., (eds), Proceedings of the 1988 Connectionist Models Summer School.

[6]   Williams, R.J. and Zipser, D. (1989) Experimental Analysis of the Real-time Recurrent Learning Algorithm. Connection Science 1(1), pp. 87-111.

[7]   Jordan, M. I. (1986) Attractor dynamics and parallelism in a connectionist sequential machine. Proceedings of the Eighth Annual Conference of the Cognitive Science Society, pp. 531-546.

[8]   Elman, J.L. (1988) Finding structure in time (CRL Tech. Rep. 8801). La Jolla, CA: University of California, San Diego, Center for Research in Language.

Parallel Processing in Neural Systems and Computers
R. Eckmiller, G. Hartmann and G. Hauske (Editors)
© Elsevier Science Publishers B.V. (North-Holland), 1990

TRAINING STRATEGIES FOR PROBABILISTIC RAMS

Denise GORSE

Department of Computer Science,
University College
London WC1E 6BT

John G.TAYLOR

Department of Mathematics,
King's College
London WC2R 2LS

The probabilistic random access memory (pRAM) is a very
flexible, hardware-realisable neural model which is able
to incorporate a significant degree of neurobiological
realism. A gradient descent training algorithm for such
units has been devised, which performs well in comparison
with conventional back-error propagation techniques. Forms
of reinforcement learning have also been considered, and
pRAM nets have been shown capable of learning associations
using a training scheme which is both computationally
inexpensive and biologically plausible.

1. INTRODUCTION

Networks of Boolean units (RAMs) were first studied by Kauffman
[1] in relation to genetic algorithms and were subsequently
extensively used by Aleksander in pattern recognition
applications [2]. It was noted that such deterministic units
could be regarded as having 'neuron-like' properties, but no
attempt was made to closely model biological neurons. Aleksander
[3] and Myers [4] extended the RAM model to that of a
'probabilistic logic node' (PLN), in which memory locations could
output 0,1 stochastically when addressed, typically utilizing
around 10 probability levels. The 'probabilistic random access
memory' of Gorse and Taylor[5] represents a similar, but neurobiologically motivated, approach in which memory contents take on
a continuous range of values in the interval [0,1]; the pRAM is a
device with intrinsically neuron-like behaviour which has already
been realised in hardware [6]. It is possible to demonstrate an
explicit equivalence [5] between the Taylor model of the noisy
neuron [7] and the equations describing a network of pRAMs, so
that the parameters $w(\underline{u},i)$, the probability of pRAM i emitting 1
on the addressing of location $\underline{u}$ may be interpreted in terms of
'neuronal' quantities (threshold functions, synaptic efficiencies
and so on). It is thus hoped ultimately to apply neurobiological
insights in the context of electronic networks.

2. GRADIENT DESCENT TRAINING SCHEME

Whilst it is natural to look at biologically realistic
(unsupervised) learning algorithms, it is also of interest to

see in what ways more conventional PDP training techniques could
be applied to pRAM networks. The training scheme outlined below
is conventional insofar as it involves a supervisor and the
reduction of an error function, but novel in its use of pRAM
spike train correlation analysis to implement gradient descent[8].

Consider an n-node pRAM net,each unit having L inputs. The aim of
the training proceedure is to produce a desired pattern of
activity $\{O_j\}$ on the n_o output nodes whenever the n_i input lines
deliver a certain pattern $\{I_k\}$. An error function is defined by

$$E = (1/2n_o) \sum_{j=1}^{n_o} (O_j - <o_j>)^2 \tag{1}$$

where $<o_j>$ is the actual firing rate of an output unit. The $nx2^L$
memory contents $w(\underline{u},i)$ (initially assigned random values in
[0,1]) are updated according to the rule

$$\triangle w(\underline{u},i) = -c.dE/dw(\underline{u},i) = (c/n_o) \sum_{j=1}^{n_o} (O_j - <o_j>).d<o_j>/dw(\underline{u},i) \tag{2}$$

If a training step consists of R presentations of the input
pattern $\{I_k\}$, it can be shown that

$$d<o_j>/dw(\underline{u},i) \doteq (R(\underline{u},i)/R^2) \sum_{r=1}^{R} X(a_i(t-r),o_j(t-r))$$

where X is a correlation function between the activities (0,1)
of a pair of nodes:

$$X(a_i(t),a_j(t)) = (\bar{a}_i\bar{a}_j + a_ia_j - a_i\bar{a}_j - \bar{a}_ia_j)(t)$$

($\bar{a}$ = 1-a) and $R(\underline{u},i)/R$ is the proportion of times location $\underline{u}$ in
the ith pRAM is addressed. If R and c are suitably chosen it can
be shown [8] that the above implements a form of gradient descent
in the space of memory contents $\{w(\underline{u},i)\}$.

The algorithm described above has been applied to various
problems, in particular that of determining the parity of an
input vector [8]. Training times for this problem compare well
with those obtained using conventional back-error propagation,
but were longer than those obtained by Aleksander and Myers for
the same topology. However the pRAM algorithm has a more rigorous
mathematical foundation than the Myers/Aleksander training
schemes, which are somewhat heuristic in nature.

3.   REINFORCEMENT LEARNING

It might be argued that the above training scheme, although it
does not require a detailed knowledge of the network connectivity
for its implementation, is prohibitively expensive in its need to
evaluate cross-correlations between a large number of output
spike trains. This is necessary in order to establish the role
played by each unit in determining the output error. An
alternative to fully supervised learning ('learning with a
teacher') is reinforcement learning ('learning with a critic'), in
which individual nodes only receive information about the quality
of performance of the network as a whole and have to discover for
themselves how to change their behaviour so as to improve this.

It is possible to adapt the ideas of Barto [9] to pRAM nets, adopting the update rule for addressed locations

$$\triangle w(\underline{u},i)(t) = (c1(r(a_i - w(\underline{u},i)) + c2.\bar{r}(\bar{a}_i - w(\underline{u},i)))(t) \quad (3)$$

where $r(t)$ is a global success/failure signal emitted with a probability dependent upon the net state at time t, and c1, c2 are constants in the interval [0,1]. The ratio c1/c2 represents the balance between reward and punishment; a non-zero value for c2 is necessary in order to prevent the system converging on false minima. According to (3), when r = 1 (success) the probability $w(\underline{u},i)$ changes so as to increase the chance of emitting the same value in the future, whilst if r = 0 (failure) the probability of emitting the other value when addressed increases.

The pRAM version of reinforcement learning differs from that of Barto in that noise is introduced at the synaptic rather than threshold level; it is well known that synaptic noise is the dominant source of stochastic behaviour in biological neurons.

4. 1-pRAM CHAINS

Properties of the gradient descent algorithm were investigated in [8] using a particularly simple network structure, consisting of N 1-pRAMs arranged such that the output of the ith pRAM is the input of the (i+1)st. The N-chain was trained to perform the identity function using batch mode training (no updates made until all members of the training set have been seen) with c = 2, terminating when the error defined by (1) fell below 0.001. Figure 1 shows A, the average number of training sets to convergence, plotted against N, the number of nodes, for (a) R = 32 and (b) R = 256. Increasing R enables a more reliable estimate of the error surface gradient, and hence decreased training times; this is particularly apparent for larger N. In order to compare the two algorithms, the reward/ penalty strategy was also applied to this problem (Figure 1(c)), with c1 = 1, c2 = 0.1, and an identical convergence criterion (using averages over 32 time steps). It can be seen that this computationally much less expensive stochastic search procedure outperforms the R = 32 gradient descent

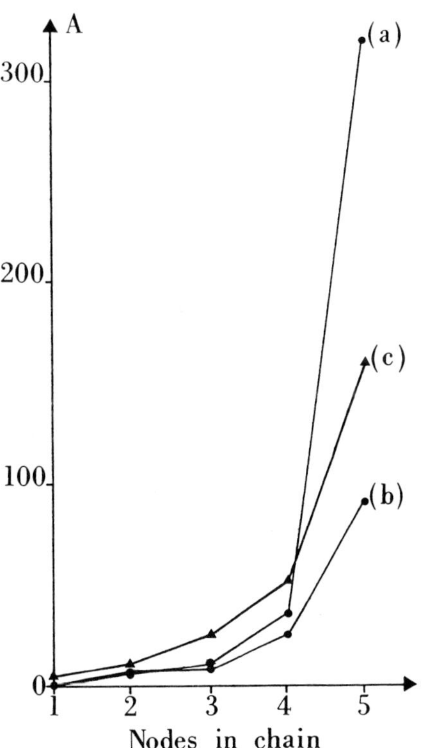

FIGURE 1
Gradient descent (a), (b) and reinforcement (c) algorithms applied to the 1-pRAM chain identity problem.

algorithm for $N > 4$, and compares well with the $R = 256$
implementation.

5. DISCUSSION

Two types of training scheme applicable to networks of
probabilistic Boolean nodes have been described. The first of
these utilizes an analysis of cross-correlated output spike trains
to implement a form of gradient descent algorithm; this method is
of very general utility, not requiring a detailed knowledge of
network connectivity, and, as shown in [8], able to deliver a
'best approximation' to a Boolean function where that function is
not exactly realisable by a given network topology. The second
type of training scheme is computationally much less expensive,
being a form of reward/penalty stochastic search in which each
node is required to discover its own role in influencing the
performance of the network as a whole. This latter training
strategy also has the advantage of a very straightforward
hardware implementation, and further work on the realisation of
such nets is in progress. A crucial problem with any proposed
training scheme is that of scaling; preliminary results have
indicated that this is particularly so for problems of high
Minsky-Papert order, such as parity. Present work aims to clarify
the situation with respect to scaling difficulties associated with
more practical problems, such as pattern recognition.

REFERENCES

[1] Kauffman, S. A., J. Theor. Biol. 22 (1969) 437.
[2] Aleksander, I., Thomas, I. V., and Bowden, P. A., Sensor
     Review (July 1984) 120.
[3] Aleksander. I., The Logic of Connectionist Systems, in:
     Aleksander, I. (ed), Neural Computing Architectures (MIT
     Press, 1989) pp. 133-155.
[4] Myers, C. E., Output Functions for Probabilistic Logic Nodes,
     in: Proceedings of the First IEE International Conference on
     Neural Networks, pp. 310-314.
[5] Gorse, D. and Taylor, J. G., Phys. Lett. A131 (1988) 326;
     Gorse, D. and Taylor, J. G., Physica D34 (1989) 90.
[6] Clarkson, T. G., Gorse, D. and Taylor, J. G., Hardware
     Realisable Models of Neural Processing, in: Proceedings of
     the First IEE International Conference on Neural Networks,
     pp. 242-246.
[7] Taylor, J. G., J. Theor. Biol. 36 (1972) 513.
[8] Gorse, D. and Taylor, J. G., A Gradient Descent Training
     Algorithm for Probabilistic Random Access Memories, King's
     College preprint, October 1989.
[9] Barto, A. G. and Jordan, M. I., Gradient Following Without
     Back-propagation in Layered Networks, in: IEEE First
     International Conference on Neural Networks 2 (1987),
     pp. 629-636.

Parallel Processing in Neural Systems and Computers
R. Eckmiller, G. Hartmann and G. Hauske (Editors)
© Elsevier Science Publishers B.V. (North-Holland), 1990

LEARNING IN NEURAL NETS BY GENETIC ALGORITHMS

Jochen HEISTERMANN

Siemens AG
Research and Development Department
Munich, Germany

For the development of parallel computer architectures it is important to get the requirements from different applications. As an example a new parallel learning algorithm for Artificial Neural Nets (ANNs) is presented: learning by evolution. The paper focusses on clarifying the basic concepts of genetic algorithms and its differences to conventional well-known learning paradigms. The inherent relationship between ANNs and the genetic theories is shown and the basic algorithm is explained.

1.INTRODUCTION

Artificial neural nets (ANNs) are networks composed of elementary units and weighted connections, which resemble simplified models of the nervous system [9]. With ANNs it seems to be possible to simulate different features of humans like learning, memory, detection of analogies or handling of similarities. Because these capabilities are difficult to achieve with conventional computer algorithms ANNs are an attractive research area opening a variety of new possibilities.

The main feature of ANNs is their capability of learning. The standard learning algorithms used in ANNs are derived from mathematical optimization theory (e.g. gradient-descent-methods [7] or simulated annealing [6]). This is due to the fact that optimization and learning are strongly related. Optimization means to find a combination of variables that maximizes a quality function. In analogy to optimization learning is the task to find an optimal configuration of the connection weights.

People involved in neural networks argument that their approach is nature oriented and animals with nervous systems do pretty well. Conventional learning algorithms usually don't have biological equivalents. It seems straightforward to use a biological approach for learning as well. In this paper such a method is discussed: learning with genetic algorithms. In the USA the work of Holland causes genetic algorithms to gain in importance as a way to solve optimization problems. Holland worked out a theoretical framework presented in [5].

All living forms are controlled by genes as basic building blocks. The genes of the fittest have more chances to be duplicated. In a highly competitive environment it is not only necessary to create well suited individuals but also fast mechanisms to adapt them to a changing environment. These individuals need so called genetic operators responsible for mixing up their genes. It is important to state that evolution of evolution mechanisms (on the meta-level) is also necessary. This chapter is the attempt to apply evolution strategies to learning in neural networks.

2. THE GENETIC APPROACH TO LEARNING

2.1 APPLYING GENETIC CONCEPTS TO ANNS

From an abstract point of view a living form can be looked at as a *genotype* or *phenotype*. The phenotype is the living form itself, while the genotype is a coding of the potential capabilities and the whole structure of the phenotype. An *instance* is defined as a list of genes with a given sequence. For our purpose the complicated coding mechanisms which decode the genotype are of no further interest, only the concept of differentiating a genotype and phenotype representation of a living form is important for the genetic approach.

An ANN is defined by the topology of the network, the activation function of the neurons and of the values of the weights [10]. The phenotype of an ANN is the ANN itself. The genotype is a coding of different features of the ANN. Which information one wants to code is dependent on the user of the genetic algorithm. Under the assumption that topology and activation function of the ANN are fixed during the learning phase, the simplest case is to code only the weights of the connections into genes, where each gene's value has a domain like $\{0,1\}$ or $\mathbb{R}$. The genetic information "survives" when individuals create offspring and the children inherit their genes from their parents. The structure of an ANN consists of a set of genes

$$(1)\ (\gamma_1,\gamma_2,...,\gamma_n),$$

where each γ_i represents one gene. An *instance* of an ANN then consists of a string of values for the genes $(\gamma_1,\gamma_2,...,\gamma_n)$. It gets these values by the function

$$(2)\ \forall\ \gamma_k \in \Xi_i\colon \lambda_i\colon \gamma_k --> a_{k_j},$$

where the task of λ_i is to assign an unique value to each gene of Ξ_i. The value of γ_k is chosen from the set $(a_{k_1},a_{k_2},...,a_{k_l})$, which is called the *set of alleles* of γ_k. While a gene can be viewed as *corresponding to the connections* of an ANN, an *allele corresponds to the weight* of a connection. Hence an instance of an ANN is represented by the string

$$(3)\ \Xi_i = (\lambda_i(\gamma_1),\lambda_i(\gamma_2),...,\lambda_i(\gamma_n)).$$

The expression of (3) is a string of alleles. The genetic algorithm is responsible for creating, deleting and ranking such strings of alleles. The structure of (3) is viewed as a genotype of an ANN, where the phenotype (further called *individual*) is simply constructed by filling in the slots of the connections with the alleles. To evaluate the goodness of such a string, a quality function is needed. This function (called μ) assigns a value $\mu(\Xi_i)$ to each individual.

The task of a genetic algorithm is to create increasingly better individuals until the quality of one individual is sufficient. The algorithm works by using so called genetic operators, which work on the structures. The basic genetic approach is the issue of chapter 2.2.

2.2 A CLOSER LOOK TO THE BASIC GENETIC OPERATORS

A detailed background of genetic operators can be acquired by reading [1], [2], [3], [4], [5], and [8]. The basic operators are recombination, selection and mutation.

A population ρ is a collection of individuals. Evolution changes the members as well as the number of members in ρ, where ρ can be expressed as

$$(4)\ \rho = (\Xi_1,\Xi_2,...,\Xi_n)$$

When the genetic algorithm starts it first creates an initial *population* of n_{init} indivi

duals with randomly chosen alleles. The task of *recombination* is the creation of a
new individual out of two individuals Ξ_i and Ξ_k of the current population. The
recombination operator $\Delta(\Xi_i,\Xi_k)$ creates a new individual Ξ_{ik} by choosing for each
gene position of Ξ_{ik} one allele either from Ξ_i or Ξ_k. In nature recombination usually
works by *crossing over*. The idea is that both gene strings of Ξ_i and Ξ_k are broken at
the same gene position γ_l. The ends of them are switched and put together again. So
$\Delta((\lambda_i(\gamma_1),\lambda_i(\gamma_2),...,\lambda_i(\gamma_n)), (\lambda_k(\gamma_1),\lambda_k(\gamma_2),..., \lambda_k(\gamma_n)))$ leads to the new individual
$\Xi_{ik} = (\lambda_i(\gamma_1),\lambda_i(\gamma_2), ... ,\lambda_i(\gamma_l),\lambda_k(\gamma_{l+1}), ... ,\lambda_k(\gamma_n))$.

By recombination new points in the search space emerge. Looking for promising
parts of the search space the bad points have to be selected out. In nature a population of some species grows until the resources of the environment force the individuals of the species to compete for food. Those who succeed live longer and produce more
offspring, therefore their genes have a higher probability to survive.

In analogy *selection* is used in genetic algorithms. Without selection, the population
would steadily increase in number. In [2] Goldberg simulates selection by calcula-
ting the ratio $\mu_i/\mu(\rho)$ where μ_i is the fitness of individual i and $\mu(\rho)$ the average fitness
of the population. Corresponding to this quotient the number of offspring of this
individual is calculated. Usually the next generation replaces the last one. This leads
to the increasing probability of inheritance of good genes, while the average fitness
rises also. This method is similar to nature conditions. Here a simpler selection
algorithm is proposed, which for tested examples leads to better results. From the
population ρ_i with size n_i two parents mate and produce a child. This process is
repeated m times yielding a population with n_i+m individuals. Then the selection
operator Ψ selects n_{i+1} points out of the population (usually the best) to generate the
next population. The better individuals have no better chance to generate offspring
than worse ones, but their children are better on the average than children of other
individuals. Since an individual is only replaced by better ones, a good individual
can't be lost by replacement. The sharp selection improves the population quickly.

Since newly created individuals have no new inheritance information, wide spread
genes have a better chance to duplicate and the number of alleles is constantly
decreasing [5]. This process results in the contraction of the population to one point,
where all individuals in the population are identical. This convergence effect is
caused by imbreeding and selection. This contraction phase is useful at the end of the
convergence process, when the population works in a very promising part of the
searchspace. In the beginning of the learning phase fast contraction of the
individuals is not useful. Diversity is necessary to examine a big part of the
searchspace. *Mutation* can be viewed as a copy error during recombination. The
mutation operator $\zeta(\Xi)$ selects a gene of Ξ and changes it by an amount ε, the
mutation variance. This happens with a mutation frequency τ. The parameters ε and
τ have a major influence on the quality of the learning algorithm.

If the mutation rate is low, the convergence process rapidly reduces the search space
spanned by the population, because the number of different alleles per gene is
steadily decreasing. So small values of ε and τ lead to a fine adjustment of the genes
to reach the optimum in the last phase of convergence. It may happen that the
population converges to a local optimum with a bad quality value. This can be stated
automatically, because in the convergence phase the number of different alleles per
gene on the average rapidly decreases while $|\mu(\Xi_{best})-\mu(\Xi_{worst})|$ also decreases. If
$\mu(\Xi_{best})$ is far from being good enough the search space of the population has to be
enlarged temporarily and therefore the possibility of finding better regions by
recombination becomes more probable again. In this context, high ε and τ values
increase the chance to create new alleles yielding better individuals. The best results
can be *achieved by using dynamically changing values for ε and τ at runtime* to
control the convergence and divergence process of the population.

We have seen the surprising effect that mutation can be used for two total different purposes. It can prevent the population from reaching local optima with a bad μ value and it can support the population in finding a local optimum with a good μ value!

3. RESULTS

Several experiments have been made with more advanced versions of the basic genetic algorithm described in this contribution. With respect to the limited space of the conference paper the reader is referred to [3], where some of the results are reported. Different standard problems for ANNs like the shifter-problem or some versions of the encoder-problem [9] were solved easily. The most interesting fact is that in [3] it was shown how to solve equation systems with ANNs. Simulations were run on a parallel computer (multiprocessor system), where experiments have been made with several populations on different processors.

The author is on the way to develop detailed theories underlying the genetic concepts. These theories should contribute to the understanding of the genetic operators.

4. SUMMARY AND CONCLUSION

In this paper it is shown how genetic algorithms can be applied to learning in ANNs. The reader gets an insight in how the ANNs have to be modelled and what the task of the genetic operators is. The purpose of the paper is to publish only the basic concept of learning with genetic algorithms in ANNs. The approach is of great practical interest, because in nature it works extremely well, whereas the standard learning techniques would hardly survive in a natural competitive environment.

ACKNOWLEDGEMENTS:
I am grateful to Sabine Canditt for carefully reviewing this report. The investigations presented in this paper are part of the project EspritII EP 2025 sponsored by the european community.

REFERENCES:
[1]   Eigen,M. and Winkler,R., Das Spiel (Piper, München, 1975).
[2]   Goldberg,D.E., Genetic algorithms in search, optimization and machine lear-
        ning (Addison-Wesley, 1989).
[3]   Heistermann,J, Parallel Algorithms for Learning in Neural Networks with
        Evolution Strategy, Parallel Computing (1989).
[4]   Hofbauer,J. and Sigmund,K., Evolutionstheorie und dynamische Systeme
        (Paul Parey, Berlin und Hamburg, 1984).
[5]   Holland,J.H., Adaption in Natural and Artificial Systems (Ann Arbor, The
        University of Michigan Press, 1975).
[6]   Kirkpatrick,S. and Gelatt,C.D. and Vecchi, M.P., Optimization by Simulated
        Annealing, Science Vol.220, S.671-680, (1983).
[7]   Neumann,K., Operations Research Verfahren Band 1-3 (Carl Hanser Verlag,
        München Wien, 1975).
[8]   Rechenberg,I., Evolutionsstrategie: Optimierung technischer Systeme nach
        Prinzipien der biologischen Evolution (Friedrich Frommann, Stuttgart-Bad
        Cannstatt, 1973).
[9]   Rumelhart,D.E. and McCLelland,J.L., Parallel Distributed Processing Vol.1
        (MIT Press, Cambridge Massachusetts, 1986).

Parallel Processing in Neural Systems and Computers
R. Eckmiller, G. Hartmann and G. Hauske (Editors)
© Elsevier Science Publishers B.V. (North-Holland), 1990

A NEURAL NETWORK WHICH ADAPTS ITS STRUCTURE TO A GIVEN SET OF PATTERNS

Stefan JOCKUSCH

Drittes Physikalisches Institut
University of Göttingen, Bürgerstr. 42-44
D-3400 Göttingen, FRG

A competitive neural network able to adapt its complexity as well as its structure to a given "environment" consisting of a set of patterns is introduced. If unknown patterns are presented, the network creates new neurons, preventing any loss of information about previously learned patterns. By adding a simple two-dimensional nearest-neighbor topology the network acquires the remarkable ability of coding the distribution of the patterns in its *structure*. In an illustrative example the network with problem-dependent structure is found to learn faster and more efficiently than a network with predefined structure.

1. INTRODUCTION

Competitive neural networks consist of a large number of neurons connected to receptors – or input lines – by links whose strength is given by real numbers called the synaptic weights. When a pattern is presented, these neurons become active according to how well their synaptic weights and the pattern match. Additionally the neurons "compete" with each other. Neurons with high activity suppress the less activated ones so that finally only the neuron is active whose synaptic weights match the pattern best. It is called the *winner*. The pattern is then said to belong to this neuron's *receptive field*. If the weights corresponding to this neurons are viewed as the output of the network, it can function as an associative memory. Learning in competitive neural networks is achieved by finding the winner neuron for a given pattern $\mathbf{x}$ and "turning" it towards $\mathbf{x}$ by a fraction ϵ of their actual Euclidean distance.

$$\mathbf{w}_{win}^{new} = \mathbf{w}_{win}^{old} + \epsilon(\mathbf{x} - \mathbf{w}_{win}^{old}). \tag{1}$$

Here $\mathbf{w}_{win}^{old}$ and $\mathbf{w}_{win}^{new}$ are the vectors of the synaptic weights before and after adjustment, respectively. If the patterns are normalized in the Euclidean sense, the winner can be determined from the inner product of the weight vectors and the pattern (see e.g. [5]):

$$\mathbf{w}_{win}^{T}\mathbf{x} = \max_{j} \mathbf{w}_{j}^{T}\mathbf{x},$$

where j runs over all weight vectors.

2. THE MORPHOGENETIC ALGORITHM

This algorithm is designed for a fixed number of neurons. As it may be time-consuming to determine the appropriate number of neurons for a given problem, I added a rule causing the insertion of new neurons into the network if unknown patterns are presented. Such a procedure was introduced by Carpenter and Grossberg [1] in their ART network. Apart from being of great practical use, similar mechanisms have been observed even in adult animals [2], so it is not inept to incorporate them in artificial neural networks.

Let us define the best match q by $q := \mathbf{w}_{win}^{T}\mathbf{x}$. If for a given pattern $\mathbf{x}$ the best match q is less than a critical match q_{crit} a new neuron is inserted whose weight vector is equal to

x. Thus, for a given set of patterns the final number of neurons depends on this threshold match. By choosing a high value of q_{crit} (close to one) the network prefers "quick" learning by inserting new neurons to "slow" learning by adaptation. From q_{crit} we can derive the minimum resolution of the pattern set in the network's final state, i.e. when presenting a pattern from the set does not lead to neuron insertion any more. If the pattern x lies on the boundary of the receptive field belonging to the weight vector w, $\mathbf{w}^T\mathbf{x} = q_{crit}$ holds. Considering normalization, it is easy to prove that the maximum error $\hat{e} = \|\mathbf{x} - \mathbf{w}\|$ is equal to $\sqrt{2(1 - q_{crit})}$.

The high learning speed achieved by setting q_{crit} close to one can lead to an overshoot in neuron insertion because the adaptation process of the weights does not have enough time to "rearrange" the weight vectors optimally. Thus the network's resolution of the pattern set is higher than that given by q_{crit}. To overcome this effect, we use a simple dynamic control of the threshold match.

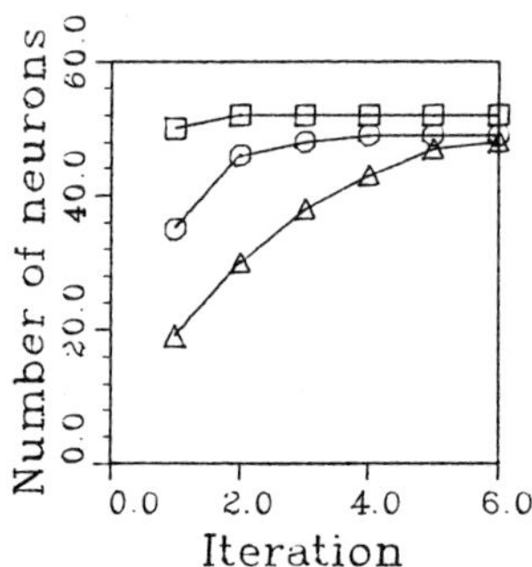

If at time t^{new} a new neuron has been inserted, q_{crit} is set to zero. Later it increases exponentially towards a maximum value $\hat{q}_{crit}$:

$$q_{crit}(t) = \hat{q}_{crit}\left[1 - \exp\left(\frac{t - t^{new}}{\tau}\right)\right]. \tag{2}$$

Thus, directly after a new neuron has been inserted, learning is restricted to adaptation for a time proportional to τ. Figure 1 shows that a larger value of τ reduces the number of neurons in the network's final state.

FIGURE 1
Number of cells as a function of time
using different values of τ=1 ($\square$), 2 ($\bigcirc$),
and 5 ($\triangle$) discrete-time units.

3. PROBLEM-DEPENDENT STRUCTURE

A very interesting extension of this algorithm consists in introducing a two-dimensional topology well known from Kohonen's "feature maps" [3]. If one arranges the neurons on a two-dimensional grid and adapts the weights of the winner neuron as well as those of its *neighbors*, after convergence neighboring cells on the grid will have neighboring receptive fields in the weight space. Consequently the neuron grid really is something like a "map" of the pattern set.

If we use the algorithm described in section 2, no *a priori* arrangement of neurons can be defined, as morphogenesis here is part of the learning process. It is an intriguing question whether we can take advantage of this fact by developing an algorithm that finds not only the optimum number of neurons but also an optimum arrangement of neurons for a given problem and resolution. To realize that some problems can be solved more efficiently by a network with suitable structure, consider the following distribution of two-dimensional patterns:

$$P(x,y) = \begin{cases} .5 & \text{if } xy \geq 0 \\ 0 & \text{otherwise} \end{cases} \tag{3}$$

It looks like a 2 × 2 chessboard consisting of two squares where patterns are equally distributed. Figure 2 shows the final state of a network with predefined rectangular structure and number of neurons trained to store this pattern set.

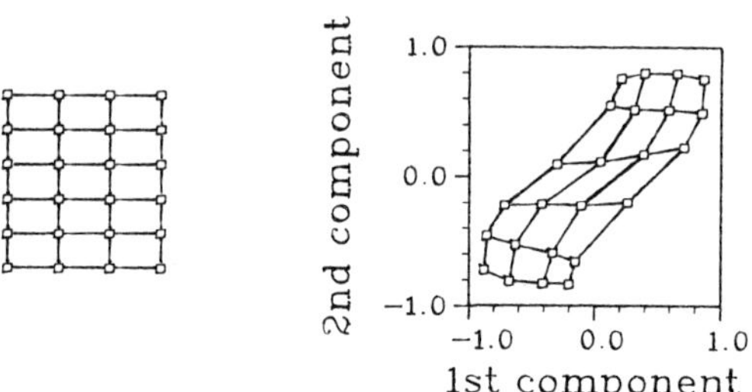

FIGURE 2

Arrangement of neurons (left) and position of weight vectors in pattern space (right) for a network with predefined structure.

Obviously this is not an optimum representation. For the morphogenetic algorithm described in section 2, the following rule was used to integrate new neurons in a two-dimensional grid: if the best match with the pattern is too small, a new neuron becomes the winner's neighbor.

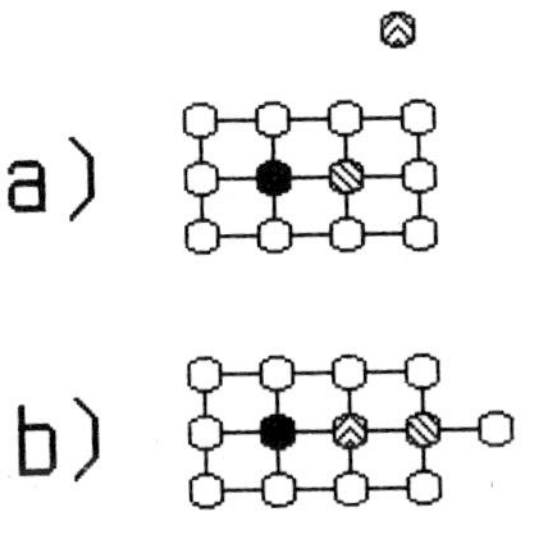

As neighboring neurons should be equidistant on the artificial "cortex", the old neighbor of the winner and its neighbors have to be shifted as shown in Figure 3. With this algorithm we get the solution shown in Figure 4; the network has a shape similar to the pattern distribution itself.

FIGURE 3

Algorithm for insertion of new neurons. One of neighbors of the winner neuron is chosen randomly (a). This neuron and its neighbors are shifted by one place and the new cell is inserted (b).

Apart from such simple problems the morphogenetic algorithm was successfully applied to a complex problem in speech research which it was originally developed for, namely, the determination of articulator movements from a continuous speech signal. As this problem requires *supervised* learning, i.e. presentation of an input and a corresponding output during training, an algorithm similar to Hecht-Nielsen's Counterpropagation algorithm [4] had to be used. For a more detailed description see [5].

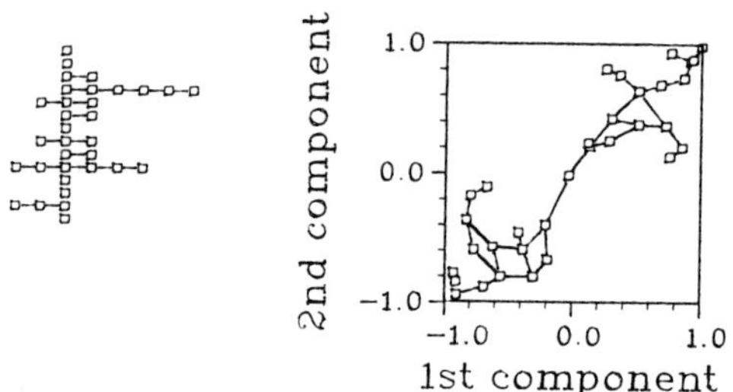

FIGURE 4

Solution found by the network with problem-dependent structure. Legend is the same as
in Figure 2.

4. CONCLUSION

The morphogenetic algorithm is capable of learning new patterns in real time without loss
of information about previously learned patterns by inserting new neurons. If the network's
growth is controlled by the parameter τ, it finds the optimum number of neurons needed·to
code a pattern distribution with a given resolution. It was shown that introducing a topology
of the neurons enables the network to acquire a structure similar to the pattern distribution.
Apart from yielding useful information about the patterns, this behavior increases learning
speed and efficiency.

Although the integration rule surely does not describe the mechanisms governing biological
network growth adequately, its performance should serve as a stimulus to intensify research
on morphogenetic learning algorithms.

ACKNOWLEDGEMENT

This work was part of project Str225/5 funded by the Deutsche Forschungsgemeinschaft
(DFG).

REFERENCES

[1] Carpenter, G. and Grossberg, S., A Massively Parallel Architecture for a Self-
Organizing Neural Pattern Recognition Machine. Computer Vision, Graphics, And
Image Processing, **37** (1987), pp. 54-115.

[2] Paton, J.A. and Nottebohm, F., Neurons Generated in the Adult Brain are Recruited
into Functional Circuits. Science **225** (1984), pp. 1046-1048.

[3] Kohonen, T., Self-Organization and Associative Memory (Springer, New York, 1987).

[4] Hecht-Nielsen, R., Counterpropagation Networks, in: Proc. IEEE First Int. Conf. Neural Networks, Vol. II (IEEE Press, New York, 1987), pp. 19-31.

[5] Jockusch, S., Panagos, G. and Strube, H.W., Anwendung neuronaler Netzwerke auf
die Schätzung von Sprachparametern, in: Fortschritte der Akustik – DAGA '89 (DPG-
GmbH, Bad Honnef, 1989), pp. 315-318.

Parallel Processing in Neural Systems and Computers
R. Eckmiller, G. Hartmann and G. Hauske (Editors)
© Elsevier Science Publishers B.V. (North-Holland), 1990

EFFICIENT LEARNING ALGORITHMS FOR SINGLE-LAYERED NEURAL NETWORKS

Nicolaos B. KARAYIANNIS and Anastasios N. VENETSANOPOULOS

Department of Electrical Engineering
University of Toronto
Toronto, Canada M5S 1A4

ABSTRACT: This paper proposes a new formulation of the training of single-layered neural networks, which provides the basis for the development of two learning algorithms with second order convergence properties. The efficiency of the proposed algorithms is experimentally verified.

1. INTRODUCTION

Neural networks is a rapidly expanding research field which recently attracted the attention of engineers and scientists as a realistic alternative towards the development of "thinking machines". This expectation is substantiated in principal by recent results in this research area. However, the transition from the dream to the reality requires an extensive amount of work on the development of efficient learning algorithms for the training of neural networks. A learning algorithm is judged on the basis of certain, rather conflicting, requirements, such as *simplicity, flexibility,* and *efficiency.* The simplicity of a learning algorithm relates to the effort required for the reproduction and application of the algorithm by an average programmer. The flexibility of a learning algorithm depends on whether the algorithm can be extended to the training of neural networks of different architectures and the effort associated with such an extension. The efficiency of a learning algorithm is evaluated with respect to the computational and time requirements for the training of a neural network and the performance of the resulting network. The delta rule is one of the contributions which have strongly influenced the neural networks research [1]. Although the delta rule is widely known from its application in adaptive filtering, its simplicity and flexibility made it a particularly attractive tool for the training of single-layered neural networks [2]. However, the learning algorithms based on the delta rule are characterized by slow convergence and, in some situations, can be trapped into local minima.

2. EFFICIENT LEARNING ALGORITHMS (ELEANNE 3 & 4)

Consider the single-layered neural network shown in Figure 1. Assuming that the key pattern $\mathbf{x}_k$ is the input of the network, the binary output of the network consists of the elements

$$y_k^o(i) = sgn(\bar{y}_k(i)) = sgn(\sum_{j=1}^{n_i} w_{ij} x_k(j)) \quad \forall \ i = 1, \ldots, n_o \tag{1}$$

where $sgn(\,.\,)$ is defined as $sgn(x) = -1$ if $x < 0$ and $sgn(x) = +1$ if $x \geq 0$. The training of neural networks is frequently based on the minimization of the objective function

$$E_k' = \frac{1}{2} \sum_{i=1}^{n_o} (y_k(i) - \hat{y}_k(i))^2 \quad \forall \ k = 1, \ldots, m \tag{2}$$

where $y_k(i)$; $i = 1, \ldots, n_o$ are the elements of the associated pattern $\mathbf{y}_k$ and $\hat{y}_k(i)$; $i = 1, \ldots, n_o$ the corresponding estimates provided by the network, given by

$$\hat{y}_k(i) = \rho(\bar{y}_k(i)) = \rho(\sum_{j=1}^{n_i} w_{ij} x_k(j)) \ ; \ i = 1, \ldots, n_o \tag{3}$$

where $\rho(\,.\,)$ is a continuous, differentiable everywhere function which approximates the signum function. In this paper, the two states of the network are chosen to be $\{-1, +1\}$. In this case, a good choice for $\rho(\,.\,)$ is the hyperbolic tangent, that is, $\rho(x) = tanh(x)$.

The estimate resulting from the minimization of (2) for all k also minimizes the function

$$E = \sum_{k=1}^{m} E_k' = \tfrac{1}{2} \sum_{k=1}^{m} \sum_{i=1}^{n_o} (y_k(i) - \hat{y}_k(i))^2 \tag{4}$$

However, it can easily be seen that $E = \sum_{i=1}^{n_o} E_i$, where

$$E_i = E_i(m) = \tfrac{1}{2} \sum_{k=1}^{m} (y_k(i) - \hat{y}_k(i))^2 \quad \forall \; i = 1, \ldots, n_o \tag{5}$$

Therefore, the minimization of (4) can alternatively be attempted by minimizing the objective function defined by (5) for any $i = 1, \ldots, n_o$. Since the estimate of each element $y_k(i)$; $k = 1, \ldots, m$ depends only on the ith row of the matrix of synaptic weights $\mathbf{W}$, each row $\mathbf{w}_i$; $i = 1, \ldots, n_o$ of $\mathbf{W}$ is estimated by minimizing the objective function (5). An efficient optimization strategy can be developed by observing that the objective function (5) is also a function of the number of associations considered. Clearly, $E_i(m-1)$ depends only on the associations $(\mathbf{y}_k, \mathbf{x}_k)$; $k = 1, \ldots, m-1$. Therefore, the network can be trained with respect to the associations $(\mathbf{y}_k, \mathbf{x}_k)$; $k = 1, \ldots, m-1$ by minimizing $E_i(m-1)$. On the other hand, the synaptic weights of the network can be updated with respect to the new association $(\mathbf{y}_m, \mathbf{x}_m)$ by minimizing $E_i(m)$. These observations suggest that the adaptation of each $\mathbf{w}_i$ be performed by sequentially considering the associations $(\mathbf{y}_k, \mathbf{x}_k)$; $k = 1, \ldots, m$. Assume that $\mathbf{w}_i$ has been updated with respect to the associations $(\mathbf{y}_k, \mathbf{x}_k)$; $k = 1, \ldots, m-1$ by minimizing $E_i(m-1)$, resulting in the estimate $\mathbf{w}_i = \mathbf{w}_i^{m-1}$. The ith row of $\mathbf{W}$ is then updated with respect to the association $(\mathbf{y}_m, \mathbf{x}_m)$ by minimizing $E_i(m)$ through the following update equation

$$\mathbf{w}_i^m = \mathbf{w}_i^{m-1} - \alpha\, \mathbf{H}_i(m)^{-1} \big|_{\mathbf{w}_i = \mathbf{w}_i^{m-1}} \; \partial E_i(m)\,/\,\partial \mathbf{w}_i \big|_{\mathbf{w}_i = \mathbf{w}_i^{m-1}} \tag{6}$$

where $\partial E_i(m)\,/\,\partial \mathbf{w}_i$ is the gradient of $E_i(m)$ with respect to $\mathbf{w}_i$, $\mathbf{H}_i(m)$ the Hessian matrix, and α is a positive real number, called here the *learning rate*.

According to the definition of $E_i(m)$ by (5), its gradient with respect to $\mathbf{w}_i$ is given by

$$\partial E_i(m)\,/\,\partial \mathbf{w}_i = - \sum_{k=1}^{m} \varepsilon_k(i)\, \mathbf{x}_k \tag{7}$$

where

$$\varepsilon_k(i) = (1 - \hat{y}_k(i))(1 + \hat{y}_k(i))(y_k(i) - \hat{y}_k(i)) \tag{8}$$

The particular form of the objective function $E_i(m)$ suggests a suitable approximation of the gradient $\partial E_i(m)\,/\,\partial \mathbf{w}_i$ which reduces further the arithmetic operations required for the adaptation of $\mathbf{w}_i$. Assume that $\mathbf{w}_i = \mathbf{w}_i^{m-1}$ minimizes $E_i(m-1)$, $\partial E_i(m-1)\,/\,\partial \mathbf{w}_i \big|_{\mathbf{w}_i = \mathbf{w}_i^{m-1}} = 0$. Then, the gradient of $E_i(m)$ with respect to $\mathbf{w}_i$ can be evaluated at $\mathbf{w}_i = \mathbf{w}_i^{m-1}$ as follows

$$\partial E_i(m)\,/\,\partial \mathbf{w}_i \big|_{\mathbf{w}_i = \mathbf{w}_i^{m-1}} = - \varepsilon_m(i)\, \mathbf{x}_m \tag{9}$$

If $\mathbf{w}_i = \mathbf{w}_i^{m-1}$ is not the exact minimum of $E_i(m-1)$, (9) is not valid. However, the current estimate $\mathbf{w}_i = \mathbf{w}_i^{m-1}$ is the result of a step towards the minimization of $E_i(m-1)$. Therefore, (9) provides a satisfactory approximation of the gradient even if $\mathbf{w}_i = \mathbf{w}_i^{m-1}$ is not the minimum of $E_i(m-1)$. Summarizing the above analysis, the update equation for each row of $\mathbf{W}$ is

$$\mathbf{w}_i^m = \mathbf{w}_i^{m-1} + \alpha\, \mathbf{P}_i(m)\, \mathbf{x}_m\, \varepsilon_i(m) \tag{10}$$

where $\mathbf{P}_i(m) = \mathbf{H}_i(m)^{-1}$ and $\varepsilon_i(m)$ are evaluated at $\mathbf{w}_i = \mathbf{w}_i^{m-1}$.

The elements of the Hessian matrix $\mathbf{H}_i(m)$ are given by $(\mathbf{H}_i(m))_{pq} = \partial^2 E_i(m)\,/\,\partial w_{ip}\,\partial w_{iq}$. In general, there is no guarantee that the Hessian matrix is positive definite and, therefore, non-singular. On the other hand, the definition of the Hessian matrix provides no indication that its inverse can be evaluated recursively. It can be shown that the Hessian matrix is of the form

$$\mathbf{H}_i(m) = \sum_{k=1}^{m} (1 + \lambda_k(i))\, c_k(i)\, \mathbf{x}_k\, \mathbf{x}_k^* \tag{11}$$

where $\lambda_k(i) = 2\,\hat{y}_k(i)(y_k(i) - \hat{y}_k(i))\,/\,(1 - \hat{y}_k(i)^2)$ and $c_k(i) = (1 - \hat{y}_k(i)^2)^2$. Assuming that $1 + \lambda_k(i) > 0 \; \forall \; k = 1, \ldots, m$, the Hessian matrix defined by (11) is positive semidefinite. However, there is no guarantee that $1 + \lambda_k(i) > 0$. It can be shown that $1 + \lambda_k(i) > 0$ only if the estimate $\hat{y}_k(i)$ provided by the network is close to the target. Since in many practical situations the algorithm is initialized randomly, there is no guarantee that the initial estimate will be close to the optimum. The above analysis indicates that the Hessian matrix (11) should be approximated by a matrix that is positive semidefinite regardless of the particular value $\hat{y}_k(i)$; $k = 1, \ldots, m$. It can be shown that the Hessian matrix can be approximated as follows

$$\mathbf{H}_i(m) \simeq \gamma \sum_{k=1}^{m} c_k(i)\, \mathbf{x}_k\, \mathbf{x}_k^* \qquad (12)$$

In (12), $\gamma = 1$ if the estimate $\hat{y}_k(i)$ is far away from the target $y_k(i)$ and $\gamma = 2$ if $\hat{y}_k(i)$ is close to the target $y_k(i)$. Since the constant γ appearing in (12) can be incorporated in the learning rate α involved in the update equation, the Hessian matrix is approximated by (12), with $\gamma = 1$.

The resulting approximation of the Hessian matrix satisfies the conditions which are necessary for the derivation of a recursive, stable algorithm. Clearly, (12) is a positive semidefinite matrix and its inverse can be evaluated recursively. Assuming that $\mathbf{P}_i(m-1) = \mathbf{H}_i(m-1)^{-1}$ is available, the evaluation of $\mathbf{P}_i(m) = \mathbf{H}_i(m)^{-1}$ can be achieved by using the matrix inversion lemma [3]. Clearly, from (12),

$$\mathbf{P}_i(m) = \mathbf{P}_i(m-1) - \delta_i(m)\, \mathbf{P}_i(m-1)\, \mathbf{x}_m\, \mathbf{x}_m^*\, \mathbf{P}_i(m-1) \qquad (13)$$

where $\delta_i(m) = c_m(i) / (1 + c_m(i)\, \mathbf{x}_m^*\, \mathbf{P}_i(m-1)\, \mathbf{x}_m)$. Each row of $\mathbf{W}$ is updated by (10), where $\varepsilon_m(i)$ is defined by (8) and $\hat{y}_m(i) = \rho(\mathbf{x}_m^*\, \mathbf{w}_i^{m-1})$. This algorithm is called ELEANNE 3.

A computationally less demanding algorithm is now derived on the basis of some simplifying, though reasonable, assumptions. The requirement for the recursive evaluation of the n_o matrices $\mathbf{P}_i(m) = \mathbf{H}_i(m)^{-1}$; $i = 1, \ldots, n_o$ is imposed by the existence of n_o different coefficients $c_k(i)$; $i = 1, \ldots, n_o$. Consider the simplifying assumption that

$$c_k(i) = \overline{c}_k(i_0) = (1 - \hat{y}_k(i_0))^2\, (1 + \hat{y}_k(i_0))^2 = (1 - \hat{y}_k(i_0)^2)^2 = \overline{c}_k \quad \forall \; i = 1, \ldots, n_o \qquad (14)$$

According to this assumption,

$$\overline{\mathbf{H}}_i(m) = \sum_{k=1}^{m} \overline{c}_k\, \mathbf{x}_k\, \mathbf{x}_k^* = \overline{\mathbf{H}}_m \quad \forall \; i = 1, \ldots, n_o \qquad (15)$$

The replacement of the n_o matrices given by (12) by the "average" matrix (15) implies that all the rows of $\mathbf{W}$ can be updated with respect to each of the m associations simultaneously. This is the major difference between the algorithm developed here and the ELEANNE 3. Each adaptation cycle of the simplified algorithm involves m sequential adaptations of all the synaptic weights of the network, each corresponding to one of the associations ($\mathbf{y}_k$, $\mathbf{x}_k$) ; $k = 1, \ldots, m$. The reliable estimation of $\overline{c}_k$ is the crucial problem resulting from the above assumption.

Assume that $\hat{y}_k(i_0)$ is a linear function of the number of adaptation cycles ν and also $\hat{y}_k(i_0) = 0$ when $\nu = 1$. In mathematical terms,

$$\hat{y}_k(i_0)^2 = \beta_k^2\, (\nu - 1)^2 \quad \forall \; k = 1, \ldots, m \qquad (16)$$

where $\beta_k > 0$ and $\beta_k\, (\nu - 1) < 1$. Since $|\hat{y}_k(i_0)| < 1$, the derivation of the estimate of $\overline{c}_k$ can be further simplified by using the well known approximation $exp(-x^2) \simeq 1 - x^2$ if $|x| < 1$. The combination of (14) and (16) indicates that $\overline{c}_k$ can be estimated during each adaptation cycle by

$$\overline{c}_k \simeq exp(-2\, \hat{y}_k(i_0)^2) = exp(-2\, \beta_k^2\, (\nu - 1)^2) = c_k(\nu) \qquad (17)$$

where ν is the number of the current adaptation cycle. The combination of (15) and (17) indicates that, during the νth adaptation cycle, the Hessian matrix can be approximated by

$$\overline{\mathbf{H}}_m(\nu) = \sum_{k=1}^{m} c_k(\nu)\, \mathbf{x}_k\, \mathbf{x}_k^* \qquad (18)$$

Therefore, $\mathbf{P}_m(\nu) = \overline{\mathbf{H}}_m(\nu)^{-1}$ can be evaluated recursively through the equation

$$\mathbf{P}_m(\nu) = \mathbf{P}_{m-1}(\nu) - \delta_m(\nu)\, \mathbf{P}_{m-1}(\nu)\, \mathbf{x}_m\, \mathbf{x}_m^*\, \mathbf{P}_{m-1}(\nu) \qquad (19)$$

where $\delta_m(\nu) = c_m(\nu) / (1 + c_m(\nu)\, \mathbf{x}_m^*\, \mathbf{P}_{m-1}(\nu)\, \mathbf{x}_m)$. During the adaptation cycle ν, the rows of the matrix of synaptic weights are updated through the following equations

$$\mathbf{w}_i^{m}(\nu) = \mathbf{w}_i^{m-1}(\nu) + \alpha\, \mathbf{P}_m(\nu)\, \mathbf{x}_m\, \varepsilon_m(i) \quad \forall \; i = 1, \ldots, m \qquad (20)$$

where $\varepsilon_m(i)$ is defined by (8) and $\hat{y}_m(i) = \rho(\mathbf{x}_m^*\, \mathbf{w}_i^{m-1}(\nu))$. The resulting algorithm is called ELEANNE 4. This algorithm can be further simplified by assuming that $\beta_k = \beta \; \forall \; k = 1, \ldots, m$ and, therefore, $c_k(\nu) = c(\nu) \; \forall \; k = 1, \ldots, m$.

3. EXPERIMENTAL RESULTS

This experiment compares the convergence of the algorithms proposed in this paper and the delta rule. A single-layered neural network with $n_i = 50$ input and $n_o = 10$ output units was trained with respect to $m = 20$ associations using the ELEANNE 3, the ELEANNE 4, and the delta rule. Figure 2 shows 20 $log_{10}(E)$ as a function of the number of adaptation cycles when the network is trained using the algorithms mentioned above. In each case, the learning rate was $\alpha = 0.1$. The parameters β_k used for the training of the network by employing the ELEANNE 4 were chosen to be $\beta_k = \beta = 0.05 \,\forall\, k$. The curves appearing in Figure 2 indicate that both algorithms proposed in this paper converge much faster than the delta rule. The superiority of these algorithms is also supported by the observation that the rate of convergence achieved by all algorithms, and the delta rule in particular, decreases considerably as the number of adaptation cycles increases. In this experiment, the major error reduction occurs during the first 100 adaptation cycles. According to Figure 2, after 100 adaptation cycles the proposed algorithms achieve an error given by 20 $log_{10}(E) \simeq -20$ compared to 20 $log_{10}(E) \simeq 10$ achieved by the delta rule. Figure 2 also indicates that the ELEANNE 3 and its simplified version, the ELEANNE 4, achieve very similar convergence. This is an experimental justification of the assumptions made for the derivation of the ELEANNE 4. On the other hand, the ELEANNE 4 requires less arithmetic operations per adaptation cycle, compared to the ELEANNE 3. Therefore, the ELEANNE 4 is a very useful tool in practical applications of neural networks.

4. CONCLUSIONS

This paper presented the derivation of two learning algorithms for single-layered neural networks. It was experimentally verified that the proposed algorithms converge much faster than the algorithm based on the delta rule. The algorithms proposed in this paper also provide the basis for the development of efficient learning algorithms for the training of multi-layered neural networks that converge faster than the well-known error back propagation algorithm [2].

ACKNOWLEDGEMENTS

The first author was supported by a University of Toronto Open Doctoral Fellowship.

REFERENCES

[1]   B. Widrow, R. G. Winter, R. A. Baxter, "Layered Neural Nets for Pattern Recognition", *IEEE Trans. on ASSP,* Vol. 36, No. 7, pp. 1109-1118, July 1988.

[2]   D. E. Rumelhart, J. L. McClelland, *Parallel Distributed Processing,* MIT Press, 1986.

[3]   B. Noble, J. W. Daniel, *Applied Linear Algebra,* Prentice-Hall, 1977.

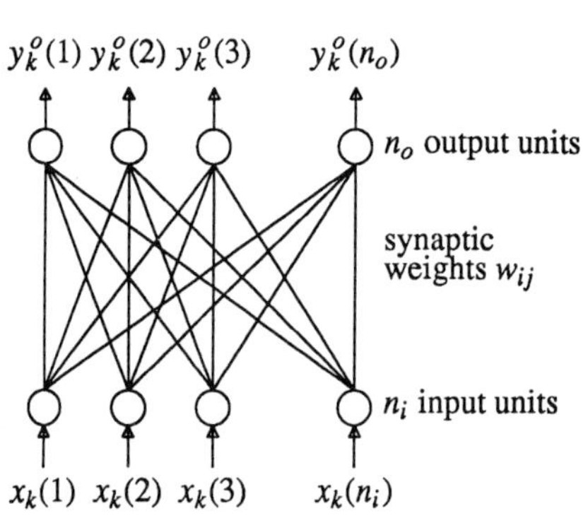

Figure 1. A single-layered neural network.

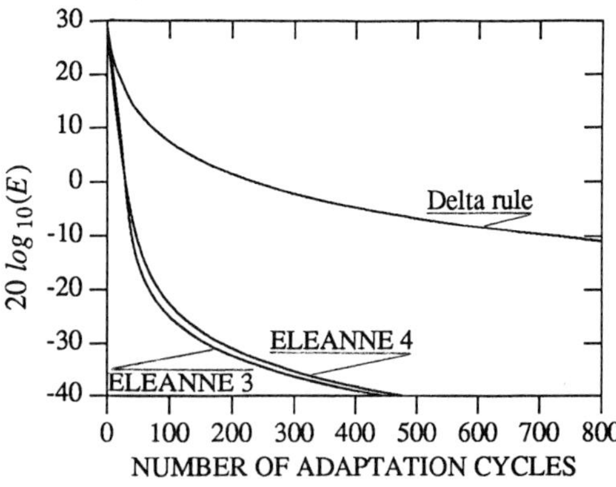

Figure 2. Training a single-layered neural network using the delta rule, the ELEANNE 3, and the ELEANNE 4.

Parallel Processing in Neural Systems and Computers
R. Eckmiller, G. Hartmann and G. Hauske (Editors)
© Elsevier Science Publishers B.V. (North-Holland), 1990

INTERNAL REPRESENTATIONS AND ASSOCIATIVE MEMORY

Teuvo Kohonen

Helsinki University of Technology
Laboratory of Computer and Information Science
Rakentajanaukio 2 C, SF-02150 Espoo, Finland

1. INTRODUCTION

The brain is no pipeline or direct stimulus-response switchboard. Any more complex behavior must be based on anticipation and planning, and to that end the brain must be able to create models of the environment and various happenings in it. One very important question therefore relates to the nature of *internal models* or *internal representations* of knowledge and awareness in the neural realms. Although it should be self-evident that thinking is principally based on *mental images* and *mental ideas*, the latter have hardly ever been considered in the theory of neural networks. It is by no means sufficient to say that the mental images are created by associative memory, because the brain has no "projection screen" or "memory register" that could receive and hold such images. The only possibility is that the mental images are formed *virtually*, in the same sense as we perceive a holographic image when we look through the hologram. It seems, however, that the principle of mental images, although so central to all thinking, has seldom even been recognized, not to talk of its application to the theory or implementations of "artificial neural networks." What people call Neural Networks are therefore only straightforward transforming or processing networks where signal patterns or activity states define the input, and other signal patterns or activity states the output, respectively; but hardly ever is it considered on what conditions *perception* actually follows from them.

As a matter of fact I started my career in "neural networks" in the late 1960s trying to publish the mathematical explanation of such virtual images in neural networks, but it was not until after many years (1971) when it was printed [1,2,3,4]. I also experienced that in order to become understood at all, I must lower the level of abstraction of my writings by a significant amount.

Although it is necessary to be aware of the above problem, even in this presentation I must now set it aside and concentrate on something more concrete that also might be called *internal representation*, namely, the internal encoding and decoding of signal patterns in the brain, which to a significant amount seems to be defined *spatially*.

Before suggesting any models or theories for the brain, one must also realize that there often prevail major discrepancies, especially between psychologists and physiologists, concerning spatial encoding of information in the brain. This seems to stem from confusion of the concepts of *activity* and *circuit*. It would be absurd to claim that there do not exist any localized circuits; certainly the anatomical structures must be anchored to fixed places, although it is also known that the anatomical structures are slowly changing during growth and development. Although the structures are then localized and carrying out specific information processing functions (analogous to the logic circuits in computers), it must also be obvious that in a massively interconnected network, the spatial distribution of activity is a dynamical entity which, being dependent on a great number of factors, is fluctuating and may not even always seem to follow the stimuli uniquely. By various imaging techniques, however, even activity states can often be seen to be localized, even to small "foci." Such activity concentrations that thus seem to correlate with various qualities of the sensory signals (although not necessarily being identical with a specific anatomical structure) may then be identifiable with an "internal representation." Let us call this somewhat abstract, although geometrically locatable entity a *"function"*, whereby we need not claim that it is a property of the tissue, as frequently falsely interpreted.

Many kinds of such functional internal representations can be found in the brain: various sensory and motor maps that may even be topographically ordered, localized responses to higher features and patterns such as the human face, foci where pain stimuli give localized responses, and perhaps most interesting of all, the "maps" in the word processing regions of the brain (the so-called Wernicke's area) where the spatial order of neural responses seems to be *categorically organized* (for reviews, cf. [5,6,7]).

2. THE SELF-ORGANIZING MAP AS A THEORETICAL PARADIGM FOR INTERNAL REPRESENTATIONS

Since the basic network structure and learning algorithm of the [Topology Preserving] Self-Organizing [Feature] Map has already been presented in numerous connections (cf., e.g., [4]), it appears superfluous to repeat its details here. The Map, as it is henceforth called for brevity, is a layered (usually two-dimensional) network or array of neural units, each one acting like a decoder to a common input signal pattern. The latter is a combination of many simultaneous real scalar signal values, so that input information is defined in a very-high-dimensional vector space. Conversely, each neural unit appears like an independent "pattern filter" for a particular signal combination, or actually for a domain of vectorial signal values in the signal space. What is essential to the Map is that these "filters" automatically tune and align themselves into the input signal space such that they sample the input space in an orderly fashion, or form some kind of "nonlinear projection" of it onto the neural network. As a matter of fact, if each "filter" were defined by a parametric vector that defines the best tuning of the filter, then it can be shown, at least in the case of very-high-dimensional input vectors, that the density of the parameter vectors tends to approximate and follow the probability density function of the input pattern vector.

It has been demonstrated by numerous computer simulations that the Map algorithm can create "internal representations" for the most different input information. The following Maps comprise a few examples: Maps for
- acoustic frequencies (tonotopic map) [8]
- phonemes of speech [9]
- colors (hue and saturation) [10]
- geographic environment (through optic and tactile signals) [11]
- taxonomically related attributes [11]
- semantic variables and relations [12]

The Maps have also been applied to practical tasks such as
- speech recognition [13,14]
- robotics [15,16,17]
- process analysis and control [18,19]
- telecommunication [20]
- imitation of handwriting [21]
  etc.

To recapitulate, it seems that the experimentally observed brain maps have now a theoretical counterpart in the artificial Map.

What is the significance of the spacial coding and topographical order of the functions, in the brain as well as in the artificial neural networks?

Concerning spatial segregation of the functions, it must first of all be realized that the biological components cannot be compared with the technical ones, in regard to their stability and linearity; as a matter of fact, the neurons trigger very nonlinearly, and their triggering threshold is variable, too. One cannot therefore rely on any proportionality or relationship between input and output signal values, whereas presence or absence of a response at a certain site would carry information more reliably. It would therefore be reasonable to assign different components to different functions. Also if different responses are spatially separated, they are not confused, which makes the overall operation more consistent.

Sometimes, especially in classification (pattern recognition), one should be able to define highly nonlinear discriminant functions. Such a function can then be implemented by many threshold units connected in parallel, each one being responsible for a different part of the dynamic range of signals; in other words, independent fitting of parts of such functional forms becomes easier if different ranges are controlled by spatially separated parameters.

It is self-evident, too, that if the different units that operate together are ordered in such a way that mutually relevant units are brought close to each other, the interconnectivity and communication costs will be minimized.

Topological closeness and connectivity of the different representations of information also solves or at least alleviates the so-called "property inheritance" problem of semantics and artificial intelligence; in other words, if a concept is defined as a set of its attributes, then only relevant attributes ought to be specified. If a concept corresponds to a geometric location in a network, then such a localized representation would naturally only be connected to other representations (attributes) that are geometrically close to it in the network, and this closeness is provided by the self-organizing map.

It can further be shown, although it cannot be justified in detail here (cf. [12]) that topological relations of representations can also lead to natural definition of abstractions and even to gradual evolution of symbolisms at increasingly higher levels of abstraction. A further result from this is that the logic contexts in which the representations appear then become defined by topological connectivities in the Maps.

3. ASSOCIATIVE MEMORY

Most attempts to model memory, especially associative memory have restricted to finding a transformation or process by which stored patterns can be recalled through *content-addressing* (on the basis of partial contents of the stored data), or by *proximity search* (whereby an erroneous key pattern is identified with the closest key stored in memory). However, prior to that one ought to understand into what form the sensory signals or other occurrences are first transformed in the neural networks, before they start leaving memory traces.

As pointed out above, it seems justified to assume that the sensory signals, and even higher concepts such as words, are handled in spatially segregated parts in the brain. Organization of the brain into areas according to sensory modalities and other major functions was already known in the nineteenth century, and more recent studies have indicated that there also exists a fine structure of the spatial organization in the local scale. It may therefore be assumed that the most predominant form of encoding of data in the brain is spatial, i.e., the activity concentrations and patterns formed of them directly carry information that is ready to be stored and recalled in memory.

The most common principles of distributed associative memory, or neural network memory may also be generally known and not need be repeated here in detail (cf., e.g., [4], [22]). Briefly, a distributed associative memory is also an array or layer of neural units, usually two-dimensional, which may have adaptive interconnections between any pair of units. These interconnections are "conditioned" such that correlation of activities on both sides of the interconnection reinforces the connectivity, and this is the principal effect from which associative memory stems. In an idealized memory the interconnectivity is complete, whereas it can be shown that sparse, randomly made connectivity may also be useful, as long as the number of connections converging on each unit is sufficient for certain statistical accuracy. It may be mentioned that recent anatomical studies have indicated that the principal cells of the mammalian cortex have 8000 input connections on the average, and with a probability exceeding 90 per cent these inputs stem from different cells, probably scattered over a large area; so the basic theoretical neural memory models presented so far may indeed be realistic.

Fig. 1 now depicts an ultimately simplified conception of the "unified" memory. It must be understood that the real brain is composed of many different regions and parts, each one with a slightly different nature and much more complicated internal structure than delineated here. It would, however, be impossible to gain any understanding of the fundamental principles unless the functions were simplified and "compartmentalized" in some way, of which Fig. 1 is an extremely crude example.

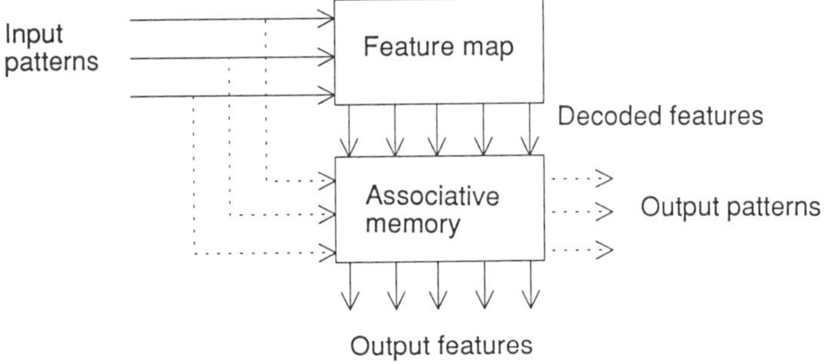

Fig. 1

4. DISCUSSION

First of all it should be self-evident that the brain is the most advanced creation of the Nature, and therefore its theory cannot be very simple. There are many complex structures in it that have evolved in the long course of phylogenesis, and being often implicit, their functions are not at all obvious to us. Naturally we may always say that we cannot understand anything unless we simplify and reduce the facts ultimately, and that we need not understand the whole brain if we are only interested in some new biologically motivated signal processing principles found in this research. The latter may indeed be the main motivation for the new technologies of Artificial Neural Networks, but one should then no longer refer to the brain, talking about "higher brain functions".

By the simple organization depicted in Fig. 1 it also seems possible to implement an associative memory that is invariant with respect to some transformation groups of patterns. Consider that the feature map, or a set of Maps can in some way be made to extract a number of different features from the primary signals, some of which are independent of translation, some of rotation, some of size, and so on. Further consider that the original input signals or images are associated and memorized as such together with all the invariant features, as delineated by the dashed arrows. Associative or content-addressable recall of the original stored patterns on the basis of their transformed or marred versions then becomes feasible, because part of the key features would remain invariant in transformations.

Finally we ought to consider whether the above simple associative memory that was devised for static signal patterns deserves to be named a genuine associative memory; in most subjective experiences, namely, associative memory is experienced as dynamical, being capable of storing sequences of patterns and other dynamical events. While there indeed exist some simple models suggested for dynamical memories, their capacities and noise tolerance, however, are not particularly good. The only principle that has any practical significance as memory is the static one referred to above. It too can be made to store or become selective to time sequences of signals in either of the following principal ways: 1. A subset of the output signals from the memory can be fed back with a constant, sufficiently long delay, whereby new input patterns become automatically associated with concomitant earlier outputs; given some pattern from this sequence as a key input, the rest of the sequence is then recalled by the feedbacks. 2. There can be plenty of signal delays of different length within the memory network itself, whereby signals presented at the inputs at different times may coincide at the inputs of some network units and become mutually associated or recalled like static signals.

The above elementary organization is only an illustrative example of the many specialized functions that different modules in a neural system, artificial or biological, may carry out. Most contemporary studies of the artificial neural networks still concentrate on single modules, and there indeed remains to be found out what the best system equations for them are. Before the neural networks can implement any more demanding tasks such as the control of autonomous, flexible robots, one also has to start implementing organizations consisting of many mutually interconnected modules, some of them operating as feature analyzers, some as memories, some as adaptive controllers of output actions, and some eventually performing still other processing functions, such as sampling of data, or control of attention.

REFERENCES

[1]   T. Kohonen, Introduction of the Principle of Virtual Images in Associative Memories, Acta Polytechn. Scandinavica, El. 29, 1971.
[2]   T. Kohonen, "A New Model for Randomly Organized Associative Memory," Int. J. Neurosc. **5**, 27-29, 1973.
[3]   T. Kohonen, "An Introduction to Neural Computing," Neural Networks **1**, 3-16, 1988.
[4]   T. Kohonen, *Self-Organization and Associative Memory*, 3rd ed. (Springer, Berlin, Heidelberg, Germany, 1989).
[5]   H. Goodglass, A. Wingfield, and M.R. Hyde, "Category Specific Dissociations in Naming and Recognition by Aphasic Patients," Cortex **22**, 87-102, 1986.

[6] A. Caramazza, "Some Aspects of Language Processing Revealed through the Analysis of Acquired Aphasia: The Lexical System," *Ann. Rev. Neurosci.* **11**, 395-421, 1988.

[7] A. Kertesz (ed.), *Localization in Neuropsychology* (Academic Press, New York, 1983).

[8] T. Kohonen, "Self-organized formation of topologically correct feature maps," *Biol. Cybern.* **43**, 59-69, 1982.

[9] T. Kohonen, K. Mäkisara, and T. Saramäki, "Phonotopic Maps - Insightful Representation of Phonological Features for Speech Recognition," *Proc. of the Seventh Int. Conf. on Pattern Recognition*, Montreal, Canada (IEEE Computer Society, Silver Spring, 1984) pp. 182-185.

[10] J. Saarinen and T. Kohonen, "Self-Organized Formation of Colour Maps in a Model Cortex," *Perception* **14**, 711-719, 1985.

[11] T. Kohonen, "Clustering, Taxonomy, and Topological Maps of Patterns," *Proc. of Sixth Int. Conf. on Pattern Recognition*, Munich, Germany, October 19-22, 1982, pp. 114-128.

[12] H. Ritter and T. Kohonen, "Self-Organizing Semantic Maps," *Biol. Cybern.* **61**, 241-254, 1989.

[13] T. Kohonen, K. Torkkola, M. Shozakai, J. Kangas, and O. Ventä, "Microprocessor Implementation of a Large Vocabulary Speech Recognizer and Phonetic Typewriter for Finnish and Japanese," *Proc. of European Conference on Speech Technology* (CEP Consultants Ltd, Edinburgh, 1987) pp. 377-380.

[14] T. Kohonen, "The 'Neural' Phonetic Typewriter," *Computer* **21**, 11-22, March 1988.

[15] H.J. Ritter and K. Schulten, "Extending Kohonen's Self-Organizing Mapping Algorithm to Learn Ballistic Movements," *NATO ASI Series*, vol. F41, 393-406, 1988.

[16] H.J. Ritter, T.M. Martinetz, and K.J. Schulten, "Topology Conserving Maps for Learning Visuo-Motor-Coordination," *Neural Networks* **2**, 159-168, 1989.

[17] D.H. Graf and W.R. LaLonde, "Neuroplanners for Hand/Eye Coordination," *Proc. IJCNN 89 Int. Joint Conf. on Neural Networks,* Washington, D.C., pp. II-543 - II-548, 1989.

[18] K.M. Marks and K.F. Goser, "Analysis of VLSI Process Data Based on Self-Organizing Feature Maps," *Proc. Neuro-Nîmes'88*, Nîmes, France, pp. 337-347, 1988.

[19] V. Tryba, K.M. Marks, U. Rückert, and K. Goser, "Selbstorganisierende Karten als lernende klassifizierende Speicher," *ITG Fachbericht* **102**, 407-419, 1988.

[20] D.S. Bradburn, "Reducing Transmission Error Effects Using a Self-Organizing Network," *Proc. IJCNN 89 Int. Joint Conf. on Neural Networks,* Washington, D.C., pp. II-531 - 537, 1989.

[21] P. Morasso, "Neural Models of Cursive Script Handwriting," *Proc. IJCNN 89 Int. Joint Conf. on Neural Networks,* Washington, D.C., pp. II-539 - II542, 1989.

[22] G.E. Hinton and J.A. Anderson (eds.), *Parallel Models of Associative Memory*, Updated Edition (Lawrence Erlbaum Ass., Hillsdale, New Jersey, 1989).

Parallel Processing in Neural Systems and Computers
R. Eckmiller, G. Hartmann and G. Hauske (Editors)
© Elsevier Science Publishers B.V. (North-Holland), 1990

HEBBIAN LEARNING OF PRINCIPAL COMPONENTS

Anders Krogh and John A. Hertz
The Niels Bohr Institute and Nordita
Blegdamsvej 17, DK-2100 Copenhagen, Denmark

ABSTRACT

Recently Oja proposed a new learning scheme called the subspace method for a network with M units to extract the principal components of the input. Here we prove Oja's conjecture: the weight vectors converge to an othonormal set spanning the subspace spanned by the M largest eigenvalues of the correlation matrix (corresponding to the first M principal components).

1. INTRODUCTION

Recently there has been a renewed interest in Hebbian learning in formal neural networks. In a classic paper Oja [1] proved that a neuron using a specific kind of Hebbian learning would find the principal component of the correlation matrix of the inputs. More recently, similar learning schemes have been proposed by Linsker [3] and Kammen and Yuille [4] for the generation of orientation–selective cells in the visual cortex.

It has also been shown [5,6] that a two–layer feed forward network with M *linear* hidden units used for autoassociation (trained on identical input and output) projects the inputs onto the subspace spanned by the M largest principal components.

Oja [2] has generalized his learning scheme to a network with M units that develops *orthogonal* weight vectors in the subspace spanned by the M largest principal components. Oja showed this with numerical results, and here we prove it mathematically.

The method of Sanger [7] is similar to Oja's, but he forces the first weight vector to be the first principal component, the second, the second principal component, etc. So it is more like applying Oja's old method [1] to the first output unit, then to the second unit in the subspace orthogonal to the first weight vector, and so on.

2. A NEURON THAT EXTRACTS THE LARGEST PRINCIPAL COMPONENT

In this section we show that the learning rule given by Oja [1] extracts the largest principal component of the inputs. The proof was given by Oja, and we only include it here for completeness and to set the scene for the more complicated proof of the next section.

The "network" consists of one unit with N input lines each with a weight factor w_j and it computes the weighted sum of inputs:

$$S = \sum_j w_j \xi_j = \mathbf{w} \cdot \boldsymbol{\xi}. \tag{1}$$

We assume the input vectors $\boldsymbol{\xi}$ are drawn from some distribution on R^N.

Oja found that by modifying the Hebb learning rule in a clever way, it is possible to make the weight vector approach a *normalized* eigenvector to the correlation matrix,

$$C_{ij} \equiv \langle \xi_i \xi_j \rangle \tag{2}$$

without having to do the normalization by hand. The modfication consists of adding a weight decay of size S^2 to the Hebbian term:

$$\dot{w}_j = \eta S(\xi_j - S w_j). \tag{3}$$

Note that (3) looks like reverse delta-rule learning; $\dot{w}$ depends on the difference between the actual *input* and the back-propagated output.

The weight changes averaged over the distribution of inputs are then

$$(1/\eta)\langle \dot{w}_i \rangle = \sum_j C_{ij} w_j - \left[\sum_{jk} w_j C_{jk} w_k \right] w_i. \tag{4}$$

At a fixed point the average change of weights is zero, so at the fixed point the left hand side is zero, and we see that $\mathbf{w}$ is an eigenvector of C with the eigenvalue

$$\lambda = \sum_{jk} w_j C_{jk} w_k. \tag{5}$$

The right hand side of (5) is equal to $\lambda \sum_j w_j^2$, so it follows that the Euclidean norm of $\mathbf{w}$ is 1.

All the eigenvectors are fixed points, and we now show that only the one(s) belonging to the largest one, $\lambda_{\max}$, is stable. Assume that $\mathbf{e}_\alpha$, $\alpha = 1 \ldots N$, is a set of orthogonal normalized eigenvectors with eigenvalues λ^α. If $\mathbf{v}$ is a small vector, we now want to find the change in $\mathbf{w}$ (from (3)) if it is initially $\mathbf{e}_\beta + \mathbf{v}$ (sum on repeated indices):

$$\begin{aligned}
\frac{1}{\eta}\dot{v}_j &= C_{jj'}(e_{\beta j'} + v_{j'}) - (e_{\beta j} + v_j)\left[(e_{\beta j'} + v_{j'})C_{j'j'''}(e_{\beta j'''} + v_{j'''})\right] \\
&= C_{jj'}v_{j'} - \lambda^\beta v_j - e_{\beta j}(2\lambda^\beta v_{j'} e_{\beta j'})
\end{aligned} \tag{6}$$

where only the first order terms in $\mathbf{v}$ are kept.

In the basis of the eigenvectors this becomes (no summation on α and β)

$$\begin{aligned}
(1/\eta)\dot{v}_\alpha &= e_{\alpha j}C_{jj'}e_{\gamma j'}v_\gamma - \lambda^\beta v_\alpha - \delta_{\alpha\beta}(2\lambda^\beta v_\gamma e_{\gamma j'}e_{\beta j'}) \\
&= [(\lambda^\alpha - \lambda^\beta) - 2\lambda^\beta \delta_{\alpha\beta}]v_\alpha.
\end{aligned} \tag{7}$$

Here we use the convention $v_\alpha \equiv \sum_j e_{\alpha j} v_j$.

For the fixed point to be stable the expression in the braces has to be negative for all α (contracting in all directions). This is the case only if λ^β is the maximal eigenvalue. If this is not the case $\lambda^\alpha - \lambda^\beta$ will be positive for some α's and $\mathbf{w}$ will "slide away" in those directions.

3. THE SUBSPACE METHOD

Recently Oja [2] proposed a way to find the principal component subspace with a network consisting of M units identical to the one described above. More specifically the network converges to weight vectors that form an othonormal basis for the subspace spanned by the M largest principal components. This claim was not proven, but supported by numerical calculations. Here we give the proof.

The units are labeled by $i = 1 \ldots M$ and the weight from input j to unit i by w_{ij}. The generalized learning rule is

$$\dot{w}_{ij} = \eta S_i(\xi_j - \sum_{i'} S_{i'} w_{i'j}). \tag{8}$$

To find the fixed points we can do as we did above when deriving (4). Assuming that $\dot{w}_{ij} = 0$ and averaging over the input patterns yields

$$0 = (1/\eta)\dot{w}_{ij} = C_{jj'}w_{ij'} - w_{i'j}(w_{ij'}C_{j'j''}w_{i'j''}) \tag{9}$$

where summation on all repeated indices is implicit. The subspace spanned by the weight vectors $\mathbf{w}_i$ is called V_0 and the orthogonal subspace $V_0^{\perp}$. We now show that V_0 is invariant under C by multiplying (9) by a vector $\mathbf{a} \in V_0^{\perp}$:

$$0 = a_j[C_{jj'}w_{ij'} - w_{i'j}(w_{ij'}C_{j'j''}w_{i'j''})] = a_jC_{jj'}w_{ij'} \tag{10}$$

so $C_{jj'}w_{ij'} \in V_0$ which implies that V_0 is spanned by eigenvectors of C.

Multiplying (9) by one of the weight vectors instead gives

$$0 = w_{i''j}[C_{jj'}w_{ij'} - w_{i'j}(w_{ij'}C_{j'j''}w_{i'j''})] = [\delta_{i'i''} - w_{i''j}w_{i'j}]w_{ij'}C_{j'j''}w_{i'j''}. \tag{11}$$

For the expression in the brackets to be zero the weight vectors have to be *orthonormal*.

This proves that the fixed points of the learning procedure gives orthonormal weights that project onto a subspace spanned by M eigenvectors of C.

We now prove that only the subspace spanned by the M largest principal components is stable. Assume that $\mathbf{w}_i$ is a fixed point and add a small perturbation $\mathbf{v}_i$, as we did in the previous section. Then to first order in the perturbation the weight changes are

$$\frac{1}{\eta}\langle \dot{v}_{ij}\rangle = C_{jj'}v_{ij'} - C_{j'j''}[v_{ij'}w_{i'j''}w_{i'j} + w_{ij'}v_{i'j''}w_{i'j} + w_{ij'}w_{i'j''}v_{i'j}]. \tag{12}$$

Parts of these vectors in input space ($\dot{\mathbf{v}}_i$) lie in V_0 and the other parts in $V_0^{\perp}$. To see if V_0 is stable we need to show that the parts in $V_0^{\perp}$ are decreased by the learning. The vectors are split up into two parts:

$$\vec{v}_i = \vec{v}_i^{\,0} + \vec{v}_i^{\,\perp}. \tag{13}$$

Because V_0 is closed under C it is easy to find the part of (12) in $V_0^{\perp}$,

$$\frac{1}{\eta}\langle \dot{v}_{ij}^{\perp}\rangle = C_{jj'}v_{ij'}^{\perp} - w_{ij'}C_{j'j''}w_{i'j''}v_{i'j}^{\perp} \tag{14}$$

where the sum is still over repeated indices. We transform this to the basis where C is diagonal and number the eigenvectors $\mathbf{e}_i$ such that $V_0 = span\{\mathbf{e}_1, \ldots, \mathbf{e}_M\}$ and the rest spans $V_0^{\perp}$. The last equation then becomes

$$\frac{1}{\eta}\langle \dot{v}_{i\alpha}^{\perp}\rangle = [\lambda^{\alpha}\delta_{ii'} - w_{ij'}C_{j'j''}w_{i'j''}]v_{i'\alpha}^{\perp}. \tag{15}$$

The last matrix in the braces $\tilde{C}_{ii'} \equiv w_{ij'}C_{j'j''}w_{i'j''}$ is just C in the basis of the weight vectors. Since the weight vectors are orthonormal $\tilde{\mathsf{C}}$ will have the eigenvalues $\lambda^1 \ldots \lambda^M$. The matrix in the braces $(\lambda^{\alpha}\delta_{ii'} - \tilde{C}_{ii'})$ will then have eigenvalues $\lambda^{\alpha} - \lambda^{\beta}$ with $\beta \leq M$. Note that $v_{i\alpha}^{\perp} = 0$ for all $\alpha \leq M$ and for $\alpha > M$ the eigenvalues will be negative if and only if

$$\lambda^{\alpha} < \lambda^{\beta} \text{ for } \alpha > M \text{ and } \beta \leq M. \tag{16}$$

When the eigenvalues are negative the perturbation will die out in $V_0^{\perp}$, so V_0 is stable.

Oja found that in V_0 rotations that keeps the weights orthonormal are allowed, so we do not expect all perturbations to die out unless there are some special directions of the weights that are preferred. To study what is going on in V_0 we look at the part of (12) that stays in V_0:

$$\frac{1}{\eta}\langle \dot{v}_{ij}^{0}\rangle = C_{jj'}v_{ij'}^{0} - C_{j'j''}[v_{ij'}^{0}w_{i'j''}w_{i'j} + w_{ij'}v_{i'j''}^{0}w_{i'j} + w_{ij'}w_{i'j''}v_{i'j}^{0}]. \tag{17}$$

In the basis of the weight vectors with $\tilde{v}_{ik} \equiv \sum_j w_{kj} v_{ij}^0$ this can be written as

$$
\begin{aligned}
\frac{1}{\eta} \langle \dot{\tilde{v}}_{ik} \rangle &= \tilde{C}_{kk'} \tilde{v}_{ik'} - \tilde{C}_{ii'} \tilde{v}_{i'k} - \tilde{C}_{kk'} \tilde{v}_{ik'} - \tilde{C}_{ik'} \tilde{v}_{kk'} \\
&= -(\tilde{C}_{ii'} \tilde{v}_{i'k} + \tilde{C}_{ik'} \tilde{v}_{kk'})
\end{aligned}
\tag{18}
$$

$\tilde{C}_{ik}$ and $\tilde{v}_{ik}$ are now considered as $M \times M$ matrices in the subspace V_0. We transform into the basis where $\tilde{C}$ is diagonal using an orthogonal transformation S_{ai}, and $\tilde{v}_{ab} \equiv \sum_{i,k} S_{ai} \tilde{v}_{ik} S_{bk}$:

$$
\frac{1}{\eta} \langle \dot{\tilde{v}}_{ab} \rangle = -\lambda^a (\tilde{v}_{ab} + \tilde{v}_{ba}).
\tag{19}
$$

Thus for every pair (a, b) there are two eigenmodes: the symmetric combination

$$
u_{ab}^s \equiv (\tilde{v}_{ab} + \tilde{v}_{ba})
\tag{20}
$$

which relaxes according to

$$
\dot{u}_{ab}^s = (\dot{v}_{ab} + \dot{v}_{ba}) = -(\lambda^a + \lambda^b) u_{ab}^s,
\tag{21}
$$

and the asymmetric combination

$$
u_{ab}^a \equiv \left(\frac{1}{\lambda^a} \tilde{v}_{ab} - \frac{1}{\lambda^b} \tilde{v}_{ba} \right)
\tag{22}
$$

which does *not* relax, because $\dot{u}_{ka}^a = 0$.

Combining (20) and (22) we see that an arbitrary infinitesimal pertubation will relax to an antisymmetric combination $\tilde{v}_{ab} = -\tilde{v}_{ba}$, that is, to a rotation in the a–b plane.

We have thus proved that for a network of M linear units and the learning rule (8) the M weight vectors converge to an orthonormal basis for the subspace spanned by the first M principal components of the input correlation matrix.

REFERENCES

[1] E. Oja, *J. Math. Biol.* **15** 267–273 (1982).

[2] E. Oja, *Int. Journ. of Neural Systems* **1** 61–68 (1989).

[3] R. Linsker, *Proc. Nat. Acad. Sci. USA* **83** 7508–7512, 8390–8394, 8779–8793 (1986).

[4] D. M. Kammen and A. L. Yuille, *Biol. Cybernetics* **59** 23-31 (1988).

[5] G. W. Cottrell, P. W. Munro, and D. Zipser, in *Advances in Cognitive Science* v.2, N. E. Sharkey, (ed.); Norwood, NJ: Abbex (1989).

[6] P. Baldi and K. Hornik, *Neural Networks* **2** 53–58 (1989).

[7] T. Sanger, MIT preprint (1989).

Parallel Processing in Neural Systems and Computers
R. Eckmiller, G. Hartmann and G. Hauske (Editors)
© Elsevier Science Publishers B.V. (North-Holland), 1990

THE ANTI–HEBB RULE DERIVED FROM INFORMATION THEORY

Hans KÜHNEL and Paul TAVAN

Physik-Department
Technische Universität München
8046 Garching, Federal Republic of Germany

Recently an "Anti–Hebb" learning rule has been proposed for lateral connections in two–layered nets of linear neurons [1]. As a result of unsupervised learning these nets perform a principal component analysis of the input information [2]. Here we show that the Anti–Hebb rule can be derived by information theoretical arguments.

1. INTRODUCTION

A crucial step in the processing of sensory information in the brain is the decomposition of the huge amount of primary sensory data into a smaller set of features which provide the input data for the following stages of information processing. If a most general decomposition of sensory information is required, the nature of the extracted features should be determined by the structure of the sensory data.

A well known method to transform correlated input data into a set of statistically independent features (or 'factors'), ordered according to decreasing information content, is principal component analysis (see e.g. [3]). In principal component analysis the data are projected onto the directions of maximum variance (in decreasing order). Oja [4] has shown that a single neuron with fixed overall synaptical strength obeying a Hebbian learning rule acts as a matched filter for the first principal component and, hence, extracts the maximum possible information from the input data. Various approaches have been attempted to extend this method as to obtain also the following principal components (see e.g. [5,6,7,8]).

Recently, Rubner et al. [1,2,9] have succeeded in designing a net of linear neurons which acts as a matched filter for all the principal components (in decreasing order). Applications to the formation of feature detectors in the visual cortex of mammals have been discussed. As a main ingredient to their neural net these authors introduced a sign–reversed "Anti"–Hebbian learning rule (see also the paper by Földiak [8] and Kohonens novelty filter [10]).

In this paper we want to show that the Anti–Hebb learning rule can be derived taking an information theoretical approach naturally arising if one tries to design or analyze neural nets (for related information theoretical approaches see e.g. [7,11,12,13]).

2. THE NETWORK MODEL

The network shown in Fig. 1 consists of two layers both with N neurons. All neurons have real, continuous-valued activities $\mathbf{i} = (i_1,..,i_j,..,i_N)$ and $\mathbf{o} = (o1,..,o_k,..o_N)$ for input and output layer, respectively, with $j,k = 1,..,N$. The two layers are completely interconnected. The weight of the connection between units j and k is denoted by w_{kj}. The set of weights connecting output unit k to all input units forms the weight vector $\mathbf{w}_k$ which is normalized to unit length $|\mathbf{w}_k|^2 = \sum_j w_{kj}^2 = 1$. The transpose of this vector is the k-th row of the weight matrix $\mathbf{W}$. The second layer has hierachically ordered lateral connections with unit k connected only to units $l < k$. The connection from l to k is named u_{kl}.

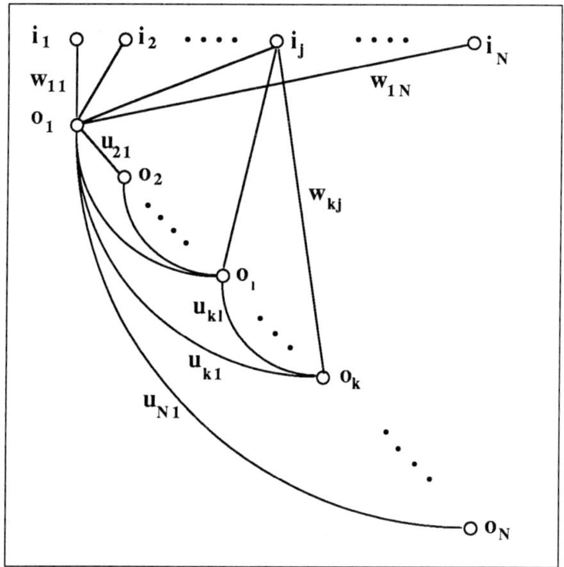

Fig.1 Scheme of the network.

For the network we assume linear neurons, i.e., the activities of units in the second layer result from the equation

$$o_k = \sum_j w_{kj} i_j + \sum_{l<k} u_{kl} o_l. \tag{1}$$

Presenting the pattern $\mathbf{p} = (p_1,...,p_N)^T$ to the network, the input units assume the values $\mathbf{i} = \mathbf{p}$ and the output units have activities $\mathbf{o} = \tilde{\mathbf{W}}\mathbf{i}$. Here, the transformation matrix $\tilde{\mathbf{W}}$ can be written as $\tilde{\mathbf{W}} = (\tilde{\mathbf{w}}_1,..,\tilde{\mathbf{w}}_k,..\tilde{\mathbf{w}}_N)^T$ with $\tilde{\mathbf{w}}_k = \mathbf{w}_k + \sum_{l<k} u_{kl}\tilde{\mathbf{w}}_l$. Note, that this implies $\det \tilde{\mathbf{W}} = \det \mathbf{W}$.

The weights between the two layers are adjusted according to a Hebbian rule $\Delta\mathbf{w}_k = \eta o_k \mathbf{i}$ with a positive learning parameter η. We assume that the input patterns have zero mean, i.e., $\langle \mathbf{p} \rangle = \mathbf{0}$. After every update the weight vektors $\hat{\mathbf{w}}_k(t+1) = \mathbf{w}_k(t) + \Delta\mathbf{w}_k$ are renormalized: $\mathbf{w}_k(t+1) = \hat{\mathbf{w}}_k(t+1)/|\hat{\mathbf{w}}_k(t+1)|$. The learning rule for the lateral connections in the second layer still remains to be specified. We intend to show that the "Anti–Hebb" learning rule $\delta u_{kl} = -\mu o_k o_l$, with positive learning parameter μ, which has been suggested by Rubner et al., can be derived from an extremum principle. The principle requires the *minimization of the mutual information* between every two output units. As a result the network performs a principal component analysis (for a mathematical analysis of the net see [2]).

3. MINIMIZATION OF MUTUAL INFORMATION

To enable an analytical treatment of the problem we consider a multivariate gaussian probability density function (PDF) of the presented patterns

$$P(\mathbf{p}) = \frac{1}{(2\pi)^{N/2}(\det \mathbf{Q}_p)^{1/2}} \exp\left(-\frac{1}{2}\mathbf{p}^T\mathbf{Q}_p^{-1}\mathbf{p}\right), \tag{2}$$

where $\mathbf{Q}_p = \left\langle \mathbf{p}\,\mathbf{p}^T\right\rangle$ is the covariance matrix of the input data. For two output units o_l, o_k, $l < k$, the mutual information is defined as (cf. [14])

$$I(o_l, o_k) = \iint do_l do_k P(o_l, o_k) \ln\left(\frac{P(o_l, o_k)}{P(o_l)P(o_k)}\right) = \left\langle \ln\left(\frac{P(o_l, o_k)}{P(o_l)P(o_k)}\right)\right\rangle. \tag{3}$$

Here, $P(o_l, o_k)$ is the joint PDF of the output units l and k, whereas $P(o_l)$ and $P(o_k)$ are the PDF's for the single units. Note, that for statistically independent output units $P(o_l, o_k) = P(o_l)P(o_k)$ and, therefore, $I(o_l, o_k) = 0$. The mutual information $I(o_l, o_k)$ depends on the connection strength u_{kl} between the two cells. Performing a gradient descent with respect to this parameter, $\partial I/\partial u_{kl} =: \langle Z_u \rangle$, we arrive at a learning rule $\Delta u_{kl} = -\mu Z_u$ for the lateral connection from unit o_l to o_k. To evaluate Z_u we first consider the PDF $P(\mathbf{o})$ of the output of the network

$$P(\mathbf{o}) = \frac{1}{(2\pi)^{N/2}(\det \mathbf{Q}_o)^{1/2}} \exp(-\frac{1}{2}\mathbf{o}^T\mathbf{Q}_o^{-1}\mathbf{o}), \tag{4}$$

with

$$\mathbf{Q}_o = \tilde{\mathbf{W}}\mathbf{Q}_p\tilde{\mathbf{W}}^T. \tag{5}$$

Note, that $\det \mathbf{Q}_o$ is independent of $u_{kl}, \forall k, l$. Integrating $P(\mathbf{o})$ over all $o_r \neq o_l, o_k$ we find

$$P(o_l, o_k) = \frac{1}{2\pi\sqrt{\det \mathbf{Q}_{lk}}} \exp\left(-\frac{1}{2}(o_l, o_k)\mathbf{Q}_{lk}^{-1}(o_l, o_k)^T\right), \tag{6}$$

and analogously

$$P(o_k) = \frac{1}{\sqrt{2\pi Q_{kk}}} \exp\left(-\frac{1}{2}\frac{o_k^2}{Q_{kk}^o}\right), \tag{7}$$

where $\mathbf{Q}_{lk}$ is the 2×2 covariance submatrix for the output units l and k and Q_{kk}^o is the k'th diagonal element of $\mathbf{Q}_o$. To evaluate the derivative $\partial I/\partial u_{kl}$ we use the following identities

$$\frac{\partial o_l}{\partial u_{kl}} = \frac{\partial Q_{ll}^o}{\partial u_{kl}} = 0, \quad \frac{\partial o_k}{\partial u_{kl}} = o_l, \quad \frac{\partial Q_{kk}^o}{\partial u_{kl}} = 2Q_{kl}^o, \quad \frac{\partial Q_{lk}^o}{\partial u_{kl}} = Q_{ll}^o \tag{8}$$

which are easily obtained using $Q_{kl}^o = \langle o_k o_l \rangle$ and Eq. (1). Straightforward calculation shows, that

$$Z_u = Z_1 + Z_2, \tag{9}$$

where

$$Z_1 = \frac{1}{Q^o_{kk}} o_l o_k \quad \text{and} \quad Z_2 = \frac{Q^o_{lk}}{Q^o_{kk}} \left(1 - \frac{o^2_k}{Q^o_{kk}}\right). \tag{10}$$

Z_1 is proportional to $o_l o_k$ and, therefore, is the "Anti–Hebb" term in the learning rule $\Delta u_{kl} = -\mu Z_u$ introduced above. For slowly varying u_{kl} the term Z_2 has no effect because its average $\langle Z_2 \rangle$ vanishes. Furthermore, Z_2 is very small in the case of a close to diagonal covariance matrix $\mathbf{Q}_o$ since it is of order $O(Q^o_{lk}/Q^o_{kk})$. Therefore, neglecting Z_2, we finally arrive at

$$\Delta u_{kl} = -\mu \frac{o_k o_l}{\langle o^2_k \rangle} \tag{11}$$

which is the desired result.

ACKNOWLEDGEMENT

The authors like to thank J. Rubner for stimulating discussions.

REFERENCES

[1] Rubner, J. and Schulten, K., Biol Cybern, in press.

[2] Rubner, J. and Tavan, P., Europhys Lett, in press.

[3] Lawley, D.N. and Maxwell, A.E., Factor Analysis as a Statistical Method (Butterworths, London, 1963).

[4] Oja, E., J Math Biology **15** (1982) 267.

[5] Oja, E., Int J Neural Systems **1** 1 (1989) 61.

[6] Baldi, P. and Hornik, K., Neural Networks **2** (1989) 53.

[7] Plumbley, M.D. and Fallside, F., An Information-Theoretic Approach to Unsupervised Connectionist Models, submitted to: Proceedings fo the Connectionist Models Summer School 1988 (Morgan-Kaufmann, San Mateo CA).

[8] Földiak, P., Adaptive Network for Optimal Linear Feature Extraction, in: Proceedings of the IJCNN, Washington 1989, I-401.

[9] Rubner, J., Schulten, K. and Tavan, P., A Self–Organizing Network for Complete Feature Extraction, this volume.

[10] Kohonen, Self–Organization and Assoziative Memory, (2nd Ed.), (Springer, Berlin, 1988).

[11] Linsker, R., IEEE Computer, March (1988) 105.

[12] Linsker, R., An Application of the Principle of Maximum Information Preservation to Linear Systems, in: Touretzky, D.S. (ed.), Advances in Neural Information Processing Systems 1, Denver 1988 (Morgan-Kaufmann, San Mateo CA, 1989).

[13] Linsker, R., How to Generate Ordered Maps by Maximizing the Mutual Information between Input and Output Signals, IBM Research Report RC 14624 (#65530) 5/22/89.

[14] Shannon, C.E. and Weaver, W., The Mathematical Theory of Communication (University of Illinois Press, Urbana, 1949).

Parallel Processing in Neural Systems and Computers
R. Eckmiller, G. Hartmann and G. Hauske (Editors)
© Elsevier Science Publishers B.V. (North-Holland), 1990

DYNAMICAL LEARNING IN NETWORKS WITH SPARSE CONNECTIVITY

Karl E. Kürten

Institut für Neuroinformatik, Ruhr-Universität Bochum,
D-4630 Bochum, BRD
and
Institut für Theoretische Physik, Universität zu Köln,
D-5000 Köln , BRD

We present a sparsely connected neural network model designed to carry out visual perception. Computer simulations show that for random as well as for structured patterns the network connectivity can be quasi-optimally adapted to the specific structure of the information the network is asked to capture. It is further demonstrated that our model substantially outperforms its fully connected counterpart.

1. INTRODUCTION

Dynamic artificial binary neural networks are connectionist models based on a microscopic description of massively parallel processing units exhibiting collective phenomena[1-6] associated with information processing and cognitive behavior. One detrimental assumption of some of the currently popular models is the full connectivity of the neuron-like elements. However, highly connected networks of respectable size suitable for such cognitive tasks as pattern recognition or image processing raise immense wiring problems in hardware realization and make real-time software simulations computationally expensive. On the other hand, full connectivity is rarely found in nature and it is well known that systems with low connectivity are often amenable to analytic study[3,5].

2. THE MODEL

The network model consists of N sparsely interconnected formal neurons which are either active ($6_i = 1$) or silent ($6_i = -1$). Each unit i is supposed to receive exactly K inputs from the *other* neurons of the network such that each unit receives a total net internal stimulus

$$h_i(t) = \sum_{j \neq i} c_{ij}\, 6_j(t) \,/\, |c|_i \qquad\qquad i = 1,\ldots,N \ , \qquad (1)$$

where the sum is performed over the arbitrary K neighbours of neuron i. The coupling coefficients c_{ij} are supposed to be real numbers and the quantity $|c|_i$ is a suitable normalization factor. The actual binary state of each cell is then determined via the deterministic threshold rule

$$6_i(t+1) = \text{sign}[h_i(t)] \qquad\qquad i = 1,\ldots,N \ . \qquad (2)$$

The network is supposed to memorize a series of $p = \alpha K$

configurations $\Pi^{(\mu)} = (\pi_1^{(\mu)},\ldots,\pi_N^{(\mu)})$ $\mu=1,\ldots,p$. In order to guarantee that the patterns are fixed points of the dynamics (2) and the information is engraved as strongly as possible, the coupling   constants   have   to   be   chosen   such   that   the pN inequalities

$$\pi_i^{(\mu)} h_i (\pi_1^{(\mu)},\ldots,\pi_N^{(\mu)}) > \kappa \qquad i=1,\ldots,N \quad \mu=1,\ldots,p \tag{3}$$

are satisfied with a polarization parameter κ chosen as large as possible. The dynamical learning process is then formulated in terms of Hebbian-like self-organization via [7]

$$c_{ij}(t+1) = c_{ij}(t) + \frac{1}{K} \sum_{\mu=1}^{p} \Theta(\kappa - R^{(\mu)}) \pi_i^{(\mu)} \pi_j^{(\mu)}. \tag{4}$$

In fact, it is well known that a network with random connectivity can store up to 2K unbiased random patterns. However, analytical studies[8] predict that depending on the initial condition two phases will arise for $\alpha > \alpha_G = 0.42$: an ordered phase, characterized by perfect recognition, and a chaotic phase, where the memory of the initial condition is totally lost. Thus, above the critical value of $\alpha_G = 0.42$ the system is not able to work as an associative memory, whereas below the critical value any macroscopic initial overlap with a memorized pattern leads to perfect recognition.

3. TRIAL AND ERROR STRATEGY

In view of technical applications, however, one would like to have systems capable of an associative recall without error for sizes of the storage capacity far beyond the critical capacity α_G = 0.42. Whenever a random choice of neighbours fails to fulfill condition (3), our simple strategy to circumvent overloading instabilities is, to switch to another random choice until the network has found its quasi-optimal connections. We allow at most 25 learning passes and stop the learning procedure when (3) is satisfied for a given minimum value of κ. Before the retrieval phase the original patterns are degraded by random noise, and presented as initial conditions for the network to recall.

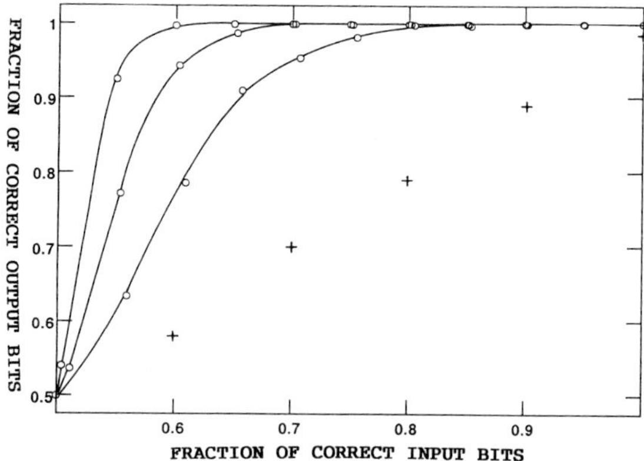

Fig.1: Mean fraction of correctly recalled bits as a function of the fraction of correct input bits for N=2500, N=625 and N=225 with K=8 and α = 0.5 (from above). Crosses show the same quantity for N=225 cells *without* our strategy.

Figure 1 shows the mean fraction of correctly retrieved bits as a function of the fraction of correct input bits for K=8 and α = 0.5 for different total numbers of cells. The final connections have been obtained after a few neighbour search trials on average. Note the strong size dependence of the mean final fraction of correctly retrieved information, clearly indicating that for sufficiently large N an initial damage of less than fifty percent leads to an essentially perfect recall. For comparison, crosses show the unsatisfactory performance of a network with N = 225 cells *without* the trial and error strategy, where the system can exhibit chaotic behavior. Furthermore, computer simulations show that also for larger values of α the trial and error strategy works successfully, though with increasing α the learning time increases rapidly making the process less efficient.

4. APPLICATION TO STRUCTURED INFORMATION

In this section we demonstrate that our network model is not restricted to capture random information, but is also able to process strongly structured patterns interpreted as different objects residing on a uniform background as in figure 2. In the case of random information we have seen that it is of crucial importance to make a 'good' choice of the neighbours with which each individual cell connects. Thus, due to strong correlations within the individual patterns, each neuron is now supposed to be *first* connected with its eight neighbours in the Moore neighbourhood, while additional suitable long-ranged neurons emerge from a trial and error scheme until a quasi-optimal network topology has been found. This picture is in accord with general ideas, in particular with Braitenberg's[9], about the interconnectivity of cortical pyramidal cells.

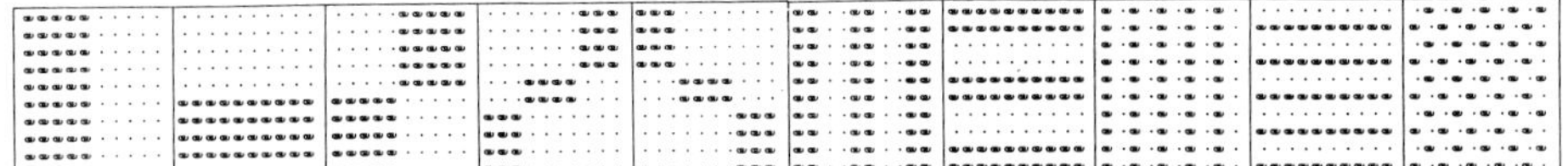

Fig.2: Original samples used for the learning process

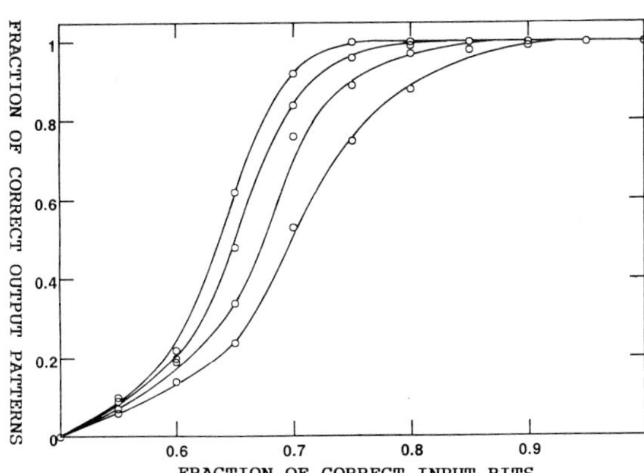

Fig. 3: Mean final fraction of perfectly recalled patterns as a function of the random fraction of correct input bits for K=50, K=40, K=30 and K=20 (from above) with N=100 cells. The data for the fully connected network are indistinguishable from K=50.

Figure 3 shows the remarkable effect that a connectivity of
fifty percent does *not* lead to a degradation of the performance
of the fully connected network. This can be explained by the view
that our synaptic plasticity scheme, as it occurs in biological
nerve nets, involves an optimization process in the sense that
information is engraved as *strongly* and *efficiently* as possible.
Finally we recover the patterns which have been degraded by non
random *specific* corruptions. Figure 4 shows the degraded patterns
presented to a network with only 30% connectivity.

Fig.4: Corrupted samples used for the recall process.

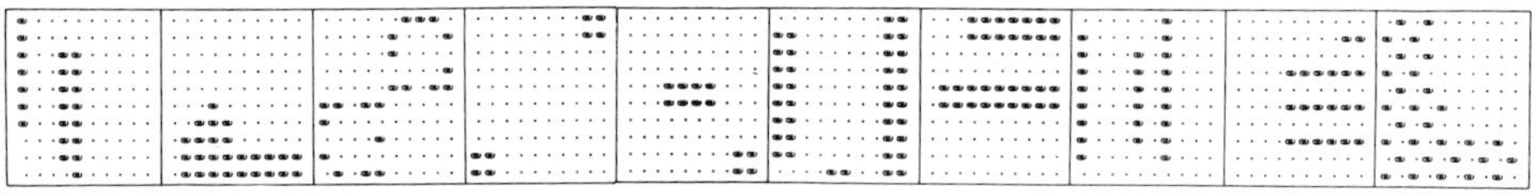

Though more than fifty percent of each 'object' has been
destroyed on average, the network is able to recall the full
information perfectly after only about two time steps. By
contrast, fully connected networks usually miss a few bits.
Apparently, sparsely connected networks, based on a geometric
and topological architecture, adapted to the structure of the
patterns, can recall them more efficiently.

5. CONCLUSION AND OUTLOOK

We have shown that the performance of sparsely connected
networks is superior to that of their fully-connected
counterparts. Moreover, compared with complete connectivity, the
storage capacity per connection is largely improved. Preliminary
work based on evolutionary optimization strategies in terms of
genetic algorithms indicates clearly that near-optimal answers
can be obtained.

ACKNOWLEDGEMENTS

The author gratefully acknowledges financial support for this
work by the German Federal Department of Research and Technology
(BMFT) under grant Nr. ITR 8800K4. It is a pleasure to thank D.
Stauffer for carefully reading the manuscript and J.W. Clark, H.
Geiger, J.L. van Hemmen, U. Keller, M.L. Ristig, W. von Seelen
and G.Senger for useful und stimulating discussions.

REFERENCES

[1] Clark, J.W., Rafelski, J. and Winston, J.V., Physics Reports
     123 4 (1985) 215-273
[2] Caianiello, E., J.Theoret. Biol. 2 (1965) 204-235
[3] Derrida B., Gardner, E. and Zippelius, A., Europhys. Lett. 4
     (1987) 167-173
[4] Hopfield J.J., Proc.Natl.Acad.Sci. 79 (1982) 2554-2558
[5] Kürten K.E., Phys.Lett.A 129 3 (1988) 157-160
[6] Kürten K.E., J. Phys. France 50 (1989) 2313-2323
[7] Minsky M.L. and Papert S., (1969) Perceptrons
     (Cambridge, MA:MIT Press)
[8] Gardner E., J.Phys.A 22 (1989) 1969-1974
[9] Braitenberg V., Lecture Notes in Biomathematics 21 (1978)
     171-188

Parallel Processing in Neural Systems and Computers
R. Eckmiller, G. Hartmann and G. Hauske (Editors)
© Elsevier Science Publishers B.V. (North-Holland), 1990

SELF-ORGANIZATION VERSUS PROGRAMMING IN MASSIVELY PARALLEL SYSTEMS: A CASE STUDY

Bernard Manderick and Frans Moyson

Artificial Intelligence Laboratory
Free University of Brussels
Brussels, Belgium

The emergence of self-organization in massively parallel systems is discussed and illustrated by an example: the adaptive response of ant colonies to their environment. Here, each individual ant has a very simple behavior. Yet, the colony as a whole exhibits "intelligent" behavior, e.g. the colony is able to discover and to exploit food sources in its environment. Also, some difficulties with self-organization in parallel systems are pointed at and indications are given of how these difficulties might be tackled.

1. INTRODUCTION

Self-organization as opposed to programming offers new perspectives for computer science in general and artificial intelligence in particular. In this paper, a massively parallel system is presented which simulates the collective behavior of ant colonies. This system is a characteristic for many massively parallel systems [10]: 1) the behavior of the system as a whole emerges from local interactions between individual component, i.e. it is an example of self-organization, 2) the system is fault-tolerant, i.e. if some of the individual components malfunction this does not necessarily affect the global behavior, and 3) the system is flexible, i.e. the system is able to adapt itself to changes in the environment. These three properties are difficult if not impossible to achieve by sequential or small-scale parallel programming. The case study has also a practical significance. In [11], it is shown how it can be used as a metaphor for solving some problems in robotics.

The paper is organized as follows. First, we place massive parallelism in the context of complex dynamical systems. Second, we describe the key concepts of complex dynamics among which the notion of self-organization and the necessary conditions for its occurrence. Third, a massively parallel algorithm for the collective behavior of ant colonies is presented. It is an example of how very simple agents can cooperate to achieve emergent functionality. Finally, the case study of the ant colonies is used to discuss some difficulties with self-organization and it is suggested how these problems might be tackled.

2. MASSIVE PARALLELISM

Parallelism gives additional computing power. One kind of parallelism, namely data level parallelism, uses a very large computational network of similar processes, each executing the same (simple) task. These processes can interact by communicating the results of their computation. Experiences with data level parallelism have shown that it is a powerful paradigm [4].

In [8], this paradigm is used to simulate the behavior of fluids. The usual approach is to solve the Navier-Stokes equations which describe the macroscopic behavior of fluids. These equations are solved using sophisticated numerical techniques, e.g. finite element methods. The data level approach does not take the Navier-Stokes equations as a starting point. Instead, it simply simulates the collisions between particles constituting the fluid. This is a description at the microscopic level. The macroscopic behavior of the fluid has to emerge from these local interactions.

This approach is characteristic in two respects for a large class of data level programs. First, it is an example of massive parallelism. The fluid is simulated by a very large number of colliding particles. These collisions are local interactions and hence intrinsically parallel. Second, the global behavior of the fluid is an emergent property of these interactions. This raises some problems since there is now a large gap between the level of programming and the level of observation and verification. The programming is situated at the level of the constituent components also called the microscopic level, while we observe and verify the macroscopic behavior of the system. In the fluids example, the program is a set of rules specifying how particles collide and the problem is to find a set of rules which shows the intended macroscopic behavior, i.e. the global behavior of fluids. This can be very hard. The relation between the micro- and the macroscopic level has to be understood better and the theory of complex dynamical systems [3,7,9] provides a natural framework to study this relation.

We want to study massively parallel systems in the broader context of the theory of complex dynamical systems. Since this theory relies on the mathematical theory of continuous dynamical systems we will use massive parallelism in the following sense: a system is *massively parallel* if it can also be modeled as a continuous and not only as a discrete system. According to this definition, the example above from fluid dynamics is massively parallel: its macroscopic behavior changes continuously.

Under certain conditions complex dynamical systems (and according to the above definition also massively parallel systems) can self-organize. Self-organization in massively parallel systems seems to offer new prospects for AI and the term complex dynamics has been coined to denote this new paradigm [10]. In the next section, we sketch the theory of complex dynamical systems. This includes the necessary conditions for self-organization to occur.

3. COMPLEX DYNAMICAL SYSTEMS AND SELF-ORGANIZATION

Complex dynamical systems and self-organization have already been studied in physics, chemistry, biology and other branches of science [3,7,9]. A well-known example from chemistry is the Belousov-Zhabotinsky reaction. In an initially homogeneous mixture of given reactants, temporal oscillations occur: the mixture changes color periodically, from red to blue and vice versa. It acts as a chemical clock. Under certain conditions, the same reaction can even produce spatial and spatio-temporal patterns on the macroscopic level. Numerous other examples of self-organizing systems can be found in [3,7,9].

A system is called *complex* if it consists of a large number of interacting entities. The interacting entities constitute the microscopic level. In physics and chemistry, the description at this level is probabilistic and the theory of stochastic processes is one major component of complex dynamics. A system can also be looked at at the macroscopic level. In this case, the system is characterized by collective variables describing the system as a whole. The change in time of these variables determine its evolution. A system is *dynamical* if this evolution can be described by a set of ordinary or partial differential equations. Dynamical systems theory, which studies these sets, is another major component of complex dynamics.

Self-organization at the macroscopic level is the manifestation of cooperative behavior or reduction of degrees of freedom at the microscopic level. Central questions in complex dynamics are: what mechanisms at the underlying level are necessary for self-organization to occur at the macro-level, and under what conditions is this behavior manifested.

These questions can be answered quite generally. The potential of self-organization is only realized when the system is constrained by its environment. Moreover, the local interactions have to obey a non-linear law; for instance they might involve feedback mechanisms. So, self-organization is a property of non-linear systems operating in non-equilibrium conditions. Self-organization as a decrease in randomness, the role of constraints and non-linearity will be illustrated by the behavior of ant colonies.

4.　THE COLLECTIVE BEHAVIOR OF ANTS

4.1. Introduction

The establishment by ant colonies of a flow between a food source and the nest occurs without central supervision. Indeed, no one ant decides where the path will be. Instead, the path emerges out of the local interaction between the ants constrained by their environment.

Initially, the ants move around randomly and independently of each other. When one ant meets a food source it will leave a trail (of pheromone) on its way. This trail will attract ants, decreasing the randomness in their motion and leading them to the food source. When these ants reach the food source they too will leave a trail behind. This will enforce the original trail. Ultimately, by chance, a trail between food source and nest will develop. This path will continually be enforced by the ants using it. However, when the food source is depleted the trail will no longer be enforced and will decay. Even when following a trail, movements of ants remain relatively random. This way ants will be able to cross interruptions in the path, and ants losing the trail can find a new food source: the randomness of their behavior is an advantage for the ants [2].

As always, self-organization results from the decrease in randomness at the microscopic level. However, some randomness persists and is responsible for the adaptive behavior of the ant colony. The feedback mechanism (the trail decreases randomness and the decreased randomness reinforces the trail) is the non-linearity responsible for the self-organization. The presence of the food source is the constraint. When it is depleted the trail disappears and the ants again start to move in all possible directions: the ant colony is again in a state of maximal randomness.

The example of the collective behavior of ants shows how something (here the making of a path between food and nest) can be done through the cooperation of many small agents (the ants) with a minimum of communication between them. These small agents do not even have to know about each other's existence. Even more, they do not know they are laying a path together. The only thing each ant does is leaving behind a trace of pheromone as soon as it has found food.

4.2. The Simulation and Experimental Results

Control over an individual ant is embodied in one function and this function is applied in parallel to all ants. It consists of a conditional with 4 clauses. First, it checks whether food has been found. If so the ant reverses its direction and looks for smell. Furthermore a counter is initialized. This counter controls for how long the ant will be able to leave a trace of pheromone behind. As soon as the counter reaches zero, no more pheromone is left behind. Second, if no food has been found the function checks for the nest. If the nest has been encountered the ant again reverses direction and looks for smell. Third, if the ant did not encounter food or nest, and if it smells nothing, it travels in a random way. If it smells something, its new direction is determined by it. And fourth, the function checks if an ant has to deposit pheromone or not.

In a typical experiment, a path between nest and (possibly) multiple food sources emerges provided enough ants are present. During an experiment one sees that in the beginning no smell is present anywhere. Slowly the ants begin to spread out all over their world. Some of them will encounter the food. This will result in some smell to be deposited (two left figures). The first ants finding the food do not know anymore where their nest is and so can not lay a path immediately reaching the nest. But they do lay a trace starting from the food. This attracts more ants to the food. More ants results in more smell being distributed. This ultimately results in smell being distributed (almost) all over the place (middle figures).

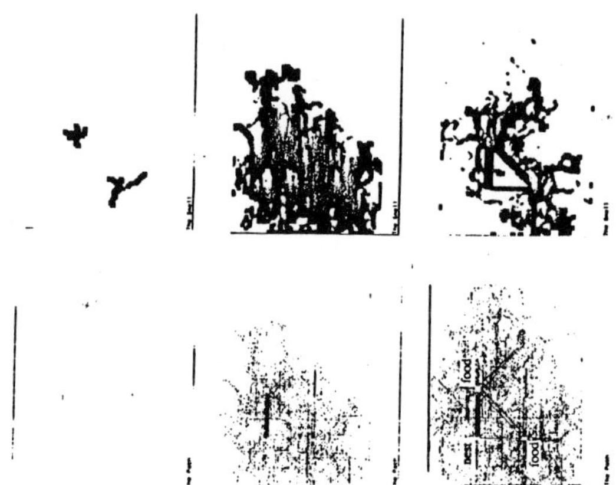

Path and Smell: These three figures depict the evolution of one experiment. Two food sources and one nest are present. In the left figure some ants have found the food. In the middle figure, there is pheromone all over the place. In the right figure paths between the nest and the two food sources are clearly visible. Note that a path between the two food sources has developed too.

As soon as one ant carrying food reaches the nest, a path exists between food and nest. Other paths exist as well. But ants that have reached the nest will turn around and go back to the food. And come back to the nest. This path between food and nest will continue to be strengthened, while the paths leading nowhere will slowly disappear, because the smell emanating from the pheromone automatically decreases. New paths leading nowhere will be generated. But they, in turn, will slowly disappear for the reason mentioned above. As a result one sees a path between the food and the nest with other trails around it. The path between food and nest is a strong one, while the others are weak ones (right figures).

The algorithm described above is an example of self-organization, i.e. the intended behavior (the discovery and exploitation of food sources) emerges from local interactions between ants. Moreover, our experiments have shown that the algorithm is 1) fault-tolerant (the behavior of the colony is unaffected by the removal or the disturbance of the behavior of some of its ants), and 2) flexible (if a food source is moved during an experiment the path is moving with it). These properties are difficult to achieve by sequential or small-scale parallel programming.

5. DIFFICULTIES WITH COMPLEX DYNAMICS

There are not only opportunities but also difficulties associated with the complex dynamics paradigm. These are illustrated by our case study. Self-organization in the ant colony occurs only for certain values of the parameters in the model. These parameters are the number of ants per unit of area and how fast the pheronome diffuses. To find these values a mathematical model for the behavior of ant colonies was used and a bifurcation analysis was performed. From this, constraints on the parameter values which lead to self-organization can be derived and this helps to find values for which self-organization does occur - see [6] for more details.

However, this approach can not be generalized easily to all massively parallel systems since the mathematical model can be too complicated and/or depend on too many parameters. Hence, we are left with a search problem in a large parameter space. Genetic algorithms might help here. The genetic algorithm is a powerful weak search method based on natural evolution [5]. Genetic algorithms have been applied with success to problems of parameter tuning - cfr. [1] for an overview. We are currently investigating this line of research.

6. CONCLUSION

The adaptive behavior of ant colonies emerging from the local interactions between the ants is a property of the colony as a whole and cannot be reduced to the behavior of individual ants: the whole is more than the sum of the parts. There is no "knowledge" necessary for creating a path available at the level of individual ants. Note also that if the colony is not large enough, no paths will result. Here, self-organization is an example of "intelligence" (the construction of a path between food and nest) emerging from non-intelligence (leaving a trace when carrying food). Furthermore, there is no a priori reason to suspect that this behavior will result. Nothing in the individual ants' behavior indicates that they have to lay a path from food to nest.

The example of ant colonies can be stated in a more general framework, namely the theory of complex dynamical systems. Just like any self-organizational behavior it is a consequence of the reduction of degrees of freedom at the micro level, i.e. the level of individual ants. To us, it seems difficult to program this flexible behavior explicitly. This way, the theory of complex dynamical systems provides a new paradigm for understanding massively parallel systems.

However, we have to face some problems. The parameter values for which self-organization occurs may be hard to find. Either because we cannot find a simple mathematical model or because there are too many parameters in the model. Genetic algorithms are of interest here.

ACKNOWLEDGEMENTS

Thanks to all people from the Dynamics Group of the VUB AI Lab for various discussions. In particular, we would like to mention Prof. Steels for providing us the opportunities to work in the exciting field of complex dynamics. Bernard Manderick is sponsored by IWONL.

REFERENCES

[1] De Jong, K., Learning with Genetic Algorithms: An Overview, in: Machine Learning, Vol. 3, Nos 2/3 (1988)

[2] Deneubourg, J.L., Pasteels, J.M, and Verhaeghe, J.C., Probabilistic Behaviour in Ants: A Strategy of Errors?, in: Journal of Theoretical Biology, Vol. 105 (1983)

[3] Haken, H., Synergetics (Springer-Verlag, Berlin, 1978)

[4] Hillis, W.D., and Steele, G.L., Data Parallel Algorithms, in: Communications of the ACM, Vol. 29, No. 12 (1986)

[5] Holland, J.H., Adaptation in Natural and Artificial Systems (University of Michigan Press, Ann Arbor, 1975)

[6] Moyson, F., and Manderick, B., The Collective Behavior of Ants: An Example of Self-Organization in Massive Parallelism, in: Proceedings of the AAAI Symposium on Parallel Models of Intelligence, Stanford (1988)

[7] Nicolis, G., and Prigogine, I., Exploring Complexity (Piper Verlag, Münich, 1987)

[8] Salem, J.B., and Wolfram, S., Thermodynamics and Hydrodynamics with Cellular Automata, (Thinking Machines Corporation, Cambridge, 1985)

[9] Serra, R., Andretta, M., Compiani, M. and Zanarini, G., Introduction to the Physics of Complex Systems: The mesoscopic approach to fluctuation, non linearity and self-organization (Pergamon Press, Oxford, 1986)

[10] Steels, L., AI and Complex Dynamics, in: Proceedings of the IFIP workshop on concepts and techniques for knowledge-based systems, (Mount Fuji 1987)

[11] Steels, L., Cooperation between distributed agents through self-organization, in: Robotics and Autonomous Systems, in print (1990)

Parallel Processing in Neural Systems and Computers
R. Eckmiller, G. Hartmann and G. Hauske (Editors)
© Elsevier Science Publishers B.V. (North-Holland), 1990

Top-Down Learning in Modular Feed-Forward Networks

Arnfried Ossen

Technische Universität Berlin, Institut für Angewandte Informatik
Franklinstr. 28/29, D-1000 Berlin 10
<ao@coma.UUCP>

Abstract — *This article proposes a means of using traditional methods of knowledge structuring, namely composition and modularization, to fight learning time complexity in* connectionist networks. *It is shown how a modular, hierarchical system of auto-associative feed-forward modules can be obtained by constraining low-level internal representations using high-level modules. Modules can learn separately and in a* top-down *manner, if some a* priori knowledge about the problem is available.

1 Introduction

Learning in *Connectionist Networks* is equivalent to finding the set of weights that enable the network to approximate its environment to an arbitrary degree. A couple of gradient-descent procedures are known [1] that are able to generate such internal representations of the environment for simple classification or function approximation tasks. At present, real-world problems, e.g. complex pattern recognition, are out of reach for these procedures. Significant progress is needed in their ability to generalize, that is, e.g. in a classifier network, to classify a new instance correctly after having seen a sufficiently large set of typical cases and in their learning-time efficiency. A promising way of reducing complexity is the use of familiar methods of knowledge structuring like composition and modularization. This can be achieved, if *a priori* knowledge about the problem is built into the network and a means of interpretation of internal representations is available.

The back-propagation algorithm (BP) [2] is a widely-used learning algorithm, that has been successfully applied to several small and medium-sized problems. BP is an interesting candidate as a learning algorithm for structured networks, because it can generate internal representations that allow the network to generalize according to the underlying regularities of the environment; in auto-associative mode this can even be achieved without an external supervisor to specify the desired activations of the output units. But there are deficiencies: Back-propagation does not scale very well with the problem size. It is approximately of order (n^3) [1], where n is the number of weights. The internal representations are distributed, i.e. concepts are not represented by single units, but by a pattern of activity over the units, giving the network inherent fault tolerance. On the other hand, interpretation of network behavior in terms of rules is limited to trivial problems. We can conclude: BP has to be enhanced in two respects: **a)** improvement of learning speed without losing generalization properties; **b)** facilitate interpretation of distributed internal representations.

2 Optimization Strategies

Le Cun [3] has pointed out that the number of training examples required for good generalization scales like the logarithm of the number of functions which a specific network architecture can implement. Assuming that a small network is less general than a bigger one, i.e. that it can implement less functions, provided it is still able to compute the desired function we can conclude: small networks yield better generalization than bigger networks, given the same amount of training data. Learning complexity per learning cycle decreases, then, simply because small networks contain less links and units. On the other hand, overall learning efficiency may get worse because of a decreasing convergence rate. Additionally, since small networks are more likely to generate compact encodings, we can also expect an improvement in interpretability of internal representations.

A couple of problem-independent strategies, also called the **minimal network** approach, are known for the construction of small nets. Either an auxiliary error term is added to the cost function in order to penalize network configurations with many active links and/or units [4], or the relevance of units with respect to the error is determined and the least relevant units are deleted [5]. The problem with minimal network construction using auxiliary error terms is the difficulty of relative weighting of terms, which may cause convergence problems. Also, it might still be very difficult to understand the emerging patterns of activity in the hidden layers.

Constraining the weight space is also possible through the use of knowledge about the problem or **a priori knowledge**. It can be used to impose equality constraints among the weights [3]. Significant reduction in learning complexity and good generalization can be obtained, but the weight sharing constraint does not help with the interpretation of internal representations.

A third approach is proposed in this article. *A priori* knowledge is used to select a constraint space for internal representations of modules. The constraints are enforced by substituting the hidden layers of *low-level* modules by a pretrained auto-associative network. After an *abstract* module has been trained to establish a constraint space, the *low-level* modules learn to develop their internal representations in terms of *high-level* input/output patterns (see figure 1). If the *low-level* modules are also auto-associative, this scheme can be extended to arbitrarily deep hierarchical, modular structures, that correspond to the bottom-up system proposed by Hinton [1]. If viewed as a perception system, *low-level* modules correspond to peripheral modules, whereas *high-level* modules implement more abstract representations. Since abstract modules are trained prior to low-level modules, the learning scheme may be called **top-down**, advancing from more abstract to more concrete.

3   A Constrained Network

The constraint space defined by an abstract module has to constrain the space of internal representations of less abstract modules to enhance generalization and convergence, but still be general enough to capture the underlying regularities of the evironment. On the other hand, it has to facilitate the encoding of external values and the decoding of adopted patterns of activity in terms of external concepts.

I have chosen a coarse coding [6] scheme of scalar values. In comparison to value-unit codings, it shows improved convergence [7]. It also provides the resolution needed for the development of internal representations. The intended scalar value can simply be represented by sampling a unimodal function centered at this value [8]. Decoding of scalars is possible via auxiliary networks that map scalar representations to an activation value or by a least square error procedure.

An auto-associative network m_1 trained for identity mapping of arbitrary scalar representations will generalize to sensible representations at the outputs if clamped to unknown input data. If back-propagated to the input units, the error can be used to modify the inputs in a way that minimizes the error at the output units. If placed in the middle of a surrounding network m_2, adopted representations at the output units of m_1 serve as encodings of internal representations of m_2, while the back-propagated errors can be used to modify the weights between the input layer of m_2 and the input layer of m_1 (see figure 1). Thus two optimizations take place: **a)** the output layer of m_1 will eventually generate almost perfect scalar representations, without any weight changes in m_1 proper; **b)** m_2 converges to its desired input/output mapping.

The top-down learning scheme was tested on an *8-4-8* encoder whose hidden layer had been replaced by a *4-4-4* module. Three simulation runs were carried out (see figure 2): **a)** standard *8-4-8* system; **b)** *8-4-4-4-8* system (three hidden layers); **c)** *8-[4-4-4]-8* system with pretrained *4-4-4* abstract module. The weights of the links of the three hidden layers are copied from a *4-4-4* system that has learned to auto-associate arbitrarily positioned coarse-coded scalars in a separate learning procedure. After being copied, the weights in the *4-4-4* module are fixed at their current

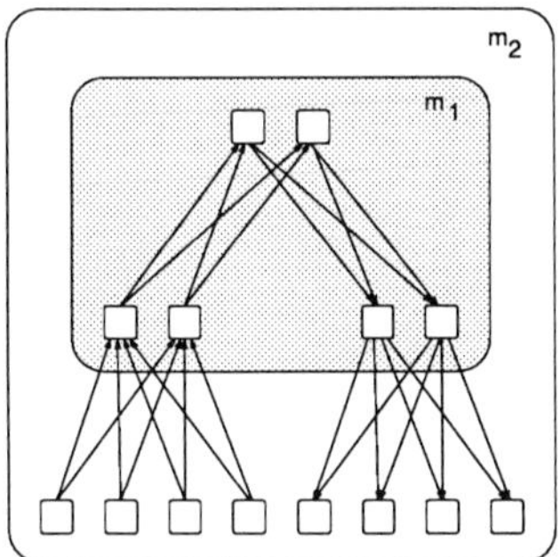

Figure 1: An auto-associative 2-2-2 module m_1 as an abstraction of the hidden layer of m_2.

values. Only weights from the input layer to the input layer of the *4-4-4* system and from the output layer of the *4-4-4* system to the output layer and the respective biases are subject to change by the back-propagation procedure. The *4-4-4* system merely serves to propagate activations and back-propagate errors, thereby constraining the internal representations of the encoder to scalar representations.

The convergence of the modular system is very similar to the original encoder and about one order of magnitude better than a system without separate training. Of course, the training effort for the pretrained module has to be taken into account too. But since back-propagation is of polynomial order, the additional complexity (40 free weights) is low in comparison to the complexity of the training cycles for the *8-4-4-4-8* (120 free weights) and *8-[4-4-4]-8* (80 free weights) systems.

This system can be applied to **nonlinear dimensionality reduction** problems [8], which turn out to be the special case of equal representations at low-level and abstract input layers. The abstract module has to be pretrained to auto-associate the chosen scalar coding, while the low-level modules auto-associate the higher dimensional data. The abstract module pressures internal patterns of activation into the desired coding, without need for special convergence strategies[1]. Figure 3 shows a nonlinear dimensionality reduction from two-dimensional data to a one-dimensional constraint using a *16-[8-8-8]-16* architecture.

4   Discussion

Gradient-descent learning procedures for *connectionist networks* can be enhanced by constraining the space of internal representations. I have shown how a constraint space can be obtained by pretrained abstract modules, if some *a priori* knowledge about the problem is available. This not only allows the required learning epochs to be reduced; a means of interpreting patterns of activity at module interfaces is also achieved. For the future, the latter may be of particular use for the construction of hybrid systems consisting of connectionist and symbolic parts, by allowing interpretation of distributed representations in terms of symbols.

A   Learning Procedure Adjustments

All simulations were carried out using the standard back-propagation algorithm with a fixed learning rate of $\eta = 0.2$ and a fixed momentum of $\alpha = 0.1$, weight update after each presentation (on-line mode), and sequential presentation of patterns. To speed up learning, the gradient was normalized[2] before updating the weight. Scalar codings were created using the second derivative of the sigmoidial function $1/(1 + e^{-tx})$ at a temperature of $t = 8.0$.

---

[1] To obtain convergence, Saund [8] has to use a simulated annealing schedule (smoothness of scalar coding) and a method of "encouraging" scalarized behavior by increasing peaks and decreasing valleys in a particular trial.

[2] This follows from the empirical observation that the product of optimal learning rate and absolute value of the gradient remains almost constant over all learning cycles, see [9].

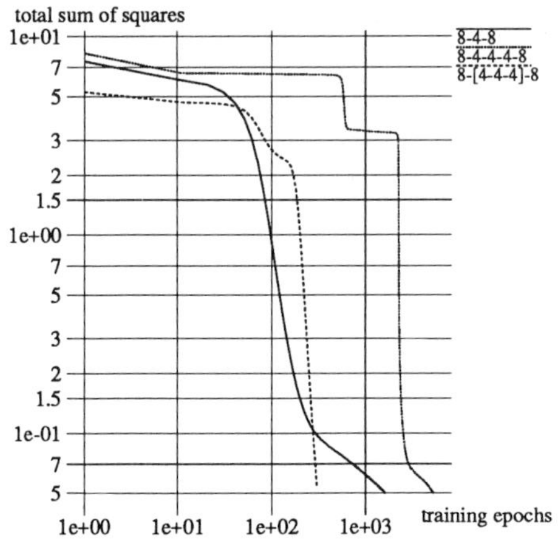

Figure 2: Convergence of *8-4-8* encoder, *8-4-4-4-8* encoder, and *8-[4-4-4]-8* encoder with pretrained abstract module.

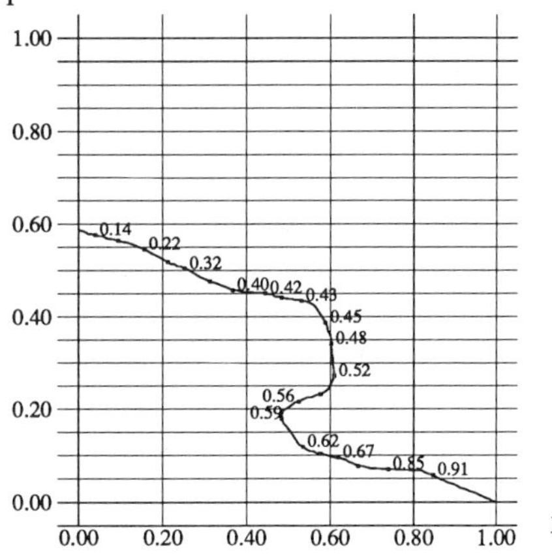

Figure 3: Nonlinear dimensionality reduction from two-dimensional curve data to one-dimensional data denoting points along the curve.

Acknowledgement

I would like to thank Albrecht Biedl for helpful comments during preparation of this article.

References

[1] Geoffrey E. Hinton. Connectionist learning procedures. Technical Report CMU-CS-87-115, Carnegie-Mellon University, Pittsburgh PA 15213, 1987.

[2] D. E. Rumelhart, G. E. Hinton, and R. J. Williams. Learning internal representations by error propagation. In *Parallel Distributed Processing: Explorations in the Microstructure of Cognition*, chapter 8. MIT Press/Bradford Books, 1986.

[3] Yann le Cun. Generalization and network design strategies. In Rolf Pfeiffer, Zoltan Schreter, Françoise Fogelman-Solié, and Luc Steels, editors, *Proceedings — Connectionism in Perspective*. Swiss Group for Artificial Intelligence and Cognititive Science (SGAICO), Elsevier Science Publishers B.V., 1989.

[4] Yves Chauvin. A back–propagation algorithm with optimal use of hidden units. In David S. Touretzky, editor, *Advances in Neural Information Processing Systems I*, pages 519–526. Morgan Kaufmann Publishers, San Mateo, California, 1989.

[5] Michael C. Mozer and Paul Smolensky. Skeletonization: A technique for trimming the fat from a network via relevance assessment. Technical Report CU-CS-421-89, University of Colorado at Boulder, Boulder, January 1989.

[6] Geoffrey E. Hinton, James L. McClelland, and David E. Rumelhart. Distributed representations. In *Parallel Distributed Processing: Explorations in the Microstructure of Cognition*, chapter 3. MIT Press/Bradford Books, 1986.

[7] Peter J. B. Hancock. Data representation in neural nets: An empirical study. In David Touretzky, Geoffrey Hinton, and Terrence Sejnowski, editors, *Proceedings of the 1988 Connectionist Models Summer School*, pages 11–20, San Mateo, CA, 1989. Morgan Kaufmann Publishers.

[8] Eric Saund. Dimensionality-reduction using connectionist networks. *IEEE Transactions on Pattern Analysis and Machine Intelligence*, 11(3):304–314, March 1989.

[9] Ralf Salomon. Adaptiv geregelte Lernrate bei Back-propagation. Technical Report 89-24, Technische Universität Berlin, 1989. Forschungsberichte des Fachbereichs Informatik.

Parallel Processing in Neural Systems and Computers
R. Eckmiller, G. Hartmann and G. Hauske (Editors)
© Elsevier Science Publishers B.V. (North-Holland), 1990

Genetic Generation of Backpropagation Trained Neural Networks

Wolfram Schiffmann   Klaus Mecklenburg

University of Koblenz

Usually Backpropagation works on a fixed network structure. In the talk we show that this procedure does not release the full power of Backpropagation. A method for automatic generation of problem specific network structures is introduced.

1 Introduction

Backpropagation (BP) is one of the best known and most successful methods to train neural networks. Because there is no feedback, the trained networks operate very rapidly. Therefore this is the only solution for realtime applications. An important drawback of BP is the fact that the structure of the network must be supplied by the developer. One has to specify the number of neurons and their distribution over several layers. Moreover, there is no systematic approach to interconnect neurons in different layers. In the next section we show by an concrete example how the learning process is influenced by the supplied net-structure.

The point in question is how to build network structures automatically. We have decided to use a genetic algorithm that generates optimal net-structures by the interplay of mutation and selection [1]. The basics of this method are presented in the third section.

The system is implemented with UNIX and X-WINDOWS and it uses RPCs of the Sun-NFS. In this way it is possible to execute independent functions of the program in parallel manner. Section four is devoted to the underlying communication mechanisms and the distribution of the required tasks over the available processing nodes.

2 Structure and Backpropagation

We have examined the learning behaviour of three different networks. Our results show that BP works best with an well structured network. The sample set for BP was generated with the structure of net $\mathcal{A}$ (Figure 1), whose connection-weights and neuron-thresholds were determined randomly. Then the input values were permuted between 0.0 and 1.0 in steps of 0.02 and the corresponding outputs were determined. All the networks are trained with this sample set and by means of BP. The parameters of the network components used were adjusted randomly and the number of iterations (epochs) was limited to 90. At each epoch the whole sample set (51^2) was presented. The I/O-behaviour of the three trained networks was compared with the sample set. The simulation showed that net $\mathcal{A}$ matched best while the other two networks are only suboptimal solutions (see Figure 2). For every input pattern of the sample set the relative error is represented by the gray level of the corresponding field. Because each field has the size of 7x7 pixels, there are 50 possible discrete values. This gray level range is scaled by the difference between the maximum error of net $\mathcal{A}$ and the maximum error of the nets $\mathcal{B}$ and $\mathcal{C}$.

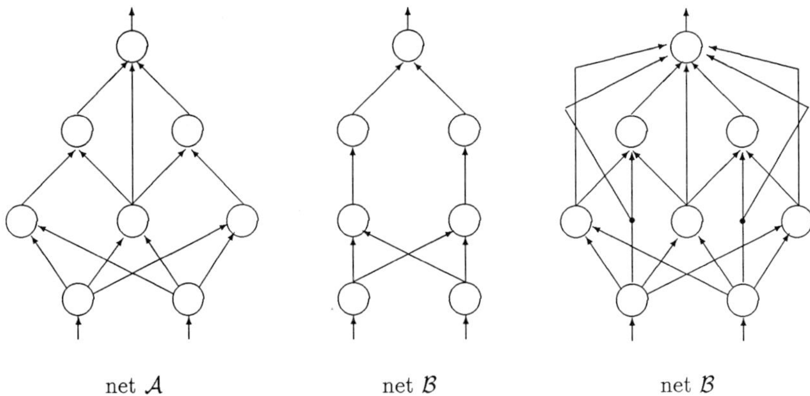

Figure 1: Structures of the used networks

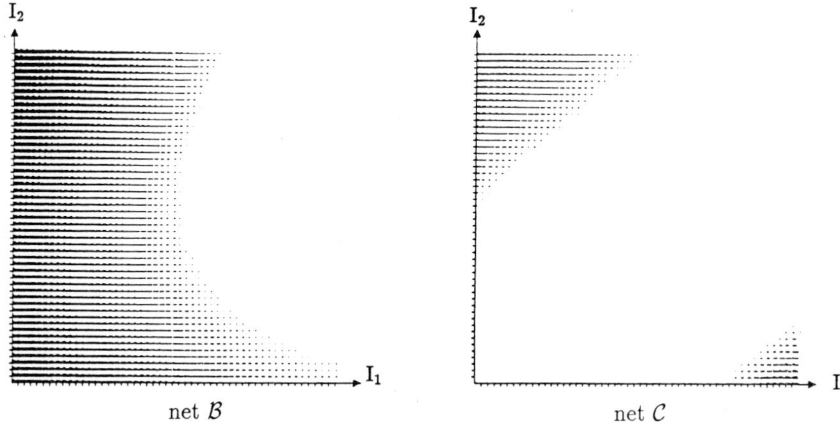

Figure 2: Relative errors over sample set

3 Basic algorithm

As we have seen in the previous example the learning behaviour of the networks is strongly influenced by the underlying structure. Therefore we have looked for a procedure to generate optimal network structures suited for BP automatically. There is no deterministic procedure which is able to predict the necessary modifications in structure by comparing the network I/O with the sample set. Our algorithm imitates the principles of biological evolution. The structure of the network is optimized by repeated cycles of mutation, BP-training and selection. The main program steps are displayed in Figure 3. The most important parts of the program will be discussed in the following (see also [3]).

3.1 Mutation

The construction plans of μ parent networks are stored in a genepool. They will be reproduced $\frac{\nu}{\mu}$ times by faulty copying of the construction plans. In this way ν child networks are generated.

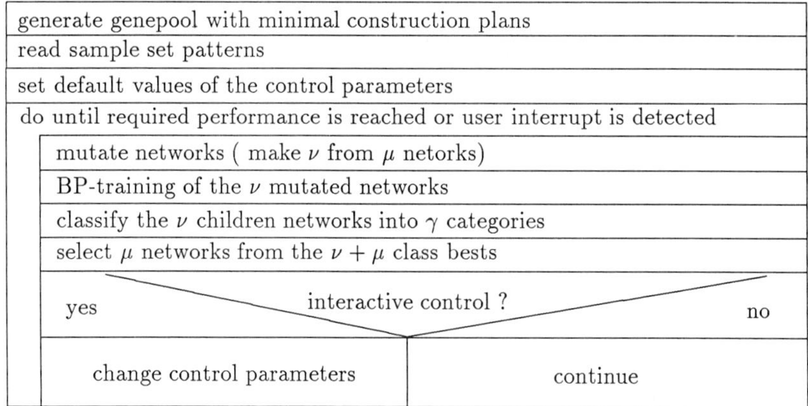

Figure 3: Raw program for genetic optimization

The values of ν and μ for a specific implementation depend on the available computing power and the mass storage capacity. The construction plan is modified by deleting or inserting one of two network components: neurons or connections.

3.2 Backpropagation

The ν mutated child networks are trained by means of BP [2]. That is equivalent to a parameter optimization of the components of the construction plans. The number of iteration steps and the learning rate are fixed by initializiation. A considerable advantage of the foregoing mutation step is that BP never gets stuck in local minima. Slowly learning networks will die out by selection.

3.3 Selection

The parent networks for the next generation are selected from the actual ν children and the μ parent networks. To prevent a high degree of structural variability, the child networks are classified in γ categories. As a feature for this classification we are using the number of neurons. We denote the network with the smallest error as the *best network*. Three cases of selection can occur:

$\gamma = \mu$ From each of the γ categories the best network is selected.

$\gamma > \mu$ From each of the γ categories the best network is marked. In a second step the μ best ones are selected from the marked ones.

$\gamma < \mu$ From each of the γ categories the best network is selected. The missing $\mu - \gamma$ networks are duplicated randomly from the former selected ones.

4 Implementation

The described algorithm was realized on a LAN with different UNIX machines, including some SUN 3/50s, SUN 3/60s, SUN 4/150s, one Celerity C2000 and one C3000. Altogether they have an considerable computing capability which is required by the time-consuming numerical BP. To reduce the overall number of processes we have joined mutation with BP. Many combined mutation-/BP-processes can run in parallel on the available computing nodes and work on different partitions of the genepool. One supervisory process which includes selection and a graphical user-interface starts the independent mutation-/BP-processes by Remote Procedure Calls (RPC). In the talk we will present an more detailed view of the user interface and our results in generating networks for character recognition. Our next goal is to increase the computing power by adding a transputer network under HELIOS and X-WINDOWS.

References

[1] Rechenberg, Ingo, Evolutionsstrategie, (problemata frommann-holzboog, 1973)

[2] McClelland, James L., Rumelhart, David E. Explorations in Parallel Distributed Processing, Vol.1-3, (MIT Press, 3rd revision, 1988)

[3] Schiffmann, Wolfram H., Selbstorganisation neuronaler Netze nach den Prinzipien der Evolution, (Fachbericht Physik Nr. 7, Universität Koblenz, 1989)

Parallel Processing in Neural Systems and Computers
R. Eckmiller, G. Hartmann and G. Hauske (Editors)
© Elsevier Science Publishers B.V. (North-Holland), 1990

TEMPORAL-DIFFERENCE-DRIVEN LEARNING IN RECURRENT NETWORKS

Jürgen Schmidhuber*

Institut für Informatik
Technische Universität München
München, Germany

The majority of learning algorithms for dynamic recurrent networks is based on gradient descent techniques. Especially the supervised learning algorithms of this kind use to depend on biologically implausible non-local computations and are computationally expensive.
There is another class of learning procedures which require to consider only the most recent unit activations, but still in principle allow to bridge arbitrary time delays during credit assignment. These algorithms can be thought of as learning by using discrepancies between successive adaptive expectations about future successes or failures. They are based on early work by Samuel, which has been extended by Holland and by Sutton, who created the expression 'Temporal Difference (TD-)Methods'. In this work we describe two learning algorithms for dynamic recurrent networks which bear strong relationships to TD-methods and which require local on-line computations only.

INTRODUCTION

In the following we focus on on-line learning networks with internal and external feedback in non-stationary environments. Internal feedback is given by recurrent connections that allow cyclic activation flow and 'short time memory'. External feedback is given by output actions of the network which may influence the state of the environment thus producing new inputs.

Recently various on-line learning algorithms for supervised and reinforcement learning in fully recurrent networks have been developed [3][4][11][10]. These algorithms are based on gradient descent techniques, which are used to minimize cumulative errors or to maximize cumulative reinforcement signals observed over time.

Gradient descent methods have to consider every past activation for computing appropriate weight changes. Is the concept of gradient descent the only one that allows spatio-temporal credit assignment in recurrent networks? We want to point out that there exists a quite different approach.

This paper is structured as follows: First the principles of non-gradient-based methods by Samuel, Holland, and Sutton are explained. Then two related but also quite different weight changing algorithms for dynamic recurrent reinforcement learning networks based on such principles are described. These algorithms consider only the most recent unit activations for learning, still they in principle allow to bridge arbitrary time delays during on-line credit assignment.

PREVIOUS WORK

Samuel's Principle. 'Samuels' Principle' [5] can be formulated as follows:
If a system is in a state which it assumes to be a bad state, but there is a transition which leads to a state assumed to be a good state, then this transition should be encouraged. Furthermore, from now on the 'bad' state also can be considered to be a good state. Transitions from good states to bad states have to be treated in an analogue fashion.

---

* Research sponsored by a scholarship from SIEMENS AG

This principle was the basic concept for learning in Samuel's successful checkers player (which also included a number of other features, like 'alpha-beta-pruning', which are not relevant for this discussion). Samuel himself states:

'We are attempting to make the score, calculated for the current board position, look like that calculated for the terminal board position of the chain of moves which most probably will occur during actual play.'

Holland's Bucket Brigade Algorithm for Classifier Systems. In this section we shortly review the main idea of Holland's learning algorithm for classifier systems [2]. Messages in form of bitstrings of size n can be placed on a global message list either by the environment or by entities called classifiers. Each classifier consists of a condition part and an action part defining a message it might send to the message list. Both parts are strings out of $\{0, 1, \_\}^n$ where the '$\_$' serves as a 'don't care' if it appears in the condition part. A non-negative real number is associated with each classifier indicating its 'strength'.

During one cycle all messages on the message list are compared with the condition parts of all classifiers of the system. Each matching classifier computes a 'bid' by multiplying its specificity (the number of non-don't cares in its condition part) with the product of its strength and a small factor. The highest bidding classifiers may place their message on the message list of the next cycle, but they have to pay with their bid which is distributed among the classifiers active during the last time step which set up the triggering conditions (this explains the name bucket brigade).

Certain messages result in an action within the environment (like moving a robot one step). Because some of these actions may be regarded as 'useful' by an external critic who can give payoff by increasing the strengths of the currently active classifiers, learning may take place. The central idea is that classifiers which are not active when the environment gives payoff but which had an important role for setting the stage for directly rewarded classifiers can earn their credit by participating in 'bucket brigade chains'.

Sutton's Temporal Difference (TD) Methods. TD-methods were introduced as a generalization of gradient descent techniques specialized for prediction [9]. Temporally successive predictions about future outcomes of some dynamical system serve for credit assignment. For the case where the dynamical system can be described as a Markov process Sutton demonstrates that TD methods can give more accurate predictions than conventional gradient descent techniques *and* require less peak computation. He also points out relationships to Samuel's work and to Holland's bucket brigade. TD methods have been successfully applied to delayed reinforcement learning [1].

ALGORITHM A

In this section we propose a combination of principles of the bucket brigade algorithm with principles of neural networks. The basic network structure is an arbitrary (possibly cyclic) directed graph, where the nodes are familiar processing units. Some units are used for input purposes, others serve as outputs and may be coupled with effectors that may change the environment, which in turn may change the current input. Thus we have external and internal feedback. The set of non-input units is partitioned into predefined 'competitive subsets'.

Notation: $x_j(t)$ is the activation of the jth unit at time t, $w_{ij}(t)$ is the weight on the directed connection from unit i to unit j at time t, $c_{ij}(t) = x_i(t-1)w_{ij}(t-1)$ is the 'contribution' of some connection between i and j at time t, $net_j(t) = \sum_i c_{ij}(t)$.

Here is a description of the algorithm's discrete time version.

1. Initialize all modifiable weights to a positive value. The activations of the input units are determined by environmental perception. All other unit activations are set equal to zero.

2. For all time ticks t:

At time t activate the input-units with values determined by the environment. For all non-input units j compute $net_j(t)$. Activate the unit k with the highest $net_k(t)$ of each competitive subset. In the simplest case: Unit j's activation $x_j(t)$ equals 1 if the non-input unit j is active, and 0 otherwise. Motoric actions based on the activations of the output units may change the environmental state *(external feedback). If unit j is active then its positive modifiable weights change according to*

$$\Delta w_{ij}(t) = -\lambda c_{ij}(t) + \frac{c_{ij}(t-1)}{net_j(t-1)} \sum_{k\,active} \lambda c_{jk}(t) + \eta c_{ij}(t)$$

Here $0 < \lambda < 1$ determines how much of its weight some particular connection has to pay to those connections that were responsible for *setting the stage* at the previous time step, even in the absence of external payoff. η is a small constant if unit j is an output unit and if there is external payoff, and η is 0 otherwise.

We get a dissipative system: 'Weight substance' enters the system in the case of payoff, flows through 'bucket brigade chains', and leaves the system through connections coming from input units.

Experiments. To demonstrate its non-linear and temporal learning capabilities the NBB has been successfully tested on an XOR-problem and on a task where the problem was to distinguish different kinds of motion across a one-dimensional 'retina' [6].

ALGORITHM B

Algorithm B may be viewed as a direct application of Sutton's TD-methods to the temporal evolution of recurrent networks. State transitions in a completely recurrent primary network are observed by a second non-recurrent 'self-supervised' adaptive network, called the critic. Differences between successive state evaluations made by the critic provide update information for the recurrent network.

The fully recurrent primary network consists of linear input units and binary probabilistic non-input units. Its state vector at time t is called $x(t)$. We consider the case where the learning phase is dividable into 'episodes'. (However, a generalization to cases where there are no externally defined episodes is possible.) An episode starts with the initialization of the system's unit activations and is finished when the final reinforcement R becomes known. Here is a description of the algorithm's discrete time version [7][8].

First all weights are randomly initialized to real values.

For all episodes:

In the beginning of each episode, at the first time tick, the activations of input units of the recurrent primary network are initialized to values determined by sensory perceptions from the environment, and the activations of hidden and output units are initialized to 0. For all following time ticks, until there is external reinforcement R (a real number) indicating failure or success :

At a given time tick t:

1. The critic (e.g. a back-propagation network with one output) receives as input the complete activation vector $x(t-1)$ of all units of the primary network. Its one-dimensional output r is interpreted as a prediction of the final reinforcement to be received in the future [1].

2. Each probabilistic non-input unit i of the primary network sums its weighted inputs, this sum is passed to the logistic function $l(x) = \frac{1}{1+e^{-x}}$ which gives the probability that the activation $x_i(t)$ becomes 1, or 0, respectively. Each unit i also stores its last activation $x_i(t-1)$. Output units may cause an action in the environment, this may lead to new activations for the input units (external feedback).

3. If there is external reinforcement R (this means the end of the current episode) then the variable r' is defined to be equal to R.

Otherwise r' is defined to be a new estimation of final reinforcement, obtained by letting the critic evaluate the new state of the primary network. In case of a linear critic $r' = x^T(t)v(t)$.

Using its static learning algorithm (e.g. the generalized delta rule) the critic associates the last activation vector $x(t-1)$ of the recurrent network with r', thus 'transporting expectation back in time' for one time step. So the critic's 'self-supervised' error is given by $r' - r$. Its weight vector is updated according to the its static learning rule; the result is the new weight vector $v(t+1)$.

4. Each directed weight $w_{ij}(t)$ from unit i to unit j of the recurrent network is immediately altered according to $\Delta w_{ij}(t) = \lambda(r' - r)x_i(t-1)x_j(t)$ (with λ being a positive constant), thus encouraging (or discouraging, respectively) the last transition.

Experiments. In [6] we described an experiment where that algorithm successfully was applied to a 'delayed XOR problem' with static inputs. A slightly modified version of the algorithm was successfully applied to a broomstick balancing task with asymmetrically scaled inputs. These tasks could not be learned without the use of hidden units. (In contrast to the work described in [1] no prewired decoder was used for the stick balancing task.)

CONCLUSION

Two algorithms for on-line credit assignment in dynamic recurrent networks have been presented. None of them uses gradient descent as the basic concept for learning (although algorithm B includes a component whose adaptation is based on local gradient descent). Both algorithms are similar in that they work on-line, only depending on computations being local in space and time. Both deal with *arbitrary* temporal delays by handing adaptation information back into time, but never 'further than one time tick' at a time. In both cases there is no need for any unit to store all past activations or to accumulate a sum of past activations [10].

The relationship of algorithm B and TD methods is obvious. But algorithm A is also a TD-like algorithm. Following Sutton's discussion of Hollands bucket brigade for classifier systems we may view the connection strengths as expectations about future successes or failures. If there is a dynamic equilibrium of weight flow through the connections then there is a match of successive expectations about the future. In that case the weights do not change, just like the weights of algorithm B.

Both algorithms represent general credit assignment schemes for neural networks. 'General' often seems to imply 'weak'. How 'weak' are the algorithms? It remains to be seen whether they can be successfully applied to more difficult control tasks.

REFERENCES

[1] A. G. Barto, R. S. Sutton, and C. W. Anderson. Neuronlike adaptive elements that can solve difficult learning control problems. *IEEE Transactions on Systems, Man, and Cybernetics*, SMC-13, 834-846, 1983.

[2] J. H. Holland. Properties of the bucket brigade. In *Proceedings of an International Conference on Genetic Algorithms, 1-7*, Hillsdale, NJ: Lawrence Erlbaum Associates, 1985.

[3] B. A. Pearlmutter. Learning state space trajectories in recurrent neural networks. Technical report, Dept. of Comp. Sci., Carnegie Mellon Univ., Pittsburgh, 1988.

[4] A. J. Robinson and F. Fallside. Static and dynamic error propagation networks with application to speech coding. *Proceedings of Neural Information Processing Systems, American Institute of Physics*, 1987.

[5] A. L. Samuel. Some studies in machine learning using the game of checkers. *IBM Journal on Research and Development*, 3, 210-229, 1959.

[6] J. H. Schmidhuber. The neural bucket brigade. In R. Pfeifer, Z. Schreter, Z. Fogelman, and L. Steels, editors, *Proceedings of the International Conference 'Connectionism in Perspective'*, *Zürich, Switzerland*, Amsterdam: Elsevier, 1988.

[7] J. H. Schmidhuber. Networks adjusting networks. In J. Kindermann and A. Linden, editors, *Proceedings of 'Distributed Adaptive Neural Information Processing'*, *St. Augustin* , 1989.

[8] J. H. Schmidhuber. Recurrent networks adjusted by adaptive critics. In *IJCNN International Joint Conference on Neural Networks, Washington*, 1990.

[9] R. S. Sutton. Learning to predict by the methods of temporal differences. *Machine Learning*, 3, 9-44, 1988.

[10] R. J. Williams. Toward a theory of reinforcement-learning connectionist systems. Technical Report NU-CCS-88-3, College of Comp. Sci., Northeastern University, Boston, MA, 1988.

[11] R. J. Williams and D. Zipser. A learning algorithm for continually running fully recurrent networks. Technical Report ICS Report 8805, Univ. of California, San Diego, La Jolla, 1988.

Parallel Processing in Neural Systems and Computers
R. Eckmiller, G. Hartmann and G. Hauske (Editors)
© Elsevier Science Publishers B.V. (North-Holland), 1990

ASSOCIATIVE MEMORY VERSUS LINEAR FILTER: DISTINCT OPERATIONAL MODES OF STABLE ARTIFICIAL NEURAL SYSTEMS

Bernd Schürmann, Ira Leuthäusser, and Jürgen Hollatz

Corporate Research & Development, ZFE IS INF 2
Siemens AG, 8000 München 83, West Germany

The behavior of a dynamically stable neural network with formal neurons of nonlinear graded response is compared with its linearized version. It is demonstrated that the linear model is incapable of storing patterns. In the nonlinear case, an approximate analytic expression for the connection weight matrix at the end of the learning phase is discussed. Relationships to other approaches are pointed out.

1. INTRODUCTION

In neural network research, two-state (spin glass like) recurrent models [1] and feed-forward algorithms [2] have received much attention.Continuous ('graded response') models have been investigated less, in particular when combined with unsupervised learning. Lately, basic research along these lines is increasing [3, 4], often relying on the pioneer work of Amari, e. g. [5]. Advantages of graded response models are their mathematical elegance and their biological plausibility [6]. If they are in addition globally stable, dynamical reliability is a further advantage. Such stable, graded response neural systems with unsupervised learning are the subject of our contribution.

2. RETROSPECTION

We summarize the model employed. The dynamics of the   activation states z_k for a fully connected network with q formal neurons is assumed to be given by [7] :

$$dz_k/dt \equiv \dot{z}_k(t) = - Cz_k + \sum_{s=1}^{q} w_{sk}(t) \, S[z_s(t)] + K_k \, . \tag{2.1}$$

The change of z_k with time originates from a passive decay term with constant C, from an internal input term consisting of a weighted sum of output signals $S(z_s)$ of units linked with node k, and from an external input to node k, being the component K_k of a pattern vector $\mathbf{K}$. As signal function we use $S(z) = \tanh(z/T)$.

The vector $\mathbf{K}$ may contain noise, i. e. it is assumed to be composed of a constant systematic ('clean') part and a rapidly varying random part:

$$\mathbf{K} = \mathbf{K}_{clean} + \mathbf{K}_{rand}. \tag{2.2}$$

For the internal and external input terms of (2.1) to be of comparable size, $\mathbf{K}_{clean}$ is assumed to scale with the number of nodes, i. e.$K_k{}^{clean} \simeq q \cdot sgn(K_k{}^{clean})$.

The network is asymptotically stable, if the weights w_{sk} obey the 'detailed balance' relation [4]

$$\mu_k \, w_{sk}(t) = \mu_s \, w_{ks}(t) \, , \tag{2.3}$$

and underly the dynamics

$$\dot{w}_{sk}(t) = - \lambda \, w_{sk}(t) + \varepsilon_{sk} \, S[z_s(t)] \, S[z_k(t)] \, , \tag{2.4}$$

with $\varepsilon_{sk} = \mu_{max(s,k)} / \mu_k$ and real positive constants μ_s, μ_k and λ. The change of the weights w_{sk} with time is caused by a 'forgetting' term with constant λ and by a Hebbian learning term. The expression

$$L(t) = \sum_{s=1}^{q} \mu_s \left[\tanh(z_s/T)\,(Cz_s - K_s) - C\,T\,\ln\cosh(z_s/T)\right] \tag{2.5}$$

$$- (1/2)\,\Sigma_{s,k}\,\mu_k\,w_{sk}\,(\tanh(z_s/T)\tanh(z_k/T) - (1/2)\,\lambda\,w_{sk}/\varepsilon_{sk})$$

is a Lyapunov function for the system, i. e. $dL/dt < 0$. Finally, we define the setting in which our investigations take place. During the learning phase, for a time interval $\delta t_\alpha^{(v)}$, in the α-th presentation cycle, sample pattern vectors $\mathbf{K}^{(v)}$ from a pattern distribution P_v are offered to the system, a new sample at each time step within $\delta t_\alpha^{(v)}$. Altogether, there are m distributions of different patterns and c cycles. The system learns the centroids of the distributions which are the clean patterns $\mathbf{K}^{(v)}{}_{\text{clean}}$ [8].

3. NONLINEAR VERSUS LINEAR DYNAMICS

Consider a time period $\delta t_\alpha^{(v)}$ when different representations of a sample vector $\mathbf{K}^{(v)}$ are presented to the system at each time step. For transparency, the indices v and α are omitted in this section. We assume for the ratio of the passive decay and the forgetting constants

$$C/\lambda \gg 1. \tag{3.1}$$

Typical values are $C \simeq 10$ and $\lambda \simeq 1$. (3.1) introduces 'short term memory' and 'long term memory' time scales C^{-1} and λ^{-1} for the activation states z_k and the connection weights w_{sk}, respectively.

3.1 Filtering Behavior with Linear Neurons

For $z_k/T \ll 1$, (2.1) and (2.4) become

$$\dot{z}_k(t) \simeq -Cz_k + \Sigma_s\,w_{sk}(t)\,[z_s(t)/T] + K_k, \tag{3.2}$$

$$\dot{w}_{sk}(t) = -\lambda\,w_{sk}(t) + \varepsilon_{sk}\,z_s(t)\,z_k(t)/T^2. \tag{3.3}$$

Near the end of the presentation period δt for samples $\mathbf{K}$, the noise in (3.2) is averaged out, and the activation states z_k become stationary. Hence,

$$z_k \simeq (1/C)\,(\Sigma_s\,w_{sk}(t)\,z_k/T + K_k^{\text{clean}}). \tag{3.4}$$

Retaining only the leading order term on the right hand side (r. h. s.), (3.4) becomes

$$z_k \simeq K_k^{\text{clean}}/C. \tag{3.5}$$

This yields for the weight matrix (3.3) a linear equation with the solution

$$w_{sk}(t) = w_{sk}(0)\,e^{-\lambda t} + [\varepsilon_{sk}\,K_s^{\text{clean}}\,K_k^{\text{clean}}/(\lambda C^2 T^2)]\,(1 - e^{-\lambda t}). \tag{3.6}$$

Inserting z_k and w_{sk} in (2.5), gives for the Lyapunov function to leading order, apart from irrelevant constant terms, a function linear with respect to the activation states:

$$L \simeq -\sum_{s=1}^{q} (\mu_s/T)\,K_s^{\text{clean}}\,z_s. \tag{3.7}$$

Hence, the Lyapunov function does not have any minima in activation state space. Consequently, the system possesses no memory. At most, it can act as a filter.

3.2 Pattern Storage with Nonlinear Neural Network Dynamics

In the strongly nonlinear limit $z_k/T \gg 1$, the Lyapunov function (2.5) becomes, up to irrelevant constant terms [4],

$$L(t) \simeq -\Sigma_s\,\mu_s\,K_s^{\text{clean}}\,\operatorname{sgn} z_s - (1/2)\Sigma_{s,k}\,\mu_k\,w_{sk}\,(\operatorname{sgn} z_s \operatorname{sgn} z_k - (1/2)\,\lambda\,w_{sk}/\varepsilon_{sk}). \tag{3.8}$$

Because of the term bilinear in $\{\operatorname{sgn} z_k\}$, this function possesses minima in activation state space, and hence the system is, in addition to filtering, capable of pattern storing. The similarity of (3.8) to Hopfield's energy function for 2-state models [9] is obvious. The term linear in $\{\operatorname{sgn} z_k\}$ constitutes an external field. This enhances pattern retrieval [10].

4. THE LEARNING MATRIX IN THE STRONGLY NONLINEAR LIMIT

4.1 An Approximate Analytic Expression

We derive an analytic result for the connection strengths at the end of the learning phase. For transparency, we make the simplifying assumption that in each cycle a, $a = 1, ..., c$, the presentation periods for samples from a specific distribution P_ν, $\nu = 1, ..., m$, are equal, i. e. $\delta t_a^{(\nu)} \equiv \delta t^{(\nu)}$. Moreover, we introduce the duration time $\tau = \Sigma_\nu \delta t^{(\nu)}$ for one learning cycle. Because of (3.1), at the end of sufficiently long time intervals $\delta t^{(\nu)}$, $\mathrm{sgn}\,(z_s) \simeq \mathrm{sgn}\,(K_s^{(\nu)\mathrm{clean}}) \equiv \xi_s^{(\nu)}$ so that

$$\dot{w}_{sk}^{(\nu)} \simeq -\lambda\, w^{(\nu)}{}_{sk} + \varepsilon_{sk}\, \xi_s^{(\nu)}\, \xi_k^{(\nu)} \;, \tag{4.1}$$

with the solution

$$w_{sk}^{(\nu)} = w_{sk}^{(\nu-1)}\, e^{-\lambda t} + (\varepsilon_{sk}/\lambda)\, \xi_s^{(\nu)}\, \xi_k^{(\nu)}\, (1 - e^{-\lambda t}), \quad \nu = 1, .., m. \tag{4.2}$$

By use of this recurrence relation, an analytic expression for the connection strengths at the end of the learning phase is obtained:

$$w_{sk}^{\mathrm{learn}}\, (w_{sk}^{(0)}, \{\mathbf{K}_{\mathrm{clean}}^{(\nu)}\};\, \lambda;\, \varepsilon_{sk};\, \{\delta t^{(\nu)}\}, \tau, c)$$

$$= w_{sk}^{(0)}\, e^{-\lambda \tau c} + \frac{1 - e^{-\lambda \tau c}}{1 - e^{-\lambda \tau}} \left\{ \sum_{\nu=1}^{m} \varepsilon_{sk}\, \xi_s^{(\nu)}\, \xi_k^{(\nu)} \;\; \frac{e^{\lambda \delta t^{(\nu)}} - 1}{\lambda}\; e^{-\lambda \sum_{\mu=\nu}^{m} \delta t^{(\mu)}} \right\}. \tag{4.3}$$

To simplify the discussion, we assume $w_{sk}^{(0)} = 0$. For $c = 1$, the factor $F(\lambda, \tau, c) \equiv [1 - \exp(-\lambda \tau c)] / [1 - \exp(-\lambda \tau)]$ becomes unity so that the expression in the brackets is the weight matrix element after the 1st cycle. It consists of a sum of m terms, one for each pattern. It is seen that the pattern learnt last is weighted the most and the one learnt first is weighted the least (cf. the factor $\exp\,[-\lambda \sum_{\mu=\nu}^{m} \delta t^{(\mu)}]$ in (4.3)). Hence, the system tends to forget its past during the course of learning. If desired, forgetting can be remedied by offering the earlier patterns for a longer period $\delta t^{(\nu)}$ than the later ones (cf. the factor $\exp\,[\lambda\, \delta t^{(\nu)}] - 1$ in (4.3). For $c \gg 1$, the front factor F becomes larger than unity, i. e. the overall strength of the learning matrix is increased as compared to $c = 1$. This may not necessarily improve the net's performance in the recall phase, since there is a delicate balance between the internal and external inputs of (2.1).

4.2. Relation to other approaches

In this subsection, two limits of (4.3) are discussed which are the Hopfield [9] and the marginalist learning schemes [11].

4.2.1. The Hopfield limit

The forgetting constant λ plays a crucial role in (4.3). If it becomes too large, the system will not learn at all. For small λ, the Hopfield limit [9] is obtained as follows: (i) put $c = 1$, (ii) put $\varepsilon_{sk} = 1$ (symmetric matrices), (iii) observe that $(\exp\,[\lambda\,\delta t^{(\nu)}] - 1)/\lambda \simeq \delta t^{(\nu)}$ for $\lambda \to 0$, (iv) $\exp\,[-\lambda \sum_{\mu=\nu}^{m} \delta t^{(\mu)}] \simeq 1$ for $\lambda \to 0$, and finally, put all $\delta t^{(\nu)}$ equal to unity (for a typical time step size of $h = 0.1$, $\delta t^{(\nu)} = 1$ means a pattern is presented 10 times).

4.2.2. The Marginalist Learning Scheme :

In the marginalist learning scheme [11], the acquisition intensity for learning patterns decreases exponentially. Consider now equal time intervals $\delta t^{(\nu)} = \delta t$. For the case $c = 1$, the marginal learning scheme is regained, if $\lambda \delta t$ is chosen equal to $(\varepsilon/2)\,(\varepsilon/N)$ for $\lambda \delta t \ll 1$ and $\varepsilon_{sk} = 1$:

$$w_{sk}^{\mathrm{learn}} = w_{sk}^{(0)}\, e^{-\lambda \tau} + \frac{\varepsilon}{N} \sum_{\nu=1}^{m} e^{-(\varepsilon^2/N)\,(m-\nu)}\, \xi_s^{(\nu)}\, \xi_k^{(\nu)} \;. \tag{4.4}$$

In this case the marginalist learning scheme has its optimal storage capacity α^{opt} at $\epsilon^{opt} = 4.09$ with $\alpha^{opt} = 0.049$.

For repetitive learning of a sequence with $m = gN$ patterns, i.e. $c \gg 1$, the marginalist learning scheme finds for the weights:

$$T_{ij} = (\epsilon/N) \, 1/\sqrt{1-\exp(-\epsilon^2 g)} \, \Sigma_{\mu=0} \exp(-\mu \, \epsilon^2/2N) \, \xi_i^{(\mu)}\xi_j^{(\mu)} \tag{4.5}$$

where $\mu = 0$ corresponds to the pattern learnt last. In the limit $\epsilon^2 g \ll 1$, the Hopfield limit is regained and hence the Hopfield storage capacity $\alpha_{Hopfield} = 0.138$. Thus, by repetitive learning with $c \gg 1$, the optimal storage capacity of the marginalist learning scheme can be enhanced up to the Hopfield limit. For eq.(4.3) this means in the limit $c \to \infty$, $\epsilon_{sk} = 1$, $\tau = m\delta t$ and $\epsilon^2 g \ll 1$

$$w_{sk}^{learn} \approx 2/(Ng\epsilon) \, \Sigma_{v=1} \, \xi_s^{(v)} \, \xi_k^{(v)}. \tag{4.6}$$

Again the Hopfield limit is found as $2/(g\epsilon)$ is constant and thus the storage capacity is $\alpha_{Hopfield} = 0.138$.

5. OUTLOOK

Insight has been gained in the operational modes of stable neural networks in dependence of the degree of nonlinearity of their signal functions. In the linear limit, the system at best can act as a filter. In contrast, in the strongly nonlinear limit, it is able to store patterns and hence possesses associative memory.

We have obtained, at the end of the learning phase, an approximate analytic expression for the learning matrix which contains other results as limiting cases. Our model does not suffer from the limitations of thermodynamic approaches. It automatically incorporates finite size effects. These enhance the storage capacity [11] and in addition wash out phase transitions. Hence, in case of overloading, the model's performance rather deteriorates gradually instead of collapsing suddenly.

ACKNOWLEDGEMENT

J. Hollatz should like to thank the Siemens AG for financial assistance.

[1] Sompolinsky, H., Physics Today (December 1988) 70.

[2] Rumelhart, D.E., Hinton, G.E. and Williams, R.J., Learning Internal Representations by Error Propagation, in: Rumelhart, D.E., McClelland, J.L., Parallel Distributed Processing, Volume 1(MIT Press, Cambridge, MA, 1986) pp. 318-362.

[3] Kosko, B., Appl. Opt. **26** (1987) 4947.

[4] Schürmann, B., Phys. Rev. **A 40** (1989) 2681.

[5] Amari, S.-I., Biol. Cybernetics **26** (1977) 175.

[6] Hopfield, J.J., Proc. Natl. Acad. Sci. USA **81** (1984) 3088.

[7] Grossberg, S., Neural Networks 1 (1988) 17.

[8] Schürmann, B., Sampling Learning, Recall and Filtering in Stable, Adaptive Neural Systems with Graded Response, Proceedings of the International Joint Conference on Neural Networks, Washington, D.C., Jan. 1990, in print.

[9] Hopfield, J.J., Proc. Natl. Acad. Sci. USA **79** (1982) 2554.

[10] Amit, D.J., Gutfreund, H. and Sompolinsky, H., Ann. Phys. [N.Y.] **173** (1987) 30.

[11] Mézard, M., Nadal, J.P. and Toulouse, G., J. Physique **47** (1986) 1457.

Parallel Processing in Neural Systems and Computers
R. Eckmiller, G. Hartmann and G. Hauske (Editors)
© Elsevier Science Publishers B.V. (North-Holland), 1990

Competitive Sequence Learning

Christoph Windheuser, Jörg Kindermann
Gesellschaft für Mathematik und Datenverarbeitung Z1.ASYMP
5205 Sankt Augustin
email: chwi@gmdzi.uucp — joerg@gmdzi.uucp

Abstract

We present a multilayer neural network which learns to classify sequential input patterns with an unsupervised competitive learning algorithm. During training the units of different layers specialize on sequences of different length. The recognition of a long sequence proceeds incrementally through recognition of its parts in lower layers. The desired temporal behavior of the units is implemented by delayed connections. The units additionally are provided with a hysteresis function acting as a temporal memory. When a sequence is partially recognized, the units generate expectations with respect to the remainder of the sequence. This allows the processing of noisy input patterns and incomplete sequences. The model worked with encouraging results for the recognition of German words.

Keywords: multilayer network, competitive learning, time invariant sequence recognition.

1 Introduction: Neural Networks for Sequence Learning.

Neural networks are good at static associative and classification tasks. As such they have also been used for inherently sequential information processing like speech recognition. The trick is to map the input signal of a time interval onto a spatial representation. This is called the time window approach.

After the first successes [2], [5] it was however soon realized that the sequential nature of signals was difficult to incorporate into plain neural networks. The time window approach imposes rigid constraints on the temporal context of the signal being analyzed. Several different approaches, such as delayed links [8], and dynamic recurrent networks [1] have been proposed to overcome these difficulties. All of these use supervised training techniques for network adaptation. This seems to be problematic, particularly in domains where it is difficult to define objective functions [3]. Another drawback is that dynamic recurrent networks and their attractors are still not very well understood [6], [10] .

Therefore it is attractive to have a network which is capable of unsupervised adaptation to sequential input, using temporal contexts. See [7] for a first attempt. In our paper we describe the design of a hierarchical, self-organizing network and its adaptation algorithm.

2 The Basic Idea.

We assume a "1-out-of-n" representation of the input patterns, i. e. at every time at most one input unit is active. Now we suppose a unit which receives input from two other units, unit 1 and unit 2. The connection to unit 1 is delayed with a delay time equal to the time of the update cycle of the units, whereas the connection to unit 2 is undelayed. When in a first time step unit 1 is active and in the next time step unit 2 is active, the unit above receives the two activations simultaneously. If this unit works like an AND gate, it recognizes a sequence of activations from units 1 and 2. Because unit 1 is not active any longer, the above unit will loose its activity in the third time step, even when unit 2 still is active. To avoid this, the above unit has a *hysteresis feature*. This means that the unit needs a lot of activation (from unit 1 *and* unit 2) to become active, but much less activation to stay active (only from unit 2). So the time unit 1 or unit 2 is active is unimportant, only the *change* of the activation from unit 1 to unit 2 is recognized by the above unit.

3　The Behavior of the Units.

Every unit receives input from units in the previous layer. The set $\mathcal{P}$ denotes the indices of all units in the previous layer sending activation to unit j. Then the total netinput of unit j is:

$$net_j = \sum_{i \in \mathcal{P}} w_{ji} a_i$$

a_i is the activation of unit i and w_{ji} is the weight of the connection from unit i to unit j.

Every unit j computes its activation a_j from the net input net_j with the *activation function*. To avoid unbounded increase of the activation, it is restricted to an interval:

$$a_j \in [0;\ 1]$$

The main task of a unit is to work like an AND gate. Only if it receives input from a delayed and an undelayed connection, it becomes active. The simplest way realizing this is using a *heavy-side function* $h(net_j)$ with:

$$h(net_j) = \begin{cases} 0 & \text{if } net_j < \theta \\ 1 & \text{if } net_j \geq \theta \end{cases}$$

θ denotes an arbitrary threshold.

One disadvantage of the heavy-side function is its discontinuity between the active and nonactive state. To work with vague or noisy input patterns, we need a continuous activation function. The activation of the unit then is equivalent to the recognition probability.

So we choose the well known *sigmoidal function* as activation function:

$$f(net_j) = \frac{1}{1 + e^{-a(net_j - b)}}$$

With the parameters a and b we determine the shape of the function: parameter a controls the steepness. With this parameter we are able to approximate the heavy-side function.

Parameter b is used to implement an important feature of our units: the hysteresis. This means that a unit will be pushed to a high level of activity only by an input greater than some upper threshold θ_U, subsequently this level is sustained by a much lower input than was necessary to trigger the increase. When the input falls down below some lower threshold θ_L, the activity decreases to zero. The activation function looks like a hysteresis loop (see figure 1).

The hysteresis function was also found in groups of neurons in the cortex [9].

We define the hysteresis activation function by setting the value of parameter b dynamicly in dependence of the netinput. Beyond this we need a *state* of a unit: it is either *active* or *inactive*.

state is defined for the values $net_j < \theta_L$ and $net_j > \theta_U$:

$$state = \begin{cases} inactive & \text{if } net_j < \theta_L \\ active & \text{if } net_j > \theta_U \end{cases}$$

For $\theta_L \leq net_j \leq \theta_U$ *state* is dependent on its previous value:

$$state = \begin{cases} inactive & \text{if previous } state \\ & \text{was } inactive \\ active & \text{if previous } state \\ & \text{was } active \end{cases}$$

Now we can define parameter b:

$$b = \begin{cases} b_1 & \text{if } state = nonactive \\ b_2 & \text{if } state = active \end{cases}$$

In our simulations we choose the following parameters:

$$\begin{aligned} a\ \ &= 20 \\ b_1\ &= 0.75 \\ b_2\ &= 0.25 \\ \theta_L\ &= 0.2 \\ \theta_U\ &= 0.8 \end{aligned}$$

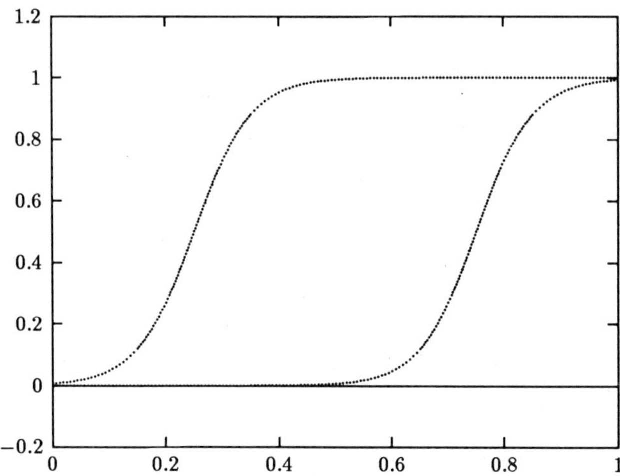

Figure 1: The hysteresis loop

4   The Architecture.

The units are grouped in layers. Every unit in one layer receives activation from all units in the previous layer and sends activation to all units in the successive layer. The first layer is the input layer and the last layer is the output layer of the network. Each unit is connected to *every* unit in the previous layer. One of these connections is *undelayed*, the others are *delayed* with a discrete delay time (multiples of the update cycle). The undelayed connection has a fixed weight (in our simulations the value 0.5), the delayed connections have modifiable weights, which are changed by the learning rule. During the training one of the delayed connections will evolve to its maximum value, whereas the other delayed connections decrease to zero. After successful training a unit will receive input only from one delayed and one undelayed connection. But this means that the unit has specialized itself to recognize a particular two step sequence.

Inside a layer the units are organized in *groups*. Each unit of a group has an undelayed connection to the same unit in the previous layer. So the number of groups in a layer is equal to the number of units in the previous layer. In the simplest architecture each group consists of only *one* unit. This means that every layer has the same number of units and every unit in a layer receives undelayed input from a different unit in the previous layer. Figure 2 shows such a model with three units per layer. If the number of units per layer is N, every layer can recognize N of $N(N-1)$ possible two step sequences. With this simple architecture it is impossible to recognize two sequences with the same second pattern, because these sequences must be recognized in the same group. Therefore the unit in this group will specialize itself to recognize the most frequent sequence.

When a group consists of more than one unit, all units have the same input connections. To avoid specialization on the same sequence, the units inside the group are in competition to each other. In that way only one unit, which receives most input, will win and will be able to change its weights towards the recognized sequence. All other units of the group decrease their activation to zero.

To determine the number of units in a group of a layer, we have to estimate the possible two step sequences with identical second patterns, witch have to be recognized in that layer. So we need some a priori knowledge about the input pattern to choose the optimal architecture.

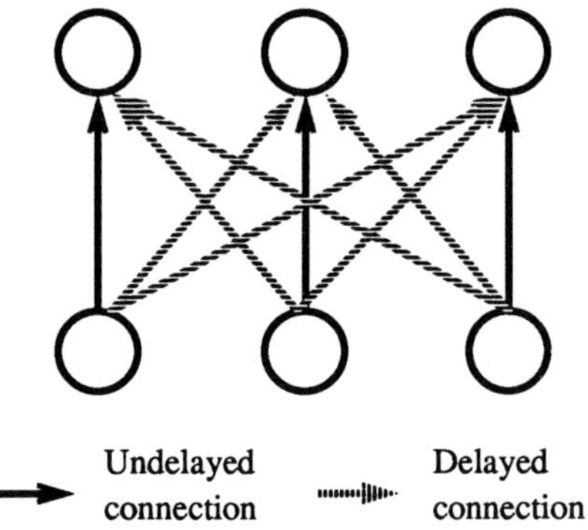

Figure 2: A simple network with 3 units per layer

5 Learning.

In our model we use the well known and biologically plausible Hebbian learning rule [4]. It says that a weight is increased only when the sending *and* the receiving unit both are active. We only modify the delayed weights, the undelayed weights are fixed.

The weights are restricted to an interval:

$$w_{ji} \in [0;\ w_{max}];\ \ w_{max} > 0 \tag{1}$$

We only use positive weights.

To avoid all weights reaching their maximum value after some training time they are in competition to each other. This is implemented by *normalizing* the weights:

$$\sum_{i \in \mathcal{P}} w_{ji} = const = w_{sum} \tag{2}$$

with:

$$w_{sum} \geq w_{max}$$

Every update cycle we increase each weight w_{ji} by a value δ_{ji} which is proportional to the product of the activation of unit i and j. To avoid violating equation (1), we have to restrict the increasing of the weights:

$$\delta_{ji} = \alpha(w_{max} - w_{ji})a_i a_j \tag{3}$$

$\alpha \in [0;\ 1]$ denotes the *learning rate*.

Because of equation (3) it is impossible for δ_{ji} to become greater than $w_{max} - w_{ji}$ and hence w_{ji} cannot become greater than w_{max}.

When we increase w_{ji} by δ_{ji}, we have to decrease all other weights to avoid violating equation (2). But because no weight can become negative, the decrease is proportional to the value of the weights. For that reason the condition $w_{sum} \geq w_{max}$ is important.

The update cycle of the weights consists of two steps: After δ_{ji} is computed by equation (3), the first step normalizes the weights:

$$w'_{jk} = w_{jk} - \delta_{ji}\frac{w_{jk}}{w_{max}} \quad\quad k \in \mathcal{P} \tag{4}$$

In the second step weight w_{ji} is increased by δ_{ji}, the other weights are left unchanged:

$$w''_{jk} = \begin{cases} w'_{jk} & \text{if } k \neq i \\ w'_{jk} + \delta_{ji} & \text{if } k = i \end{cases} \quad\quad k \in \mathcal{P} \tag{5}$$

6 Simulation Results.

We built a model to recognize German words consisting of the six letters: *a, e, n, o, t, u*. In the first layer – the input layer – six units represent the six letters. In the second layer we recognize sequences of two letters. The delay time of the delayed connections is equal to one update cycle. Because in our pattern set the maximum number of two letter sequences with identical second letter is three, every group in layer two consists of three units. So the total number of units in layer two is 18. In Layer three the delay time is equal to two update cycles. The units in this layer are able to recognize sequences consisting of two two letter sequences recognized in layer two. Every group consists of two units because we want to learn words with te same two last letters. Layer three is the output layer. Therefore our model is able to recognize words consisting of four letters which are presented successively in the input layer.

To reach maximum stability, we set $w_{max} = w_{sum} = 0.5$. So when a weight reaches its maximum value of 0.5, all other weights of the unit decrease to zero. The starting point of the weights is $\frac{w_{sum}}{N}$, N denotes the number of units in the layer. The learning rate α is set to 1.0.

This model works with encouraging results. After twenty presentations of twelve German words consisting of four letters, the weights of layer two evolved to their final values. After thirty further presentations, the weights of the units in layer three reached their final values. Twelve of the units in this layer reached their maximum activation when the particular word was presented in the input layer. The other units never had any activation. In addition, the trained network was able to recognize time stretched sequences. When it had learned the word *"anne"*, it also recognized the word *"aannnnee"*. This could be a disadvantage in some tasks, but it is well suited for speech recognition.

In another experiment we set $w_{max} = w_{sum} = 0.75$. With this parameters one weight per unit was able to reach 0.75. This means after the appearance of pattern one of the learned sequence, the unit reached an activation of 0.5 *without* the presentation of pattern two. This can be interpreted as an *expectation* of the second pattern. Therefore it was not necessary to present the second pattern with full activation to recognized the sequence quite well. Because this works in every layer, the network was able to recognize a word even when only the beginning of the word was presented clearly.

7 Conclusion.

We have presented a new approach to sequence recognition by unsupervised learning. When the number of units is restricted, only the most frequent sequences will be learned. Learning and recognition is hierarchically organized, so it is rather efficient, because a unit representing a micro sequence in a lower layer is used by several units in a higher layer, representing longer sequences. After successful training the network is able to recognize time stretched and noisy sequences. When the input sequence is ambiguous, the network will choose that sequence which appears more frequently in the training set. Problems remain with the restricted network architecture, leading to a combinatorial explosion of unit numbers in higher layers. Future research has to modify the learning rule to allow adaptive undelayed connections. This would increase the flexibility of adaptation to different sequences in the training set.

Acknowledgement: The preparation of this paper was supported in part by grant ITR 8800 L7 from the German Federal Ministry for Scientific Research and Technology (BMFT).

References

[1] S. Anderson, J. Merrill, and R. Port. Dynamic speech categorization with recurrent networks. In D. Touretzky, G. Hinton, and T. Sejnowski, editors, *Proceedings of the 1988 Connectionist Models Summer School*, pages 398–406, Morgan Kaufmann, Palo Alto, June 1988.

[2] H. Bourlard and C. J. Wellekens. Multilayer perceptrons and automatic speech recognition. In M. Caudill and C. Butler, editors, *Proceedings of IEEE First International Conference on*

Neural Networks, San Diego, pages 407–416, IEEE, 1987.

[3] J. B. Hampshire and A. H. Waibel. *A Novel Objective Function for Improved Phoneme Recognition Using Time-Delay Neural Networks*. Technical Report CMU-CS-89-118, Computer Science Department, Carnegie Mellon University, Pittsburgh, PA, March 1989.

[4] Donald O. Hebb. *The Organization of Behavior*. Wiley, New York, 1949.

[5] T. Kohonen. The 'neural' phonetic typewriter. *IEEE Computer*, 21:11–22, 1988.

[6] B. A. Pearlmutter. *Learning State Space Trajectories in Recurrent Neural Networks*. Technical Report CMU-CS-88-191, Carnegie Mellon University, 1988.

[7] V. V. Tolat and A. M. Peterson. A self-organizing neural network for classifying sequences. In *Proceedings of the First International Joint Conference on Neural Networks, Washington, DC*, pages II 561 – II 568, IEEE, IEEE TAB Neural Network Committee, San Diego, 1989.

[8] A. H. Waibel. Modular construction of time-delay neural networks for speech recognition. *Neural Computation*, 1:39–46, 1989.

[9] H.R. Wilson and J.D. Cowan. A mathematical theory of functional dynamics of cortical and thalamic nervous tissue. *Kybernetik*, 13:55–80, 1973.

[10] M. Zak. Terminal attractors in neural networks. *Neural Networks*, 2(4):259–274, 1989.

Section 6
Evaluation of
Artificial Neural Networks

Parallel Processing in Neural Systems and Computers
R. Eckmiller, G. Hartmann and G. Hauske (Editors)
© Elsevier Science Publishers B.V. (North-Holland), 1990

225

IDEAL NEURONS FOR NEURAL COMPUTERS

Igor Aleksander, FEng.

Department of Electrical Engineering
Imperial College of Science Technology and Medicine
London SW7 2BT
e-mail:UMEDE01@uk.ac.imperial.cc.vaxa

ABSTRACT

The building of neural computers based on biological models of the neuron is fashionable but has never been properly justified as being optimal in an engineering sense (least cost for the required function). There are many alternatives properly founded in engineering practice and logic[1]. This paper seeks to define an ideal specification (the Ideal Artificial Neuron or IAN) based on engineering considerations. For the first time the design of a G-ram is disclosed which is an embodyment of an IAN. As the input size of the IAN grows the cost of its embodyment grows exponentially. It is proposed that the specification of an IAN device may be approximated by the use of logic decomposition techniques. These are based on the G-ram. The tradeoff between cost and performance of decomposed forms is briefly discussed, resulting in the conclusion that the conventional neuron based on the McCulloch and Pitts formulation is at the lowest end of the performance range.

1. INTRODUCTION

The common feature of the "neuron" in most current work on artificial neural networks is an n-input device with input values labelled:
$$i_1, i_2, i_3, \dots , i_n$$
where such values could be real or binary numbers. We restate the familiar function where each input i_j is multiplied by some "synaptic" weight w_j, and the entire neuron reacts by producing an output which is a function f(S) of the sum of products :
$$S = \sum i_j\, w_j$$
Hopfield[2] used the McCulloch and Pitts formulation in which i_j is binary, w_j has real values and f(S) is binary in the sense that f(S) assumes one value if S exceeds some threshold, and another if it does not. He showed that a *fully connected network* of K neurons with each neuron connected through n=K-1 inputs to the outputs of all the other neurons but not its own, has an important "emergent" property of forming a content addressable memory. We shall first state this property formally and then show that there exists an ideal neural function, quite distinct from the above, which has perfect content-addressing performance. We shall then show that this leads to a family of functions for which a cost is associated with content addressing performance. The conventional neural formulation is seen to be at the low-cost, poor performance end of this family.

2.   THE IDEAL NEURON

A binary net (such as the Hopfield model described above) can, at any one time, be in a state *s* taken from a set **S** with a total of 2^K possible states, one for each binary pattern of K variables. **T,** a subset of **S**, is defined as a training set. The "neurons" must adapt their input/output functions so as to maintain each state of **T** stable in the net with time. The property of *conent addressability* is present if, starting the net in state *s* (from set **S**), the net goes through a set of state changes ending in the element of **T** which is most similar to *s*. ("Similar" refers to a measure of the proportion of K binary points in which two patterns agree). We note that in Hopfield nets, content addressability is not guaranteed for two reasons: first, there exist stable states which are outside **T** (these are known as "false minima") and second, **T** itself cannot be arbitrarily determined due to the limited variability of the McCulloch and Pitts function (called "linear separability").

This work is supported by the UK Science and Engineering Research Council.

An Ideal Artificial Neuron (IAN) may be defined as having the following properties: 1) It must record the appropriate response to all the patterns in **T** during a learning period, and subsequently be capable of producing these responses at its output when addressed by all such patterns. 2) Given an input due to a state not in **T**, the IAN must produce a response as the most similar pattern in **T**. 3) States which lead to input patterns that do not have a clear, single, most similiar, element of **T** cause the output to be arbitrarily 0 or 1 with equal probability (the *u* state).

We assume that IANs are placed into a Hopfield-like net which is fully connected as described above, and which is asynchronous in the sense that only one neuron can attempt to change its output at any one time, and all neurons will have made such an attempt in W time units. In consequence to the definition of the IAN, a network starting in a state *s* with a clear "nearest neighbour" *s'* in **T** will enter *s'* within time W. A state without a clear nearest neighbour in **T** will cause the net neurons to select their outputs arbitrarily until a state in **T** is reached.

As an aside, this perspective shows two ways in which the topology of a Hopfield net is arbitrarily restricted. First, a synchronous net in which all elements are allowed to attempt to fire simultaneously (at the arrival of a "clocking" pulse) will enter the nearest state in **T** in one step rather than W time units. Second, the "blindness" of a neuron to its own output is unnecessary and merely leads to anomalies with respect to an ideal behaviour (to be discussed elsewhere).

3.   THE G-RAM

It is quite possible to make an electronic IAN by the following modification of known random-access-memory (RAM) neuron models such as the Probabilistic Logic Node (PLN) [3] (this is true of more sophisticated models auch as the pRAM[4] or the n-state PLN (nPLN)[5], but this is beyond the scope of this paper). Instead of the usual "read" and "use" phases in such devices, a "spread" phase is added during which additional storage locations in the RAM are set according to properties (2) and (3) of the IAN discussed above. We call the physical embodyment of the IAN the G-RAM or "generalising random-access memory".

The G-RAM has K address inputs and hence 2^K storage locations. Each storage location contains B bits. In the style of pRAMs and nPLNs, the 2^B possible messages stored at each location are ordered to represent discrete firing probabilities of generating a 0 or a 1 at the output terminal of the device. These probabilities are set during a training period as in Myers, 1989[4]. There are interconnections between the storage locations so that a location address Aj can be influenced by locations with addresses Aj1 to AjK each of which differs from Aj by exactly one bit, the bit being different for each of the K adresses. This influence is determined by a *spreading algorithm* which is applied during the "spreading" phase. A specific example of such a design is given below.

Let B=2 and say that Pj and Qj are the two bits of the storage location addressed by input address Aj. Only three of the four encodings of these two bits are used:

Pj,Qj=00 causes the overall output to be 0,
Pj,Qj=11 causes the overall output to be 1,
Pj,Qj=01 causes the overall output to be 0 and 1 with equal probability.

The logic of the spreading algorithm is:

$$\sim Qj = \sim Pj \cdot Qj \cdot \sum_{j=1}^{K} \left(\sim Pji \cdot \sim Qji \cdot \prod_{l \neq j} (\sim Pjl \cdot Qjl) \right)$$

$$Pj = \sim Pj \cdot Qj \cdot \sum_{j=1}^{K} \left(Pji \cdot Qji \cdot \prod_{l \neq j} (\sim Pjl \cdot Qjl) \right)$$

Here $\sum$ refers to a multiple OR operation and $\prod$ to a multiple AND, • being AND and ~ being NOT.

This is clearly implementable with conventional digital techniques. Clocking arrangements have not been included in this arrangement, but naturally, they have to exist in order to achieve spreading in synchronism to all nearest neighbours only. It may be shown that the above procedure converges to a stable internal state for the contents of the memory locations.

4.   THE COST OF G-RAM: NEED FOR DECOMPOSITION

The cost of G-ram increases exponentially with K as total memory $= KB2^K$. This mitigates against G-ram being used directly as IANs, but suggests that the IAN be decomposed into G-ram circuitry. There are several ways of doing this including methods based on pyramids and WISARD discriminators[7]. For the purposes of making clear what is meant by "a decomposition" we shall pursue an example based on the latter of these two. This form of decomposition is shown in fig. 1.

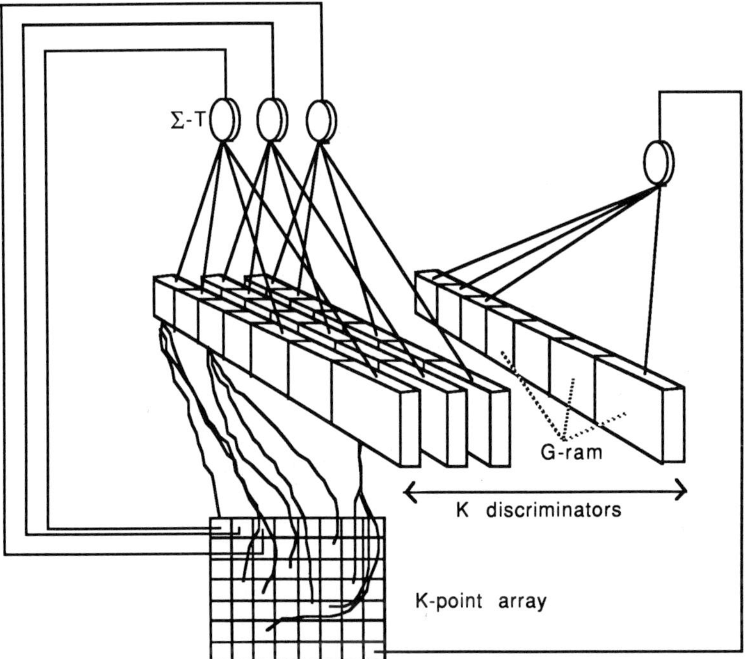

Fig 1. A decomposition into G-ram discriminators.

Here the K-input is covered by K/N N-input G-rams where the IAN would use just one. These G-ram are organised into a "discriminator" in which the total output of the G-rams is summed and thresholded before being fed back to the input. An alternative is not to threshold but to feed back an overall probability which is related to the sum of the firing probabilities of the G-rams that support an output. The cost of the system is

$$K(K/N)B2^N$$

Putting some figures to this: K=100, B=2, N=4 gives a cost of 3.2 x 10^5 bits. This should be compared to the gigantic 10^{32} bits of the un-decomposed G-ram implementation with K=100, and B=2.

5.   SOME NOTES ON PERFORMANCE

Clearly, decomposition leads to a deterioration of performance with respect to an IAN. In a short paper of this kind a complete analysis of performance will not be presented. Indeed, the problem-dependence of such a consideration makes it fit for a deep study in its own right. This is left for a longer version. However, here it is possible to describe the character of this deterioration in an empirical manner. The form that the deterioration takes is that of overgeneralisation. This leads to the creation of stability for untrained states . In the IAN, the same states would lead to one of the trained states. The limit occurs when the output of a discriminator becomes dependent only on the corresponding input and independent of all the others. At this point any pattern would remain stable in the net.   As N is reduced from a maximum of K, it can be shown that the above trivial limit case becomes more likely (the generalisation of G-ram discriminators follows the same pattern as in the WISARD discriminator which is covered in Aleksander and Morton[7]). This has implications of major significance which are elaborated below.

6. RELATIONSHIP TO CONVENTIONAL MODELS

If N is unity, the G-ram discriminator output sum takes the form:

$$\sum Ij(Wj)(1\text{-}Ij)(Wj')$$

Wj is the content of the storage location of the jth G-ram addressed by an input (Ij) value of 1, while Wj' is the content of the address for the input value of 0. Were one to make Wj'=0 the formulation becomes the conventional one of the McCulloch and Pitts model and that assumed in much of the connectionist paradigm.   However, as has been argued above, the performance of N=1 systems is the most likely to provide false minima in the content-addressable feedback structure.

7. CONCLUSIONS

The main thrust of this paper is to show that there is an engineering tradeoff between performance and cost in neural nets used for content-addressing tasks.   An Ideal Artificial Neuron has been defined which gives   perfect content-addressable performance.   The G-ram is a silicon-implementable memory in which the training information can be spread to addresses related to unseen inputs and which is an embodyment of the IAN. However , direct implementations of IANs in G-ram are shown to be costly in terms of required memory and to the end of reducing the cost, a decomposition of the IAN in terms of G-rams is proposed. It is argued that as levels of decomposition reduce the costs, the performance of the system deteriorates through the creation of unwanted stable states. This provides for choice in the cost/performance continuum. It is also shown that the McCulloch and Pitts formulation is at the lowest performance end of this continuum casting doubt on the justification for building neural nets using models derived from this formulation.

8. REFERENCES

1. Aleksander, I. (1989): *Neural Computing Architectures* , London, North Oxford Academic; Cambridge, Mass, MIT Press.
2. Hopfield, J.J.(1982): Neural networks and physical systems with emergent collective properties, *Proc. Nat. Acad. Sci. USA,* **79,** 2554-2558
3. Aleksander I.(1989): The logic of connectionist systems. In (Aleksander, ed.)*Neural Computing Architectures* , London, North Oxford Academic; Cambridge, Mass, MIT Press.
4. Myers, C (1989): Output functions for probabilistic logic nodes. *Proc IEE First Conference on Artificial Neural Networks*, London.
5. Gorse, D. and Taylor, J.G.(1989): Phys. Lett. A, **131,** 326-332.
6. Aleksander, I., and Morton, H.B.(1990):   An overview of weightless neural nets. *Proc IJCNN 90*, Washington.
7 Aleksander, I., and Morton, H.B.(1989) : *An introduction to neural computing*(chapter 5). Chapman and Hall (UK), MIT press (USA).

Parallel Processing in Neural Systems and Computers
R. Eckmiller, G. Hartmann and G. Hauske (Editors)
© Elsevier Science Publishers B.V. (North-Holland), 1990

ON THE BIOLOGICALLY MOTIVATED DERIVATION OF KOHONEN'S SELF-ORGANIZING FEATURE MAPS

Reinhard Acker, Andreas Kurz[+]

Department of Control Theory and Robotics
Technical University of Darmstadt, Germany

Kohonen gives a "Shortcut Algorithm" characterizing the behaviour of his self-organization feature maps. He derived this algorithm from biological motivated differential equations. This paper describes this derivation shortly and reports on simulations of the authors which show that self-organizing is a property of the Shortcut Algorithm but not a property of the biological justificated equations.

1. INTRODUCTION

Kohonen [1] describes artificial idealized neurons justifying their properties by results of biophysical research. The input-output behaviour of these artificial neurons is linear between two saturation limits and their adaptation feature responds to a Hebbian learning rule which is extended by a "forgetting term". Kohonen puts many of these neurons together obtaining a neuronal network which he calls "self-organizing feature map" [1][2]. The one-dimensional case of a self-organizing feature map is shown in figure 1.

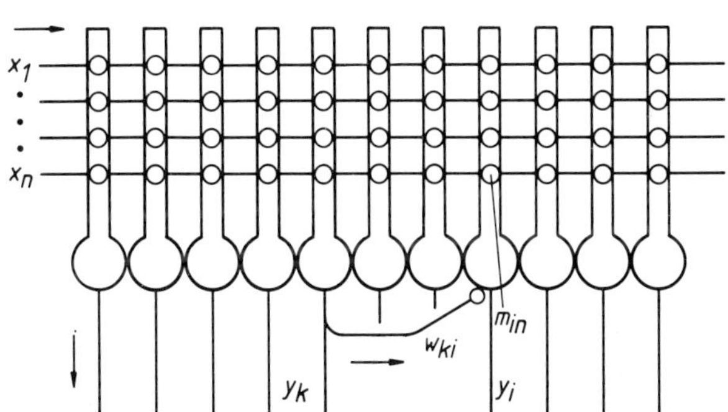

FIGURE 1
One-dimensional self-organizing feature map

The external input components x_j of the network are connected in parallel to all the neurons via synapses weighted by coupling coefficients m_{ij}. The output y_i of each neuron i is linked to its neighbours in the lateral direction. This feedback connections are weighted by coupling coefficients w_{ki} which are a function of the

---

[+] supported by Ministry for Research and Technology (BMFT) under grant ITR 8800 B/5

distance between two neurons ("Mexican hat" function, figure 2).

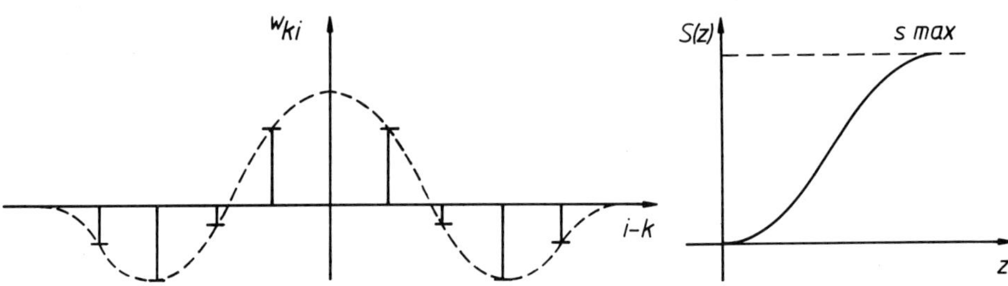

FIGURE 2
Mexican hat function

FIGURE 3
Saturation function

The behaviour of a network with p neurons and with an external input of n components can be described according to Kohonen by a system of differential equations which Kohonen approximates by the following system of discrete equations:

$$y_i(k) = S\left(\sum_{j=1}^{n} m_{ij}(k)x_j(k) + \sum_{j=1}^{p} w_{ij}y_i(k-1) \right) \quad i=1\ldots p \quad (1a)$$

$$m_{ij}(k+1) = m_{ij}(k) + a\left(y_i(k)x_j(k) - B(y_i(k))m_{ij}(k) \right)$$
$$i=1\ldots p; \; j=1\ldots n \quad (1b)$$

with:
 S : saturation function (figure 3)
 B : nonlinear scalar "forgetting" function
 a : learning factor

The first set of equations (1a) describes the input-output behaviour of the network, the second set (1b) describes the adaptation of the input synapses.

2. THE DEVELOPMENT OF ACTIVITY CLUSTERS

Kohonen [1] examined the behaviour caused by the equations (1a) in detail. His simulations showed that the behaviour has the following features: If the parameters of the mexican hat function are choosen suitably, the development of just one coherent area of neurons can be observed the output of which is the maximum value smax of the saturation function S (activity cluster). Each other neuron is inactive (output = 0) after the transient of the network. Close to the center of this activity cluster a neuron c is found for which holds:

$$\sum_{j=1}^{n} m_{cj}(k)x_j(k) = \max_{i=1..p} \sum_{j=1}^{n} m_{ij}(k)x_j(k)$$
$$\text{(max. scalar product)} \quad (2)$$

This features could be essentially confirmed in our own simulations. Furtheron, we could confirm Kohonen's observation that an activity cluster which has been established is in a stable state where it cannot be manipulated by changing the input signal. By

reason of this property of the activity clusters, Kohonen introduced the additional assumption (A1) that the network is reset periodically caused by a (not further specified) external influence.
The saturation limits of the saturation function (figure 2) stabilize the output activities y_i to either a high value (smax) or to
zero. Therefore neglecting the transition phase, the equations (1)
can be simplified as follows when $B(0)$ and $B(smax)$ are choosen
suitably and when the assumption (A1) is respected:

$$m_{ij}(k+1) = m_{ij}(k) + a(x_j(k) - m_{ij}(k)) \quad \text{for } i \in N_{c,r}$$
$$m_{ij}(k+1) = m_{ij}(k) \quad \text{for } i \notin N_{c,r}$$
$$i=1...p; \; j=1...n \quad (3)$$

whereby for the neuron c (the center of the activity cluster)
equation (2) holds and $N_{c,r}$ is the set of neurons in an environment of the neuron c (activity cluster) characterized by
the radius r.

Since the complex development process of the activity clusters is
idealized in these equations, the behaviour of the network can be
calculated much faster as it would be possible according to the
equations (1).

3. THE SHORTCUT ALGORITHM

Kohonen modified the equations of section 2 further. He exchanged
the condition (2) (determination of the center of the activity
cluster) in these equations for the following one:

$$\sum_{j=1}^{n} (m_{cj}(k) - x_j(k))^2 = \min_{i=1..p} \sum_{j=1}^{n} (m_{ij}(k) - x_j(k))^2$$
$$(\text{min. Euclidean distance}) \quad (4)$$

In addition, he stipulates that the the learning factor a and the
radius of the activity centers are functions which decrease mono-
tonically in time (assumption A2). The computing rules which are
defined by the equations (3) with condition (4) and by the assumption (A2) are called by Kohonen "Shortcut Algorithm". By simulation and – for a simple one-dimensional case – by theoretical conclusion he showed that a process called "self-organization" takes
place when the behaviour of the network is computed according to
the Shortcut Algorithm. In this process Kohonen distinguishes two
phases. In the first one, an "ordering process" is observed where
neurons in topological neighbourhood become sensitized to input
signals which are topological neigbours in the input signal space.
The second phase is a "convergence phase" where the network adapts
to the probability density of the input signal occurance.

4. SELF-ORGANIZATION AND THE BIOLOGICAL MOTIVATED EQUATIONS

The main property of the Shortcut Algorithm is the self-organization. On the other hand according to Kohonen the Shortcut Algorithm is equivalent to the equations (3) (with the condition (2)
for computing the center neuron). These equations on their turn
are derived from the biological motivated equations (1). Therefore

we should expect that the self-organization is not only a property
of the Shortcut Algorithm but also a property of the equations (3)
and (1). We simulated both the equations (2),(3) (with respect to
assumption (A2)) and the equations (1) (with respect to assumption
(A1)). However, in both cases we couldn't observe the ordering
process which is crucial for the self-organization. Thus the
transition from the condition (2) (maximal scalar product) to the
condition (4) (euclidean distance) is essential for the
self-organization property. For backing up this conclusion we
changed the equations (1a) in such a way that the center of the
activity clusters are established near that neuron for which
equation (4) holds. The simulations show, that the desired
ordering process takes place then. Thus it can be verified also on
this low level of simulation that the transition from the scalar
product to the euclidean distance is decisive for the
self-organizing effect. Kohonen's argument for the transition to
the euclidean distance is that the vectors of the synaptic coupling factors m_{ij} would be normalized to a specific length when the
learning of the network proceeds, independent of the input signals
which occur in this process. Then, he further argues, the
condition (2) (maximal scalar product) and the condition (4)
(minimal euclidean distance) are equivalent and could be exchanged
[1]. But, neither we could derive such a normalizing effect by
theoretical reasoning nor we could observe it in our simulations.

5. BIOLOGICAL INTERPRETATION OF THE SHORTCUT ALGORITHM

If we don't want to treat the Shortcut Algorithm only as a special
method of cluster analyzing, the question arises how an artificial
neuronal network must be built up which behaves in conformity with
the Shortcut Algorithm and which can be justificated by biological
facts. Since the condition (4) is fundamental for the desired
self-organization property, we must take into account this
condition for answering this question. Here we consider two
possibilities: Firstly, we could demand that the input signals and
the vectors of coupling factors are always normalized. This would
mean, that the information is contained in the direction of the
input vectors only. Since the condition (4) is equivalent to
condition (2) in this case, we can then use Kohonens network unchanged to establish the self-organizing feature maps. However, it
is the question then which biological mechanism could provide for
the normalization of the input vectors. Secondly, we could stipu-
late a complex neuron which has the features for computing an
output signal which grows if the euclidean distance between the
input vector and the vector of synaptic coupling factors gets
smaller. However, the biological justification of such a complex
neuron seems to be difficult. So a biological interpretation of
the Shortcut Algorithm is an open problem in our opinion.

REFERENCES

[1] Kohonen, T., Self-Organization and Associative Memory (Berlin,
    Heidelberg, New York, London, Paris. Tokyo, 1987).
[2] Kohonen, T., The "Neural" Phonetic Typewriter, Computer
    (March, 1988) pp. 11-24.

All-or-None Connections or Variable Weights?

A. DE CALLATAY

IBM European Systems Research Institute
A.K. Watson IEC, La Hulpe, Belgium

Abstract. *Parallel computers made of recording content addressable memories (RCAM) have been suggested as a central element of robot controllers (Callataÿ, 1969, 1986, 1988). These controllers learn invariant rules describing the actions and find how to achieve goals. They have a neural mapping (Callataÿ, 1989a, 1989b).*

RCAMs are neuron layers in which synapses are all-or-none switches. This paper compares RCAMs with other neural networks. RCAMs differ from analog neural networks by the use of timers, classifiers and latches. RCAMs are efficient for learning sparse relations in a large knowledge base, for logic deduction, for reasoning by analogy, and for integration in complex systems. The paper also describes hybrid robot controllers integrating different neural networks.

Introduction. The recording content addressable memories (RCAM) defined here are neural networks built with neurons having switches instead of synapses with a variable weight. RCAMs are the main components of *Symbolic Neural Networks* (SNN). RCAMs differ from one-layer analog neural networks (ANN) by having classifiers, timers and latches. An example of a robot controller including RCAMs is briefly described. The following functions of SNNs, ANNs, dynamic networks and hybrid systems are discussed: exact and approximate recognition, impulse density, representation, classification, association, adaptation, computation speed, learning and preservation of invariant rules, consistency, problem solving, reasoning by analogy, discriminative learning, noise production and elimination, capacity, redundancies, robustness, and integration.

Recording content addressable memory. RCAM is made of cells connected to input wires (ON or OFF) by latches. The latches are switches which are initially open. They learn by being irreversibly closed at once, and by subsequently acquiring a modifiable positive weight. Each cell computes a weighted sum of the impulses received, and is proportionally activated. The group of cells is controlled by a classifier allowing no more than one cell to be ON. RCAM output is a cable made of the cell output wires. This cable is an input to several other RCAMs or to set-points of servo-mechanisms. RCAM input is read and output is issued when external synchronization signals are received from central timers.

Document retrieval algorithm. RCAM recognition is described by an algorithm of document retrieval in which keywords in the query and in the document description have weights. The evaluation of a document is computed by a sum for each keyword of the products of the weights in the query and the document, or by a more complex algorithm. Redundant or synonymous keywords can be used. The selected documents are sorted and the best one is chosen. Alternative methods take several first documents and give an output weight to each selected one. Categorizers to authorize memorization are more strict than those for recognition. RCAMs are normally used with a low impulse density (say one wire ON per 1000 or 10,000). Density is further reduced by enabling few RCAMs. RCAMs have neural mappings no worse than adaptive neural networks. A wire ON carries a "burst" of neural impulses (Callataÿ, 1986).

Symbolic/analog computations. A representation in a neural layer occurs when the cell activations normally produced by a sensory situation are generated in another way. This representation is an intention when the system can find the robot's sequence of muscular actions generating this situation in the physical world. Conscious content is intentional and seems to be representable by symbols. The connectionist rule of the radical behaviorists: *stimulus → response* has been enhanced in the cognivitists' formula: *stimulus & internal state → intention → response*. We must represent intentions and will them before acting. I suggest that most vertebrates also use this formula. The first phase selecting an intention is adaptive and can be implemented with a simple perceptron because the stimuli are discriminated elsewhere. The second phase is a resolution analogous to problem solving in logic programming.

Complex devices for finding invariant rules. After each action, the controller adds in its memory an invariant transition rule: *do actions C when body is in posture A and has to go in posture B after a time T.* These transition rules use static symbols to store dynamic events. Each rule is associated with a timer or a clock to take dynamics into account. The rules translate repetitive physical laws in naive physics. The cerebrum symbolic command is followed by a cerebellar corrective action to skillfully stop the movement and keep equilibrium. To offset the neural computational delay, the cerebellar cortex predicts the subsequent situation, and provides feedback signals with a negative delay to the cerebellar nuclei.

Memorization in RCAM. The invariant rules are learned when a *memorization* signal is delivered. The latches between the active input wires and the output cells are closed. The learning uses Hebb's method in an all-or-none mode. The decision to store a document differs from the recognition decision. A new input/output association is stored in the cell having recognized the input pattern or in another cell. A controllable assignment of the memorizing cell is needed to avoid the *duplication problem* defined by Minsky (1986).

Choice of the memorizing cell. Simple simulations have shown that random occurrences can be classified in well-spread symbolic domain ranges. The algorithms are discussed in Callataÿ (1986). They take into account the similarity of recognition, the number of closed latches per cell, and the saturation of the cell group. In some cases, the result is known when learning takes place (cerebellum). Piaget's sensory-motor scheme takes advantage of information external to the RCAM (expected goal, recognition by high levels). Other algorithms for similarity-based reasoning have been developed (Stanfill and Waltz, 1987).

Learning. Learning is a misleading word in the multidisciplinary environment of brain research. For computer scientists, it may mean the discovery of a generalized rule, or the automated design of device parameters (e.g. the weights of a neural network). Genetic learning is computerized design using the methods of natural selection.

For psychologists, learning requires a body (e.g. a robot) acting and sensing. Learning refers only to the real-time memory changes produced by experience. For brain modeling here, the interesting learning is psychological. RCAMs learn the invariant rules predicting the outcome of a body movement. RCAMs irreversibly add the rules corresponding to the experiments (Callataÿ, 1986). Experiments providing rich information are found by Piaget's sensory-motor scheme.

Symbolic neural networks have a set of predefined possible decisions (one per neuron). Rewards change perceptron choices by modifying weights, i.e. some previous information is forgotten and something new about the current event is memorized (what neural networks learn is still controversial). In RCAMs, reinforcement is always maximum, without forgetting.

Content Addressable Memories (CAMs) and Associative Memories (ASMs). Computer CAMs are devices made of AND circuits storing and recognizing any content. They have been used in brain models (Callataÿ, 1969), but were unable to find similarities. After a period, an RCAM always finds a command in function of the activation of the closed switches. RCAMs are called *"recording"* CAMs because storing a pattern in a cell is a controllable, clear-cut and at-once operation. Finding similarities is not incompatible with the expression "content addressable". ASMs are networks in which some input nodes are held activated. ASMs converge to a state described by a pattern, thus they retrieve the missing data of a pattern. RCAMs recognize the best match with the presented pattern, but not the missing input data and therefore are not *associative.* One output symbol in a CAM with OR exit circuits can correspond to several input patterns. The missing data are ambiguous and cannot be retrieved without a complex perception. CAM data are recognized within a fixed time interval. Associative memories, such as Hopfield's networks, may need a variable time to converge to an equilibrium state. It is not clear how the output of associative memories can be used for further processing, except as an input to a CAM. RCAM classifiers are useful to prevent combinatorial explosion.

Logical consistency. On the one hand, unlike CAMs, RCAMs do not always logically fire rules (Callataÿ, 1988) because:
- Without rules perfectly matching, RCAMs decide at best (by analogy).
- With two input cables (a,b,..) and (c,d,..), a cell learning *a & c* and *b & d* stores also *a & d.* By storing examples, a cell builds a generalized rule for all patterns producing an action.
- RCAMs cannot learn exclusive OR. This needs several layers, with a first layer sparsely connected. Learning in the first layer does not seem useful.
- Quantification is not compatible with categorization. First order problems are solved with short term memories added to RCAMs (Callataÿ, 1988). Negation cannot be directly represented.

On the other hand, RCAMs preserve invariant rules. The irreversible latches implement the single assignment rule giving correct logic computation in pure Lisp and Prolog.

Computational methods by inversion. Knowledge is represented by formulas whose procedural interpretation permits computation. Performance improvement is mainly achieved by inversion. When the inverse of $y = f(x)$ is $x = g(y)$, $f()$ and $g()$ describe arithmetic operations. Matrix inversion is the method for linear equations. Inversion is also used for symbolic computations:
- *Conversion by tables:* Binary search (requiring a sorted table) is better than iterative browsing. A sort on the second column builds an inverted table finding an argument index. In connectionist hardware, each index is represented by an activated wire, and each argument by a connected wire in another domain. Inversion is done by changing the impulse direction.
- *Document retrieval:* The direct access finds the keywords describing a document. The disk-based computers use inverted files storing the lists of documents using each keyword. A search merges these lists and sorts the documents having queried keywords. In connectionist hardware, each RCAM cell directly computes a weighted sum of matching keywords.
- *Production rules:* RETE networks are the inverse of a set of production rules (Forgy, 1982). RETE networks should be compiled, and their use as brain models is therefore unlikely

(Anderson, 1987). RCAMs permit implementing some RETE networks in symbolic neural networks (Callataÿ, 1988).

Neural network paradigm shift. The organization of executable programs depends on the hardware. The simple pattern matching in a neural network is a complex operation in a von Neumann processor. A paradigm shift brought by neural architecture is a change in the relative complexity of operations, specially noticeable for parallel processing. Symbolic network rules are natural declarations like those used in expert systems.

Problem solving by networks. Circular directed network processor systems (DNPS) are symbolic neural networks for which the following theorem applies: given present facts and a goal, the solution found by a DNPS made of CAMs is true (Callataÿ, 1988). The theorem can be extended: In the same way that the equations of relativity become those of classical mechanics when speeds are slow, DNPS systems with RCAMs are shown to be logical resolvers in exceptional cases: i.e. when all needed rules are known, are still valid after hierarchical decomposition and categorization, are based on predictable events in which the relevant input/output data have been recorded, and are such that a solution can be found in one network sweep (Callataÿ, 1988). In the same way that relativity gives one speed for light, a DNPS always finds an action by analogy within a fixed reaction time, as required for an animal. DNPS operations never halt. A DNPS problem is solved by analogy during one processing loop, whatever the data.

A definite advantage of symbolic versus analog neural networks is this capability of generating new correct reasoning paths. In the 1960s, it was found that the dynamic circuits were inefficient for learning and problem solving. Perceptrons were static and limited. Later the AI approach (as seen in expert systems) was found brittle, slow and unreliable with large knowledge bases. Breaking the logical iron collar was advocated by Minsky (1975). Adaptive neural networks have thrown away the benefits of deductive logic. Symbolic neural networks may reconcile the problem solving ability of symbolic representations with the robustness of neural networks.

Noise in pattern recognition. The first quality of a clerk is to not hinder the work of others. In a symbolic neural network, the classifiers prevent most of the watching components from intervening. Therefore the number of circulating logical impulses is kept very low. Similarly in AI, programs work better when only a small part of the whole knowledge is kept instantiated at a time (say one concept per 100,000, i.e. 100,000 per 10 billion cells in human brain). After a conflict set resolution, anything concerning the set not selected becomes noise. This is why classification (winner takes all) is efficient. Suppressing the low frequency transmissions may improve the discriminatory performances of adaptive neural networks. Distributed memories remain robust after local changes, and resistant to local noise (Rumelhart and McClelland, 1986). Such robustness has not been shown if all synapses are randomly modified by normal metabolism. All-or-none dendritic spine resorption (latch closure) may persist for a lifetime (Callataÿ, 1989b).

Capacity. In symbolic networks, the number of input categories is limited by the lifetime, independent of the captors. A human lifetime lasts for 10 billion quarter seconds, and rules may be added to the memory for each period. A rule is used only when its condition pattern recurs or is recognized as similar, but a repetition does not give the opportunity to learn. Preventing learning is a major problem in symbolic neural networks.

A million different patterns are coded with one active wire in each of 3 groups of 100 wires or by one active wire per 1,000,000. In this last system, the synapses on the cells issuing these wires have few chances to be used for more than one event type. It was computed that up to 6% of latches can be used before their double use becomes confusing (Callataÿ, 1986). Using analog instead of all-or-none devices provides more powerful recognition, but this may not be relevant if the problem is to reduce large amounts of sensory data to useful subsets. Animals have many inputs and few outputs.

Grand-mother neurons. The name *"grand-mother"* cell invites misconceptions. There is a different grand-mother neuron for each decision. Each neuron, unlike a grand-mother, has only one idea (one reaction type to events). Many top neurons are simultaneously active in every area of the top realms: one decides a move to target, an other locates the target, an other focuses on the target angle with the body, an other cares about the ground configuration on which to run, an other manages the body position and equilibrium state, etc... Grand-mother neurons have speculative neural mappings (Callataÿ, 1989b).

Redundancy and robustness. An RCAM domain is determined by the inputs and outputs of the group and by the sensors and actuators eventually connected. The size is dependent on the number of categories in each intermediate level. RCAMs may have a fully or partially redundant domain of connections. Contradictory orders produced by redundant or almost-redundant grand-mothers are normally suppressed by a subsequent categorizer.

Distributed memory can mean the coding of a datum in many components, different data coded in the same component, and/or the redundant storage of the same datum. When a circuit codes different data, modification of one datum may modify other independent information. When each part of a code has a single meaning (structural coding), modification or addition by learning is safe and allows generalization. Systematic redundancy makes a system very reliable. If a neuron is destroyed, redundant neurons will take the decision, and what had been learned by this neuron will be learned by an other. A destruction of all redundant nodes will normally not be noticed in the external behavior, as an observer (or consciousness) does not know whether the person has forgotten, or has never learned, an event.

Feature extractors. Feature extractors (FE) should scan only what could make sense in our world. In symbolic networks, the output of FE dynamic processing is subsequently classified into symbols. Only a tiny amount of relevantly correlated information is extracted from the sensors (compare what is noticed in a briefly presented picture with the information captured by a photography having the discriminative power of the retina).

Learning or wired-in FEs? Nature has probably not built a general purpose FE, because this needs more parts than a specialized one (say a billion times more). Specialized FEs do not learn by definition. Can we design interesting learning FEs somewhere between the general purpose ones and the specialized ones? This design should solve the following problems: How can dynamic networks learn? Until otherwise shown, perceptrons and adaptive neural networks are static (Minsky and Papert, 1969; Smolensky, 1988). How can the experiments of early life show which features are interesting to extract? The reward signal used by natural selection is survival, thus this signal is delayed for a lifetime. How many experimental trials are required for this learning? Natural selection took a few billion years. Specialized FEs can use all their circuits because nothing is in reserve to be possibly selected for learning. FEs represent a few percent of the neural mass in man, but may compute more than the cortices, due to better opportunities for analog computation and parallelism.

Preservation of knowledge. FEs are dynamic networks described by differential non-linear equations transforming sensory data (Koch and Poggio, 1987). Servo-mechanisms and FEs can have adjusting circuits to provide an output independent of changing internal conditions. Besides this self-tuning, FEs should not learn during the lifetime to preserve the logical consistency of the invariant rules learned on-line. Interposing an adaptive connectionist layer between the FEs and the symbolic neural layers jeopardizes the logical consistency of the stored rules and does not seem necessary. General purpose learning symbolic processors (cerebral cortical columns) may be used everywhere as all logic resolvers have a common processing principle. Cerebral columns differs in their sizes, I/O connections and acquired knowledge.

Conclusions. The comments given here concern only a controller for a robot environment. Reasons were given to discard some known systems in favor of a hybrid system in which dynamic networks of non-learning feature extractors and self-tuned servo-mechanisms are controlled by symbolic neural networks. The intentions are chosen by perceptrons using already discriminated data. The following are examples of suggested hybrid neural networks. A speculative cerebellum model (Callataÿ, 1986) uses an input vector shaped by classifiers (e.g. one wire active per 10). It has three layers: (1) A non-learning layer of cells with sparse connections (4 per cell). (2) A fully connected RCAM layer. A teacher gives afterwards which symbol should have been produced (the cerebellum is supposed to predict the subsequent body state, known 100 ms later). (3) A perceptron layer, rewarded to learn the best reaction for each situation (e.g. to produce movements ending motionless or balanced). The cerebrum model does not need teachers, but is much more complex (Callataÿ, 1986). In both models, adaptation is provided by the competitive evaluation of classifiers and by perceptrons in layers remote from the logic resolver.

References

Anderson J.R. (1987) Skill acquisition: Compilation of weak-method problem solutions, *Psychological review*, Vol. 94, No. 2, 192-210.

deCallataÿ A. (1969) Brain Model with Periodic Processing. *Curr. Mod. Biol.*, 2, pp. 307-319.

deCallataÿ A. (1986) *Natural and Artificial Intelligence: Processor Systems Compared to the Human Brain*, North Holland (Elsevier), Amsterdam.

deCallataÿ A. (1988) Logic programs directly processed in a network of content addressable memories, *Future Generations Computer Systems*, 4, 117-131.

deCallataÿ A. (1989a) Can Artificial Intelligence help in finding how brains may work? in *Brain Dynamics: Progress and Perspectives*, eds. E. Basar, T. Bulloch, Springer, Heidelberg, pp. 214-232.

deCallataÿ A. (1989b) Biological aspects of neural networks, in *Artificial Neural Networks*, Presses Polytechniques Romandes, pp. 1-15.

Forgy C. L. (1982), RETE: a fast algorithm for the many pattern/many object pattern match problem, *Artificial Intelligence J.*, 19, 17-37.

Koch C. and Poggio T. (1987) Biophysics of computation: neurons, synapses and membranes, in *Synaptic Function*, eds: G. Edelman, W.E. Gall and W.M. Cowan, pp. 637-698, Wiley, New York.

Minsky M. and Papert S. (1969) *Perceptrons*, The MIT Press, Cambridge, Massachusetts.

Minsky M. (1975) A Framework for Representing Knowledge, in *The Psychology of Computer Vision*, Winston P.H. ed., McGraw Hill, New York, pp. 211-277.

Minsky M. (1986) *The Society of Mind*, Simon and Schuster, New York

Rumelhart D.E. and McClelland J.L. and the PDP group (1986), *Parallel Distributed Processing: Explorations in the Microstructure of Cognition, Vol. 1 and 2*, M.I.T, Cambridge, Massachusetts.

Smolensky P. (1988) On the proper treatment of connectionism, *Behavioral and Brain Science*, 11, 1-74

Stanfill C. and Waltz D. (1987) Toward memory-based reasoning, *CACM* 29, 12, 1213-1228.

Parallel Processing in Neural Systems and Computers
R. Eckmiller, G. Hartmann and G. Hauske (Editors)
© Elsevier Science Publishers B.V. (North-Holland), 1990

LEARNING TO PREDICT THE CONSEQUENCES OF ONE OWN'S ACTIONS

Federico Cecconi   Domenico Parisi
Institute of Psychology
National Research Council
Rome, Italy

Abstract

Neural networks with recurrent memory units were assumed to be contained in small "animals" moving in a 2-dimensional world with "food" elements. The networks were first taught to predict the next angle and distance of food given the present angle and distance and the animal's current move as input. These predicting networks were then taught to approach food, i.e. to select a sequence of moves that was more likely to bring the animal to a food element. The prior learning of a prediction ability greatly increases what the network learns during the successive teaching of food approaching behavior in comparison with networks that learn to approach food without prior prediction learning.

Introduction

In order to move efficiently in an environment organisms must have some map of the environment. For example, if an animal is to find food it must be able to approach the food and therefore it must know how the position of the food changes relative to the animal's position. This knowledge represents a very simple map of the geometrical properties of the environment. Such a map can be used to develop efficient sequences of actions that can guide the animal to where the food is located. In the present research we use a neural network approach to investigate the question of whether learning to predict how the position of the food varies with the animal's actions represents a favourable pre-condition to learning to approach and eat food. The research was inspired by Patarnello and Carnevali[1], who had little animals learn efficient food getting strategies using simulated annealing in networks with logical units. Another related study is Booker's[2], who constructed a classifier system (Holland[3]) that builds an internal model of its environment to locate food and avoid noxious stimulation.

The general design of the experiment is a 2x2 matrix with two variables: learning or not learning to predict the position of the food and learning or not learning to approach the food. There are 12 networks in each of the four conditions.

Learning to predict food's position

The world in which our "creatures" learn how to predict the position of the food is a grid of 10x10 square cells. Each world hosts only one creature. The creature at each particular time occupies one cell. The creature faces in one particular direction and has a repertory of 4 possible actions: (1) move forward one cell in the facing direction; (2) turn 90° right; (3) turn 90° left; (4) do nothing.

The environment contains a single food element which occupies one cell. The creature receives information about the food's position in terms of the food's distance from the animal and the angle formed by the line connecting the creature and the food with the animal's facing direction.

We assume that our little creature contains a neural network with a layer of 4 input units, a layer of 4 output units, and an intermediate layer of 7 hidden units. In addition, the network has a layer of 7 recurrent memory units[4, 5]. These units copy the state of the hidden units at one cycle and re-input this state to the hidden units at the successive cycle. Each hidden unit sends a connection of fixed weight of 1 to its corresponding memory unit, while every memory unit sends connections with learnable weights to all hidden units.

There are two sets of input units, each set including two units. The first set encodes the action that the creature is currently doing. The other set encodes information about the food. The output units also are divided into two sets of two units. The first set represents the next action that the creature decides to do. The second set represents the creature's prediction of the location of the food (distance and angle) when the current action is completed.

In each cycle the network is given, as input, knowledge about the position of the food and knowledge about the action that the creature is doing, and it gives as output a decision about the next action and a prediction about the location of the food after the current action. In the successive cycle the action which has been decided by the network in the preceding cycle is given back to the network as input. There is an algorithm that computes the correct location of the food after each action of the creature and this information is used both as teaching input for the network's prediction units in one cycle and as input to the network in the successive cycle. The network is successfully taught by backpropagation to give correct predictions about locations of food.

Learning to approach food

In the second part of the experiment our goal was to train a network to approach and eat food, i.e. to develop strategies (= sequences of actions) that were likely to bring the creature to a food cell so as to eat the food. The same network architecture of the preceding experiment was used. However, the "worlds" were larger (80x80 cells) and they contained many food elements (35% of cells on average were filled with food). A creature was permitted to roam in its environment, selecting actions at will. Whenever it stepped on a food cell, it "ate" the food, which consequently disappeared from the world. Eating ability was measured as number of food units eaten during a lifetime.

Teaching to eat was realized by constructing an algorithm called SEQUENCE which did the following. Whenever the creature happened, by chance, to step on a food cell, its spontaneous behavior was stopped. The algorithm recorded the sequence of the last 8 actions that had led the creature to the food cell and then it re-positioned the creature in the cell where it had been 8 actions back. The creature was restarted again but this time backpropagation was applied. The network received a teaching input on the output units that code the action selected by the animal. (No teaching input was supplied to the other output units coding the prediction.) As teaching input the corresponding action previously selected at that point by the animal was used, this having been a successful action that led the animal to the food.

Results and discussion

As already mentioned, we had four conditions. 12 networks were trained to predict and then they were also taught to approach food. Another set of 12 networks was only trained to predict and a third set only to approach food. A fourth final set was taught nothing. As a final test of performance we measured both the prediction error (PE) and the quantity of food eaten by each network in a test world. We conducted an analysis of variance on the data obtaining significant results ($p < .01$) and subsequent pair comparisons showed that all differences between pairs of conditions are statistically significant.

Let us consider prediction ability first. Unsurprisingly, the prediction error is quite high in the animals that have not been explicitly taught to predict (PE=29.1). However, there appears to be a better prediction ability in the animals that are taught to approach food (PE=19.9). Learning how to approach food can produce some ability to predict where the food will be next, even in the absence of any teaching of such a predictive ability.

The prediction error is of course much lower in the animals that are taught to predict. But the relationship of food approaching ability to the ability to predict emerges here too since the animals that are subsequently taught to approach food have half the prediction (PE=1.4) error of the animals that do not undergo such teaching (PE=3.3).

Turning to the other performance measure, eating ability, we should note that the animals that are not taught either to predict or to eat do not show any such ability. Their eating behavior (35.9%) is at chance level (35%). However, in this case too it is interesting to note that there is some positive transfer from prediction ability to the ability to approach food. The animals that are taught to predict the position of the food show an eating performance (38.4%) which is above the chance level of the control animals (35.9%).

Quite understandably, the ability to approach and eat food emerges more clearly in the animals that are explicitly taught to eat. But the most interesting result is obtained here. The animals that are taught to eat but have not previously acquired a prediction ability eat 43.2% of the available food, which is not a particulary exciting performance given 35% as the chance level. One must consider, however, that these animals are taught sequences of actions that have led them to the food but are not necessarily the best, most efficient, sequences. For example, one such sequence may include sub-sequences of 4 right turns that do not bring the animal any nearer to the food. What is interesting however is that this same sub-optimal teaching leads to an almost perfect eating performance (97.9% of food units eaten) if it is addressed to animals that have already acquired an ability to predict where the food is relative to where the animal is.

We conclude that possessing a map of the environment, in the sense of being able to predict how the food changes its position relative to the animal, increases considerably how much is learned when the animal is trained to approach and find food. Learning to predict in this sense is like learning the direct kinematics from actions to results. The knowledge so acquired is then applied to learning an inverse kinematics from results (goals) to actions. We have obtained similar results for animals that do not move but have a 2-segment arm with which they must learn move in order to reach a fixed object. Here too, learning to predict the position of the "hand" (end-point of the arm) helps in learning to position the "hand" where the object is located (Parisi and Cecconi[6]). We believe that learning to produce desired effects with one's own behavior may be generally facilitated by learning to predict the effects of this behavior and that this might explain so-called exploratory and play behavior of young individuals in primate species (including humans), which can be interpreted as behavior functionally generated in order to learn to predict its effects.

<u>References</u>

1. S.Patarnello and P.Carnevali, <u>Neural networks for behavioral experiments: training an individual to find food in an external environment</u>. Rome IBM Research Center, 1988.

2. L.B.Booker, "Classifier systems that learn internal worlds models," <u>Machine Learning</u>, Vol.3, pp.161-192, October 1988.

3. J.H.Holland, "Adaptation". In R.Rosen and F.M.Snell (eds.), <u>Progress in theoretical biology</u> (Vol.4). New York: Academic Press, 1976.

4. M.I.Jordan, <u>The learning of representations for sequential performances</u>. Unpublished doctoral dissertation, University of California, San Diego, 1985.

5. J.Elman, <u>Finding structure in time</u>. Institute of Cognitive Science, University of California, San Diego, 1988.

6. Parisi, D. and Cecconi, F. <u>Predicting hand movements and reaching for objects</u>. Institute of Psychology, National Research Council, 1989.

Parallel Processing in Neural Systems and Computers
R. Eckmiller, G. Hartmann and G. Hauske (Editors)
© Elsevier Science Publishers B.V. (North-Holland), 1990

VARIOUS ROLES FOR INHIBITORY INTERNEURONS
IN THE FORMATION OF ASSOCIATIVE MEMORY

Ingolf E. DAMMASCH and Joachim R. WOLFF

Department of Anatomy
University of Göttingen
Kreuzbergring 36
D-3400 Göttingen, F.R.Germany

It is suggested that a mathematical model for associative memory
should include inhibitory units if it is to realize the process of
learning and recalling in a way similar to neurobiology. The effects
of inhibition are exploited (1) for balancing an initial connectivity, (2) for focussing the effects of afferent input to the network,
and (3) for realizing "Hebbian" plasticity with neuronal properties.

1. INTRODUCTION

Hopfield [1] showed that networks composed of those logical neurons introduced
by McCulloch and Pitts [2] may work as associative memories. Following the
spin-glass paradigm (e.g. [3]), the storage of patterns is distributed over the
whole synaptic connectivity, and the recall consists of a relaxation into a
basin of attraction that is closest to a given input pattern. A pattern is
defined by the vector of 0/1-states of the network's neurons, and it is learned
by the so-called Hebbian rule [4] modifying the synaptic weights between two
neurons according to their correlated activation. The Hopfield model contains
several mathematical assumptions, e.g. about network dynamics, connectivity
values, and the learning rule. Apparently, it is useful to investigate how
these formalisms can be realized by the brain in accordance with neurobiological observations.

Investigations on neural networks may have two aims, (i) elucidating how the
brain may work, and (ii) finding new ways to build computers suitable for
parallel tasks. As expressed e.g. by Hartstein and Koch [5], there is an
interest in local learning rules in order to build silicon technologies that
learn by example.

2. NETWORK DYNAMICS: THE RECALLING PROCESS

The dynamic behavior of the neural network is defined by

$$z(k,t+dt) = \begin{cases} 1, & \text{if } \sum_{j=1}^{N} c(k,j)*z(j,t) = MP(k,t) \geq \theta \\ 0, & \text{otherwise} \end{cases}, \quad k \in \{1,..,N\}$$

where $z(k,t) \in \{0,1\}$ is the state of neuron k at time t, N is the number of
neurons in the network, and $c(k,j)$ is the synaptic weight by which neuron j
influences neuron k. If the summed influence - the membrane potential $MP(k,t)$ -
exceeds a threshold θ, then the state of neuron k in the next time instant will
be 1 ("active"). The N neurons may perform this update in a random order (time
horizon $dt = 1/N$) or all at one time ($dt = 1$). So far, although simplified,

the process seems plausible in terms of electrophysiological evidence.

The coefficients $c(k,j)$ have to be calculated in a learning process as described in the next chapter. Hopfield's model starts with $c(k,j) = 0$ and proceeds by adding and subtracting synaptic increments in each learning step, leading to positive and negative synaptic weights. This implies that each neuron can exert excitatory and inhibitory influences simultaneously on other neurons. In contrast, neurobiology suggests that most neurons rather have an either-or character. To avoid this problem, a basic initial connectivity is introduced with increments being added or subtracted during the learning process. To maintain the functional behavior, the abundand basic excitation has to be balanced in a kind of threshold adjustment. This is done by inhibitory neurons. In the most general case, the learning process is started with a balanced but otherwise randomly connected network consisting of NE excitatory and (N-NE) inhibitory neurons. When θ is set to 0, as done by Hopfield without loss of generality, "balanced" means

$$\sum_{j=NE+1}^{N} c(k,j) = - \sum_{j=1}^{NE} c(k,j) \ , \quad k \in \{1,..,N\} \ , \quad \text{initially, and}$$

$$MP(k,t) = \sum_{j=1}^{NE} c(k,j)*z(j,t) + \sum_{j=NE+1}^{N} c(k,j)*z(j,t)$$

again defines the dynamic. This implies that the (excitatory) pattern activity projects to a population of inhibitors, which feed back on the neurons of the pattern. While this may work with satisfying accuracy for large numbers of neurons, distinct patterns, and few disturbances, it cannot generally guaranteed to be correct in the sense of relaxing from an arbitrary input pattern towards the "right", i.e. closest pattern being stored.

The right pattern classification can be imposed by homogeneity conditions, introduced by Dammasch [6], as follows:

$$\sum_{j=NE+1}^{N} c(k,j) = - NE*c' \ , \quad k \in \{1,..,NE\} \ , \quad \text{initially, and}$$

$$MP(k,t) = \sum_{j=1}^{NE} c(k,j)*z(j,t) + \sum_{j=NE+1}^{N} c(k,j)*(\sum_{j=1}^{NE} z(j,t)/NE \ )$$

implying that the initial excitatory connectivity $c(k,j) = c'$, $k,j \in \{1,..,NE\}$, is exactly balanced by a population of inhibitors whose effect is proportional to the pattern's activity. In this case, there has to be a complete excitatory connectivity before and after learning, in the sense that all possible synapses are realized and initially have the same weight. If this constraint is weakened, other conditions must be strengthened: For inhomogeneous networks with $c(k,j) \in \{0,c'\}$ one needs

$$c(k,NE+j) = - c(k,j) \ , \quad j \in \{1,..,NE\} \ , \quad k \in \{1,..,NE\} \ , \quad \text{initially, and}$$

$$MP(k,t) = \sum_{j=1}^{NE} c(k,j)*z(j,t) + \sum_{j=NE+1}^{N} c(k,j)*z(j-NE,t)$$

With these conditions, a more flexible topology can be allowed, i.e. there may be "holes" in the initial connectivity, or defects in the connectivity modelling "lesions" after learning. In the extreme case, for each excitatory neuron E1 there has to be an inhibitory neuron I1 that initially has the same projec-

tions to other excitatory neurons E2. I1 must have the same functional behavior
as E1, so it has to be triggered by E1 directly, or by the same afferent input
that triggers E1.

```
                    ( E1 ) ——————————————→ ( E2 )
afferent  =>               \             ↗
input     =>                 ↘        ↗
                            ( I1 )
```

While it may appear trivial to expand the network structure in the described
way, this only holds as long as just the initial connectivity is considered.
During the learning process and for pattern storage, the inhibitory interneurons play a further role.

3. NETWORK PLASTICITY: THE LEARNING PROCESS

In order to function as an associative memory, the network's synapses have to
have the right weights, leading to the right classification dynamic in the
above-mentioned sense. Neurons with correlated activity in the patterns must
have a stronger excitatory influence on each other than uncorrelated ones. Local rules that calculate the necessary synaptic modifications by comparing pre-
and postsynaptic activity are usually called Hebbian learning rules. This is
due to one sentence in Hebb's book [4] that may also apply to the phenomenon of
conditioning, cf. Carpenter [7]. This kind of plasticity poses some problems:
How is a pattern presented and permanent reinforcement avoided? How does a
synapse "know" that it is responsible for triggering the postsynaptic neuron,
and how does it calculate the correlation? Together with a formula like

$$c(i,j) = \overline{\sum_{p}} \; (2*z(i,p)-1)*(2*z(j,p)-1) \; , \quad i,j \in \{1,..,NE\} \; , \quad p \in \{patterns\} \; ,$$

Hopfield [1] states that "the Hebbian property need not reside in single synapses; small groups of cells which produce such a net effect would suffice". A
similar approach was presented by Dammasch [6], neurobiologically based and
using the compensation theory for synaptogenesis as introduced by Wolff [8,9]:

A pattern presentation is modelled as excitatory afference to the neurons
concerned in the network, acting over a period of time that is morphogenetical-
ly relevant but limited. Neurons with correlated activity ($z(i,p) = z(j,p) = 1$)
stabilize their mutual influence via hyperexcitation and overcompensation,
reduction of existing synapses and formation of new ones. Synapses between
neurons that are both inactive ($z(i,p) = z(j,p) = 0$) remain unchanged.
Between uncorrelated neurons ($z(i,p) \neq z(j,p)$), a reduction of synapses results.

Inhibitory interneurons are utilized twofold in this process: First, feedforward inhibition focuses the afference on the relevant neurons and avoids
side-effects, i.e. it keeps the rest of the network unaffected by the afference. Second, all neurons lose some inhibitory input elements while compensating the overall loss of excitatory synapses during morphogenesis (cf. [6],
[9]), i.e. the desired plasticity emerges as a net effect.

4. RESULTS AND CONCLUSIONS

Computer simulations modelling a network architecture that includes inhibitory
interneurons show that the classification behavior was kept at the same level
of accuracy as e.g. for homogeneous networks [6], if some possible synapses

were not realized in the initial connectivity and were not developed during learning. This means that the compensation algorithm works satisfactorily even with a non-homogeneous topology ("holes") in the network connectivity. Classification performance became slightly worse due to resulting disturbances in the threshold balance, when synapses were "removed" ($c(i,j)$ set to 0) after the learning process.

The role of the inhibitors appears plausible, since GABAergic neurons have been identified as short-axon interneurons; see Mugnaini and Oertel [10]. In summary, on a neurobiologically verifyable set of conditions for network architecture, dynamics and plasticity, even dilute and nonsymmetric networks as analysed by Derrida et al. [11] can function as associative memories. While the emergence of overall activity balance and morphogenetic stability could already be demonstrated by Dammasch et al. [9], some questions remain to be answered about the emergence and the stabilization of the correct, i.e initially balanced level of a specified feed-forward inhibition.

REFERENCES

[1] Hopfield JJ, Proc Natl Acad Sci USA 79 (1982) 2554
[2] McCulloch WS and Pitts WH, Bull Math Biophys 5 (1943) 115
[3] Kinzel W, Z Physik B 60 (1985) 205
[4] Hebb DO, The organization of behavior (Wiley, New York, 1949)
[5] Hartstein A and Koch RH, Neural Networks 2 (1989) 395
[6] Dammasch IE, Structural realization of a Hebb-type learning rule, in: Cotterill RMJ (ed) Models of brain function (Cambridge University Press, 1989) in print
[7] Carpenter G, Neural Networks 2 (1989) 243
[8] Wolff JR, Some morphogenetic aspects of the development of the central nervous system, in: Immelmann K et al. (eds) Behavioral development (Cambridge University Press, 1981) 164-190
[9] Dammasch IE, Wagner GP and Wolff JR, Biol Cybern 54 (1986) 211
[10] Mugnaini E and Oertel WH, An atlas of the distribution of GABAergic neurons and terminals in the rat CNS as revealed by GAD immunohisto-chemistry, in: Björklund A, Hökfelt T (eds) Handbook of chemical neuro-anatomy, Vol.4, Part 1, Ch.10 (Elsevier, Amsterdam, 1985) 436-608
[11] Derrida B, Gardner E and Zippelius A, Europhys Lett 4,2 (1987) 167

Parallel Processing in Neural Systems and Computers
R. Eckmiller, G. Hartmann and G. Hauske (Editors)
© Elsevier Science Publishers B.V. (North-Holland), 1990

SOLUBLE LOW-ACTIVITY NETWORKS WITH NONLINEAR SYNAPSES

Harald ENGLISCH

Sektion Informatik
Karl-Marx-Universität
DDR-7010 Leipzig, GDR

J. Leo VAN HEMMEN

Physik-Department
Technische Universität München
D-8046 Garching (bei München), FRG

The dynamics of low-activity networks with nonlinear, e.g. clipped, synapses is studied in the limit where the number q of stored patterns is proportional to the connectivity C per neuron, i.e. $q=\alpha C$, and $C\to\infty$. The nonlinearity is described by a synaptic function Φ; e.g. for clipping we have $\Phi(x)=sgn(x)$. We consider two cases: (i) C=N, as in the original Little-Hopfield model, with parallel dynamics for a single time step, and (ii) C≪N, a strongly diluted system. In both cases, a signal-to-noise ratio analysis shows that there exists a critical pattern-to-connectivity ratio $\alpha_c=(2/\pi)\ \langle x\Phi(x)\rangle^2/\langle\Phi(x)^2\rangle$ such that beyond α_c no pattern is stable. Here the angular brackets indicate a Gaussian average.

1. ASSOCIATIVE MEMORY FOR LOW-ACTIVITY PATTERNS

The nonlinear Little-Hopfield model with dynamics
$$S_i(t+1)=sgn(h_i(t)), \quad h_i(t)=\sum_{j\neq i} J_{ij}S_j(t) \tag{1}$$
stores q N-bit patterns $\xi_\gamma=\{\xi_{i\gamma};\ 1\leq i\leq N\}$ in the synapses
$$J_{ij}=N^{-1}Q(\xi_i,\xi_j), \quad \xi_i=\{\xi_{i\gamma};\ 1\leq\gamma\leq q\},\ \xi_{i\gamma}\in\{-1,\ 1\} \tag{2}$$
only with the help of the information available to neuron i and j (locality). The synaptic kernel $Q(\xi_i,\xi_j)=Q(\xi_j,\xi_i)$ is some function defined on $\mathbb{R}^q\times\mathbb{R}^q$ [1]. If the ξ_γ are random (independent, identically distributed) with mean $E(\xi_{i\gamma})=0$, then the inner-product models
$$Q(\xi_i,\xi_j)=\sqrt{q}\ \Phi(\xi_i\cdot\xi_j/\sqrt{q}) \tag{3}$$
with an odd function $\Phi(x)=-\Phi(-x)$ provide an important subclass, where the dot denotes the scalar product. The original

Little-Hopfield model corresponds to $\Phi(x)=x$ and is therefore called linear.

Parga and Virasoro [2] proposed a variant of the Little-Hopfield model for the storage of correlated patterns. Applied to random patterns with mean $E(\xi_{i\gamma})=a\neq0$ [3, 4] it yields

$$J_{ij}=(1+\sum_{\gamma}(\xi_{i\gamma}-a)(\xi_{j\gamma}-a)/(1-a^2))/N. \tag{4}$$

Since we are interested in the storage of patterns with different means $E(\xi_{i\gamma})=a_\gamma$ as in [5] we generalize (4) to

$$J_{ij}=\sum_{(\gamma)} v_\gamma(\xi_i)v_\gamma(\xi_j) \tag{5}$$

with $v_\gamma(\xi_i)=(\xi_{i\gamma}-a_\gamma)/\sqrt{1-a_\gamma^2}$ and $v_\emptyset(\xi_i)=1$. The symbol $\sum_{(\gamma)}$ denotes a sum with respect to all indices γ and the empty set $\emptyset$. Equation (5) may be reinterpreted by saying that we sum over the patterns and the progenitor (in the sense of [2]) $v_\emptyset(\xi_i)=1$. In the special case where all a_γ are equal to zero we get the Little-Hopfield model which stores the q patterns ξ_γ, $1\leq\gamma\leq q$, together with the trivial pattern $\xi_\emptyset=(1,1,\ldots,1)$.

The ideas leading to equations (3) and (5) suggest that the appropriate inner-product model for the storage of patterns with mean $a_\gamma\neq0$ should be either

$$Q_i(\xi_i,\xi_j)=\sqrt{q}\ \Phi(\sum_{(\gamma)} v_\gamma(\xi_i)v_\gamma(\xi_j)/\sqrt{q}) \tag{6}$$

or

$$Q_e(\xi_i,\xi_j)=\sqrt{q}\ \Phi(\sum_\gamma v_\gamma(\xi_i)v_\gamma(\xi_j)/\sqrt{q})+\langle x\Phi(x)\rangle. \tag{7}$$

The size $\langle x\Phi(x)\rangle=\int x\Phi(x)e^{-x^2}dx/\sqrt{2\pi}$ of the constant term in Q_e (the kernel with the constant excluded from Φ) originates from the condition $E(Q_e)=E(Q_i)=\langle x\Phi(x)\rangle$ [6: Appendix].

For hardware realizations the first variant Q_i (with the constant term included in Φ) with $\Phi(x)=sgn(x)$ is preferable since Q_i has only two values with equal absolute value $\sqrt{q}$ in contrast to the kernel $Q_e\in\{-\sqrt{q}+\sqrt{2/\pi},\sqrt{q}+\sqrt{2/\pi}\}$.

However formula (1) with J_{ij} given by Q_e can be rewritten to a simple expression similar to that in [4]

$$S_i(t+1)=sgn(\sum_{j\neq i}\tilde{J}_{ij}S_j(t)+\langle x\Phi(x)\rangle E(S(t))), \tag{8}$$

where $E(S(t))$ is the mean $\sum_j S_j(t)/N$ and $\tilde{J}_{ij}=$

$\sqrt{q}\ \Phi(\sum_\gamma v_\gamma(\xi_i)v_\gamma(\xi_j)/\sqrt{q})/N$ corresponds to the first term in Q_e.

In the next section the retrieval quality $m_\gamma(t)=$

$\sum_i S_i(t)v_\gamma(\xi_i)/\sqrt{1-a_\gamma^2}N$ (the normalization $1/\sqrt{1-a_\gamma^2}N$ guarantees

$m_\gamma(t)\in[-1,1]$; for $a_\gamma=0$ this definition coincides with the usual one $\Sigma S_i(t)\xi_{i\gamma}/N$; the use of $v_\gamma(\xi_i)$ instead of $\xi_{i\gamma}$ implies, that for $a_\gamma\neq0$ and $N\to\infty$ a pattern $S(t)$ with $E(S(t))\neq0$ being independent of ξ_γ also yields $m_\gamma(t)=0$) will be calculated for the dilute model proposed by Derrida e.a. [7]: $J_{ij}=Q(\xi_i,\xi_j)/C$ with probability C/N and $J_{ij}=0$ otherwise, where the dilution is given by an independent process destroying the symmetry $J_{ij}=J_{ji}$. The retrieval quality $m_\gamma(t)$ can be calculated rigorously as long as

$$C^{2t-2}\ll N, \tag{9}$$

e.g. for $t=1$ in the parallely operating Little-Hopfield model. Since the kernels Q_i and Q_e do not fulfil the invariance condition $Q(\xi_i\circ\eta,\xi_j\circ\eta)=Q(\xi_i,\xi_j)$ with $(\xi_i\circ\eta)_\gamma=\xi_{i\gamma}\cdot\eta_\gamma$, $\eta_\gamma\in\{-1,1\}$, the methods of [1] cannot be directly applied. Nevertheless an analysis of nonlinear networks for low-activity patterns in the limit $t\to\infty$ will be presented elsewhere.

2. THE PERFORMANCE OF THE NONLINEAR MODEL

Provided (9) is valid and $S(0)$ is correlated only with ξ_γ the retrieval quality is given by

$$m(t+1)=erf((E(S(t))+(1-a_\gamma)m_\gamma(t))\langle x\Phi(x)\rangle/\sqrt{2\alpha\langle\Phi(x)^2\rangle})/2+$$

$$erf((-E(S(t))+(1+a_\gamma)m_\gamma(t))\langle x\Phi(x)\rangle/\sqrt{2\alpha\langle\Phi(x)^2\rangle})/2 \tag{10}$$

for both Q_i and Q_e, where $erf(x)=2\int_0^x e^{-y^2}dy/\sqrt{\pi}$ is the error function [4]. Let us assume that $S_i(0)=\xi_{i\gamma}$ is valid with a probability independent of whether $\xi_{i\gamma}=1$ or $\xi_{i\gamma}=-1$, i.e. with a probability $(1+m_\gamma(0))/2$. Then for all t it holds that $E(S(t))=a_\gamma m_\gamma(t)$, and (10) reduces to

$$m_\gamma(t+1)=erf(m_\gamma(t)/\sqrt{2\alpha_\Phi}) \tag{11}$$

with a rescaled ratio

$$\alpha_\Phi=\alpha\langle\Phi(x)^2\rangle/\langle x\Phi(x)\rangle^2. \tag{12}$$

The stable fix-points of (11) are decreasing with respect to α_Φ with $m_\gamma(\infty)=1$ for $\alpha_\Phi=0$, $1>m_\gamma(\infty)>0$ for $0<\alpha_\Phi<2/\pi\simeq0.637$ and $m_\gamma(\infty)=0$ for $\alpha_\Phi\geq2/\pi$. It follows that for clipped synapses ($\Phi(x)=sgn(x)$) the critical ratio is $(2/\pi)^2$. Due to Schwarz inequality applied to the Hilbert space $L^2(\mathbb{R}, e^{-x^2/2}dx/\sqrt{2\pi})$ the effective ratio α_Φ is always less than α. Both ratios coincide only for linear synapses $\Phi(x)=cx$.

A similar result can be derived for the serially updating network

$$S_i(t+1/N)=sgn(h_i(t)), \quad S_j(t+1/N)=S_j(t) \text{ for } j\neq i \tag{13}$$

as long as $C^{2t}\ll N$, where the time step $1/N$ is chosen in order to ensure that after one time unit N neuron values are updated as in the parallely working network. In the limit $N\to\infty$ (11) has to be replaced by a differential equation [7]

$$\dot{m}_\gamma(t)=erf(m_\gamma(t)/\sqrt{2\alpha_\Phi})-m_\gamma(t) \tag{14}$$

which possesses the same fix-points as (11).

One can try to improve the performance of the network by the introduction of pattern-dependent weights ε_γ [8] in

$$Q_i(\xi_i,\xi_j)=\sqrt{q}\;\Phi(\sum_{(\gamma)}\varepsilon_\gamma v_\gamma(\xi_i)v_\gamma(\xi_j)/\sqrt{q}). \tag{15}$$

A simple criterion for the optimization of ε_γ would be to maximize $\min_\gamma m_\gamma(t)$ under the assumption $m_\gamma(0)=$const. Since equation (11) is independent of α_γ the optimal solution is the trivial one $\varepsilon_\gamma=$const. This simple answer is not longer correct for the dynamics [4]

$$S_i(t+1)=sgn(\sum_{j\neq i}J_{ij}(S_j(t)-E(S(t)))) \tag{16}$$

leading to a better retrieval quality according to

$$m_\gamma(t+1)=erf(m_\gamma(t)/\sqrt{2\alpha_\Phi(1-E(S(t))^2}). \tag{17}$$

Here the optimal ε_γ depend on α_γ, the constant initial retrieval quality and the number of time steps t.

ACKNOWLEDGEMENTS

We thank the SFB 123 and the NTZ at the Universities of Heidelberg and Leipzig, resp., for the financial support.

REFERENCES

[1] van Hemmen, J.L., Grensing, D., Huber, A. and Kühn, R., J. Stat. Phys. 50 (1988) 231.

[2] Parga, N. and Virasoro, M.A., J. Physique 47 (1986) 1857.

[3] Bös, S., Diploma thesis (1988) Univ. Heidelberg (FRG).

[4] Englisch, H. and Böhlau, P., Optimization of Strongly Diluted Networks, Preprint NTZ KMU Leipzig, submitted to Proc. "Neuroinformatik", Eberswalde (GDR) 1989.

[5] Englisch, H. and Pastur, L.A., Spectral Analytic Approach to the ADALINE Learning for Neural Networks, Preprint NTZ KMU Leipzig, submitted to Proc. "Neurocomputers and Attention", Moscow (USSR) 1989.

[6] Englisch, H., Engel, A., Schütte, A. and Stcherbina, M.V., Improved Retrieval in Nets of Formal Neurons with Thresholds and Non-Linear Synapses, Preprint NTZ KMU Leipzig, submitted to Proc. "Mathematical Psychology", Berlin (GDR) 1988.

[7] Derrida, B., Gardner, E. and Zippelius, A., Europhys. Lett. 4 (1987) 187.

[8] van Hemmen, J.L. and Zagrebnov, V.A., J. Phys. A 20 (1987) 3989.

Parallel Processing in Neural Systems and Computers
R. Eckmiller, G. Hartmann and G. Hauske (Editors)
Elsevier Science Publishers B.V. (North-Holland), 1990

IMPLEMENTATION OF FUZZY PRODUCTION SYSTEMS WITH NEURAL NETWORKS

Wolfgang Eppler

Institut für Rechnerentwurf und Fehlertoleranz
(Prof. Dr.-Ing. Detlef Schmid)
Universität Karlsruhe
Karlsruhe, West Germany

The power of neural nets comes from features like massive parallelism and self-organization, i.e. they are not programmed. Simulations show that learning in a 'blank' network is very time consuming. Therefore this paper proposes a method for the prestructuring of neural nets. It seems to be quite natural using production systems with fuzzy logic to achieve a prestructured network with a better starting point for learning a special task.

1. INTRODUCTION

In this paper a neural net will be regarded as a controller accepting patterns at the input and determining actions at the output. There is no processing of linguistic knowledge, i.e. facts like "A has property B" cannot be handled with this approach if B can't be derived from the input patterns. A further, but not severe limitation is the use of a local representation. At the moment there is no possibility known representing rules in a distributed manner. First approaches to handle rule based systems in a distributed environment show, that there always is a bijective mapping between the local and the distributed representation of objects or facts [1]. Therefore every local rule based system can be transformed into a distributed one with much more redundancy and internal fault tolerance.

1.1. Motivation to prestructuring

Up to now any hard learning in neural nets is extremely time consuming, e.g. a net with nine neurons and four hidden units needs several hundreds to more than thousand learning cycles, each of them consisting of the whole presentation of all learning patterns. It´s difficult to determine the huge set of test patterns in trigger networks. But the crucial point is the limitation of the network, i.e. determining the breadth and depth of the network used according to the problem to be solved.

It is not very feasible to cope with complex tasks in a manner like this, i.e. starting learning in a totally unstructured and "blank" network. In nature the evolution provides for "prestructured" organisms at birth. In technical systems a similar state could be achieved by "logical prelearning" or "prestructuring".

1.2. Possibilities of prestructuring

There are two possibilities of prestructuring:
  - Partitioning of tasks into subtasks, which can be learned easier: The size of the nets used for those subtasks is determined in the old manner by trial and error. Afterwards the solution to these subtasks are combined and integrated in one huge network.
  - Mapping of conventional methods into the structure of neural nets. There are two further possibilities:
    • The mapping of algorithms written in a procedural programming language.
    This method seems to be unsuited to a network, in which further improvements should be possible. Little local changes can distort the whole algorithm and it is not possible to achieve a better solution step by step.
    • The mapping of rule based systems into neural nets.
    It would be desirable having a language like PROLOG or an expert system shell like KEE with a compiler performing the translation to neural nets. Up to now many problems like the variable binding or the representation of facts are still unsolved (despite [2, 3, 4]!).

1.3. Prestructuring with standard logic

Rules like "if ($\sim$AB + A$\sim$C) then D" can be implemented easily in neural nets - not necessarily in a one-level form, but always in a multi-level form with weights $\in \{-1, 0, 1\}$. There are some approaches using rules in this manner or in an extended form with semantic nets [5]. They all show a big disadvantage that the net isn't

able to learn. But in a complex or smoothly changing environment the rules aren´t quite accurate to the task being solved. Further learning may change the rules partly and modify the net towards a better execution of the task.
-> Is it now possible to interpret the modified net as a rule based system?

2.   PRESTRUCTURING WITH FUZZY LOGIC

2.1.   Two levels of description

In a prestructured network with a threshold output function we have ideal values, because they are elements of a discrete range (weights $\in \{-1, 0, 1\}$ and output activities $\in \{0, 1\}$). In this case we can regard a neuron as a *set of input patterns*. The elements of this set lead to an activation of the corresponding neuron. After a learning phase the weights are in the interval [-1, 1] (limited by a "decay" term which decreases the weights in time) and the output activities are in [0, 1]. Even in this case it is possible looking at a neuron as a pattern set, but with "fuzzy" borders. The output activity o_{rij} of a neuron defines the membership of each input pattern (the vector o_{ri-1}, which consists of the output activities o_{ri-1k}) to the pattern set. There are now two levels of description: the rule level, which is static, and the dynamic level. On the *rule level* a neuron is described as a feature or a set of binary patterns, the on-set of this neuron. The patterns are composed of other, 'previous' features. This pattern set often is written as a logical combination of these previous features. It is also possible

using a weighted feature vector $\begin{pmatrix} w_1 \\ .. \\ w_n \end{pmatrix}_\theta$. The weights have only to be added and compared to the bias θ to

show the possible feature combinations producing the on-set of the neuron.

In the running mode these weighted vectors are overlaid by the current input vector (normally the components are multiplied). On this *dynamic level* the membership of the input pattern to a neuron is computed by the output function of a neuron.

2.2.   Fuzzy logical interpretation of neural nets

According to the fuzzy logic of Zadeh [6] we can describe neuron N as a fuzzy set with the elements $o_{r,i-1}$ and membership values μ_N of these elements to set N:

$$N = \sum_r \mu_N(o_{r,i-1}) / o_{r,i-1} = \sum_r o_{rij} / o_{r,i-1}$$

with   r :   number of input patterns
       i :   layer i
       j :   column j
       o_{rij} :   output activity of neuron N
       o_{ri-1} :   input pattern r = vector of output activities.

o_{rij} is interpreted as the membership function $\mu_n(o_{r,i-1})$ and is being computed by

$o_{rij} = \sigma(\sum_k w_{ijk} o_{ri-1k} - \theta_{ij})$. (Here θ_{ij} is the bias of neuron N_{ij} and w_{ijk} is the weight between neuron N_{i-1k} and neuron N_{ij}).

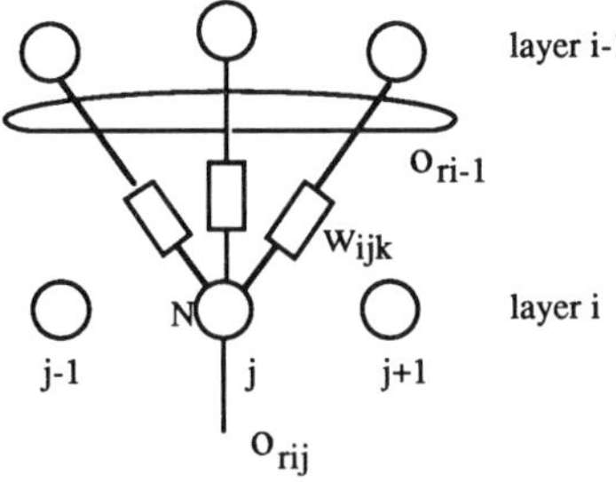

Fig.1 Description of a neuron as fuzzy set

The same two levels of description are important, if we use a different way of looking at the significance of a simple neuron: the production rule "if AB + ~BC then D" defines the fuzzy set and neuron D in a static manner. Nothing is said what happens in the running mode if A only partly is active, i.e. μ_A(input pattern x_0) < 1. In this case a new fuzzy set B' is created with the same elements used as B, but with a reduced membership function

$$\mu_{B'}(y) = \min (\mu_A(x_o), \mu_B (y)) \qquad \text{for all } y \in B, \text{ or}$$

$$\mu_{B'}(y) = \mu_A(x_o) * \mu_B (y) \qquad \text{for all } y \in B.$$

The second equation can be transformed easily into a neural net, if $\mu_A(x)$ is the output activity of neuron A and $\mu_B(y)$ is the weight between neurons A and B. The first equation also can be used if the input function of a neuron is changed to $net_i = \sum \min(o_j, w_{ij}) - \theta_i$, which shows moreover computing advantages.

2.3. The conclusion of a rule

In the rule "if A then B" B may be a fuzzy set, too. B contains elements with a specified membership to B (i.e. $B = \{\mu_B(x) / x + \mu_B(y) / y\}$). The same elements may have a different membership to an other fuzzy set D at the same time. E.g. there is an additional rule "if C then D", with $D = \{\mu_D(x) / x + \mu_D(y) / y\}$. Then we can regard the elements x and y as new sets composed of the previous sets B and D. The elements of these new sets are the patterns composed of the features B and D and belonging to the on-sets of the neurons B and D. The patterns can be described by the vectors

$$x = \begin{pmatrix} \mu_B(x) \\ \mu_D(x) \end{pmatrix}_{\theta 1}, \qquad y = \begin{pmatrix} \mu_B(y) \\ \mu_D(y) \end{pmatrix}_{\theta 2}$$

or, in a more logical notation: x = f(B, D), e.g. x = B + ~D

2.4. The capability of fuzzy logic

Fuzzy logic combined with neural nets is applicable to both the prestructuring and the interpretation of a neural net.
prestructuring:
    a) Input patterns with their membership to a linguistic variable are given, weights are sought.
       E.g.   if x = big **then** y = ...   with big = {0.1/150 + 0.5/180 + 0.9/210}
            If the values 150, 180 and 210 are coded badly, we have to use an onehot decoding with

$$\mu_A(x_r) = o_r = \sigma(\sum w x_r - \theta) = w .$$

    b) Rules with feature combinations as conditions are given, weights are sought.
       E.g.   if x = big + small * thick **then** y = heavy
            if y = heavy **then** z = ...

$$\text{There exist many solutions, e.g. } \begin{pmatrix} 0.8 \\ 0.4 \\ 0.4 \end{pmatrix}_{0.6}$$

interpretation:
    A network with weights and biases is given,

    a) for each neuron the pattern set with its membership function μ is sought,
    b) the feature combination each neuron is composed of is sought.

$$\text{(weights and biases) -> weighted feature vector } \begin{pmatrix} w_1 \\ ... \\ w_n \end{pmatrix}_{\theta} \text{ -> logical form, e.g. } M_1 + M_2 \cdot M_3$$

3. EXAMPLE

To demonstrate a small example a former paper of Kickert and Lemke [7] is chosen. They controlled a warm water plant with five fuzzy rules R1, ... , R5 of the form like:

    if   T = "very small" **and** dT = "small"
    then dF1 = " small",   with   T : temperature error
                              dT : change in error
                           dF1 : change in hot water flow.

"small" is a linguistic variable [6], i.e. a fuzzy set with the membership funktion $\mu(x) = (1 + (3 (x - 1)) 2)-1$. The other linguistic variables are defined similarly (see Fig. 2).

The implementation into a neural net is to be carried out as follows: variable T consists of a range of values, say 0 till 9. These numbers are coded and presented to n input neurons. The next layer consists of neurons representing the linguistic variables "not small", "small", "very small", "medium" and "big". They receive their input patterns from the two independent domains T and dT. The input weights correspond to the membership values of the linguistic variables. The following layer defines the logical combinations of the conditions of the rules. The conclusions of the rules are also fuzzy sets and their elements are located in the next layer, with their membership values as input weights. These elements may be mapped to an arbitrary coding in the last layer of neurons.

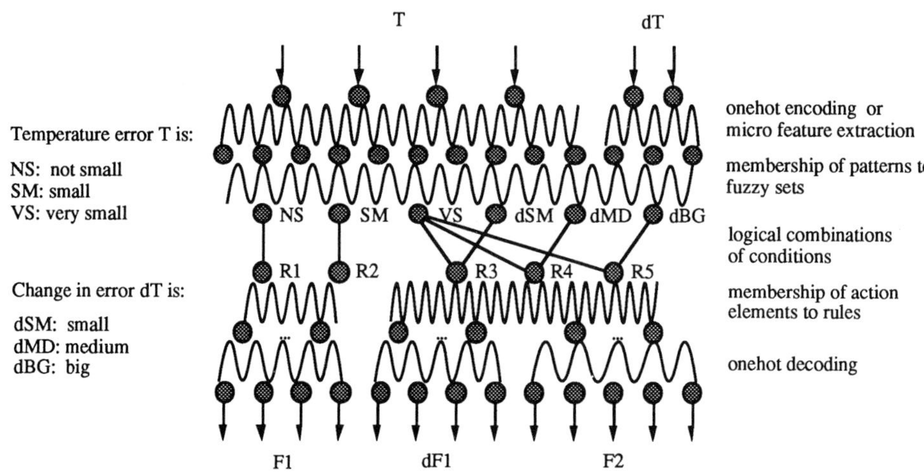

Fig.2 A small production system in a neural net

This implementation can be performed automatically with a compiler. After the phase of prestructuring this neural net behaves like a conventional rule based system. In a following learning phase weights are changed and therefore rules are modified. Simulations showed better results when every rule was represented by a *collection* of neurons rather than by a single one. In this way not only rules are modified but new rules represented with smaller collections emerge from the original collections.

4.   CONCLUSIONS

Prestructured networks which were changed by learning can be interpreted with production rules in fuzzy logic. It was showed that prestructuring itself may be done with rules formulated in fuzzy logic. Up to now this is only applicable in controller-like systems with no unification or storage of linguistic knowledge. Such extensions towards general expert systems (with the ability of learning!) will be future work.

REFERENCES

[1]   Touretzky, D.S. and Hinton, G.E., Distributed Connectionist Production System, Cognitive Science 12 (1988), pp.423-466
[2]   Smolensky, P., On variable binding and the representation of the symbolic structures in connectionist systems, CU-CS-355-87 (1987), pp.1-68
[3]   Feldman, J.A. and Ballard, D.H., Connectionist Models and Their Properties, Cognitive Science 6 (1982), pp.205-254
[4]   Rumelhart, D.E. and McClelland, J.L., Parallel Distributed Processing, Vol.1,2 (1986)
[5]   Shastri, L., A Connectionist Approach to Knowledge Representation and Limited Inference, Cognitive Science 12 (1988), pp.331-392
[6]   Zadeh, L.A., Outline of a New Approach to the Analysis of Complex Systems and Decision Processes, IEEE Transact. on Syst., Man and Cyb. No.1 (1973), pp.28-44
[7]   Kickert, W.J.M. and Lemke, H.R.v.N., Application of a Fuzzy Controller in a Warm Water Plant, Automatica Vol.12 (1976), pp.301-308

Parallel Processing in Neural Systems and Computers
R. Eckmiller, G. Hartmann and G. Hauske (Editors)
© Elsevier Science Publishers B.V. (North-Holland), 1990

A NEURAL NETWORK FOR JOB SEQUENCING

Luyuan Fang and William H. Wilson
Department of Computer Science
Flinders University of South Australia
Bedford Park, S.A. 5042, AUSTRALIA

Tao Li
Department of Computer Science
Monash University
Clayton, VIC 3168, AUSTRALIA

This paper presents the design of a competition based neural network for sequencing unit time jobs. The problem is polynomial time solvable. The neural network always finds optimal solutions. The approach taken is conceptually simple and often results in good quality optimization neural networks.

1. INTRODUCTION

Artificial neural networks have been used in solving many combinatorial optimizations problems since the seminal work of Hopfield and Tank [1, 2, 3, 4, 9]. These networks have produced promising results. Some examples can be found in [6, 7].

In the conventional approach of optimization neural network design, an energy function is first specified for a given problem. Equations of motion for neurons are then derived from the energy function. Although elegant, this approach is not always practical. It is sometimes difficult to find a good energy function. On the other hand, some researchers have adopted a heuristic approach to the design of neural networks for various applications. One interesting approach is the *competition-based* neural network [8]. In this approach, neurons are allowed to compete to become active under certain constraints. The activation rules are designed using heuristics. This approach has produced competitive results as compared to heuristic algorithms.

In this paper we present neural network implementations for a simple optimization problem which is not NP-complete. Tank and Hopfield [5] also solved a few simple optimization problems like this using neural networks. Competition based networks are used in our research. Although the neural network is based on local representation, it also exhibits globally verifiable behaviors such as convergence property. The neural network presented here is *guaranteed* to produce optimal solutions if the resource requirements in each row (and in each column) are all different.

2. NEURAL NETWORK FOR JOB SEQUENCING

This is a problem for which polynomial time algorithms exist. We use this as an example to show that neural networks which always yield optimal solutions for computationally tractable problems can often be found.

We are given n jobs and each job J_i is associated with an integer deadline $d_i \geq 0$ and a profit $p_i \geq 0$. A profit p_i is earned iff the job J_i is completed before its deadline. Each job requires a unit time to process and only one machine is available for processing jobs. A feasible solution is a subset of jobs, each of which can be completed by its deadline. The sum of the profits earned is the value of the solution. An optimal solution is a feasible solution with maximum value (total profit). A greedy heuristic algorithm can be found in [5] which always finds an optimal solution for each set of jobs. The algorithm first sorts the jobs by their profits and schedules them according to that order.

The neural network for job sequencing is organized as a $n \times n$ array of neurons, where n is the number of jobs. If the maximum deadline among the set of jobs is less than n, say m, then we only need an array of n rows and m columns. Each column represents a time slot and each row represents a job. Thus the variable x_{ij} denotes the value of a neuron which represents the assignment of job i to time slot j. There is a connection between a neuron x_{ij} to every other neuron on row i and on column j. These connections all have the same weight, -1. This is actually the competitive geometry for the 1-out-of-n constraint.

3. FORMULATING EQUATIONS OF MOTION

A simple form of constraint, 1-out-of-n, is used in the job sequencing neural network. We use the following equation of motion for 1-out-of-n constraint with cost coefficients.

$$\frac{dx_i}{dt} = [(1 - \epsilon) - \sum_{j \neq i}^{n} x_j]c_i, \tag{1}$$

where ϵ is a small constant near 0 and c_i's are the cost coefficients.

Proposition 1: The equation of motion defined above guarantees to enter a global stable state in each run and the global stable state entered must have one variable assume the value 1 and rest 0 if $c_i's$ are all different and if the variables are initialized to 0.

Due to page limits, the proof is not presented here. We notice that at most one neuron is allowed to have the value 1 on each row and on each column in a feasible solution since each job is assigned to at most one time slot and each time slot can be assigned at most one job. So we use one equation to test the row assignment condition and another to test the column assignment condition. These equations are similar to the 1-out-of-n equation except that the costs are all set to 1. The row condition equation for x_{ij} is
$$sum1 = 1 - \epsilon - \sum_{k \neq j} x_{ik}^m 1,$$
and the column condition equation for x_{ij} is
$$sum2 = 1 - \epsilon - \sum_{k \neq i} x_{kj}^m 2.$$
Typically, $m1 < m2$. In fact, x_{ab}^m in the sums can be replaced by the threshold function used for 1-out-of-n equations if appropriate threshold values for the two conditions can be readily found. The choice of $m1 < m2$ implies that the row condition of a neuron will become negative before its column condition becomes negative.

The equations of motion is formed by taking into account the above two conditions and is given below.

$$\frac{dx_{ij}}{dt} = \Psi(sum1, sum2, A + (j + 1)p_i/B). \tag{2}$$

and Ψ is defined by

$$\Psi(s_1, s_2, P) = \begin{cases} -P, & if \ s_1 \leq 0 \ or \ s_2 \leq 0 \\ P, & if \ s_1 > 0 \ and \ s_2 > 0 \end{cases}$$

Notice that the term $A + (j + 1)p_i/B$ increases as the time slot j increases. This means that the network tries to sequence each job as close to its deadline as possible. In addition, the increase of activation for x_{ij} is proportional to the profit p_i of job i. Hence, jobs with high profits compete strongly and are more likely to be sequenced. For any variable $x_{i,j}$ such that $j > d_i$, $x_{i,j}$ is always 0 because it makes no sense to sequence a job after its deadline.

Although not immediately obvious, this strategy has the same effect as the sequential greedy algorithm. The circuit realization of a typical neuron in the job sequencing network is shown in Figure 2.

4. NETWORK CONVERGENCE

The conditions $sum1$ and $sum2$ guarantee that $dx_{a,b}/dt$ can only change sign once if the neurons are initialized to different values or if the profits are all different (as discussed in the convergence property study of the generalized assignment). Hence, the system always converges to some stable state. The system generates the same results as the greedy sequential algorithm if costs are different or if the neurons are initialized to different values. Due to page limit, the proof of the proposition below is omitted here.

Proposition 2: If the profits of the jobs are different, the neural network always finds solutions with the same total profits as the greedy sequential algorithm. Hence, it always finds optimal solutions.

5. SIMULATION RESULTS

The job sequencing neural network has been tested by a large number of data sets. In the simulation, 4 groups of data were used, with each group containing 10 sets of data. The numbers of jobs the four groups were 7, 10, 15, and 20 respectively. The data were randomly generated. The profits were bounded within a large interval and the deadlines were bounded by the number of jobs.

From our simulation experiments, a very important observation is that the number of steps required a simulation run is independent of the network size. It depends on the costs of the jobs to a certain degree. The numerical requirement for simulation to converge to correct solutions is very low. Hence, each simulation run can be executed quickly. Normally, no more than a few thousand steps are need in each run.

For all sets of test data, each of which contains different profits for the jobs, the neural network produced exactly the same results as the greedy algorithm (that is, optimal solutions). A few sets of data contain profits which are the same for several jobs. For such data sets, if the neurons are initialized with the same value, then some jobs would be sequenced in the same time slot, resulting in infeasible solutions.

6. SUMMARY

A neural network has been designed to show the feasibility of this approach. This neural network is guaranteed to produce optimal solutions. This demonstrates the feasibility of competition based neural network for combinatorial optimization. In addition, the simulation can be finished in almost constant time independent of the network size if performed on a massively parallel machine.

ACKNOWLEDGEMENTS

The support of the Australian Research Council under the Research Grant for Parallel Symbolic Processing is acknowledged.

REFERENCES

[1] Hopfield,J.J., "Neural networks and physical systems with emergent collective computational abilities", *Proc. Natl. Acad. Sci. USA*, vol.79, pp.2553-2558, Apr. 1982.

[2] Hopfield,J.J., "Neurons with graded response have collective computational properties like those of two-state neurons", *Proc. Nat. Acad. Sci. USA*, vol.81, pp.3088-3092, May, 1984.

[3] Hopfield, J. J., and Tank, D. W., "'Neural' computation of decisions in Optimization problems", *Biological Cybernetics*, vol. 52, pp.141-152. Berlin: Springer-Verlag. 1985.

[4] Hopfield,J.J and Tank,D.W, "Computing with Neural Circuits: A Model", *Science*, Vol.233, No.4767, pp.625-633, Aug.1986.

[5] Horowitz, E. and Sahni, S. *Fundamentals of Computer Algorithms*, Computer Science Press, 1978.

[6] Peterson, C. and Anderson, J.R. "Neural Networks and NP-complete Optimization Problems: A Performance Study on the Graph Bisection Problem", *Complex Systems* 2, pp.59-89, 1988.

[7] Peterson, C. and Soderberg, B. "A New Method for Mapping Optimization Problems onto Neural Networks", Int. Journal of Neural Systems, 1(3), 1989, pp.3-22.

[8] Reggia, J. "Methods for deriving competitive activation mechanisms", *Proc. IJCNN 1989*, June, 1989, Washington D.C., pp357-363.

[9] Tank, D. W., and Hopfield, J. J., "Simple 'Neural' optimization networks: An A/D converter, signal decision circuit, and a linear programming circuit", *IEEE Transactions on Circuit Systems*, vol. CAS-33, pp.533-541, May 1986.

Parallel Processing in Neural Systems and Computers
R. Eckmiller, G. Hartmann and G. Hauske (Editors)
© Elsevier Science Publishers B.V. (North-Holland), 1990

INCREASING THE STORAGE CAPACITY AND SHAPING THE BASINS OF ATTRACTION OF NEURAL NETS

Shlomo Geva and Joaquin Sitte
Faculty of Information Technology
Queensland University of Technology
GPO Box 2434 Brisbane Q 4001 Australia

We reformulate the autoassociative network as a two layer network with feedback, which includes the higher order correlation and the pseudo-inverse networks as special cases. From an analysis of the dynamics we conclude that cooperation between the stored prototype vectors is the main factor limiting the network's performance as an associative memory. We suggest a simple scheme for improving the performance by reducing the cooperation. Exhaustive scans of the state space of small networks support our conclusions.

1 Introduction

The number, shapes and sizes of the basins of attraction of a neural network determine its performance as an associative memory. In the auto-associative network [1] of N neurons, with a Hebbian correlation storage rule (also known as the Hopfield net), one has little control over the shapes and sizes of the basins of attraction. Also, associative recall tends to break down when more than $0.15N$ random prototype vectors are stored [2]. A learning rule based on the pseudo-inverse matrix [4] significantly improves the network's behaviour. The basins are nearly of the same size with the prototypes close to their centers. The pseudo-inverse method guarantees the recall of N prototype vectors, but the associativity is almost lost above $0.5N$ [3]

Higher order correlation rules [5] [6] [7] [8] have been proposed for improving storage capacity and associativity. The difficulty with higher order correlation rules is the combinatorial growth in the number of synaptic matrix elements needed.

Here, we present an autoassociative network which formally contains the previously mentioned models as special cases, and improves their performance by relatively simple means.

2 A two layer network

We use a two layer feedback network, which is based on Psaltis's [5] formulation of the autoassociative network for optical implementation. The two layers are: input layer and memory layer. The neurons in the input layer are two-state neurons corresponding to the N components of the input vectors. The states of the input layer belong to the hypercube $\{-1, 1\}^N$. Each neuron in the memory layer represents one of the m memorized prototype vectors. The neurons in the input layer are connected to each neuron in the memory layer and vice versa. No connections exist between the neurons in a layer. There are two unidirectional synapses between a neuron in the input layer and a neuron in the memory layer. Their synaptic strengths may be different.

The network evolves in a parallel synchronized fashion. The state $\vec{s}$ of the input layer produces an excitation of the memory neurons, which is defined as a function of the input state and the prototype vector:

$$v_\mu = F_\mu(\vec{M}^\mu, \vec{s}) \qquad \mu = 1..m \tag{1}$$

The memory neurons feed back to the input layer producing excitations:

$$h_i = \sum_{\mu=1}^{m} v_\mu M_i^\mu \tag{2}$$

The process performed by all neurons at the input layer can be summarized in vector notation:

$$\vec{h}(\vec{s}) = \mathbf{M}\vec{v}(\vec{s}) \tag{3}$$

where $\mathbf{M}$ is the matrix whose columns are the prototype vectors. The excitation of the input layer is a linear combination of the memorized prototype vectors.

The new state of the input layer is the squashed excitation vector:

$$s_i(t+1) = \begin{cases} s_i(t) & \text{if } h_i(t) = 0 \\ sign(h_i(t)) & \text{otherwise} \end{cases} \tag{4}$$

For associative recall the input layer is set to an initial state. The memory layer neurons evaluate their excitation via the synapses from the input layer. The input layer, in turn, is excited by the memory layer and determines its new state. The process repeats until a stable state is reached at the input layer.

3   Relation to existing models

Although our two layer network has some similarity with Carpenter and Grossberg's ART networks [9] it lacks their elaborate control mechanisms. We limit our attention to the simpler autoassociative nets.

If we chose the excitation function F_μ of the neurons in the memory layer to be the dot product (overlap) of the input vector with the corresponding prototype vector

$$v_\mu = \vec{M}^\mu \vec{s} \tag{5}$$

the network essentially becomes a Hopfield network

$$\begin{aligned} \vec{h}(\vec{s}) &= \mathbf{M}\vec{v}(\vec{s}) \tag{6} \\ &= \mathbf{M}\mathbf{M}^T\vec{s} \tag{7} \\ &= \mathbf{J}\vec{s} \tag{8} \end{aligned}$$

where the synaptic matrix $\mathbf{J}$ is defined by the Hebbian correlation rule:

$$\mathbf{J} = \mathbf{M}\mathbf{M}^T \tag{9}$$

Unlike the Hopfield model we do not set the diagonal elements of the synaptic matrix to zero. The non-zero diagonal introduces self- excitation which hinders the relaxation of the network towards its energy minima. It turns out that self-excitation has little effect in most of the cases discussed below.

It is also a straight forward matter to cast the higher order correlations model or the pseudo inverse model into the two layer network form.

Higher order correlations are introduced into a single layer network, by means of multi-dimensional synaptic matrices:

$$J_{ij\ldots n} = \sum_{\mu=1}^{m} M_i^\mu M_j^\mu \ldots M_n^\mu \tag{10}$$

The excitation of a neuron is defined as:

$$h_i(\vec{s}) = \sum_{j,k,\ldots} J_{ijk\ldots} s_j s_k \ldots \tag{11}$$

By substituting (10) into (11), rearranging the order of summations, and expanding this expression for a given order of correlations, say q, we derive the following expression for the excitation:

$$h_i(\vec{s}) = \sum_{\mu=1}^{m} (\vec{M}^\mu \vec{s})^q M_i^\mu \tag{12}$$

and the excitation vector:

$$\vec{h}(\vec{s}) = \mathbf{M}\vec{o^q} \tag{13}$$

where $\vec{o^q}$ is the vector whose components are the overlaps of the prototype vectors with the input vector, raised to the power q. Therefore higher order correlations can be introduced in our two layer model by setting:

$$\vec{v}(\vec{s}) = \vec{o^q} \tag{14}$$

Several orders of correlation can be simultaneously considered by making $\vec{v}$ a polynomial of the overlap. There is no need for storing multidimensional synaptic matrices.

The pseudo-inverse method, defines the synaptic matrix by:

$$\mathbf{J} = \mathbf{M}\mathbf{M}^+ \tag{15}$$

where $\mathbf{M}^+$ is the pseudo inverse of $\mathbf{M}$. We shall refer to the rows of the pseudo-inverse matrix $\mathbf{M}$ as the *pseudo-prototypes*. The excitation is calculated in the same way as in the Hopfield network. By defining the vector $\vec{o}^+$, whose components are the overlaps of the input vector $\vec{s}$ with each of the pseudo- prototype vectors we write:

$$\vec{h}(\vec{s}) = \mathbf{M}(\mathbf{M}^+\vec{s}) \tag{16}$$
$$= \mathbf{M}\vec{o^+} \tag{17}$$

The pseudo-inverse model is obtained by setting

$$\vec{v}(\vec{s}) = \vec{o^+} \tag{18}$$

4 Cooperation and competition of prototypes

The two layer formulation suggests a simple explanation of the improved storage capacity of higher order correlation networks. According to (3), each memory neuron tries to enforce its prototype on the input layer, in proportion to its excitation. If the input layer is set to a state in the vicinity of a prototype vector, the feedback signal from that prototype may dominate the excitation of the input layer. The new state of the input layer will then be equal to the prototype vector. In that case we can say that competition determines the result. However, when the input vector is at comparable distances from several prototype vectors, the feedback signals are also comparable and may enforce a compromise at the input layer. When this occurs one typically refers to the state as a noise or spurious state, and we can also say that cooperation determines the result. This is sometimes a useful property, but too much cooperation renders the network useless as an associative memory.

Cooperation is much more likely to happen with first order correlations (linear excitation), than with second order correlation. Because of the faster decay of the quadratic excitation, more of the distant prototypes are needed to dominate a closer one. As the order of correlation is increased the memory layer excitation becomes a steeper function of the Hamming distance, therefore cooperation diminishes in the vicinity of the prototypes. States close to a prototype will be pulled towards it. Ultimately, high orders of correlations approach a winner-take-all situation, where the prototype with the greatest overlap with the input vector always enforces itself on the input layer.

When the excitation is simply a power of the overlap, the function is either symmetric or antisymmetric around zero. Consequently, prototypes having a large negative overlap with the input state contribute strongly to the excitation of the input layer. Far away prototypes produce undesired cooperation. A simple solution is to set the memory excitation to zero for negative overlaps. Prototypes beyond a distance of $N/2$ do not participate in determining the new state of the input layer. When there is no prototype within a distance of $N/2$ the network simply does not associate the input state with any prototype.

This procedure can be applied to any level of correlations desired, and of course allows one to use any polynomial or power series as the memory excitation function. In particular, the exponential function already exhibits the desired behaviour of steep decrease and of taking small values for negative overlaps. Contributions from far away states rapidly become negligible.

To illustrate this behaviour we did exhaustive scans of the state space of a network with 16 neurons in the input layer, and sets of 32 random memory prototypes. Fig.1 shows a comparison of the average basin shape for several functions, with and without cut-off. A clear improvement in the shape of the basins is evident. The number of prototypes which are recallable also increases for shorter cut-off distances, as shown in Fig.2. This improvement is at the expense of reducing the range of the basins.

As an example, reducing the cut-off distance to 4 allowed us to recall 32 random prototypes even with the linear excitation. The average fractions of first, second and third nearest neighbours included in basins were 0.98, 0.93 and 0.72 respectively.

The exponential excitation function not only provided the best basins (Fig.1), but also associated 96% of all states with prototypes. This can be linked to the fact that the exponential function decays smoothly and does not require a cut-off. By comparison the linear and quadratic excitations never exceeded 50% regardless of the cut-off.

The same line of reasoning leads to an explanation of the improved performance of the pseudo inverse method for calculating the synaptic matrix. When the prototype vectors are mutually orthogonal, up to N prototype vectors can always be recalled. Even with first order correlations their basins remain fairly large for up to $N/2$ stored prototypes. The reason is that for a state of the input layer that is in the vicinity of a prototype, due to the orthogonality, the excitation of all the other prototype neurons will be close to zero. Effectively there is little cooperation.

In broad terms, the pseudo prototypes generated by the pseudo inverse matrix *distort* the image of the input layer state, in such a way, that when it is in the vicinity of a prototype, the excitation of the other prototypes is reduced. When prototypes are input they actually look orthogonal to the memory layer neurons.

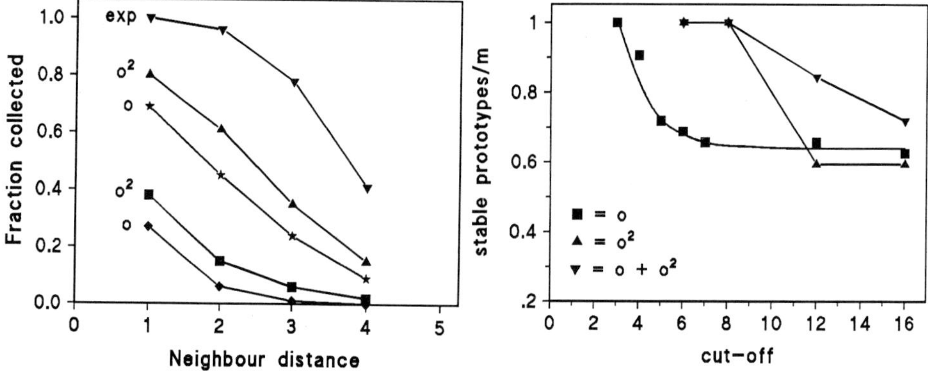

Figure 1: Average shape of the basins for a cut-off at 8, and without cut-off (two bottom curves).

Figure 2: Fraction of prototypes than can be recalled vs. the cut-off distance

References

1. S. Amari & K. Maginu , Neural Networks **1** (1988) 63
2. J.J. Hopfield , Proc. Nat. Acad. Sci. USA, **79** (1982) 2445
3. I. Kanter & H. Sampolinsky, Phys. Rev. A, **35** (1987) 380
4. L. Personnaz, I. Guyon & G. Dreyfus, J. Physique Lett. **46** (1985) L-359
5. D. Psaltis & C.H. Park, AIP Conf. Proc. 151,"Neural Networks for Computing" (Snowbird, Utah) (1986), 370
6. H.H. Chen, Y.C. Lee, G.Z. Sun, H.Y. Lee, T. Maxwell & C.L. Giles, AIP Conf. Proc. 151, "Neural Networks for Computing" (Snowbird, Utah) (1986), 87
7. P. Baldi & S.S. Venkatesh, Phys. Rev. Lett., **58** (1987)913
8. M.L. Minsky & S. Papert, Perceptrons, MIT Press (1969)
9. G.A. Carpenter & S. Grossberg, Applied Optics **26** (1987) 4919

Parallel Processing in Neural Systems and Computers
R. Eckmiller, G. Hartmann and G. Hauske (Editors)
© Elsevier Science Publishers B.V. (North-Holland), 1990

Drift and Diffusion in Backpropagation Networks

G. Radons, H.G. Schuster and D. Werner
Institut für Theoretische Physik, Universität Kiel, D-2300 Kiel, FRG

It is shown that that escape from local minima in the parameter space of Backpropagation networks is possible through chaotic diffusion if the network is trained by periodic pattern presentation, or via induced parameter fluctuations if patterns are presented in a random fashion. In the latter case and for small leaning rates we find that learning of the network is gouverned by a Fokker–Planck equation, describing drift and diffusion in parameter space. These results suggest how for perfectly trainable networks parameters can be made to converge to globally optimal values corresponding to an errorfree implementation of the desired input–output relations. For cases where perfect learning is impossible we demonstrate the usefulness of a simulated annealing-like procedure to reach the minimal error state.

Backpropagation (BP) [1] is currently the most prominent neural network and has proven to be able to e.g. convert text to phonems [2], to predict chaotic time series [3], to predict the secondary structure of proteins [4] or to learn how to drive a car [5].

BP is an algorithm which iteratively adjusts internal parameters $\vec{w} = (w_1, w_2, \cdots, w_N)$, the "synaptic strengths" and thresholds of a multilayer network, to implement relations between input patterns $\vec{\xi}^\mu = (\xi_1^\mu, \xi_2^\mu, \cdots, \xi_M^\mu), \mu = 1, 2, \cdots$ and L-dimensional target outputs $\vec{f}(\vec{\xi}^\mu)$. The number of patterns can be finite or infinite. If an input pattern $\vec{\xi}^\mu$ is fed into the network, it produces an output $\vec{g}(\vec{\xi}^\mu; \vec{w})$ which is in general different from the desired output $\vec{f}(\vec{\xi}^\mu)$. The task consists in finding network parameters $\vec{w}^*$ which minimize the *averaged error*

$$E(\vec{w}) = \frac{1}{2} \sum_\mu \left[\vec{f}(\vec{\xi}^\mu) - \vec{g}(\vec{\xi}^\mu; \vec{w}) \right]^2 \tag{1}$$

between actually produced outputs $\vec{g}(\vec{\xi}^\mu; \vec{w})$ and target outputs $\vec{f}(\vec{\xi}^\mu)$. Whithin BP this problem is tackled by performing gradient descent steps on the surfaces of *individual errors* $E^\mu(\vec{w}) = \frac{1}{2}[\vec{f}(\vec{\xi}^\mu) - \vec{g}(\vec{\xi}^\mu; \vec{w})]^2$, i.e. iteratively changing $\vec{w}$ according to

$$\vec{w}(t + 1) = \vec{w}(t) - \eta \, \vec{\nabla} E^\mu(\vec{w}(t)) \tag{2}$$

where η is the learning rate. Often an additional term the so called momentum term is included in (2) to improve the performance of Backpropagation [1]. In the following this term is omitted for simplicity but can easily taken into account. A special choice of the nonlinear function $\vec{g}(\vec{\xi}^\mu; \vec{w})$ in Backpropagation networks allows for an elegant implementation of the update steps (2) by back-propagating the error through the layers of the network. The following discussion and conclusions, however, are largely independent of the form of $\vec{g}$ and therefore apply also to more general nonlinear optimization problems.

Note that a trajectory $\vec{w}(t), t = 0, 1, 2, \cdots$ generated by system (2) depends in an essential way on how the patterns $\vec{\xi}^\mu$ are presented. For instance if the $\vec{\xi}^\mu$ are presented periodically, i.e. $\vec{\xi}(t) = \vec{\xi}(t + p)$ with p being the number of patterns, the force $\vec{F}^\mu(\vec{w}) \equiv -\vec{\nabla} E^\mu(\vec{w}) = -\vec{\nabla}_w E(\vec{\xi}^\mu; \vec{w})$ is explicitly time dependent with $\vec{F}^\mu(\vec{w}) = \vec{F}(\vec{\xi}(t); \vec{w}) = \vec{F}(\vec{\xi}(t + p); \vec{w})$. Thus in this case Eq. (2) describes a discrete periodically driven nonlinear dynamical system. Such systems are known to exhibit chaotic behaviour and one wonders if chaos may help the system to escape from local minima in the error landscape $E(\vec{w})$. The problem of getting stuck in local minima is well known for gradient descent algorithms but can indeed be avoided via chaotic diffusion resulting from periodic pattern presentation in Eq. (2). This is demonstrated in Fig. 1 for a simple example which nevertheless exhibits the typical features of the full many-dimensional problem. In this example we consider only one "neuron" with one-dimensional input ξ and target output $f(\xi)$ and the nonlinear output function g coincides with the transfer function of a single neural unit $g(\xi; \vec{w}) = \sigma(w_1 \xi + w_2)$ where σ is the

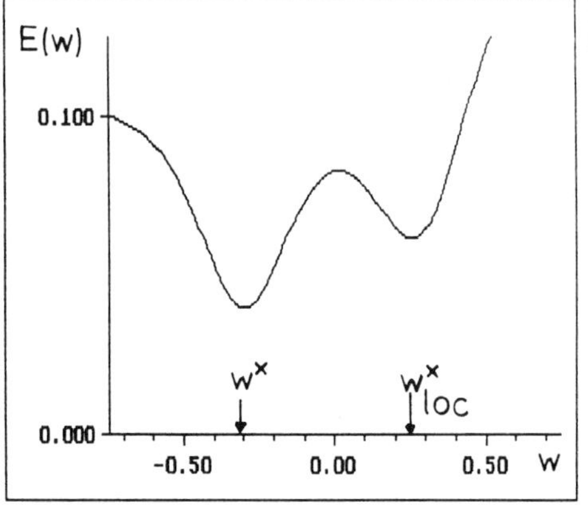
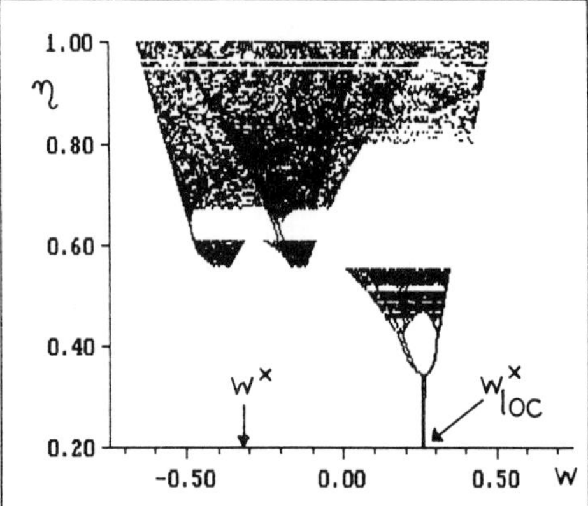

Figure 1: a) Error landscape $E(w)$ of Eq. (1). b) Attractors of the BP algorithm Eq. (2) for periodic pattern presentation for different learning rates η: Escape from local minimum w^*_{loc} via chaotic diffusion

typical sigmoid function $\sigma(x) = \frac{1}{2}(1 + \tanh x)$ used in Backpropagation networks. The task of the "network" consist in learning of only two patterns ξ^1 and ξ^2 but with only one free parameter w since we fix w_1 at a constant value $w_1 = a$ and try to optimize the "threshold" parameter " $w_2 = w$. The resulting one-dimensional error landscape is given by

$$E(w) = \sum_{\mu=1,2} E^\mu(w) = \sum_{\mu=1,2} \frac{1}{2}[f(\xi^\mu) - \sigma(a\xi^\mu + w)]^2 \tag{3}$$

and is depicted in Fig. 1a. Here the errors E^μ could be made zero individually but not the total or averaged error E. This, of course, stems from the fact that both conditions $f(\xi^\mu) = \sigma(a\xi^\mu + w), \mu = 1, 2$ cannot in general be fulfilled simultaneously by only one parameter w. This competition between individual patterns leads to local minima in $E(w)$. The main point we want to demonstrate in Fig. 1 is that periodic pattern presentation can result in self-generated stochasticity making the system to leave local minima of the error surface. In Fig. 1b this is done by plotting the asymptotic distribution of $w(t)$ for $t \to \infty$ and for different values of the leaning rate η of Eq. (3). For alternating presentation of the two patterns ξ^1 and ξ^2 we find for small η a period 2 attractor (a fix point in the Poincaré section at multiples of the period) i.e. $w(t)$ asymptotically jumps back and forth between two values w^*_1 and w^*_2 in the local minimum of $E(w)$. As η is increased one finds the typical period doubling sequence resulting in chaotic behaviour for large enough values of η, which finally drives the system out of the local minimum at w^*_{loc}. Note, however, that not even the mean value $\overline{w}$ of the asymptotically reached parameter values does in general coincide with the parameter w^* of a local or global minimum of $E(w)$ ($\overline{w} = \frac{1}{2}(\vec{w}^*_1 + \vec{w}^*_2)$ in the above case and for the period 2 attractor). This statement is easily checked analytically near a parabolic minimum of the error surface. This point is overcome with stochastic pattern presentation (Chapt. 11 of [1]) which is treated in the following.

For stochastic pattern presentation the update steps in Eq. (2) are performed by changing the individual error functions E^μ in every step in a random fashion. This turns the corresponding force $\vec{F}(\vec{\xi}(t); \vec{w}) = -\vec{\nabla}_w E(\vec{\xi}(t); \vec{w})$ into a stochastic force. Since now every realization of the random sequence $\vec{\xi}(t), t = 0, 1, 2, \cdots$ gives rise to a different solution $w(t)$ of Eq. (1), only probabilistic statements can be made.

The central quantity corresponding to the stochastic process $\vec{w}(t + 1) = \vec{w}(t) + \eta\vec{F}(\vec{\xi}(t)\,\vec{w}(t))$ is the probability density $P(\vec{w}, t)$ of finding the system at time t in state $\vec{w}$ of the parameter space. It is straight forward to show that the time evolution of $P(\vec{w}, t)$ is governed by a Chapman-Kolmogorov equation of the form

$$P(\vec{w}, t + 1) = \int T(\vec{w}, \vec{w}') \, P(\vec{w}', t) \, d\vec{w}' \tag{4}$$

where the transition matrix is given by

$$T(\vec{w}, \vec{w}') = \left\langle \delta(\vec{w} - \vec{w}' - \eta \, \vec{F}(\vec{\xi}; \vec{w}')) \right\rangle \tag{5}$$

and $<\cdots>$ denotes the average over all patterns $\vec{\xi}^\mu$ i.e. $<f(\vec{\xi})> = \int d\vec{\xi}\, f(\vec{\xi})\rho(\vec{\xi})$ with the distribution function $\rho(\vec{\xi})$ which may be continous or discrete and is normalized to one. To proceed we write the r.h.s. of (4) formally as a power series in the learning rate η and use that $P(\vec{w}, t)$ vanishes at infinity. For simplicity we give the result in one dimension

$$P(w, t+1) - P(w, t) = \sum_{l=1}^{\infty} \frac{(-\eta)^l}{l!} \frac{\partial^{(l)}}{\partial w^{(l)}} \left[\left\langle F^l(\xi; w) \right\rangle P(w, t) \right] \tag{6}$$

Eq. (6) is a Kramers-Moyal like expansion [7] which is fully equivalent to (4) but equally intractable. For small learning rates η, however, it is justified to truncate the r.h.s. of Eq. (6) after the second term and to replace the difference on the l.h.s. by a differential. We obtain for the general case a multidimensional Fokker-Planck Equation (FPE).

$$\frac{\partial P(\vec{w}, t)}{\partial t} = -\eta \sum_l \frac{\partial}{\partial w_l} [F_l(\vec{w}) \, P(\vec{w}, t)] + \frac{\eta^2}{2} \sum_{kl} \frac{\partial^2}{\partial w_k \partial w_l} [D_{kl}(\vec{w}) \, P(\vec{w}, t)] \tag{7}$$

with drift coefficients $\eta F_l(\vec{w})$ which are simply the components of the average force $\vec{F}(\vec{w}) = -\vec{\nabla} E(\vec{w})$ resulting from the averaged error surface Eq. (1) as expected and a diffusion tensor $\eta^2 D_{kl}(\vec{w})$ with $D_{kl}(\vec{w}) = <F_k(\vec{\xi}; \vec{w}) F_l(\vec{\xi}; \vec{w})>$. We are mainly interested in stationary solutions $P^*(\vec{w}, t+1) = P^*(\vec{w}, t) = P^*(\vec{w})$ of Eq. (4) i.e. the asymptotic distribution of network parameters after long times. It is known that stationary solutions of (4) can be nowhere differentiable e.g. fractal measures, whereas stationary states obtained from approximation (7) are differentiable. In the following we demonstrate the usefulness of (7) for practical purposes. In one dimension the stationary solution $P^*(\vec{w})$ of Eq. (7) is easily obtained by integration [7]

$$P^*(w) = \frac{\mathcal{N}}{D(w)} \exp\left(\frac{2}{\eta} \int^w \frac{F(w')}{D(w')} dw'\right) \tag{8}$$

where $\mathcal{N}$ is a normalization constant and the parameter dependent diffusion tensor is reduced to $D(w) = <F^2(\xi, w)>$. This solution is compared in Fig. 2a with results from numerical simulations of Eq. (4) for the simple system (3). It is observed that the numerically obtained distribution $P(w, t)$ indeed evolves towards the stationary solution (8) of the FPE (7). Multidimensional generalizations of solution (8) can be found under certain conditions [7], which will be considered elsewhere.

There is an interesting aspect of Eq. (8) with respect to perfectly trainable networks. These networks are characterized by the fact that there exist paramters $\vec{w}^*$ such that *all* individual errors E^μ are minimal at w^* namely zero. Since the E^μ are differentiable functions of $\vec{w}$ all individual forces $\vec{F}(\vec{\xi}^\mu; \vec{w})$ must also vanish at $\vec{w}^*$. This means that $\vec{F}(\vec{w}) = <\vec{F}(\vec{\xi}^\mu; \vec{w})>$ as well as $D_{kl} = \left\langle F_k(\vec{\xi}; \vec{w}) F_l(\vec{\xi}; \vec{w}) \right\rangle$ is zero at $\vec{w} = \vec{w}^*$. In (8) this leads to a logarithmic divergence of the integral at $\vec{w}^*$ because $\frac{D(w)}{F(w)} \to 0$ for $\vec{w} \to \vec{w}^*$. As a consequence $P^*(w)$ diverges at $\vec{w} = \vec{w}^*$ i.e. becomes a δ-distribution. From this we cannot conclude, however, that the total probability is concentrated in the δ-distribution. If one could show that there are no local minima other than possibly degenerate absolute minima, then the drift term in Eq. (8) would drive all probability into absolute minima of $E(\vec{w})$ and with probability one parameters would take their optimal values. Though this often happens in perfectly trainable networks, we cannot exclude the occurence of local minima even in this case.

To drive the system from these local minima into absolute minima is, however, easily accomplished by choosing high learning rates η. This causes strong fluctuations of $\vec{w}$ in local minima but not at absolute minima where in perfectly trainable networks the fluctuations are zero. Thus trajectories $\vec{w}(t)$ are "heated up" everywhere in parameter space except at absolute minima where they "freeze" into their optimal values.

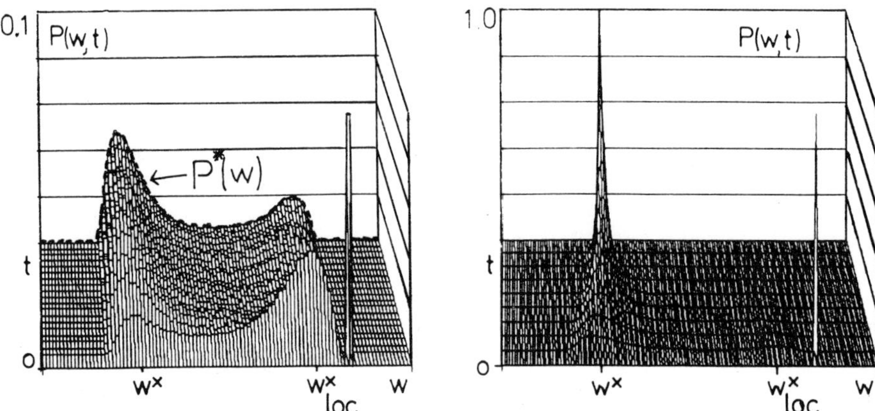

Figure 2: Evolution of the probability $P(w, t)$ to its equilibrium $P^*(w)$, a) for fixed $\eta = 0.5$, the dashed line is $P^*(w)$ of Eq. (8), b) for exponentially decreasing $\eta(t)$: All samples converge to the global minimum $\vec{w}^*$ (Note: Same initial condition as in a), different P-scale).

This picture is confirmed directly by calculating the induced parameter fluctuations from Eq. (2) near a minimum $\vec{w}^*$ of $E(\vec{w})$ in the quadratic approximation for $\eta \ll 1$

$$\overline{\vec{w}^2} - \overline{\vec{w}}^2 = \frac{\eta}{2} \langle \vec{F}(\vec{\xi}; \vec{w}^*)^T < \vec{\nabla} \otimes \vec{F}(\vec{\xi}; \vec{w}^*) >^{-1} \vec{F}(\vec{\xi}; \vec{w}^*) \rangle \tag{9}$$

Comparing this in the one-dimensional case with the coordinate fluctuation of an overdamped particle in a quadratic potential in a heat bath of temperature T shows that $\frac{\eta}{2}<F^2>$ takes the role of the temperature, which vanishes at the absolute minima in perfectly trainable networks. This analogy suggests that in networks where perfect learning is impossible a simulated annealing-like procedure [8] is appropriate to find global minima of the error surface. For BP nets this simply means decreasing η in the course of time. In Fig. 2b we demonstrate the usefulness of this procedure again for our simple example Eq. (3) with η decreasing exponentially in time. We see that all samples $w(t)$ finally relax into the global minimum at w^* although there was initially a high probability of finding them in the local minimum at w^*_{loc} . From the above general considerations and also from numerical simulations it follows that the same picture holds also for higher dimensional problems i.e. full networks with many parameters providing us with an optimized strategy of finding optimal parameters in Backpropagation nets.

One of us (D.W.) acknowledges support from the SFB 185 "Nichtlineare Dynamik"

References

[1] D. Rumelhart and J. McClelland. *Parallel Distributed Processing.* Volume 1, MIT Press, Cambridge, MA, 1986.

[2] C.R. Rosenberg and T.J. Sejnowski. *Complex Systems*, 1:145, 1987.

[3] A.S. Lapedes and R. Farber. Technical Report LA-UR-87, Los Alamos National Laboratory, 1987. submitted to Proc. IEEE.

[4] N. Qian and T.J. Sejnowski. *J.Mol.Biol.*, 202:865, 1988.

[5] D.A. Pomerleau. In D.S. Touretzky, editor, *Advances in Neural Information Processing Systems 1*, Morgan Kaufmann Publishers, San Mateo, CA, 1989.

[6] D. Werner, G. Radons, and H.G. Schuster. to be published.

[7] C.W. Gardiner. *Handbook of Stochastic Methods.* Springer Verlag, Berlin, 1983.

[8] S. Kirkpatrick, C.P. Gelatt.Jr., and M.P. Vecci. *Science*, 220:671, 1983.

Parallel Processing in Neural Systems and Computers
R. Eckmiller, G. Hartmann and G. Hauske (Editors)
© Elsevier Science Publishers B.V. (North-Holland), 1990

Solving Constraint Problems using Feedback Neural Networks

H. Nikolaus Schaller

Lehrstuhl für Datenverarbeitung
Technische Universität München
München, FRG.

A device called m-of-n-flop is presented which behaves like a multistable flip-flop. Stability and parameter selection is discussed. How to interconnect several m-of-n-flops to solve constraint satisfaction problems like coloring a map or placing queens on a chess board will be shown.

1. INTRODUCTION

A typical example of constraint satisfaction problems is to select colors for a map. The constraints are that adjacent countries are colored differently. It is easy to formulate constraints valid solutions must conform to, but it is difficult to find them. Systematically trying all possible assignments (e.g. using PROLOG) is only practicable for small parameter sets.
The neural network approach is to represent possible assignments by neuron states. The switching of states is controlled so, that a finally reached stable state fulfills all constraints. Designing a constraint problem solver is done by first designing data storage structures and then programming the constraints.

2. THE M-OF-N-FLOP

2.1. Topology

We consider an m-of-n-flop (Fig. 1) which is an extension of the n-flop presented in [1]. An m-of-n-flop consists of n binary neurons (states 0 and 1). The neurons are input adders (Σ) combined with threshold-zero elements (T) as defined below. The circuit is to be designed that exactly m neurons become active if the network state stabilizes. This is accomplished by selecting appropriate connection weights.

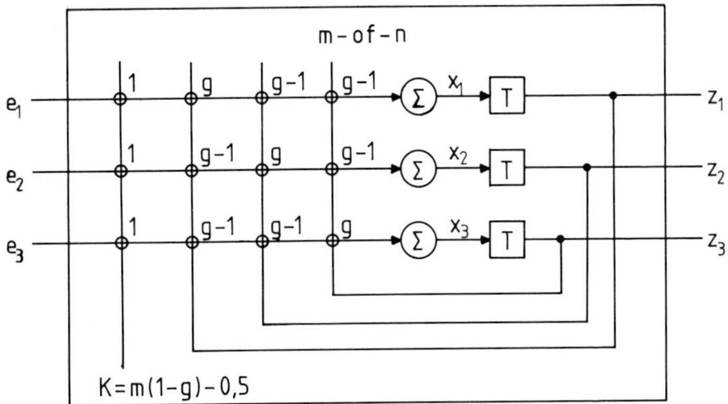

Fig. 1: Scheme of a m-of-n-flop (n=3)

2.2. Updating rule

The continuous repetition of the following steps is assumed. A neuron j is randomly selected from the pool of n neurons. The activation x_j of this neuron is (all sums are taken from $i = 1$ to n):

$$x_j = \Sigma\, W_{ij} z_i + K + e_j$$

with synapses weights defined as $W_{ii}=g$ and $W_{ij}=g\text{-}1$. The state of neuron i is z_i. The value $K=m(1\text{-}g)\text{-}0.5$ is a constant excitement which will control the number of active neurons. The external input e_j will be discussed below. The new state z_j results from applying a threshold function

$$
\begin{aligned}
z_j &:= 0 &&\text{for } x_j < 0 \\
z_j &:= \text{indeterminate} &&\text{for } x_j = 0 \quad (\text{see 2.5}) \\
z_j &:= 1 &&\text{for } x_j > 0.
\end{aligned}
$$

Another representation using the above definitions and designating the number of excessively active neurons $m' = (\Sigma\, z_i) \text{ - } m$ is

$$x_j = z_j \text{ - } 0.5 + m'(g\text{-}1) + e_j.$$

The switching of neurons has been randomly serialized to avoid the problems imposed by oscillations and metastable behaviour.

2.3. Selecting a new state

The following cases have to be considered. The selected neuron can be active or inactive and m' can be positive, zero or negative. Depending on this, a new state has to be selected to move the system in the direction of the stable state $m' = 0$ (if possible). The following table summarizes the required new state of neuron j

z_j	$m' < 0$	$m' = 0$	$m' > 0$	
0	1	0	0	new z_j
1	1	1	0.	

2.4. Limits for g

Depending on the required new state z_j, x_j must be positive or negative. This leads to six inequalities for g. Selecting the most restrictive condition ($z_j=1$ and $m' = 1$) results in:

$$g < 0.5$$

This means that the direct feedback can be even positive. A nonzero feedback is an extension to the basic Hopfield type networks [2, 3].

2.5. Controlling the m-of-n-flop

An m-of-n-flop can be seen as a data register based on an m-of-n-code. Applying sufficiently large negative values to the input of an active neuron will turn it off, while positive values will turn it on. The thresholds are

$$
\begin{aligned}
e_j &< \text{-0.5-}m'(g\text{-}1) &&\text{to turn } z_j \text{ off} \\
e_j &> 0.5\text{-}m'(g\text{-}1) &&\text{to turn } z_j \text{ on.}
\end{aligned}
$$

This level is constant if the circuit is initially in the stable state m' = 0. After some random selection steps another neuron with e_j = 0 will be switched on or off to fulfill the m-of-n-condition. The case x_j = 0 can be avoided if the input values are integers (as in section 3) and g is irrational. The proof is not presented here.

3. PROGRAMMING OF CONSTRAINTS

3.1. Connecting m-of-n-flops

How to connect two m-of-n-flops is shown in Fig. 2. If the interconnection weights are either 0 (no connection) or -1 (connection), the input values e_j will be zero or negative integer values. What does a connection from neuron output A to input B mean? If z_A = 1, there will be a negative input to x_B which tries to turn off z_B. Because of this, a connection means "if neuron A is active, turn off neuron B". If all constraints programmed by such connections are satisfied, all e_j are zero which is a stable state.

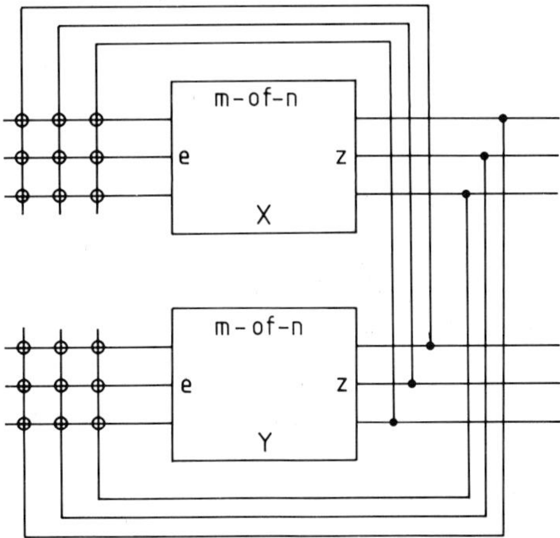

Fig. 2: Connecting two m-of-n-flops (n = 3)

3.2. Connecting 1-of-n-flops for numerical relations

If the neurons of two 1-of-n-flops represent the integers from 1 to n, relations like X = Y and X > Y can be easily formulated by decomposing the relation into rules like "if X equals 3, then set Y not equal 4 and vice versa". This is shown in Fig. 3 where each line represents two mutual connections with weight -1.

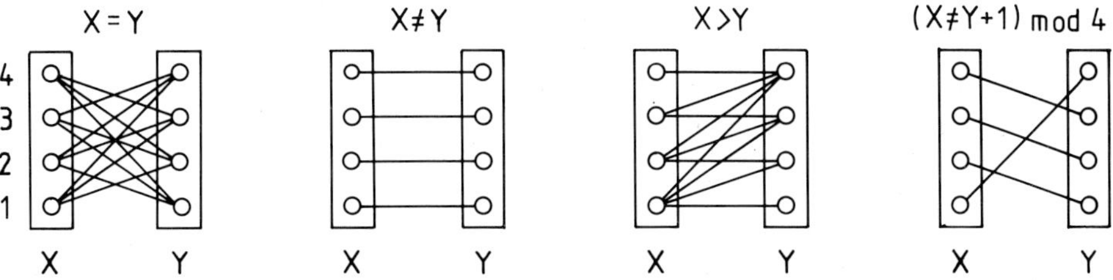

Fig. 3: Programming of relations

3.3. Examples

In the map coloring problem [1] different colors have to be assigned to adjacent countries. This can be programmed by using one 1-of-n-flop per country where n is the number of colors. Each possible color is associated with one neuron. The constraints are $X_a \neq X_b$ for all pairs (a, b) of adjacent countries.
A permutation matrix [2] (e.g. for machine assignments) can be generated by a set of n columns of 1-of-n-flops representing row assignments. The value (selected row) of each n-flop must be different from all the others.
Very similar is the problem of placing n queens on a n^2 chess board [1, 4] where additional connections have to be added for the diagonal lines.
Note that the permutations and the queens problem can be represented either by n 1-of-n-flops or one n-of-n^2-flop with external self-connections.

4. COMPARISION WITH RESULTS FROM OTHER AUTHORS

- Shackleford [1] (as does Hopfield [2, 3]) uses the sigmoid function $z_i = (1 + e^{-Gx_j})^{-1}$ instead of threshold elements. He discusses the influence of the gain G and states some vague rules about G and K: "... lower bound on G for a 4-flop is about 4. ... Acceptable values of K for a 4-flop range from about 0.75 to 1.25.". This paper shows that the same performance can be reached with G being infinite and K can be fixed by the designer to control the number of active neurons. The global activation term of [1] turns out to be a direct feedback g = 1 which must be reduced for G being increased to infinity.
- Hopfield describes a similar system [3] but explicitly excludes direct feedback even in his global inhibition term which has the purpose to enforce the 1-of-n-solution. The result is an indifferent equilibrium when all state variables are zero.
- Adorf [4] solves the 1-of-n-constraint by adding a guard neuron (GDS network) to the basic n-flop which becomes active only if *no* neuron is active ($\Sigma z_i = 0$). It will activate some arbitrarily selected neurons to kick the n-flop out of the 0-of-n-state. This is equivalent to a $\Sigma z_i > 0$ constraint.

5. ACKNOWLEDGEMENT

I would like to thank Prof. J. Swoboda and my colleagues for making this contribution possible.

6. REFERENCES

[1]   J.B. Shackleford, Neural Data Structures: Programming with Neurons (HP-Journal, 69-78, June 1989)

[2]   J.J. Hopfield, D.W. Tank, "Neural" Computation of Decisions in Optimization Problems (Biol. Cybern. 52, 141-152, 1985)

[3]   J.J. Hopfield, Collective Computation, Content-Addressable Memory, And Optimization Problems (Complexity in Information Theory, Ed.: Y.S.Abu-Mostafa, Springer-Verlag, 99-114, 1989)

[4]   H.-M. Adorf, Connectionism and Neural Networks (Knowledge Based Systems in Astronomy, Eds.: F.Murtagh & A.Heck, Springer-Verlag, 215-245, 1989)

Parallel Processing in Neural Systems and Computers
R. Eckmiller, G. Hartmann and G. Hauske (Editors)
Elsevier Science Publishers B.V. (North-Holland), 1990

THE EXACT EVALUATION OF MEMORY CAPACITY FOR NEW STOCHASTIC VERSIONS OF THE HOPFIELD NET

Peter WHITTLE*

Statistical Laboratory
University of Cambridge
Cambridge, England

1. INTRODUCTION: THE ANTIPHON

A task of great interest is the economical design of a processor which operates reliably, despite being constructed of unreliable components. The simplest case of a processor is a memory; to achieve results for this case would yield valuable insight. The aim is then to achieve a memory of maximal *capacity* (i.e. nats of information reliably stored per component) rather than the other abilities which have been sought in the neural net literature, such as generalisation, association or learning. However, a system which copes with unreliable elements will acquire some of these qualities.

A device proposed in Whittle [1] is the *antiphon*. This was based on the analogy of circulating a message around a noisy communication loop, and functioned by reverberating a pattern of excitation between two sets of nodes. There are M α-nodes and N β-nodes, denoted respectively by $\alpha(j)$ and $\beta(k)$ ($j = 1, 2, \ldots, M$; $k = 1, 2, \ldots, N$). For reasons which will transpire, these could loosely be called *processing nodes* and *sensory nodes* respectively. At time t the α-nodes hold a pattern $\xi(t) = (\xi(j, t))$ where the elements $\xi(j, t)$ adopt values 0 or 1; these patterns are the possible memory values. The node $\beta(k)$ receives an input

$$u(k, t) = \sum_{j} \xi(j, t)c(j, k) \tag{1}$$

where $c(j, k)$ is the capacity of the arc between $\alpha(j)$ and $\beta(k)$, assumed to be 0 or 1. The β-node is the unreliable element; it yields an output with probability distribution

$$P(y(k, t) = y | u(k, t) = u) = p(y|u) \tag{2}$$

independent for different k conditional on the inputs u. From these outputs an input

$$x(j, t + 1) = \sum_{k} c(j, k)y(k, t) \tag{3}$$

---

*This work was carried out during the author's tenure of a Senior Fellowship awarded by the Science and Engineering Research Council, U.K.

is formed, which is then normalised to

$$\xi(j, t+1) = f(x(j, t+1)) \tag{4}$$

where $f(x)$ is a threshold response function, taking values 0 or 1 according as $x < h$ or $x \geq h$.

One can regard this model as a communication loop in which the message ξ is coded into u (the $c(j, k)$ representing the letters of the codewords), so yielding a channel output y from which ξ is inferred by the decoding steps (3) and (4). One can also regard it as an alternation between reliable local storage (in the α-nodes) and unreliable distributed storage (in the β-nodes), storage in the α-nodes being possible for only part of the time. The 'biological' interpretation (although no biological fidelity is claimed) is that a memory is held by allowing the abstract memory held in the processing nodes to evoke its sensory counterpart; this then to re-evoke the abstract memory, etc.

If the β-response (2) is faithful, so that $y(k, t) = u(k, t)$, then elimination of x, y and u from the relations above yields

$$\xi(j, t+1) = f(\sum_k w(j, k)\xi(j, t)) \qquad (w(j, k) = \sum_i c(j, i)c(k, i)). \tag{5}$$

Relation (5) is the dynamic relation of a deterministic Hopfield net. The antiphon is thus a stochastic version of the Hopfield net, with an expanded set of variables and recursions. It differs, however, from the stochasticisation which has been most studied hitherto: the spin-glass model.

2. RELIABILITY AND CAPACITY

We shall take N, the number of unreliable elements, as defining the *size* of the system. Let us define the error probability ϵ as $P(\xi(t+1) \neq \xi(t))$, this taking account of the randomness of the network $(c(j, k))$ as well as of the β-output y. We shall say that a given mode of operation is *reliable* if $\epsilon \to 0$ as $N \to \infty$. Suppose this procedure distinguishes $R(N)$ memory values for a system of size N, and so has a memory size of $\log R(N)$ nats. We shall say the system stores information at *rate K* if $(1/N)\log R(N) \to K$ as $N \to \infty$; the *capacity* of the system is the supremum of reliable rates.

For the spin-glass model it is usually claimed (see [2], [3], [4]) that $R(N) \approx 0.15N$, which would correspond to zero capacity. Obviously $R(N)$ must be exponentially large in N if positive capacity is to be realised. This is considered to be impossible for the spin-glass model, principally because excitation states which correspond to simultaneous evocation of more than one 'basic' memory trace (corresponding to $\xi(j, t)$'s being non-zero for more than one value of j, in our model) are considered 'spurious'.

3.   CAPACITY ASSERTIONS FOR THE ANTIPHON

Capacities are evaluated over a class of random networks. We suppose that the $c(j,k)$ independently take values 0 and 1 with probabilities $\phi = 1 - \theta$ and θ. Let $C(s,\theta)$ be the capacity for a given value of θ under s-fold excitation, so that the possible memory traces correspond to configurations in which exactly s of the ξ-components are non-zero. Define

$$\bar{p}_i(y) = \sum_u \binom{s-1}{u} \phi^{s-u-1} \theta^u p(y|u+i), \quad \bar{p}(y) = \sum_u \binom{s}{u} \phi^{s-u} \theta^u p(y|u).$$

Then it can be shown that

$$C(s,\theta) = s \sup_{z \geq 1} \sum_y [\theta \bar{p}_1(y) \log z - \bar{p}(y) \log(\phi + \theta z^y)]. \tag{6}$$

In the binary case, when y takes only the values 0 or 1, this reduces to

$$C(s,\theta) = s(-\bar{p} \log \bar{p} + \phi \bar{p}_0 \log \bar{p}_0 + \theta \bar{p}_1 \log \bar{p}_1) \tag{7}$$

where $\bar{p} = 1 - \bar{q} = \bar{p}(1)$, etc.

Suppose that, instead of the antiphon decoding rules (3) and (4), one used optimal inference methods on the 'observations' y to estimate ξ. It can be shown that the capacity evaluation corresponding to (6) then becomes

$$C'(s,\theta) = H[\bar{p}(\cdot)] - \sum_u \binom{s}{u} \phi^{s-u} \theta^u H[p(\cdot|u)] \tag{8}$$

where $H[p(\cdot)]$ is the Shannon information $-\sum_y p(y) \log p(y)$ associated with a distribution $p(\cdot)$.

Comparison of (6) and (8) is most striking in the binary case with $s = 1$, when we have

$$C(1,\theta) = -\bar{p} \log \bar{p} + \phi p_0 \log p_0 + \theta p_1 \log p_1 \tag{9}$$

$$C'(1,\theta) = C(1,\theta) - \bar{q} \log \bar{q} + \phi q_0 \log q_0 + \theta q_1 \log q_1. \tag{10}$$

The effect of antiphon decoding is then to strip all the q-terms from (10), leaving only the p-terms (9).

These capacities are all realised with M exponentially large in N, indeed necessarily of order $\exp[C(s,\theta)N/s]$. It is only by letting s grow with N that M can be made of order N, which is what one would wish if N is truly to be a measure of size. If s grows then θ must also decrease, as s^{-1}, if the 'background noise' provided by other α-nodes is not to swamp the detection of activity of a prescribed α-node. Suppose then that $\theta = \lambda/s$. One finds that the limiting capacity for large s is

$$C_\infty(\lambda) = \lambda \sup_{z \geq 1} E_u \sum_y [y P(y|u+1) \log z + (1 - z^y)p(y|u)]$$

where u is here to be regarded as the background noise, a Poisson variable with expectation λ. A positive capacity can also be realised with M proportional to N and

$$s \propto N/\log N.$$

REFERENCES

[1] Whittle, P., The antiphon; a device for reliable memory from unreliable components, Proc. Roy. Soc. A 423 (1989) 201.

[2] Amit, D.J., Gutfreund, H. and Sompolinsky, H., Storing infinite numbers of patterns in a spin-glass model of neural networks, Phys. Rev. Letters 55 (1985) 1530.

[3] McEliece, R.J., Posner, E.C., Rodemich, E.R. and Venkatesh, S.S., The capacity of the Hopfield associative memory, IEEE Trans. Inf. Th. IT-33 (1987) 461.

[4] Newman, C.M., Memory capacity in neural network models: rigorous lower bounds, Neural Networks 1 (1988) 223.

Parallel Processing in Neural Systems and Computers
R. Eckmiller, G. Hartmann and G. Hauske (Editors)
Elsevier Science Publishers B.V. (North-Holland), 1990

Sparse coding: the link between low level vision and associative memory *

Christoph ZETZSCHE

Lehrstuhl für Nachrichtentechnik
Technische Universität München
Arcisstr. 21, D–8000 München 2, Fed. Rep. of Germany

Low level visual processing provides reduced entropy of the internal representations of natural images. However, there seems to be no obvious advantage for the biological system, since it can neither use it to construct a more compact image code nor to save transmission capacity. The efficiency of associative memories, on the other hand, is assumed to depend on a sparse distribution of input activity, but no appropriate coding procedures are known. Both problems may be solved if seen in connection: The afferent information processing provides an adequate code for associative capabilities of higher memory stages.

1. Introduction

The different levels of information processing in the brain are usually covered by different research areas, each with its typical problems, methods, and concepts. In this paper we show that two unsolved problems from different research areas, namely the apparently useless entropy reduction in low level vision and the need for sparse coding in associative memories, have an important aspect in common.

2. Low level vision

Recent statistical investigations on signal processing properties of low level vision have revealed that these structures transform the statistical dependencies in natural images into a highly uneven, zero peaked activity distribution, i.e. a low entropy of the internal image representations is achieved [1–4]. The entropy reduction takes place in two steps. Within object areas there is a high probability of neighbouring points having the same luminance. This statistical dependency is an isotropic property, i.e. it is independent of direction. It is removed on the retinal level by isotropic lateral inhibition. However, this operation can not remove statistical dependencies between points of the remaining object contours. These anisotropic, i.e. orientation specific dependencies are reduced on the cortical level by application of orientation specific filters. In the ideal case of a locally straight contour only one of these filters will give a response, while all others show zero activity. As a consequence, the activity distributions in the visual system will become more and more concentrated around zero activity from stage to stage (Fig.1).

This entropy reduction seems to offer an explanation for structures found in low level vision: They reduce the information that is transmitted to higher stages. However, the low entropy caused by an uneven activity distribution is no advantage by itself. In technical systems it is used to assign code words of different length to activity levels (e.g. by Huffman coding). If seldom occuring levels are assigned to long words and more frequently occuring to short ones, the necessary storage size can be reduced or more information can be sent over a limited capacity transmission line.

However, such a type of encoding does not fit into any current theory of brain function, where it is generally believed that amplitudes are encoded by spike frequencies, and that

---

* Research supported by European Community, project BRAIN–ST2J–0418–C (EDB).

274   *C. Zetzsche*

each neural "transmission line" is associated with a given position in the visual field. Hence we are faced with the problem that in spite of the effort spent to reduce statistical dependencies there seems to be no obvious usefulness for the system.

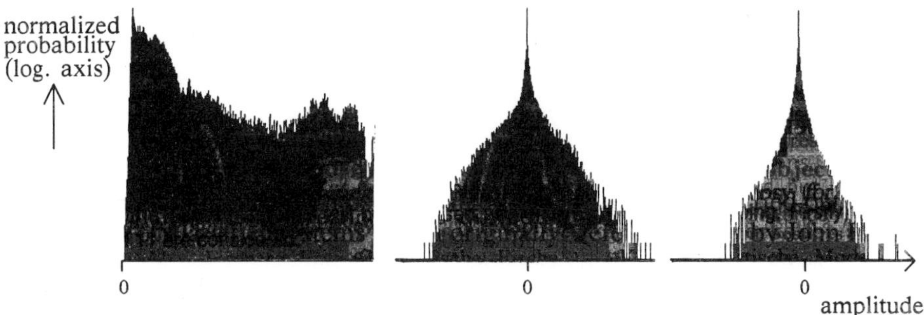

FIGURE 1: Simulated activity distributions. a) input , b) retina, c) cortex.

3. Associative memories

The various modifications of the basic associative memory scheme include linear and nonlinear networks, static and dynamic processing, continuous and finite state networks [5,6]. An important aspect of all memories is their storage capacity. Linear systems theory predicts that in a NxN matrix no more than N linear independent vectors can be stored [6]. Recent statistical approaches indicate that even more vectors can be stored if they are subject to a particular constraint: They have to show a "sparse" distribution of activity, i.e. the vectors have to contain only a few active units [7-11]. Binary autoassociative matrix memories achieve their optimal performance if the matrix density is 50% and if the number of nonzero values in the input vectors is logarithmically related to the vector size [7,8]. Sparsification schemes based on increasing vector size will therefore run into exponential problems. So far, the advantages of sparse coding have been verified only by simulations with random input vectors or with sparsification schemes based on combinatorial considerations rather than on neurophysiological evidence [7,9,11]. To our knowledge, there exist no suggestions which brain structures provide the necessary encoding stages and which type of sparse coding is used in the biological system.

4. A shortcut: direct combination of low level vision and associative memory

4.1 Simplified model

For our simulations we used a straightforward combination of size and orientation specific preprocessing and a binary autoassociative matrix memory. It has to be expected that this simplified model will not achieve optimal performance, since it does not take into account the neural machinery between early processing stages and higher memory centers. However, it will provide information about which image structures can be successfully sparsified by the first stages, and which will need a more complex processing. This, in turn, can lead to a hypothesis about the operations of the intermediate stages.

Four 512*512 pictures containing a wide variety of pictorial information were selected for the simulations. Each of these pictures was splitted into 256 sub-pictures of size 32 * 32. The resulting 1024 sub-pictures were filtered by an orthogonal set of orientation selective bandpass filters with octave bandwidth, 3 different radial center frequencies and 6 orientations. All filter outputs were optimally sub-sampled and strongly amplitude quantized in order to generate a binary vector of length 2440 for the representation of each sub-picture. Due to this strong quantization the image information is slightly degraded (see Fig. 3a for a "perfect" reconstruction).

4.2 Simulation results

The binary vectors are successively loaded into the memory. After each vector is loaded the matrix density (Fig.2a) and the cumulated average recall error (Fig.2b) are evaluated. The optimum 50% density is reached with 700 vectors stored. Recall error is rising fast, reaching the 10% level with 350 vectors stored. A reconstructed picture after storage of all 1024 vectors shows that the main errors arise from the low frequency range (Fig.3b). This is caused by a broader activity distribution within low frequency signals.

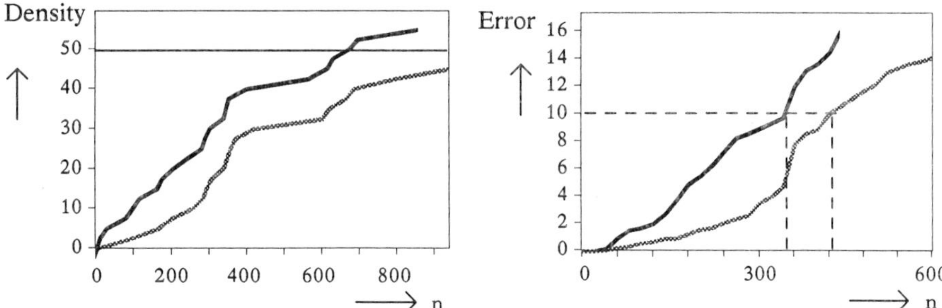

FIGURE 2: Matrix density and average recall error vs. number of stored vectors. Black lines: full frequency range. Grey lines: without low frequencies.

Therefore we tested a modified scheme where these filter outputs are not stored into memory. This reduced the vector size by 20% from 2440 to 1920. The development of matrix density is now significantly slower, and the 50% optimum load level is not reached. The increase in recall error is also reduced. Up to 300 vectors can be stored with less than 3% recall error. After this point, however, there is a steep increase, so the 10% error level is reached with as few as 420 vectors stored. The pictures reconstructed after storage of 1024 vectors look significantly better then before (Fig.3c), but it should be kept in mind that the 20% low frequency information is not influenced by the memory.

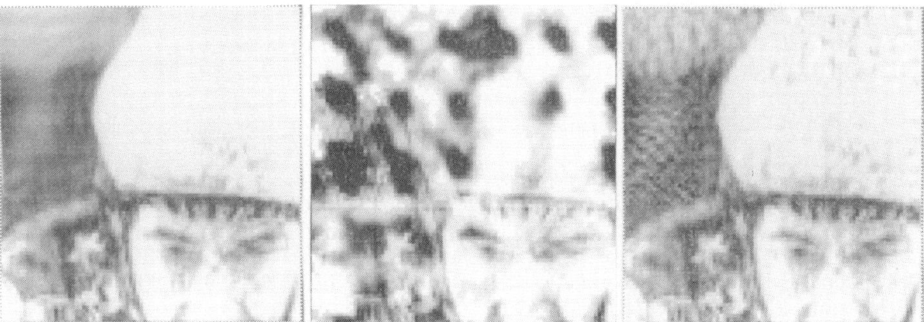

FIGURE 3: Reconstructed picture. a) correct recall b) full frequency range, 1024 vectors stored, c) no low frequency influence, 1024 vectors stored.

5. Discussion

As has been expected, the simplified model does not reach the optimal performance, but the analysis of its shortcomings provides important insights into the function of the more complex intermediate processing stages:

We have observed a fast increase in error rate associated with loading pictures containing textured areas. These vectors have a high percentage of active values, and even one single nonsparse vector can destroy the matrix structure. It is interesting to note that the visual system seems to avoid such a structural overload. Textures are perceptually treated different from other patterns. There is no representation of the microstruture, rather they are encoded as homogenous areas charcterized by a few statistical attributes [12].

Evidence concerning the relevance of low frequency information is contradictory. Although threshold measurements of human vision indicate its unimportance, low frequency components have to be represented with maximum precision in image coding [2]. Taking into account the limited dynamic range of a single cortical neuron the latter would imply an encoding by a group of neurons. This offers the possibility of further sparsification, but the increase of vector size may lead to combinatorial problems.

The degree of sparsity obtained in our experiments is not sufficient. For the vector sizes used the optimal values should be of the order of 0.5%. We have reached 8.7% with, and 3.6% without the low frequency information. This seems to be the limit that can be reached with linear filter operations. There is strong information–theoretical and neurophysiological evidence that the next level of processing contains highly nonlinear detectors which respond no longer to lines and edges but only to the even more rare two–dimensional signal variations like corners, line–ends, etc. [13,14].

It should be noted that an optimal capacity of associative memories is easily achieved by encoding schemes based on purely formal assumptions. However, these will be of limited use for the understanding and description of brain functions, unless they can be clearly linked to neurophysiological structures and perceptual processes. As yet, there exists no complete theory of how an optimal sparse encoding can be produced by the nervous system. The properties of visual information processing can be used as a basis for the development of such a theory.

References

[1]   Field, D. J.,   J. Opt. Soc. Am. A4(12) (1987) 2379–2394
[2]   Watson, A. B.,   J. Opt. Soc. Am. A4(12) (1987) 2401–2417.
[3]   Zetzsche, C. and Schönecker, W.,   Perception 16(2) (1987) 229.
[4]   Daugman, J. G.,   IEEE Trans. ASSP-36(7) (1988) 1169–1179.
[5]   Lippmann, R. P., IEEE ASSP Magazine April (1987) 4–22.
[6]   Kohonen, T.,   Self–organization and associative memory (Springer, Berlin, New York , 1988)
[7]   Palm, G.,   Biol. Cybern. 36 (1980) 19–31.
[8]   Willshaw, D.J., Holography, associative memory, and inductive generalization, in: Hinton, G.E. and Anderson, J.A., (eds.), Parallel models of associative memory (Lawrence Erlbaum Assoc.,Hillsdale, New Jersey, 1981) pp. 83–104.
[9]   Baum, E. B., Moody, J. and Wilczek, F.,   Biol. Cybern. 59 (1988) 217–228.
[10] Bottini, S.,   Biol. Cybern. 59 (1988) 151–159.
[11] Marcinowski, M., Coding problems and associative memories. (In German), diploma thesis (Univ. Tübingen, 1987)
[12] Julesz, B., Nature 290 (1981) 91–97.
[13] Marko, H., IEEE Trans. SMC-4 (1974) 34–39.
[14] Zetzsche, C. and Barth, E., Fundamental limits of linear filters in the visual processing of two–dimensional signals. Vision Res. (1990) in print.

Parallel Processing in Neural Systems and Computers
R. Eckmiller, G. Hartmann and G. Hauske (Editors)
© Elsevier Science Publishers B.V. (North-Holland), 1990

The Convergence of Parallel Boltzmann Machines

Patrick J. Zwietering and Emile H.L. Aarts

Eindhoven University of Technology
Department of Mathematics and
Computing Science, P.O. Box 513,
5600 MB Eindhoven, The Netherlands

Philips Research Laboratories
P.O. Box 80.000, 5600 JA
Eindhoven, The Netherlands

We discuss the main results obtained in a study of a mathematical model of synchronously parallel Boltzmann machines. We present supporting evidence for the conjecture that a synchronously parallel Boltzmann machine maximizes a consensus function that consists of a weighted sum of the regular consensus function and a pseudo consensus function. The weighting is determined by the fraction of units that can change their states simultaneously.

Keywords: Boltzmann machines, neural networks, synchronous parallelism, combinatorial optimization, simulated annealing.

1   Introduction and Notation

A Boltzmann machine [1] can be viewed as a probabilistic neural network. Here we consider a special class of such devices in which the units (neurons) respond simultaneously in synchronised groups. A so called *synchronously parallel Boltzmann machine* (SPBM) $\mathcal{B} = (\mathcal{N}, \mathcal{P}^{(\tau)})$, is given by:

(i) a pseudograph $\mathcal{N} = (\mathcal{U}, \mathcal{C})$, denoting the underlying network of the Boltzmann machine, where $\mathcal{U}$ denotes the finite set of units and $\mathcal{C} \subseteq \mathcal{U} \times \mathcal{U}$ specifies the connection pattern of the network. With each connection $\{u, v\} \in \mathcal{C}$ a connection strength $s_{\{u,v\}} \in \mathbb{R}$ is associated. The configuration ($\equiv$ global state) of a SPBM is denoted by k, a $|\mathcal{U}|$-dimensional vector whose components $k(\cdot) \in \{0, 1\}$ indicate the states of the individual units; $\mathcal{R}$ is the set of configurations. The *regular consensus function* $C_k : \mathcal{R} \to \mathbb{R}$ is defined as

$$C_k = \sum_{\{u,v\} \in \mathcal{C}} s_{\{u,v\}} k(u) k(v). \tag{1}$$

The objective of a Boltzmann machine is to reach a configuration with maximal consensus. This is accomplished by using:

(ii) a transition mechanism $\mathcal{P}^{(\tau)} = (G, A^{(\tau)})$, consisting of the following two steps. Firstly a subset of units is generated that may propose a state transition, with $G(\mathcal{U}_s)$ denoting the probability of generating the set $\mathcal{U}_s$. Secondly each unit in the subset evaluates a state transition based on the current states of all other units, using the well-known probabilistic acceptance criterion (cf. [2]). These evaluations are performed in parallel. The acceptance probability $A_{kl}^{(\tau)}(\mathcal{U}_s)$ expresses the probability of accepting all the (unit) state transitions required to transform k into l, given that the set of units $\mathcal{U}_s$ is allowed to make a transition and τ is the current value of a control parameter.

Hence, due to our synchronous approach, we can associate with $\mathcal{P}^{(\tau)}$ a transition matrix $P^{(\tau)}$, given by

$$P_{kl}^{(\tau)} = \sum_{\mathcal{U}_s \subseteq \mathcal{U}} G(\mathcal{U}_s) A_{kl}^{(\tau)}(\mathcal{U}_s). \tag{2}$$

If G and $A^{(\tau)}$ are both stochastic, $P^{(\tau)}$ is, and hence it has a corresponding homogeneous Markov chain (for fixed τ). Under some mild conditions on G and $A^{(\tau)}$ this Markov chain can be shown to be irreducible and aperiodic and therefore has a unique stationary distribution $q^{(\tau)}$. We now say that the SPBM *functions properly* if the objective is asymptotically guaranteed, i.e. if $\lim_{\tau \to \infty} q^{(\tau)} = r$, where r is a distribution over $\mathcal{R}_{opt}$, the set of configurations with maximal consensus.

The subject of this paper is a study about the effect of the amount of parallelism applied in the transition mechanism, upon the functioning of a SPBM. Before dealing with this subject, we first need to be able to distinguish between different modes of parallelism. The amount of parallelism applied in the transition mechanism of a SPBM is entirely determined by G, because we assume that the evaluations in the 'acceptance-phase' are performed in parallel. In [1] four different modes of parallelism were introduced, which can be subdivided as shown in figure 1. A SPBM is called sequential if always exactly one unit is generated,

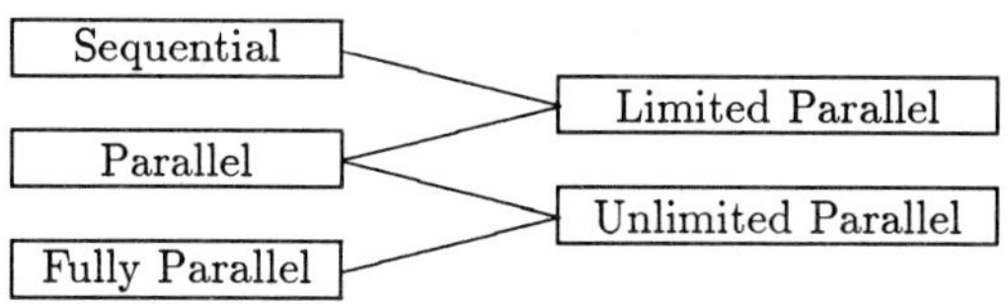

Figure 1: Parallelism in Synchronously Parallel Boltzmann machines.

it applies limited parallelism if never two connected units are in the generated subset and applies full parallelism if always all units are generated. From [1] we know that a SPBM applying limited parallelism functions properly, whereas in case of full parallelism this is generally not true. There is however a large open gap between these two extremes. In order to fill this gap we propose to describe the amount of parallelism in a SPBM with a single parameter $\alpha \in (0, 1]$, denoting the average fraction of the units that are allowed to make a transition. With $\mathcal{B}(\alpha)$ we denote a SPBM which is fixed except for the amount of parallelism. The following proposition is based upon experimental evidence.

Proposition 1 *If a SPBM $\mathcal{B}(\alpha)$ functions properly and $\hat{\alpha} \leq \alpha$ then $\mathcal{B}(\hat{\alpha})$ also functions properly.*

This is equivalent to the following.

Proposition 2 *For each SPBM $\mathcal{B}(\cdot)$ a critical value α_c exists, such that $\mathcal{B}(\alpha)$ functions properly if and only if $\alpha < \alpha_c$.*

Using that a SPBM $\mathcal{B}(\alpha)$ functions properly if it applies limited parallelism, we find that Proposition 2 implies $\alpha_c > 1/m$, where m is the chromatic number of the graph $\mathcal{N}$. In the next section we present mathematical evidence to support the propositions. For rigorous proofs of the presented theorems the reader is referred to [3].

2   Main Results

The problem of proving something about the behaviour of $q^{(\tau)}$ is that there does not exist an explicit analytical expression for $q^{(\tau)}$, except for those modes of parallelism mentioned in [1]. We therefore tried to calculate $\lim_{\tau\to\infty} q^{(\tau)}$ without such an expression, the result being the subject of Conjecture 1. But first we need the following preliminaries.

- k_u is the configuration that is obtained from k by changing the state of unit u.

- $\Delta C_k(u) = C_{k_u} - C_k$.

- $U_{kl} = \{u \in \mathcal{U} \mid k(u) \neq l(u)\}$ is the set of units that should change their states in order to transform configuration k into l.

Next the acceptance probability is given by (see [1], [3]):

$$A_{kl}^{(\tau)}(\mathcal{U}_s) = \begin{cases} \prod_{u\in\mathcal{U}_{kl}} [1 + \exp(-\tau\Delta C_k(u))]^{-1} \prod_{u\in\mathcal{U}_s\backslash\mathcal{U}_{kl}} [1 + \exp(+\tau\Delta C_k(u))]^{-1} & \text{if } \mathcal{U}_{kl} \subseteq \mathcal{U}_s \\ 0 & \text{otherwise.} \end{cases} \tag{3}$$

It can be shown (cf. [3]) that for $\mathcal{U}_{kl} \subseteq \mathcal{U}_s$ (3) is equivalent to:

$$A_{kl}^{(\tau)}(\mathcal{U}_s) = \exp(\tau D_{kl}) \left/ \sum_{m\in\mathcal{R}_k(\mathcal{U}_s)} \exp(\tau D_{km}) \right. , \tag{4}$$

where $\mathcal{R}_k(\mathcal{U}_s) = \{m \in \mathcal{R} \mid \mathcal{U}_{km} \subseteq \mathcal{U}_s\}$ denotes the set of all configurations that can be reached from k by a state transition of one or more units in $\mathcal{U}_s$, and D is the matrix whose elements are given by:

$$D_{kl} = 2 \sum_{\{u,v\}\in\mathcal{C}} s_{\{u,v\}}k(u)l(v) + \sum_{u\in\mathcal{U}} s_{\{u,u\}}|k(u) - l(u)|. \tag{5}$$

Now, let the configuration $k_{\mathcal{U}_s}^* \in \mathcal{R}_k(\mathcal{U}_s)$ be defined such that for all $l \in \mathcal{R}_k(\mathcal{U}_s)$ we have $D_{kl} \leq D_{kk_{\mathcal{U}_s}^*}$. From (4) it is clear that at large values of τ, $k_{\mathcal{U}_s}^*$ has a large probability of being selected from $\mathcal{R}_k(\mathcal{U}_s)$. Hence, it is very probable to observe that the SPBM is in a configuration with a large value of the *generalised pseudo consensus function* $S_k(\mathcal{U}_s) \equiv D_{kk_{\mathcal{U}_s}^*}$. If we also account for the generation probability, we come to the following:

Conjecture 1 *A SPBM $\mathcal{B}$ converges asymptotically to a distribution over the set $\widetilde{\mathcal{R}}_{opt}$ for which the extended consensus function given by $\tilde{C}_k = \sum\limits_{\mathcal{U}_s\subseteq\mathcal{U}} G(\mathcal{U}_s)S_k(\mathcal{U}_s)$ is maximal.*

Conjecture 1 is supported by the following theorems.

Theorem 1 *For a SPBM applying limited parallelism, Conjecture 1 is true.*

Theorem 2 *For a SPBM applying full parallelism, Conjecture 1 is true.*

In order to relate Conjecture 1 to the Propositions 1 and 2 we define the notion of a fair SPBM, by adding a very 'natural' condition for the generation probability: A SPBM is called *fair* if the probability that a unit is allowed to make a transition is equal for all units. One can easily verify that this probability then equals the average fraction α of the units that are allowed to make a transition. We now have

Theorem 3 *For a fair SPBM $\mathcal{B}(\alpha)$ the extended consensus function satisfies*

$$\tilde{C}_k = (1 - \alpha)S_k(\emptyset) + \alpha S_k(\mathcal{U}). \tag{6}$$

Note that Theorem 3 implies that the extended consensus function consists of a weighted contribution of $S_k(\emptyset) = 2C_k$ and $S_k = S_k(\mathcal{U})$.

Theorem 4 *Assume Conjecture 1 is true and let α_c be defined as*

$$\alpha_c = \min \left\{ \frac{2C_{opt} - 2C_k}{S_k - 2C_k} \,\middle|\, k \in \mathcal{R} \setminus \mathcal{R}_{opt}, \, S_k \geq 2C_{opt} \right\}. \tag{7}$$

Then Proposition 2 is true.

It is not hard to derive lower and upper bounds for α_c. Let 0 denote the configuration with all units "off", define $S_{opt} = \max \{S_k \mid k \in \mathcal{R}\}$, $C_{nop} = \max \{C_k \mid k \in \mathcal{R} \setminus \mathcal{R}_{opt}\}$ and assume $S_{opt} \geq 2C_{opt} > 0$, then we have (cf. [3]):

$$\frac{2C_{opt} - 2C_{nop}}{S_{opt} - 2C_{nop}} \leq \alpha_c \leq \frac{2C_{opt}}{S_0}. \tag{8}$$

In some rare cases these bounds can coincide (see [3]), but experiments indicate that in general the upper bound is much sharper than the lower bound. Since it is often impossible to calculate α_c given by (7), we therefore use the upper bound as an estimate for α_c. Further research might yield a better estimate.

3   Discussion

We have conjectured that synchronously parallel Boltzmann machines maximize a consensus function that is a weighted sum of the regular consensus function (which is maximized by a sequential Boltzmann machine) and a pseudo consensus function (which is maximized by a fully parallel Boltzmann machine). The weighting is determined by the amount of parallelism, i.e. the fraction of units that are allowed to propose a state transition simultaneously. This fraction can be easily related to the problem instance at hand.

We have implemented simulations of synchronously parallel Boltzmann machines designed for a number of problems. In [3] the results for the independent set problem and the max-cut problem are discussed, whereas in [4] the results for the 0-1 knapsack problem are shown. All these results provide numerical evidence for the validity of the conjecture mentioned above. The practical use of an analysis is in the existence of a simple measure that determines an upper bound on the amount of parallelism in a synchronously parallel Boltzmann machine, that guarantees convergence to the optimum of the regular consensus function.

References

[1] AARTS, E.H.L. AND KORST, J.H.M., *Simulated Annealing and Boltzmann Machines* (Wiley, Chichester, 1989).

[2] HINTON, G.E., SEJNOWSKI, T.J. AND ACKLEY, D.H., Boltzmann machines: constraint satisfaction networks that learn, Carnegie-Mellon Univ. (1984) *Technical Report CMU-CS-84-119*.

[3] ZWIETERING, P.J. AND AARTS, E.H.L., Synchronously Parallel Boltzmann machines: a Mathematical Model, Eindhoven Univ. of Technology (1989) *Memorandum COSOR 89-21*.

[4] ZWIETERING, P.J. AND AARTS, E.H.L., A Note on the Convergence of a Synchronously Parallel Boltzmann machine for the Knapsack Problem, Eindhoven Univ. of Technology (1989) *Memorandum COSOR 89-28*.

Section 7
Hardware and Software Simulators
for Neural Networks

Parallel Processing in Neural Systems and Computers
R. Eckmiller, G. Hartmann and G. Hauske (Editors)
© Elsevier Science Publishers B.V. (North-Holland), 1990

A TRANSPUTER BASED GENERAL SIMULATOR FOR CONNECTIONIST MODELS

Hans Peter Ernst, Branislav Mokry and Zoltan Schreter

AI-Lab, Institute for Informatics
University of Zurich
Switzerland

We have developed a transputer based connectionist simulator, with the help of which the experimentator defines his network models using a set of simulation specific C functions and prefabricated general constructs for the layout of the desired network topology. The implementation of parallelism itself is straightforward and simple, providing the user with the desired flexibility and demanding only a few necessary concessions to the aspects of parallel programming.

1. INTRODUCTION

Transputer systems offer a cheap, simple and efficient approach to parallel processing. Therefore they are often considered ideal candidates for the simulation of connectionist models, which are implicitly of parallel nature. A general simulator for such models should allow the user the definition of arbitrary network topologies and learning rules. These demands can be satisfied by providing a framework of functions and data structures, which serve as a flexible toolbox for the construction of such simulations. Of course this framework should be designed in such a way, that the user may concentrate on the design of his network models, without paying much attention to, say, the system user interface, or, as in our case, the parallel nature of the implementation.

Led by these assumptions, we have developed a simulator for a system consisting of four transputers using the C programming language , in which the experimentator defines his network models with the help of a set of simulation specific C functions and prefabricated general constructs for the layout of the desired network topology. The implementation of parallelism itself is straightforward and simple, providing the user with the desired flexibility and demanding only a few necessary concessions to the aspects of parallel programming.

2. HARDWARE

There are many so called transputer boards available for use in microcomputer systems, most offering different configurations of up to four of the INMOS T414 and T800 transputers. We have chosen the Levco TransLink II card, a NuBus-compatible card for the Mac II, that can be configured with one to four transputer modules of different memory sizes and types. For the setup of the link topology, the card uses the C004-chip, another INMOS product. The host controls the board by means of standard Macintosh driver calls.

The version of the board we use is equipped with four T800-20 transputers, each delivering about 1.5 MFLOPS and 10 million instructions per second. The maximum theoretical data transfer rate between the T800's is about 2 MByte/s, per link, per direction.

3. LANGUAGE

In deciding which language to use for the system development, we had to choose between occam and C. The advantages of occam for use in transputer programming are well known. Transputers, after all, can be seen as hardware implementations of the occam model. Though there exist neural network simulators based on occam (Koikkalainen & Oja [1]), the language does not provide the means for creating the abstractions of the problem domain desired in a general simulator, like records, lists, dynamic data structures and the like. Worse, our design is based to a large degree on the use of mapping functions, dynamic memory allocation and higher level data structures. These possibilities are supported by the C language of course, but the price which is paid for them is a relatively poor handling of parallel processing by C library routines.

This conflict was solved by making the observation that the basic system design did not need any parallel tasks on the transputers themselves. The transputers are used as a *multicomputer* system, without any regard to their *multitasking* abilities.

The decision for C has resulted in a shorter development time and also reduces the time people need to get accustomed to working with the simulator. Another positive effect is the portability of the system, which could be easily adapted for use on a C-programmable accelerator board, with only a few modifications to existing simulation models. At this point it should be noted, that the actual end user interface is programmed as an independent Macintosh application and could therefore be used as a front end to different 'simulator boards'.

4. COMPONENTS PROVIDED FOR NETWORK MODELS

Each simulation is built on three basic abstractions, the *unit*, the *link* and the *set*.

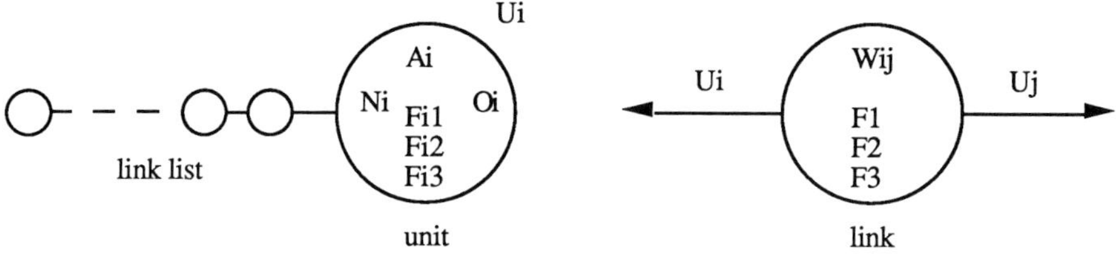

FIGURE 1a) The basic unit   FIGURE 1b) The basic link

A unit (figure 1a) is defined by six status parameters, namely Output (O_i), Activation (A_i), Netinput (N_i) which have special meanings and three free parameters Free1 (F_i1), Free2 (F_i2) and Free3 (F_i3) which can be used as necessary. Connections between units are assigned to the 'receiving' unit and are held in a link list.

A link (figure 1b) has four status parameters, of which only one, Weight (W_{ij}) has a special meaning, the other three being free parameters, similarly as in the case of units. Each link points to the two units which it connects.

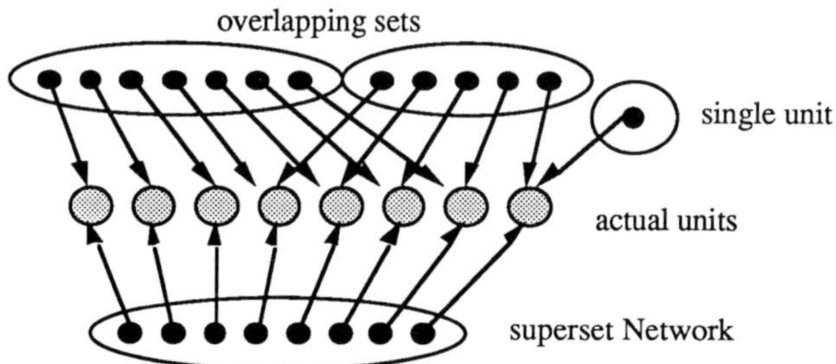

FIGURE 2. Set concept

The set concept (figure 2) is used to define groups of units which have to be connected to each other or which may be updated in parallel. Each simulation model consists of a superset Network, which contains all of the existing units. All the other sets are subsets of Network. By realizing sets as sets of pointers to units, arbitrarily complicated groupings of units can be defined.

5. IMPLEMENTATION

The central problem in running a network simulation on several transputers consists in the *distribution* of the workload on the available processors. We have chosen a very simple approach, by dividing a network evenly in a round robin fashion among the transputers (figure 3). This way we can be sure that the workload is distributed almost optimally. For simplicity and because the amount of memory required by units is small compared to that used by links, units exist redundantly on all four transputers. Those units assigned to the specific transputer are called active units, the others virtual units. The links leading to a unit, it's link list, are created only on it's 'home' transputer.

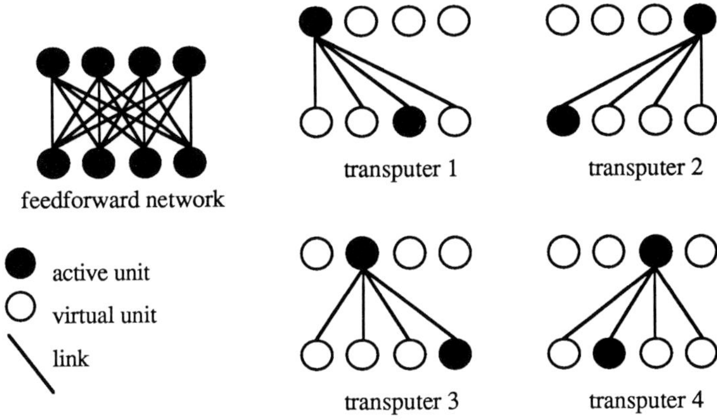

FIGURE 3. Allocation of units and links

Using this distribution concept, unit updates, which do not depend on each other, can be computed in parallel. The only thing that one has to make sure is, that the necessary nonlocal data, usually the output values of other units, are made available when necessary. This can be done, by copying the desired unit parameters from the active units into the respective virtual units whenever an update of a layer of units depending on these values is made.

To guarantee that this concept remains transparent with regard to the user code, a set of mapping functions is provided, which apply so called unit or link procedures to all units of a set or all links between two sets. There exist two classes of such functions, basically one for active units and one for active and virtual units. The first are used for parallel processing, the second for communication purposes like the synchronisation of virtual units or for initialization tasks, which have to be performed identically on all four transputers.

6. A SIMULATION EXAMPLE

The following code segments for a simple pattern associator network should give an idea of the programming style supported by the simulator. The network in question consists of an input and an output layer, which are defined in **Build** as InputSet and OutputSet. The network learns using the delta rule.

```
Build()
{ CreateNetwork(NoUnits);
  InputSet = MakeSet(NoInputs, "Input");
  OutputSet = MakeSet(NoOutputs, "Output");
  ConnectSets(InputSet, OutputSet, LearnFlag, LowerWeightLimit, UpperWeightLimit); }
```

Run()
{ SetNextPattern(InputSet, Output);
  SetNextPattern(OutputSet, Free2);   ... the Free2 parameter holds the unit target activity
  TraverseUnits(OutputSet, ComputeActivation);
  TraverseUnits(OutputSet, SetErrorSignal);
  TraverseLinks(InputSet, OutputSet, ModifyWeights); }

ComputeActivation(unit)
{ unit->Output = 1.0 / (1.0 + exp(-WeightedSum(unit))); }

SetErrorSignal(unit)
{ unit->Free1 = unit->Free2 - unit->Output; ...the Free1 parameter holds the unit error signal }

ModifyWeights(link)
{ link->Weight += LearningRate * link->From->Output * link->To->Free1; }

Build and **Run** are basic functions which have to be included in every simulation file. **Build** is used for the definition of the network sets and their connectivity. **Build** is also, among other things, used for the setup of commands and parameters which can be called or acessed from the user interface. **Run** describes the actions which have to be executed during a simulation cycle. **ComputeActivation, SetErrorSignal** and **ModifyWeights** illustrate the concept of the unit procedure and are used to compute the sigmoid output update function, the error signal of a unit and the weight change of a link.

7. PERFORMANCE

It is well known that parallel processing is most efficient when used for solving large problems. Not surprisingly we have found this to be true with our simulator as well, as can be seen in figure 4 where CUPS stands for connection updates per second during backpropagation learning.

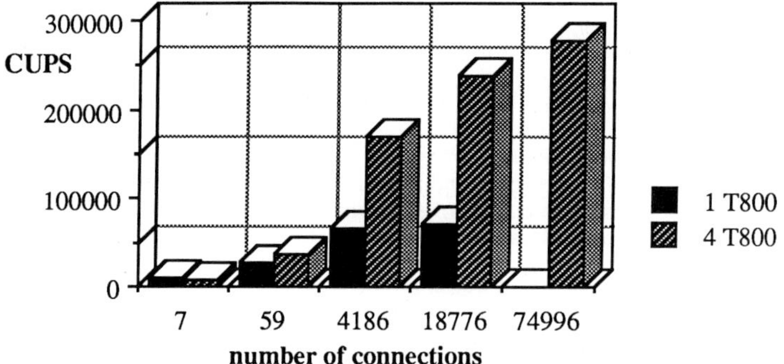

FIGURE 4. Performance

It should also be mentioned that when comparing the simulator's performance with several different network simulators running on a IBM PS2-80 system without arithmetic coprocessor, we generally found a speed advantage on the order of two magnitudes of the transputer based system.

REFERENCE

[1] Koikkalainen, P., Oja, E. (1988). Specification and implementation environment for neural networks using communicating sequential processes. *Proceedings of the IEEE International Conference on Neural Networks, San Diego July 24-27, 1988,* I-533-540.

Parallel Processing in Neural Systems and Computers
R. Eckmiller, G. Hartmann and G. Hauske (Editors)
© Elsevier Science Publishers B.V. (North-Holland), 1990

Neural Pascal
A Language for Neural Network Programming

H.Peter Gumm
Dept. of Computer Science
SUNY at New Paltz
New Paltz, N.Y. 12561
gummp@snynewba.bitnet

Ferdinand B. Hergert
Siemens – Corporate Research
and Development – ZFE IS INF2
D-8000 Munich 83
hergert@ztivax.uucp

Abstract
Neural Pascal is an extension of object-oriented Pascal, designed to allow easy specification and simulation of neural networks. Syntactically, just a handful of extensions to Pascal had to be added. Mainly they are syntax for the declaration of neurons, links and nets; commands to build (and change) the net topology; generators to iterate through all elements of a linear datatype.

1. Introduction

Neural network models are frequently visualized as graphs with nodes representing neurons and edges representing synapses between the neurons. Subclasses of neurons are distinguished by their functions or by their location within the network. Often these subclasses are arranged in layers with different layers containing neurons of different functionality. The algorithms driving such a network will take advantage of the structuring of the net, or of the properties of neurons when performing calculations or when updating states of neurons.

It seems highly desirable to translate this view of a network into an executable program as directly as possible with the actual translation of the code into memory accesses and pointer references shielded from the concern of the programmer. There is no data structure present in the repertoire of imperative languages, such as Pascal, that would be suitable to represent this kind of graph-like structure as needed for neural networks.

On the other hand we did not want to design a new language from scratch but rather extend a widely used language to simplify the use of the new features. In the design of Neural Pascal it turned out, fortunately, that we could get by with only few new data structures and a handful of associated new commands. These data structures and commands were built on top of Turbo Pascal 5.5 which is an extension to standard Pascal mainly in its provision of object-oriented features. The functionality of NP's new data structures fits in nicely with an object- oriented approach. The new data types are objects with their private data and methods, and with the capability of inheriting properties and methods to subtypes.

Neuro Pascal is currently realized as a preprocessor that translates NP-code into Pascal source code, which is then compiled using Turbo-Pascal's efficient compiler. Program development takes place in a comfortable and intuitive integrated development environment that is similar in appearance and functionality to the development environment provided by Turbo Pascal.

2. The Language

Links and Ports:

Our underlying concept of a neural network is that of a directed graph whose nodes may be grouped into nodes of different types. A connection between nodes is called a 'link'. A link is a directed edge between two nodes. The main purpose of a link is to provide an access from one node to another provided there is a link connecting the two. Just as any other Pascal object, links are typed, so different link types may be required when connecting different kinds of neurons.

Links of the same type leading into one neuron are collected together into what is called 'ports' (see figure 1). A neuron which is the target for several links from other neurons will have to provide one or more ports to receive these links. By grouping incoming links into ports two aims are achieved: Firstly, strong typing is preserved by requiring links of different types to come in through separate ports , and secondly, network algorithms can appropriately handle the different kinds of connections or access exactly those nodes that are connected to a given neuron through a particular port. Ac-

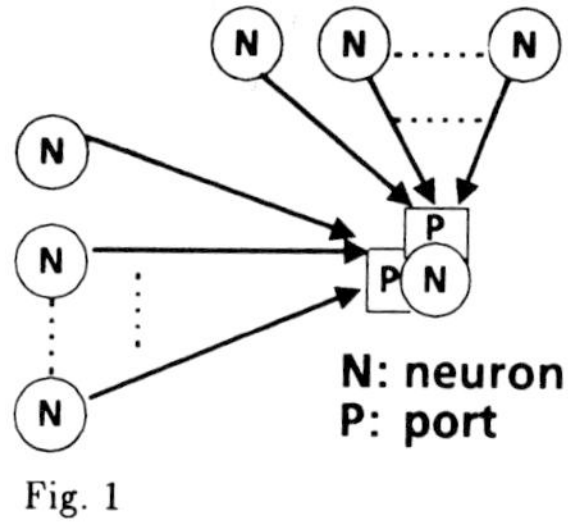

Fig. 1

cess through a link is provided from the target neuron to the source neuron of the link. The reason for this lies in the fact that in updating a neuron N, all those neurons that are directly linked to N have to be consulted, i.e. for every link L arriving at N we have to access the node at the source of L. For the same reason, the type of the neuron at the source of a link is part of the type of the link.

```
TYPE
   inLink  = LINK FROM input
     weight : real
     BEGIN
       weight := random - 0.5
     END;
```
Fig. 2

Figure 2 gives an example of the definition of a link (inLink) that has as source an object (neuron) of type input. This link type has its own data field weight and as a matter of illustration we provided inLinks with an additional initialization code, which is automatically executed when such a link is created.

Figure 3 shows how neurons can be provided with appropriate ports for the links to hook into. This is done using NPs type constructor PORT. In this example the neuron of type hidden possesses two ports: port entry for links of type inLink, and port lateral to receive links of type rigid.

Creating a net topology:

A complete network will consist of many different kinds of neurons, grouped somehow according to their function in the net and connections between some of these neurons. In order to connect two neurons with a link, a SEW statement is supplied. SEW will create a link between the source neuron and the target neuron and hook the link into the designated port of the target neuron. The link type is inferred from the type of the port. At the same time the initialization code for the created link is executed.

```
hidden = OBJECT
   activation : real;
   entry    : PORT OF inLink;
   lateral : PORT OF rigid;
END;
```
Fig. 3

The network topology can thus be created dynamically. Accordingly, we provide the CUT statement for removing links again. CUT causes a link to be removed. Just as with SEW, the user need not worry about storage allocation and reclamation. This is handled by the system. The syntax of SEW and CUT is given by:

```
SEW <source neuron> TO <port id> OF <target neuron>
CUT <source neuron> FROM <port id> OF <target neuron>
```

A network can now easily be modeled as an object consisting of groups of neurons. Even though the network topology can be changed dynamically, this will not be needed in many applications and it should be generally recommended to build the network topology by supplying the network object with an appropriate initialization procedure.

Accessing Neurons Through Links:

The purpose of links is to provide access paths to neurons. Given a link `li`, its source is referred to as "`li.?`". Typically, the same kind of processing is done with all links arriving at a given port. (This must be a rationale for creating an appropriate selection of ports). NP provides an iterator to do the same kind of processing with all links arriving at a given port: `ALL <linkId> TO <port> DO <statement>` will execute the `<statement>` with `<linkId>` running through all links arriving at `<port>`.

Suppose for instance that hidden neurons have been defined with a private method "Fire", this method could now be defined as given in figure 4. An important point to observe is that this method is attached to just a neuron and does not need to know anything about the the topology of the network of which this neuron is a part.

A slight variation of the ALL statement is the WITH ALL statement :

```
WITH ALL <linkId> TO <port> DO <statement>,
```

which is equivalent to

```
ALL <linkId> TO <port> DO WITH <linkId> DO <statement>.
```

```
PROCEDURE hidden.Fire;
VAR
  sum : real;
BEGIN
  sum := 0;
  ALL li TO entry DO
    sum := sum + li.weight *
                   li.?.activation;
  ALL li TO lateral DO
    sum := sum - li.?.activation;
END;  { hidden.Fire }
```
Fig. 4

The order of processing the links in an ALL-statement is unspecified. The linkId need not be declared; its type is inferred from the type of the `<port>`. Essentially an ALL statement sets up a local block with the `<linkID>` having only the body of the ALL statement in its scope.

Inheritance:

In object oriented Pascal new types of objects can be defined by extending existing objects by new fields or methods. Since the instances of a type are called a "class" in object oriented parlance, the instances of the so extended type are said to form a "subclass". A class may have several different subclasses, and eventually the subclass hierarchy will form a tree. Since subclasses inherit all properties of their ancestor classes, objects of subclasses may appear at any place where an object of the parent class is expected.

```
TYPE
  genericLink = LINK FROM neuron;

  dendrite = LINK(genericLink)
    weight : real;
    BEGIN {init. code}
      weight := random - 0.5
    END;
```
Fig. 5

In NP, links are objects too, with the type of the neuron, where the link originates, is part of its type. Fig. 5 shows the definition of a generic link (`genericLink`) and a subclass (`dendrite`) derived from it.

3. The language environment

NP development takes place in an integrated environment with menu bars, popup windows, pulldown menus, quite similar to the environment Turbo Pascal 5.5 provides. The main menu bar provides choices for syntax checking or translating the original program into Turbo Pascal. The choice "Compile" creates an executable file by invoking the command line Turbo Pascal compiler. Some menu choices activate further pulldown menus for dealing with file handling or even to provide operating system shells, or to set switches and toggles. The environment is meant to be selfexplanatory to programmers familiar with Turbo Pascal or similar environments. The "translate" menu item shows the source code of the intermediate Turbo Pascal program in an editor window.

4. Planned Extensions

In the current version of NP links are always directed. If symmetrical links are required, we must use two links, one in either direction. The only problem with this approach is that data cannot be aliased between the two links. In the next version facilities for the creation of bidirectional links will be added.

References

[1] (Borland International) *Turbo-Pascal 5.5. Object Oriented Programming Guide.* Scotts Valley, Ca. 1989.

[2] H.P. Gumm, F.B. Hergert, *Neural Pascal (NP)*, Siemens Technical Report, INF2-ANN-5-89, 1989.

Parallel Processing in Neural Systems and Computers
R. Eckmiller, G. Hartmann and G. Hauske (Editors)
© Elsevier Science Publishers B.V. (North-Holland), 1990

MODELING OF NEURONAL SYSTEMS ON TRANSPUTER NETWORKS[#]

M. Migliore[+], G.F. Ayala[^] and S.L. Fornili[*][+]

[+]Institute for Interdisciplinary Applications of Physics, Nat. Res. Council
Via Archirafi 36, I-90123 Palermo, Italy.
[^]Chair of Psycophysiology, University of Palermo,
Via Pascoli 8, I-90143 Palermo, Italy.
[*]Physics Department, University of Palermo,
Via Archirafi 36, I-90123 Palermo, Italy.

We report findings on parallel simulations of neuronal systems on high-performance and yet cost-effective Transputer networks of various sizes. Our data indicate that high efficiency of these arrays can be obtained. The computing power can be easily increased by addition of Transputer-based modules.

1. INTRODUCTION

The computing power requirements is perhaps the most important factor to be considered in planning computer simulations of neuronal systems. The complexity of the differential equations, the use of small integration steps to obtain an acceptable accuracy, and the size of the neuronal model to appropriately investigate a given problem, almost inevitably lead to CPU-bounded computer programs.

During the last few years a parallel processing approach [1] has been taken exploring the potentialities of new computer architectures. Parallel simulations of neuronal systems have been efficiently conducted on such computers [2]. In this work we extend this approach by simulating the neuronal behaviour on computer systems based on Transputers [3]. Their great versatility and potentiality have been already shown by efficiently running different computationally intensive applications on rather small and economical arrays [4]-[9]. Our aim is to show that, without any special program sophistication, it is easily possible to achieve and maintain very high efficiency and speed-up factors adding more processors modules even with small neuronal systems.

2. COMPUTATIONAL PROCEDURE

The neuronal systems we use as a test in this work are composed by 2, 6 and 48 neurons arranged in a linear array as shown in Fig.1.

Each neuron includes one somatic compartment and six segments connected to form dendritic and axonal trees. Segments are composed by six compartments. Each compartments communicates with adjacent neighbours and is supposed small enough to be considered isopotential. To each axonal, somatic or dendritic compartment correspond first-order ordinary differential equations describing the passive cable properties, the active calcium and potassium related conductances in the dendritic region, and the Hodgkin-Huxley ionic currents in the axonal region. The exponential integration scheme used in this work, and the underlying theory and mathematical details can be found in Ref.10. However, our implementation can be easily adapted to other integration schemes or different biological models as long as compartmentalized neuronal systems are used.

[#]The present work has been carried out at IAIF-CNR and supported also by local MPI 60% and CRRN-SM local fundings.

M. Migliore et al.

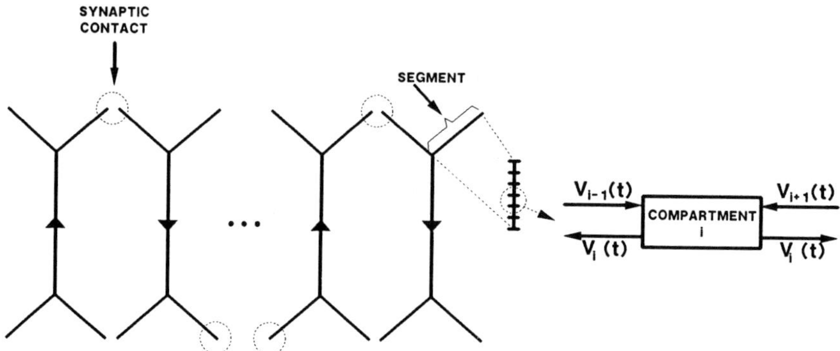

FIGURE 1

The neuronal systems used in the simulations. Each neuron is composed by 1 somatic, 18 axonal and 18 dentritic compartments. Each compartment communicate with adjacent neighbours. $V_k(t)$ represents the membrane potential of compartment k at time step t.

The Transputer-based system we use in these simulations is composed by a *master* Transputer module with 2-Mbyte of RAM and one T800-20 Transputer, hosted by a PC-XT/AT interface board, and a TRANSTECH TSMB16 board featuring 16 modules - the *workers* - each with one T800-20 and 1-Mbyte RAM. A double linked ring topology has been chosen for interprocessors connection, as shown schematically in Fig.2, in which the master Transputer provides data routing along the ring and I/O connection with the monitor and filing system of the host PC. We implemented networks of 4, 8, 12 and 16 Transputers by programming a C004 link switch controlled by a configuration program [11].

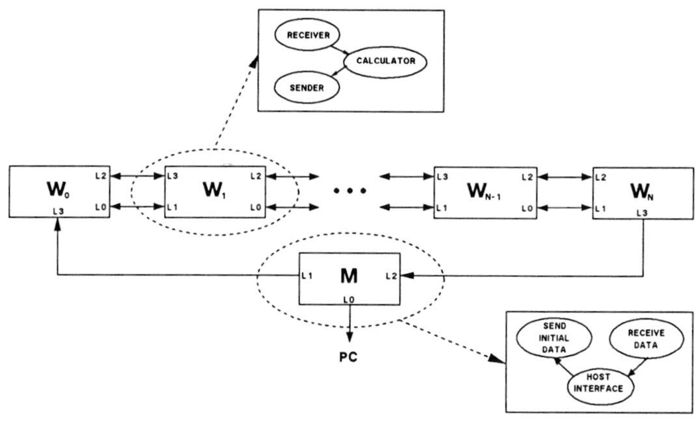

FIGURE 2

Interprocessor connection topology and process allocation on the master (*M*) and worker (*W*) Transputers.

The neuronal system to be simulated is divided in subsystems and each of them is assigned to one processor. Our Occam2 [12] concurrent program running on each Transputer has been implemented using INMOS TDS 700D development system [13], by merging and adapting the DENDR51 and AXON04 FORTRAN programs described in Ref.9. At the beginning of each simulation, a copy of this template program is loaded on each worker by the master together with data relative to the subsystem that has been assigned to it. This worker program, schematically shown in the exploded view of Fig.2, consists of three main processes - *calculator*, *sender* and *receiver*. To calculate the membrane potential relative to a given compartment, the *calculator* process needs information about the adjacent compartments at each time step. As long as the latter compartments belong to the same subsystem, this information can be found on the same Transputer. Otherwise, the *sender* and *receiver* processes exchange information with adjacent processors. These processes are also involved when - every 500 time steps in the present application - data are collected by the master from all the workers for storage and/or on-line display.

3. RESULTS AND DISCUSSION

In Fig.3, we show the speed-up factor for neuronal systems and Transputer networks of different sizes and the 16-processor network efficiency, ϵ, which is usually defined by $\epsilon = (T_1/P)/T_P$, where P is the number of processors, and T_1 and T_P the execution times on a 1- and P-processor systems, respectively.

As one can see from Fig.3, the efficiency of the 16-Transputer network we use is 65% for the very small 2 neurons system. However, the efficiency rapidly increases by increasing the neuronal system size, being 80% for 6 neurons and 95% for 48 neurons.

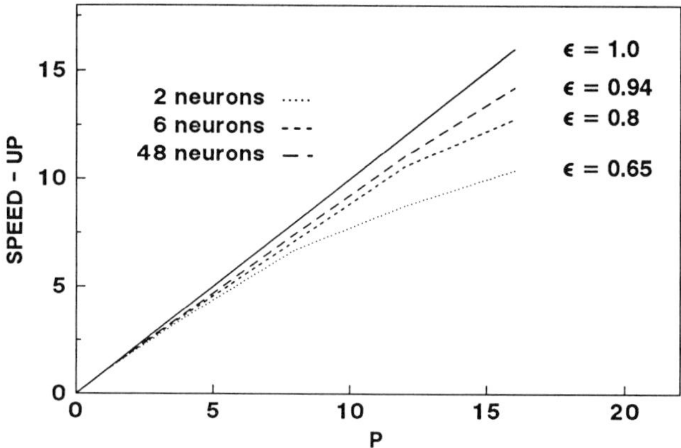

FIGURE 3

Speed-up curves for simulations of 2, 6 and 48 neurons on Transputer networks including up to 16 processors. The network efficiency, ϵ, is also shown for each neuronal system simulated on the 16-Transputer array.

Performance degradation is a common and important problem to be considered in any parallel processing approach, and can be mainly induced by communication overhead or load unbalancing. Although membrane potentials of each compartment can be computed concurrently, this process requires information from neighbouring compartments. This implies some synchronization of the time evolution of the entire neuronal system at each time step. Further, in the case of an arbitrary neuronal system, good load balancing cannot be obtained simply dividing the neural system into subsystems composed by an equal number of compartments. Indeed, computation for the axonal and somatic regions is several times slower than for dendrites, due to the Hodgkin-Huxley equations. Thus, a simple preliminary trial-and-error procedure could be needed to balance the processors load, which depends on the equations used to model the neuron, the integration method, the neuronal system morphology and processor network size. However, once the appropriate subsystems are found, it is no longer necessary to adjust them as long as the simulation conditions are unchanged. In the compartmentalized neuron models, communication between adjacent compartments cannot be avoided or drastically reduced without affecting the simulation accuracy. The communication overhead is, however, independent of the size of the neuronal system. Therefore, its total fractional relevance in terms of CPU-time for a given processor network topology decreases with larger neuronal systems. In our simulations, communication overhead has only a minor role, because the ring topology does not need a router, that is an expensive process in terms of CPU-time, as long as the neuronal system could be divided in such a way that subsystems do not have more than four point of contact with neighbors. In this case, the four links of each Transputer and their flexibility provides all the necessary direct connections to avoid a router process.

Simulating the 48 neuron system, we could assign an integer number of neurons to each Transputer for all networks. In this almost ideal case the computational load is perfectly balanced among *workers*. The deviation from the theoretical limit is caused by the remaining sequential parts of the program that run on the *master* and by the communications between processors that occurs each time data have to be gathered to the master. Simulating 6 neurons, additional efficiency degradation is caused by some communication overhead due to inter-processor communication at each time step. Finally, with only 2 neurons there is also a problem with load balancing, since the computational cost of a single axonal compartment become a significant fraction of the total simulation time if too many processors, with respect to the neuronal system size, are present in the network.

4. CONCLUSIONS

We presented and discussed results of an implementation of a concurrent program to simulate neuronal systems using a compartmentalized neuron model, on cost-effective and modular Transputer networks. The neuronal system was divided into smaller subsystems that are processed in parallel on the Transputer array. High efficiency and speed-up factors are obtained even with neuronal system composed by only 6 neurons (222 compartments in our case). Simulation of larger neuronal systems (e.g. 48 neurons involving 1776 compartments) result in 95% efficiency using a 16-processor network.

ACKNOWLEDGEMENTS

We wish to thank Dr. F. Bruge' and Dr. V. Martorana for fruitful discussions, and Mr. S. Pappalardo, Mr. M. Lapis and Mr. A. La Gattuta for technical assistance.

REFERENCES

[1] G. Fox, M. Johnson, G. Lyzenga, S. Otto, J. Salmon, D. Walker, *Solving Problems on Concurrent Processors* vol.1, Englewood Cliffs, NJ: Prentice Hall, 1988.
[2] M.E. Nelson, W. Furmanski and J.M. Bower, "Simulating neurons and networks on parallel computers", in *Methods in neuronal modeling: from synapses to networks*, T.J. Sejnowski and T.A. Poggio, Eds. Cambridge, MA: The MIT Press, 1989.
[3] INMOS, *The Transputer Data Book*. Bristol: INMOS, 1988.
[4] D. Fincham "Parallel Computers and Molecular Simulation", *Molec. Simul.* vol.1, 1-45, 1987.
[5] C.R. Askew, D. B. Carpenter, J.T. Chalker, A.J.G. Hey, M. Moore, D.A. Nicole and D.J. Pritchard, "MC simulation on Transputer Arrays", *Par. Comp.* vol.6, 247-258, 1988.
[6] F. Bruge', V. Martorana, and S.L. Fornili "Concurrent statistical simulation on a Transputer-based systems", in *Proceedinds of CONPAR88*, C. R. Jesshope and K. D. Reinartz, Eds. Cambridge: Cambridge University Press, 1989.
[7] F. Bruge', P. L. San Biagio, S. L. Fornili "Transputer-based upgrading of a laser photon correlator", *Rev. Sci. Instrum.* vol.60, 222-225, 1989.
[8] F. Bruge', P. L. San Biagio, S. L. Fornili "New photon correlator design based on Transputer array concurrency", *Rev. Sci. Instrum.*, in press, 1989.
[9] M. Migliore, V. Martorana, F. Sciortino, "An Algorithm to find all paths between two nodes in a graph", *J. Comp. Phys.*, in press, 1989.
[10] R. J. MacGregor, *Neural and Brain Modeling*. London: Academic Press, 1987.
[11] Courtesy of Dr. F.Bruge'.
[12] INMOS, *Occam2 Reference Manual*. Hertfordshire, U.K.: Prentice-Hall, 1986.
[13] INMOS, *Transputer Development System*. Hertfordshire, U.K.: Prentice-Hall, 1988.

Parallel Processing in Neural Systems and Computers
R. Eckmiller, G. Hartmann and G. Hauske (Editors)
© Elsevier Science Publishers B.V. (North-Holland), 1990

A Development Tool for Neural Networks Simulations on Transputers

Markus Miksa

Department of Mathematics / Computer Science
University of Marburg
3500 Marburg
Federal Republic of Germany

In this paper a new Artificial Neural Network Editor (ANNE) is presented. The architecture of this new neural network development-tool is discussed. The application area is outlined. ANNE supports the object oriented development of any interactive neural network simulation based on transputer networks of any size. ANNE's data structures correspond to objects which may be spread over the whole transputer net in whatever way and which are addressed by specific messages. This architecture allows development and test of programs on a standard PC. Later they are simply downloaded and run on the transputer hardware. ANNE offers libraries for menu driven graphical user interfaces. It is possible to generate arbitrarily connected neural networks. The ultimate goal of this new simulation development system, is to reduce the time needed to construct new neural network applications.

1. INTRODUCTION

Simulation tools allowing easy change of network parameters and rapid turn around are crucial in the process of developing neural networks. A high degree of parallelism is a key feature of efficient natural neural networks. Evidently this will also be a key point designing powerful hard- and software to simulate artificial neural network. Conventional sequential hardware is not enough powerful for neural network simulations.

There are two ways to achieve the desired performance with parallel machines:
- Special purpose computers along with inflexible user interfaces.
- General purpose architectures programmed with parallel computer-languages. Only the The second alternative is flexible enough to implement different connectionism models of any kind. Transputer-Systems are parallel general purpose computers. They are in common use and are well suited to implement neural networks. Although they realise parallelism more coarse grained as natural neural nets do. The problem with transputer systems is the difficulty to program them. The expertise to program parallel systems is rare and the software-development tools for transputer-systems are still clumsy. Researchers interested only in speeding up their simulations are faced with a lot of software engeneering problems they never expected. Much time and effort is needed to implement a simulation program on a transputer based system. These problems are in contrast to the possible advantage achieved by shorter runtime particulary if the program is not used in an industrial context but for scientific investigation.

Considering these circumstances, ANNE has been developed. ANNE supports the development of simulation programs for neural networks on transputer based systems with an arbitrary degree of parallelism.

The main features of ANNE are:
1. Support of user-interaction, supporting an interactive, graphical menu interface.
2. Hiding the details of parallel programming by an object oriented programming paradigma.

It is possible to use the same source code for both the transputer based implementation and the stand-alone version on the host. Thus an application may be developed and tested with any well known comfortable programming environment on the host system. Having finished the development process, the program my be transfered to the transputer net without any complication. A very important advantage of ANNE is the size-invariance of the transputer net. It may be enlarged without the necessity to recompile already existing programs.

This paper presents the following topics:
First the requirements of general neural network simulations are discussed. Then the special approach of a transputer net is explained. The next chapter describes briefly the architecture of ANNE. Finally it is explained how programs are developped using ANNE.

2. THE PURPOSE OF NEURAL NETWORK SIMULATIONS

Two tasks of a neural network simulation are distinguished:
a) Network construction including initialisation of activities and weights.
b) Execution of network functions including learning.

The intended network-paradigma is outlined coarsely by the program design. A set of parameters is used to tune the model's behavior. Depending on the chosen model the most interesting parameters are:
- The number of Units.
- The number of Layers.
- Lateral and vertical connections.
- Activation- and output values and functions.
- Learning rules.
- Synchronous or asychronous net update ... [1]

Only the opportunity of interactive modification of these parameters allows the exploration of the network behavior.

3. DECOMPOSITION OF NEURAL NETWORK PROGRAMS ON PARALLEL TRANSPUTER SYSTEMS

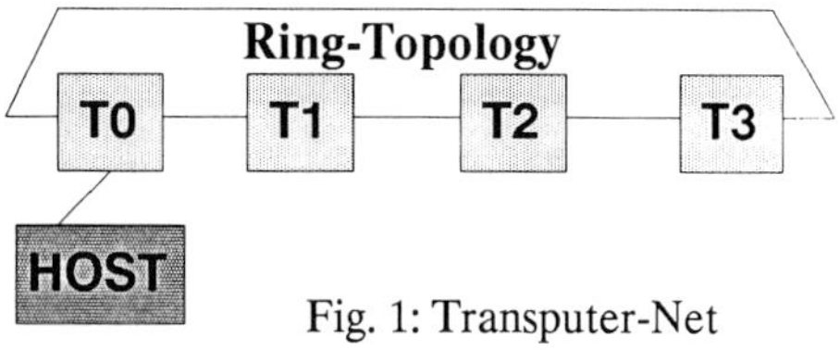

The horsework of any simulation program is the continuous computation of the unit netto input. Therefore the neural net performance is measured in million interconnects per second (MICPS). The transputer T800 is a powerful (10 MIPS) microprocessor with an integrated floating point unit (1,5 MFLOPS) and 4 serial links (10-20 MBit/s). It is

Fig. 1: Transputer-Net

well suited to work in a multiprocessor system according the Multiple Instruction, Multiple Data (MIMD) principle. Furthermore, it provides a local memory of approximately 1 MB, depending on the manufacturer. The processor nodes may be connected in a hypercube, ring, torus or any other useful interconnection topology. Data I/O is usualy done by a link

connected standard Host. ANNE runs on a personal computer using MS-DOS.
The aim of the problem decomposition is to minimize communication overhead and therefor
to optimize the system performance. A simple und appropriate decomposition is data sharing:
In a simulation with X Units, each of N transputers is responsible for X / N Units. To reduce
communication cost during
update cycles, every transputer's
local memory is provided by all
necessary information, e.g. the
activity vector, required for an
independent update of all it's
units [2].

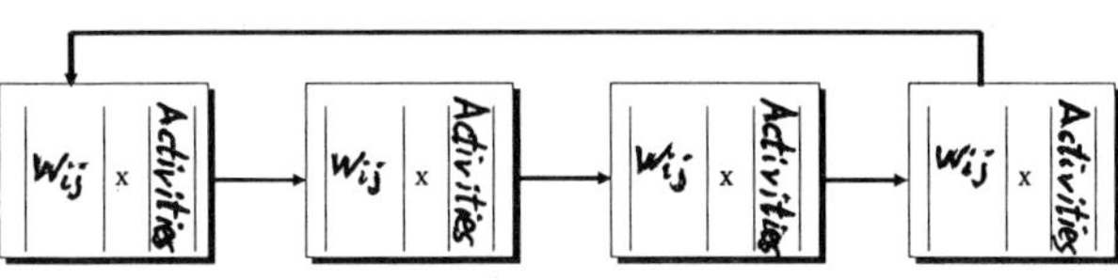

Fig. 2: Transputer Communication

Unit up dates may work
completely independently and in
parallel on each single transputer. Communication operations occur only after a completed
cycle.Their only task is to update the common activity vector.

4. PROBLEMS IN PROGRAMMING TRANSPUTER-SYSTEMS

Some serious problems complicate the programming of parallel transputer systems:

-Data routing: In processor networks of different sizes and different topologies data have to be
routed between arbitrary chosen nodes.
-Data consistensy: Application data are distributed over the transputer net. Some of them are
shared by all nodes, others are restricted to certain processors. Display data are only needed
on the host. The interactive modifications have to be passed to all concerning processors.
-Program development: The turn around time in usual programming environments for
transputers is much longer than for conventional personal computers. Sometimes a new
computer language has to be learned, only for a special transputer application.   [3]

5. THE ARCHITECTURE OF ANNE

Hard- und Softwarerequirements:

ANNE was developed for PC based transputer systems. The application language is C, as well
on the host as on the transputer net. Transputer communication is supported by the
reconfigurable runtime system EXPRESS (a trademark of Definicon coroperation) and is
independent from a certain transputer system topology.

Graphical user-interface
Paradise:

Data-I/O on the PC is
supported by a set of
library functions. For
data presentation and
mouse oriented
manipulation of
connections and activies
a hinton display is
offered. Graphical
dialog-boxes,
file-selectors, window
oriented text I/O and
easily programmed

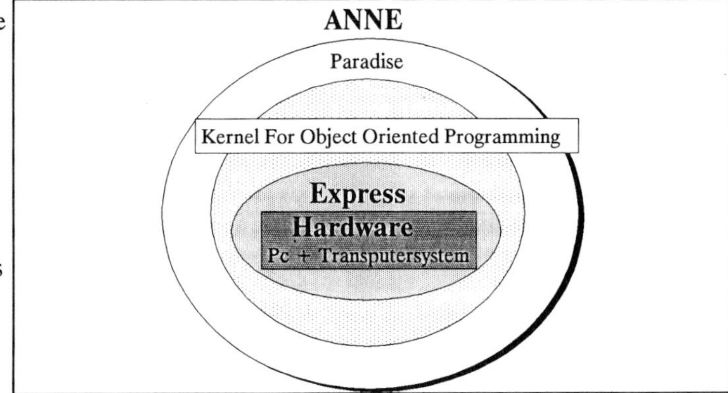

Fig. 3: Architecture of ANNE

menu functions facilitate the implementation of I/O menus for parameter manipulation.

The core of ANNE is an object oriented communication system hiding the parallelism and communication completely:
The most important point of programming interactive neural net applications for transputer systems is the fact that two classes of data have to be handeled:

A) SHARED: Shared Data e.g. the amount of layers and Units ..., are common to all nodes and have to be kept consistent on all nodes during runtime, specially after user input.
B) LOCAL : Other data have to be hold locally at certain processors. These might be groups of units and their connections. These data require too much system memory to be loc all nodes.

These facts and the required reconfigurability of the necessary data structures, lead to the request for complete data encapsulation. Together with the message passing concept of the transputer hardware a completely object oriented program-design is derived.

All user defined data structures are layed out as objects, for example "ANN" the neural Net, "LAYER" a data structure keeping all informations concerning a Layer of Units ... (s. Fig. 4). Objects of the same class share a set of functions, called methods. These functions are selected by sending messages to the objects. Each object knows it's dataclass and the processor or the processorgroup where it is allocated. Therefore after object allocation the programmer must not bother about details concerning the physical locations of data distributed on the transputer system. The object oriented message-passing system knows where messages have to be passed. The destination of a message may be the host a single transputer or a certain group of transputers if their data class is SHARED.

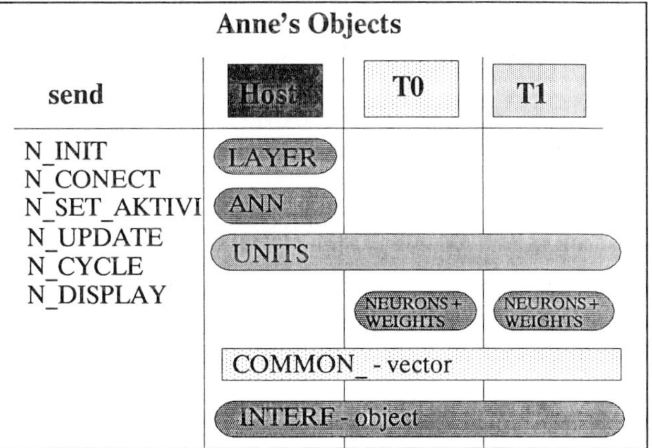

Fig. 4: ANNE's Object Oriented Program Design

Summing up it may be said ANNE reduces the effort for developing neural networks on transputers drastically, because it offers for example a library of objects and methods for common tasks.

REFERENCES

[1] Rumelhart D.E.and McClelland J.L. (ed.). Parallel Distributed Processing. Vol. I + II. MIT Press, 1986.

[2] Ghosh J. and Hwang K. (1989) Mapping neural networks onto message-passing multicomputers. Journal of Parallel and Distributed Computing, 6(2): 291-330.

[3] Büttner W. (1989) Grundlagenforschung und Softwareentwicklung im Siemensproject "Neurodemonstrato". In: W.Brauer (ed.) Wissensbasierte Systeme, Springer Heidelberg.

Parallel Processing in Neural Systems and Computers
R. Eckmiller, G. Hartmann and G. Hauske (Editors)
© Elsevier Science Publishers B.V. (North-Holland), 1990

A SOFTWARE ENVIRONMENT FOR FLEXIBLE AND RAPID PROTOTYPING OF NEURAL NETWORK APPLICATIONS.

Jos A. G. Nijhuis, Poul E. de Haan, Lambert Spaanenburg and Frank Warkowski

Institute for Microelectronics Stuttgart
Allmandring 30a
D–7000 Stuttgart 80, West Germany

The NNSIM neural network simulation environment provides for flexible and fast prototyping of neural networks in mixed applications. The flexibility of the simulator is demonstrated in two examples. The first experiment shows Carver Mead's analog silicon retina whereas the second one presents a mixed neural–digital image processing system.

1. INTRODUCTION

Neural computing has gained increasing interest over the last years, but the number of applications is still quite limited. This is mainly caused by the difficulty in probing architectures that exploit the advantages of neural computing. In the search for suitable hardware implementations a flexible simulation environment is required to accommodate for "step–by–step" concept evaluation [1]. Most of the currently available neural network simulators have a limited scope: (a) the number of neural network structures that can be simulated is limited to a few standard topologies [2] and/or (b) the inclusion of device characteristics is cumbersome [3,4]. The NNSIM simulation environment [5] answers these demands in offering the neural engineer parametrizable models valid on a wide range of abstractions. Therefore a typical description extends over several abstraction layers (conceptual upto electrical) and modes (neural, analog and digital). This allows for "in–application" simulation. The NNSIM flexible tooling set supports next to automatic and interactive also interruptive neural design evaluation. This accelerates design iterations where the optimal neural architecture still must be determined.

Section 2 introduces the NNSIM simulation environment together with the definition, construction and evaluation of neural applications. Section 3 presents two experiments illustrating the wide variety of neural designs handled by NNSIM. Finally in section 4 the conclusions are summarized.

2. THE NNSIM ENVIRONMENT

The architecture of NNSIM is shown in figure 1. The network handler implements the physical layer of the neural network simulator whereas the respective procedural interfaces offer a conceptual layer to the designer. This layered approach results in a flexible simulation environments whose functionality can be changed incrementally by adding new functions to the procedural interface. System designs can be written as C–programs, to which the neural part can be added using C–function calls.
The network information can be saved for off–line inspection. Such information is also applied to migrate the simulated neural model into an actual realization as semi–custom IC. Overall this data transparency makes for a truly open CAD environment.

All functions in the procedural–interface support the on–line incremental creation and modification of a neural design. Simulations can be interrupted and continued at any time allowing the designer to make changes without having to start all over again. The current NNSIM release offers over 110 functions including error reporting, file read/write and graphic routines.

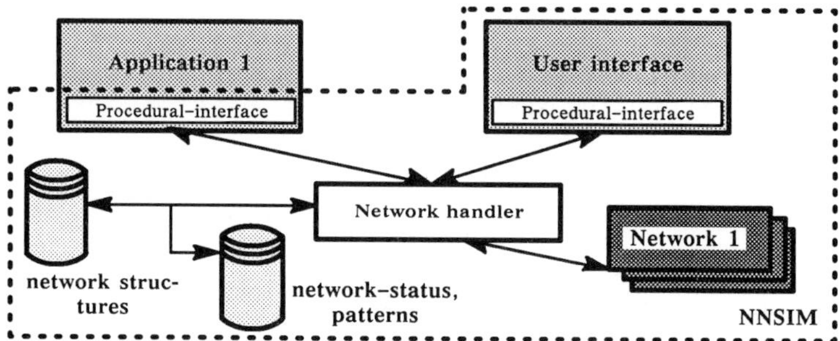

FIGURE 1 The NNSIM architecture

Most neural network simulators separate the construction phase, during which a neural network description is translated into an internal data structure, from the simulation phase. Each small design/simulation change urges to build the complete internal data structure anew. Hence design iterations can not be interactive. With NNSIM the definition of a neural topology is an integral part of the environment: network construction and simulation are indistinguishable operations on the same database. During a simulation session the designer can add, delete or insert user-enforced malfunctions in any desired order, so that various neural design realizations can be tested or a failure analysis can be made [6].

3. SOME EXPERIMENTS

In the first experiment the operation of Carver Mead's silicon retina [7] is investigated. This neural network contains a large number of analog non-linear elements. The topology of a 20 by 20 neuron retina is shown in figure 2 together with a part of the NNSIM code used to generate the hexagonal structure.


```
/*..create neurons*/
for(k=1;k< =20;k+ +)
    neuron_$add(0,(k-1)*20+ 1,k%2+ 1,2*k-1,1,20,&d,&status);
/*..add connections*/
for(k=1;k< =20;k+ +)
{   for(i= (k-1)*20+ 1;i< =k*20;i+ +)
    {   if(i<k*20)
        {   connection_$add(i,l+ 1,0,0,1,0.0,1,&status);
            ......
            ......
            ......
            ......
```

FIGURE 2 The hexagonal topology of the silicon retina

The silicon retina as presented by Mead performs a low-pass spatial filtering on the incoming image. The simulation results presented in figure 3 are based on a variant that is discussed in [8]. Each 3D-plot shows the output levels of the neurons of a 50 by 50 neuron retina. The original image has very sharp transitions, represented by the steep edges in figure 3A, as only binary pixel values are used. During the spatial filtering process the transitions are smoothed as can be seen in figure 3B and 3C. The simulated network contained 2500 neurons and about 7500 connections. Creating a network of this size takes about 20 seconds on an APOLLO DN3500 whereas each simulation cycle takes about 7 seconds.

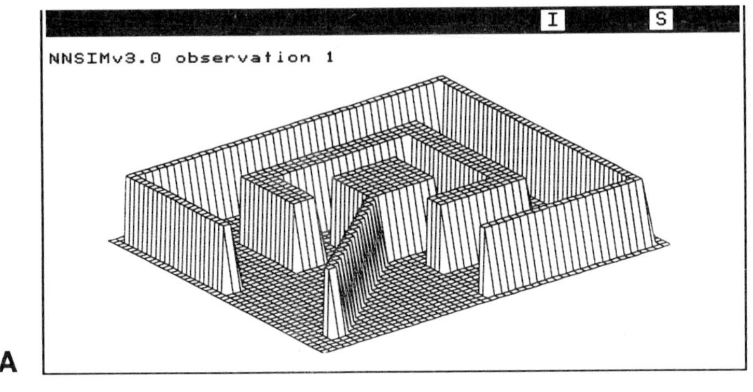

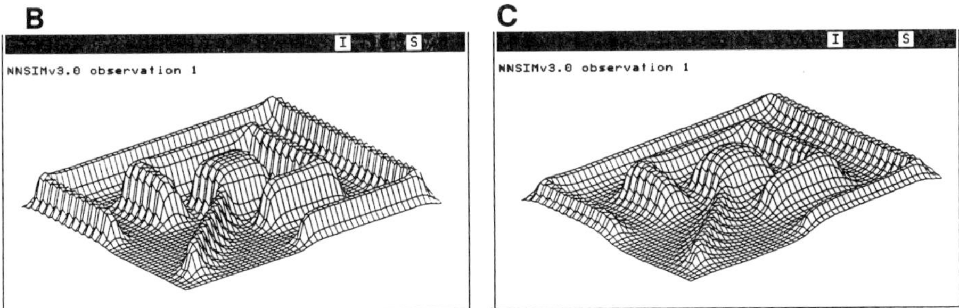

FIGURE 3 The operation of the silicon retina as simulated by NNSIM.
(a=original image,b=after 5 iterations,c=after 70 iterations)

The next experiment involves a mixture of digital and neural hardware [9]. Figure 4 shows the architecture of an image processing system and the corresponding graphic screen of NNSIM. In this experiment the neural–digital image processor is used to detect whether or not the image contains a rectangle. Two neural networks are used to detect horizontal and vertical oriented line activity on the input sensor. A third neural network will enable the digital part only when enough horizontal and vertical activity is present for the possible occurrence of a rectangle.

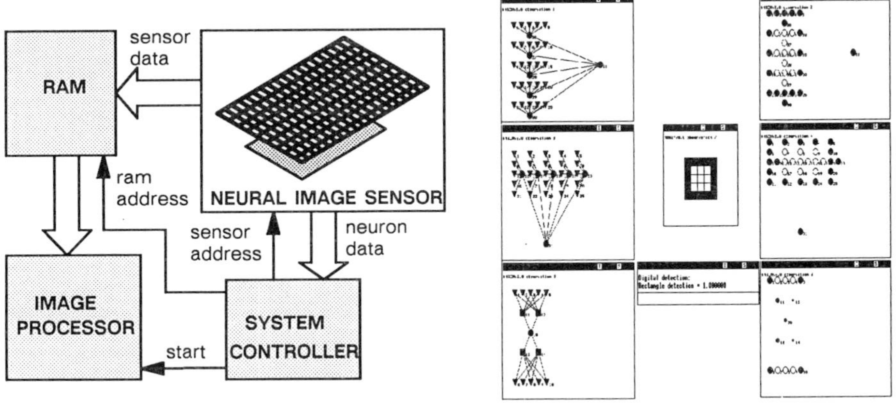

FIGURE 4 Mixed neural–digital image processing system

The system controller part uses the outputs of the neural network to decide which part of the sensor input should be loaded into the RAM for inspection by the image processor.
So the 'slower' digital image processor will only be activated when the neural part decides on a possible occurrence of a rectangle and even then neural data is used to reduce the amount of sensor data that should be processed. This mixed neural–digital approach benefits from the speed of the neural network part and the accuracy of the digital image processor, so that critical images can not only be recognized but also interpreted correctly in real–time.

4. CONCLUSIONS

The neural network simulation environment NNSIM provides means for an efficient, flexible and fast prototyping of neural network applications. NNSIM is developed to handle simulations of neural hardware at various description levels together with non–neural system parts. Besides automatic simulation NNSIM offers interactive and interruptive modes that are specially suited to investigate the influence of different neural network topologies and parameter settings on the expected input data response.

ACKNOWLEDGEMENTS

This work has been supported by Daimler Benz AG and the Bundesministerium für Forschung und Technologie under contract "TV 8926 3". Furthermore the cooperation within the DFG project "Photo–diode array processor" and the ESPRIT/BRA project NERVES is acknowledged.

REFERENCES

[1] DARPA neural network study, *AFCEA International Press, Fairfax VA*, 1988

[2] McClellands, J. and Rumelhart, D., "Explorations in parallel distributed processing", *MIT Press, Cambridge MA*, 1988

[3] Feldman, J.A., Fanty, M.A., Goddard, N.H. and Lynne, K.J., "Computing with Structured Connectionist Networks", *Communications of the AMC*, **31–2**, 1988, 170–187

[4] Paik, E., Gungner, D. and Skrzypek, J., "UCLA SFINX – A Neural Network Simulation Environment", *IEEE First International Conference on Neural Networks, San Diego CA*, 1988, 367–375

[5] Nijhuis, J.A.G., Spaanenburg, L. and Warkowski, F., "Structure and Application of NNSIM: a General–Purpose Neural Network SIMulator", *Microprocessing and Microprogramming*, **27**, August 1989, 189–194

[6] Nijhuis, J.A.G. and Spaanenburg, L., "Fault tolerance of neural associative memories", *IEE Proceedings E*, **136–5**, September 1989, 389–394

[7] Mead, C., "Analog VLSI and Neural Systems", *Addison–Wesley, Reading MA*, 1989

[8] Hutchinson, J.M. and Koch, C., "Simple Analog and Hybrid Networks for Surface Interpolation", in Denker, J.S., (ed), Neural Networks for Computing, *American Institute of Physics, Snowbird UT*, **151**, 1986, 235–240

[9] Nijhuis, J.A.G., Spaanenburg, L., Warkowski, F., Höfflinger, B., Darianian, M., Martini, Ch., Waldschmidt, K. and Kollbach, D., "DUITS: Driver–User Interface for Tablet Symbols", *Proceedings 2nd PROMETHEUS workshop, Stockholm, Sweden*, **1**, October 1989, 350–358

Parallel Processing in Neural Systems and Computers
R. Eckmiller, G. Hartmann and G. Hauske (Editors)
© Elsevier Science Publishers B.V. (North-Holland), 1990

SYSTOLIC SIMULATION OF MULTILAYER, FEEDFORWARD NEURAL NETWORKS

Nikolay PETKOV[*]

University of Erlangen-Nürnberg,
Institute of Informatics - IMMD (III)
Martensstraße 3,
D-8520 Erlangen, West Germany

ABSTRACT: It is shown how two different systolic algorithms for matrix-by-vector multiplication can be alternately used to simulate a multilayer, feedforward, artificial neural network. Similar algorithms are given for the error back propagation and weight correction. A systolic neural network simulator based exclusively on these algorithms has been implemented on a ring of transputers.

1. INTRODUCTION

Figure 1 shows a paradigm of a three-layer, feedforward, artificial neural network. The mechanisms for simulating such a network in both the working and the learning phase have been extensively studied and the reader is referred to [1] for a model description, application examples, and further references. Less attention has been paid to the parallel computer implementation, mainly because model issues have been found more interesting, and because implementation was thought to be straightforward and architecture dependent.

The implementation of this model on a parallel computer is, however, not a trivial task, if the efficiency is considered. The network communication pattern leads to non-scalable mappings of the network onto any target multiprocessor. Even small changes in the number of neurons of one of the layers, e.g. the hidden one, may lead to major changes in the way in which load is distributed among the processors and data is moved around. Instead of using a straightforward mapping of neurons onto processors, we identify those parts of the simulation problem which are computationally intensive and apply to them algorithms which have already proved efficient. In particular, the computation of the activation values of the neurons of the hidden layer and output layer can be reduced to two matrix-by-vector multiplications. In the following, it is shown how two different systolic algorithms for matrix-by-vector multiplication which were previously used for iterative solution of systems of linear equations (pp. 74-77, 105-106 in [2]) can be used to realize the feedforward, neural network model. As in the solution of systems of linear equations referred to, the alternate use of both algorithms leads to highly efficient execution in which the output of the first algorithm matches the form in which the input for the second algorithm is required. Similar algorithms are given for the learning phase in which the error is backpropagated and the synaptic strengths are corrected.

The paper is organized as follows. The feedforward, neural network model is considered briefly in Section 2 mainly for reasons of terminology and denotions. In Section 3, systolic algorithms are given for the efficient realization of this model. Implementation issues are considered in Section 4.

---

[*]This work was supported by a research award from the Alexander von Humboldt Foundation.

2. SIMULATION MECHANISM

In the following, we consider a three-layer network with n, m, and l neurons in the first, second, and third layers, respectively. Let x_i, i = 1, 2, ... , n, denote the values output by the neurons of the first (input) layer. The values output by the neurons of the second, so called hidden layer are determined by computing first the respective activation values

$$\underline{y}_i := \sum_{j=1}^{n} a_{ij}x_j , \qquad i = 1, 2, ... , m, \tag{1}$$

where a_{ij} denotes the "synaptic strength" of the connection from neuron j of the first layer to neuron i of the second layer. The output values are next computed from the activation values as follows (The latter are underlined for clarity.):

$$y_i := f(\underline{y}_i - \Theta_i) , \qquad i = 1, 2, ... , m, \tag{2}$$

where Θ_i is the bias of the neuron with number i and f is a non-linear, non-decreasing, differentiable activation function with $f(-\infty) = 0$, $f(0) = 1/2$, and $f(\infty) = 1$.

Similarly, the activation values $\underline{z}_i$ and output values z_i, i = 1, ... , l, of the neurons of third layer are computed from the output values of the second layer:

$$\underline{z}_i := \sum_{j=1}^{m} b_{ij}y_j , \qquad i = 1, 2, ... , l, \tag{3}$$

$$z_i := f(\underline{z}_i - \Phi_i) , \qquad i = 1, 2, ... , l. \tag{4}$$

In the learning phase, an input vector **x** is presented to the network and an output vector **z** is computed first as in the working phase. This output vector is compared with a target output vector **t**, and if different, an error vector $\delta\mathbf{z}$ is computed as follows:

$$\delta z_i := f'(\underline{z}_i - \Phi_i)(t_i - z_i) , \qquad i = 1, ... , l. \tag{5}$$

To decrease the error, the weights a_{ij} and b_{ij} are corrected as follows:

$$b_{ij} := b_{ij} + q\,\delta z_i y_j , \qquad i = 1, ... , l, \quad j = 1, ... , m, \tag{6}$$

where q is a constant called the learning rate. To correct the weights a_{ij}, an error vector for the second layer is computed first by backpropagating the output error $\delta\mathbf{z}$

$$\delta y_i := f'(\underline{y}_i - \Theta_i) \sum_{j=1}^{l} b_{ji}\,\delta z_j , \qquad i = 1, ... , m \tag{7}$$

and then a_{ij} are computed as follows:

$$a_{ij} := a_{ij} + q\,\delta y_i x_j , \qquad i = 1, ... , n; \quad j = 1, ... , m. \tag{8}$$

3. SYSTOLIC SIMULATION ALGORITHM

First, we consider the feedforward propagation in the working phase (1-4). The main computational effort is due to the multiplications (1) and (3) of the matrices $\mathbf{A} = [a_{ij}]$, i = 1, ... , m, j = 1, ... , n, and $\mathbf{B} = [b_{ij}]$, i = 1, ... , l, j = 1, ... , m, with the vectors $\mathbf{x} = (x_1, x_2, ... , x_n)$ and $\mathbf{y} = (y_1, y_2, ... , y_m)$, respectively. For these operations, we propose to use the systolic algorithms shown in Figure 2 and Figure 3, respectively. For simplicity, only a relatively small example is considered here (n = 4, m = 3, l = 3). We use a cellular automata representation of systolic algorithms in which the cells are interconnected via delay elements (denoted by small black boxes) to avoid broadcasting. The cell function is specified by a procedure which is executed once in a clock period. The data input is described graphically by representing the time period at which a data unit is input by the distance of this data unit to the respective input of the array. More details about this form of representation of parallel systolic algorithms and many examples can be found in [2].

In the first algorithm, the components of the input vector **x** come from the left, the matrix **A** is input row-wise from below, and the components of the vector **y** are accumulated in the respective cells (see pp.76 in [2] for further details). After a cell completes all accumulations which belong to the first algorithm, it evaluates the function f for the accumulated value thus implementing (2). The biases Θ_j, i = 1, ... , m, can be taken into account either by loading them into the respective cells before the accumulation begins (as shown in Figure 2) or by considering them as a zeroth column of the matrix **A** and providing a zeroth component x_0 = 1 of the input vector **x**.

In the second algorithm, the matrix **B** is input column-wise and the components z_i, i = 1, ... , l, of the output vector **z** are computed while moving in the pipe of cells. The biases Φ_i, i = 1, ... , l, are input from the left as starting values for this process (see pp.75 in [2] for detailed description). When bleeding z_i, i = 1, ... , l, from the right array end, the host evaluates the function f for the outcoming values thus implementing (4).

The learning phase starts with computing an output vector **z** for an input vector **x** just as in the working phase. While bleeding z_i and computing z_i, the host computes also the components of the error vector δz according to (5). Next, δz is fed into the array to correct the weight matrix **B** and to compute the error vector δy for the hidden layer according to (6) and (7), respectively. These computations can be implemented by an algorithm whose data flow is similar to the one shown in Figure 3. It differs in feeding δz_i instead of Φ_i and in the cell function to be used. The latter is shown in Figure 4. Last, the components of the weight matrix **A** are corrected according to (8) using an algorithm which differs from the algorithm shown in Figure 2 only by the cell function used, Figure 5.

4. CONCLUDING REMARKS

A high efficiency is achieved by the algorithms presented, because at each stage the data is output in a form matching the input requirements of the next stage. Generalizations for any (odd) number of layers are straightforward. There is no communication bottleneck between the array and the host, since data is transferred at the same rate at which it is processed. The algorithms can be directly mapped onto a pipeline of processors and we have implemented them on a ring of transputers (The pipeline is closed to a ring by the host which communicates to both array ends). The matrices **A** and **B** are not fed into the array by the host but reside row-wise and column-wise, respectively, in the nodes of the target transputer array. Usually, the number of neurons m in the hidden layer will be much greater than the number of processing nodes p in the target system. The algorithms are readily modified for this case: instead single components of **y**, **A** and **B**, segments of m/p components of **y** and the respective columns of **A** and rows of **B** should be used.

There is one particular point of practical importance. The cell functions used in the different algorithms will generally require different execution times. This may give rise to an efficiency loss when an algorithm with slow cell function (large granularity) follows an algorithm with fast cell function (small granularity). This problem has been solved by a further modification of the algorithms: instead of components of the vectors **x** and **z** and the matrices **A** and **B**, segments of **x** and **z** and blocks of **A** and **B** should be used. The granularity of the algorithms can thus be changed by changing the respective segment and block sizes.

REFERENCES:

[1] D. E. Rumelhart, J. L. McClelland, and the PDP Res. Group: *"Parallel Distributed Processing"*, vol. 1: *"Foundations"* (Cambridge, Mass.: MIT Press, 1988) Chapter 8, pp. 318-362
[2] N. Petkov: *"Systolische Algorithmen und Arrays"* (Berlin: Akademie-Verlag, 1989)

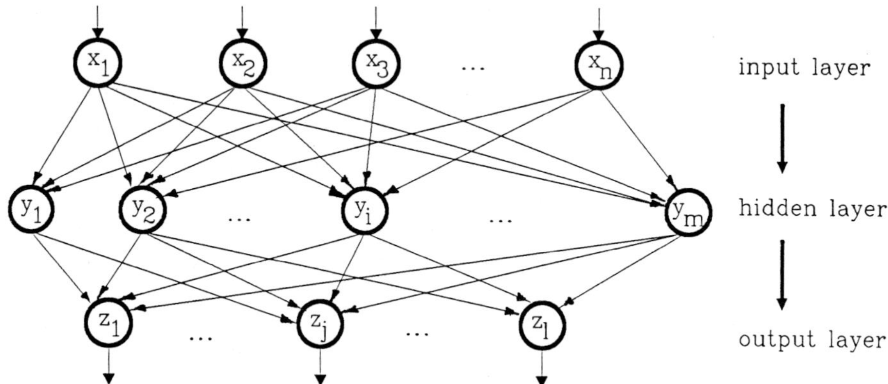

Fig.1 A three—layer, feedforward neural network

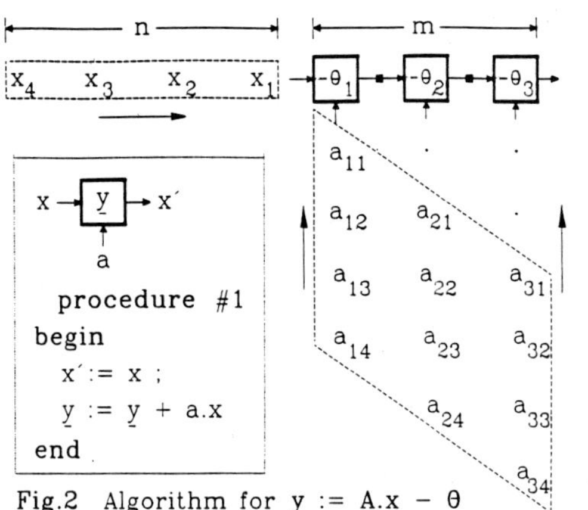

Fig.2   Algorithm for $\underline{y} := A.x - \theta$

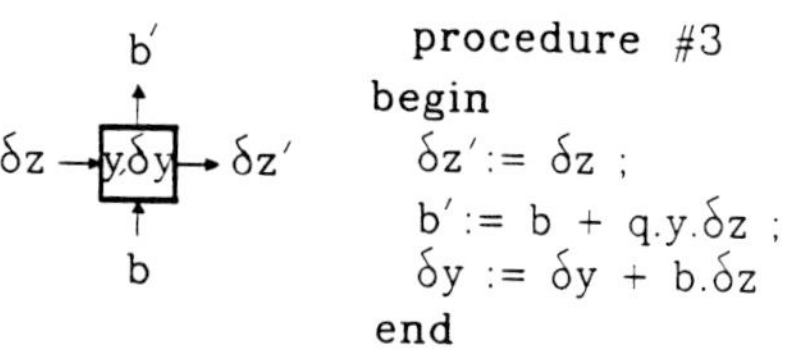

Fig.4 Cell for the correction of B and the computation of δy

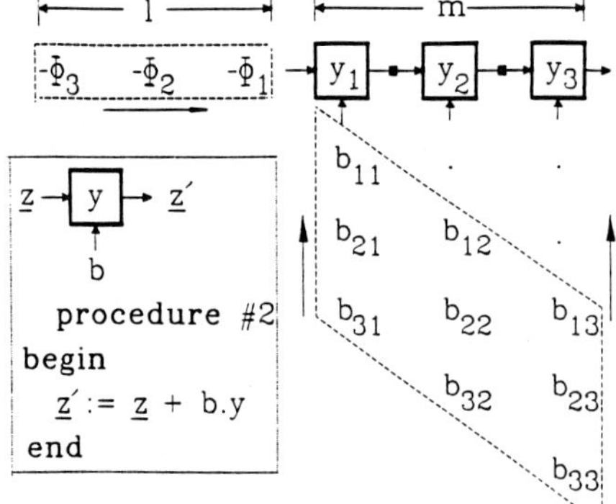

Fig.3 Algorithm for $\underline{z} := B.y - \Phi$

Fig.5 Cell for the correction of A

Parallel Processing in Neural Systems and Computers
R. Eckmiller, G. Hartmann and G. Hauske (Editors)
© Elsevier Science Publishers B.V. (North-Holland), 1990

Artificial Neural Network for Real-Time Task Allocation in Fault-Tolerant, Distributed Processing System

Peter W. Protzel*

Institute for Computer Applications in Science and Engineering
Mail Stop 132C, NASA Langley Research Center
Hampton, Virginia 23665, USA (e-mail: protzel@icase.edu)

Abstract The failure of a processor in a fault-tolerant, real-time distributed processing system requires a fast and reliable mechanism to reallocate the tasks among the remaining processors. This paper describes how a special type of Artificial Neural Network (ANN) based on the method of Hopfield and Tank can be used to generate such a task allocation by observing certain constraints and by balancing the load of the processors. Since the ANN is a critical component in this application, the fault-tolerance of the ANN itself is investigated by injecting simulated faults and observing the performance degradation. It is shown that the fault-tolerance exhibited by the ANN is fundamentally different from conventional systems in that faults act like additional constraints of the problem, but do not impair the convergence to a valid solution.

1. Introduction

Finding the best allocation of tasks among the processing elements is a common problem in distributed systems, which can be usually formulated as a constrained optimization problem [1]. Conventional solution methods are often computationally too expensive when a real-time system requires a very fast solution. This paper concentrates on a special type of distributed system that operates under hard real-time constraints and has to meet very high reliability requirements. An example of such a system is the Software-Implemented Fault-Tolerance (SIFT) computer used by NASA as an experimental vehicle for fault-tolerant systems research [3]. The SIFT architecture can accommodate up to eight processors in a fully distributed configuration with a point-to-point communication link between every pair of processors. The system achieves an extreme fault-tolerance by its capability of detecting and isolating possible hardware faults, which requires a reconfiguration of the system and a reallocation of the tasks among the remaining processors. Thus, it is not the initial task allocation, but the reallocation of tasks after a processor failure, that is time-critical and has to be performed by a highly reliable mechanism. The use of look-up tables has the disadvantage that the number of combinations of tasks and processors is very large for even moderately sized systems [1].

In this paper, we investigate the application of a certain type of Artificial Neural Network (ANN) to generate new task-allocations after a processor failure. The ANN is based on the model of Hopfield and Tank (H&T, 1985 [2]) who first demonstrated the use of a simple nonlinear system, based on analog electronics, to obtain approximate solutions to certain optimization problems. In previous work, we focused on the fault-tolerance and performance characteristics of these 'optimization networks', and argued that they are especially suited to solve special purpose, real-time control problems because of their potential benefits of high speed, fault-tolerance, and low weight and power consumption when implemented in hardware [4, 5]. The task allocation considered here is such a real-time problem and the next section describes the details of the

*   This research was supported by the National Aeronautics and Space Administration under NASA Contract No. NAS1–18605 while the author was in residence at ICASE.

system and the mapping of the problem onto the ANN. An example of the operation of the network and performance characteristics including the fault-tolerance of the ANN are presented in section three.

2. ANN for Task Allocation and Load Balancing

The exemplary distributed system considered here resembles a simplified version of the SIFT computer and is based on a model described in [1], in which a conventional heuristic algorithm is used to solve the problem. In section three, we use this algorithm as a benchmark to assess the ANN-performance. The system consists of m identical processors and executes n tasks. Each task is replicated into r clones that are executed by different processors and submitted to a majority voter in order to detect and to mask possible hardware failures. Assuming periodic real-time tasks for a typical flight control system, the number of instructions per execution of task j, the frequency of execution and the execution rate of the processor determine the load that a certain task places on a processor, which is called the utilization u_j of task j. A particular allocation can be described by a variable V_{ij} with $V_{ij}=1$ if task j is scheduled on processor i and $V_{ij}=0$ otherwise. Then $p_i = \sum_j u_j V_{ij}$ represents the utilization of processor i under the allocation V_{ij}. The task allocation has to observe the constraint that each task has to be executed by exactly r different processors and should be done in a way that achieves an optimal load balancing among the processors. According to [1], minimizing the sum of all p_i^2 also minimizes the statistical variance of the p_i's, which is a measure of the imbalance of the processor utilizations. A load balancing is desired for this application because an imbalance potentially decreases the reliability of the system for a number of reasons discussed in [1]. We further assume that there are enough processors to accommodate a (balanced) assignment without capacity or scheduling violations.

A detailed description of Hopfield and Tank's original approach of using an ANN to obtain solutions for optimization problems can be found in [2] and will not be explained here because of space limitations, but we will give all the problem-specific information necessary to recreate our results. The task allocation problem is represented by a network consisting of a two dimensional array of m×n 'neurons' or elements, in which the output V_{ij} of an element is bounded between 0 and 1 and corresponds to the 'hypothesis' that task j is assigned to processor i. In order to map the task allocation problem onto the network, it has to be expressed as a function whose minima correspond to (local) solutions of the problem. The quantity

$$E^\star = a \sum_{j=1}^{n} \left(\sum_{i=1}^{m} V_{ij} - r \right)^2 + b \sum_{i=1}^{m} \sum_{j=1}^{n} V_{ij} \left(1 - V_{ij} \right) + c \sum_{i=1}^{m} \left(\sum_{j=1}^{n} u_j V_{ij} \right)^2 \tag{1}$$

is such a function where the first term has a minimum if the constraint is met (i.e. each task is executed by exactly r processors), the second term forces the outputs to converge to either 0 or 1, and the third term represents the cost-function to be minimized. Mapping (1) onto the energy function of the ANN yields the values for the interconnections and the external current required to simulate the time evolution of the network [2]. We used Hopfield and Tank's original transfer function and gain, and the results described in the next section are based on the parameter values a=150, b=10, and c=700.

3. Operation, Performance, and Fault-Tolerance of the ANN

Figure 1 illustrates the operation of an ANN by showing the 'neuron' output after initialization and convergence with the size of the squares corresponding to the output value. Initial values are $V_{ij}=0.5+\delta$ with small, uniform noise $-10^{-7} \leq \delta \leq 10^{-7}$. The task utilizations are randomly generated from a uniform distribution between 0.01 and 0.1. Since the ANN should be used as a fast mechanism for producing a new task allocation in case of a processor failure, the network has to be provided with the information of which processor has failed and has to implement this information as an additional constraint. The unavailability of a processor k can be represented by enforcing $V_{kj}=0$ for all j, either by external currents of sufficient strength or by switches

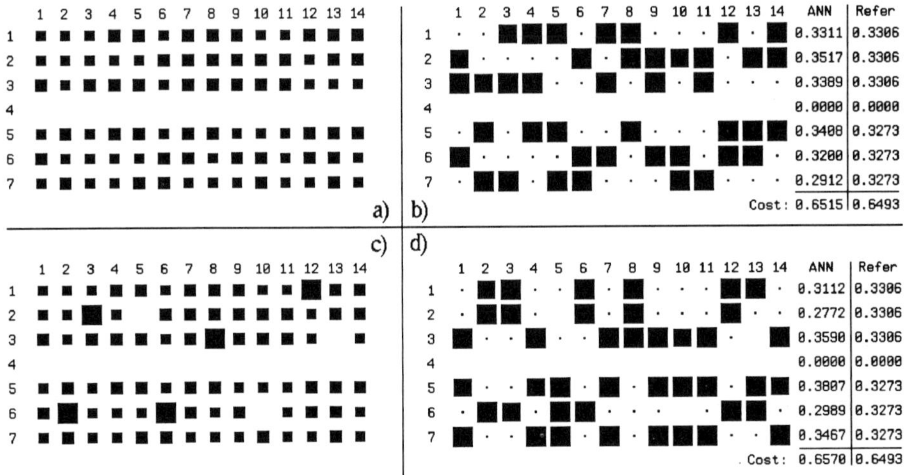

Figure 1. Illustration of the operation and convergence of a network generating a task allocation after the failure of processor #4 (m=7, n=14, r=3): a) initialization of the network (no faults), b) solution after convergence with resulting processor utilizations in comparison to the reference algorithm, c) initialization (five stuck-at-1 and three stuck-at-0 faults injected into the network), d) solution after convergence under the presence of the injected faults.

connecting the output to ground potential (Figure 1a,c). The latter method has the advantage that a possible "stuck-at-1" hardware fault of an element, in which the output is permanently pulled to the highest potential, is 'overwritten' by the external switch. This is important because the ANN becomes a critical component of the system, and its performance in the presence of hardware faults within the ANN itself is of some concern. The most likely failure modes are "stuck-at-1" and "stuck-at-0" faults of the operational amplifiers that represent the 'neurons' in a hardware implementation. In order to study this behavior, we subjected the (simulated) ANN to a varying number of simulated stuck-at-1 and stuck-at-0 faults in randomly selected locations. This is illustrated in Figure 1c where the fault locations are clearly recognizable after the initialization of the network. Interestingly, we could confirm our previous results [5] that the faults do not impair the convergence but act as additional constraints of the problem. This becomes also evident by the equivalence of stuck-at-0 faults and the way the unavailability of a processor is enforced. The outputs of the 'faulty' ANN after convergence are shown in Figure 1d, and it can be seen how the new configuration takes the faults into account with a solution that is only slightly worse than in the fault-free case.

The performance of the ANN is also illustrated in Figure 1b,d, which lists the processor utilizations resulting from the ANN solution in comparison with a simple, heuristic reference algorithm [1]. As can be seen from the cost values, which are the sum of the squares of the processor utilizations, the ANN is outperformed by the algorithm, although the difference of the cost values is only of the order of one percent. Since the performance of the ANN varies considerably for different random initializations and different input data, it is necessary to consider a sufficient number of instances to allow a statistically relevant assessment of the average performance. We used a comparative performance measure defined in [4], which relates the solution (cost) of the ANN c_{ann} to the cost obtained by the reference algorithm c_{ref} and to the average of cost of a purely random solution c_{ran}. The *solution quality* q is defined as $q=(c_{ran}-c_{ann})/(c_{ran}-c_{ref})$ and has the value q=1 if $c_{ann}=c_{ref}$ and the value q=0 if $c_{ann}=c_{ran}$. Thus, the normalized solution quality can be averaged over different problem instances, is independent of the problem size, and even allows a comparison of the ANN performance for different types of optimization problems [4].

The solution quality q is used in Figure 2 to demonstrate the performance degradation of an ANN with 192 elements in the presence of up to 8 'injected' stuck-at-0 or stuck-at-1 faults.

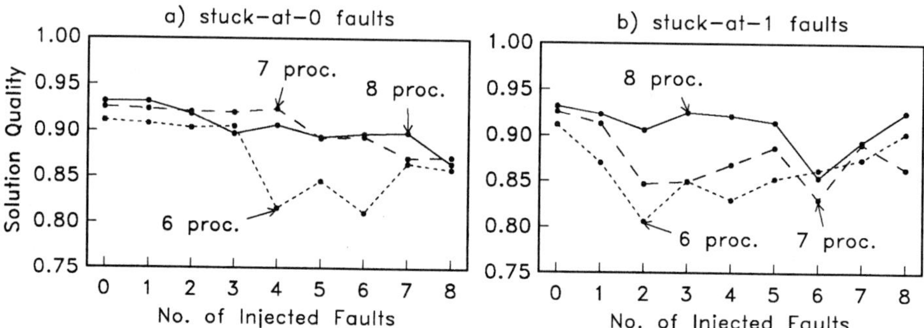

Figure 2. Performance degradation of the ANN allocating n=24 triplicated tasks (r=3) to m=8, 7, and 6 processors. (Each data point represents the average solution quality for 7 different data sets and 7 random initializations for each set.)

The number of processors refers to the remaining number of available processors in the system. For example, if the distributed system consists initially of 8 processors, then m=7 refers to the operation of the network after a failure of one processor with the corresponding row of the ANN output switched to zero (cf. Figure 1). The results in Figure 2 confirm the qualitative observation in Figure 1 that the ANN exhibits an extreme fault-tolerance compared to conventional systems. Since the faults are randomly located and act as additional constraints of the problem, it is possible that one or more faults accidently 'dictate' a better solution than the network would have found without faults. This explains the occasional *performance increase* after fault-injection and the nonmonotonic characteristic of the performance degradation, which is, of course, only possible because of the 'suboptimal' performance of the ANN in the fault-free case. It is also important to note that none of the simulations converged to an invalid solution or to a solution that violates the capacity constraint $p_i<1$, although the latter was not explicitly enforced.

In summary, we think that, for this type of ANN, there exist applications like the one described in this paper that can take advantage of the speed and fault-tolerance of future hardware implementations, although, in most cases, the actual performance of the ANN does not reach the performance of the best available, conventional optimization algorithm [4]. Unlike in conventional systems, where fault-tolerance is a calculated design goal and can only be achieved by using redundant resources, this type of ANN exhibits an *inherent* fault-tolerance that results directly from the functional characteristics of the nonlinear dynamical system.

References

[1] Bannister, J. A., and Trivedi, K. S. Task allocation in fault-tolerant distributed systems. In *Hard Real-Time Systems (Tutorial)*, J. A. Stankovic and K. Ramamritham, Eds. IEEE Computer Society Press, 1988, pp. 256–272.

[2] Hopfield, J. J., and Tank, D. W. "Neural" computation of decisions in optimization problems. *Biological Cybernetics 52* (1985), 141–152.

[3] Palumbo, D. L., and Butler, R. W. A performance evaluation of the software-implemented fault-tolerance computer. *J. Guidance 9*, 2 (March-April 1986), 175–180.

[4] Protzel, P. W. Comparative performance measure for neural networks solving optimization problems. In *Proceedings of the International Joint Conference on Neural Networks IJCNN-90, Washington, D.C.* (January 1990).

[5] Protzel, P. W., and Arras, M. K. Fault-tolerance of optimization networks: Treating faults as additional constraints. In *Proceedings of the International Joint Conference on Neural Networks IJCNN-90, Washington, D.C.* (January 1990).

Parallel Processing in Neural Systems and Computers
R. Eckmiller, G. Hartmann and G. Hauske (Editors)
© Elsevier Science Publishers B.V. (North-Holland), 1990

311

Distributed Processing Hardware for Realization of Artificial Neural Networks

Peter Richert[*], Gunther Hess[*], Bedrich Hosticka[**], Martin Kesper[**], and Markus Schwarz[*]

[*] Fraunhofer Institute of Microelectronic Circuits and Systems, Finkenstraße 61, D-4100 Duisburg, Federal Republic of Germany

[**] Fachgebiet Elektronische Bauelemente und Schaltungen, University of Duisburg, P.O. Box 101629, D-4100 Duisburg, Federal Republic of Germany

In this contribution we present an advanced concept of neural hardware that realizes two important features: 1. each neuron possesses membrane potential, learnable synaptic weigths and delays, threshold, and transfer function, all variable, and 2. each neuron possesses its own communication hardware to realize global communications through fault-tolerant interconnection multiplexing.

1. INTRODUCTION

In recent years a considerable effort has developed to build VLSI hardware for artificial neural networks [1]. Up to now most of the proposed and realized networks are based on very simple models of biological neuron, such as Mc Culloch-Pitts model [2]. There have been some attempts to combine the pulse coding scheme of biological neurons with VLSI technology [3, 4, 5] but most of them stress not biological relevance but technical advantages like small size, low power dissipation and ease of design.

In the work described in this contribution we also employ pulse coding for hardware neurons. Our approach, however, uses asynchronous information processing and hence it differs fundamentally from standard self-organizing mechanisms used in todays neural networks. Learning rules can be embedded into neural networks by special use of "learning neurons" thus needing no external analytical learning.

Our aim is to integrate this new type of neurons into VLSI hardware thus being able to build networks containing large number of neurons. This is not just problem of fabricating neurons and synapses, but also problem of connecting the chips together and realizing communications. We prefer to separate the switching communication task from the neuron processing but integrate both together on single chip. As we will show our solution enables not only connection and communication between adjacent chips, but global communication within the neural network is also possible. This has been achieved by introducing a communication processor, called "switch box", which enables separate processing of communication signals.

2. CONCEPT DESCRIPTION

The block diagram of the proposed neural hardware is shown in Fig. 1. It is apparent that the actual neural processor is strictly separated from the communication hardware, i.e. switch box, as mentioned above.

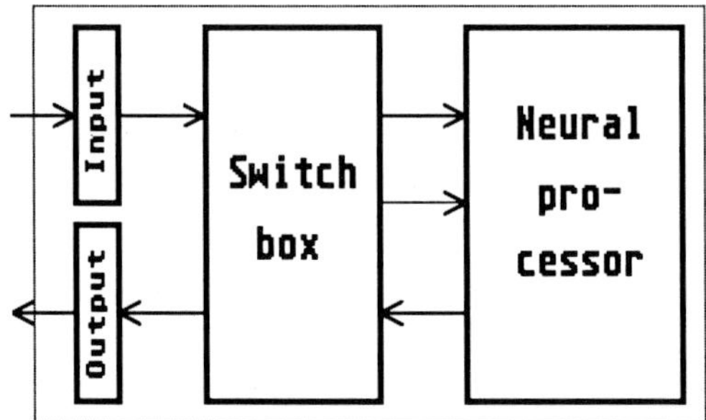

Fig. 1: Block diagram of the proposed neural hardware

All received and transmitted messages must pass through the switch box. Each neuron has its own global address to ensure reception of personal messages. At the same time the switch box can pass along messages from other neurons without processing and thus increase routing capability. Further, the switch box can generate addresses for messages to be transmitted by the neuron. In case a personal message has been received, the incoming control and data signals are sent to the neural processor. The control signals contain synapse or neuron parameters and the data contains the neural information. The message format consisting of the start header, neuron address, control, and data is depicted in Fig. 2.

Fig. 2: Message format

3. SWITCH BOX

The block diagram of the switch box is illustrated in Fig. 3. The switch box checks addresses of all incoming messages and compares them against the addresses of the receiving neuron and its neighbours stored in the content-addressable memory (CAM).

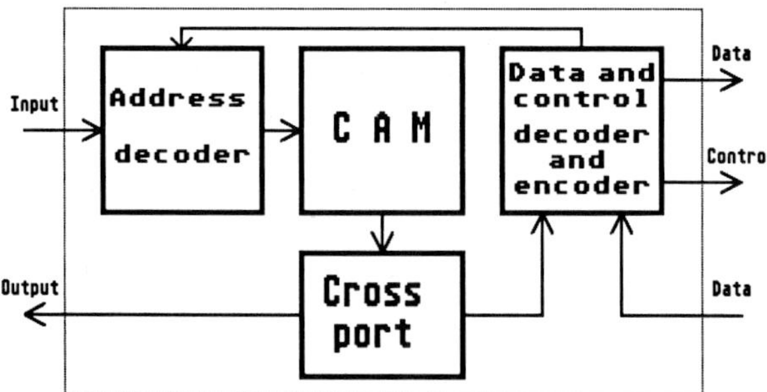

Fig. 3: Block diagram of the switch box

Personal messages, i.e. incoming messages containing the address of the receiving neuron, are sent via the cross port to the neural processor. Other messages are simply passed along to other neurons, again through the cross port, till they have reached the final destination, i.e. the destination neuron. This bypassing enables the neurons to communicate globally though they are connected physically only with their nearest neighbours, either in rectangular or hexagonal arrays. The direction of passing messages, i.e. north, south, east, west, etc., is determined by the address hits scored in the CAM.

Finally, the switch box can also generate addresses of neurons to which the messages including data from the transmitting neuron should be sent to. These addresses define the neuron clusters, to which a single neuron broadcasts identical messages, thus effectively reducing broadcasting transmission requirements. By changing these addresses we can effectively redefine the topology of the neural network.

4. NEURAL PROCESSOR

The block diagram of the neural processor can be found in Fig. 4. The data are actually only binary "ones" and "zeros" thus imitating neuro-biological pulse trains.

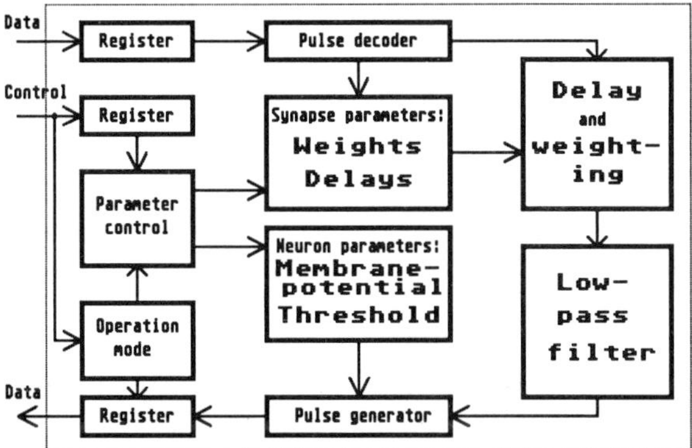

Fig. 4: Block diagram of the neural processor

The neural information is contained in the pulse period, which means that the pulse frequency and phase carry the information. The pulse frequency (or firing rate) lies in the range from less than 1 Hz up to a maximum of 500 Hz, i.e. much lower than the system clock rate, which is in excess of 10 MHz for a realization in CMOS technology. The propagation and machine delays are thus much shorter than the minimum pulse period and do not affect the neural processing. As the messages run at a very high clock rate their time constants are much smaller than the time constants of the neurons and thus the neurons appear to operate in asynchronous mode. This allows processing of messages upon arrival, a correspondence to data-flow concepts.

The synapses process received excitatory and inhibitory data signals which are individually weighted and delayed. All weights and delays are programmed under control of learning algorithms that can be implemented either externally or even internally, i.e. in the neural processor itself. In case external learning is employed, the learned parameters can be adapted externally, i.e. under control of an external host computer, and then be sent to each neuron individually. If internal learning is used, i.e. the neuron realizes internally the delta learning rule, the neural processor must be set into the internal learning mode. Its output can be then used for weight or delay adaption of synapses of other neurons.

The neuron parameters are membrane potential, transfer function, threshold below which the neuron ceases firing, and operational mode, that means external or internal learning. There is no special mode for associative recall (i.e. operation with fixed parameters) because this can be simply achieved by disabling all learning functions of the neuron.

5. CONCLUSIONS

One of the main issues in the development of neural processing hardware is realization of high connectivity. Today planar monolithic integration technologies allow only implementation of mainly two-dimensional metal interconnection layers. Besides, the chip pins are located only at the perimeter thus effectively limiting input / output capabilities.

The concept presented in this contribution places high emphasis on solving communication problems. At the same time it allows implementation of neurons that are close to their biological originals, including variable synapses, synaptical delays, neuron membrane potential, transfer function, and neuron threshold. Learning can be external or internal. The interconnection problem was solved by introducing an elaborate communication system to allow passing messages through the neural network. This measure reduces interconnection hardware and pin-count in realization on monolithic chips because the interconnection can be easily time-multiplexed. The technique also allows selfcontrolled bypassing of defect neurons if desired and thus increases fault tolerance.

ACKNOWLEDGEMENTS

This work was supported in part by the Ministry of Science and Research of North-Rhine-Westphalia under research grant awarded jointly to Prof. B. Hosticka, University of Duisburg, and Prof. R. Eckmiller, University of Düsseldorf.

REFERENCES

[1] C. Mead, "Analog VLSI and Neural Systems", Adison-Wesley, Reading, 1989.

[2] W.S. Mc Culloch and W.H. Pitts, "A Logical Calculus of the Ideas Immanent in Neuron Activity", Bull. Math. Biophys., Vol. 5, pp. 115 - 133.

[3] N.E. Cotter, K. Smith, M. Gasper, "A Pulse Width Modulation Design Approach and Programmable Logic for Artificial Neural Networks", in: Proc. of 5th MIT Conf. Advanced Research in VLSI, MIT, Cambridge, pp. 1 - 15, 1988.

[4] R. Eckmiller, "Electronic Simulation of the Vertebrate Retina", IEEE Tran. BME-22, pp. 305 - 311, 1975.

[5] A. Hamilton, A.F. Murray, L. Tarassenko, "Programmable Analog Pulse-Firing Neural Networks", in: D.S. Touretzky, ed., Advances in Neural Inf. Proc. Systems I, Morgan Kaufmann, San Mateo, pp. 671 - 677, 1989.

Parallel Processing in Neural Systems and Computers
R. Eckmiller, G. Hartmann and G. Hauske (Editors)
Elsevier Science Publishers B.V. (North-Holland), 1990

MULTIPROCESSOR SIMULATION OF A SELF-ORGANIZING NEURAL NETWORK ON NERV

Rainer STOTZKA, Reiner HAUSER and Reinhard MÄNNER

Physics Institute, University of Heidelberg
Philosophenweg 12, D-6900 Heidelberg,West Germany
stotzka@dhdphy5.bitnet

This paper describes the neural network multiprocessor simulation system NERV. It is based on the VME bus which has been extended by a broadcast and a global max finder feature. In addition, the parallel simulation of a self-organizing multi-layer neural network is discussed which develops Mexican Hat and orientation-sensitive cells.

1. INTRODUCTION

Current neural network theories are not yet developed sufficiently to allow to select a certain model and to choose appropriate parameters for a specific application. The development of neural networks for practical applications as well as basic research in this field therefore require computer simulation. Due to the high number of parameters, the simulation of neural networks is computationally extremely demanding [2]. Many computations are therefore done on parallel computers of the MIMD [9, 10] or SIMD type [11]. This paper describes a very flexible simulation system, the NERV multiprocessor [2, 7], an integrated hardware and software system which provides extreme computing power. We implemented a self-organizing neural network on a small prototype to test its suitability. Below, we describe its hardware and software structure. We discuss the neural network model which has been simulated and the way it has been implemented as a parallel program. Simulation results are presented.

2. THE NERV SYSTEM

The NERV system consists of a multiprocessor that is controlled by a host computer. The multiprocessor is assembled of a high number ($\leq$320) of processors operating in parallel (see Fig. 2). They are controlled by a master processor which also provides the link to the host. All processors of the NERV system, i.e. the host, the master, and the processing elements, use the 32 bit microprocessor 68020.

The multiprocessor is based on the VME bus standard. However, a few special bus lines are used for the implementation of broadcast transfers and the global max finder logic. Broadcasts are required to transfer changes in neuron states to all processors. The max finder supports the simulation of asynchronous models which use global criteria for the selection of a single neuron for updating. Selection and broadcast transfers can both be done in a single bus cycle. It has been shown previously [2] that by using these bus extensions, the communication time between the processing elements (PEs) is less than the calculation time, so that hundreds of processors can be used on a single bus.

A single VME crate may contain up to 20 multiprocessor boards and a single master processor. Each multiprocessor board consists of up to 16 PEs. PEs only provide the most basic features in order to get the maximum packing density. One PE consists of a CPU, 1 MByte of static memory, and some random logic implemented in PALs. All local memories are mapped into the address space of the master processor and can be accessed directly. The CPU of a PE, however, can communicate with other PEs and the master processor only via broadcast transfers. The PEs are used to execute all computationally demanding tasks, e.g. parallel learning and network dynamic algorithms.

The master processor, a commercial VME board, is responsible for synchronizing the parallel processors, for the communication between the host computer and the parallel processors, for buffering often used algorithms and data, and for preprocessing and formatting of data during I/O.

Currently, the host computer is a Macintosh II which offers a friendly graphical user interface. It is used as a program development system for the multiprocessor, as well as for preparing and presenting

This work has been supported by the Interdisziplinäres Zentrum für Wissenschaftliches Rechnen Heidelberg input and output data during the simulation by means of commercially available software. To provide a standardized environment, a similar user interface under UNIX and X-Windows is under development.

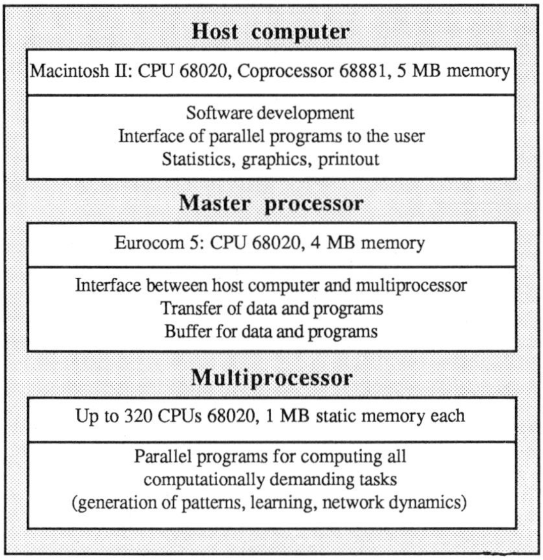

Like the NERV hardware, the simulation software [7] is also structured into three parts that are executed on the host computer, the master processor, and the multiprocessor (Fig. 1). The implementation language Modula-2 has been chosen, since it supports modular programming and provides strong type checking across module boundaries. From the users point of view, the distributed processor architecture is transparent to a very high degree. An interactive simulation control is implemented in which the user can easily set up a neural network, adjust its parameters, and get first results very fast.

Host computer
Macintosh II: CPU 68020, Coprocessor 68881, 5 MB memory
Software development Interface of parallel programs to the user Statistics, graphics, printout

Master processor
Eurocom 5: CPU 68020, 4 MB memory
Interface between host computer and multiprocessor Transfer of data and programs Buffer for data and programs

Multiprocessor
Up to 320 CPUs 68020, 1 MB static memory each
Parallel programs for computing all computationally demanding tasks (generation of patterns, learning, network dynamics)

FIGURE 1

Hardware and software structure of the NERV system

3. THE SELF-ORGANIZING NEURAL NETWORK

As a first test implementation of a neural network, we chose a multi-layer network that has been shown to have self-organizing features comparable to the mammalian visual system. In the layers of the retina and the visual cortex Hubel and Wiesel [4] found many different specialized cells which are capable to perform e.g. Laplace filtering for edge detection (ON- or OFF-center neurons) or show orientation sensitiveness (half- and stripe-cells). Linsker [5, 6] described a feedforward neural network with similar features. His model is also able to develop feature detecting cells only in response to the presentation of random noise input.

The Linsker network consists of several two-dimensional layers. The neurons in each layer are placed on a regular grid. The first layer is the input layer and represents the receptor cells. There exist only feedforward synaptic connections between adjacent layers and no lateral connections within one layer. Each cell of a layer has N synaptic bonds, which are randomly distributed according to a density distribution, e.g. Gaussian, within a neighborhood (with radius r) of the opposite cell in the previous layer. The size of this receptive field grows from layer to layer. This can be described by r_k/r_{k-1}, where k denotes the actual layer and k-1 the layer before. The dynamics of the network can be formulated as follows: the inputs of the cells in layer 1 are random noise signals. At each time step, new uncorrelated signals are presented. A neuron of layer k calculates his activity according to a linear update rule. Then the synapses c_i, i=1,...,N, change their weights according to a Hebb-type learning rule. The initial values of the synapses are randomly chosen. Synaptic bonds changed by Hebbian learning rules normally increase or decrease without limit. We therefore limit the synaptic values to -0.5 and 0.5.

The self-organization of the synapses in the receptive fields of these cells can be described as a minimization of an energy function:

$$ E = -k_1 \sum_{j=1}^{N} c_j - \frac{1}{2} \sum_{i=1}^{N} \sum_{j=1}^{N} \left(Q_{ij}^{(k-1)} + k_2 \right) c_j c_i \quad , \quad c_i \in [-\tfrac{1}{2}, \tfrac{1}{2}] $$

$Q_{ij}^{(k-1)}$ is the covariance of the activity of neurons i and j in the previous layer and depends only on the distance between these neurons. If in layer 1 uncorrelated signals are chosen, $Q_{ij}^{(1)}=0$ if i≠j. Normally the covariance function has a Gaussian shape in layer 2 and a Mexican Hat shape in layer 3 whose parameters depend on r_k/r_{k-1} and on the constants k_1 and k_2. The direct simulation of the maturation of feature detecting cells in a Linsker network is computationally very demanding. Instead, we minimize the energy function to calculate stable cells. We apply the Hopfield model, since our energy function has the same structure as the energy function used by Hopfield [3]. The synapses of one neuron of the Linsker model can therefore be represented by binary Hopfield neurons [8] with values [-0.5,0.5]. For the model described we analyzed the effects of different parameter settings on the feature detecting properties of the cells obtained by the self-organizing process.

4. IMPLEMENTATION

To minimize the energy function of the Linsker model a Hopfield model with a random iteration [1] was implemented on a NERV prototype system with 3 PEs. A copy of the states of the N Hopfield neurons - which are equivalent to the synapses of the Linsker model - is stored in the local memory of every processor. It also stores a copy of the network dynamic program. This program is executed on all processors without any synchronization. Each PE selects a neuron at random and computes its local field

$$h(i) = k_1 + \sum_{j=1}^{N} \left(Q_{ij}^{(k-1)} + k_2 \right) c_i \; .$$

The values of the covariance function are taken from a look-up table, so that the N^2 synaptic weights of the Hopfield model don't have to be stored. A threshold function is used to determine the new neuron state as -0.5 or 0.5. If the neuron state has been changed, it is updated via broadcast in the local memory of all other processors. Then another neuron is chosen until convergence of this process (Fig. 2).

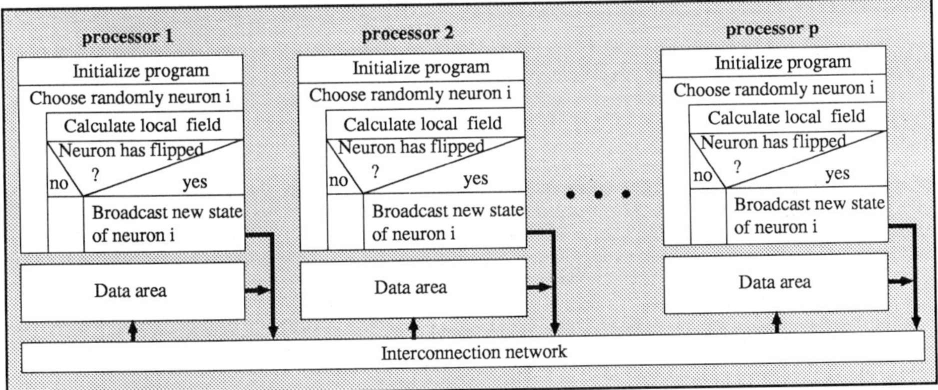

FIGURE 2
The parallel implementation of the energy minimization using the Hopfield model.

The parallel implementation of the simulation described has some important features. The first one is its efficiency. The computation of the local field of a neuron requires N integer additions and multiplications. Only then it is decided whether a neuron update via broadcast is required. The communication time is therefore negligible, so that a high number of processors could be used concurrently. Secondly, all processors operate without any synchronization independently of each other and at maximum speed. If processors are added, the system performance increases linearly without any changes to hardware or software. Accordingly, processor failures result in a graceful degradation of the system performance only. Even in cases where wrong data are broadcasted, the system would still converge to the correct result. Moreover, the program can be developed easily since it is a sequential program and the parallel structure of the NERV hardware does not have to be taken into account.

5. RESULTS

A network with 1000 Linsker synapses, i.e. 1000 Hopfield neurons, was simulated on the NERV system. The synapses were Gaussian distributed in the receptive field. A Mexican Hat form was taken as the covariance function, which is equivalent to simulating the maturation of a cell in layer 4 of the Linsker model. Small fluctuations in density lead to asymmetric cell types, as described in [8]. Random values were chosen as the initial states of the synapses. During 10000 iteration steps feature detecting cells emerged (Fig. 3). The total computation time was ≈5 min. A simulation run on a single processor system showed that 30000 iteration steps are needed until the cell is stable.

It is important to note that the number of neurons which can be simulated by the NERV system depends essentially on the size of the available memory. Usually, the memory requirements are determined by the size of the synaptic matrix that has to be stored. In the application described above, however, the synapses can be obtained on-line and very fast via table-lookup. Thus, most of the main memory is available for storage of neuron states and coordinates. In the current development status it is possible to minimize the Linsker energy function for 20000 neurons. A simulation run with >8000 neurons has been tested successfully. It required ≈10 hours on a 3 processor machine.

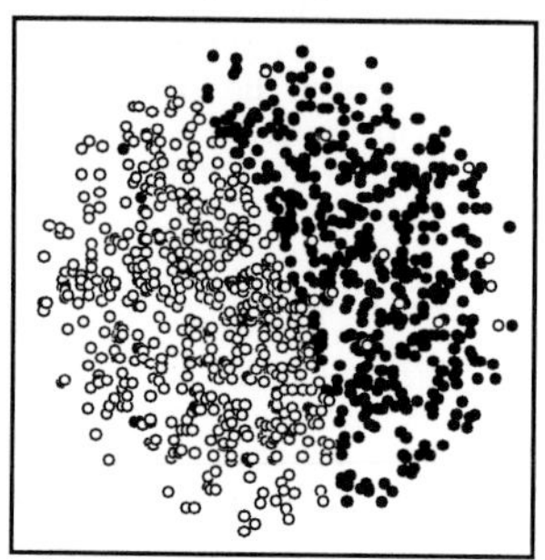 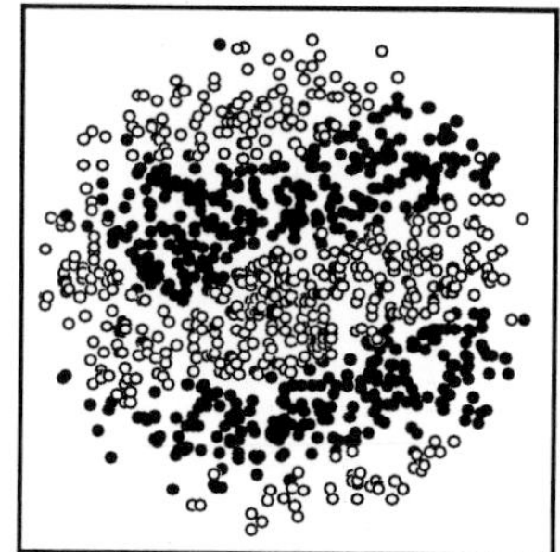

FIGURE 3.a   FIGURE 3.b

Maturated receptive fields of layer 4. The synapses in the receptive fields are Gaussian distributed. They developed to -0.5 (bright circles) or 0.5 (dark circles). {k_1=0, k_2=0, 3.a: r_4/r_3=1.5, 3.b: r_4/r_3=4}

6. CONCLUSIONS

A multiprocessor system for the simulation of neural networks, NERV, has been developed. Whereas the system is able to use up to 320 processors and a maximum of 320 MBytes of main memory. The development of system and application software has been done up to now with a small prototype only. As a realistic test application we chose the simulation of a multi-layered feedforward network. From this applications we can draw the following conclusions. The basic software of the NERV system, i.e., the operating system, the multiprocessor-host communication system, the user interface, and the library containing standard application modules allowed a very fast implementation of the test application. We were able to develop the application software and to obtain the results presented above within one day. We could verify that the computing power increased linearly with the number of processors. We also verified that the system operates fault tolerant.

The next steps are 1) to increase the number of available processors to 64 to provide a higher computing power, 2) to enhance to user interface, e.g. to provide a batch processing feature for production runs, 3) to install the system under a UNIX/X-Windows environment to provide a standard environment.

REFERENCES

[1]   Fogelman-Soulie F., Weisbuch G., Random Iterations of Threshold Networks and Associative Memory, SIAM J. of Computing 16, 1, Feb. 1987, 203-219

[2]   Hauser R., Horner H., Makhaniok M., Männer R., Architectural Considerations for NERV - a General Purpose Neural Network Simulation System, Proc. WOPPLOT 89, Wildbad Kreuth, Germany, 1989, in print

[3]   Hopfield J. J., Neural Networks and Physical Systems with Emergent Collective Computational Abilities, Proc. Nat'l Acad. Sci. USA 79, April 1982, 2554-2558

[4]   Hubel D. H., Eye, Brain and Vision, The Scientific American Lib., New York, 1988

[5]   Linsker R., From Basic Network Principles to Neural Architecture, Proc. Nat'l Acad. Sci. USA 83, Oct.-Nov. 1986, 7508-7512, 8390-8394, 8779-8783.

[6]   Linsker R., Self-Organization in a Perceptual Network, IEEE Computer 21, 3, 1988, 105-117

[7]   Männer R., Horner H., Hauser R., Genthner A., Multiprocessor Simulation of Neural Networks with NERV, Proc. Supercomputing ´89, Reno, NV, 1989, 457-465

[8]   Stotzka R., Männer R., Self-Organization of a Multilayered Feedforward Neural Network, Proc. Int´l Joint Conf. on Neural Networks, Washington, DC, 1990, in print

[9]   Forrest B.M., Roweth D., Stroud N., Wallace D.J., Wilson G.V., Implementing Neural Network Models on Parallel Computers, Comp. J. 30, 5, 1987, 413-419

[10] Kuczewski R.M., Myers M.H., Crawford W.J., Neurocomputer Workstations and Processors: Approaches and Application, Proc. IEEE 1st Int´l Conf. on Neural Networks III, 1987, 487-426

[11] Hillis D.W., The Connection Machine, The MIT Press, Cambridge, MA, 1985

Parallel Processing in Neural Systems and Computers
R. Eckmiller, G. Hartmann and G. Hauske (Editors)
© Elsevier Science Publishers B.V. (North-Holland), 1990

IC^3 : A NEURAL ASIC FOR REAL–TIME PROTOTYPING

Frank Warkowski, Lambert Spaanenburg, and Jos A. G. Nijhuis

Institute for Microelectronics Stuttgart
Allmandring 30a
D–7000 Stuttgart 80, West Germany

The architecture and application of a master–slice concept for rapid prototyping of neural networks is presented. After simulation with the NNSIM neural network simulation environment a netlist specifies the interconnection on a master of neural building blocks, which can be personalized to various architectures with metalization, while maintaining a range of options for mask– and software programmability.

1. INTRODUCTION

The current revival of interest in neural networks brought some new application ideas, but efficiency in realization still falls short of its biological counterpart. With the advent of submicron silicon technologies the gap may be closed, which leaves the problem of implementing such systems. The paper addresses this problem by showing a smooth migration path from the neural simulation model to the actual silicon.

Analog realizations of neural networks are nothing else than a large, nonlinear analog computer. Application of analog computers stagnated because of low integration level, accuracy and stability problems (long term and over temperature) as well as programmability and testing problems. The benefit however is the small amount of silicon elements required to build neurons and synapses, such that even with present–day technology non–trivial single–chip neural networks can be created.

In contrast digital realizations tend to be larger but by the lack of accuracy and stability problems also a higher scale of integration can be used, while programmability and testability are seemingly trivial problems. Though current digital realizations are comparatively small, they lend themselves well for size increase as a result of decreased technological detail or increased level of integration.

The concept proposed here starts from a simulation environment [Nijh89], that supplies a netlist and a pattern file. Next to that it assumes a Sea–Of–Gates array (the Gate Forest [Beu88]) to allow for short turn–around on fabrication. A library of neural functional cells allows the netlist to be mapped on the master–slice using conventional CAD software. The pattern file serves for hardwired programming. The master–slice topology is chosen such that a further mapping into a full–custom design (eventually on wafer–scale) is feasible.

The electronic concept of this IC^3 (IMS Collective Computation Circuit) has been influenced by ideas in literature about time–coded information representation in neural network realizations [Mur89], [Hir89], [Tom89]. Drawbacks of these ideas (f. e. bus of multiple, synchronous clocks) are avoided here. A novel means to enhance the testability eliminates the occurrence of long scanchains and therefore simplifies chip test.

2. THE NEURAL NETWORK STRUCTURE

The neural network prototype development starts with the use of the NNSIM simulation environment. This allows the neural system to be analyzed within its application. The level of abstraction covers the range from conceptual to electrical/physical, but especially where sensory peripherals are used a more realistic testing–ground is still required. Such a testing–ground will be PC–based, see figure 1. From the NNSIM netlist silicon prototypes need to be generated, that will be controlled by the PC to provide a real–time test. Later on the same NNSIM netlist is to be used to generate the final full–custom silicon neural network.

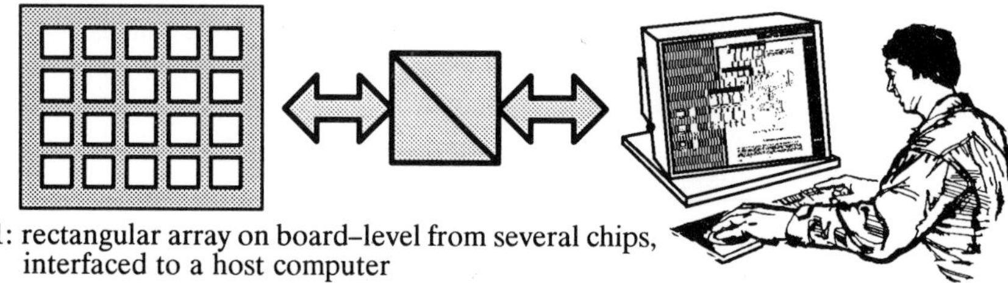

Figure 1: rectangular array on board–level from several chips, interfaced to a host computer

Within this concept a master–slice chip is required with off–chip learning, offering still a large degree of mask– and software programmability. This master–slice is based on the digital neural model shown in figure 2 and uses pulse logic. The synapses can be set in the neutral, normal, transparent and test mode. Sign control is applied to differentiate between excitatory and inhibitory dendrites. All functional cells are based on 4–bit units, which can be tapped to smaller widths or concatenated to larger widths.

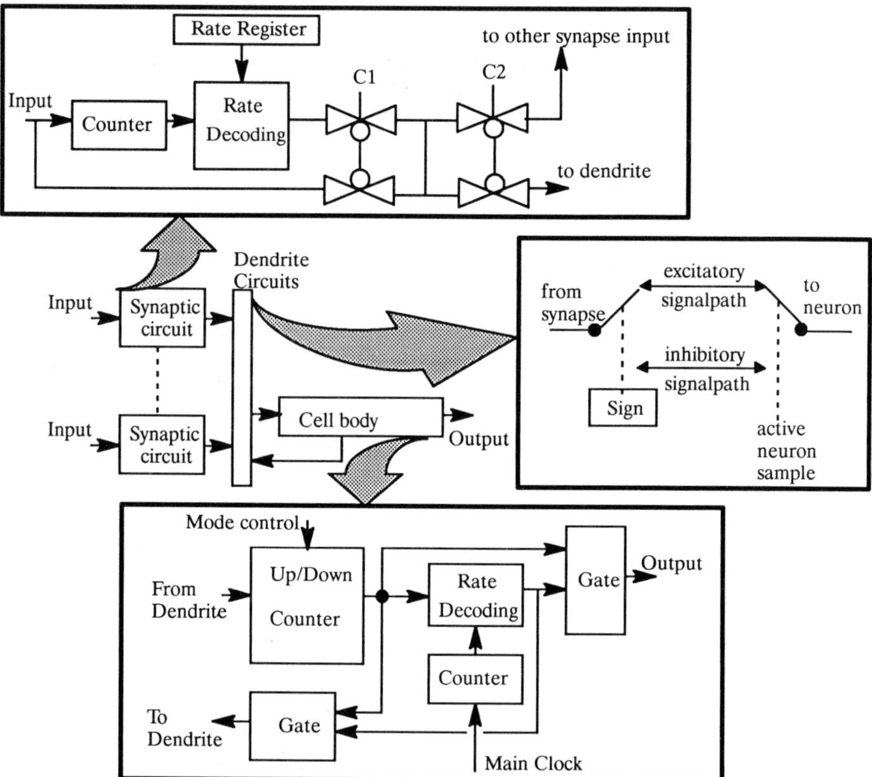

Figure 2:   The single neuron and its components.

3. NETWORK TOPOLOGY

The array shape on system level is not fixed, therefore on chip level the shape should be such, that circular, rectangular and hexagonal array shapes are possible with a minimum of unused neurons at the system border. A problem for the universality of a neuron master slice is the difference in granularity between the neural network and the Sea–Of–Gates target. The Gate Forest master is orthogonal and has different granularity in horizontal and vertical direction. To achieve a greater flexibility, neuron circuits are split into synaptic, dendritic and soma parts as building blocks. They are connected both by explicit metal wiring as by software controllable switches.

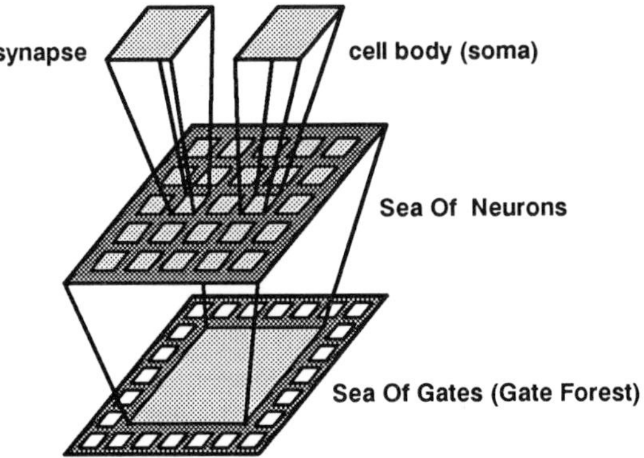

Figure 3: Flexibility is achieved in a two step mapping process: The neural network consisting of connected synaptic, dendritic and soma circuits is mapped on a Sea Of Neurons, which itself is implemented on a Sea Of Gates array

Definition of connectivity is done in two hierarchical steps (Figure 3). Number and distribution of the building blocks is fixed by definition and fabrication of the master. Interconnection of these blocks is defined during personalization of the prefabricated masters with metal1 and metal2. In operation, switches in the building blocks allow software controlled fine tuning of the interconnection topology. A number of connections will be accommodated by setting a synapse circuit into the transparent mode. As the Gate Forest template supports all types of integrated digital logic, some area will not be usable for synapses and neurons. This area is of course minimized in constructing the library of functional cells, but some area loss will be unavoidable. This area is used for explicit wiring.

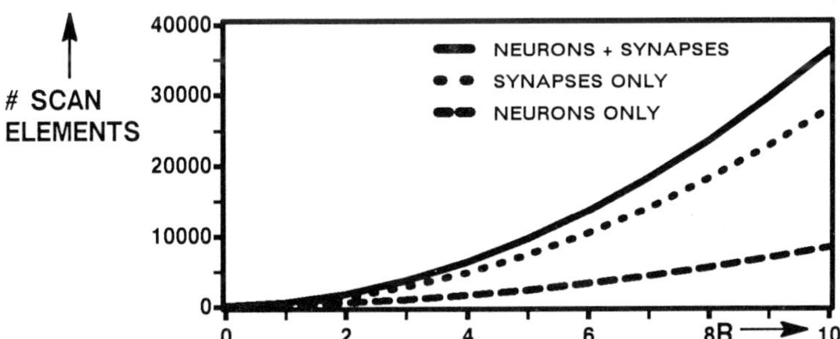

Figure 4: Scanpath length of a hexagonal neural network with ring count R.

A last point of interest is testability. A neural network with redundant capacity can be trained to be fault–tolerant; however for any fabricated chip the presence of functionality must still be ascertained [War89]. Using conventional design–for–testability techniques does not bring a solution due to the large amount of memory locations. As shows from figure 4 the main contribution stems from the synapses. By setting the synapses in the built–in test mode an asynchronous scanpath is created that allows for a single–cycle flush of the entire scanchain. This leaves only the neurons for the synchronous scanpath, reducing actual testtime to 20%.

4. DISCUSSION AND APPLICATION

A chip architecture is presented for rapid prototyping of digital neural networks. This architecture and the principle of master personalization has been chosen to allow a high degree of flexibility in realization of neural networks which have been developed and specified with the NNSIM simulation environment [Nijh89]. The hardware implementation with IC's is necessary either where a neural network is part of a real–time signal processing system or in small systems where a microprogrammed neural network part is far beyond the limits for size or power of the entire system. The table below shows the properties of the chip proposed.

	IC^3		IC^3
Status	chip proposed	Neuron Timing	synchronous
Connectivity	full on–chip, certain radius only off–chip	Synapse States	digital, n x 4 bit
Programmability	electrical, digital	Neuron States	digital, n x 4 bit
Learning	off–chip/host	Neuron Number	19/chip
Parallelism	serial per neuron, all neurons in parallel	Synapse Number	19/neuron or 19x19/chip
Circuit Technology	2μm CMOS, IMS Gate Forest semi custom gate array	Interconnects	parallel, digital 1 bit pulse coded
Input Timing	asynchronous		

ACKNOWLEDGEMENTS

This work has been supported by Daimler Benz AG and the Bundesministerium für Forschung und Technologie under contract "TV 8926 3". Stimulating discussions with Prof. Eckmiller (Uni Düsseldorf), Prof. Höfflinger (IMS) and Prof. Waldschmidt (Uni Frankfurt) are gratefully acknowledged.

REFERENCES

[Beu88] Beunder, M., Kernhof, J. and Hoefflinger, B., "The CMOS GATE FOREST: An Efficient and Flexible High Performance ASIC Design Environment", IEEE Journal of Solid State Circuits 23, pp.387–399

[Hir89] Hirai, Y., Kamada, K., Yamada, M. and Ooyama, M., "A Digital Neuro–Chip with Unlimited Connectability for Large Scale Neural Networks", Proc. of the IJCNN 1989 (Washington D.C., June 1989) pp.II.163 — II.169

[Mur89] Murray, A., Hamilton, A. and Tarassenko, L., "Fully Programmable Analogue VLSI Devices for the Implementation of Neural Networks", Proc. of the Journées d'Électronique (Lausanne, October 1989) pp.265–278

[Nijh89] Nijhuis, J.A.G., Spaanenburg, L. and Warkowski, F., "Structure and Application of NNSIM: a General–Purpose Neural Network SIMulator", Microprocessing and Microprogramming 27, pp.189–194

[Tom89] Tomberg, J., Ritoniemi, T., Kaski, K. and Tenhunen, H., "Fully Digital neural Network Implementation Based on Pulse Density Modulation", Proc. CICC'89 (San Diego, May 1989) pp.12.7.1 – 12.7.4

[War89] Warkowski, F., Leenstra, J., Nijhuis, J.A.G. and Spaanenburg, L., "Issues in the test of artificial neural networks", Proc. ICCD'89 (Boston, October 1989) pp.487 – 490

Parallel Processing in Neural Systems and Computers
R. Eckmiller, G. Hartmann and G. Hauske (Editors)
© Elsevier Science Publishers B.V. (North-Holland), 1990

NEW CONCEPTS FOR INFORMATION PROCESSING IN CONNECTIONISTIC SYSTEMS

T. Waschulzik*, H. Geiger, M. Arnoldi, D. Böller, A. Nischwitz, W. Brauer°

Kratzer Automatisierung GmbH   °Institut für Informatik
Unterschleißheim/München   der Technischen Universität München

1. INTRODUCTION

A connectionistic model is presented in which frequency–coded McCulloch Pits neurons are used; learning is achieved by an enhanced Hebb rule (Singer–modified). The internal connectivity scheme is similar to Kohonen net's. Then enhancements of this model are presented which serve to successfully deal with tasks of object recognition in the realm of technical automation. For these enhancements the algorithm's strict locality was one of the deciding criteria.

The neuron model is expanded by an adjustment formed by the opposing effect of adaptation and sensitization. Through this adaptation the neurons are able to dynamically adjust to changes of the variable boundary conditions. It will be shown that shortterm learning can be used very advantageously to enhance the network's processing capabilities. Besides the conventional synapses with additive components synapses with multiplicative influences are used. If additionally a structure is superimposed on the distribution of synapses, connectionistic systems which can adapt to the problem at hand without learning result. For the associative storage of mappings a interconnection scheme is presented that independenly searchs for internal representation using given boundary conditions.

Using a network for the problem–oriented forming of detectors as an example, a concept for a universally interactive hierarchical network model with feedback over more than one level is presented.

* Supported in part by the German Ministry of Science (BMFT) under
  contract 413–5839–ITR8800 E3.

2. MODEL

2.1. Basic Model

The basic model is moreless the generally used model. The main contribution are the additional features.

We use processing units in our investigations whose activity ($a_i(t)$) is described by an integer. The neuron's activity depends on the net input $net_i(t)$. For the damping of oscillations the net input of the last step can be taken into account.
The neuron's output $o_i(t)$ is determined with the help of a linear threshold–function (simple approximation of the sigma–function), depending on the momentary net depolarisation.
Two different algorithms are used for learning, the modified Hebb–rule for *auto associations* and forced learning for *hetero associations*.

2.2. The Adjustment

The adjustment ($adj_i(t)$) enables the neuron to change its behavior, depending on its history [1,2]. It is modeled by the two antagonistic processes of adaptation and sensitization. The neuron's sensitization (or adaption) depends on its momentary acitivity and a given norm value for activity and strength of its actual sensitization (or adaptation).

2.3. Short–term Learning

When constructing a neural network for information processing, one often faces the problem of wanting to dynamically and reversibly adapt the interconnection strength to the problem at hand. This is e.g. necessary, when processing a picture with various objects if one wants to strengthen the coupling of neurons belonging to the same object [3,4].
Another application is the modified *winner takes all*–mechanism. Compared to the commonly applied version [5], one gets a considerably quicker converging algorithm if the inhibitory synapses between the competing neurons are modified as follows:

> Increase the inhibitory effect of synapses whose presynaptic activity is greater than the postsynaptic one, and diminish the inhibitory effect of all other synapses.

2.4. Multiplying Synapses and Dendritic–Tree–Structures

Often one not only wants to dynamically change the coupling strength of individual synapses, but also the connecting structures' topology. This is impossible to do with the standard synapses with additive influence on the net–input, this calls for different types of synapses. In order not to lose the powerful capabilities of the hitherto used models its seems feasible to use different types of synapses in one network. One example thereof are the *SIGMA–PI–UNITS* [6].

If one allows besides the simple combination of additive or multiplicative synapses more complex patterns of connectivity, it follows that one must impose a structure on the set of synapses. A synapse is now not only described by the presynaptic element, the coupling strength and the postsynaptic element, but additionally by the type of functional interdependence and a parameter pointing to the synapses' position within the superimposed structure.
A tree structure is used for the patterning of the synapses, thereby guaranteeing the structures' high flexibility. In analogy to biology we call these structures dendritic trees. The computation of the state of activity is achieved via the computation of the depolarisation at every point of the dendritic tree. The roots' depolarisation is defined as the net–input.

At the moment it is investigated, how far problems of the invariant pattern recognition and stereo vision can be solved with this new tool. In contrast to other network models the well–known XOR–problem, for example, can be solved without using hidden layers:

3. NEW CONCEPTS FOR USING STRUCTURED NETWORKS

3.1 Internal Representation

In the following a pattern of connectivity will be explained with which connectionistic systems can learn projections of strongly linearly depented patterns [2]. This makes it possible to self–organize such projections without using the very time–intensive back propagation method [6]. The projection is divided into two substeps. The input patterns are projected onto an internal representation and these are then – as step two – projected onto the output pattern. This requires an associative memory that – starting with one particular pattern – generates new activity patterns that are statistically distributed throughout the whole state space. If a new pattern ist to be learned, it is presented to the system; the memory settles into a definite state.
A special structure of connection now tests whether this internal state is already associated whith an other stored pattern. In this case the internal representation is disturbed in such a way that a radically different state is achieved and the testing cycle is repeated until a unique internal representation is found [7].

3.2 Feedback Over More Than One Hierarchical Layer

One often encounters the problem of wanting to influence information processing on a distinct lower level with the information gathered on a higher hierarchical level. So a Network which detects objects from visual patterns has been built, that finds its feature detectors in an optimal way. When the patterns are taught, the network can select the feature detectors in such a way, that only that detectors are created which allow discrimination of the taught patterns. In that network feedback over more than one hierarchical layer is necessary.

The retinotopic feature detectors take their input from a layer with a second order Laplace filter. The set of neurons which learns to react to features on the same retinotopic field are coupled with a winner takes all (wta) mechanism. From these neurons the neurons on the next hierarchical layer gather their input and so on. From the highest level of feature detectors an associative memory with internal representation collects information. If an error appears, i.e. two different patterns cannot be distinguished on the highest feature detector level, an error signal is generated via veto neurons. This signal which has to feed back over more than one hierarchical layer modifies the wta–mechanism in such a way, that exactly these neurons learn, which contain a feature in their receptive field with which the actual pattern can be discriminated from all other previously taught patterns. This selection and all other processes run only with information which is locally available to the neuron and is only influenced by the connectivity pattern and the previously described common algorithms [8].

4. METHODS

We created an extensive comfortable development system (NETUSE) consisting of the three parts NBS, MERGE and NTS. NBS contains the network description language and its compiler. During the implementation of NBS the modularity and the mutual independence of the different language elements was strongly emphasized. This modularity makes it possible to include standard libraries and to incorporate subnet–definitions into a more complex network and enables us to define even complex networks simply, compactly and accurately.

MERGE merges already trained networks. The merging in this universal approach is only possible because of the strictly local algorithms used. The third part, NTS, is used for investigating and manipulating networks. It runs in interactive mode as well as in batch mode. The NETUSE system is momentarily used with the operating systems VMS and SCO/XENIX.

Since definiton of dendritic tree structures is a relatively new concept in neural networks, we give a short description of how this is done in our system.

The position within the tree is described by a character string that defines the path from the root to this node. Synapses are treated as branches of the first degree. A neuron's state of activity is recursively computed by its dendritic tree structure:

Abbreviations:

p   = string defining a position in the dendritic tree
o   = concatenation of strings, ""'' is the empty string
string (i)   = string corresponding to the integer i
$depo_{n,p}$ (t) = depolarisation of the neuron n on position p at time t
$k_{n,p}$ (t)   = coupling strength from synapses of neuron n on position p
                at time t
präs (n,p)   = presynaptic neuron of a synapses from neuron n at position
$v_{n,p}$ = degree of branching of neuron n on position p

The definition of the structure:

net input :

$$net_i (t) = depo_{i,""} (t)$$

branching :

$$depo_{n,p} (t) = \sum_{i = 1}^{v_{n,p}} depo_{n, (p\ o\ string\ (i)\ o\ "\_")} (t)$$

internal muliplicative synapses :

$$depo_{n,p} (t) = depo_{n, (p\ o\ "1\_")} * (k_{n,p} (t) * out_{präs\ (n,p)} (t))$$

internal additive synapses:

$$depo_{n,p} (t) = depo_{n, (p\ o\ "1\_")} + (k_{n,p} (t) * out_{präs\ (n,p)} (t))$$

additive synapses at a leaf :

$$depo_{n,p} (t) = (k_{n,p} (t) * out_{präs\ (n,p)} (t))$$

5. DISCUSSION

It has been shown that using strictly local algorithms and an enhanced neuron model a powerful tool for the development of connectionistic models can be built. This has far–reaching advantages for the practical use of such systems. A connectionistic system can adjust to changing environmental conditions (e.g. in picture analysis: change of illumination or camera properties). Aided by short–term learning neurons can form groups and compete as a group with others. Multiplicative synapses combined with dendrite–tree–structures make it possible for the topology of the connectivity structure to dynamically adjust to specialized tasks. By the use of appropriate connectivities a network chooses an internal representation for the projection even of strictly linear–dependent patterns. Networks with feedback reaching across several hierarchical levels can – dependent on the results of the processing – very efficiently recognize features relevant to the recognition task.

The algorithms applied can easily be transported to massively parallel hardware; trained subnets can be incorporated into new tasks or into the development of more complex systems. As is common practice in other, complex systems, design and test can be made in a strongly modular fashion. It is our opinion that further development of connectionstic systems does not involve new and complex rules of learning, but rather a continued development of connectionistic structures and a permanent and critical test of the neuron model.

REFERENCES

[1] Waschulzik, T., Optische Mustererkennung in neuronalen Architekturen (Dipl.–Arb., TU München, 1988)

[2] Böller, D., Realisierung eines hierarchischen neuronalen Netzwerks zur assoziativen Erkennung von Objekten in natürlicher Umgebung (Dipl.–Arb., TU München, 1988)

[3] Nischwitz, A., Objekterkennung und Kamerasteuerung mit neuronalen Netzwerken (Dipl.Arb., TU München, 1989)

[4] Von der Malsburg, C & Bienenstock, E., Statistical Coding and Short–term Synaptic Plasticity, A Scheme for Knowledge Representation in the Brain, in: Bienenstock, E, Fogelman Soulié & Weisbuch, G. (eds.). Disordered systems and biological organization, NATO ASI Series F20, (Springer, Berlin, 1986)

[5] Großberg, S. & Ellias, S. A., Biological Cybernetics 20 (1975) pp 69–98.

[6] Rummelhart, D. E., Hinton, G. E. & McClelland, J. L., A General Framework for Parallel Distributed Processing, in: Rummelhart, D. E. & McClelland, J. L. (eds.), ParallelDistributed Processing, Vol. 1, (MIT press, 1986).

[7] Ackley, D. H., Hinton, T. J. & Sejnowski, Cognitive Science 9 (1985) pp 147–169.

[8] Arnoldi, M. , Mustererkennung mit Hilfe selbstorganisierender adaptiver Mechanismen (Dipl.–Arb., TU München, 1989).

––––––––––––––––

Thomas Waschulzik, Dr. Hans Geiger, Martin Arnoldi, Dagmar Böller, Alfred Nischwitz, c/o Kratzer Automatisierung GmbH, Maxfeldhof 5–6, D–8044 Unterschleißheim/München, Tel.: (089) 310 20 37, Telex: 521 68 43, Telefax: (089) 317 22 46. Prof. Dr. W. Brauer, c/o Technische Universität München, Institut für Informatik, Arcisstraße 21, D–8000 München 2, Tel.: (089)2105–2401, Telex: tume d 05–22 854, Telefax: (089)2105–8207.

Section 8
Neural Networks
for Visual
Pattern Recognition

Parallel Processing in Neural Systems and Computers
R. Eckmiller, G. Hartmann and G. Hauske (Editors)
© Elsevier Science Publishers B.V. (North-Holland), 1990

RECOGNITION OF NEW SPATIO-TEMPORAL PATTERNS BY ADAPTIVE JUNCTION

Yoshiaki Ajioka†, Yuichiro Anzai‡ and Hideo Aiso‡

†Department of Computer Science
‡Department of Electrical Engineering
Keio University§
Yokohama, JAPAN

Adaptive Junction is a neural network that learns to discriminate signals moving along either a vertical or horizontal direction in a two-dimensional plane. In this paper, we show how new spatio-temporal patterns can be recognized by Adaptive Junction networks that were trained to recognize simple patterns.

1.INTRODUCTION

In previous work[1] described a neural network called Adaptive Junction that learns to discriminate simple spatio-temporal patterns moving along either a vertical or horizontal direction in a two-dimensional plane. However, those results were restricted to the successful discrimination of trained examples rather than novel input patterns. In this paper, we provide new results on how Adaptive Junction, trained with simple patterns, can discriminate new spatio-temporal patterns that were not learned previously.

2.NETWORK

Adaptive Junction networks have three layers: (1) the sensory layer (S-layer), (2) the response layer (R-layer), (3) the teaching layer (T-layer). The S-layer consists of sensor neurons (S-neurons) that accept input patterns. The R-layer consists of response neurons (R-neurons) that are classified into two types, "discrimination output" neurons (O-neurons) and "feature detection" neurons (F-neurons). O-neurons represent discrimination results, and F-neurons represent whether or not input signals are useful for discrimination. Finally, the T-layer consists of teacher neurons (T-neurons) that supply teacher signals used for supervised learning. Those neurons are connected in the networks as depicted in Figure 1. Note that the networks do not have any hidden units. Adaptive Junction's neuron models, the learning rules and the number of R-neurons needed for discrimination are described in [1], [2] and [3].

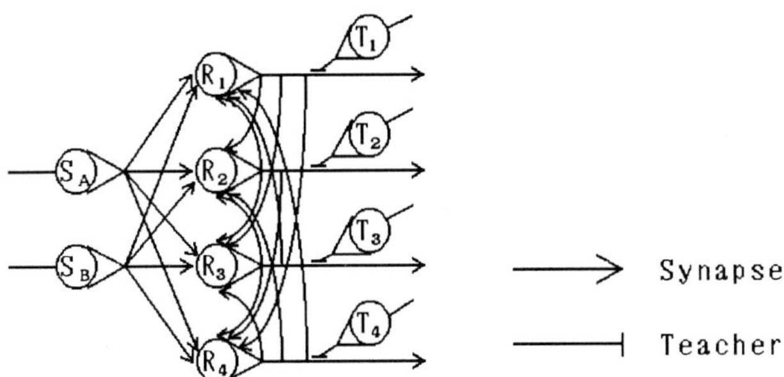

Figure 1
A network of Adaptive Junction.

§3-14-1 Hiyoshi, Kohoku-ku, Yokohama 223 JAPAN

3.DESCRIPTION OF THE PROBLEM

The aim of this paper is to described how Adaptive Junction can flexibly recognized new spatio-temporal patterns that were not learned previously. Below, we describe what is meant by a spatio-temporal pattern.

Suppose that nine S-neurons, named from A to I, are laid out along two-dimensions as shown in Figure 2. A signal is made to appear and to move in some directions by firing only one S-neuron at a time, as shown in Figure 3 (for pattern-ABC). In all, there are twelve possible patterns (e.g., pattern-IHG, pattern-BEH, pattern-GDA). The task of the network is to discriminate the four directions in which the signals moved. Simulation results for this problem are described in [1], [2] and [3].

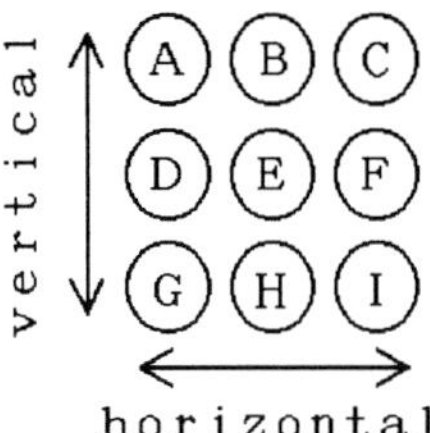

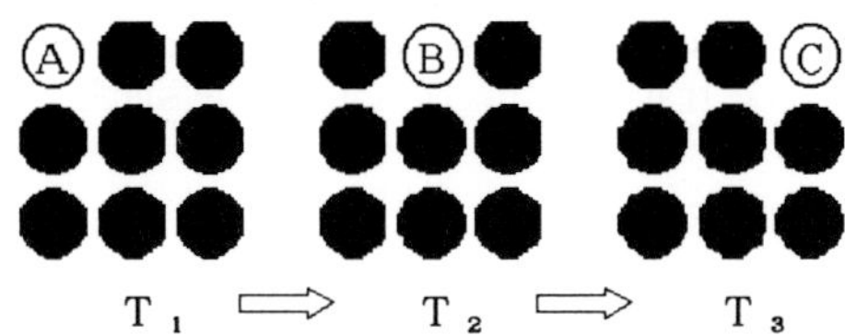

<table>
<tr><td align="center">Figure 2</td><td align="center">Figure 3</td></tr>
<tr><td>S-neurons laid out along
two-dimensions.</td><td>Two-dimensional spatio-temporal pattern-ABC: A white
circle corresponds to an excited neuron, and a black circle
to an inhibited neuron.</td></tr>
</table>

4.RECOGNITION OF NEW PATTERNS

In this section, we describe the simulation results for recognition of two-dimensional spatio-temporal patterns that were not learned previously. Before running the simulations, the network was trained by using the twelve kinds of simple patterns described before.

The input procedure and response of the network are as follows. First, a spatio-temporal pattern is given to the network by presenting each element comprising the pattern in turn. For example, in the spatio-temporal pattern-ABC, first element-A is fed into the network, then element-B, and finally element-C. Note that Adaptive Junction has an additional network that learns the context of each element, that is, what elements were input just before each element (see [2] for details). In response, Adaptive Junction recognizes the direction that is represented by the spatio-temporal pattern. For example, the direction of pattern-ABC is to the right.

Once the network has learned all of the simple spatio-temporal patterns for a direction, we expect that the network should be able to recognize the direction of more complex patterns. That is , the network is taught to recognize pattern-ABC, pattern-DEF, and pattern-GHI, which all represent the movement of a signal to the right. We expect that Adaptive Junction should now be able to recognize that, say, pattern-ABF or pattern-DHF both represent movement to the right, even though the network was not trained on those specific patterns.

The simulation results are shown in Figure 4. Each graph shows the temporal activation of a neuron. The first row of these graphs for the R-layer shows the directions, and the other rows show the detected feature. The columns correspond to up, down, left and right, respectively.

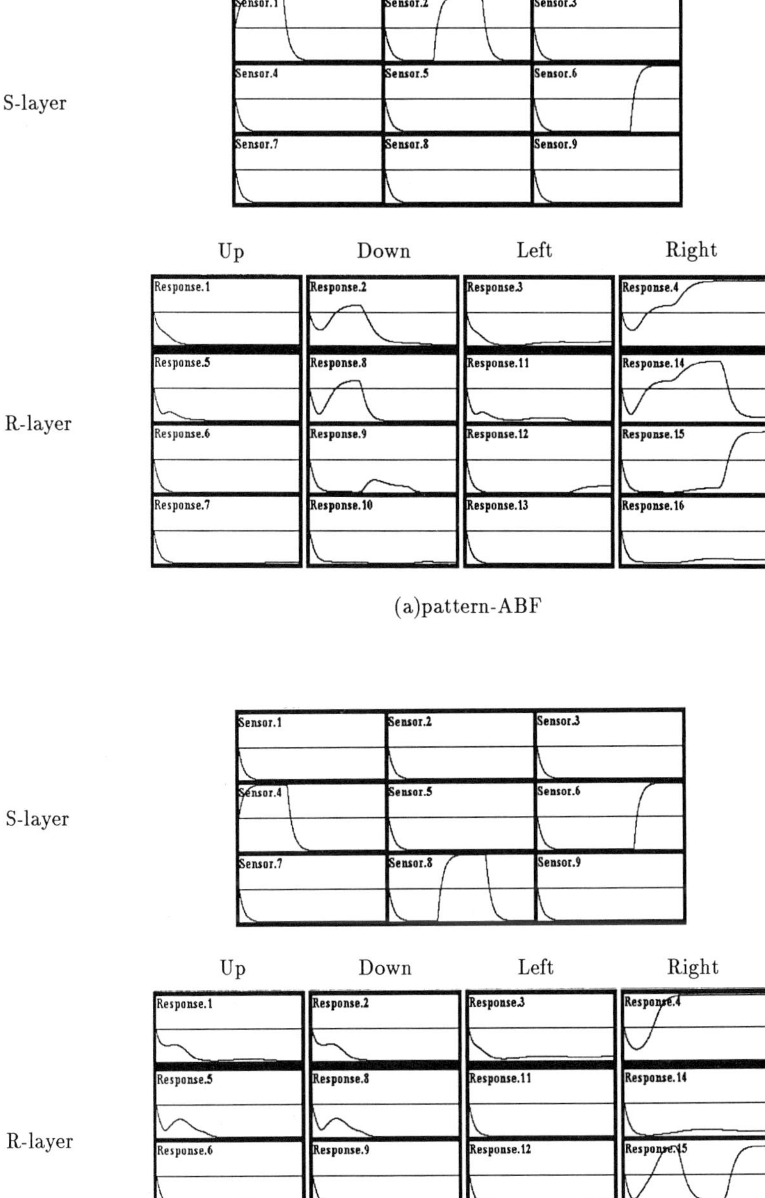

(a)pattern-ABF

(b)pattern-DHF

Figure 4

Results of recognition of new spatio-temporal patterns. (See text for details.)

5.DISCUSSION

The mechanisms of discriminating learning patterns and recognizing new patterns can be summerized as follows:

- Discriminating learning patterns (see [1] and [3])

  1. If Adaptive Junction can decide the directions of spatio-temporal input patterns using the first elements of these patterns, it outputs the directions.
  2. Otherwise, Adaptive Junction outputs all the directions of the spatio-temporal patterns whose first element is equal to that of the present input pattern.
  3. Because the Adaptive Junction network presented here can decide the directions using the second elements, it outputs the correct directions of input patterns.
  4. When the third elements are input to the network, the outputs of this network remain because of cooperation between the O-neuron that indicates the direction of the present input pattern and F-neuron that indicates the present input pattern.

- Recognizing new patterns

  1. When the first elements are input to the network, Adaptive Junction performs in the same fashion as it dose when discriminates patterns.
  2. When the second and third elements, which are different from the training patterns, are input, Adaptive Junction cannot remember the directions as was described above because the F-neurons that were expected to remain activated for the present input patterns cannot be excited by these newly input elements.
  3. However, if there are F-neurons that indicate training patterns whose directions are the same as the present directions, and whose second and third elements are equal to the second and third ones of the input patterns, such F-neurons take the place of F-neurons that cannot remain activated.

6.CONCLUSION

From the above simulation results, even though networks of Adaptive Junction have not enough memory elements such as recurrent networks[4][5][6], Adaptive Junction can recognize spatio-temporal patterns using dynamics of the networks.

REFERENCES

[1] Ajioka, Y., Ishikawa, M., Akamatsu, M. and Anzai, Y., The mechanisms of learning on Adaptive Junction, Symposium on Paradigms of Learning and Their Applications, Proc. of IPS JAPAN Symposium, (JAPAN, Vol. 89, No. 4 1989), pp. 11-20.

[2] Ajioka, Y., Anzai, Y. and Aiso, H., The discrimination of rotating patterns and the local representation on Adaptive Junction, IEICE Technical Report, NC89, (JAPAN, 1989), in press.

[3] Ajioka, Y., Anzai, Y. and Aiso, H., Adaptive Junction: A spatio-temporal neuron model, IJCNN-90 Washington D.C., in press.

[4] Jordan, M.I., Generic Constraints on Underspecified Target Trajectories, Proc. of IJCNN-89, (Vol. 1 1989), pp. 217-225.

[5] Smith, A.W. and Zipser, D., Encoding sequential structure: experience with real-time recurrent learning algorithm, Proc. of IJCNN-89, (Vol. 1 1989), pp. 645-648.

[6] Williams, R.J. and Zipser, D., A learning algorithm for continually running fully recurrent neural networks, ICS Report 8805, UCSD, (California, U.S.A. 1988).

Parallel Processing in Neural Systems and Computers
R. Eckmiller, G. Hartmann and G. Hauske (Editors)
© Elsevier Science Publishers B.V. (North-Holland), 1990

Hierarchical Spin Model for Stereo Interpretation Using Phase Sensitive Detectors[†]

R. Divko and K. Schulten*
Physik-Department
Technische Universität München, D-8046 Garching

The property of physical spin systems to assume long range order is exploited to establish global depth interpretation of real scene stereo pairs. We choose spins to code for local disparity between corresponding pixels in both pictures [1]. The energy of the spin system entails a field contribution, which incorporates the information on the stereo images, and an attractive interaction between alligned neighboring feature spins, which implements the continuity propertiy of disparities. The energy function is chosen in a way that the ground state of the feature spin system – found by simulated annealing [2] – corresponds to the best global depth interpretation of the stereo pair. Disparity detectors, based on relative phase measurements between left and right picture, yield information on the local field.

1. Introduction

Our brain is effortlessly inferring depth from the two slightly different images of the world formed by our eyes. The difference of the position of objects in both images is called disparity and can be used to obtain a depth map of the observed scene. Current computer algorithms solving the stereo correspondence problem are based on two different approaches. Feature based models use specific features as matching primitives, e.g. zero crossings, proposed by Marr and Poggio [3]. Correspondence-less algorithms are not based on matching features but use cross-correlation between parts of each image. For example algorithms proposed by Jenkin [4] and Sanger [5] locally compute disparity from complex phase difference obtained from the Gabor transformed [6] images. Another classification scheme of algorithms evolves from the question, wether the algorithms are local [4,5] or involve cooperation between neighboring features [3]. We showed previously [1], that with cooperative spin algorithms constraints obtained from 'apriori' knowledge on the images, e.g. continuity of depth, can easily be implemented. In the following we propose a hierarchical stochastic algorithm analog to physical spin systems with interaction between neighboring features and a field interaction based on the local measurements of phase detecting filters. The algorithm proposed belongs to a class of regularization methods [7] known as Markov random fields [8].

2. Feature Spins

In physics a spin system is characterized by spin variables s_{ij}, which can assume a certain set of values, and by the energy function of the system. In the two-dimensional Ising model the spin variables take the values ± 1 and the energy of the system is

$$E = -J \sum_{<(i,j),(k,l)>} s_{ij}\, s_{kl} - \sum_{(i,j)} h_{ij}\, s_{ij} \tag{1}$$

where the brackets indicate summation over next neighbors. In the ferromagnetic case ($J > 0$)

[†]This work has been supported by BMFT ITR-8800-G9

*Department of Physics and Beckman Institute, University of Illinois, Urbana, IL, 61801 USA

the first term gives a negative contribution for equal neighboring spins and, therefore, establishes a tendency for an alignment of the spins. The second term describes the interaction of the spins with a local magnetic field h_{ij} tending to align the spins locally with the field.

To use spin systems for pattern recognition, features, e.g. the disparity between corresponding pixels, are coded in spins. The pattern to be processed is incorporated in the field. Global 'apriori' knowledge, like continuity constraints, on the pattern to be interpreted is implemented in the interaction between the feature spins. The energy of the feature spin system is chosen such, that the ground state, i.e. the state of the feature spins with the lowest total energy, corresponds to the best global interpretation of the pattern. It is achieved by the standard method of Monte Carlo annealing [2].

3. Phase Detecting Filters

The external field serves to communicate the pattern to be processed to the system of feature spins. In our model we use the output of filters capable of computing local phase differences between both images of the stereo pair [4, 5]. These filters yield confident disparity information only on sparsely distributed regions of the image.

Disparity detecting filters are based on the property of the Fourier transform, that two functions globaly shifted against each other by the ammount $\delta(x)$ yield a phase factor of $e^{i\omega\,\delta(x)}$, where ω is the frequency. This property can be generalized to the case of a Gabor transform, i.e. a Fourier transform weighted with a Gaussian, if the width of the Gaussian is smaller than the relative shift. If $l(x)$ and $r(x)$ are the left and right image repectively, with a local relative disparity $\delta(x)$ between them, the Gabor transforms $L(x,\omega)$ and $R(x,\omega)$ of both pictures are locally related to each other:

$$R(x,\omega) \approx \exp^{i\omega\delta(x)} L(x,\omega).$$

Therefore $\delta(x) = \frac{1}{\omega}arg(\frac{R(x,\omega)}{L(x,\omega)})$ is a measure for the local disparity. The value of $c = min(|\frac{R}{L}|, |\frac{L}{R}|)$ is a confidence measure. Values not equal to one are indicating a false disparity value.

4. A Hierarchical Algorithm for Stereo Vision

Because the disparity filters yield confident results only if the maximum disparity to detect is smaller than the wavelength of the Gabor filter, a hierarchical process from larger to smaller wavelengths is used. The number of waves within the Gaussian is kept constant. Beginning with a sufficiently large wavelength the resulting disparities together with their confidence values are presented to a spin system to yield global disparity interpretation. At each wavelength the left image is shifted according to the disparity map, obtained by the spin system. This modified image is used to compute a disparity map at a smaller value of the wavelength.

5. The spin system

At each level of the hierarchy the following spin system is used. The disparity spins s_{ij} at lattice sites (i,j) can take integer values ($s_{ij} \in \{0, \pm 1, \ldots\}$). The energy function of the disparity spin system is $E = E_i + E_f$ with

$$E_i = J_i \sum_{<(i,j),(k,l)>} (s_{ij} - s_{kl})^2$$
$$E_f = J_f \sum_{(i,j)} c_{ij}(s_{ij} - f_{ij})^2. \tag{2}$$

The interaction energy E_i between the disparity spins induces an alignment between neighboring spins. The field energy E_f incorporates the disparities f_{ij} from the detectors and their confidences c_{ij}. E_f yields a low energy contribution for spins, well aligned with the field f_{ij}. For high values of the confidences c_{ij}, the field contribution becomes more important.

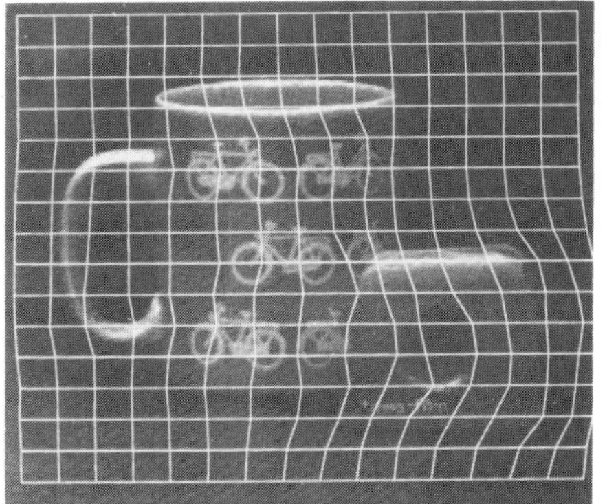

Fig. 1: Cup with tape holder - corresponding pixels in left and right image are indicated by the superimposed grid.

6. Results

The result for a 256×220 stereo image of a cup and a tape holder in front of it is presented in figure 1. The grid marks a subset of corresponding pixels in the left and in the right image as obtained from the algorithm. The energy weights used are $J_e = 0.25$ and $J_f = 1$ and the wavelengths have been lowered from 64 to 4 pixel in steps of a factor of 2. The disparity map coded as an intesity map is shown in Figure 2. The brighter the map the larger the disparity. Maximum disparity in the stereo image is about 23 pixels.

7. Conclusion

The hierarchical spin system proposed yields a disparity map for real stereo images. It can be extended to other features than the intensity of the images. With a combination of several features, e.g. intensities, edges and lines, as input to the disparity sensitive filters the algorithm can yield a better confidence of the disparity interpretation [9].

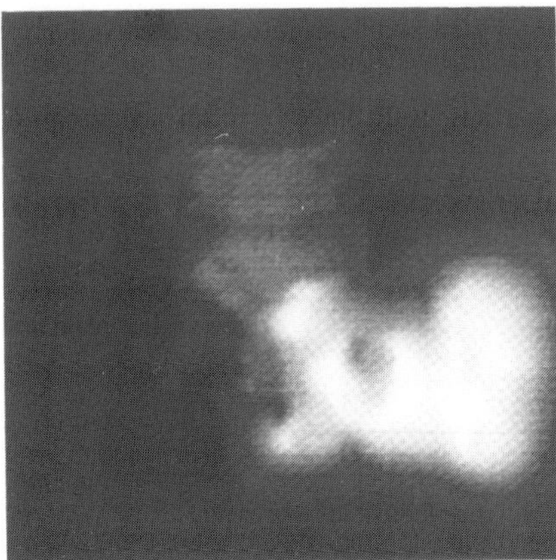

Fig. 2: Density coded disparity map for cup with tape holder.

REFERENCES

[1] Divko R, Schulten K (1986), *Stochastic Spin Models for Pattern Recognition*, in *Neural Networks for Computing*, Ed. Denker JS, 129-134, American Institute of Physics Publication

[2] Kirkpatrick J, Gelatt CD, Vecchi MP (1983), Science, 220:671

[3] Marr D, Poggio T (1979), *A computational theory of human stereo vision*, Proc R Soc Lond B 204:301-328

[4] Jenkin M, Jepson AD, (1988), *The measurement of binocular disparity*, In *Computational processes in human vision: An interdisciplinatory perspective*, Ed. Phylyshin Z, 69-98, Ablex Press, Norwood, NJ

[5] Sanger TD, (1988), *Stereo Disparity Computation Using Gabor Filters*, Biol Cybern 59:405-418

[6] Gabor D (1946) *Theory of communication*, J Inst Elec Eng 93 (III):429-457

[7] Poggio T, Torre V, Koch C (1985), *Computational Vision and Regularization Theory*, Nature 317:314-319

[8] Geman S, Geman S (1984), *Stochastic Relaxation, Gibbs Distribution, and the Bayesian Restoration of Images*, IEEE PAMI-6:721-741

[9] Divko R, Schulten K (1989), in preparation

Parallel Processing in Neural Systems and Computers
R. Eckmiller, G. Hartmann and G. Hauske (Editors)
© Elsevier Science Publishers B.V. (North-Holland), 1990

A SHIFT INVARIANT NETWORK UTILIZING THE GRT

M. Fang and G. Häusler

Department of Optics, Physics Institute
University of Erlangen
D-8520 Erlangen, FRG

A feedback network is described that restores pictures independent of their location in space. The network combines a space invariant transform with binary input and output, and a special coupling matrix. This space invariant transform, the Generalized Rapid Transform (GRT), can be implemented within the network.

1. INTRODUCTION

General associative networks for both binary and continuous valued input vectors that use Hebb's rule have been described and analyzed by Kohonen [1], Nakano [2], Amari [3], and by Hopfield [4]. The feedback network consists of a coupling matrix which carries the information about the learned patterns and a nonlinear mapping function, as shown in fig. 1.

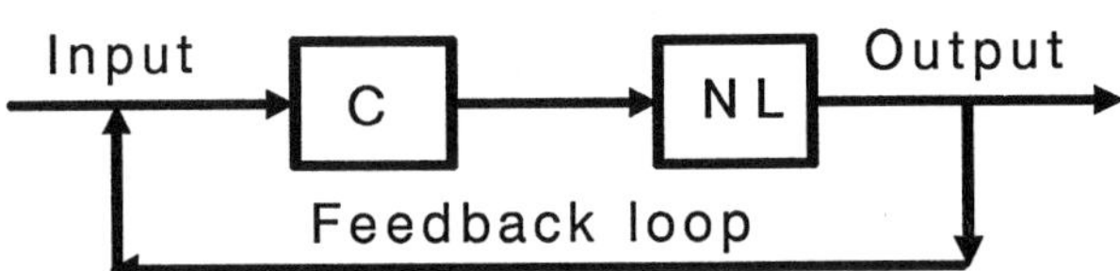

FIGURE 1
Block diagram of the feedback network

The network can be characterized by the following algorithm:

$$V_i(k+1) = NL\ (\ \sum_{j=1}^{N} C_{ij}V_j(k)\ )\tag{1}$$

where $\mathbf{V}$ is the k-th iteration of a bipolar binary vector $(-1,1)$ of length N. NL is a nonlinear mapping function, for example, the signum function. C_{ij} is the element of the coupling matrix with size N x N. The system can learn Q input patterns $\mathbf{V}^q$, $q=1,\cdots,Q$, by the following construction of the coupling matrix:

$$C_{ij} = \sum_{q=1}^{Q} V_i^q V_j^q \qquad \text{for } i \neq j\tag{2a}$$

$$C_{ii} = 0\tag{2b}$$

For the application of this above mentioned model the following features have to be considered:

(1) The model is shift variant. If the input pattern is disturbed
by noise, the system can restore it even with a noise level of
50% [5]. However, even small shifts of a few pixels usually cause
failure. This is due to the fact that recognition is essentially
based on the inner product of the stored vectors and the input
vector [6].

(2) The network can learn and recognize about 0.15N uncorrelated
binary patterns. However, in general, patterns are correlated,
hence costly orthogonalization procedures are necessary.

In the following we want to demonstrate a shift-invariant binary
associative feedback network for binary valued input vectors that
can overcome some of the above mentioned problems.

2. Basic idea

The first problem (shift variance) can be solved by using a shift
invariant transform which can be easily implemented within the
network. The principle is sketched in fig. 2.

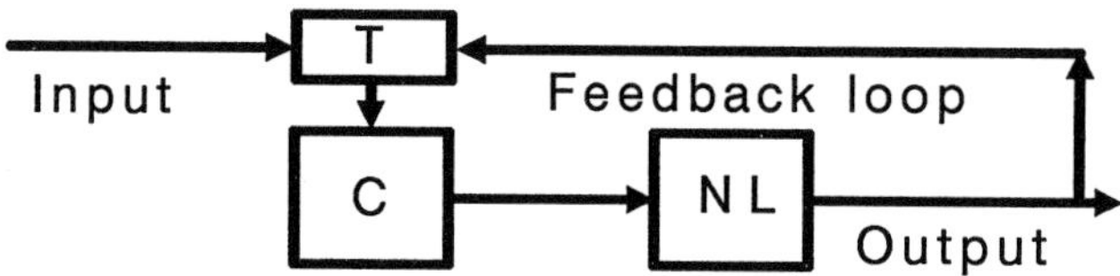

FIGURE 2
Block diagram of a shift invariant feedback network

The series of the operators is: space invariant transform T,
coupling matrix C_{ij} and nonlinearity NL. The coupling matrix is
different from the rule, described in eq.(2). It is now trained
with the original patterns V^q _and_ the corresponding transform
patterns $\tilde{V}^q$:

$$C_{ij} = \sum_{q=1}^{Q} V_i^q \cdot \tilde{V}_j^q \tag{3}$$

The system equation can be written as

$$V_i(k+1) = NL\{ \sum_{j=1}^{N} C_{ij}\tilde{V}_j(k) \} = NL\{ \sum_{q=1}^{Q} [\sum_{j=1}^{N} \tilde{V}_j^q \tilde{V}_j^q(k)] V_i^q \} \tag{4}$$

The [...]-term is the inner product of the transformed learned
vector V^q and the transformed input vector. For uncorrelated
learned vectors this term is large only for that input vector
close to a learned vector. Hence, summation over q and clipping
yields essentially that learned vector which is most similar to
the input vector.

However, most patterns and their transform patterns are correlated. In order to avoid orthogonalization procedures we use the
method of Kinzel [5], which is based on the destruction of
synapses. The coupling matrix can be constructed in this case
with the following rule:

$$C_{ij}= V_i^1 \cdot \tilde{V}_j^1, \text{ and } C_{ij}= 0, \text{ if } C_{ij}V_i^q \cdot \tilde{V}_j^q \lessgtr 0 \text{ for any } q>1 \qquad (5)$$

The invariance of our network depends on the invariance of the transform T. In general, one can achieve all desirable invariant properties, if a suitable transform can be found. For achieving the shift invariance one can use the Generalized Rapid Transform defined by Wagh and Kanetkar [7] that can be easily implemented within the networks. In the following section we briefly review this transform.

3. Generalized Rapid Transform (GRT)

The Generalized Rapid Transform [7] (GRT) is similar to the Rapid Transform (RT) introduced by Reitboeck and Brody [8]. The signal flow graph of both transforms is shown in fig. 3a, left side. Wagh and Kanetkar generalized the rapid transform by using any two commutative operators instead of the two special commutative operators "$a_i + a_j$" and "$|a_i - a_j|$" applied to two pixels a_i and a_j of the input pattern in the RT. Two special commutative operators for binary input and output may be $\max(a_i,a_j)$ and $\min(a_i,a_j)$ as shown in fig. 3a. Such a transform with two commutative operators shows the desired cyclic shift invariance [7]. This class of transforms has a computational complexity of $O(NlogN)$; N is the length of the input vector.

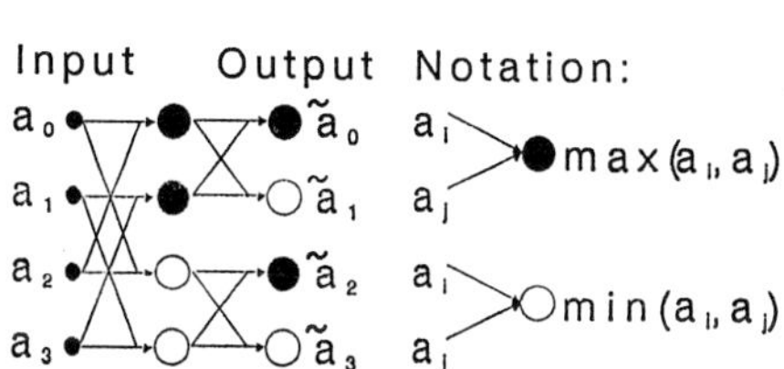

FIGURE 3a	FIGURE 3b
The signal flow diagram of GRT for an input vector with four pixels	A feed forward network with ln(N) layers for the implementation of the GRT.

For our purpose, the GRT shows one more interesting property. It can be easily implemented by a feed forward network. This is illustrated in fig. 3b. The weights of the interconnections are all set to 1. The thresholds of the nonlinear operators F1 and F2 have to be set to small negative (0-d) and positive (0+d) values as shown in fig. 3b.

4. Numerical Experiments

In this section we investigate our network numerically. As an input pattern we choose the characters A, B and C. The binary patterns with two values of "+1" and "-1" are represented with

bright and dark areas, respectively. For matrix training we used
eq. (5). We perturbed these three patterns in the following way:

- At first patterns were shifted upwards by two pixels.
- a "cross" pattern was subtracted from the shifted patterns, as
  an example of a deterministic perturbation.
- Then random noise was added.

Our model can restore the perturbed patterns within a few
iterations as shown in fig. 4.

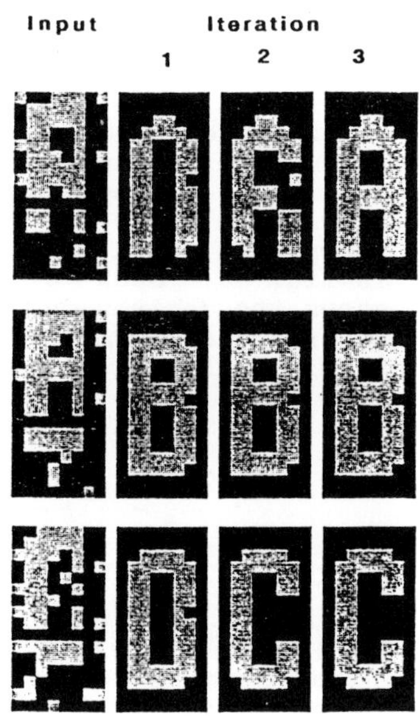

FIGURE 4
Associative restoration of disturbed characters A, B, C.

5. CONCLUSIONS

The proposed feedback network shows space invariance and auto
associativity. The space invariant transform used can easily be
implemented within a multi-step feed forward network.

REFERENCES

[1] T. Kohonen, IEEE Trans. Computer C-21 (1972) 353.
[2] K. Nakano, IEEE Trans. SMC-2 (1972) 380.
[3] S. Amari, IEEE Trans. Computer C-21 (1972) 1197.
[4] J.J. Hopfield, Proc. Natl. Acad. Sci. USA 79 (1982) 2554.
[5] W. Kinzel, Zeitschrift für Physik B, Condensed Matter 60
    (1985) 205.
[6] N.H. Farhat et al. Appl. Opt. 24 (1985) 1469.
[7] M.D. Wagh and S.V. Kanetkar, IEEE Trans. ASSP 25 (1977) 203.
[8] H. Reitboeck and T.P. Brody, Information and Control 15
    (1969) 130.

Parallel Processing in Neural Systems and Computers
R. Eckmiller, G. Hartmann and G. Hauske (Editors)
© Elsevier Science Publishers B.V. (North-Holland), 1990

APPLYING THE ART1 ARCHITECTURE TO A PATTERN RECOGNITION TASK

Edson C. de B. C. Filho and David L. Bisset
Electronic Enginnering Laboratories
University of Kent
Canterbury
Kent, CT2 7NT

This paper shows how the pattern recognition abilities of the *Adaptive Resonance Theory* (ART) architectures can be realised through a number of minor modifications to its implementation. It presents results for a postcode recognition application. Parameters for the assessment of its self-organisation abilities are also explored, as are the effects of various system parameters on recognition performance.

1. INTRODUCTION

The adaptive resonance theory uses a system of self stabilising patterns to represent the classes to be recognised. These patterns are automatically generated during the learning process and are not supplied by the user. The output from the system is in the form of recognition categories which are also selected by the learning system and are outside the normal control of the user, this necessitates some form of post processing to assign classes to the recognition categories. The ART system contains built in constraints to prevent an explosion of the number of recognition categories. ART was introduced by Grossberg[5]. Currently, there are two different ART architectures, ART1[3] and ART2[2]. The ART1 architecture is characterised by using binary input patterns, the ART2 architecture is characterised by using analogue input patterns.

In order to investigate ART architectures in a recognition environment, a real recognition task has been chosen. The task consists of recognising machine generated postcodes extracted from envelops by a solid state camera. Each pattern has been coded as 384 bit binary pattern and normalised to fall centrally on a matrix of 16×24 points. For each class there are three hundred characters of differing fonts and styles. Because the input patterns are binary, the ART1 architecture is suitable for use on this task.

For all the experiments in this paper, the networks are trained with one set of patterns and recalled with a different set of patterns. No attempt is made to improve the quality of patterns through pre-processing, nor is any classifier employed to further process the recognition categories, except for a simple scheme to assign them to classes. For this reason the network's skill at self-organisation, generalisation and discrimination is put to the test, in a realistic recognition task.

2. ART1 ARCHITECTURE

The ART1 architecture has been proposed by Carpenter and Grossberg[3]. ART1 self-organises input patterns into recognition categories while maintaining a balancing between

The authors would like to acknowledgement the support of the Brazilian Research Council(CNPq) and the University of Kent in carrying out this research.

the properties of *Plasticity* (discrimination) and *Stability* (generalisation). These properties are represented in ART1 as follows: the plasticity property through the system's ability to creates new recognition categories when unfamiliar patterns stimulate the network, and the stability property through its ability to clusters similar patterns to the same recognition category. Similarity is determined by the degree of matching between the new pattern and previously stored patterns (prototypes). The matching process is controlled by a rule of similarity, called the vigilance rule.

The ART1 architecture is shown in Figure 1. There are two well defined processes the *bottom-up process* often referred to as an adaptive filtering or contrast–enhancement process, which provides the pattern Y, and the *top–down process* which performs template matching and serves to stabilise the learning, this produces the match pattern X^*. As soon as the *Input Pattern I* is clamped at the input buffer, it activates the F_1 layer producing the pattern X, identical to I. The pattern X is multiplied (LTM gated) by the weights at each F_2 node. This produces at the F_2 *layer* the contrast–enhanced pattern Y. The pattern Y is the result of a *Winner–Take–All Process*, where only one neuron from F_2 can be the winner. This node produces an output value of one. The winner neuron in F_2, is also called the resonant neuron.

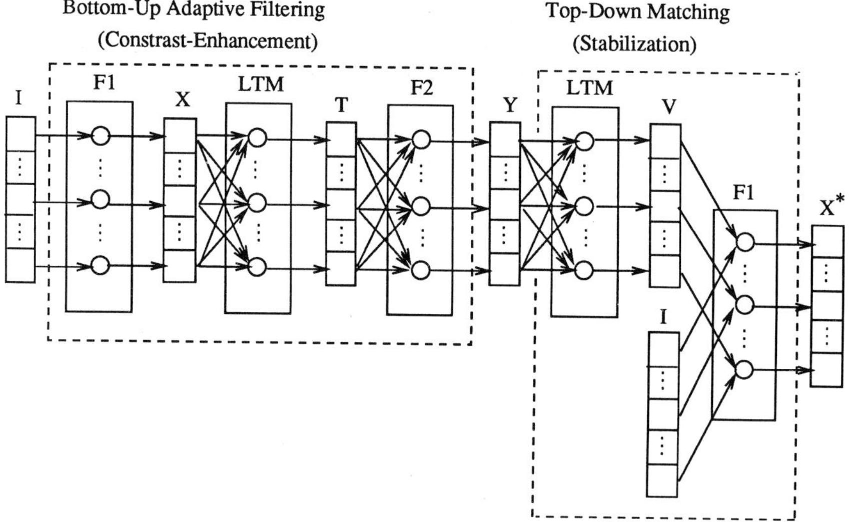

Figure 1 Art1 Architecture

At this point the bottom–up process finishes and the top–down process starts. The pattern Y activates the top–down LTM (binary valued weights) which produces the pattern V. The pattern V is known as the *Top–Down expectation* or critical feature, but is commonly referred to as the prototype pattern for that recognition category. The prototype V and the input pattern I activate F_1 where the matching process occurs, this is simply an *AND* function, this produces the *match pattern X^**. The complete process composed of bottom–up and top–down activations is called a *trial*.

The match pattern X^* indicates the similarity of the input pattern and the prototype pattern. To the match pattern, X^*, a rule of similarity is applied that decides whether the resonant category is stable or whether another recognition category must be activated. There are a number of different possibilities at this point in the process, either a further recognition category will be found to match, in which case the system stabilises, or no categories will match. In this case a node at F_2 that has not yet been assigned to a recognition category will be assigned to this new category.

The pattern matching between I and X^* is performed according to the following rule, known as the reset rule:

$$\frac{|X^*|}{|I|} \geq \rho$$

where $|X^*|$ is the number of ones in X^*; $|I|$ is the number of ones in I; ρ is a given *Vigilance Parameter* which is set by the user. If this inequality is true then the recognition category is deemed to be stable, if not the F_2 node is suppressed and the search for a stable recognition category continues.

After a stable category has been chosen the network learns the input pattern through the adaptation of bottom–up and top–down weights (LTM). There are a number of different ways to implement the learning rules presented by Carpenter and Grossberg[3], in this paper the fast learning rule has been employed. this results in binary weight values for the top–down weights.

3. ART1 NETWORK AS PATTERN RECOGNITION MACHINE

These processes give rise to a number of problems if the ART system is to be used in a traditional pattern recognition role. The system is designed to constantly reorganise itself at the presentation of each new pattern, this can lead to the corruption of the stored prototypes when an incorrect classification is made, and presents difficulties in extracting results from the system in terms of a fixed set of recognition classes. The implementation of ART1 used in this paper, (i.e. with the fast learning rule), means that the LTM values are not incrementally adjusted but forced to new values for each pattern presented, this results in severe corruption of the stored prototypes if poor quality or bad patterns are classified. (This is an acute problem in the early stages of self-organisation). This problem does not arise if the incremental version of ART1 is used since the absorption of patterns into the prototypes is slower, and thus poor or incorrect patterns are less likely to cause corruption.

In order to solve these problems the ART1 network is modified to have two different phases. The *Self-organisation Phase* where the network decides how to assign input patterns to recognition categories, and the *Recalling Phase* where the network looks for a recognition category corresponding to the input pattern.

3.1. Self-Organising Phase

This phase consists of presenting the training patterns to the system and allowing it to self–organise its outputs. The network realises an ideal self–organisation when for all class of input patterns C_i, all patterns of class C_i fall in the same stable category v_j and patterns of other classes do not fall in v_j. This ideal case will almost never occur in real problems due to the large number of classes involved and the wide variety of patterns found in the same class. By controlling the vigilance, or matching parameter, it is possible to obtain different assignments of classes to recognition categories. A low vigilance value tends to cluster more than one class into the same recognition category, and a high vigilance value tends to split one class between more than one recognition category. In the extreme case a high vigilance will cause each input pattern to occupy a different recognition category, and a very low vigilance value will cause all input patterns to occupy the same recognition category. The penalty paid for higher vigilance parameters and larger numbers of recognition categories, or prototypes, is a larger amount of memory required to store them (since each category requires a binary weight at each of the F_1 nodes). The optimum vigilance value will depend on the quality of the input patterns and the implementation resources.

There are a pair of parameters of the learning process that can be monitored to discover the progress of categorisation. These parameters are important because of the lack of control available to the user in a self-organising system. The first of these monitoring parameters is the number of stable categories, or prototypes, expressed as a percentage of the number of training patterns, this gives the user the ability to forecast the size of networks with a given set of parameters. For high vigilance values this percentage will be large. The second of these monitoring parameters is the number of stable categories which represent more than one class, (i.e. patterns from more than one class were used in the formation of the recognition category). This parameter is expressed as a percentage of the total number of stable categories, and is called the percentage of intersections (i.e. the percentage of recognition categories that contain more than one class).

3.2.Recalling Phase

The recalling phase consists of leaving the network to reach a stable recognition category and then applying a *Discrimination Rule* to the stable category in order to discover the fixed class. During this phase the learning rules are turned off. The discrimination rule can be either probabilistic or deterministic. A simple example of a probabilistic decision would be to output probabilities for the stable recognition category for each of the classes, these probabilities would be based on the numbers of training patterns of each class that formed the recognition category. A simple example of a deterministic decision would be to take a majority decision in favour of the class that contributed most to the formation of the activated recognition category.

The vigilance parameter is just as important in the recalling phase as on the self-organising phase. A high vigilance value means that the network will be more selective in stabilising the match between prototypes and input patterns, which can in turn cause higher reject rates. Previous work by the authors has shown that recognition performance can be considerably improved if the vigilance value is lowered during the recalling phase[1].

4. CHARACTER RECOGNITION

In order to analyse the self-organisation and recalling of patterns in a real task, the task of recognising machine printed postcode numerals is used. The analysis of performance is carried out in terms of monitoring the recognition rates, the error rates and the reject rates for different values of the learning and recalling vigilance parameters, and for different numbers of patterns per class in the training set. The pattern recognition task has been implemented on the *Rochester Connectionist Simulator*[4].

All input patterns used in the recalling phase are different from those used in the learning phase. Figure 2 shows some typical patterns from each of the classes. Figure 3 gives some indication of the variation of the patterns within individual classes.

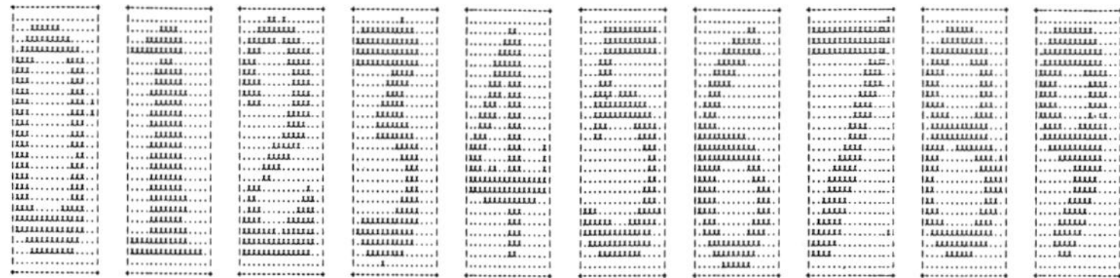

Figure 2: Samples of Numeric Patterns

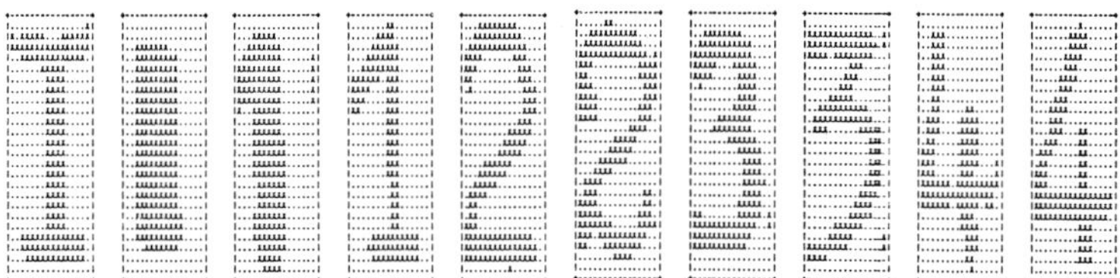

Figure 3: Samples of Different Numeric Fonts

4.1. Self-Organisation Experiments

The analysis of the self-organisation has been carried out by concentrating on the three monitoring variables, as described above, analysed with respect to the number of patterns in the training sets and the learning vigilance parameter.

Figure 4.a and Figure 4.b show the percentage of stable categories and intersections, respectively, obtained from the average of ten experiments, varying the learning vigilance parameter from 0.1 to 0.9, for different sizes of training set.

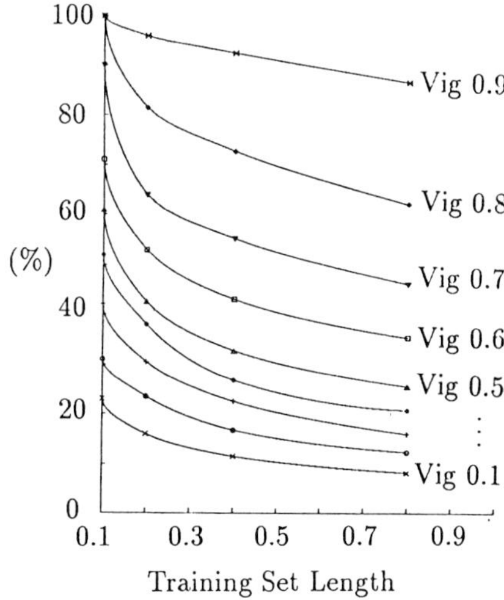

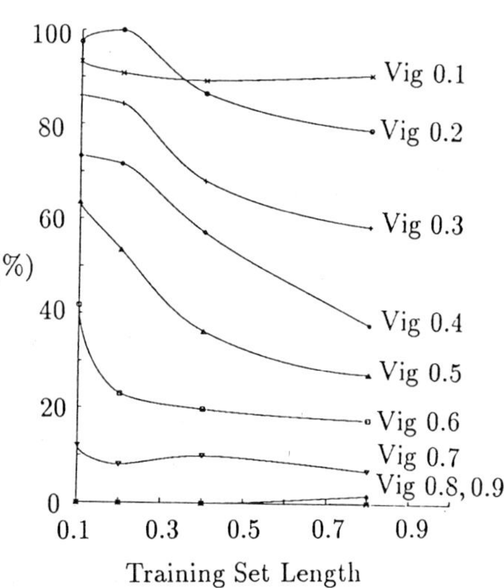

Figure 4.a: Percentage of Stable Categories Figure 4.b: Percentage of Intersections

The percentage of stable categories gives some indication of the generalising properties of the network, since it shows how many input patterns have been clustered into each recognition category. The percentage of intersections needs to be as low as possible, and gives some indication of the discrimination ability of the network. Clearly for vigilance values of 0.8 and training set lengths of less than 5 patterns this is zero, which is ideal. From these results it can be seen that that ideal assignment of one pattern class to one recognition category has not been found. However the learning vigilance value of 0.8 shows a good balance between generalisation and discrimination.

4.2. Recalling Experiments

In order to analyse the recalling phase the percentage of correct, error and reject responses and the number of trials are monitored, while varying the number of training patterns and the learning and recalling vigilance values.

Figure 5 shows the percentage of correct, error and reject responses for learning vigilances of 0.6 and 0.9 varying the recalling vigilance from 0.1 to 0.9. For each point one hundred recalling patterns of each class have been used, this gives a total of one thousand patterns for each point.

As can be seen the higher learning vigilance gives a much better performance for smaller training sets, but this also generates more recognition categories. Clearly lowering the recalling vigilance strongly affects the ability of the recognition system. Preliminary tests indicate that although the recognition performance appears not to degrade with very low recalling vigilance values, the ability of the network to reject patterns that do not fall in the input pattern classes is severely affected. Thus to achieve a given reject performance for unknown patterns it may be necessary use higher recalling vigilance values than these graphs indicate.

The results clearly show that for a learning vigilance of 0.9, a much improved response can be obtained if the recalling vigilance is lowered to 0.6. If the recognition response with the learning vigilance set to 0.6, and 8 training patterns per class, is acceptable, then from Figure 4 it can be seen that there will be a considerable decrease in the amount of memory required to store the prototypes. In this particular example there will be a saving of some 50%, without a significant reduction in the recognition rate (although to achieve such a result is is necessary to have a lower recalling vigilance which may have other effects as noted above).

Figure 6 shows the average number of trials for the previous recognition experiments, this indicates that for low recalling vigilance values the ART system is working with direct access to the recognition categories, and no searches are being invoked.

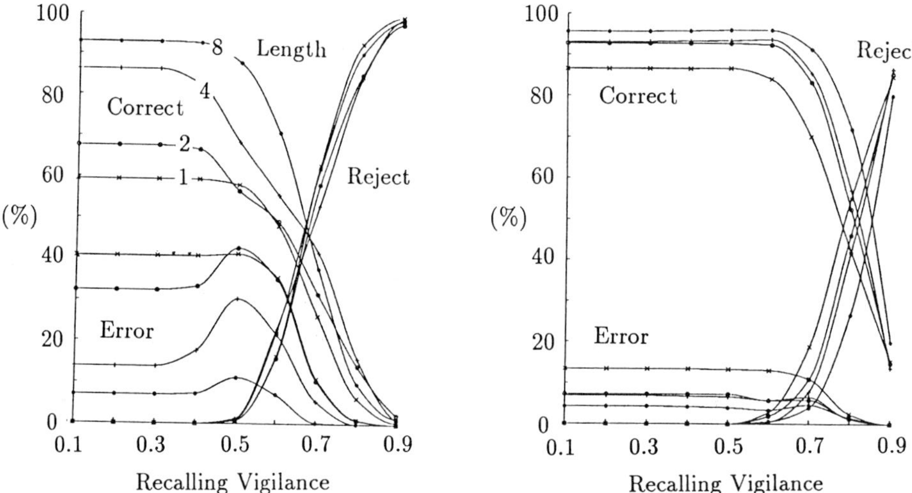

Figure 5.a: Learning Vigilance 0.6   Figure 5.b: Learning Vigilance 0.9

Figure 5: Recognition Performances

5. CONCLUSION

The paper has shown how the ART1 network can be used in a pattern recognition task to perform as well as many other neural network algorithms on the same task [1]. The modifications required to achieve this performance are as follows:

- The operation of the network is divided into two phases, a self-organising phase and a recalling phase.

- The value of the vigilance parameter is set separately for these two phases.

- During the self-organising phase statistics about the assignment of classes to recognition categories (prototypes) is gathered and this is used to form a discrimination rule to allow results from the network to be translated into a recognition class.

By applying these modifications to the ART1 system an improved recognition response has been shown in a real recognition problem. It has also been possible to investigate the de-

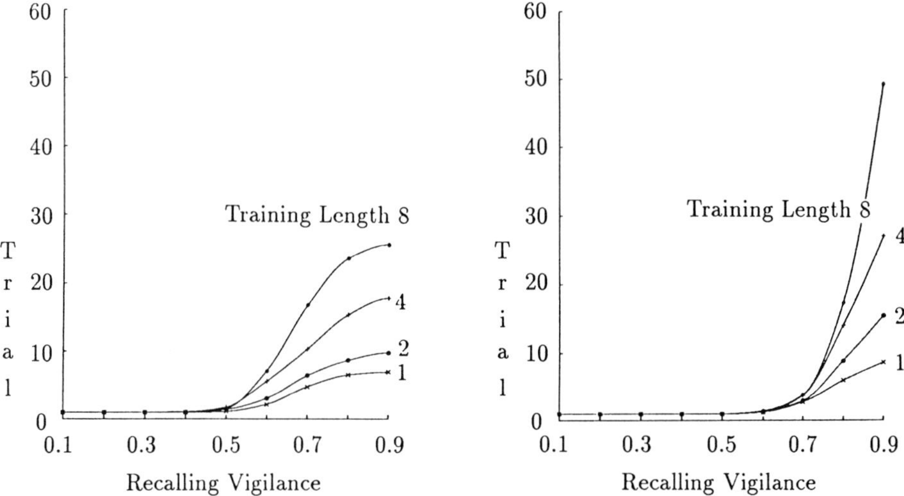

Figure 6.a: Learning Vigilance 0.6   Figure 6.b: Learning Vigilance 0.9

Figure 6: Number of Trials

pendence of performance on various parameters of the system. This has been achieved by monitoring a number of system parameters, which are able to show the self-organisation ability of the network, and allow optimal system parameters to be selected. It has been shown that high rates of recognition are achievable provided that large numbers of prototypes are acceptable. It has also been demonstrated that for a small sacrifice in recognition performance large savings can be made in the amount of memory used to store the prototype patterns by lowering the learning vigilance and adopting the modifications indicated in this paper.

References

[1] D L Bisset, E C D B C Filho, and M C Fairhurst. A comparative study of neural network structures for practical application in a pattern recognition enviroment. In *First IEE International Conference on Artificial Neural Networks*, pages 378–382, London, UK, October 1989. IEE.

[2] G A Carpenter and S Grossberg. Art2: Self-organization of stable category recognition codes for analog input patterns. *Applied Optics*, 26(23):4919–4930, 1987.

[3] G A Carpenter and S Grossberg. A massively parallel architecture for a self-organizing neural pattern recognition machine. *Computer Vision, Graphics, and Image Processing*, 37:54–115, 1987.

[4] J A Feldman, M A Fanty, and N H Goddard. Computing with structured neural networks. *Computer*, 21(3):91–102, 1988.

[5] S Grossberg. Adaptive pattern classification and universal recoding: II feedback, expectation, olfaction, illusions. *Biological Cybernetics*, 23:187–202, 1976.

Parallel Processing in Neural Systems and Computers
R. Eckmiller, G. Hartmann and G. Hauske (Editors)
© Elsevier Science Publishers B.V. (North-Holland), 1990

NEURAL NETWORK MODELS FOR VISUAL PATTERN RECOGNITION

Kunihiko FUKUSHIMA

Department of Biophysical Engineering
Faculty of Engineering Science
Osaka University
Toyonaka, Osaka 560, Japan

Modeling neural networks is useful in understanding the mechanism of the brain, and also in obtaining design principles for new information processors. With this approach, various models of visual pattern recognition has been developed: for example, a "neocognitron" for deformation-invariant pattern-recognition; a selective attention model, which has the function of not only pattern-recognition but also associative recall, segmentation and restoration of patterns; and so on.

1. INTRODUCTION

Modeling neural networks is a promising approach not only to understanding the mechanism of the brain, but also to obtaining design principles for new information processors. In the modeling approach, we study how to interconnect neurons to synthesize a brain model, or a network with the same functions and abilities as the brain.

When synthesizing a model, we try to follow physiological evidence as faithfully as possible. For parts that are not yet clear, we construct a hypothesis and synthesize a model that follows the hypothesis. We then analyze or simulate the behavior of the model, and compare it with that of the brain. If we find any discrepancy in behavior of the model and the brain, we change the initial hypothesis and modify the model. We then test the behavior of the model again. We repeat this procedure until the model behaves in the same way as the brain. Although we must still verify the validity of the model by physiological experiment, it is probable that the brain uses the same mechanism as the model, because both respond in the same way. The relationship between modeling neural networks and neurophysiology resembles that between theoretical physics and experimental physics.

Modeling neural networks is useful not only in explaining the brain itself, but also in engineering. It brings the results of neurophysiological and psychological research to engineering in the most direct way possible. Once we complete a model, we can easily see the essential algorithms of information-processing in the brain, because unessential factors in the biological brain are discarded in the model. We can use the algorithms directly as design principles for new information-processors.

Many models have been constructed along these lines. Some models of visual information processing proposed by the author are introduced in this article.

2. FEATURE EXTRACTION BY MULTILAYERED NETWORKS

Hubel and Wiesel hypothesized that the cells in the visual area are connected in a hierarchical manner [1]. In the 1960's and early 1970's, various kinds of model of the visual system have been constructed following this hypothesis.

The author also proposed such a model [2],[3]. It is a multi-layered network
consisting of a cascade connection of many layers of cells. The input layer U_0
consists of photoreceptors, and the succeeding layers consist of analog-
threshold elements. Cells of the first layer U_1 correspond to the retinal
ganglion cells, and extract the contrast component of the input pattern. Cells
of layer U_2, U_3 and U_4 correspond to simple cells, complex cells and
hypercomplex cells, respectively. The input pattern is analysed into line
segments of different orientation in layer U_2. The curvature of the input
pattern is extracted in layer U_5 in the highest stage.

Several models for extracting binocular disparity have also been designed
following similar ideas [4],[5].

3. COGNITRON: A SELF-ORGANIZING MULTILAYERED NETWORK

A famous example of a network which can learn to recognize patterns is the
three-layered "perceptron" [6]. Although the perceptron has a three-layered
hierarchical structure, it is only in the highest stage that the cells have
variable input connections. The connections to the cells in other layers are
all fixed and unmodifiable.

It is well known that the capability of a hierarchical multilayered network to
process information is greatly improved by increasing the number of layers in
the network. In order to give full play to the capability of a multilayered
network, however, the input connections of the cells must be modifiable not
only in the highest stage, but also in the intermediate lower stages. Various
studies have been made to find learning procedures which can be used
effectively for the self-organization of multilayered networks.

An example of supervised learning (or learning-with-a-teacher) useful for such
a purpose is the "back propagation" [7].

In the model called "cognitron" [8], which the author proposed earlier,
unsupervised learning (or learning-without-a-teacher) is used. Self-
organization of the cognitron is performed on the following principle: Among
the cells situated in a certain small area (called a "competition" area), only
the one responding most strongly has its input connections reinforced. The
amount of reinforcement of each input connection to this maximum-output cell is
proportional to the intensity of the response of the cell from which the
relevant connection leads. This principle is applied to both excitatory and
inhibitory connections.

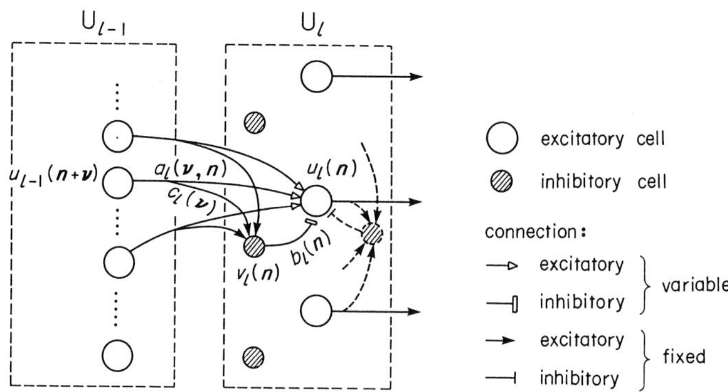

Fig. 1. The structure of the cognitron (modified from [8]). The
interconnection between two adjoining layers is shown.

Because of the "winner-take-all" nature of this principle, the formation of more than one cell to extract the same feature does not occur, and "division of labor" among the cells occurs automatically. With this principle, the network also exhibits self-repair.

The cognitron consists of a number of neural layers of similar structure, connected in cascade. Each layer has a number of excitatory cells and the same number of inhibitory cells. Figure 1 shows interconnection between two adjoining layers. After the training (or learning) is complete, each excitatory cell in the network acquires the ability to extract a feature of the stimulus presented during the training period. Through the excitatory connections, the cell receives signals indicating the existence of the feature to be extracted. If an irrelevant feature is presented, the inhibitory signal via the inhibitory cell becomes stronger than the direct excitatory signals.

4. NEOCOGNITRON: DEFORMATION-INVARIANT RECOGNITION OF VISUAL PATTERNS

Although the cognitron can acquire the ability to recognize patterns by unsupervised learning, the cognitron, like many other models, does not have the ability to recognize shifted or distorted patterns correctly.

The neocognitron [9],[10] can recognize stimulus patterns correctly, even if the patterns are shifted or distorted. It can acquire this ability either by unsupervised learning or by supervised learning. Since the neocognitron has the function of generalization, it is not necessary to teach all the deformed versions of the training patterns. It even comes to recognize correctly a pattern which has not been presented before, provided it resembles one of the training patterns.

The neocognitron, like the cognitron, is a hierarchical multilayered network and consists of a cascade of many layers of cells. The initial stage of the network, called U_0, is the input layer and consists of a two-dimensional array of receptor cells. Each of the succeeding stages has a layer of "S-cells" followed by a layer of "C-cells". Thus, in the whole network, layers of S-cells and C-cells are arranged alternately.

The S-cells are feature-extracting cells, and somewhat resemble the simple cells of the visual cortex. They have variable input connections, and can acquire the ability to extract features by learning (or training) in the same way as the excitatory cells of the cognitron. After learning is finished, an S-cell is activated only when a particular feature is presented in a certain position in the input layer. The features extracted by the S-cells are determined during the learning process. Generally speaking, local features, such as a line at a particular orientation, are extracted in the lower stages. More global features, such as part of a training pattern, are extracted in the higher stages.

C-cells somewhat resemble complex cells. They are put into the network to allow for positional error in the features of the stimulus. The connections from S-cells to C-cells are fixed and invariable. Each C-cell receives signals from a group of S-cells which extract the same feature, but from slightly different positions. The C-cell is activated if at least one of these S-cells is active. Even if the stimulus feature is shifted and another S-cell is activated instead of the first one, the same C-cell keeps responding.

In the entire network, with its alternating layers of S-cells and C-cells, the process of feature-extraction by the S-cells and the toleration of a shift by the C-cells is repeated, as shown in Fig. 2. During this process, local features extracted in lower stages are gradually integrated into more global features. Since changes in relative position of local features are tolerated by the C-cells, S-cells in the succeeding stage can extract even deformed global features. Tolerating positional errors a little at a time at each stage, rather than all in one step, is important for recognizing deformed patterns.

The layer of C-cells at the highest stage works as the recognition layer: the response of the cells in this layer shows the final result of pattern recognition by the neocognitron. Each cell integrates all the information of the input pattern. Only one cell, corresponding to the category of the input pattern, is activated in this stage.

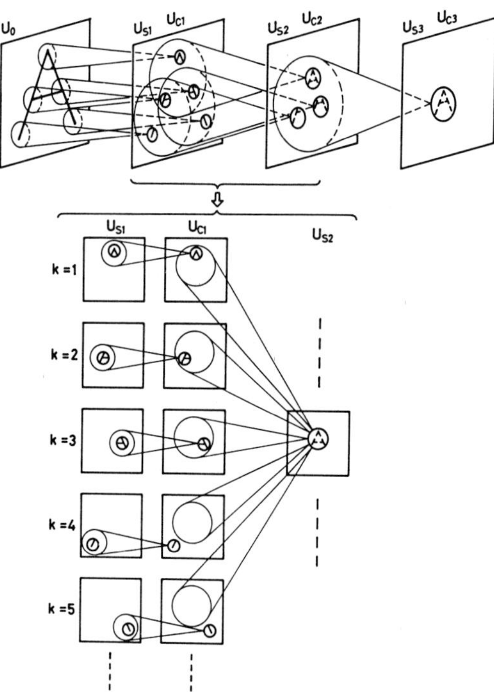

Fig. 2. Illustration of the process of pattern recognition in the neocognitron. (Modified from [9]).

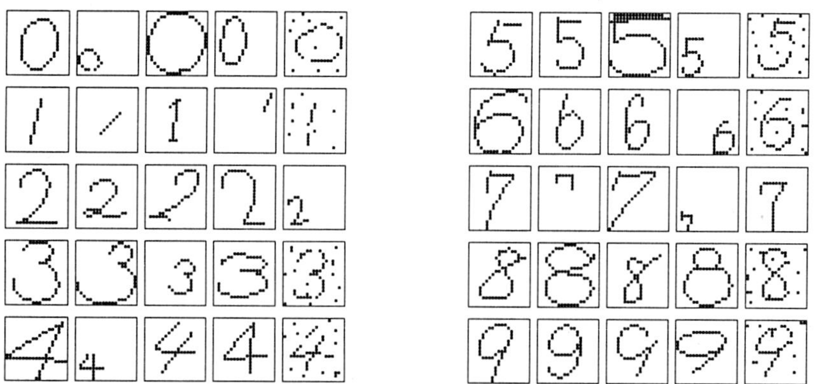

Fig. 3. Some examples of deformed numeral patterns which the neocognitron has recognize correctly [10].

In order to demonstrate the ability of the neocognitron, the author and his group have designed a system [10] which recognizes hand-written numerals from "0" to "9". It has been implemented on several kinds of computers: from a micro-computer [11] to a parallel computer [12].

Figure 3 shows some examples of deformed input patterns which the system has recognized correctly. As can be seen from the figure, the neocognitron recognizes input patterns correctly even if they are shifted, changed in size, or deformed. Even though the input patterns have some parts missing or are contaminated by noise, the neocognitron recognizes them correctly.

5. SELECTIVE ATTENTION MODEL

Although the neocognitron has a considerable ability to recognize deformed patterns, it does not always recognize patterns correctly, when two or more patterns are presented simultaneously. In order to increase the ability of the neocognitron, backward (i.e., top-down) connections were added to the conventional neocognitron which had only forward (i.e., bottom-up) connections.

The new model obtained in this way has the function of selective attention in visual pattern recognition [13],[14]. At each stage of the hierarchical network of the model, there are several kinds of cells. Figure 4 illustrates how the cells, represented by circles, are connected with each other. Although the figure shows only one of each kind of cell in each stage, there are actually numerous cells in a two-dimensional array.

In this hierarchy, the forward signals manage the function of pattern recognition. The forward paths in the network have almost the same structure and function as the neocognitron. The lowest stage of the forward paths is the input layer U_{C0}, to which stimulus patterns are presented. The response of the recognition layer U_{C3} at the highest stage shows the final result of pattern recognition of the network.

The cells in the backward paths are arranged in the network making a mirror image of the cells in the forward paths. The same is true also between the forward and the backward connections.

The output of the recognition layer is returned to the lower stages through the backward paths. The forward and backward signals interact. The forward signals gate backward signal flow, and, at the same time, the backward signals facilitate forward signal flow. The backward signals manage the function of selective attention, pattern-segmentation and associative recall. The lowest stage of the backward paths is the "recall" layer W_{C0}, in which the result of

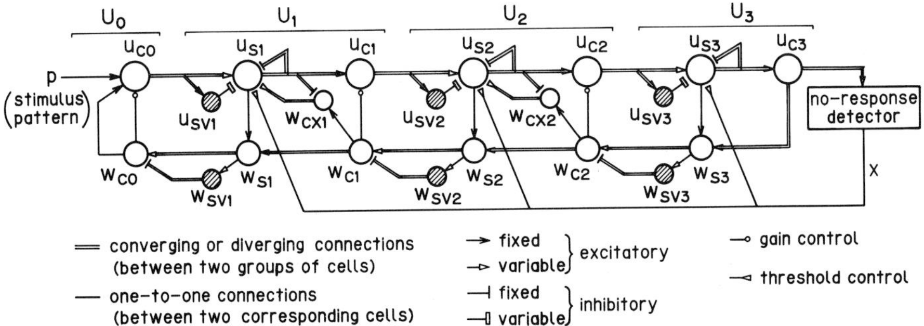

Fig. 4. Hierarchical structure of the interconnections between different kinds of cell [14].

associative recall appears. We can also interpret the output of the recall layer as the result of segmentation. The response of the recall-layer W_{C0} is fed back positively to the input layer U_{C0}.

Figure 5 summarizes the abilities of the model. When a composite figure consisting of two patterns or more is presented to the model which has finished learning, the model selectively focuses its attention on one pattern after another, segments each pattern from the others, and recognizes it separately. The model also has the function of associative recall. Even if noise or defects affect the stimulus pattern, the model can recognize it and recall the complete pattern from which the noise has been eliminated and defects corrected. Perfect recall does not require that the stimulus pattern be identical in shape to the training pattern. A deformed pattern, or one changed in size can be recognized correctly and the missing portions restored.

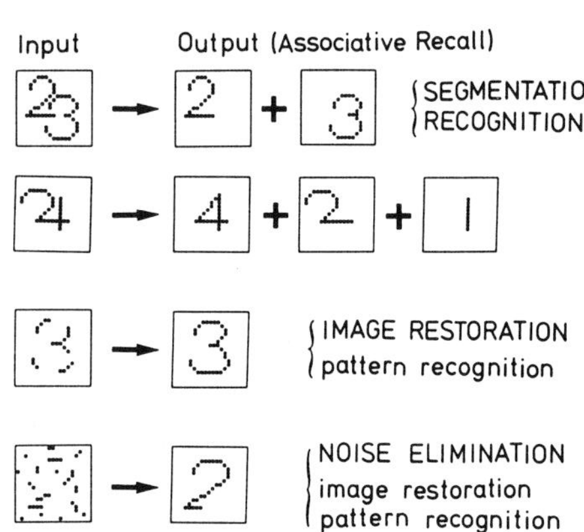

Fig. 5. Summary of the abilities of the selective attention model.

ACKNOWLEDGEMENTS

The author would like to thank S. Miyake, T. Ito and M. Fujii of NHK Science and Technical Research Laboratories for their help on various aspects of this work. This work was supported in part by a Bioscience Grant for International Joint Research Project from the NEDO, Japan.

REFERENCES

[1] Hubel, D.H. and Wiesel, T.N.: J. Neurophysiol. **28** (1965) 229-289.
[2] Fukushima, K.: Kybernetik **7** (1970) 153-160.
[3] Fukushima, K.: NHK Technical Journal **23** (1971) 351-367 (in Japanese).
[4] Hirai, Y. and Fukushima, K.: Biol. Cybernetics **18** (1975) 19-29.
[5] Fujii, M. and Fukushima, K.: Tech. Report, IEICE Japan, **MBE88**-148 (1989) (in Japanese).
[6] Rosenblatt, R., Principles of Neurodynamics, (Spartan Books, Washington, DC, 1962)
[7] Rumelhart, D.E., Hinton, G.E. and Williams, R.J., Learning Internal Representations by Error Propagation, in: Rumelhart, D.J., McClelland, J.L. (eds.), Parallel Distributed Processing, vol. 1, (MIT Press, Cambridge, MA, 1986) pp. 318-362.
[8] Fukushima, K.: Biol. Cybernetics **20** (1975) 121-136.
[9] Fukushima, K.: Biol. Cybernetics **36** (1980) 193-202.
[10] Fukushima, K.: Neural Networks **1** (1988) 119-130.
[11] Fukushima, K., Miyake, S., Ito, T. and Kouno, T.: Trans. Inf. Process. Soc. Japan **28** (1987) 627-635 (in Japanese).
[12] Ito, T., Fukushima, K. and Miyake, S.: "Examination of implementing a neural network on a parallel computer -- Neocognitron on NCUBE", Trans. IEICE Japan (1990) in press (in Japanese).
[13] Fukushima, K.: Biol. Cybernetics **55** (1986) 5-15.
[14] Fukushima, K.: Applied Optics **26** (1987) 4985-4992.

Parallel Processing in Neural Systems and Computers
R. Eckmiller, G. Hartmann and G. Hauske (Editors)
© Elsevier Science Publishers B.V. (North-Holland), 1990

ΣΠ-NETWORKS FOR MOTION AND INVARIANT FORM ANALYSES

Helmut GLÜNDER

Institut für Medizinische Psychologie
der Ludwig-Maximilians-Universität
Gœthestraße 31, D-8000 München 2
Federal Republic of Germany

Spatiotemporal relations that are represented by properly chosen multilinear forms of signal values are shown to be useful for form-independent motion analyses and for the extraction of form descriptors that are invariant under geometric transformations. Both tasks are performed by identical parallel computing ΣΠ-networks.

1. INTRODUCTION

The extraction of geometric relations between two or more points of a pictorial pattern plays a fundamental role in the evaluation of form-related pattern properties that are invariant under geometric transformations. Since any kind of statements about the presence or the degree of such relations involve decisions, nonlinear combinations of the signal values at two or more sites of a static image are inevitable. Multilinear terms, i.e. products of signal values that are taken from an image, represent convenient measures of spatial coincidences (Figure 1a). Motion analysis, on the other hand, additionally requires the evaluation of temporal relations, i.e., between those values of a continuously time-varying image that are found at the same location but are separated by one or more time intervals (Figure 1b). Together with a system for the analysis of spatial relations, spatiotemporal coincidence detection (Figure 2), i.e., motion analysis becomes feasible. It is remarkable that, from this point of view, the gradient and the correlative approaches to motion analysis do not differ.

For both applications, pooling of defined sets of multilinear terms, i.e., the computation of multilinear forms, is requisite if invariance of the form descriptors under geometric transformations, or independence of the velocity estimates from object form are demanded. These conditions are dual and, as will be shown, can be established by a common computing structure that is itself based on this dualism, in so far, as biological processing is concerned: Firstly, there is no need for invariant features if there is no variance, i.e., neither motion nor movement. Secondly, there is no self-organization of the highly specific motion analyzer networks if they are to function dependent on object form, because great numbers of equal stimuli are needed for the formation of suitable networks.

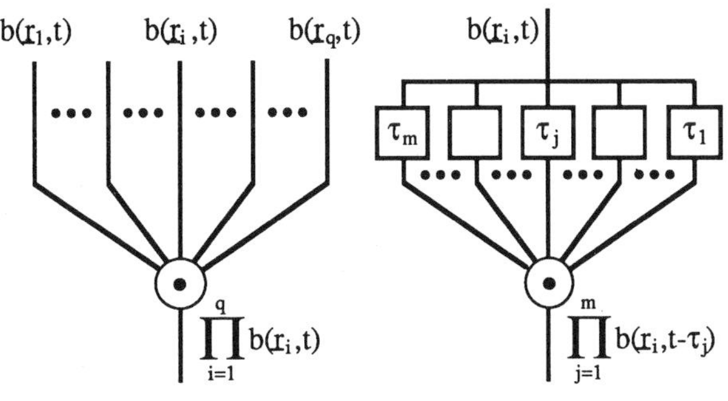

FIGURE 1a
Detection of spatial relations by simultaneous multiplication of q signal values

FIGURE 1b
Temporal coincidence detector multiplying m signal values sampled at different times at the same location

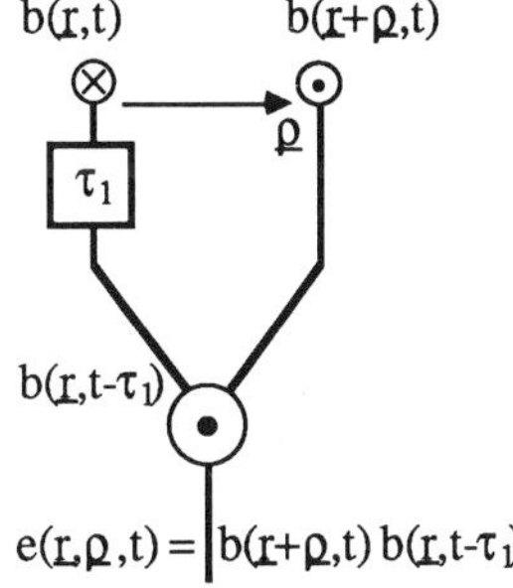

FIGURE 3
Simplest processing unit for the detection of spatiotemporal relations

2.   BASIC COINCIDENCE DETECTORS

Figure 1a,b shows schematics of multiplying computing units for the extraction of spatial and temporal relations from a continuously time-varying image $b(\underline{r},t)$, with $\underline{r}=(x,y)^T$, which result in q- and m-linear terms respectively. In Figure 2 a general circuit for the analysis of spatiotemporal relations is depicted which leads to qm-linear terms. Although terms of orders qm>2 may be needed to solve special tasks in motion and invariant form analysis, the simplest spatiotemporal detector of order qm=2, with $\tau_{11}=\tau_1$=const and τ_{21}=0, is used in what follows. Such a unit represents a so-called ρ-tuned velocity detector with its tuning velocity defined by $\underline{v}_{tun}=(\underline{r}_2-\underline{r}_1)/\tau_1=:\underline{\rho}/\tau_1$ (Figure 3). In other words, this bilocal detector delivers a coincidence signal $e(\underline{r},\underline{\rho},t)$, if a bright spot moves from point $\underline{r}$ straight to point $\underline{r}+\underline{\rho}$ with the constant speed $|\underline{v}_{tun}|$.

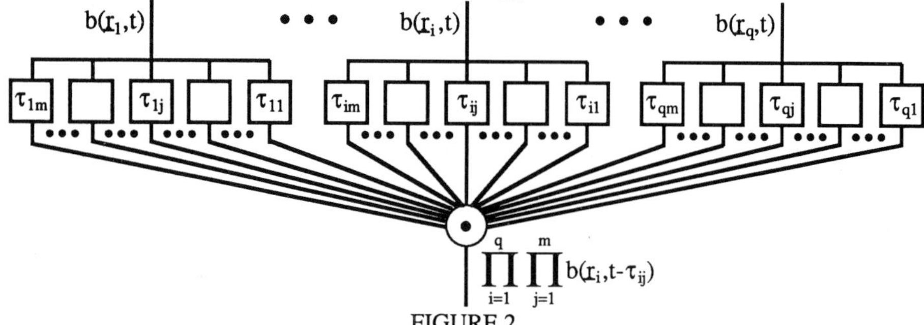

FIGURE 2

General analyzer unit for spatiotemporal relations of time-varying images producing qm-linear terms

3.   MOTION ANALYSIS

If a single detector unit of the type shown in Figure 3 is confronted with motion of more complex but still rigid patterns $b(\underline{r})$, it will in general give rise to form-related deviations from the ideal velocity estimate. This is due to the well-known correspondence and window problems (see for instance [1]). They can be eased or even eliminated if extended arrays of elementary detectors are applied in combination with suitable schemes for the pooling of their coincidence signals. For the analysis of frontoparallel translations, for instance, a spectrum of differently tuned units must be isotropically arrranged around every location of the image representation and the coincidence signals of all units with identical vectors $\underline{\rho}$ must be summed. Thereby, the modified autocorrelation function (MACF) of second order is computed, with each of its coefficients representing a bilinear form.

$$^2k_{\rho d}(\underline{\rho},t) = \iint b(\underline{r} + \underline{\rho}, t)\, b(\underline{r}, t - \tau_1)\, d\underline{r} \qquad \text{for real-valued } b(\underline{r},t) \qquad (1)$$

Form-independent velocity estimates $\underline{v}_d(t)=\underline{\rho}_d(t)/\tau_1$ can be obtained from the MACF by evaluating the coordinates $\underline{\rho}_d(t)$ of its absolute maximum which is always defined if the pattern is nonperiodic.

In order to analyze expansions or frontoparallel rotations, the input signals to each detector unit are taken from one of the straight lines of a line-bundle, or from one of the circles of a set of concentric circles respectively. In the first case, summation is applied to all bilinear terms that stem from units whose input positions are related by the same scale factor s with respect to the point of intersection of the line-bundle. In the second case, all terms are summed that result from units whose input positions are related by the same rotation angle ψ with respect to the centre of the circles. A universal formulation of such systems is given by the generalized MACF of order q which also allows for any kinds of impulse responses h(t), and not only of the ideal delay $h_d(t)=\delta(t-\tau_1)$.

$$^qk_{zh}(\underline{z}_1...\underline{z}_i...\underline{z}_q, h_1...h_i...h_q, t) = \iint \prod_{i=1}^{q} [b(\mathcal{T}(\underline{z}_i)\{\underline{r}\},t) * h_i(t)\, \delta(\underline{r})]\, d\underline{r} \qquad \text{for real-valued } b(\underline{r}) \qquad (2)$$

Operator $\mathcal{T}(\underline{z}_i)$ denotes mappings $\underline{r} \to \underline{r}'$ defined by parameter vectors $\underline{z}_i$. It was shown recently [2] that form-independent velocity estimates, e.g. of frontoparallel translations $\underline{v}_c(t)=\underline{\rho}_c(t)/\tau_h$, can also be obtained from systems comprising bilocal detectors with lowpass filters of arbitrary characteristic, if the unequivocally defined centroid location $\underline{\rho}_c(t)$ of the generalized MACF is used in conjunction with a characteristic time constant τ_h of the filter. Since the evaluation of centroids relies on integral operations, it is less susceptible to noise and less costly than maximum detection.

Owing to the specific network structure, i.e., to the geometric arrangement of the detector units in the plane of analysis, each analyzer system 'views' a specific aspect of object motion. Consequently, certain kinds of compound motion are decomposed according to the implemented systems. Cycloidic motion of points on a rolling wheel, for instance, can be decomposed into their translatory and rotatory components. This selectivity allows one to deal with motion of nonrigid objects as well [2]: Each system reacts to motion of moderately deforming patterns by adequate deviations of its analysis from the one that would have been caused by the corresponding rigid motion process. The fact that velocity estimates are independent of the object form is essentially due to the spatial integration of motion information. Globally acting systems, however, can only analyze motion of a single object correctly and thus, among other reasons, local autocorrelative networks are required. They are feasible, if the regions of analysis allow at least two applications of the considered transformation parameter and if the overlap between neighbouring regions measures its single application. In order to obtain regional or global MACFs, the locally computed results need simply be added. This grouping especially makes sense if it depends on size and location of the object [2,3,4].

Higher order analyses (cf. Figure 2) become necessary if analyzer systems are to detect not only specific types of motion but velocity profiles, such as those caused by motion or movement along the line of sight. In this case, scale or expansion factors are a hyperbolic function of time which may be analyzed by a third order system (qm=3) built from elementary detectors incorporating two different delay times $\tau_{11}>\tau_{21}\neq0$ and $\tau_{31}=0$. However, this approach to the analysis of depth motion depends strongly on object size. One can cope with this problem by using a two-stage design of second order systems which computes the reciprocal of the 'time to contact' [5].

4. INVARIANT FORM ANALYSIS

If motion analyzers of the type described by equation (2) are confronted with stimuli that remain static for periods longer than the maximum duration of its impulse responses $h_i(t)$, their reaction is described by the generalized autocorrelation function (ACF) of order q

$$^qk_z(\underline{z}_1\ldots\underline{z}_i\ldots\underline{z}_q) = \iint\ \prod_{i=1}^{q} b(\mathcal{T}(\underline{z}_i)\{\underline{r}\})\,d\underline{r} \qquad \text{for real-valued } b(\underline{r})\,, \tag{3}$$

with each of its coefficients representing a q-linear form. Generalized ACFs that depend on a single type of transformation are inherently invariant under this very transformation. This holds true for every single autocorrelation coefficient, since all q-linear terms that stem from image points that are related by the same set of transformation parameters $\underline{z}_1\ldots\underline{z}_i\ldots\underline{z}_q$ are summed. The invariance of autocorrelative features is exclusively due to this spatial averaging. Hence, the extent of feature invariance for a given type of geometric transformation depends on the number of q-linear terms that contribute to a coefficient. For certain groups of geometric transformations it is even possible to derive form descriptors from the corresponding generalized ACF that are multiple invariant, e.g. under the whole group of similitudes [6]. Such features describe patterns by their general symmetries, self-congruences and self-similarities, i.e., by categories known from Gestalt-psychology.

Even if all coefficients, e.g. of the translatory second order ACF $^2k_\rho(\underline{\rho})$, are available, they do not unequivocally determine the form of a pattern, i.e., its exact reconstruction is not possible. This situation is due to the loss of information about the relative positions between the two-point relations of a pattern (loss of coherence). The translatory third order ACF $^3k_\rho(\underline{\rho}_1,\underline{\rho}_2)$, with $\underline{\rho}_3=\underline{0}$, however, perfectly defines patterns of finite extent, except for their shift position (invariance), and thus allows for the unequivocal reconstruction of their form [7,8,9]. Comparable considerations hold for the generalized triple ACF $^3k_{s\psi}(\underline{z}_1,\underline{z}_2)$, with $\underline{z}_i=(s_i,\psi_i)^T$ and $\underline{z}_3=\underline{0}$, which depends on scale factors s and rotation angles ψ. This becomes immediately clear if one imagines an image representation in polar coordinates with a logarithmically graded radial axis, for which both types of transformations are changed into pure translations. Pattern description by this ACF is rotation as well as scale invariant with respect to the considered fixed point. The reconstruction of the pattern form is unequivocally possible but will not include size and angular position. It remains to be mentioned that pattern reconstruction in both cases is essentially based on all those autocorrelation coefficients that depend on the parameter vector by which the two most distant points of a pattern are related. Therefore, local analyzers, even if the results of many of them are combined, cannot provide perfect invariant pattern descriptions (cf. the similar conclusions in [9]).

5.   COMPUTING ELEMENTS FOR MULTILINEAR FORMS

The direct computation of a q-linear form is performed by a $\Sigma\Pi$ computing element which contains coincidence detectors and sums their q-linear terms. Hence, each coefficient of a generalized MACF can be calculated by a single $\Sigma\Pi$ element, provided the input signals are properly grouped. A polynomial approximation, using threshold logic elements (TLUs), results in cross correlation coefficients of the time-varying image and spatiotemporal masks that consist of q nonzero values. These masks are the counterparts of the q signal values which contribute to each q-linear term. The computed coefficients are nonlinearly weighted and summed according to the same schemes that serve for the production of the corresponding q-linear forms. A necessary condition for their approximation is a nonlinearity of polynomial degree $p \geq q$. The most severe problem associated with TLU-approaches is the fact that the desired q-linear terms are contained in sums with many other polynomial terms. The latter are either redundant, if, e.g. all q signals are nonzero, or they may distort the result, if, e.g. one of them is zero: $(1+1+1)^3 = (1.5+1.5+0)^3$. Furthermore, processing must cope with a large signal dynamic and provide sufficient accuracy in order to evaluate the contribution of q-linear terms to the polynomials. Obviously, the situation is aggravated after pooling.

Since networks of orders $q > 3$ do not considerably improve the performance but – owing to the exponential decrease of the probability for coincidences –, have little chance to self-organize, the optimum order is estimated to be $2 \leq q \leq 3$. 'TLU-neurons' that perform cross correlations by linearly weighted synaptic transmission of signal values should therefore receive only about three input signals which is inconsistent with neurobiology. A direct evaluation of q-linear forms by TLUs, i.e., through masks of large q in conjunction with polynomial degrees $2 \leq p \leq 3$, is in general not feasible for the pooling schemes considered here. '$\Sigma\Pi$-neurons' perform coincidence detection by nonlinear dendritic interaction of neighbouring synaptic inputs and produce outputs that essentially are sums of coincidence signals. It is conjectured that coincidence detection in real neurons is based on dendritic conductance changes that depend in a highly nonlinear way on the membrane potential [10]. Synaptic modification, i.e., self-structuring, is assumed to depend on the success of the coincidence [11]. Thus, learning and processing are based on the same local mechanism. However, a global rule is additionally needed for the self-organization of the highly specific pooling structure.

ACKNOWLEDGEMENTS

This paper was prepared and written while on leave at the 'Ecole Nationale Supérieure des Télécom. de Bretagne'. The support of the 'Dépt. Math. et Syst. de Com.' is acknowledged.

REFERENCES

[1]   Poggio, T.; Woodward, Y. and Torre, V., Optical flow: computational properties and networks, biological and analog, in: Durbin, R.; Miall, C. and Mitchison, G. (eds.), *The Computing Neuron* (Addison-Wesley, Wokingham, 1989) pp.355-370.
[2]   Glünder, H., Correlative velocity estimation: visual motion analysis, independent of object form, in arrays of velocity-tuned bilocal detectors, J. Opt. Soc. Am. A **7** (1990) 317.
[3]   Bülthoff, H.; Little, J. and Poggio, T., A parallel algorithm for real-time computation of optical flow, Nature **337** (1989) 549.
[4]   Yuille, A.L. and Grzywacz, N.M., A computational theory for the perception of coherent visual motion, Nature **333** (1988) 71.
[5]   Glünder, H., A dualistic view of motion and invariant shape analysis, in: Simon, J.C. (ed.), *From Pixels to Features* (Elsevier, Amsterdam, 1989) pp.323-332.
[6]   Glünder, H., Invariant description of pictorial patterns via generalized autocorrelation functions, in: Meyer- Ebrecht, D. (ed.), *ASST '87* (Springer, Berlin, 1987) pp.84-87.
[7]   Lohmann, A.W. and Wirnitzer, B., Triple correlations, Proc. IEEE **72** (1984) 889.
[8]   McLaughlin, J.A. and Raviv, J., Nth-order autocorrelations in pattern recognition, Information and Control **12** (1968) 121.
[9]   Minsky, M. and Papert, S., *Perceptrons* (The MIT Press, Cambridge/MA, [1]1969, [2]1988)
[10] Dingledine, R., N-methyl-aspartate activates voltage-dependent calcium conductance in rat hippocampal pyramidal cells, J. Physiol. **343** (1983) 385.
[11] Kelso, S.R.; Ganong, A.H. and Brown, T.H., Hebbian synapses in hippocampus, Proc. Natl. Acad. Sci. USA **83** (1986) 5326.

Parallel Processing in Neural Systems and Computers
R. Eckmiller, G. Hartmann and G. Hauske (Editors)
© Elsevier Science Publishers B.V. (North-Holland), 1990

SELF ORGANIZATION OF A NETWORK
LINKING FEATURES BY SYNCHRONIZATION[*]

Georg HARTMANN and Siegbert DRÜE

Fachbereich Elektrotechnik
Universität Paderborn
Pohlweg 47-49, D4790 Paderborn, West-Germany

In a two dimensional network of neurons with oriented
receptive fields, we could show that subsets of neurons,
responding to continuous contours, learn to synchronize
their action potentials. A distributed mechanism provides
wide-range synchronization on the base of self organizing
local interconnections. So an exploding number of meaningful combinations of neurons, representing continuous contours, can be synchronized by a constant number of interconnections per neuron.

1. INTRODUCTION

Several groups have proposed models for feature linking by temporal
codes [1], [2], [3], and neurophysiological experiments [3], [4],
[5] are supporting this concept. Synchronization mechanisms have
been described by [3] and by the authors. Synchronity is a good
label for neurons with matching features, this temporal label can
easily be detected, and networks are able to learn synchronization.

2. A CHAIN OF SYNCHRONIZED NEURONS

Our model neurons are of the well known type, described by French
and Stein [6]. To explain the synchronizing mechanism, we restrict
our two dimensional network to a chain for a moment (fig. 1). The
neurons are driven by afferent signals at feeding inputs. In addition each neuron receives signals from its next neighbours (fig. 1)
at trigger inputs. A neuron may be stimulated by a single spike at
a trigger input if its membrane potential is close to the threshold. A neuron, however, will not be activated by trigger signals
only. This is due to a very short time constant at the triggering
inputs, preventing significant temporal integration.

Suppose, all neurons of the chain in fig. 1 receive feeding input.
Neuron 1 will trigger neuron 2, and after a short delay neuron 2
will send a spike to neuron 3 etc. (fig. 2a). The total delay may
be reduced, if neuron 6 happens to fire first (fig. 2b). But these
examples do not describe the real situation. A neuron may fire
input driven, it may fire due to stimulation by its neighbour, or
it may be in its absolute refractory period during stimulation. In
fig. 2c, neurons 3, 6 and 10 are firing input driven, all the other
neurons are firing due to stimulation. Neuron 3 triggers its neighbours 2 and 4, neuron 2 triggers neuron 1, and neuron 4 would like
to trigger neuron 5. But neuron 5 has been triggered just before by

---

[*] Supported by Ministry for Res. and Techn., Grant No. ITR 8800 D/0

neuron 6 and so it is in its absolute refractory period.

In a chain with n neurons, i neurons will fire input driven and s=n-i will fire due to stimulation. From each of the i input driven neurons, two wave fronts of triggering signals are starting, traveling up and down the chain. A wave front stops, as soon as it collides with another front, travelling in the opposite direction (fig. 2c). The mean number of neurons, triggered by one wave front is n/2i, which is obviously independent of the number n of neurons in the chain. This number n/2i multiplied by the delay between triggering signal and output spike, is a good measure for the time interval, within which all neurons of a chain are firing. A detailed description of the synchronization mechanism and of the parameters of the model neurons is given in [8].

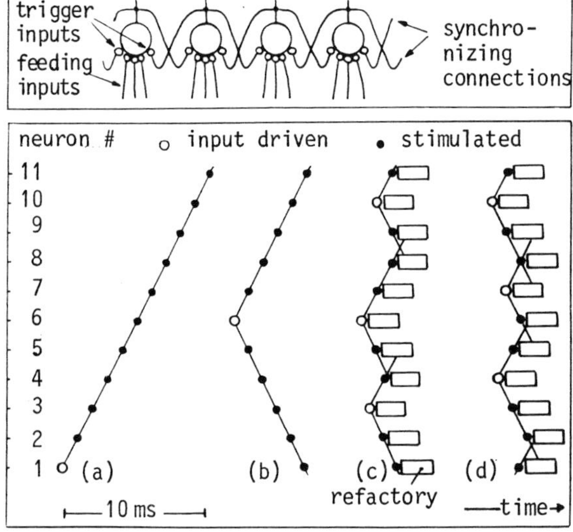

Fig. 1: The synchronizing connections in a chain of neurons.

Fig. 2: The synchronization mechanism in a chain of neurons. Without input driven activity of the stimulated neurons, delay would cumulate proportional to the length of the chain (a, b). With input driven activity, wave fronts start from different points of the chain and stop at refractory neurons (c, d).

3. FEATURE LINKING IN A TWO DIMENSIONAL VISUAL FIELD

We simulated a two dimensional network, representing a small visual field. This visual field was subdivided into 16x16 subfields, arranged in a hexagonal grid (fig. 3). For each subfield there is a complete set of detector neurons with oriented receptive fields. These detectors are described more precisely in previous publications (Hartmann [7]). In our simulations we have simplified the feeding inputs and we have also reduced the complexity of the detector set by omitting detectors with highly curved receptive fields. We have added, however, the interconnections, necessary for synchronization.

Suppose a bright line is running through subfield A, B, C, D, E, F, G in fig. 3. The line shall fit to the receptive fields of neuron 1, 2, 3, 4, 5, 6, and 7, so that these neurons are excited. As in our chain configuration, neuron 1 and 2, 2 and 3, and all the other adjacent pairs shall be mutually interconnected. As we have seen in the last chapter, all the neurons of this chain will synchronize their spikes.

Now we change the input pattern to a slightly different contour, encoded by the neurons 1, 2, 8, 9, 10, and 7 (fig. 3). In this case neuron 2 must be connected with neuron 8 instead of neuron 3, and similarly neuron 7 with neuron 10 instead of neuron 6. There may be other lines, requiring connections between neuron 2 and 11 or 12 or 13. Generally spoken, each neuron in a subfield must be mutually connected with two groups of fitting neurons in two adjacent subfields.

As all neurons with "fitting" receptive fields are mutually linked now, also those neurons will always be connected which are activated by any arbitrary continuous line. But the above discussed chain is just a subset of the complete interconnection, and so we only have to prove, that the additional connections will not disturb the synchronization of the chain. We have to discuss two cases. Firstly, a neuron with full interconnection will not only send triggering spikes to its active neighbours in the chain, but also to inactive fitting neighbours. But a triggering spike can not excite a neuron without sufficient feeding input. Secondly, a neuron with full interconnection can not only receive triggering spikes from its neighbours, but also from its neighbours outside the chain. These outside neighbours, however, are not active and so they will not influence synchronization.

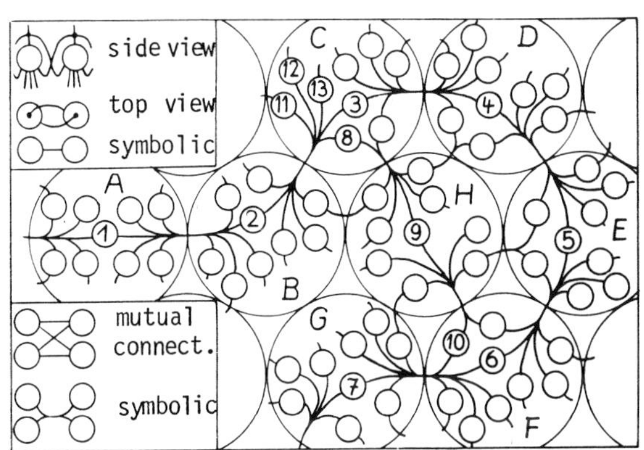

Fig. 3: The synchronizing connections in a two dimensional assembly of neurons. Only part of neurones is shown for clearness. Also for clearness, the two connections between a pair of neurons are symbolized by one (box top left). Similarly mutual connections between groups of neurons are symbolized (box bottom left).

In simulations we have presented different continuous lines activating up to 137 neurons. Synchronization was as good as in the case of simple chains, and was not significantly influenced by spontaneous activity of "outside neighbours".

4. SELF ORGANIZATION OF SYNCHRONIZATION

This synchronous response to continuous contours is due to the special interconnection scheme between trigger inputs of matching neurons. Simulations show that these interconnections can be organized by unsupervised learning. We presented arbitrary continuous contours to the network, and allowed the weights at the trigger inputs to change according to a Hebbian-like rule. The general idea is to start with a complete mutual interconnection between all the trigger inputs of neurons in adjacent subfields and to increase the weights of simultaneously active pairs. As the synchronizing interconnections are exclusively local, we restricted the network to a group of seven subfields (fig. 3). The resulting weights were copied to the other subfields at the end of the learning period.

Now we presented arbitrary continuous contours, running through the group of seven subfields. Let the pair of neurons i and j with matching receptive fields simultaneously be activated by a contour, and let neuron j send a spike to all its neighbours, then weight w_{ij} at the trigger input of neuron i will be increased by Δw_{ij} if neuron i is supraliminal.

If only continuous contours are presented, this learning rule is sufficient. If also arbitrary textures are presented, there will be almost no response of the detector neurons. A very small fraction of textures will fulfil the input condition of one detector neuron, and only an extremely low fraction may happen to active a non-matching pair of neighbouring neurons, causing erroneous learning. These errors are corrected by a very small term in the learning rule, decreasing all weights w_{ij} at subliminal neurons i, receiving a spike from a neighbour j. We can write

$$\Delta w_{ij} \text{ (per spike from j at i)} = \begin{cases} p \text{ if } w_{ij} \leq W \text{ and i supraliminal} \\ -q \text{ if } w_{ij} \geq q \text{ and i subliminal} \end{cases}$$

with $q \ll p$ and W as an upper limit for w_{ij}. The positive increments $\Delta w_{ij} = p$ were adjusted to provide quick learning, and to change a weight to its maximum W by some hundred spikes from a matching neighbour. In a biological system quick learning is necessary, if synchronization shall be learnt by "looking around in a structured environment". The negative increments were adjusted to destroy erroneous weights very slowly and to preserve correctly learnt weights of matching pairs. Learning was switched off at the end of a period with high plasticity. There are first hints that neurons with complex receptive fields may be organized on the base of synchronization mechanism.

REFERENCES

[1]   von der Malsburg, C.: The correlation theory of brainfunction. Internal report 81-2, Dpt. Neurobiology, Max Planck Institute for Biophysical Chemistry (1981)
[2]   Koenderinck, J.: The concept of local sign. In A. J. van Doorn et al., Eds. Limits in Perception, VNU Sci. Press. 495-549 (1984)
[3]   Eckhorn, R. et al.: Feature linking via stimulus-evoked oscillations: Experimental results from cat visual cortex and funcional implications from a network model. Proc. IJCNN89, IEEE, 1.723-1.730 (1989)
[4]   Freeman, W. J.: Mass action in the nervous system. Academic Press New York (1975)
[5]   Gray, C. M., Singer, W.: Stimulus specific neuronal oscillations in the cat visual cortex: a cortical functional unit. Soc. Neurosc. abstr. 404.3 (1987)
[6]   French, A. S., Stein, R. B.: A flexible neuronal analog using integrated circuits. IEEE Trans. Biomed. Eng., 17, 248-253 (1970)
[7]   Hartmann, G.: Processing of continuous lines and edges by the visual system. Biol. Cybern. 47, 43-50 (1983)

Parallel Processing in Neural Systems and Computers
R. Eckmiller, G. Hartmann and G. Hauske (Editors)
© Elsevier Science Publishers B.V. (North-Holland), 1990

A SELF-ORGANIZING NETWORK FOR COMPLETE FEATURE EXTRACTION

Jeanne RUBNER, Klaus SCHULTEN* and Paul TAVAN

Physik-Department
Technische Universität München
8046 Garching, Federal Republic of Germany

We describe a two-layered network of linear neurons that organizes itself as to extract the maximal amount of information contained in a set of presented patterns. The weights between layers obey a Hebbian rule, whereas the lateral, hierarchically organized weights within the output layer follow an Anti-Hebbian rule. For a proper choice of the learning parameters, this rule forces the activities of the output units to become uncorrelated and the lateral weights to vanish. The weights between the two layers converge to the eigenvectors of the covariance matrix of input patterns, i.e. the network performs a principal component analysis of the input information. Consequently the output units become detectors of orthogonal features, similar to ones found in the brain of mammals.

1. INTRODUCTION

Although part of the synaptic connections in the brain is genetically specified, postnatal visual input plays an essential role in the organization, birth and death of synapses. One expects that local rules, like Hebb's rule [1], govern the postnatal organization of the brain and the formation of feature detectors or visual filters. These expectations raise the general question how a sensory system, in response to input information, can organize itself according to local rules so as to form feature detectors which encode mutually independent aspects of the information contained in patterns presented to it.

In the following section we describe a simple, two-layered neural network as a model for such a system. We sketch the mathematical properties of this model and present results of simulations yielding visual filters that respond to patterns of varying orientations and spatial frequencies.

2. THE NETWORK MODEL

The network consists of an input and an output layer with N_i and N_o neurons, respectively. The units exhibit real, continuous-valued activities $\mathbf{i} = (i_1, .., i_{N_i})$ and $\mathbf{o} = (o_1, .., o_{N_o})$. The two layers are completely interconnected, and the weight of the connection between input unit j and output unit m is denoted by w_{jm}. The set of weights connecting an output unit m to all input units forms the weight vector $\mathbf{w}_m$, the transpose of which is the m-th row of the weight matrix $\mathbf{W}$. The set of N_π presented patterns is denoted by $\{\mathbf{p}^\pi = (p_1^\pi, .., p_{N_i}^\pi), \pi = 1, \ldots, N_\pi\}$. The activities of the input units correspond to the presented patterns, i.e., $\mathbf{i} = \mathbf{p}^\pi$, the activities of the output units are the sums of the inputs weighted by the synaptic strengths, i.e., $\mathbf{o}^\pi = \mathbf{W}\,\mathbf{p}^\pi$.

The weights between the two layers are adjusted upon presentation of an input pattern $\mathbf{p}^\pi$ according to a Hebbian rule, i.e., $\Delta\mathbf{w}_m = \eta\,\mathbf{p}^\pi o_m^\pi$ with positive η [1]. As suggested in Ref. [4], we choose the pattern set such that $\langle \mathbf{p}^\pi \rangle = 0$. Here, the brackets $\langle \ldots \rangle$ denote the average over the set of patterns.

If the network contains a single output unit, the Hebbian rule and an Euclidean normalization of weights after every update, i.e., $\sum_i w_{i1}^2 = 1$, render weights which characterize the direction of maximal variance of the pattern set [4], [5]. Equivalently, the weightvector $\mathbf{w}_1$ converges

---

*present adress: Department of Physics, University of Illinois, Urbana, Ill 61801, USA

to the eigenvector with the largest eigenvalue of the covariance matrix $\mathbf{C}$ of the pattern set, whose elements are given by are given by $C_{jk} = \langle p_j^\pi p_k^\pi \rangle$. Diagonalizing a covariance matrix corresponds to the statistical technique of principal component analysis [6]. Thus, a Hebbian learning rule for normalized weights yields the first principal component of the input data set. Consequently, the output unit corresponds to a visual filter extracting the most important feature contained in the set of presented patterns.

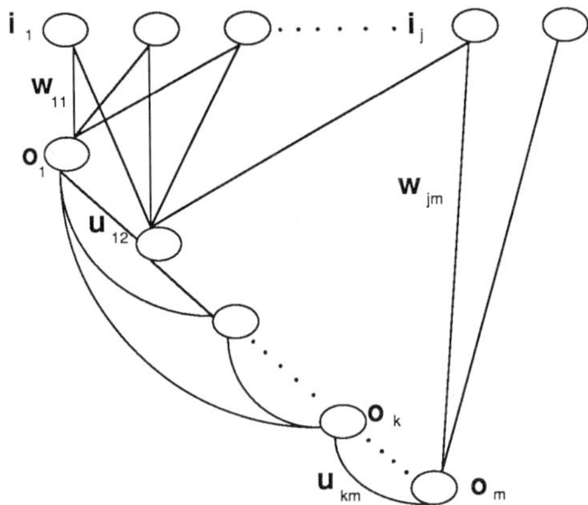

Fig.1 Scheme of the proposed network.

However, a single unit only processes a fraction of the total information contained in a pattern. In order to transmit the complete information between the two layers, more output cells and a different learning rule are required. As a result of an unsupervised learning process, the weight vectors of these units should represent the remaining principal components, i.e., the remaining eigenvectors of $\mathbf{C}$. For that purpose we assume the existence of lateral, hierarchically organized connections with weights u_{lm} between the output units. Then the activity of the m-th output cell is given by $o_m^\pi = \mathbf{w}_m \cdot \mathbf{p}^\pi + \sum_{l<m} u_{lm} \mathbf{w}_l \cdot \mathbf{p}^\pi$. Figure 1 shows a scheme of the network. We propose that these lateral weights adapt themselves according to an *anti-Hebbian* rule: the change of the lateral synaptic weight u_{lm} between two output units l and m is negatively proportional to the product of pre- and postsynaptic activities, namely

$$\Delta u_{lm} = -\mu \, o_l^\pi \, o_m^\pi, \tag{1}$$

where μ is a positive learning parameter. This strictly local rule forces the lateral weights to vanish and the activities of the output cells to become uncorrelated. Correspondingly, the weight vectors $\mathbf{w}_m$ converge to the eigenvectors $\mathbf{c}_\alpha$ of the covariance matrix $\mathbf{C}$. As a result, the output units correspond to analyzers of mutually orthogonal features that extract the directions of diminishing variance of the input patterns.

3. MATHEMATICAL ANALYSIS

Consider first the case of N_i input and $N_o = 2$ output cells, connected by the lateral weight $u \equiv u_{12}$. For slow weight changes, average products of pre- and postsynaptic activities can be used in the learning rules. Then it is possible to expand the weight vectors in terms of eigenvectors $\mathbf{c}_\alpha$ of $\mathbf{C}$ and to derive differential equations for the expansion coefficients $d_{1\alpha}$, $d_{2\alpha}$ and for the lateral connection u, namely

$$\dot{d}_{m\alpha} = -d_{m\alpha} + \frac{(1 + \eta\lambda_\alpha)d_{m\alpha} + \delta_{m2}\,\eta\,u\,\lambda_\alpha\,d_{1\alpha}}{\sqrt{\sum_\beta \left[(1 + \eta\lambda_\beta)d_{m\beta} + \delta_{m2}\,\eta\,u\,\lambda_\beta\,d_{1\beta}\right]^2}} \equiv f_{d_{m\alpha}}(d_{m\beta}, u), \tag{2}$$

and

$$\dot{u} = -\mu \sum_\beta \lambda_\beta d_{1\beta} d_{2\beta} - \mu u \sum_\beta \lambda_\beta d_{1\beta}^2 \equiv f_u(d_{m\beta}, u), \tag{3}$$

where the dot denotes the time derivative and λ_β the β-th eigenvalue of $\mathbf{C}$ ($\lambda_\beta \geq \lambda_{\beta+1}$).

Carrying out a linear stability analysis for this system of coupled differential equations [3], one finds upper and lower limits for the learning parameter μ,

$$\frac{2}{\lambda_1} > \mu > z^{(2)} = \frac{\eta \Delta \lambda}{\lambda_1 (1 + \eta \lambda_2)}. \tag{4}$$

In order to get analytical results for the case of more than two output units, we assume that $n - 1$ of the total N_o weight vectors have already converged to the first $n - 1$ eigenvectors of $\mathbf{C}$ and that the lateral weights between the corresponding output units have vanished. (Note that a corresponding procedure is not necessary in practice, since our simulations have shown that the weight vectors converge simultaneously). Then, for reasons of symmetry, only those variables u_{mn} and $d_{n\alpha}$ are coupled for which $m = \alpha < n$. Requiring a stable fixpoint for the corresponding differential equations, again yields a lower limit for μ, namely

$$\mu > z^{(n)} = \frac{\eta(\lambda_1 - \lambda_n)}{\lambda_1 (1 + \eta \lambda_n)}. \tag{5}$$

We have checked these analytical results by applying the proposed learning scheme to one-dimensional, random patterns with nearest-neighbor correlations. This corresponds to diagonalizing the tight-binding Hamiltonian of a linear chain of atoms, the eigenvalues and eigenvectors of which are well-known. Thus, we can compute $d_{m\alpha}(t)$ as well as the upper and lower limits for μ. The numerical behavior of the weight vectors and lateral weights is in excellent agreement with the analytically predicted behavior [3].

4. ORIENTATION AND SPATIAL FREQUENCY SELECTIVE CELLS

In the following, we examine which are the essential features of spatially varying patterns and compare the receptive fields obtained by our learning scheme with the ones of simple cells, feature detectors selective to edges or bars, which represent the first stage of spatial information processing in the primary visual cortex [7].

For this purpose, we consider a rectangular lattice of $N_i \times N_i'$ sensory input units representing the receptive field of N_o output units, with $N_o \leq N_i N_i'$. We generate two-dimensional patterns of varying intensity by first selecting random numbers from the interval $[-1, +1]$. Then, in order to introduce information about the topological structure of the receptive field, the random input intensities are correlated, e.g., with their nearest neighbors in both directions. We assume vanishing boundary conditions. Note, that this averaging of neighboring signals corresponds to introducing an additional layer with random activities and with fixed and restricted connections to the input layer.

Receptive fields of simple cells in cat striate cortex can be described by Gabor functions [8], which consist of an oscillatory part, namely a sinusoidal plane wave and a Gaussian, exponentially decaying part. In analogy to the one-dimensional case of random patterns, one expects receptive fields corresponding to the eigenfunctions of a tight binding lattice of atoms, i.e., sinusoidal plane waves with vanishing boundary conditions. In order to implement an exponential decay, we scale the weights between layers by a two-dimensional Gaussian distribution centered at the lattice location $(N_i/2, N_i'/2)$ with widths in x- and y-directions σ_1 and σ_2. The non-homogeneous distribution of weights between layers could correspond to a higher density of nearby input cells.

If the Gaussian distribution is not rotationnally symmetric (i.e., if $\sigma_1/\sigma_2 \neq 1$), degeneracy of eigenvalues is broken and mixing of eigenfunctions does not occur. However, the orientation of

Fig.2 From left to right and top to bottom: contour plots of receptive fields
of output units 1-8 in the case of a square lattice of 20 × 20 input units.

receptive fields is pre–determined, due to imposed symmetry axes. Figure 2 displays contour
plots of the receptive fields of the first eight output cells after 10000 learning cycles (from
left to right and top to bottom). Solid lines correspond to positive, dashed lines to negative
synaptic weights. The input lattice was square, with 20 × 20 units and the parameters of
the Gaussian distribution of weights were $\sigma_1 = 11$ and $\sigma_2 = 14$. Learning parameters η and
μ were equal to 0.05 and 0.1, respectively. Due to the non symmetric Gaussian distribution
of weights, all units have slightly elongated receptive fields. The first unit corresponds to
a simple cell with all-inhibitory synaptic weights. The receptive fields of the second and
third units display an excitatory and an inhibitory region and resemble simple cells, selective
to edges of a fixed orientation. The fourth and sixth units have receptive fields with two
zero-crossings, corresponding to simple cells, selective to bars of a fixed orientation. The
seventh unit is as well orientation selective, with four alternating excitatory and inhibitory
regions. This unit would maximally respond to two parallel lines or bars with fixed distance
and orientation. All the described units have receptive fields that resemble recorded receptive
fields of simple cells in the primary visual cortex [8]. Up to now, there has not been any
experimental evidence for receptive fields of the type of the fifth and eighth units, displaying
four and six lobes. However, if the scheme of spatial information processing in terms of a
local Fourier analysis is correct, such receptive fields might exist in the visual cortex.

ACKNOWLEDGEMENTS

This work has been supported by the Bundesministerium für Forschung und Technologie
(ITR-8800-G9) and the Deutsche Forschungsgemeinschaft SFB 143-C1.

REFERENCES

[1] Hebb, D.O., The Organization of Behavior (Wiley, New York, 1949).
[2] Rubner, J. and Schulten, K., Biol Cybern, in press.
[3] Rubner, J. and Tavan, P., Europhys Lett, in press.
[4] Oja, E., J Math Biology 15 (1982) 267.
[5] Linsker, R., IEEE Computer March (1988) 105.
[6] Lawley, D.N. and Maxwell, A.E., Factor Analysis as a Statistical Method (Butterworths,
    London, 1963).
[7] Hubel, D.H. and Wiesel, T.N., J Physiol 160 (1962) 106.
[8] Jones, J.P. and Palmer, L.A., J Neurophysiol 58 (1987) 1187.

Parallel Processing in Neural Systems and Computers
R. Eckmiller, G. Hartmann and G. Hauske (Editors)
© Elsevier Science Publishers B.V. (North-Holland), 1990

A Self-Organising Neural Net for Depth Movement Analysis

Fritz Seytter

Department of Artificial Intelligence†
University of Edinburgh
Edinburgh, Scotland

Depth movement is a stimulus that is common as well as important for all visually orienting creatures and widely neglected in computational models of vision.

Some animals like cats, and humans, are not born with a working visual system and learn to see when they first receive visual stimuli.

A system has been devised and simulated which is able to develop into a working analysing device for depth movement. It consists of a layered network of 'neurons' with linear and nonlinear connections (Σ-Π Units), involving time delays.

The structure is at first indifferent to the stimulus to be recognised. Simple local and spread mechanisms of the net, associated with biological properties, lead to the formation of detectors sensitive to depth movement. Important for this is the typical sequence of scaling rates produced by a linear depth movement which can be used to measure the 'Time to Contact' of an approaching object.

1 Introduction

1.1 Predetermined versus self-organising systems

In a creature that is able to perceive and react perfectly from the moment of birth all neural connections must be determined genetically. But the high number of connections, often of neurons spatially far apart in the brain tissue, suggest that this is not the case. It is more likely that a general structure is encoded genetically, defining which part of the brain is connected with which other part via axons, how far the dendrites reach out etc., rather than every single synaptic connection. It is the probability for connections to the vicinity and to remote areas of the brain that is predetermined. This prestructured brain must be trained by stimuli from the outside world in order to function correctly. For biological evidence supporting this view see [2] and [3].

1.2 The optical flow field originated by depth movement

The movement of an object in a (3-d) scene will cause a pattern of local velocities in the (2-d) projection on the retina, called optical flow field. The mathematical properties of the optical flow field have been analysed thoroughly in many publications, e.g. [4].

Consider a flat, textured object, orientated perpendicularly to the direction it is seen, which moves with constant speed towards the observer. The projection of the object will be expanding as the object is looming. The local velocities point away from a focus of expansion and their magnitude increases linearly with the distance from the centre.

For a linear depth movement the rate of expansion grows in time according to a hyperbolic law. Independent of actual speed or size of the object, the rate of expansion contains information about the time remaining until collision (Time to Contact) [4].

---

† The work leading to the results presented was performed at the Technische Universität München, W-Germany in partial fulfilment of the requirements for the degree of a Dipl.Ing. [1].

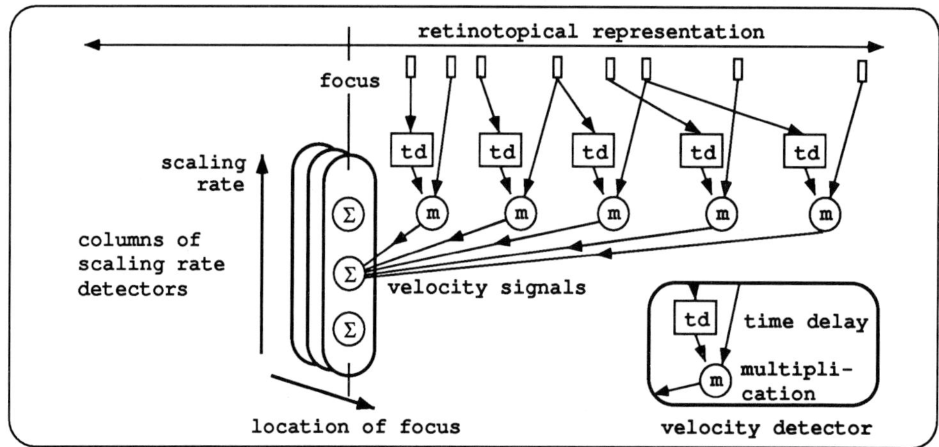

Figure 1: Wiring of a scaling rate detector.

2   A System for Analysing Depth Movement

2.1   Velocity detectors

To be able to analyse the optical flow field we need detectors for local velocities in a retinotopical representation. A neural setup with the ability to extract local velocities was described by Hassenstein and Reichardt in 1956 [5], a simplified variant of which is used here. The signals from two points in a retinotopical representation are combined nonlinearly (by multiplication), one of them delayed by a fixed amount of time, thus forming a detector for a certain velocity and a certain direction of movement. In the model presented here it is assumed that the delay time is the same for all detectors. The velocity detectors are tuned to different velocities by the detector width.

This simple setup for the velocity detectors requires pre-processed input with a low density of excitation. Otherwise the detectors are swamped by false combinations that excite corresponding inputs by chance. Edge extraction proved to reduce the density of excitation sufficiently.

2.2   Connecting velocity detectors to detect scaling rate

In order to analyse the optical flow field originated by depth movement, we need to collect signals from velocity detectors at particular locations, with particular tuned velocities: those that respond when a certain scaling rate occurs with respect to a certain focus of expansion. For a geometrically undistorted retinotopical representation the parameters for the motion detectors can be easily evaluated [6]. The resulting setup is shown in Figure 1, which is a cross section through a scaling rate detector. The first stage shows the velocity detectors, which can serve several scaling rate detectors at different locations. Then there are columns of scaling rate detectors (summation) which react to certain Time to Contact values of depth movement with its focus of expansion at a specific location. The selection of velocity detectors for 3 different, successive scaling rate detectors is shown in Figure 2a.

If an object approaches, the scaling rate detectors of a column associated with a certain focus location will react successively, giving an estimate for the Time to Contact and the direction from where the object approaches.

2.3   Making the system adaptive

If the retinotopical representation is geometrically distorted, for example due to a fovea, it is impossible to wire the analysing systems according to simple geometrical rules as we did for the ideal

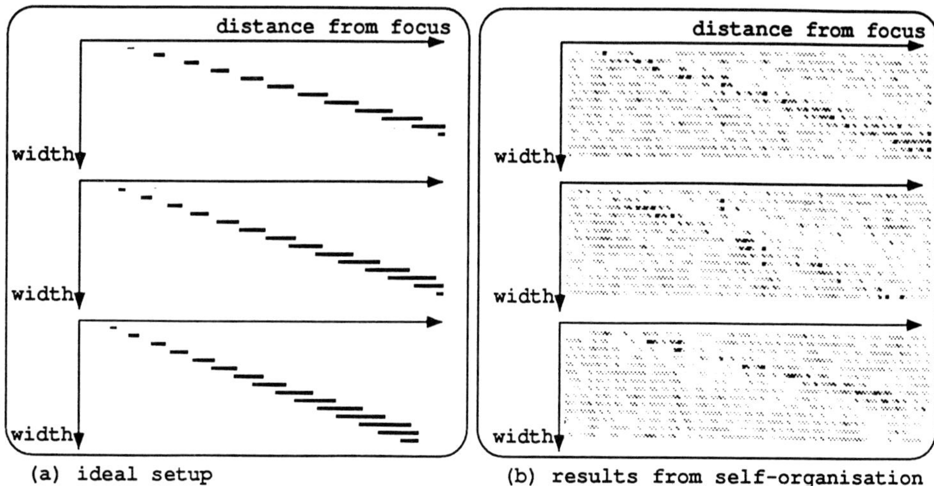

Figure 2: The selection of velocity detectors to form scaling rate detectors.

setup of Figure 1. Rather than analysing the geometrical mapping of the camera system exactly one could use a self-structuring mechanism that automatically chooses the right subset of velocity detectors to form a scaling rate detector.

Such a system has been implemented in two versions: one learns the correct mapping in a supervised way, and the other relies on the regularities in the depth movement stimuli to structure itself out of purely local rules. The setup for both versions of the system is as follows:

- A scaling rate detector forms a weighted sum of a great number of randomly chosen velocity detectors within a certain area of the visual field. For each location in the image there exist a number of such summation neurons.

- The weight of a connection is adjustable, following a local synaptic (Hebbian) rule [7] which strengthens a weight when the velocity detector and the scaling rate detector are active at the same time and weakens it when only one of them is active. Hebbian rules are widely accepted as one of the mechanism responsible for learning in biological nervous systems.

- The overall activity is controlled by measuring the average over all weights, which is then used to adjust the strengthening and weakening rates of the Hebbian rule.

2.3.1 Supervised learning

Under supervised learning, a depth movement sequence is given as input. The scaling rate detector which should respond at a certain time (i.e. to a certain scaling rate) is activated from outside, whereas all others are suppressed. After offering enough training sequences the system will have changed its weights so that it responds like the ideal, pre-wired setup. The only difference is that this adaptive solution can deal with any homeomorphic distortion of the retinal image.

2.3.2 Unsupervised learning, self-organisation

For a self-structuring system it cannot be known in advance which output will encode which feature, unless the structure already encodes neighbourhood relationships [8] which are reflected by the input. Here, temporal neighbourhood is an important feature which should be reflected by structural neighbourhood in the system. One way to achieve this is lateral influence among the scaling rate detectors. Since it is necessary to map temporal neighbourhoods onto the spatial structure of the system, time delays are required in the interactions if we assume that the Hebbian rules only work for concurrent events. The mechanism used here is the following:

- The neuron which accumulates the highest input signal becomes active, all others in the same column [1] are suppressed, in a 'winner-take-all' fashion.
- The weights are updated according to the Hebbian Rule. Weights from signals that lead to the excitation of the active neuron are strengthened, ones that tried to excite others are weakened.
- Once a neuron has fired, it is exhausted and cannot be excited again for a certain recovery time. This prevents one neuron from becoming dominant over the others.

In this system the temporal neighbourhood will not be reflected by spatial neighbourhood, that is, it develops scaling rate detectors which are randomly distributed in the column as shown in Figure 2b. At the start of the learning phase there is no information in the system about the structure to be developed. However in the end we obtain a set of detectors for scaling rate, after offering enough training sequences. This shows clearly that the system is self-structuring.

3   Conclusion

- It has been possible to create a working system for depth movement analysis out of a 'homogeneous' pre-structure by self-organisation.
- The system developed is capable of becoming sensitive to the stimulus it has been trained on, using simple, biologically suggestive mechanisms.
- Temporal neighbourhood is mapped onto spatial structure using time delays and recovery time, mechanisms that can also be found in real neurons.

References

[1] Seytter F., *System zur Analyse des "Optical Flow" bei Tiefenbewegungen*, (Diplomarbeit TU München, Lehrstuhl für Nachrichtentechnik, 1989)

[2] Hubel D.H., Wiesel T.N., *Receptive fields of cells in striate cortex of very young, visually inexperienced kittens*, J.Neurophysiol. 26 (1963) 994-1002

[3] Singer W. et al., *Restriction of visual experience to a single orientation affects the organization of orientation columns in cat visual cortex: a study with Deoxyglucose*, Expl. Brain Res. 41 (1981) 199-215

[4] Lee D.N., *The optic flow field: The foundation of vision*, Phil. Trans.Roy.Soc.Lond.B, 290 (1980) 169-179

[5] Hassenstein B., Reichardt W., *Systemtheoretische Analyse der Zeit-, Reihenfolgen- und Vorzeichenauswertung bei der Bewegungsrezeption des Rüsselkäfers 'Chlorophanus'*, Z. Naturforschg. 11b (1956) 513-524

[6] Glünder H.,*Invariante Bildbeschreibung mit Hilfe von Autovergleichs- Funktionen*, (Dissertation TU München, Fakultät für Elektrotechnik und Informationstechnik, 1988)

[7] Hebb D.O, *The organization of behaviour*, (Wiley, New York, 1949)

[8] Ritter H., *Selbstorganisierende neuronale Karten* (Dissertation TU München, Fakultät für Physik, 1988)

---

[1]In the present implementation it is assumed that the spatial neighbourhood relations (columns) have already been established. See [8] for the self-structuring of spatial maps.

Section 9
Neural Networks
for Auditory
Pattern Recognition

Parallel Processing in Neural Systems and Computers
R. Eckmiller, G. Hartmann and G. Hauske (Editors)
© Elsevier Science Publishers B.V. (North-Holland), 1990

SEQUENCE ANALYSIS IN FEEDBACK MULTILAYER PERCEPTRONS

Hans-Ulrich BAUER* and Theo GEISEL

Institut für Theoretische Physik and SFB Nichtlineare Dynamik
Universität Frankfurt, D-6000 Frankfurt/Main 11, Fed. Rep. of Germany

Neural style speech recognition requires sequence analyzing neural networks
which provide certain invariance properties. These include time warping invariance and robustness with regard to coarticulation of sequence elements (phonemes). In this contribution we show that multilayer perceptrons with feedback
from the output layer to the input layer meet these requirements. We discuss
a learning rule and the stability of the resulting multistable states. Several test
examples are given, including a threestage network which detects words in a
sequence of letters.

1. INTRODUCTION

In many perception problems the time variation in the continuous input stream contains
information, which has to be taken into account for successful recognition. In the visual system this is the case for the detection of motion, a problem with a rather simple, nevertheless
very hard to detect time variation [1-3]. The most prominent recognition problem which
requires temporal as well as spacual analysis of the input stream is speech recognition.
Since human speech recognition performance is far beyond that of any artificial system, a
lot of work has been focused on the application of neural networks to this problem [4-10].
Besides the enormous (combinatorial) number of possible sequences the networks have to
cope with time warping of the input sequence and with coarticulation of the constituents
of the sequence. A review of neural nets for speech recognition is given in Ref. [11], other
work on sequence analysis with neural networks includes [12-15].

In this contribution we describe the application of multilayer perceptrons with feedback
from the output to the input layer to the analysis of input sequences (Fig. 1). This
kind of network provides time warping invariance as a built-in feature. It allows for rapid
online dismission of redundant data because only preprocessed information is stored in the
feedback loop. Feedback multilayer perceptrons (FMLPs) can be combined hierarchically
in order to process increasingly long sequences, which change on increasingly slow time
scales. In the following sections we briefly describe an appropriate learning rule, discuss the
stability properties of the resulting multistable states, and present several examples, which
demonstrate the invariance with respect to time warping and coarticulation. Finally, we
show that a combination of three FMLPs can detect words in a random sequence of letters,

---

* Supported by the Deutsche Forschungsgemeinschaft (Grant Ge 385/7-1).

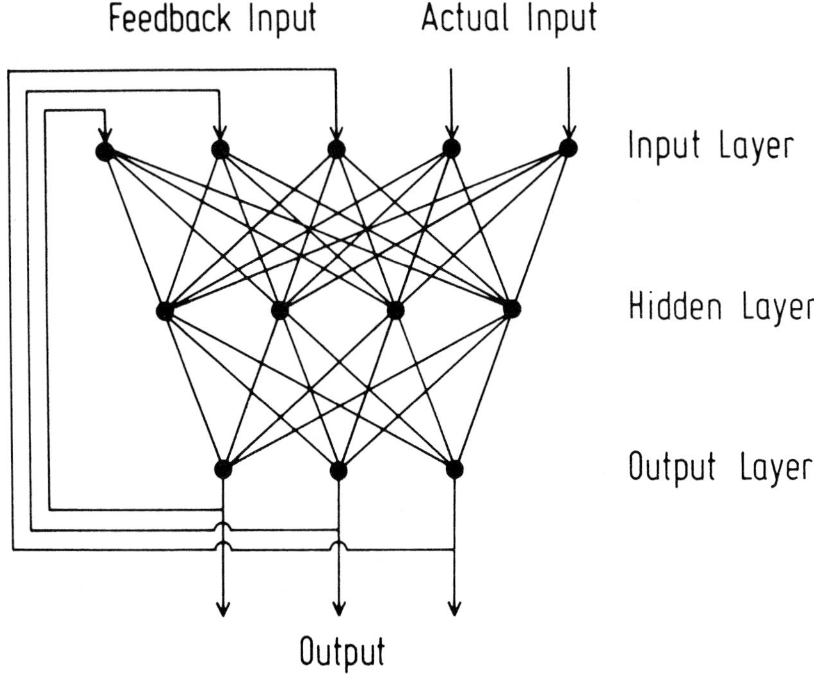

FIGURE 1
Basic topology of a feedback multilayer perceptron (FMLP).

each of which is presented for a random amount of time. More details on some of the results
can be found elsewhere [16].

2. OPERATION OF THE FMLP

2.1 "OPEN LOOP" LEARNING RULE

In contrast to an approach by Almeida [17] and Pineda [18] the feedback in our networks
does not serve to improve the performance but to provide a short term memory. Therefore,
one particular input pattern is not mapped onto one particular output pattern, but onto one
pattern out of several, depending on the history. So Almeida's and Pineda's learning rule
is not applicable in this case. Instead we train the FMLPs by standard conjugate gradient
backpropagation, and apply the desired teacher signal in place of the actual output signal
at the feedback input nodes. The effect of this substitution is like opening the feedback
loop during learning, hence the term "open loop learning". Similar ideas seem to have been
used by other authors too [19-21].

Unfortunately, in contrast to Almeida's and Pineda's learning scheme, stability of the resulting multistable states is not guaranteed. However, the following trick helps to improve
stability. In MLPs the teacher signal may not contain zeroes or ones, because this would
drive the weights beyond any limit. Instead some semibinary values like 0.1 or 0.9 are used.
Linear stability analysis of FMLPs about the multistable states reveals that choosing the
semibinary representation closer to 0 resp. 1 tends to decrease the eigenvalues of the corre-

sponding Jacobian. With all eigenvalues eventually obeying $| \lambda | < 1$ the particular state is stable.

2.2 TEST OF THE STABILITY PROPERTIES

In order to test this analytical stability argument we trained several networks on the same processing task (analyzing a sequence of 2-bit patterns). Slowly decreasing the distance ϵ between the teacher values and 0 or 1 and retraining the nets at each ϵ-value led to a sequence of network realizations. Analyzing the stability properties of each pattern in each net showed, that the eigenvalues of the Jacobians indeed decreased (and stability increased) with decreasing ϵ. Below some threshold ϵ all patterns turned out to be stable.

2.3 COARTICULATION

Having smooth instead of sharp transitions between the letters constitutes the problem of coarticulation. Even after training with sharp transitions only the FMLPs manage to identify a substantial percentage of smooth transitions. Preliminary results indicate, that additional training of a few smooth transitions allows for the transition time to be varied in a wide range without decreasing the detection performance.

3. APPLICATION EXAMPLE

Last we discuss a simple test problem and show, how it can be solved by a combination of three FMLPs. The input sequence consists of a (random) stream of letters ("r", "s", "t", "p", "o", "e"). Each letter is presented for a random amount of time, between 3 and 10 network time steps, in order to mimic time warping. In the tests conducted so far, transitions between the letters were sharp. The network has to detect online the occurence of the words "post", "pest", "rost", "rest", "rose", "pose" in the input stream. Input as well as ouput are represented as bit strings (letters plus blank as 7 4-bit strings, words plus blank as 7 3-bit strings). The task is distributed between three FMLPs which operate in sequence. The first subnet detects transitions bewteen two letters and ouputs 15 different 6-bit strings ("syllables"). The second net detects transitions between two and three letter segments, the last net serves as a filter which keeps the last word alive at its output until the next complete word occurs. The performance of the overall network is shown in Table 1. Each column corresponds to one time step. The four different entries per column indicate the actual input and the ouput of the three subnets. N denotes the blank. The first completed word, pest, is kept alive at the output of the third net until the next complete word (rest) occurs. Tests which ran over as many as 2000 time steps resulted in the same perfect performance as shown in Table 1.

REFERENCES

[1] W.S. Stornetta, T. Hogg, B.A. Huberman, Neural Inf. Proc. Sys., ed. D. Anderson, AIP, New York, (1988) 750.
[2] H. T. Wang, B. Mathur, C. Koch, Neural Computation **1** (1989) 92.
[3] H.-U. Bauer, T. Geisel, Int. Journal of Neural Systems, in print (1989).
[4] B. Gold, R.P. Lippmann, M.L. Malpass, Proc. of First Int'l Conf. on Neural Networks, IEEE (1987) IV-427.
[5] B. Kämmerer, W. Küpper, Proc. of nEuro 88, Ed. L. Personnaz, G. Dreyfus, IDSET Paris (1989) 531.
[6] A.J.R. Robinson, F. Fallside, Proc. of nEuro 88, Ed. L. Personnaz, G. Dreyfus, IDSET Paris (1989) 541.

TABLE 1

Performance of three-stage FMLP for the word detection problem.

```
actual input first net:    p    p    p    e    e    e    s    s
output first net:          p    p    p    pe   pe   pe   es   es
output second net:         N    N    N    pe   pe   pe   pes  pes
output third net:          N    N    N    N    N    N    N    N


s    s    s    s    s    s    s    t     t     t     t     t     r    r
es   es   es   es   es   es   es   st    st    st    st    st    r    r
pes  pes  pes  pes  pes  pes  pes  pest  pest  pest  pest  pest  N    N
N    N    N    N    N    N    N    pest  pest  pest  pest  pest  pest pest


r     e     e     e     e     s     s     s     s     s     t     t     t     t
r     re    re    re    re    es    es    es    es    es    st    st    st    st
N     re    re    re    re    res   res   res   res   res   rest  rest  rest  rest
pest  pest  pest  pest  pest  pest  pest  pest  pest  pest  rest  rest  rest  rest


r     r     r     r     r     r     r     r     o     o     o     o     s     s
r     r     r     r     r     r     r     r     ro    ro    ro    ro    os    os
N     N     N     N     N     N     N     N     ro    ro    ro    ro    ros   ros
rest  rest  rest  rest  rest  rest  rest  rest  rest  rest  rest  rest  rest  rest


s     s     r     r     r     r     r     r     o     o     o     o     o     o
os    os    r     r     r     r     r     r     ro    ro    ro    ro    ro    ro
ros   ros   N     N     N     N     N     N     ro    ro    ro    ro    ro    ro
rest  rest  rest  rest  rest  rest  rest  rest  rest  rest  rest  rest  rest  rest


o     o     s     s     s     s     s     s     s     s     e     e     e     e
ro    ro    os    os    os    os    os    os    os    os    se    se    se    se
ro    ro    ros   ros   ros   ros   ros   ros   ros   ros   rose  rose  rose  rose
rest  rest  rest  rest  rest  rest  rest  rest  rest  rest  rose  rose  rose  rose
```


[7] S. Shamma, Proc. First Int'l Conf. on Neural Networks, IEEE (1987) IV-397.

[8] M.A. Cohen, S. Grossberg, D. Storck, Proc.1.Int'l Conf.on N.N., IEEE (1987) IV-443.

[9] A. Waibel, Neural Computation 1 (1989) 39.

[10] R.L. Watrous, L. Shastri, Proc. 1.Int'l Conf. on N. Networks, IEEE (1987) IV-381.

[11] R.P. Lippmann, Neural Computation 1 (1989) 1.

[12] D.W. Tank, J.J. Hopfield, Proc. Natl. Acad. Sci. USA 84 (1987) 1896.

[13] H. Sompolinsky, I. Kanter, Phys. Rev. Lett. 57(22) (1986) 2861.

[14] U. Riedel, R. Kühn, J.L. van Hemmen, Phys. Rev A 38(2) (1988) 1105.

[15] J. Buhmann, K. Schulten, Europhys. Lett. 4(10) (1987) 1205.

[16] H.-U. Bauer, T. Geisel, Phys. Rev. A, in print (1989).

[17] L.B. Almeida, Neural Computing, Ed. I. Aleksander, Kogan Page, London, (1989).

[18] F. J. Pineda, Phys. Rev. Lett. 59 (19) (1987) 2229.

[19] P.J. Werbos, Neural Networks 1 (1988) 339.

[20] M.I. Jordan, ICS Rep. 8604 (Inst. for Cognitive Science, UC San Diego) (1986).

[21] J.L. Elman, CRL Tech. Rep. (Center for Research in Language, UC San Diego) (1988).

Parallel Processing in Neural Systems and Computers
R. Eckmiller, G. Hartmann and G. Hauske (Editors)
© Elsevier Science Publishers B.V. (North-Holland), 1990

A NEURAL NET FOR RECOGNITION AND STORING OF SPOKEN WORDS

Holger BEHME

Drittes Physikalisches Institut
Universität Göttingen
Göttingen, FRG

A neural net for speech processing is described. It
is able to build up a representation of the input space,
the 'phonotopic map'. With this map as a basis, word recognition can be done as well as storing and reproduction
of words.

1. INTRODUCTION

Speech recognition is one of the promising applications of
neural networks. Here a network shall be described which, based
on Kohonen's "Feature Map" [1], represents spoken words as
special sequences of cells or cell activities. This network can
serve as a basis for word recognition as well as for some kind
of "word-normalisation". The network has been designed and
trained to form a "phonotopic map", where every phoneme has its
special neurons assigned to it. A word, as a sequence of
phonemes, is translated into a sequence of neurons. This
sequence is subject to further processing, e.g., recognition
algorithms.

2. THE PHONOTOPIC MAP

The speech data used in our experiments was produced by the
following process: spoken words, sampled with 10 kHz, were
Fourier transformed (length 512 samples with 300-sample Hamming
window, successive frames 10 ms apart), converted into a 257-
channel power spectrum and then transformed into 37-channel Bark
spectra according to Traunmüller's formula [2]:

$$z \text{ (Bark)} = \frac{26.81 \ f}{1960 \ Hz + f} - 0.53 \ .$$

Actually, the details of this preprocessing are not very important; in former experiments we used 16-channel Bark spectra.
Even some power law (square roots, for example) may be applied.
The phonotopic map will adapt to a very wide variety of
preprocessing methods, as can be understood by looking at the
adaptation rules below.

The speech data is regarded as a sequence of vectors $e(t)$ (in
the 37-dimensional real vector space). The network consists of N
neurons (in our case N = 384 = 16x24), each one connected by to
all 37 input channels by synapses. These synapses are regarded
as a vector in Euclidean space, too, and are called the "representation vector" $r(j)$ of neuron j. The neurons are arranged in
a two-dimensional lattice, where each neuron has its fixed
neighbours.

During the training, the input vectors e(t) of a large set of
many different words are presented one after another (in this
stage, the sequence of the input vectors is not important). For
every e(t), the following steps are carried out:
1. Find that neuron i, whose representation vector r(i) is
   closest (with respect to Euclidean metric) to the e(t).
2. Change r(i) according to r(i) --> r(i) + c(e(t)-r(i)).
3. Change the r(j) of the neighbours of i according to
   r(j) --> r(j) + h(i,j)c(e(t)-r(j)).
Here c is a convergence parameter slowly decreasing to zero with
time, and h(i,j) is a function being nonzero only in the neighbourhood of i and decreasing with the lattice distance towards j.
   At the beginning, the r(j) are randomly distributed; after
convergence 3 they form the phonotopic map: each neuron has
adapted its r(j) to represent a special part of the input space
- a special class of input vectors or, as one can say, a special
"sound". Moreover, lattice neighbours represent similar sounds,
lattice distance now being a kind of "similarity measure". If
the Bark spectra of some word are fed into the network and step
1 above is applied, a sequence of neurons i(t) is obtained as a
representation of the input sequence e(t). (Remember that e(t)
is a 37-dimensional real vector while i(t) is only one integer!)

3. WORD RECOGNITION

   In the above way, a representation of a word is found as a
sequence of integers. (The sequences may have different lengths
for different words or even for different realisations of the
same word.) We now tried a very simple recognition method: for
each word it was counted, how often each cell appeared in the
sequence i(t). This transforms the word into a pattern of N
(=384) integer values (most of which are zero). These patterns
were presented to a two-layered perceptron. Notice that by
transforming the sequence into a static pattern, the temporal
information is discarded. However, the recognition worked very
well.
   As speech data, the ten digits were used (spoken in German by
one speaker). Each of the digits was spoken 12 times in
different ways (different speed, emphasis etc.). First of all, a
phonotopic map was trained with these data. Then the words were
transformed into static patterns as described above, getting a
total of 120 patterns, one dozen for each digit.
   In the first experiment, the perceptron was trained with all
patterns. After 50 training cycles we checked the recognition
rate, which was 117 out of 120 (97.5%).
   In further experiments we gave the perceptron only 110
patterns as training data and left one version of each digit to
test the generalisation facilities of our perceptron. Again, for
the recognition of the training data we got error rates of about
2%. The recognition rate for the unknown words varied between
70% and 100%. This variation in recognition rate reflects the
vari- ation between the words. (If the perceptron has never seen
a very exotic version of a word, it will have difficulties to
recognize it. If, on the other hand, the unknown version is an
average version, it will be easy to recognize.)

   The perceptron recognition was only a very crude approach, but
even this crude approach worked. Also a recognition by Euclidean
distance was tried and produced similar results. When using more
words and more speakers, we cannot afford anymore throwing away

the temporal information contained in the sequence i(t). Below, an extended network shall be described, which deals with this information in a special way.

4. STORING AND REPRODUCING WORDS

The problem we considered was to find a way of storing the sequence i(t) in the network, including its temporal structure. With such a feature, a reproduction of something close to the original Bark spectra sequence could be obtained by using the representation vectors of the i(t). But not the reproduction as such is the real intention: if the network can "remember" an ideal form of a word, it could compare a somehow disturbed version with its memory and remove some of the disturbances.

How can the sequence be stored in the network without adding any neurons? - The idea is to build up synaptic connections between the neurons (not only to the 37 input channels, as needed for the phonotopic map). Each neuron i should learn which neurons j are to follow in the sequence. Therefore some changes have to be made:
A cell activity is introduced, which is increased when the neuron is selected in the sequence or by excitation through synapses. To be selected in the sequence means that the neuron has to represent the input in this time cycle, i.e. its representation vector is closest to the present input vector. The activity decreases by a constant factor (<1) every time cycle. No special nonlinearity is used, but the activity is bounded between 0 and an upper limit. All synapses start at zero strength, and only those synapses are increased which connect a neuron to its successors in the sequence. The connections also lead to higher order successors, thus removing ambiguities in the sequence. In every learning step, activity is fed to the actual neuron i(t), the new activities via synapses are calculated and the synapses leading from formerly excited neurons to the 'present' neuron are strengthened according to the cell activities. In detail, the steps performed in each time cycle (for every input vector) are:
1. Select a representing neuron i(t): during learning, apply step 1 of the phonotopic map; during reproduction, choose the neuron with maximum activity.
2. Calculate synapses: during learning, increase the strengths of all synapses leading from those neurons which have been representing neurons during the last T (e.g. 10) time cycles to the presently selected neuron i:
   for j=i(t-1),..,i(t-T) set s(j->i) = s(j->i) + e a(j) with s(j->i) being the strength of the synapse from j to i, a(j) being the activity of neuron j and e a small positive number like 0.01.
   If the representing neuron has already been representing in the preceding cycle, decrease all synapses leading off this neuron: for all k set s(i->k) = f s(i->k) with f below 1, e.g. 0.9. This holds during learning; during reproduction, on the other hand, synapses are increased in such a case, but are restored as soon as the sequence leaves this neuron. This is called the 'impatience rule': one can think of the neuron as to become impatient to continue the sequence. This rule is necessary for correct reproduction of those parts of the sequence where one neuron is selected more than once in succession.

3. Calculate new activities:
     (i)   increase the activity of the representing neuron by a
           constant amount;
     (ii) excite all neurons through their mutual synapses;
     (iii) decrease all activities by a constant factor (e.g. 0.3).
     Always account for of the limitations in cell activity.

The reproduction is started by giving activity to the first
few neurons of the sequence; then the maximum of activity will
travel through the network following the trace of the sequence.
It will stop for a while at cells with weak synapses until the
'impatience' becomes large enough, thus reproducing repetitions
in the sequence, and it will run safely over "crossings" of the
sequence with itself because the synapses look more than one
time cycle into the future. Thus the sequence $i(t)$ is
reconstructed.

With this network, experiments have been carried out. It is
indeed able to store words, even if the sequences have "crossings" or other adversities. The quality of the reconstructed
Bark spectra depends on the number of cells in the phonotopic
map. Further experiments showed that the network can be used,
for example, to change a different sequence $i(t)$ into that one
it had learned, if the differences are not too large. Details
about the network and the phonotopic map can be found in [4].

5. CONCLUSION

The phonotopic map, a vector quantisation of the input space
(Bark spectra), yields a basis for word recognition by neural
networks. Even very simple recognition algorithms (two-layered
perceptrons) are able to work well with the help of the map. The
map can be extended by internal synapses to support more
sophisticated use of the temporal structure of the sequence.

ACKNOWLEDGEMENT
This work has been supported by the Federal Ministry of Research
and Technology (BMFT), project ITR 8800 C8.

REFERENCES

[1] T. Kohonen, Self-Organisation and Associative Memory
    (Springer, Heidelberg, 1988)

[2] H. Traunmüller and F. Lacerda, Speech Communication 6
    (1987) 143-157.

[3] H. Ritter and K. Schulten, Biological Cybernetics 71
    (1988) 59-71.

[4] H. Behme and T. Gramss, Speicherung und Reproduktion ge-
    sprochener Worte mittels neuronaler Netzwerke, in: Fort-
    schritte der Akustik - DAGA '89 (DPG-GmbH, Bad Honnef, 1989)
    pp. 319-322.

Parallel Processing in Neural Systems and Computers
R. Eckmiller, G. Hartmann and G. Hauske (Editors)
© Elsevier Science Publishers B.V. (North-Holland), 1990

ADAPTIVE RESONANCE THEORY: NEURAL NETWORK ARCHITECTURES FOR SELF-ORGANIZING PATTERN RECOGNITION

Gail A. Carpenter and Stephen Grossberg

Center for Adaptive Systems
Boston University
111 Cummington Street
Boston, MA 02215 USA†

ABSTRACT

Adaptive resonance architectures are neural networks that self-organize stable pattern recognition codes in real-time in response to arbitrary sequences of analog or binary input patterns. This article outlines properties of 3 generations of these networks: ART 1, ART 2, and ART 3. In ART architectures, top-down learned expectation and matching mechanisms are critical in self-stabilizing the code learning process. A parallel search scheme updates itself adaptively as the learning process unfolds, and realizes a form of real-time hypothesis discovery, testing, learning, and recognition. After learning self-stabilizes, the search process is automatically disengaged. Thereafter input patterns directly access their recognition codes without any search. Thus recognition time for familiar inputs does not increase with the complexity of the learned code. A novel input pattern can directly access a category if it shares invariant properties with the set of familiar exemplars of that category. A parameter called the attentional vigilance parameter determines how fine the categories will be. If vigilance increases (decreases) due to environmental feedback, then the system automatically searches for and learns finer (coarser) recognition categories. Gain control parameters enable the architecture to suppress noise up to a prescribed level. The architecture's global design enables it to learn effectively despite the high degree of nonlinearity of such mechanisms.

1. INTRODUCTION

Adaptive Resonance Theory (ART) architectures are neural networks that carry out stable self-organization of recognition codes for arbitrary sequences of input patterns. Adaptive Resonance Theory first emerged from an analysis of the instabilities inherent in feedforward adaptive coding structures (Grossberg, 1976a, 1976b). More recent work has led to the development of two classes of ART neural network architectures, specified as systems of differential equations. The first class, ART 1, self-organizes recognition categories for arbitrary sequences of binary input patterns (Carpenter and Grossberg, 1987a). A second class, ART 2, does the same for either binary or analog inputs (Carpenter and Grossberg, 1987b). A third class, ART 3, includes a model of the chemical synapse that solves computational problems of ART systems embedded in network hierarchies, where there can, in general, be either fast or slow learning and distributed or compressed code representations (Carpenter and Grossberg, 1990). In particular, ART incorporates a search mechanism that serves at least three distinct functions: to correct erroneous category choices, to learn from reinforcement feedback, and to respond to changing input patterns.

The main elements of a typical ART 1 module are illustrated in Figure 1. F_1 and F_2 are fields of network nodes. An input is initially represented as a pattern of activity across the nodes, or feature detectors, of field F_1. The pattern of activity across F_2 corresponds to the category representation. Because patterns of activity in both fields may persist after input offset yet may also be quickly inhibited, these patterns are called short term memory, or STM, representations. The two fields, linked both bottom-up and top-down by adaptive

† Acknowledgements: This research was supported in part by the Air Force Office of Scientific Research (AFOSR F49620-86-C-0037 and AFOSR F49620-87-C-0018), the Army Research Office (ARO DAAL-03-88-K-0088), British Petroleum, DARPA, and the National Science Foundation (NSF DMS-86-11959 and IRI-87-16960). We wish to thank Diana Meyers, Cynthia Suchta, and Carol Yanakakis for their valuable assistance in the preparation of the manuscript.

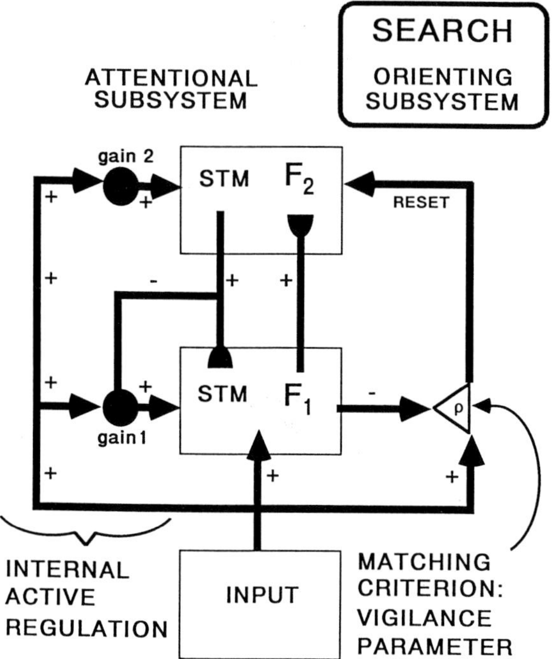

Figure 1. Typical ART 1 neural network module (Carpenter and Grossberg, 1987a).

filters, constitute the Attentional Subsystem. Because the connection weights defining the adaptive filters may be modified by inputs and may persist for very long times after input offset, these connection weights are called long term memory, or LTM, variables. An auxiliary Orienting Subsystem becomes active during search.

2. AN ART SEARCH CYCLE

Figure 2 illustrates a typical ART search cycle. An input pattern **I** registers itself as a pattern **X** of activity across F_1 (Figure 2a). The F_1 output signal vector **S** is then transmitted through the multiple converging and diverging weighted adaptive filter pathways emanating from F_1, sending a net input signal vector **T** to F_2. The internal competitive dynamics of F_2 contrast-enhance **T**. The F_2 activity vector **Y** therefore registers a compressed representation of the filtered $F_1 \rightarrow F_2$ input and corresponds to a category representation for the input active at F_1. Vector **Y** generates a signal vector **U** that is sent top-down through the second adaptive filter, giving rise to a net top-down signal vector **V** to F_1 (Figure 2b). F_1 now receives two input vectors, **I** and **V**. An ART system is designed to carry out a matching process whereby the original activity pattern **X** due to input pattern **I** may be modified by the *template pattern* **V** that is associated with the current active category. If **I** and **V** are not sufficiently similar according to a matching criterion established by a dimensionless *vigilance parameter* ρ, a reset signal quickly and enduringly shuts off the active category representation (Figure 2c), allowing a new category to become active. Search ensues (Figure 2d) until either an adequate match is made or a new category is established.

3. ART 2: THREE-LAYER COMPETITIVE FIELDS

Figure 3 shows the principal elements of a typical ART 2 module. It shares many characteristics of the ART 1 module, having both an input representation field F_1 and a category representation field F_2, as well as Attentional and Orienting Subsystems. Figure 3 also illustrates one of the main differences between the examples of ART 1 and ART 2 modules so far explicitly developed; namely, the ART 2 examples all have three processing layers within the F_1 field. These three processing layers allow the ART 2 system to stably

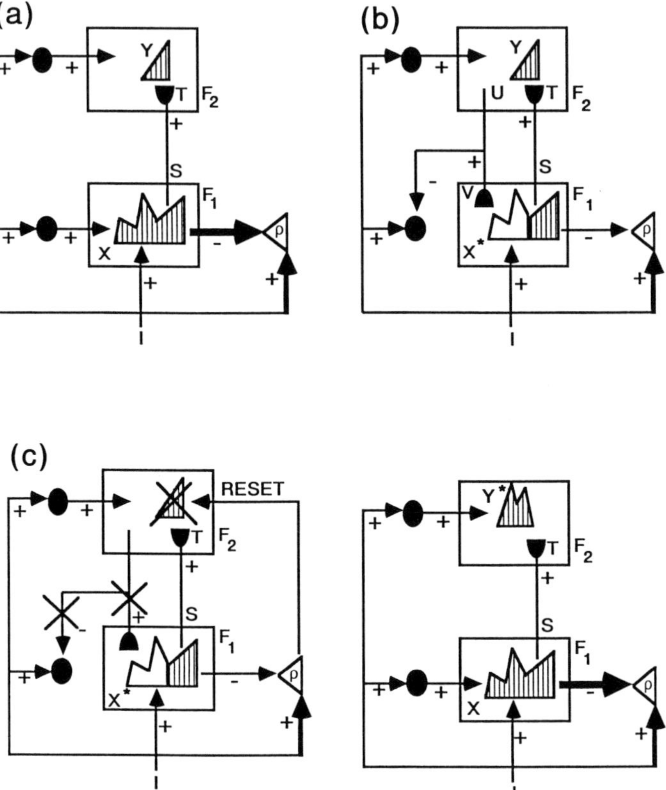

Figure 2. ART search cycle (Carpenter and Grossberg, 1987a).

categorize sequences of analog input patterns that can, in general, be arbitrarily close to one another. Unlike in models such as back propagation, this category learning process is stable even in the fast learning situation, in which the LTM variables are allowed to go to equilibrium on each learning trial. In Figure 3, one F_1 layer reads in the bottom-up input, one layer reads in the top-down filtered input from F_2, and a middle layer matches patterns from the top and bottom layers before sending a composite pattern back through the F_1 feedback loop. Both F_1 and F_2 are shunting competitive networks that contrast-enhance and normalize their activation patterns (Grossberg, 1982).

4. SEARCH IN AN ART HIERARCHY

We now consider the problem of implementing parallel search among the distributed codes of a hierarchical ART system. Assume that a top-down/bottom-up mismatch has occurred somewhere in the system. How can a reset signal search the hierarchy in such a way that an appropriate new category is selected? The search scheme for ART 1 and ART 2 modules incorporates an asymmetry in the design of levels F_1 and F_2 that is inappropriate for ART hierarchies whose fields are homologous. The ART 3 search mechanism eliminates that asymmetry.

The computational requirements of the ART search process can be fulfilled by formal properties of neurotransmitters (Figure 5), if these properties are appropriately embedded in the total architecture model. In particular, the ART 3 search equations incorporate the dynamics of production and release of a chemical transmitter substance; the inactivation of transmitter at postsynaptic binding sites; and the modulation of these processes via a nonspecific control signal. The net effect of these transmitter processes is to alter the ionic permeability at the postsynaptic membrane site, thus effecting excitation or inhibition of the

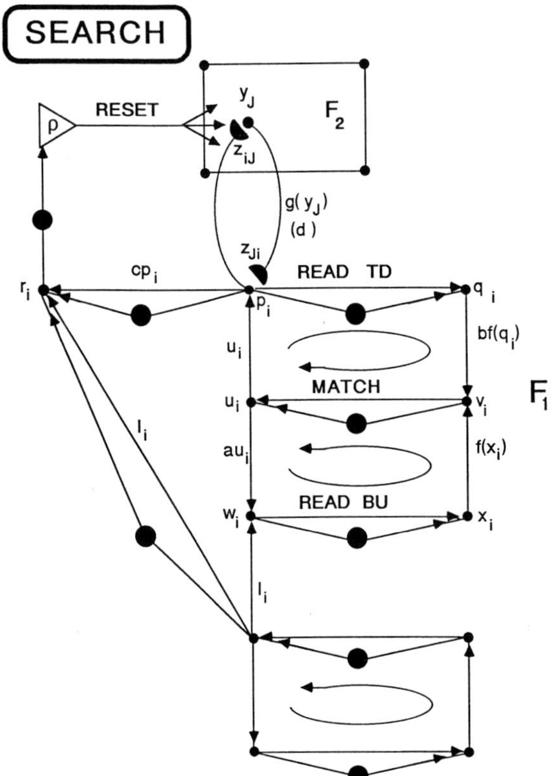

Figure 3. Typical ART 2 neural network module, with three-layer F_1 field (Carpenter and Grossberg, 1987b). Large filled circles are gain control nuclei that nonspecifically inhibit target nodes in proportion to the Euclidean norm of activity in their source fields.

postsynaptic cell.

The notation to describe these transmitter properties is summarized in Figure 6, for a synapse between the ith presynaptic node and the jth postsynaptic node. The presynaptic signal, or action potential, S_i arrives at a synapse whose adaptive weight, or long term memory trace, is denoted z_{ij}. The variable z_{ij} is identified with the maximum amount of available transmitter. When the transmitter at this synapse is fully accumulated, the amount of transmitter u_{ij} available for release is equal to z_{ij}. When a signal S_i arrives, transmitter is typically released. The variable v_{ij} denotes the amount of transmitter released into the extracellular space, a fraction of which is assumed to be bound at the postsynaptic cell surface and the remainder rendered ineffective in the extracellular space. Finally, x_j denotes the activity, or membrane potential, of the postsynaptic cell.

5. SUMMARY OF SYSTEM DYNAMICS DURING A MISMATCH-RESET CYCLE

Figure 7 summarizes system dynamics of the ART 3 search model during a single input presentation. Initially the transmitted signal pattern $\mathbf{S} \cdot \mathbf{u}_j$, as well as the postsynaptic activity x_j, are proportional to the weighted signal pattern $\mathbf{S} \cdot \mathbf{z}_j$ of the linear filter. The postsynaptic activity pattern is then contrast-enhanced, due to the internal competitive dynamics of the target field. By hypothesis, the transmitter release rate is greatly amplified in proportion to the level of postsynaptic activity. A subsequent reset signal selectively inactivates those pathways that caused an error. Following the reset wave, the new signal $\mathbf{S} \cdot \mathbf{u}_j$ is no longer proportional to $\mathbf{S} \cdot \mathbf{z}_j$ but is, rather, biased against the previously active

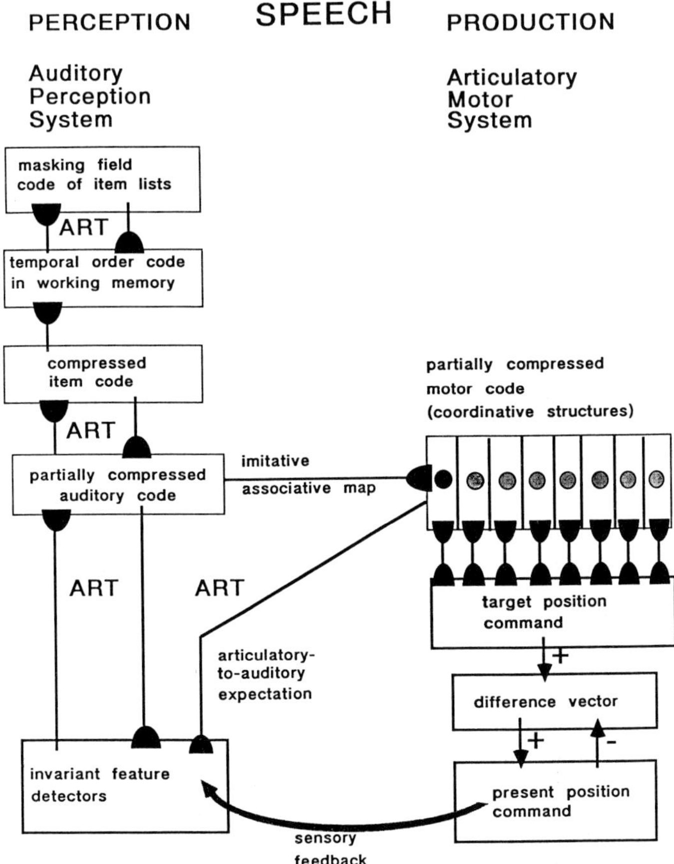

Figure 4. Neural network model of speech production and perception (Cohen, Grossberg, and Stork, 1988).

representation due to transmitter depletion at those sites. A series of reset events ensue, until an adequate match or a new category is found. Learning occurs on a time scale that is long relative to that of the search process.

6. AUTOMATIC STM RESET BY REAL-TIME INPUT SEQUENCES OR BY REINFORCEMENT

The ART 3 architecture serves other functions as well as implementing the mismatch-reset-search cycle. In particular it allows an ART system to dispense with additional processes to reset STM at onset or offset of an input pattern. The representation of input patterns as a sequence, $I_1, I_2, I_3, \ldots$, corresponds to the assumption that each input is constant for a fixed time interval. In practice, an input vector $I(t)$ may vary continuously through time. The input need never be constant over an interval, and there may be no temporal marker to signal offset or onset of "an input pattern" per se. Furthermore, feedback loops within a field or between two fields can maintain large amplitude activity even when $I(t) = 0$. Adaptive resonance develops only when activity patterns across fields are amplified by such feedback loops and remain stable for a sufficiently long time to enable adaptive weight changes to occur (Grossberg, 1976b, 1982). In particular, no reset waves are triggered during a resonant event.

The ART reset system functionally defines the onset of a "new" input as a time when the orienting subsystem emits a reset wave. This occurs, for example, in the ART 2 module

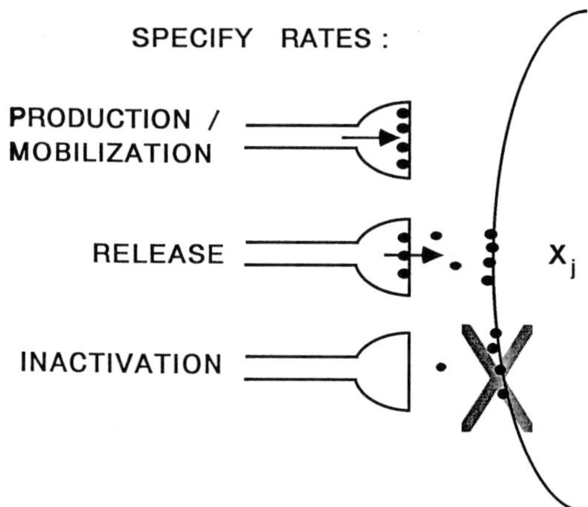

Figure 5. The ART search model specifies rate equations for transmitter production, release, and inactivation.

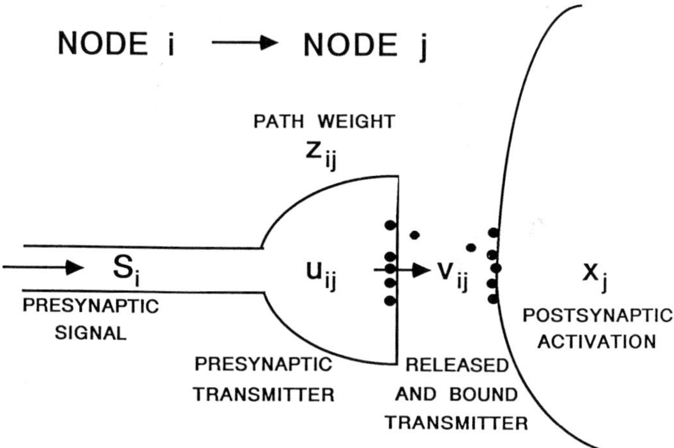

Figure 6. Notation for the ART chemical synapse.

(Figure 3) when the angle between the vectors $\mathbf{I}(t)$ and $\mathbf{p}(t)$ becomes so large that the norm of $\mathbf{r}(t)$ falls below the vigilance level $\rho(t)$, thereby triggering a search for a new category representation. This is called an *input reset* event, to distinguish it from a *mismatch reset* event, which occurs while the bottom-up input remains nearly constant over a time interval but mismatches the top-down expectation that it has elicited from level F_2 (Figure 2).

The mechanisms described thusfar for STM reset are part of the recognition learning circuit of ART 3. Recognition learning is, however, only one of several processes whereby an intelligent system can learn a correct solution to a problem. We have called Recognition, Reinforcement, and Recall the "3 R's" of neural network learning (Carpenter and Grossberg, 1988). Various types of reaction to reinforcement feedback may be useful in applications. For example, a change in vigilance alters the overall sensitivity of the system to pattern differences; a shift in attention and the reset of active features can help to overcome prior coding biases that may be maladaptive in novel contexts.

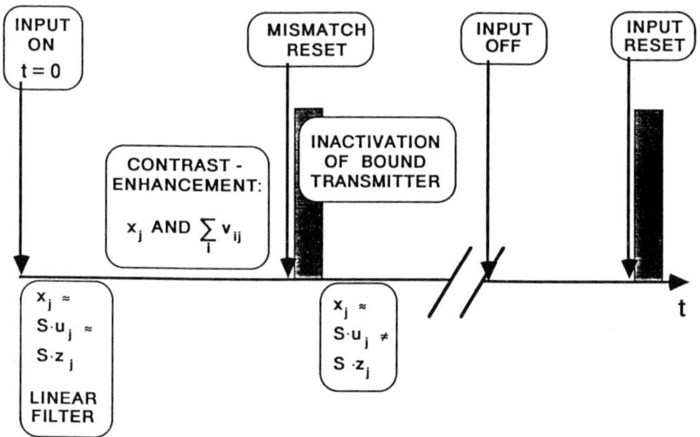

Figure 7. ART Search Hypotheses 1 and 2 implement computations to carry out search in an ART system. Input reset employs the same mechanisms as mismatch reset, initiating search when the input pattern changes significantly.

REFERENCES

Carpenter. G.A. and Grossberg, S. (1987a). A massively parallel architecture for a self-organizing neural pattern recognition machine. *Computer Vision. Graphics, and Image Processing*, **37**, 54–115.

Carpenter, G.A. and Grossberg, S. (1987b). ART 2: Self-organization of stable category recognition codes for analog input patterns. *Applied Optics*, **26**, 4919–4930.

Carpenter. G.A. and Grossberg, S. (1988). The ART of adaptive pattern recognition by a self-organizing neural network. *Computer*: Special issue on Artificial Neural Systems, **21**, 77–88.

Carpenter, G.A. and Grossberg. S. (1990). ART 3: Hierarchical search using chemical transmitters in self-organizing pattern recognition architectures. *Neural Networks*, in press.

Cohen, M.A., Grossberg. S., and Stork, D. (1988). Speech perception and production by a self-organizing neural network. In Y. C. Lee (Ed.). **Evolution, learning, cognition, and advanced architectures**. Hong Kong: World Scientific Publishers. 217–231.

Grossberg, S. (1976a). Adaptive pattern classification and universal recoding. I: Parallel development and coding of neural feature detectors. *Biological Cybernetics*. **23**, 121–134.

Grossberg. S. (1976b). Adaptive pattern classification and universal recoding, II: Feedback, expectation, olfaction. and illusions. *Biological Cybernetics*. **23**. 187–202.

Grossberg. S. (1982). **Studies of mind and brain: Neural principles of learning, perception, development, cognition, and motor control**. Boston: Reidel Press.

Grossberg, S. (Ed.) (1988). **Neural networks and natural intelligence**. Cambridge, MA: MIT Press.

Parallel Processing in Neural Systems and Computers
R. Eckmiller, G. Hartmann and G. Hauske (Editors)
© Elsevier Science Publishers B.V. (North-Holland), 1990

WORD RECOGNITION WITH A RECURRENT NEURAL NETWORK

Frank KOWALEWSKI and Hans Werner STRUBE

Universität Göttingen
Drittes Physikalisches Institut
Bürgerstr. 42-44
D-3400 Göttingen, F. R. Germany

A word recognition system using a recurrent perceptron with one
hidden layer is presented. The net is robust against velocity
variations in the speech data. It learns by error back propagation
without considering past signals. A one-speaker vocabulary of 13
words could be learned and recognized perfectly.

1. INTRODUCTION

In speech recognition, there has been increasing interest in neural networks
during the last few years. Although conventional methods have reached high
performance, neural networks may offer an advantage since they are readily
implemented on parallel computers and algorithmically simple.

In the literature neural nets are described that work on representations of
whole words (thus transforming time into a spatial coordinate, e.g. [1]) and
others that process parts of the speech data in a sequential manner (e.g.
[2]). The disadvantage of whole-word representations, e.g. based on a two-
dimensional spectrogram, is that they are very different for speech data with
differences in the time domain (e.g. velocity variations) and require the
exact determination of start and end points of a word. This is avoided by
time-sequential representations. The simplest net that works on such
representations maps parts of a word onto a particular word code (which is
constant over the whole word). Thereafter the code is time integrated (see
Figure 1).

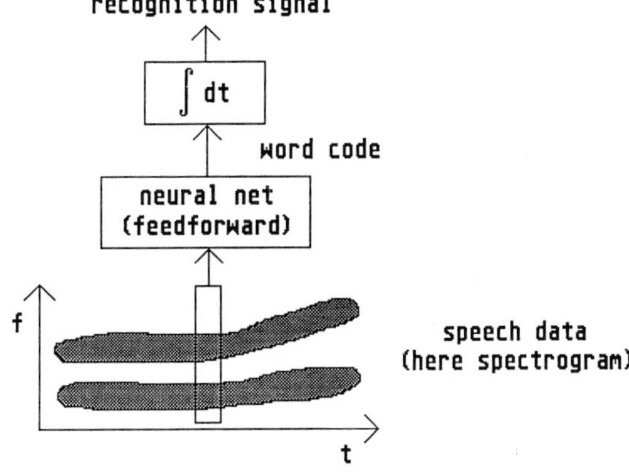

FIGURE 1
Simple time-sequential net for speech recoginition (see [3] for phoneme
recognition using this principle).

If the word parts are single analysis frames, only the frequencies of occurrence of these parts are essential. The net cannot discriminate words that differ only in the sequential order of their parts. This is avoided either by broader multi-frame parts or by recurrent nets.
Another difficulty is posed by temporal variations, since usual techniques such as dynamic programming are not applicable here. Thus the net itself should be robust against such variations.

In the following a recurrent net is presented which shows this robustness. As learning rule, error back propagation [4] is employed.

2. THE NET

It is known that 1-layer perceptrons (input layer not counted) can perform only linearly separable maps [4]. On the contrary, 2-layer nets are able to approximate arbitrary continuous mappings (on compact sets) with arbitrary accuracy [5], [6]. Therefore in the following a 2-layer net is chosen.

There are different possibilities for feedback signals in such networks:
  i) activities of hidden units,
 ii) activities of output units (see Figure 2).

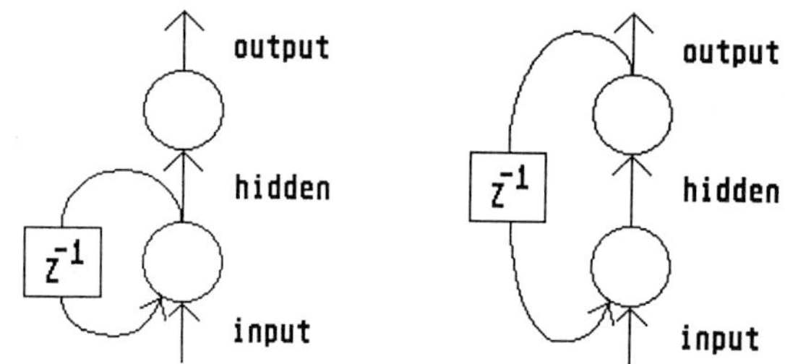

FIGURE 2
Activities of hidden (left) and output (right) units as feedback signals (circles represent whole layers).

The latter possibility has the advantage that delayed signals need not be considered for learning , as required for the first possibility [4]. This means, the error at the output is determined only by the difference between output and target and not by some error signal that is back-propagated along the feedback connection. Thus learning takes place in the corresponding feedforward net of Figure 3.

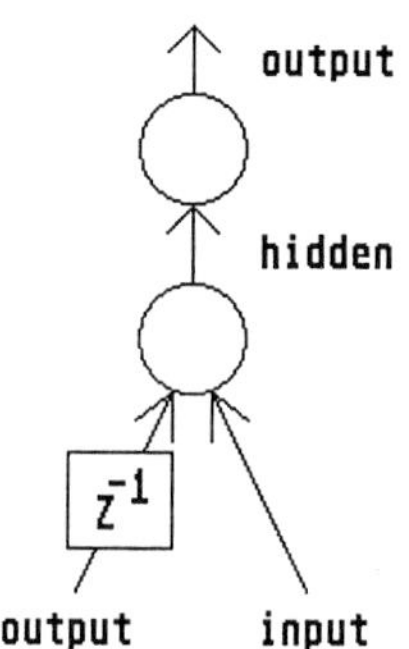

FIGURE 3
Learning net that corresponds to Figure 2 right.

Simulations with artificial data showed that this net is able to distinguish sequences that differ only in their sequential order. Besides, it generalizes to sequences that are presented with greater velocities than the learnt ones. This was tested and found to work up to a factor of 3.

3. SPEECH RECOGNITION EXPERIMENTS

The preprocessing of the speech signals is shown in Figure 4.

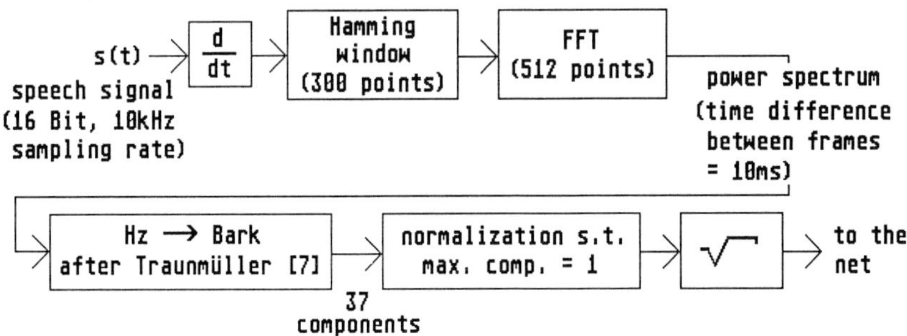

FIGURE 4
Preprocessing of speech signals.

The net itself consisted of units which are given by:

output of i-th unit: $\qquad o_i = f(s_i), \qquad s_i = \sum w_{ij} i_j + \theta_i,$

with the activation function: $\qquad f(x) = 1/(1 + \exp(-x)) - 1/2,$

where w_{ij} denotes the weight from input i_j to the i-th unit and θ_i is the threshold of the i-th unit. Thresholds were also adapted.

The input signals for the net were 2 adjacent time frames outputted from the preprocessor. Targets for the output units were given by a 1 out of n code. Recognition was done with the time-integrated output signals (see Figure 5).

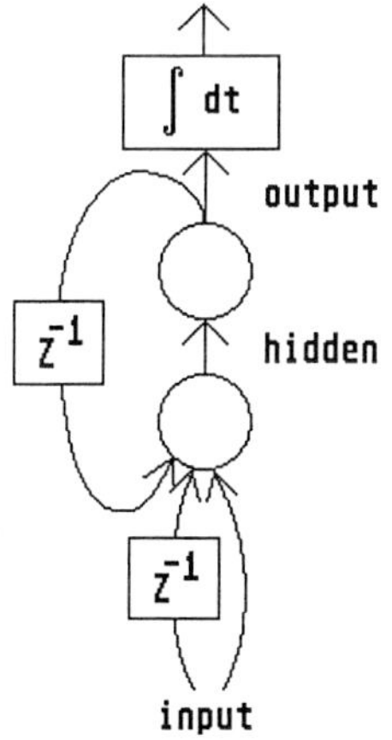

FIGURE 5
Network topology for speech recognition experiments.

In the learning phase, the weights were updated after the presentation of a whole learning sequence. The learning rate was set to some value r divided by the length of the sequence to be learnt. In the beginning r was 0.5. As soon as the recognition rate (for the learning data) decreased, r was set to 0.2. Learning was stopped when the recognition rate for the learning data reached 100% or was not increased over the last 100 learning sweeps (through the whole set of learning data).

The learnt vocabulary consisted of the 13 German numerals 'Null' (0) to 'Zwölf' (12). The net learned 6 versions of each word. A further version was processed in the test phase. All words were spoken by one male speaker. The recognition rates achieved with the learning and test data for different numbers of hidden units are summarized in Table 1.

TABLE 1

Recognition rates

number of hidden units	recognition rate for learning data	recognition rate for test data	learning sweeps (through the whole set of learning data)
20	100%	100%	310
15	100%	100%	300
10	100%	92%	260
5	100%	100%	90
3	85%	77%	70 + 100

4. CONCLUSION

A dynamical, recurrent net that can recognize sequences was presented. Sequences which differ only in their velocities are recognized as one and the same by the network. This is required in speech recognition.

In the learning process, only the instantaneous error signal was used and no back propagation done over the feedback connections. In comparison to usual back propagation in recurrent nets, this is a very time and memory saving feature of the algorithm. Besides, the given rule is biologically more plausible than the usual back propagation for recurrent nets.

Speaker dependent word recognition was done with this net. The 13 words-vocabulary could be recognized perfectly.

ACKNOWLEDGEMENTS

This work was funded by the Federal Ministry for Research and Technology (BMFT), project ITR 8800 C 8.

REFERENCES

[1] Peeling, S.M. and Moore, R.K., Speech Commun. 7 (1988) 403.

[2] Rohwer, R., Renals, S. and Terry, M., ICASSP 88 (1988) 426.

[3] Waibel, A., Hanazawa, T., Hinton, G., Shikano, K. and Lang, K., ICASSP 88 (1988) 107.

[4] Rumelhart, D.E., Hinton, G.E. and Williams, R.J., Learning Internal Representations by Error Propagation, in: Rumelhart, D.E., McClelland, J.M. et al., (eds.), Parallel Distributed Processing: Explorations in the Microstructure of Cognition, Vol. 1 (MIT Press, Cambridge (Mass.), 1986) pp. 318–362.

[5] Funahashi, K., Neural Networks 2 (1989) 183.

[6] Hornik, K., Stinchcombe, M. and White, H., Neural Networks 2 (1989) 359.

[7] Traunmüller, H. and Lacerda, F., Speech Commun. 6 (1987) 143.

Parallel Processing in Neural Systems and Computers
R. Eckmiller, G. Hartmann and G. Hauske (Editors)
© Elsevier Science Publishers B.V. (North-Holland), 1990

A FEEDBACK NETWORK FOR TEMPORAL PATTERN RECOGNITION

Teresa B. LUDERMIR[*]

Neural Systems Engineering Group
Imperial College, London SW7 2BT, England
email: JANET tbl%winge@sig.ee.ic.ac.uk

The aim of this paper is to show that the response of a neural network with feedback carries information about the order of appearance of its input patterns. They are capable of recognising some temporal patterns independently of their initial states even in the presence of input distortions. The model employed is an artificial neural network based on RAM memory devices as digital neurons (Aleksander [1]). The architecture has been used successfully in many sequence recognition problems. Some of them are presented here.

1. INTRODUCTION

Time is the essence of many pattern recognition tasks, e.g. speech recognition, motion detection and signature verification. However, connectionist learning algorithms to date are not well-suited for dealing with time-varying input patterns. In this paper it is shown that RAM networks are able to store sequential information from their input patterns and they operate efficiently on some temporal pattern recognition tasks.
Some weighted-connectionist systems have been implemented to deal with temporal recognition. Most of them use a buffer to hold the n most recent elements of the input sequences. Some drawbacks of this approach are: 1) the buffer must be sufficient in size to accommodate the longest possible input sequence; 2) by making a great deal of information simultaneously available, much computation is required each time; 3) each element of the buffer must be connected to a higher layer of the network. In consequence a large number of training examples must be used. Others systems are based on the fact that, in a connectionist network, the connections from one set of units to another implement a mapping. This approach is rather inflexible in that the mapping function used to construct the temporal context is predetermined and fixed. As a result, the representation of a temporal context must be made sufficiently rich to accommodate a wide variety of tasks.
The model used here is based on a different model called the RAM neuron model. The RAM model is based on the simple operations of a look-up table which is best implemented by random access memory (RAM) and where the knowledge is directly "stored" in the memory (the look-up tables) of the nodes during learning. Some advantages of this model are: (1) it is straightforward to implement in hardware; (2) learning is not unreasonably slow and (3) error-correction requires only a global success signal. An example of such a model is the pattern recognition device called WISARD as in Aleksander [2].

2. SEQUENTIAL BEHAVIOUR OF RAM NETWORKS WITH FEEDBACK

The type of network used in this work consists of a layer of identical RAM type digital neurons, where each of them has n-address terminals, i of them connected to an external matrix of binary elements and f of them connected to the output terminal of others neurons through clocked delay units ($n=i+f$). The RAM type digital neuron is represented in figure 2.1. The structure of a SDNN (Sequence Digital Neural Network) is represented in figure 2.2 where binary vectors x(t), r(t), r(t-1) and d(t) represent respectively at time t, the state of input matrix, the response, the delayed response and the 'desired' response of the neurons. The input and feedback connections are randomly generated. The SDNN is trained to anticipate its inputs, i.e. $d(t)=x(t+\alpha)$, such that $r(t)=x(t+\alpha)$ during test phase. During the training phase the net is fed with $x \in A$ (where A is the training set) with one (or more) RAM(s) in the write mode, and the memory position is changed $M_i[A_x(t),A_r(t)]=d_i(t)$ (A_x and A_r are input and feedback component respectively) for all RAMs in the write mode.

Supported by CNPq (Brazilian Research Council) grant no. 20.3296/86-CC

In order to justify the use of a single layer network(SLN) in sequence discrimination it will be shown that the response of SLN carries information about the order of appearance of its input patterns; that is, each output of the network depends not only on the present input but also on the previous ones.

Theorem : r(t) is function of x(1) x(2) ... x(t).

Proof : Let x = x(1) x(2) ... x(a) be the input sequence and r = r(1) r(2) ... r(a) be the corresponding response sequence of the network. Obviously, in general, x(i)'s do not relate themselves, since each sequence x is simply an element of the free monoid generated by the input symbols. Then, it is necessary to show only that r(t), 1<t<a is function of x(1) x(2) ... x(t) which gives to each r(t) of r the status of sequentiality that we are looking for. There follows a proof by induction on the length of r. r(1) = (x(1), r(0)) by definition. Suppose now that r(t) = (x(t) ... x(1), r(t-1)), then by definition r(t+1) = (x(t+1), r(t)), which gives us r(t+1) = (x(t+1) x(t) ... x(1), r(t)). Thus r(t) is function of all inputs until the instant t, for any t, as stated. Observe also that r(t) is function of r(t-1), which characterises networks with feedback.

It was shown above that the response of a net with feedback carries information about the order of appearance of its input patterns. But feedback nets are not the only way to deal with this problem. It is possible to use a feedforward RAM-net with more inputs at each time. The advantage of feedforward RAM-nets is that they are less sensitive to input errors, although it would be necessary to hold and process more information each time.

From the study of the stability properties of feedback networks in Fernandes [4] it is known that such networks are able to recover from input errors naturally. They are able to recover from input errors independently of their initial states. It is also known that the stability of the network is responsible for the increase of generalisation. Thus it has direct influence on pattern discrimination and identification. It is possible to control the stability of the network in at least two different ways: by altering memory contents and by the influence of the feedback connection as in Ludermir [6].

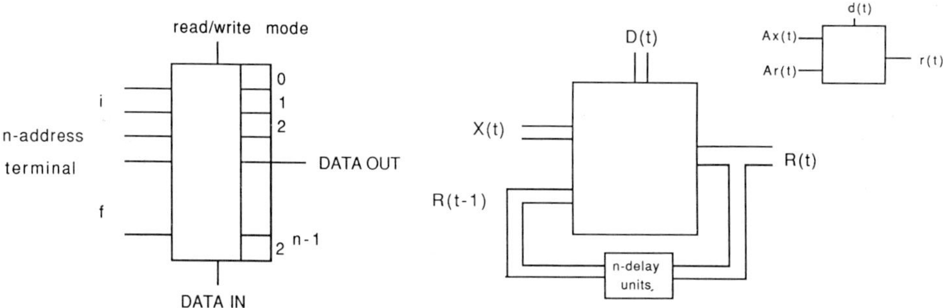

Figure 2.1 RAM type digital neuron   Figure 2.2 Sequential Digital Neural Net

3. EVALUATION OF THE FEEDBACK RAM-NETWORK

In this section three different types of classifiers are presented all based on the RAM-network described in the last section. The use of different classifiers gives a better insight into the system and shows the importance of the RAM-network itself. Also the choice of the classifier depends on the desirable application one has in mind. With all classifiers we did two kinds of experiments. The first one was to distinguish between different geometric forms, such as triangles, squares and circles. The second one was to recognize regular languages. Due to lack of space, only the results of the second set of experiments one presented here. The others results are presented in Ludermir [5]. The network used in the examples in this paper was composed of 32 RAMs with four inputs each, two from the input patterns and two from the feedback state. And the languages used were $L_1 = \{x \mid x = a^i b^j c^k d^l, i,j,k,l \geq 0\}$ and L_2 is the complement: $L_2 = U - L_1$, where U is the Universal set.

PROBABILISTIC CLASSIFIER

A classifier based on the probability $p(x \in L_k/r)$ that the input sequence $x \in L_k$ given that the response r occurred when x was fed into the net was used. Although ideally $p(x \in L_k/r) = 1$ for all symbols in the sequence $x \in L_k$, this does not generally happen. What happens is that $p(x \in L_k/r)$ changes randomly near 1 for sequences x in L_k and near 0 for sequences x not in L_k. In consequence, a measure is required which considers $p(x \in L_k/r)$ for all input symbols in the same sequence x. The measure that is being used is the continuous average

$$S_k(t) = a^{-1} \sum_{t=1}^{a} p(x \in L_k / r(t))$$

where $k = 1,2$ are the different languages. This measure ensures $0 \leq S_k \leq 1$. Then, for sequences belonging to the language L_k the sum $S_k(t)$ should be close to one, while for sequences not belonging to the language L_k the sum $S_k(t)$ should be close to zero.
In the example in figure 3.1 we show the results obtained with the probabilistic classifier using the languages L_1 and L_2 as defined above. This classifier is effective for distinguishing different geometric forms (Ludermir[5]). In the examples with regular languages, however, the discrimination is not so effective. One of the reasons for this is that more sequential information needs to be processed. As a result the use of a buffer classifier is proposed in such cases.

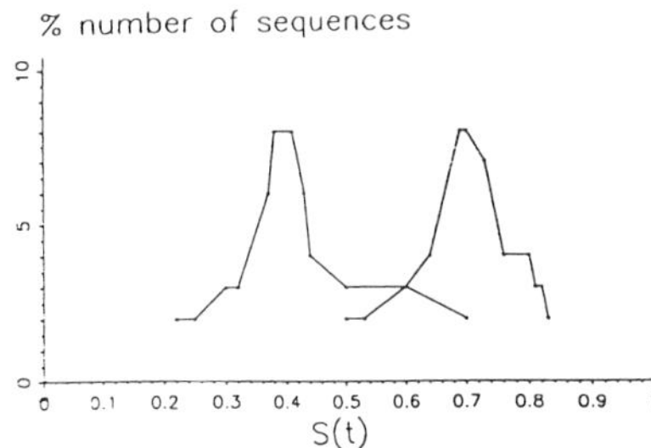

Figure 3.1 Probabilistic Classifier.   Figure 3.2 Buffer Classifier

BUFFER CLASSIFIER

The SDNN used in this work is blind to small differences in patterns. And also the feedback state does not contain enough information to recognise those small differences. In order to store more information a buffer is introduced in the probabilistic classifier. The buffer stores the time of occurrence of the states. Then the only difference between the classifiers is that here it is necessary to store the times of occurrence of the states.
It is not likely that the states in the sequences x of any class appear in the same order of the states in the training sequences for this class. The definition of a window, in which the states can occur, is needed. If the size of the window is the length of the sequence, this classifier becomes the probabilistic classifier. The sum is defined as:

$$S_k(t) = a^{-1} \sum_{t=1}^{a} p(x \in L_k / r(t'))$$

where $t' = t - t''$, $t'' = 0,1,\ldots,ws$ with ws being the window size and $k = 1,2$ ($0 \leq S_k \leq 1$ is conserved).
The results in figure 3.2 were obtained with this classifier using L_1 and L_2.
Better results were achieved with the buffer classifier than with probabilistic classifier but it usually requires the processing of much more information and the system is more complex and also takes longer to run. Despite theses drawbacks, this classifier has two properties in common with any model of temporal pattern recognition. Firstly, some memory of the input history is required. And secondly a function must be specified to combine the current memory (or the temporal context) and the current input to form a new temporal context.

FINAL STATE CLASSIFIER

In this model ideas from Formal Language Theory are used to recognise temporal patterns. It is desirable to characterise the (class of) languages(s) that can be recognised by a specific neural net. At this point only finite state languages have been used.
One way to recognise temporal patterns using neural networks is by using the final state of the network; finite automata work in the same way. Instead of probabilistic discriminations we have a set of final states.
Although the temporal transformation $f: x \to r$ examined here is a finite state representation, an analysis of the generalisation of the network is needed. The generalisation in this new classifier can be large because of the stability properties inherent in networks. After the network has been trained a finite-state structure is generated inside the network. Then the network is fed with some patterns and the final state of the network for each pattern is stored. The patterns are recognised when their final states are one of the final states stored.
The results in figure 3.3 were obtained with the final state classifier using L_1 and L_2.
As can be seen, in the examples above the results with this classifier are worse than with the previous classifiers. This is because of the large generalisation. The network recognises all sequences in the language it has been trained on but it also recognises some of the sequences

which do not belong to the language (over-generalisation). Also the network can not recognise all the finite state languages because the net is being trained only to predict the next symbol of the input sequence and there are finite state languages that do more than predict the next symbol(state). The choice of training strategy depends on the desirable application. Unfortunately, there is not a developed theory in which, for a certain kind of problem, a good training strategy can be determined.

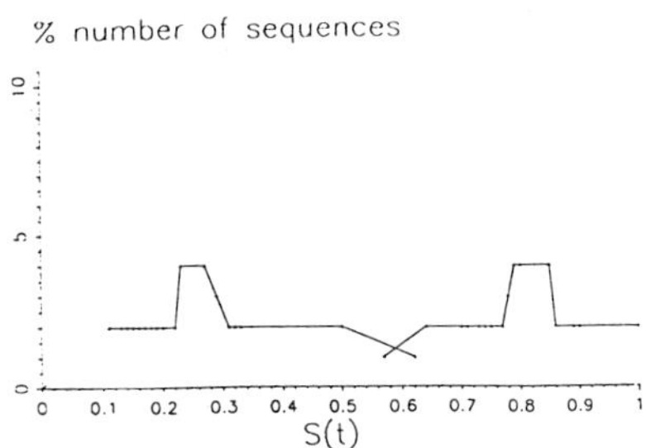

Figure 3.3 Final State Classifier.

Other motivations for this new classifier are: 1) it is very easy to know the total generalisation of the network by regular expressions; and 2) it is one way to solve problems such as parity with a single layer network with feedback.

The way that the output information of a net is processed makes a difference when working with different pattern recognition problems. Different classifiers are necessary to deal with different kinds of problems. And in order to characterise the generalisation of RAM-network, the analysis of the influence of different training strategies, classifiers, net configuration, etc are indispensable.

4. CONCLUSIONS

A methodology has been presented for Temporal Pattern Recognition. The main ideas behind this methodology are: (1) the use of a feedback RAM- network and (2) the use of different classifiers. Three different classifiers (probabilistic, buffer and final states) have been presented.

The main strengths of the method are that: (1) the response of a RAM-network with feedback carries information about the order of appearance of its input patterns and (2) the RAM-network is capable of recognising patterns independently of its initial states even in the presence of input distortions.

Many aspects of this methodology remain to be investigated since alternative training strategies were not explored. In addition probabilistic logic neurons (PLN) defined by Aleksander [3] could be adopted instead of RAM. The PLN can avoid knowledge being overwritten in the training phase and introduces some non determinism into the system.

One problem that is being analysed by the author is the the power of these feedback networks (RAM, PLN) in the recognition of languages in the Chomsky hierarchy. Different classifiers are being used in order to obtain a formal characterisation of the languages recognised by these feedback machines.

Application of this methodology to real-world problems will probably require the development of additional sequence transformations.

REFERENCES

[1] Aleksander, I. & Stonham, T.J.: "A Guide to Pattern Recognition using Random Access Memories. IEE J Comp & Digital Tech 2(1), 29-40, 1979.
[2] Aleksander, I., Thomas, W.V. & Bowden, P.A.: "WISARD, a radical step forward in image recognition". Sensor Review 4(3), pp 120-124, July 1984.
[3] Aleksander, I.: "The logic of connectionist system" in R. Eckmiller, Chr. v. d. Malsburg, eds. Neural Computers. Berlin: Springer-Verlag, 189-197, 1988.
[4] Fernandes, C.G.:"Stability Properties Inherent to Digital Neural Networks", COGNITIVA 85, Paris, 1985.
[5] Ludermir, T.B.:"A Feedback RAM-Network for Temporal Pattern Recognition", Neural Systems Eng Report, Imperial College, Dept of Electrical Eng 1989.
[6] Ludermir, T.B.:"Stability and Temporal Pattern Recognition", in Proceeding IJCNN-90, Washington.

Parallel Processing in Neural Systems and Computers
R. Eckmiller, G. Hartmann and G. Hauske (Editors)
© Elsevier Science Publishers B.V. (North-Holland), 1990

SPECTRAL PROCESSING OF HARMONIC COMPLEX TONES AND PITCH BY PDP NETWORKS

R.W. Ward TOMLINSON and W. TREURNIET

Zoologisches Institut,
Technische Hochschule Darmstadt,
Darmstadt, Federal Republic of Germany

Communications Research Center,
Department of Communications,
Ottawa, Canada.

1. INTRODUCTION

Sounds with harmonic line spectra (harmonic complex tones) give rise to a sensation of pitch which is the same as that of a sinusoidal tone with the frequency of the lowest component of the spectrum [1]. This frequency is also referred to as the fundamental component of the complex tone. The same pitch is present when the fundamental component has been filtered out of the stimulus, and arises even when two successive components are presented separately to each ear. This shows that some central nervous system processing of complex tones is required for missing fundamental perception [3]. Monkeys can perform tasks with harmonic complex tones that indicate an ability to hear missing fundamental pitches [7]. However, no neuron in monkey auditory cortex could be found that responded to complex tones solely on the basis of their pitch. A cochleotopic organization (frequency map) was observed instead [6]. The lack of pitch selectivity in the single neuron responses suggests that complex tone analysis might be accomplished by a process which is distributed across the population of cortical neurons. Could complex tones and pitch be encoded and processed by a distributed representation? Does missing fundamental pitch emerge from a parallel distributed processing (PDP) system which represents harmonic complex tones? One possibility for reconstruction of a missing fundamental is suggested by the phenomenon of pattern completion. When a neural network is presented with a fragment of a previously learned input, the network may still respond as though the entire input had been presented. Dramatic examples of pattern completion in video images of faces have been shown by Kohonen and coworkers [4]. If complex tone pitch is processed in this way, presentation of fragmentary complex tones (missing fundamentals) may still give rise to the appropriate pitch due to pattern completion. These questions are pursued using recent implementations of PDP networks.

2. METHODS

Three experiments were performed using a three-layer back propagation network with 270 input units, 100 hidden units and 270 output units. The networks were trained with complex tone spectra, and generate at the output either the same complex tone pattern, or the pure tone with the frequency of the fundamental component. The networks were then tested with complex tones missing their fundamentals to see if they could produce the appropriate complex tone (complete with fundamental), or a pure tone with the frequency of the missing fundamental.

The networks were not presented with line spectra, but rather the activity pattern in the cochlea that line spectra would produce. These "excitation patterns" are derived from results of psychophysical experiments, and were calculated using the algorithm of Moore and Glasberg [5]. The generated patterns (see Fig. 1) have a multipeaked profile, each peak representing a component of the spectrum. The peaks appear closer together at higher frequency due to the logarithmic-like scale in the cochlea. The pattern was presented to the array of 270 input units, each unit corresponding to one frequency channel in the auditory periphery. The output of the network corresponds to a neuronal representation of the input spectra, and is also spread across an array of 270 units. Two basic training paradigms were used: (1) autoassociation, or pairing of complex tone inputs with identical target outputs, and (2) "pitch association", or pairing of complex tone inputs with a pure tone at the frequency of the fundamental.

3. RESULTS AND DISCUSSION

In the first experiment, the network was trained using the autoassociation paradigm, with 386 input-output pairs whose fundamentals covered the range from 120 to 6500 Hz in 60 equi-logarithmic steps. The spectra of the tones included the fundamental frequency, and had up to nine successive components. Each component had an amplitude equivalent to 50 decibels sound pressure level.

The network was trained until the levels at each output unit came to within 3 decibels of the target output. One sample input pattern is shown in Fig. 1a. The network was then tested with novel complex tones - missing their fundamental components - to see if the outputs would have the appropriate fundamentals (Fig. 1b). The output patterns from the network not only had no fundamental components, but were almost identical to the input patterns. It can be shown analytically [6] that such networks, when trained with a sufficient variety and number of input examples, will produce identical outputs even for inputs not in the original training set. This indicates that for training sets with small numbers of input-output pairs, this identity mapping should fail, and the missing fundamental may emerge at the output through the phenomenon of pattern completion.

The second experiment used only 10 pairs in the training set. The training criterion was the same as for the previous experiment. Fig. 1c shows that it is possible for a complex tone missing its fundamental to cause the output of a pattern which includes the fundamental.

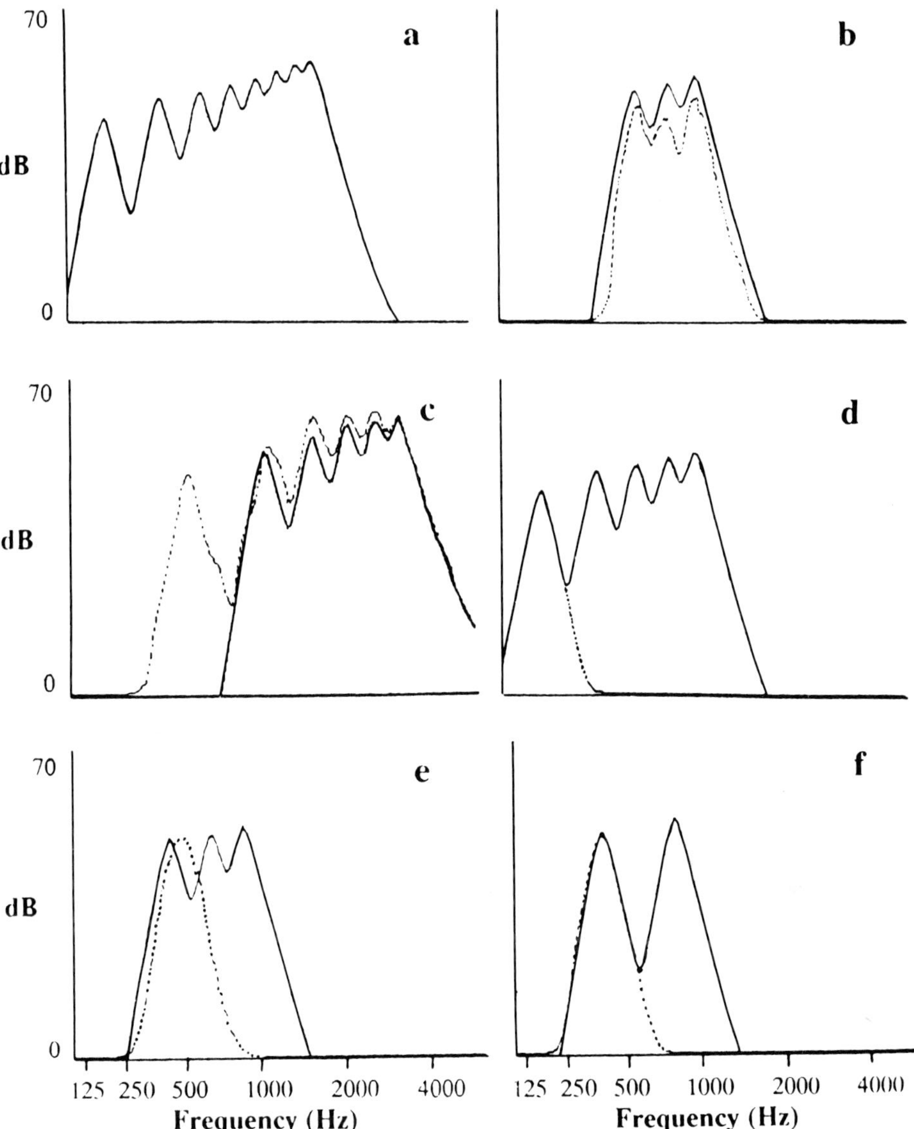

FIGURE 1. Input and output patterns from three neural networks. Vertical axes show amplitude of signal in decibels, the horizontal axis shows frequency in Hz. All fundamental frequencies are 215 Hz. Input and output patterns have solid and dotted curves respectively (a) input tone with components 1-8 (component 1 is the fundamental); (b) input and output tones from 1st network: the input tone has components 3-5; (c) input and output tones from 2nd network showing reconstruction of missing fundamental: the input tone has components 2-6; (d) input and output patterns from 3rd network: input tone has components 1-5; (e) input and output patterns from 3rd network: input tone has components 5-7; (f) two tone input to 3rd network: 400, 800 Hz.

However, this is not a robust phenomenon, and for 7 of the 10 tones from the original training set, the fundamental was either absent or unrecognizable.

The third experiment employed the "pitch association" paradigm, using 319 input-output pairs to train the network. An example of an input-output pair from the training set is shown in Fig. 1d, which displays the input complex tone, and the output "pitch" (a pure tone at the fundamental frequency). This network was only trained with complex tones that possessed a fundamental, with up to a maximum of nine successive components. When novel inputs lacking fundamentals were presented (Fig. 1e), the network produced a pure tone with the frequency of the lowest component. The outcome was similar when two single frequencies were presented at the input (Fig. 1f). In this case the output pure tone had a peak at the lower frequency input tone. These results indicate that the network used the low frequency edge of the stimulus to generate the output pure tone patterns and not the spacing pattern of the input harmonics.

In summary then, the second simulation showed that missing fundamental reconstruction through pattern completion can be performed by neural networks that process complex tones, it is possible only under restrictive conditions. The third simulation had the implicit assumption that pitch judgements are based on learned associations between complex tone spectra and the fundamental frequency. The results of this simulation indicate that such a system will be strongly influenced by complex tone features other than the spacing of successive components, for example the overall spectral shape (timbre). Psychophysical evidence [2] suggests that learned associations can affect pitch perception, so the possibility still exists that this kind of processing has a modulatory effect on pitch perception in real neural systems.

ACKNOWLEDGEMENTS

This work was supported by the Canadian Medical Research Council.

REFERENCES

[1] Boer, E. de, Handb. Sens. Physiol. V, part 3, (1976) 479.
[2] Divenyi, P.L., J. Acoust. Soc. Amer. 66 (1979) 1210.
[3] Houtsma, A.J.M., and Goldstein, J.L., J. Acoust. Soc. Amer. 51 (1972) 520.
[4] Kohonen, T., Reuhkala, E., Makisara, K., and Vainio, L., Biol. Cybern. 22 (1976) 159.
[5] Moore, B.J.C., and Glasberg, B.R., Hearing Res. 28 (1987) 209.
[6] Tomlinson, R.W.W., Ph.D Thesis, University of British Columbia. (1989)
[7] Tomlinson, R.W.W., and Schwarz, D.W.F., J. Acoust. Soc. Amer. 84 (1988) 560.

Section 10
Neural Networks
for Motor Control

Parallel Processing in Neural Systems and Computers
R. Eckmiller, G. Hartmann and G. Hauske (Editors)
© Elsevier Science Publishers B.V. (North-Holland), 1990

RESISTIVE NETWORK APPROACH FOR OBSTACLE AVOIDANCE IN TRAJECTORY PLANNING

Jürgen BECKMANN

Department of Biophysics
Universitätsstraße 1
Heinrich-Heine-Universität Düsseldorf (FRG)

This paper considers the problem of planning a trajectory between two points in a two-dimensional plane with obstacles. Starting from a diffusion model (Ritter, 1988 [2]), a modified electrical model is presented and interpreted with the statistics of random walks [3]. This model directly leads to a concept for a possible hardware implementation as a resistive network, which would fully exploit the parallelity of the underlying computations. Moreover, an optical mapping of the workspace onto photosensitive connections in the network could realize a straightforward and fast method of data acquisition.

1. INTRODUCTION

When animals or robots operate in space, they have to plan their motions. The basic task to move the whole body or parts of it to a given target position must be fulfilled under many constraints, e.g. avoidance of collisions with environmental obstacles, restrictions in the degrees of freedom of the body like joint angle limits, limitations in the available accelerations and forces etc..

This paper concentrates on the geometrical problem of finding a path that connects two given points A, B in a plane and by-passes obstacles. This problem is nontrivial due to arbitrary forms (e.g. mazes) and possible movements of the obstacles as well as the demand that the connecting path should be as short as possible (cf. Schwartz [1]). The solution presented here relies on computations that can be done in parallel by a large number of locally connected simple processing elements. This is a typical task for a neural network, and an equivalent hardware implementation will be suggested.

2. SOLUTION ALGORITHM

H. Ritter explains in his PhD thesis [2] a physical analogy that helps to solve the problem of obstacle avoidance. He considers the situation of a plane with obstacles in which a connecting path from point $\underline{r}_A$ to a target point $\underline{r}_B$ is being looked for. Moreover, this path should be as short as possible.

Regarding $\underline{r}_B$ as the source of an imaginary chemical substance that diffuses into space, decays with rate γ , and is absorbed by the obstacles, Ritter derives the following boundary problem for a "concentration" u:

$$\partial_t u(\underline{r}, t) = D \Delta u(\underline{r}, t) - \gamma u(\underline{r}, t) \tag{1}$$

$$u(\underline{r}_B, t) = 1, \qquad u(\underline{r}_{obstacle}, t) = 0, \qquad u(\underline{r}, 0) = \delta(\underline{r} - \underline{r}_B)$$

After simulating the diffusion process for some time t_0 (e.g. until the substance reaches $\underline{r}_A$), a gradient ascent in $u(\underline{r}, t_0)$ leads from $\underline{r}_A$ to $\underline{r}_B$. Ritter extends his diffusion model successfully to more abstract situations, in which the possible states of a system are represented by the nodes of a graph and two such nodes are connected by a directed edge, if a transition between the corresponding states is possible.

A disadvantage of Ritter's algorithm is that it does not say how to measure the positions of the obstacles and how to transform them into the diffusion plane. Moreover, it needs a lot of computing time as long as it is implemented on sequential computers. Therefore an equivalent model will be proposed that directly leads to a hardware implementation.

Consider the case that the workspace is represented by a plane P of an electrically conductive material with conductivity $g(\underline{r})$. The subset $R \subseteq P$ of positions occupied by obstacles shall be cut out of the plane, i.e. $g(\underline{r}) = 0$ for $\underline{r} \in R$. If in this configuration two subareas (or points) A, B of P are clamped to different voltages u_A, u_B, a stationary current will evolve. The basic idea is to exploit the fact that all currents flow along connecting paths from B to A which automatically by-pass obstacles. Furthermore the strongest current should flow along a relatively short path. To treat this concept mathematically, the resulting voltages and currents must be considered.

Let $u(\underline{r})$ be the voltage at point $\underline{r} \in P$, then $\underline{i} = -g(\underline{r})\nabla u(\underline{r})$ is the current density at $\underline{r}$, and Kirchhoff's law requires that div $\underline{i} = 0$. This leads to the following Dirichlet/von Neumann boundary problem:

$$0 = \mathrm{div}(g\nabla u) \qquad \text{for} \qquad \underline{r} \in P\backslash R \qquad\qquad (2)$$

$$u(\underline{r}) = u_A \qquad \text{for} \quad \underline{r} \in A, \qquad u(\underline{r}) = u_B \qquad \text{for} \quad \underline{r} \in B,$$

$$\nabla u \cdot \underline{n} = 0 \qquad \text{at the boundary of obstacles } (\underline{n} = \text{normal vector of that boundary})$$

In the case of a locally constant conductivity g the differential equation (2) corresponds to the stationary version of eq. (1) for $\gamma = 0$. But the boundary conditions are somewhat different in both cases. Again the chosen path from A to B shall follow a gradient ascent in u.

For an alternative interpretation of the selected path and with regard to the hardware implementation eq. (2) will now be formulated in a discretizised version, i.e. for a resistive network. In such a network every node X is connected to his neighbouring nodes Y via conductivities g_{XY}. The current i_{XY} from X to Y is

$$i_{XY} = g_{XY}(u_Y - u_X)$$

and after the summation over all neighbours Y Kirchhoff's law yields:

$$0 = \sum_Y i_{XY} = \sum_Y g_{XY}(u_Y - u_X)$$

$$\implies \quad u_X = \sum_Y \frac{g_{XY}}{\sum_{\tilde{Y}} g_{X\tilde{Y}}} \cdot u_Y =: \sum_Y w_{XY} u_Y \qquad\qquad (3)$$

Eq. (3) expresses that u must be a harmonic function, i.e. the u-value of every point X is the weighted average of the u-values of his neighbours. The boundary conditions simply are

$$u_X = u_A \qquad \text{for} \quad X \in A, \qquad u_X = u_B \qquad \text{for} \quad X \in B.$$

Doyle and Snell [3] consider random walkers on a similar structured network, who pass from X to Y with probability w_{XY} (notice that $\sum_Y w_{XY} = 1$). They define $\tilde{u}_X$ as the probability that such a random walker who starts at point X reaches B before he reaches A. Then it immediately follows that $\tilde{u}$ is a harmonic function, i.e. eq. (3) holds for $\tilde{u}$, too. As the random walker never reaches B before A if he already starts at some $X \in A$, $\tilde{u}_X = 0$ for $X \in A$. Similarly, $\tilde{u}_X = 1$ for $X \in B$.

Because of the uniqueness principle for the solutions of Dirichlet boundary problems the functions u and $\tilde{u}$ must be identical if $u_A = 0$ and $u_B = 1$. Therefore the voltage u in the resistive network may be considered as the result of what lots of random walkers found out

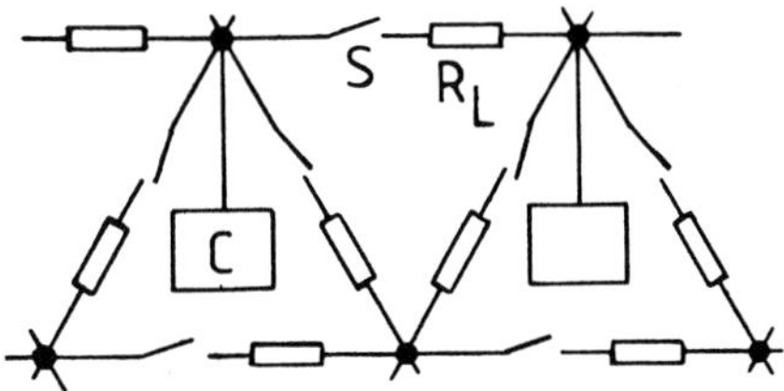

Figure 1: Resistive network: basic cells

about the ways from A to B, and gradient ascent in u follows a path that makes it more and more probable to reach B by a continuation with simple random walk only. This path is in general not identical with the geometrically shortest connection, but rather a compromise between the demands "go a short way" and "keep far away from obstacles". This seems to be a reasonable guideline for the planning of movement trajectories.

3. SUGGESTED HARDWARE IMPLEMENTATION

The concept of a resistive networks directly allows a hardware implementation. For the use under real-time conditions e.g. in robotics, an integrated circiut with the following features is proposed:

Firstly, it suffices to have a network in which neighbouring nodes are either connected via constant conductivities g or disconnected. The choice between these two states should be possible from outside the network by programmable switches (S in fig. 1). Moreover, for each connection there is another switch which is light dependent, e.g. it closes if illumination is above a certain threshold. In fig. 1 this switch is simply approximated by a photosensitive resistor R_L, that is assumed to toggle between a finite and an infinite value depending on the level of illumination.

Provided the workspace is illuminated in such a way that obstacles are dark (below threshold) and accesable area is light, an optical projection of it directly onto the resistive network automatically transforms regions of obstacles in isolated patches of the network. This would be the most elegant and fast method to measure and represent the obstacle positions at the same time.

For the complete function of the network each node needs a simple circuit (C in fig. 1) that is adressable from outside and allows to choose one of the following functions:
  - clamp the corresponding node to one of the two voltages $u = 0$ or $u = u_0$
  - put the voltage of the corresponding node onto a signal line in order to make it externally measurable
  - alter the state of the switches S.

A typical procedure to use the chip would then consist of the following steps:
  1. Map the obstacles optically onto the network.
  2. In case set further "virtual" obstacles via the switches S.
  3. Clamp the target point B to voltage $u = 0$.
  4. Clamp position A to voltage $u = u_0$; the stationary voltage distribution will evolve.
  5. Measure the node-voltages around the current position A.
  6. Take the node with the voltage nearest to 0 as the new position $\tilde{A}$.
  7. Restart at step 4. or 5. with $A = \tilde{A}$ until the target is reached, i.e. $\tilde{A} = B$.

Of course a lot of variants are possible, e.g. to clamp a different point or even the whole bondary of the network to u_0, to use optical filters in order to influence illumination levels etc.. Moreover, it might be necessary to take the voltage differences around point A only as a kind of cost function grating the possible step directions, and to decide about the actually done step after the consideration of further constraints.

4. DISCUSSION AND CONCLUSIONS

Comparing the resistive network approach with Ritter's path finding algorithm, the latter turns out to be more general and also applicable in higher dimensional abstract spaces. However, it is only available as software on conventional computers. A resistive network in contrast would directly use physical laws to do its computations and therefore fully exploit the parallelity of the problem. Together with the optical mapping of the workspace onto the functional structure of the network this would constitute an extremely fast computing tool. Though it was not actually built, the work on similar VLSI resistive networks done by C. Mead [4] demonstrates the general feasibility of this proposal.

Besides obstacle avoidance in robotics, other application areas of the network might arise in any case, where the possibility to connect two points of a two-dimensional plane shall be examined.

ACKNOWLEDGEMENTS

This work was supported in part by grants from Chamber of Commerce (IHK) Düsseldorf and Ministry for Research and Technology (BMFT) under grant No. ITR 8800F6.

REFERENCES

[1] Schwartz, J.T., Sahir, M., Hopcroft, J., Planning, Geometry, and Complexity of Robot Motion (Ablex Publ. Corp., Norwood, NJ, 1984)

[2] Ritter, H., Selbstorganisierende Neuronale Karten, PhD thesis (TU München, FRG, 1988)

[3] Doyle. P.G., Snell, J.L., Random Walks and Electric Networks (The Mathematical Association of America, 1984)

[4] Mead, C., Analog VLSI and Neural Systems (Addison-Wesley, 1989)

Parallel Processing in Neural Systems and Computers
R. Eckmiller, G. Hartmann and G. Hauske (Editors)
© Elsevier Science Publishers B.V. (North-Holland), 1990

A SIMPLE NETWORK CONTROLLING THE MOVEMENT OF A THREE JOINT PLANAR MANIPULATOR

Holk Cruse and Michael Brüwer

Department of Biol. Cybernetics
Fac. of Biology
University of Bielefeld
Bielefeld, Fed. Rep. Germany

In an earlier investigation (Cruse and Brüwer (1)) an algorithmic model was proposed which describes targeting movements of a human arm when restricted to a horizontal plane. As three joints at shoulder, elbow and wrist are allowed to move, the system is redundant. A network model is discussed here in order to replace this algorithmic model. The network solves the static problem, i.e. it provides the joint angles which the arm has to adopt in order to reach a given point in the workspace.

1. INTRODUCTION

The control of the movement of a multijointed manipulator in general includes strongly non-linear operations (Benati, Gaglio, Morasso, Tagliasco and Zaccaria (2)). The control algorithm has to cope with additional problems when the manipulator has more joints than necessary for a given task, i.e. when it has extra degrees of freedom. The human arm provides such a redundant manipulator.

To obtain the simplest version of a redundant arm, in earlier experimental investigations (Cruse (3), Cruse and Brüwer (1)) the number of degrees of freedom of the human arm was reduced to three, namely shoulder, elbow and wrist joint, while allowing the arm to move in a two-dimensional, horizontal plane. The axes of rotation of the three joints were perpendicular to the plane. The position of the endeffector, the tip of a pointer attached to the palm, is determined by two cartesian coordinates in the horizontal plane. Therefore, two joints were sufficient to move the endeffector in the workspace. The existence of the third joint produced an additional degree of freedom and therefore made the system redundant (Fig 1a). Thus a given point in the two dimensional workspace can be reached by a number of different combinations of joint angles of the manipulator.

The question arises of how the control system selects one of this infinite number of possible positions when trying to reach a given point. To solve this static problem, the following hypothesis was proposed (Cruse (1)). To each joint a cost function is attached which defines a cost value for each joint angle. The cost functions show a minimum at about the middle of the angle range of the joint and the cost values increase to either of the extreme angles. The total cost of a manipulator position is described as the sum of the actual cost values of all joints. When reaching to a given point in the workspace, according to this hypothesis that manipulator position is selected out of the geometrically possible positions which shows the minimum total cost value. In this way the number of degrees of freedom of the system is reduced and thus the redundancy problem can be solved.

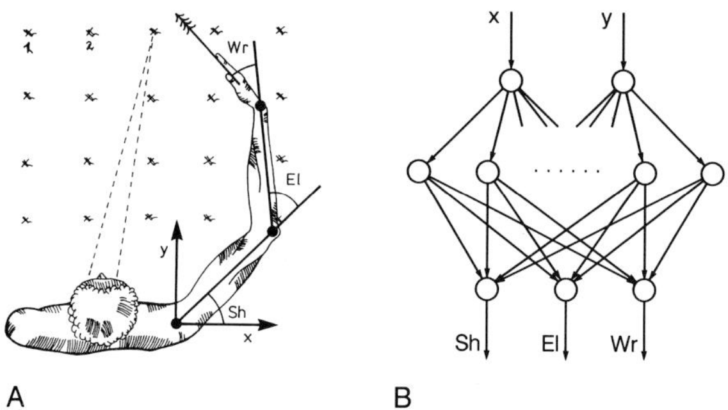

Fig. 1. (a) Top view of the experimental arrangement as seen by a medieval artist. The arm is moved in a horizontal plane with the coordinates x and y. The origin is at the shoulder joint. The three joint angles at shoulder (Sh), elbow (El) and wrist (Wr) determine the position of the tip of the arm. (b) The architecture of the network. For explanation see text.

2. THE NETWORK MODEL

The model consists of three layers of neurons (Fig 1b), the input layer, an intermediate "hidden" layer and the output layer. The network models were simulated using the software of McClelland and Rumelhart (4). The input layer consists of two neurons the excitation of which corresponds to the value of the cartesian workspace coordinates x and y. The intermediate layer consists of 20 hidden units. The output layer contains three units which give the angle values of the three joints. Only feedforward connections and only connections between neurons of directly neighbouring layers are allowed. The excitation of each neuron was between 0 and 1 with a resting value of 0.5. To train the network we followed the error back propagation rule with a learning rate of 0.05 and a momentum of 0.9. For one learning cycle (epoch) 32 input patterns were used which correspond to the x-y coordinates of 32 points in the workspace (see below, Fig 2b). The corresponding training patterns for the output, i.e. the angle tripels, were taken from the experimental results of one subject which made targeting movements to each of the 32 points.

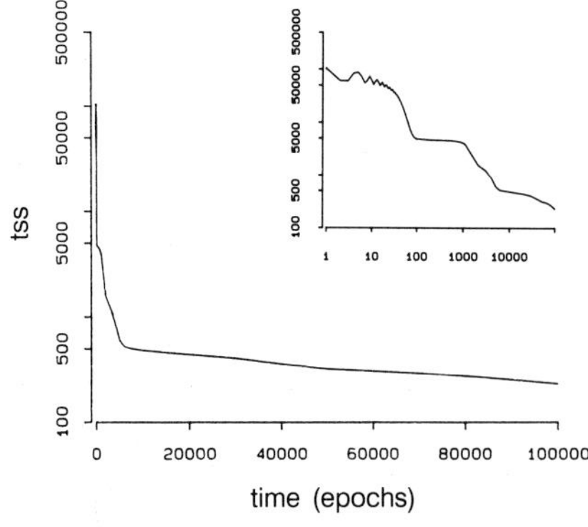

Fig. 2. Learning curve for the network shown in Fig 1b. Abscissa is the number of learning cycles (epochs). In one epoch each of the 32 target points (see Fig. 3a) was presented once. Ordinate is the total sum of squared errors (tss) in one epoch. This error is the deviation of the angle values produced by the network from the angle values obtained during experiments with the human subject. The inset figure shows the same learning curve with a logarithmic abscissa.

Fig 2 shows the progress of learning the network made during the presentation
of 100.000 cycles. The deviation between angles values calculated by the
network and those assumed by the subject was then in the order of 1 degree per
joint. In Fig 3a the x-y coordinates of these points are compared with the
positions the network actually pointed to. As the network calculates angles and
not x-y coordinates, the latter were determined from the angle values by a
separate calculation using the corresponding trigonometric formulas. The mean
Euclidean distance was 1.1 (+/- 0.8) cm for the 32 trained points (mean +/-
S.D.).

In order to test the capacity of the network to respond to untrained target
points, the response of the network to 117 new target points was calculated.
These untrained points lie on a 9 x 13 grid in the same area as the 32 trained
points. As was shown in Fig 3a for the 32 target points, in Fig 3b the
deviation for the 117 untrained points is shown. The mean Euclidean distance is
0.9 (+/- 0.5) cm for the untrained points.

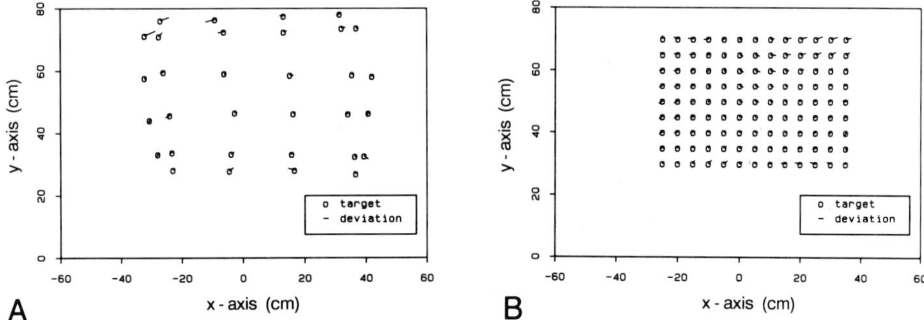

Fig 3. (a) The mean deviation from each of the 32 target points after 100 000
epochs. The circles show the position of the target points in workspace
coordinates. The length and direction of the attached bar shows the mean
deviation. (b) The deviation for 117 untrained target points shown in the same
way.

3. DISCUSSION

Models on the basis of neuronal networks for the control of a manipulator have
already been presented by several authors. Kohonen (5) and Ritter and Schulten
(6) provide systems which are based on the Kohonen algorithm (7). Another model
given by Josin (8) uses the error back propagation procedure of McClelland and
Rumelhart (4). All these models concern the non redundant arm. Here a simple
model is proposed which is able to solve the redundancy problem for the static
situation. The model obtains x-y coordinate values as input as does the model
of Josin (8). Another version of the model not described here in detail obtains
the input information via a retina-like input layer which is similar to the
approach of Kohonen (5,7) and Ritter and Schulten (6).

The redundancy problem was solved by an earlier algorithmic model by the
application of cost functions attached to each joint (see Introduction). The
model proposed here solves the problem on the basis of distributed networks.
This occurs in such a way that during the training session the weighting
factors in the network connections were learned. However, no obvious explicit
representation of the cost functions was found in the values of the weighting

factors. The network was trained using the data obtained from a human subject and thus used an external teacher. However, as the data of this subject could be well described by three cost functions, the network could have also be trained with practically the same result by implementing these cost functions into the training program.

The network in its actual form cannot deal with three properties found in the human experiments. First, one type of experiment performed by Cruse and Brüwer (1) was to start the movement with the arm in an "uncomfortable" position, i.e. a position which does not obey the minimum cost principle. These experiments cannot be modelled by this network. For this purpose the network had to be enlarged by input units which monitor the actual value of the three joint angles. Second, our earlier algorithmic model in addition to the minimum cost strategy and the mass-spring strategy contains a third strategy which also is not yet implemented in the network model. This so-called pseudoinverse control has the effect that during the movement the incremental changes of the three angles dS, dE and dW obey the rule that the sum of the squared angle changes assumes a minimum. This strategy is also important when modelling movements starting with uncomfortable arm positions.

However, two disadvantages of the algorithmic model are not found in the network model. In the algorithmic model the necessary linearisation required an iterative method of calculation. Higher exactness then requires smaller iteration steps and thus larger computing time. In the parallel network computing time is extremely short when implemented on a real parallel system. Higher exactness is only a question of the resolution of the individual units or could be increased by increasing the number of units but does not necessarily influence the computing time. Another disadvantage of the algorithmic model is the appearence of so-called singularities. As described by Cruse (9) a singularity occurs in the algorithmic model when the wrist angle obtains a value of zero. In this case it is not possible for the model to calculate the incremental angle values for the subsequent movement step. For the network model this problem does not exist because the angle values are not calculated but represented in the combination of weighting factors of the different synapses and thus correspond to a kind of distributed look-up table. This property is also of practical importance for the control of artificial manipulators as these also have to deal with the problem of singularities.

REFERENCES

(1)   Cruse,H., Brüwer,M., Biol.Cybern. 57, 137-144 (1987)
(2)   Benati,M., Gaglia,S., Morasso,P., Tagliasco,V., Zaccaria,R., Biol.Cybern. 38, 125-140 (1980)
(3)   Cruse,H., Biol.Cybern. 54, 125-132 (1986)
(4)   McClelland, J.L., Rumelhart, D.E., Explorations in parallel distributed processing. (MIT Press, Cambridge, Mass. 1988)
(5)   Kohonen, T., Proceedings of the sixth international conference on pattern recognition, pp. 114-128 (IEEE Computer Society, Silver Spring, MD)
(6)   Ritter, H.J., Schulten, K.J., in: R.Eckmiller, C. von der Malsburg (eds.) Neural computers, pp. 393-406, (Springer, Heidelberg 1987)
(7)   Kohonen,T., Biol.Cybern. 43, 59-69 (1982)
(8)   Josin,G., Biol.Cybern. 59, 283-290 (1988)
(9)   Cruse,H., in: L. Personnaz, G. Dreyfus (eds.) Neural networks from models to applications. pp. 71-77. (IDSET, Paris 1989)

Parallel Processing in Neural Systems and Computers
R. Eckmiller, G. Hartmann and G Hauske (Editors)
© Elsevier Science Publishers B.V. (North-Holland), 1990

OPTIMAL SOLUTION OF UNDERDETERMINED LINEAR MATRIX EQUATIONS IN NEURAL NETWORKS

Wolfgang J. DAUNICHT and Heinz WERNTGES

Abt. Biokybernetik, Institut fuer Physikalische Biologie
Heinrich-Heine-Universitaet Duessseldorf (FRG)

Abstract: Based on a pair of optimization criteria which are evolutionary advantageous a learning scheme is proposed that allows to solve underdetermined linear matrix equations and to optimize the solution. The learning scheme does not only yield a theoretically remarkable result but is also feasible in biological neural networks.

1. INTRODUCTION

Finding the solution of underdetermined matrix equations is a frequent problem in biological systems. For example, reflexes such as the vestibulo-ocular reflex (Daunicht 1988a) can often be linearized by considering small signal changes. Then the system may be described as a matrix equation of the type

$$A \, X \, B = C.$$

Here the output of a sensory system (B) is fed into a neural network (X) and the network's output is fed into a motor system (A), so that the action of the total system can be characterized as a reflex (C). Usually the number of signals obtained from the sensory system exceeds the number of degrees of freedom of the input signal, so that the the matrix B has more rows than columns. On the other hand, the number of actuators in the motor system is usually higher than the number of degrees of freedom on the motor system, so that the matrix A has more columns than rows. Therefore there usually exists an infinite number of networks which are able to mediate the reflex. In other words, the matrix equation describing the reflex is underdetermined. Physiological observations, however, show that biological neural networks seem to be capable not only to solve such an ill-posed problem, but also to find a unique solution in a given species. Therefore it is tempting to assume, that the uniqueness of the solution is based on an optimization process and that the optimization refers to some meaningful criterion. In the present paper a pair of optimization criteria is proposed that provide evolutional advantage to an animal, and a learning scheme is outlined that uses very little global information to find an optimal solution to the underdetermined matrix equation.

2. THE OPTIMIZATION CRITERIA

In biological systems noise seems to play an important role. Therefore it can always be assumed that in general all signals are contaminated with noise. Since noise in sensory signals destroys valuable information it seems advantageous for an animal to minimize noise on sensory signals. In other words, the neural network should be optimized in such a way that it appears to 'reconstruct' as good as possible the uncontaminated signal by exploiting the redundancy of the sensory system. The mapping implemented in the network can then be described as yielding the same output, whether operating on the contaminated vector or on the optimal estimation of the uncontaminated vector. It is known from the theory of generalized inverses (cf. Albert 1972) that such a mapping can be characterized by

$$X = X \, BB^+,$$

where the superscript $^+$ denotes the generalized inverse of a matrix. Therefore it seems a reasonable optimization criterion to minimize the deviation d_B of the actual network from an ideal one

$$d_B = \|X - X \, BB^+\|.$$

However, there is still an infinite number of networks which satisfy the matrix equation and minimize d_B. Therefore a second optimization criterion is required. Concerning the motor system it seems correspondingly advantageous to optimize the economy of movement generation. For example, it seems useful to minimize the cocontraction in the muscles by avoiding to generate motor commands that unnecessarily increase the force in 'antagonistic' muscles simultaneously. The mapping implemented in the network can then be described as yielding the minimal image vector. Such a mapping is known to meet the condition

$$X = A^+A \, X.$$

Similarly, it seems reasonable to minimize the deviation d_A

$$d_A = \|X - A^+A \, X\|.$$

This pair of criteria indeed leads to a unique network matrix X satisfying the matrix equation. It can be shown using the definition of the generalized inverse that for the unique matrix X given by

$$X = A^+CB^+$$

both deviation d_B and d_A vanish. A matrix of this form is known to be the best approximate solution of an overdetermined matrix equation (Penrose 1965) and the minimum norm solution of an underdetermined matrix equation (Kohonen 1984). The minimum norm solution is defined by minimizing the Euclidean matrix norm of X.

3. THE LEARNING SCHEME

In order to propose a learning scheme that minimizes the criteria noted above it is sufficient to ensure that the Euclidean matrix norm of a the synaptic weights of a two-layer feedforward network is minimized. Such a learning scheme is outlined in Fig. 1. The network described by the network X is assumed to consist of two layers of linear neurons with a single layer of synapses in between, connecting all first layer neurons to all second layer neurons. Each single synapse is assumed to be subdivided into two parts with long term memory and short term memory. Both subdivisions contribute to the weight of the synapse, i.e. the signals weighted by the subdivisions add up. The short term memory can be converted into long term memory, whereby the occurrence of the conversion depends on a scalar - or even binary - signal, that is homogeneously distributed to all synapses. The short term subdivision is subject to small stochastic changes of either sign with Gaussian distribution function around zero. It may be noted that random search has already been used successfully to solve control tasks (Barto et al. 1983). However, here it is essential that the distribution function be Gaussian. Further it is proposed, that the long term memory is subject to slow homogeneous decay or 'forgetting' (Werntges and Daunicht 1988) where the amount of decrease is only proportional to the long term weight. Both the amplitude of stochastic short term variation and the long term forgetting coefficient may decrease in time.

The network (X) is functionally located between the input mapping (B) and the output mapping (A), thus mediating the total mapping of the system. The total mapping of the system is now to be compared with the desired mapping (C); this comparison is made on the basis of the operation of the system on some input vectors. It is assumed, that the environment or some random process provides input signals evenly distributed over the whole input space. The difference between output vectors of the system and the desired output is evaluated by some independent critic. It suffices to produce a binary signal such as improvement or deterioriation, which is used to decide on the occurrence of the conversion between the synaptic subdivisions. The proposed mechanism may be varied and possibly accelerated by taking the amount of improvement and deterioriation into account, e.g. by modifying the amplitude of the stochastic matrix variations.

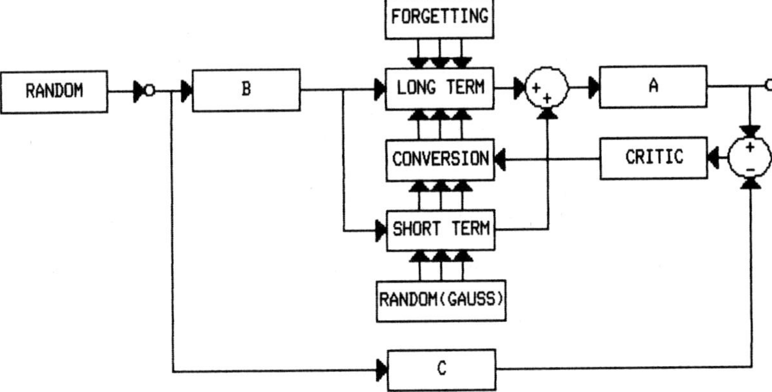

Fig. 1. Learning scheme for the optimal solution of underdetermined matrix equations

It can be shown that the mean of a matrix that is subject to such learning mechanism can be made arbitrarily close to the minimal norm solution of the underdetermined matrix equation. The proof may be outlined as follows. The matrix X is considered as a vector in the space of the matrix elements. The solutions of the matrix equation form an affined subspace of the element space. As the random matrix variations are isotropicly distributed in the element space (because of the Gaussian distribution), the decision mechanism ensures, that the mean of the actual matrices gets arbitrarily close to the solution subspace, and that the mean of the variations has no component parallel to the solution subspace. The forgetting mechanism has two effects, first it draws the actual matrix slightly from the submanifold towards zero, an effect that can be made arbitrarily small by reducing the coefficient of forgetting, and secondly it moves - as the only force - the actual matrix vector parallel to the solution subspace and it attracts it towards the unique vector within the solution subspace that minimizes the Euclidean vector norm and thereby the Euclidean matrix norm as well.

The learning scheme described in the present paper presents a mechanism that is biologically feasible and may be in fact working in the selforganization of the vestibulo-ocular reflex (Daunicht 1988 a,b). It can be expected that there is a tradeoff in accuracy vs. velocity of the optimization. Consistently it has been observed, that the vestibulo-ocular reflex in the dark is always less than perfectly compensating; the gain of the vestibulo-ocular reflex therefore provides a measure that may allow to assess the most advantageous balance that has evolved in natural systems between learning and forgetting.

4. ACKNOWLEDGEMENTS

Thanks are due to Juergen Beckmann for valuable discussions. Supported by Der Bundesminister für Forschung und Technologie (BMFT) under grant No. ITR 8800 F6

5. REFERENCES

1. Albert A. (1972) Regression and the Moore-Penrose pseudoinverse. Academic Press, New York

2. Barto A.G., Sutton R., and Anderson C.W. (1983) Neuronlike adaptive elements that cam solve difficult learning conrol problems. IEEE Trans. SMC-13, 834-846

3. Daunicht W.J. (1988a) A biophysical approach to the spatial function of eye movements, extraocular proprioception and the vestibulo-ocular reflex. Biol. Cybern. 58: 225-233

4. Daunicht W.J. (1988b) Neural networks mediating linearizable dynamic redundant sensori-motor reflexes characterized by minimum of hermitian norm.

5. Kohonen T. (1984) Self-organization and associative memory. Springer, Berlin

6. Penrose A. (1956) On best approximation solutions of linear matrix equations. Proc. Camb. Philos. Soc. 52: 17-19.

7. Werntges H. and Daunicht W.J. (1988) Effects of forgetting on the self-optimization of redundant sensory-motor control networks. Abstr. 1st. Ann. INNS Meeting, Boston, 1988, 366

Parallel Processing in Neural Systems and Computers
R. Eckmiller, G. Hartmann and G. Hauske (Editors)
© Elsevier Science Publishers B.V. (North-Holland), 1990

INVERSE KINEMATICS WITH OBSTACLE AVOIDANCE IMPLEMENTED AS A DEFANET

Wolfgang J. DAUNICHT, Martin LADES, Heinz WERNTGES and Rolf ECKMILLER

Abt. Biokybernetik, Institut fuer Physikalische Biologie
Heinrich-Heine-Universitaet Duessseldorf (FRG)

Abstract: Although it is desirable to use the redundancy of a manipulator to avoid obstacles, it seems difficult to choose the right configurations in real time, even if signals on the actual positions of the obstacles are available. To this end an neural net approach is used to approximate a high-dimensional nonlinear continuous function by a deterministic four layer feedforward network (DEFAnet). The feasibility of this approach is demonstrated by applying it to a planar redundant manipulator avoiding two obstacles with one degree of freedom, i.e. a moving window.

1. INTRODUCTION

The solution of the inverse kinematics of a manipulator is a problem, that is the more time consuming, the more complex the manipulator is. Additional computer time is needed when additional problems have to be solved such as resolving the redundancy of a manipulator or when obstacles have to be taken into account and to be avoided by choosing appropriate configurations. In fact, the task may become so difficult, that it is not or not easily solvable by conventional algorithmic calculation in real time. At this point it seems interesting to compare biological solutions of that problem. Obviously animals use learning in two respects: first to aquire the knowledge of how to implement the required function, and second, to use memory when the same or a similar situation is encountered, rather than to recalculate the desired outputs every time.
It seems that neural networks are suitable to implement conveniently inverse manipulator kinematics, provided they are capable of approximating any given continuous real function from $\mathcal{E}^n$ to $\mathcal{R}^m$ (cf. Daunicht 1989). The present paper proposes an approach to the implementation of an inverse kinematic by means of a feedforward network according to the DEFAnet (DEterministic Function Approximating network) concept (Daunicht 1990). Based upon the choice of a certain interpolation rule, this concept allows to determine the topology of the network required to approximate any function and allows to find the weights of the synapses by direct calculation as well as by training. The DEFAnet is then demonstrated to be capable of implementing global obstacle avoiding inverse kinematics of a redundant planar manipulator with any desired accuracy.

2. CONCEPT OF DEFANET

The concept of DEFAnet is based on the construction of a grid in the n-dimensional hypercube by subdividing it into n-dimensional rectangular cells by means of l_ν hyperplanes orthogonal to the ν-th axis ($\nu = 1, \ldots, n$), and the assumption that the goal function f is defined and

known at least at the corners of each cell, the grid points. The choice of the interpolation
rule that is to be implemented by the network is given by the condition, that all first partial
derivatives are constant inside a cell and along the edge of a cell.

The neural network is constructed in such a way as to follow the interpolation rule - given any
combination of function values at the corners of each cell - only by modifying the synaptic
weights of the last layer. To this end it is required to have all kinds of products of relative
input signals (taken to the power of 0 or 1) available, so that appropriate linear combinations
of them may be formed. If this is to be achieved by a network with neurons summing rather
than multiplying their inputs, a 4-layer feedforward network is required. Unfortunately, using neurons with limited non-negative output (such as natural neurons), it is not possible to
generate a product directly using logarithmic output functions, summation, and an exponential function, as the logarithm of numbers less than one would generate negative signals of
unlimited amplitude. However, the use of output functions that can be described as shifted
logarithmic functions yields linear combinations of products that can still be employed while
following the interpolation rule. The structure of a DEFAnet is illustrated by an example
(Fig.1).

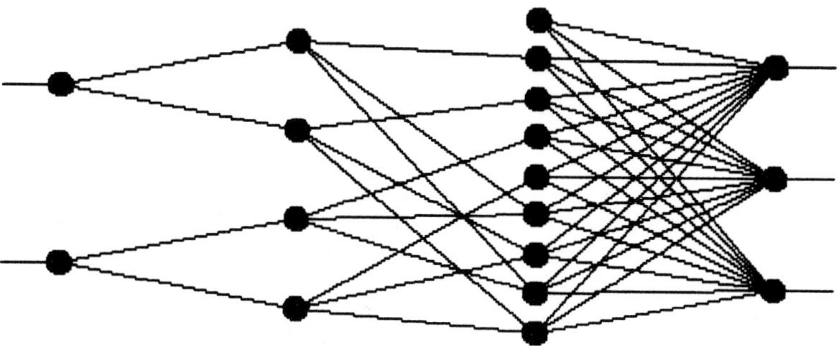

Fig. 1. Example of a DEFAnet with $n = 2, l_1 = 3, l_2 = 3$, and $m = 3$

The first layer consists of n pure fan-out units. The neurons of the second layer have
monotonous output functions which increase logarithmically over the range between two adjacent grid hyperplanes from 0 to 1. The ranges of such increase correspond to the sizes and
locations of the grid cells. To subdivide a hypercube into $\prod_{\nu=1}^{n}(l_\nu - 1)$ cells requires

$$z_{(1-2)} = \sum_{\nu=1}^{n}(l_\nu - 1)$$

synapses as well as neurons in the second layer, as each second layer neuron receives input
from only one first layer neuron. It suffices to let the synapses be excitatory, non-plastic and
their weight be 1.

The output functions of the third layer neurons increase proportional to the exponential
function and saturate at 1. To provide all required linear combinations of products the third
layer consists of $\prod_{\nu=1}^{n} l_\nu$ neurons. The number of required non-zero synaptic connections
between the second and third layer is

$$z_{(2-3)} = \sum_{\nu=1}^{n}\frac{l_\nu - 1}{l_\nu}\prod_{\rho=1}^{n} l_\rho$$

Here the synapses are excitatory and non-plastic; their weights depend on the third layer neuron to which they belong, but not on the goal function.

The output function of the only fourth layer neuron is linear and not limited. If vector rather than scalar functions are considered, i.e. the output dimension m is greater than 1, this is the only layer to increase the number of units. The number of synapses is

$$z_{(3-4)} = m \prod_{\rho=1}^{n} l_\rho$$

as all third layer neurons are connected to all fourth layer neurons. The synapses between the third and the fourth layer are considered to be plastic and to have either sign. Given the output functions of the second and third layer neurons, the values of the last synaptic layer can be determined by n-fold application of a set of recursive formulae. The size of a DEFAnet and its topology of connections is completely determined by the dimensions of the input and output signals and by the resolution of the grid along each axis.

3. INVERSE KINEMATICS BY DEFANET

The manipulator employed is a planar chain of four links with four rotational joints (Eckmiller 1988). The redundancy of such a chain can be exploited to choose configurations that avoid obstacles where necessary and possible. Obstacles may have an arbitrary shape; in the present example the obstacle consisted of a vertical window that could move (with one degree of freedom) in vertical direction. The movement of the window was assumed to be sensed and communicated to the network. This signal and the coordinates of the desired tool position (finger tip position) form the input signals to the network. The output signals consist of the commands that control the position of the four joints. In particular, the output signals and thus the configuration may have to change, when the tool position is constant and the obstacle moves (Fig.2).

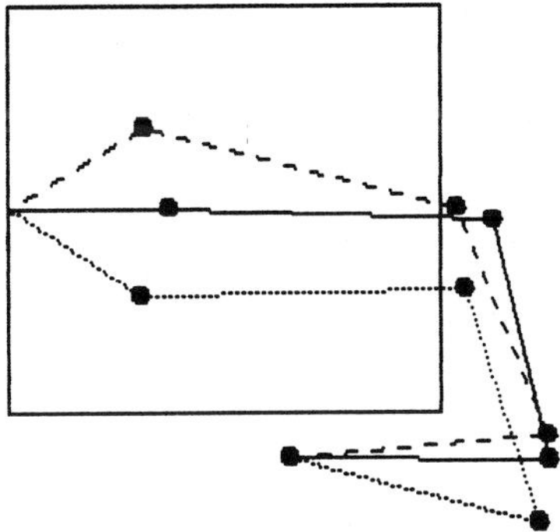

Fig. 2. Changing configuration of the four-joint machine with constant tool position

In the present work the network is trained by the delta rule (Widrow and Hoff, 1960), i.e. using a teacher. Due to the fact that only the last synaptic layer of DEFAnet is plastic, the learning process is rapidly converging, backpropagation of errors is not required. The teacher is obtained by minimizing a cost function by means of the simplex algorithm (Nedler and

Mead, 1965). The cost function employed here takes a number of constraints into account, such as not to exceed 90 degrees of deviation from an arbitrary 'resting angle' of joint rotation, and the position of the obstacles. Obstacle avoidance is achieved by means of 'Coulomb forces' between locally discrete 'charges of equal sign' on the manipulator and the obstacles. In order to avoid local minima inside the obstacles, the cost function is repeatedly calculated while moving the charges from their 'center of gravity' to the surface of the obstacle. Repeatedly minimizing such a cost function is very time consuming and far beyond realization in real time. However, storing the result of such optimization processes in a neural network - preferably implemented on a parallel computer - makes it possible to perform elegant avoidance of moving obstacles in real time, as the computation time of a feedforward network such as DEFAnet depends only on the number of consecutive layers. It may be mentioned, that the proposed procedures can also be applied to manipulators and obstacles with higher degrees of freedom, in particular by calculating high-dimensional 'centers of gravity' as a starting points for the establishment of the cost function. In addition, the proposed network approach allows to integrate sensors of different modality for the solution of complicated tasks.

4. ACKNOWLEDGEMENTS

Supported by Der Bundesminister für Forschung und Technologie (BMFT) under grant No. ITR 8800 F6

5. REFERENCES

1. Daunicht, W.J. (1989) Control of manipulators by neural networks. IEE Proc. 136: 395-399

2. Daunicht, W.J. (1990) DEFAnet - a deterministic approach to function approximation. IJCNN 1990, Washington (in press)

3. Eckmiller, R. Concept of a 4-joint machine with neural net control for the generation of 2-dimensional trajectories. Abstr. 1. Ann. Meet., Boston, 1988, 334

4. Nedler, J.A. and Mead, (1965) A simplex method for function minimization. Computer J. 7: 308

5. Widrow, G. and Hoff, M.E. (1960) Adaptive switching circuits. Institute of Radio Engineers, Western Electronic Show and Convention, Convention Record, Part 4, 96-104

Parallel Processing in Neural Systems and Computers
R. Eckmiller, G. Hartmann and G. Hauske (Editors)
© Elsevier Science Publishers B.V. (North-Holland), 1990

Parallel Neural Net Path-Planner on
Hypercube and Transputer

Alex W. Ho and Geoffrey C. Fox

Caltech Concurrent Computation Program
206-49, California Institute of Technology,
Pasadena, CA 91125, USA

Abstract

The task of finding a "good" path is of prime importance to any automated vehicle navigation controller. This task has been formulated as problems in computational geometry, which is commonly reduced to a graph search problem. We report here the preliminary results of a new approach that finds a "good", if not optimal, path dynamically and in real-time by casting the task as a classification problem [Ho:89a.] The path-planner implements a back-propagation model for classification. The multi-scale representational strategy used for mapping the problem domain onto the input space of the back-propagation model ensures the applicability of the trained network to instances of the problem, which differ in the distribution of obstacles and the size of the problem domain. Software simulations based on the parallel computer EXPRESS operating system run, without modification of code, in Transputer-based systems as well as hypercube concurrent processor, with satisfactory efficiency.

1. Introduction

Advances in robotics technology have stimulated interest in the research on autonomous learning systems. One application of an autonomous learning system is a mobile automaton [Nilsson:80] which learns from experience, and responses intelligently to unknown environment. Besides important issues such as visual and tactile sensing, spatial data handling, scenery analysis, and motor control, one other critical issue facing the development of a system with autonomous navigation capability is how to make the system plans a "good" course when it is given a high level specification of a desired target position. In this case the ability of both efficient and effective motion-planning becomes essential.

A prototype motion-planning problem for a single vehicle can be stated as follows:

> Given a partial model of the world (W), the starting (Z_s) and destined (Z_d) vehicle positions in physical space and a description of the characteristics and positions of obstacles, an objective of a navigation system is to plan a continuous collision-free path that leads a vehicle (V) from Z_s to Z_d, if such a path exists.

In this paper, the physical space is interpreted mathematically as a 2 or 3 dimensional Euclidean space. For simplicity the problem statement does not have the explicit notion of velocity and time. In other words, kinematics and dynamics are not considered in this paper although our techniques can be extended to include these.

In many circumstances, optimal paths are not required. It is often more important to obtain a "good" (i.e., nearly but not precisely optimal) path quickly than to devote precious computational resources to find the exact solution. In fact the input data is often imprecise (e.g. the exact nature of the terrain is not known) and the notion of a precise optimal path undefined. Several new optimization techniques such as simulated annealing, neural networks, elastic networks and genetic algorithms have been devised for such approximate optimization problems [Kirkpatrick:83, Hopfield:85, Fox:88kk, and Simic:89.] We will explore this problem using neural networks here.

2. Motivations

There is a category of problems that artificial neural-network algorithms, used as computation tools, are expected to be efficient in solving. These are the problems that human can handle easily and efficiently. Obvious examples are pattern and speech recognition. Another note-worthy example is playing chess. It has been a long-standing speculation that a good chess player can recognize patterns of the chess board and command a good move or counter-move efficiently, as opposes to a computer chess program which has to construct and evaluate, by using some intelligent heuristic tree-search techniques, a huge game tree of legal moves and counter-moves of the current board, iteratively until a fixed depth is reached. In the same vein the problem of motion-planning also fits well into this category.

A string-based neural formulation using the Hopfield and Tank model to find an optimal path has been studied by [Fox:89aa] and [Wong:89.] We describe here how the problem of motion-planning can be formulated as a classification problem in which the classifier is an implementation of a supervised learning neural model. An autonomous neural navigator which implements the back-propagation model and finds a "good" path, if not optimal, for the problem dynamically is presented.

3. A Supervised-Learning Neural Formulation

The objective of the simulation is to train an autonomous vehicle navigator with a human supervisor, so that the navigator can plan "good" collision-free motions for a vehicle in an unknown environment with randomly or structurally-positioned obstacles. The navigator was taught with a number of input terrains. Based on the training the navigator uses the experiences to plan best route. The planning is dynamical in the sense that the navigator plans the immediate next move for a vehicle at each discrete time-step. In other words, the navigator can cope with slowly evolving problem domain.

We assumed sensor information provides a planar view of the distribution of obstacles in the problem domain. The planar view is discretized into a rectangular grid of square cells. The sub-space occupied by an obstacle, including its boundary and interior, in the discretized domain is represented by placing a circle or a union of circles to cover that sub-space. Each of the circles has the size of a cell. In addition, the center of the circles coincide with the center of mass of the corresponding occupied cells. This scheme provides a representation that approximates both concave and convex obstacles consistently.

A vehicle is represented by a small circle with its center coincident with the center of mass of a cell. This vehicle representation is orientation-invariant. Therefore, only translation, but not rotation, is considered in the problem. For simplicity motor control of a vehicle is assumed to constrain the vehicle to translate in one of the five admissible directions (left, right, up, diagonal to the left, and to the right.) In one discrete time-step a vehicle moving in any of these five directions will position the vehicle in the corresponding neighboring cell. Although the current constraint imposed on admissible directions forbids vehicle roll-back it will become evident that extension of our formulation to motion in any direction is simple and straight-forward.

Traffic regulation imposes that any motion of a vehicle that results in a collision with an obstacle or part of an obstacle is considered "illegal". All moves that do not result in violation of a traffic regulation are called feasible moves. Depending on the optimality criteria one feasible move can be more preferable than the others.

Multi-scale technique is naturally and universally applicable to spatial problem solving in different fields. An example is in animal vision. The distribution of photo-receptors on the retina has been found to have a diminishing density away from the fovea. This shows that biological photo-receptors have different spatial scaling. Based on the same principle a multi-scale representation of a time-slice instance of the problem domain, including the vehicle, the distribution of obstacles, and the target position, was developed. For experimental purposes only a 4-scale representation scheme was used in our simulations. After discretizing the problem domain into an M × N grid, where M is in the horizontal and N is in the vertical dimension, the grid is mapped onto 36 neurons, each of which encodes the porosity of the obstacle distribution in 9 general directions with respect to the vehicle position. The 9 general directions were chosen for this preliminary study because it was easy to implement. Neurons of the finest scale encodes detailed information of the nearest neighbor of the vehicle in the grid space. Further away from the vehicle each neuron encodes information of a broader area. The further away from the vehicle the coarser are the encoded information. This sort of representation is natural because near field or local information is crucial for the immediate motion of the vehicle, while hierarchical far field information is required for global planning. By choosing an exponential function for the diminishing photo-receptor density away from the center of attention the coarsest neuron can encode information for a much bigger grid.

Any instance of the problem is treated as a pattern and represented by 36 neurons. In the situation of using the grandmother cell approach to encode the admissible directions as distinct classes, five output neurons are needed. A teacher of the path-planner encodes his/her decision of motion for a given instance (which is a pattern) by turning on the appropriate output neuron that corresponds to the desired move. The casting of the problem of path-planning into a classification problem is now completed.

4. Simulation Results

Parallel implementation of back-propagation model for character recognition on hypercubes based on a character decomposition technique has been discussed by [Ho:88c.] Our current implementation is based on the decomposition of the set of training patterns. Essentially, each processor of an NCUBE hypercube or a Transputer-based concurrent processor is responsible for only a small subset of the set of training samples. Using Parasoft EXPRESS as the communication software the same code runs in NCUBE as well as in a Transputer-based concurrent processor.

The training set we used consists of 184 patterns, each is represented by 36 neurons (47 × 24 grid.) The choice of using 184 patterns is arbitrary. Our first goal is to teach the navigator enough basic knowledge upon which it can generalize, not just memorize, to cope with most situations. We used up to 64-node NCUBE hypercube and 4-node Topologix Transputer system for our simulations. A training set of 184 patterns becomes a small problem for the case of 64 processors because each processor is then responsible for performing computations for at most 3 patterns. This is evidented in Fig. 1 and 2. Fig. 1 shows the timing results which are normalized to the time required to compute one iteration on a Sun SparcStation. The timing results of the NCUBE and Topologix runs are based on the epoch and a hybrid updated rule. The hybrid rule, in its limits, reduces either to the generally used epoch update or pattern update in back-propagation models. The hybrid update rule is efficient for parallel implementation and has a good convergence rate. Fig. 2 is a comparison of the efficiency of the parallel epoch update rule and the hybrid rule. It is obvious that epoch update requires less communication among processors and has extremely high efficiency for hypercubes of low dimension. The per iteration timing comparison is less favorable for hybrid update. However, the better convergence rate still makes it the preferred choice for this simulation.

The learning histories of the back-propagation path-planner for various parameter values are displayed in Fig. 3 to 4. Average error is defined as the average of the total quadratic error at the output neurons per pattern. The learning behavior of the planner at a fixed learning rate $\eta=0.1$ and momentum $\alpha=0.8$ for 4 different randomly generated initial states were investigated (both η and α are as defined in [Rumelhart:86].) Results suggest that the error landscape for this problem is quite smooth. Fig. 3 demonstrates that the momentum term at values of 0.2, 0.4, 0.6 and 0.8 produced roughly similar learning history, for $\eta=0.1$. Fig. 4 illustrations the dramatic effect of tuning the learning rate η for a fixed $\alpha=0.8$. At small η the planner's learning curve is very smooth but it learns slowly. Increasing the learning rate also increases the possibility of instability. For this particular problem and network configuration η higher than 0.15 indicates slow learning and big oscillations.

5. Performance of Neural Planner

The trained neural planner was put to test by submitting to it unlearned scenarios in the form of 47×24 grids and, to show the scale-invariant capability of the network, 105×53 grids. Fig. 5 and 6 are test cases of using randomly distributed obstacles, while Fig. 7 is a case of structured obstacles. The planner performed satisfactorily by planning "good" paths for each case.

6. Conclusions

Current study indicates that the back-propagation model can be efficiently implemented on distributed memory MIMD computers such as hypercube and Transputer-based systems. Also, the problem of path-planning can be approached from a different perspective, namely, a classification problem. Back-propagation path-planner performed satisfactorily in our preliminary study of simple scenarios. However, in order to claim the usefulness of the approach in practice would require a lot more research work.

Acknowledgements

This work is based on research supported by the Department of Energy grant DE-FG03-85ER25009, and by the Program Manager of Joint Tactical Fusion Office.

References

[Fox:88kk]   Geoffrey Fox, Wojtek Furmanski, Alex Ho, Jeff Koller, Petar Simic, and Issac Wong, "Neural Networks and Dynamic Complex Systems", contribution to the 1989 SCS Eastern Conference, Tampa, Florida, (March 1989,) Technical Report C^3P-695, California Institute of Technology, 1988.

[Fox:89aa]   Fox, Geoffrey C. and Gurewitz, Eitan and Wong, Yiu-fai, "A Neural Network Approach to Multi-vehicle Navigation", contribution to the 1989 SPIE Conference, Philadelphia, Pennsylvania, (Nov. 1989,) Technical Report C^3P-833, California Institute of Technology, 1989.

[Ho:88c]   Ho, Alex W. and Furmanski, W., "Pattern Recognition using Neural Networks in Hypercubes," in the Proceedings of the Third Conference on Hypercube Concurrent Computers and Applications, **Vol. 2**, (ed.) Geoffrey C. Fox, ACM Press, New York, 1011-1021 (1988), Technical Report C^3P-528, California Institute of Technology, 1988.

[Ho:89a]   Ho, Alex W., "A Back-propagation Navigation Controller for Land and Space Vehicles," Technical Report C^3P-735, California Institute of Technology, April 1989.

[Hopfield:85]   Hopfield, J. J. and Tank, D.,"Neural Computation of Decisions in Optimization Problems," *Biol. Cybernetics* **52**, 141-152 (1985.)

[Kirkpatrick:83]   Kirkpatrick, S., Gelatt, C. D., and Vecchi, M. P., "Optimization by Simulated Annealing," *Science* **220**, 671 (1983.)

[Nilsson:80]   Nilsson N.J., *Principles of Artificial Intelligence*, 1980.

[Rumelhart:86]   D.E. Rumelhart, G.E. Hinton, and R.J. Williams, "Learning Internal Representations by Error Propagation" in D.E. Rumelhart & J.L. McClelland (Eds.), "Parallel Distributed Processing: Explorations in the Microstructure of Cognition. Vol. 1: Foundations." MIT Press, 318-364 (1986.)

[Simic:89]   Simic, P., "Statistical Mechanics as the Underlying Theory of 'Elastic' and 'Neural' Optimizers," *NETWORK: Computation in Neural Systems* **1**, 1-15 (1990), Technical Report CALT-68-1556, C^3P-787, California Institute of Technology, May 1989.

[Wong:89]   Wong, Issac and Fox, Geoffrey C., "Use of neural networks for path planning," Technical Report C^3P-784, California Institute of Technology, May 1989.

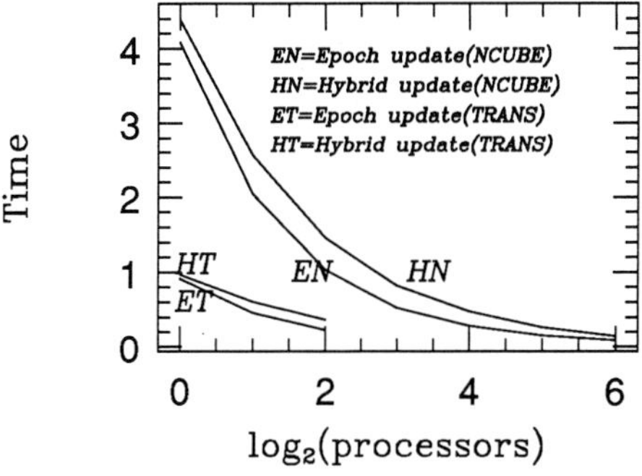

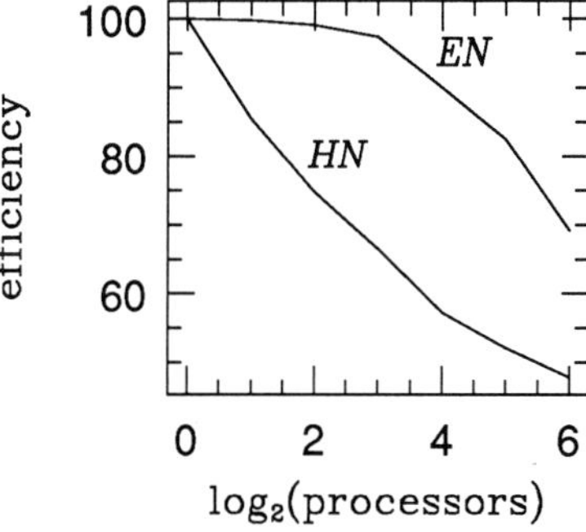

FIGURE 1
Timing results on NCUBE and Transputer

FIGURE 2
Efficiency results on NCUBE hypercube

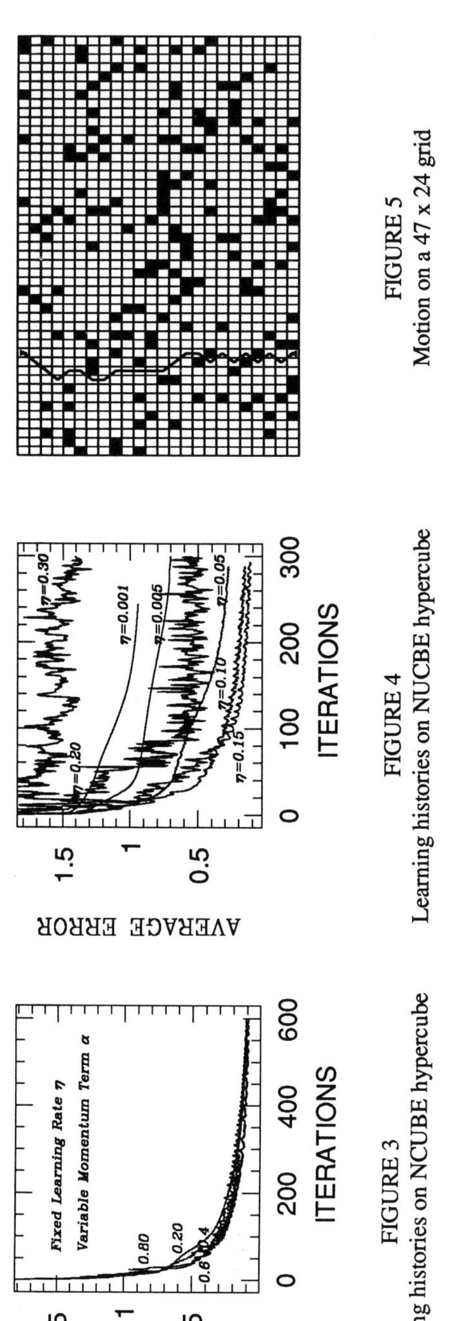

FIGURE 5

Motion on a 47 x 24 grid

FIGURE 4

Learning histories on NUCBE hypercube

FIGURE 3

Learning histories on NCUBE hypercube

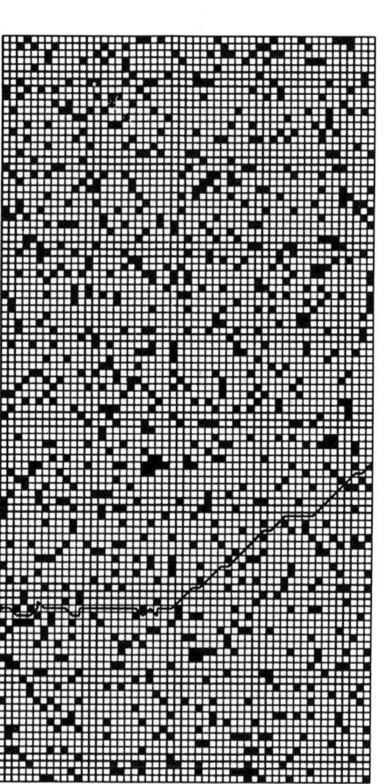

FIGURE 7

A 105 x 53 grid with structured obsctacles

FIGURE 6

A 105 x 53 grid with randomly distributed obstacles

Parallel Processing in Neural Systems and Computers
R. Eckmiller, G. Hartmann and G. Hauske (Editors)
© Elsevier Science Publishers B.V. (North-Holland), 1990

A CONTROL CONCEPT FOR A MULTIFINGERED ROBOT GRIPPER BASED ON NEURON-LIKE ASSOCIATIVE MEMORIES

Michael HORMEL[+]

Department of Control Systems Theory and Robotics
Technical University of Darmstadt

An alternative approach for the control of different
phases of grasping movements with a multifingered robot
gripper based on associative memories is presented in
this paper. The experimental setup for the multitransputer
based gripper control system is outlined.

1. INTRODUCTION

At the present time industrial robots are used for relief of human
workers from handling heavy loads and for automation of unwhole-
some workshop places. Endowed with exchangeable toolsets these
robots can easily be used for a large number of different tasks.
However when used for tasks which require exact positioning and
orienting of a grasped object most of the robots are too clumsy.
In some extreme cases it will be necessary to adjust the position
of all joints of a six degree of freedom robot in order to achieve
a small correction of the position and orientation of the robot
gripper or a grasped object. On the other hand multifingered robot
grippers exist which allow to change orientation and position of
the object directly without affecting the robot position. When
using conventional, mathematical methods for the control of such a
multifingered gripper one has to handle a large amount of computation, which has to be carried out on-line and in real-time. An
alternative approach for a control system based on theories of
neural information processing is presented throughout this paper.

2. CONTROL OF GRASPING MOVEMENTS WITH NEURON-LIKE ASSOCIATIVE MEMORIES

Actually neural information processing systems like the brain use
a combination of computation and recall of learned and stored in-
put/situation - output/reaction relationships. So part of the necessary control computation is substituted by fast, input-data
driven (associative) access to an appropriately organized memory.
Our alternative approach to multifinger gripper control is just to
imitate this procedure.

The associative memory system AMS [1] which will be used in the

---

[+] M. Hormel, Department of Control Systems Theory and Robotics,
            TU Darmstadt, Schlossgraben 1, 6100 Darmstadt, FRG
  This work was sponsored by the German Ministry for Research
and Technology (BMFT) under grant ITR 8800 B/5

gripper control system is a simulation of a neuron-like associative memory as proposed by J.S.Albus as a model for information
processing in the human cerebellar cortex [2]. AMS is extremly
useful for storing a nonlinear input-output relationship by simply
mapping n-dimensional stimulus patterns onto arbitrary m-dimensional output patterns and for a fast recall of the stored information. The operation of the system depends on stimulus driven selection of ρ cells out of a much larger number of p possible
weight cells.

During recall the memory response is calculated as the average of
the values of the activated cells. In doing so the response for
untrained stimulus patterns in the neighbourhood of trained points
is the result of an automatic multidimensional interpolation over
the output values of the trained points. During training the
values of the activated weight cells are updated by a delta-rule.
In the gripper control system the AMS can be applied e.g. for
storing the inverse kinematics of the fingers and so skipping the
need for intensive computation during the backward transformation.

In general a grasping operation can be characterized by the four
phases approach, contact, grasping, and handling. Each of these
phases can be controlled by an AMS-block which becomes active depending on the current state of the operation. During approach,
which is controlled by the robot control system only, the geometrical data of the target are used for preshaping the robot
gripper i.e. for opening the hand wide enough not to collide with
the target. In order to achieve this an AMS-block can be configured which employs the desired opening radius and the object
height as stimulus and provides the joint positions for the fingers as response.

Having reached the desired position the hand is closed until all
fingers detect a contact with the target. The detection can be
implemented by a continuous supervision of the joint forces. In
order to avoid the occurence of too high forces a very low stiffness will be chosen for the fingers in this phase.

In general it is possible that different strategies for grasping
can be activated during approach as well as during contact phases
depending on size, form and other features of the target. These
strategies include e.g. grasping with fingertips only, selecting
special contact points, etc.

During grasping the fingers execute coordinated movements in order
to compensate the internal forces of the grasped object to ensure
a stable grip. The coordination can be achieved either by using a
hierarchical approach with a coordinator specifying goals for
subsequent AMS-blocks or by providing informations about planned
actions of one single finger for all other fingers in the case of
a non-hierarchical control structure. The stiffness will be increased during this phase to exert higher forces on the target.

Finally in the handling phase the object can be manipulated by
coordinated movements of the fingers. Again stiffness will be increased to ensure the stability and to avoid sliding of the target.

3. GLOBAL SYSTEM STRUCTURE

As a testbed for the application of an associative gripper control a system consisting of a picture processing unit with two cameras, a robot control system and a robot carrying the multifingered gripper is considered (Fig. 1). The picture processing unit detects and identifies the target, computes object position and orientation in world coordinates and/or other geometrical data required for moving the robot over the target and for opening the fingers for grasping. The data are then sent to the robot control system and to the gripper control for task execution.

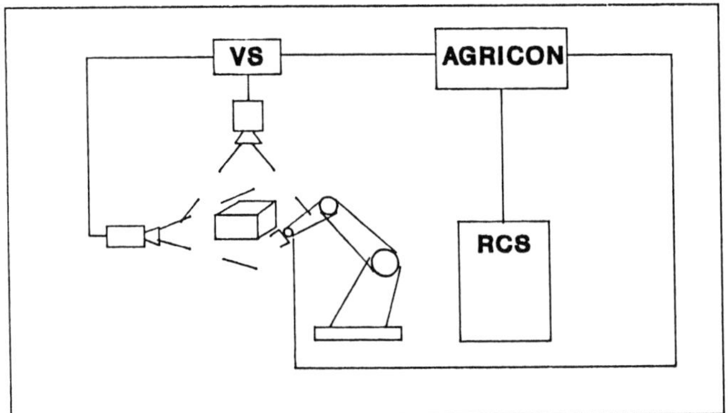

FIGURE 1
Testbed for the associative gripper control system
VS: Vision System, RCS: Robot Control System
AGRICON: Associative GRIpper CONtrol

A prototype of the multifingered gripper is currently under development [3]. The gripper as a whole consists of three fingers with three rotatorial joints each. Fig. 2 shows the basic arrangement of joints for one finger. The joints are actuated by external DC-motors via harmonic-drive gears and bowden wires as shown in Fig. 3.

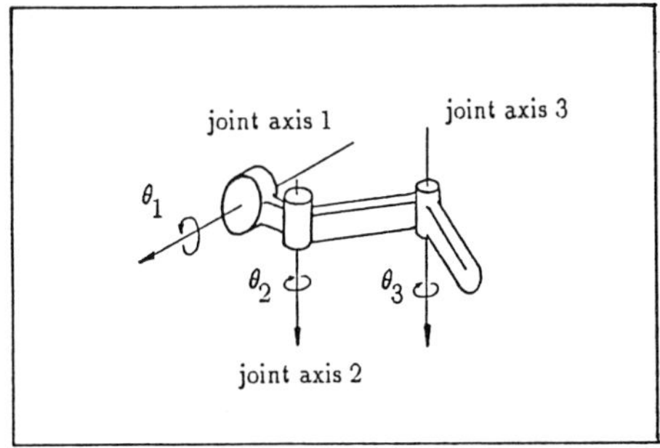

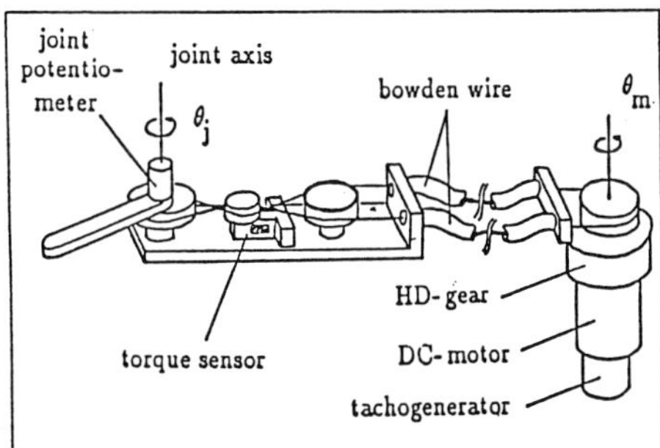

FIGURE 2
Basic joint arrangement

FIGURE 3
Mechanical joint structure

The drives are equiped with tacho-generators for generating a
velocity signal. Joint position is measured by a high precision
potentiometer attached directly to the joint axis whereas joint
torque is measured by a tension differential type torque sensor
near the joint axis. At the moment the computer hardware for the
gripper control system is based on an IBM-AT compatible computer
with 386/387 processors. For the application of associative
memories for gripper control a multiprocessor system based on
transputers is taken into consideration. Herein a single
transputer will be responsible for the movement of one single
finger.

4. STATUS

In parallel to the development of the multifingered gripper some
improvements to the associative memory system AMS have been made
concerning an information dependent variable generalization. Several simulation experiments for learning the inverse kinematics
of a simple robot arm and for moving the robot arm along a given
path using this information have been carried out successfully.
The current work is concerned with implementing an environment for
the simulation of grasping movements with multifingered grippers
of variable configuration.

5. CONCLUSION

A brief description of an alternative approach to algorithmic
multifinger gripper control has been outlined. Different phases of
a grasping movement have been presented wich shall be controlled
by learning control loops with associative memories. A rough description of a multifinger gripper hardware and the experimental
setup for learning to grasp different objects has been given. The
computer hardware will be based on a multitransputer system with
an IBM-AT compatible computer as host-system and for executing the
real-time control tasks.

REFERENCES

[1] Ersue E., Militzer J., Implementation of a Neuron-like
    Associative Memory System for Control Applications,
    Proceedings of the 2nd IASTED Conference on MINI- and
    MICROcomputer Applications MIMI'82, March 1982, Davos,
    Switzerland
[2] Albus J.S., A New Approach to Manipulator Control: The
    Cerebellar Model Articulation Controller (CMAC),
    Transactions of the ASME, series G, Vol. 97, No. 3
[3] Paetsch W., Kaneko M., A Three Fingered, Multijointed
    Gripper as a Working Tool for Experimenal Use,
    Proceedings of the 2nd EIOOT Symposium, 1989, Toulouse,
    France

Parallel Processing in Neural Systems and Computers
R. Eckmiller, G. Hartmann and G. Hauske (Editors)
© Elsevier Science Publishers B.V. (North-Holland), 1990

LEARNING OF VISUOMOTOR-COORDINATION OF A ROBOT ARM WITH REDUNDANT DEGREES OF FREEDOM

Thomas MARTINETZ, Helge RITTER and Klaus SCHULTEN

Beckman Institute for Advanced Science and Technology
and Department of Physics
University of Illinois at Urbana-Champaign
405 North Mathews Ave., Urbana, IL 61801, USA

An improved version of an earlier introduced algorithm for learning of visuomotor-coordination has been successfully applied to a simulated robot arm system with five degrees of freedom, two of which are redundant. The learning algorithm was not affected by this redundancy, and with the improved version the robot arm system is able to reduce its positioning error to about 5% after only 200 learning steps, and finally to about 0.3% of its linear dimensions after 6000 learning steps. Learning proceeds without the need of an external teacher by a sequence of trial movements using input signals from a pair of cameras. The topology conserving map used for the representation of the input-output transformation leads to an automatic resolution of the redundancy of the arm as it tries to minimize the variation of the joint angles over the work space. The improved learning algorithm incorporates some immediate feedback, so that now the robot arm is able not only to adapt to slowly occuring miscalibrations, but also can compensate for sudden changes in its geometry, such as picking up e.g. a tool.

1. INTRODUCTION

Compared to present-day robots, biological motor control systems excel with an enormous degree of flexibility. This flexibility is to a large extent due to two major factors. The first is the use of highly developed sensors, most notably vision, capable to monitor a multitude of different aspects of ongoing movements simultaneously. The second factor is the presence of a high amount of redundancy in the muskulo-skeletal system, offering for most movement goals a wide range of alternative realizations. Both strengths are enhanced further by the high degree of adaptability that neural systems exhibit for both, sensory perception and selection and control of movements.

There have been many attempts to capture these remarkable features in biologically inspired neural network models, see e.g. [1-5]. Our own work has previously considered the problem of adaptive visuo-motor coordination for a robot, using a non-redundant (simulated) three-link robot arm ([3,5]). In this contribution, we want to focus on the additional issue of controlling a redundant arm and to resolve its redundancy by the use of a topology conserving map imposing a "smoothness constraint" on the arm configurations. In addition, we present an improved version of the learning rule used in [3,5], which incorporates a more accurate visual feedback.

For a redundant robot arm, specification of a target location does not yet uniquely fix the configuration of the arm, but instead allows a wide range of possibilities still compatible with the given end effector location. The general problem of redundancy resolution is then to choose one of the infinity of possible configurations such that some overall cost-function or some performance measure is extremalized. Frequently, however, the precise form of the performance measure is of secondary interest, and serves merely as a tool to impose a smoothness constraint upon the system, i.e. to ensure that the selected configurations for two nearby end effector locations differ from each other as little as possible.

However, this is precisely the kind of task solved "naturally" by topology conserving maps. If such map is used to represent the mapping from task space to joint coordinates, neighboring task coordinates will activate neighboring nodes in the network for output. Due to the natural tendency of the map to smooth out any unnecessary variations in the outputs of neighboring nodes, any redundant degrees of freedom will be used by the map to make the variation between joint configurations for neighboring target points as small as possible. This method differs from previous approaches, like the pseudo-inverse technique, in that the optimization of smoothness comes "for free" by the natural learning dynamics of the network and does not require any auxiliary computations beyond the usual adaptation steps.

2. THE MODEL

In Fig.1 we see the robot system consisting of a robot arm of five degrees of freedom and a pair of cameras providing the spatial information about the location of the object the robot arm shall reach for. For each trial movement, the target location is chosen randomly within the work space. Each target within the three-dimensional work space corresponds to a pair of two-dimensional vectors, namely the locations of the images of the object on the two camera "retinas". As described in more detail in [3,5], it is possible to combine both locations to a four-dimensional vector $\mathbf{u}$ which then carries the visual information necessary for the network to extract the spatial position of the object. These four-dimensional vectors form the input signals for a vector-quantization network of Kohonen-type [6]. This network is able to adaptively discretize the relevant three-dimensional submanifold within the four-dimensional input space. The highly nonlinear transformation $\vec{\theta}(\mathbf{u})$ from camera input to joint angles $\vec{\theta}$ is then linearized in the vicinity of each discretization point. Below we describe how the local linear transformations are learnt from trial movements simultaneously with the adaptive discretization.

To each discretization point $\mathbf{s}$ corresponds a formal neuron of the Kohonen-network; the neuron associates a location $\mathbf{w_s}$ within the four-dimensional input space with a five-dimensional vector $\vec{\theta}_\mathbf{s}$ of joint angles and a 5×4 matrix $\mathbf{A_s}$. Angles $\vec{\theta}_\mathbf{s}$ and matrix $\mathbf{A_s}$ are used to position the end effector at the given target, if the corresponding input was closer to the discretization point $\mathbf{w_s}$ than to any other $\mathbf{w_r}$, $\mathbf{r} \neq \mathbf{s}$. The discretisation occurs in a topologically ordered manner, which means that neighboring neurons of the Kohonen-network are associated with neighboring subsets of the input space.

The positioning of the end effector consists of two phases, namely a gross-positioning, and a subsequent, usually small correction. For the gross-positioning the robot arm system uses the five-dimensional vector $\vec{\theta}_\mathbf{s}$, each element of which determines one joint angle. After this gross movement the position of the end effector in both camera "retinas" will be denoted by a four-dimensional vector $\mathbf{v}_i$, which is close to the four-dimensional retinal target location $\mathbf{u}$. Departing from the ansatz in [3,5], we now use the linear correction term

$$\Delta\vec{\theta} = \mathbf{A_s}(\mathbf{u} - \mathbf{v}_i) \tag{1}$$

for the final corrective joint movement. Here $\mathbf{A_s}$ is the Jacobian of the transformation $\vec{\theta}(\mathbf{u})$ at the discretisation point $\mathbf{s}$ and $\Delta\vec{\theta}$ is usually sufficient to correct the error of the gross movement very precisely. The final position of the end effector is seen by the cameras at $\mathbf{v}_f$.

For the learning step of $\mathbf{A_s}$ we use a linear error correction rule of Widrow-Hoff-type [7] which minimizes the quadratic error

$$E = \frac{1}{2}(\Delta\vec{\theta} - \mathbf{A_s}\Delta\mathbf{v})^2 \tag{2}$$

with $\Delta\mathbf{v} = \mathbf{v}_f - \mathbf{v}_i$ by using steepest descent. Together with (1) and by choosing the optimal step size we obtain as an improved estimate $\mathbf{A}^*$ for $\mathbf{A_s}$

$$\mathbf{A}^* = \mathbf{A_s} + ||\Delta\mathbf{v}||^{-2} \cdot \mathbf{A_s}(\mathbf{u} - \mathbf{v}_f)\Delta\mathbf{v}^T. \tag{3}$$

Because of the topologically correct assignment of the neurons to the inputs neighboring neurons of $\mathbf{s}$ have to learn similar output values $\vec{\theta}_\mathbf{r}$ and $\mathbf{A_r}$. Therefore, $\mathbf{A}^*$ is used to improve the Jacobians $\mathbf{A_r}$ for all neurons $\mathbf{r}$ in a whole neighborhood of $\mathbf{s}$. This neighborhood is defined by a function $h_\mathbf{rs}$ which is unity at $\mathbf{r} = \mathbf{s}$ and decays to zero as $\mathbf{r}$ moves away from $\mathbf{s}$. The resulting adjustments are

$$\mathbf{A_r}^{new} = \mathbf{A_r}^{old} + \epsilon h_\mathbf{rs}(\mathbf{A}^* - \mathbf{A_r}^{old}), \tag{4}$$

and provide an enormous increase in speed and stability of the convergence of the learning algorithm ([5]).

With the improved Jacobian $\mathbf{A_s}$ we are now also able to improve $\vec{\theta}_\mathbf{s}$ because $\vec{\theta}_{correct} - \vec{\theta}_\mathbf{s} = \mathbf{A_s}(\mathbf{w_s} - \mathbf{v}_i)$ so that we obtain as an improved estimate θ^* for $\vec{\theta}_\mathbf{s}$

$$\vec{\theta}^* = \vec{\theta}_\mathbf{s} + \mathbf{A_s}^{new}(\mathbf{w_s} - \mathbf{u}). \tag{5}$$

Like the Jacobians, the joint angles $\vec{\theta}_\mathbf{r}$ of neighboring units share their learning steps through

$$\vec{\theta}_\mathbf{r}^{new} = \vec{\theta}_\mathbf{r}^{old} + \epsilon h_\mathbf{rs}(\vec{\theta}^* - \vec{\theta}_\mathbf{r}^{old}). \tag{6}$$

3. SIMULATION RESULTS

In the following we describe the results of a simulation using the same parameter values for ϵ and $h_\mathbf{rs}$ and the same network topology (i.e. a three-dimensional $7 \times 12 \times 4$ lattice of 336 neurons) as in ([3]). However, in contrast to this work we now employ the improved adaptation rules (1), (3), (5) and the robot arm with five degrees of freedom shown in Fig.1. The initial output values for each neuron were chosen at random, but subject to the constraint that the resulting shape of the robot arm was convex in all cases.

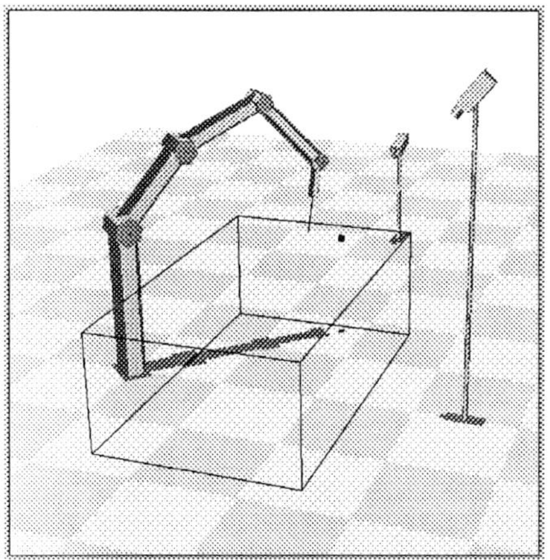

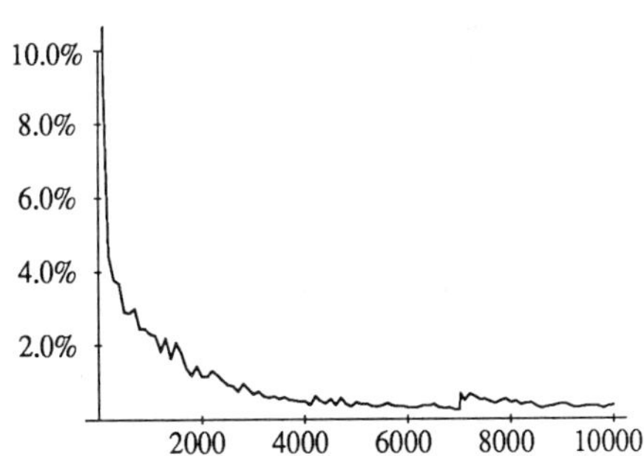

FIG.1: The robot arm and the two cameras.	FIG.2: Positioning error versus number of trials.

Figure 2 shows the development of the remaining positioning error after each trial. The error decays very rapidly to below one percent of its initial value. Although the number of links is now larger (5 vs. 3), the decay of the error is faster than in [3,5], which clearly shows the superiority of the new learning rule over that employed in [3,5]. This stems mainly from basing the corrective fine movement on the actual distance $\mathbf{u} - \mathbf{v}_i$ remaining after the preceeding gross movement, instead of using the distance $\mathbf{u} - \mathbf{w_s}$ between the target and the closest discretization point. This provides the system with an accurate feedback about its residual error, and enables the fine movement to compensate even for sudden and unexpected changes in the geometry of the arm. To demonstrate this, we suddenly increased after 7000 trials the length of the last arm segment by about 5% of its linear dimension, a situation which could result from picking up e.g. a wrench. As we see from Fig.2, even this perturbation causes only a small increase in error, which rapidly returns to its baseline value.

Due to the redundancy of the arm, each target location is compatible with a continuous subset of different arm postures. Initially, as a result of the random initialization of the network, mutually close target locations may give rise to very different arm postures selected for reaching. As learning proceeds, those posture differences not contributing to end effector location get more and more wiped out and postures which only vary smoothly over the target space emerge. An example is provided by Fig.3, which shows successive postures traversed if the end effector follows a path along a diagonal of the workspace.

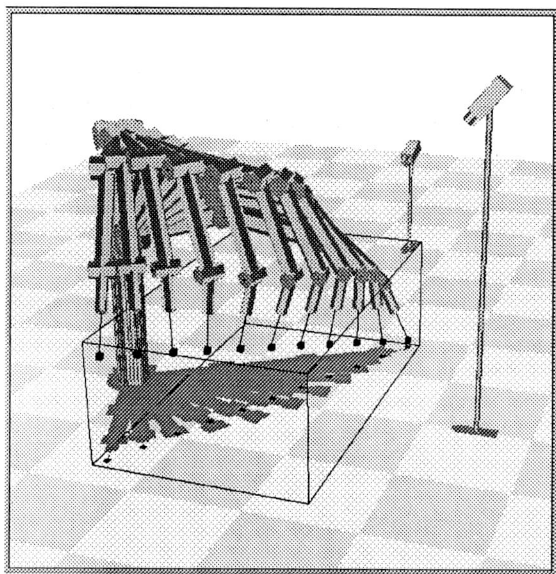

FIG.3: Stroboscopic rendering of successive robot arm postures on traversing a trajectory in the workspace after learning. Due to the redundancy of the arm, each trajectory point can be reached by an infinitude of different postures. Although the system has never obtained any explicit information on how to resolve this redundancy, the topology conserving property of the map used for the learning algorithm has forced a solution, leading to a particularly smooth movement.

ACKNOWLEDGEMENT

This work has been supported by the German Ministry of Science and Technology (ITR-8800-G9) and by the Volkswagen Foundation.

REFERENCES

[1] Eckmiller R., Beckmann J., Werntges H., Lades M. (1989) Neural Kinematics Net for a Redundant Robot ARM. IJCNN-89, Vol.II:333-340.

[2] Kuperstein M., Rubinstein J. (1989) Implementation of an Adaptive Neural Controller for Sensory-Motor Coordination. IEEE Control Systems Magazine, Vol.9, No.3, pp. 25-30.

[3] Martinetz T., Ritter H., Schulten K. (1989) 3D-Neural Net for Learning Visuomotor-Coordination of a Robot arm. IJCNN-89, Conference Proceedings, Washington 1989, Vol.II:351-356.

[4] Miller W. T. (1989) Real-Time Application of Neural Networks for Sensor-Based Control of Robots with Vision. IEEE Transactions on Systems, Man, and Cybernetics, Vol.19, No.4, pp 825-831.

[5] Ritter H., Martinetz T., Schulten K. (1988) Topology-Conserving Maps for Learning Visuomotor-Coordination. Neural Networks 2, pp. 159-168.

[6] Kohonen T. (1982) Self-organized Formation of Topologically Correct Feature Maps. Biological Cybernetics 43:59-69.

[7] Widrow B., Hoff M.E. (1960) Adaptive switching circuits, WESCON Convention Record, part IV pp. 96-104.

Parallel Processing in Neural Systems and Computers
R. Eckmiller, G. Hartmann and G. Hauske (Editors)
© Elsevier Science Publishers B.V. (North-Holland), 1990

DELTA RULE-BASED NEURAL NETWORKS MAY BE APPLICABLE TO A CONTROL TASK THAT DOES NOT PROVIDE A "TEACHER"

Heinz W. WERNTGES

Department of Biophysics
Heinrich-Heine-Universität
D-4000 Düsseldorf 1, West Germany

A calculation method is discussed that adapts delta rule-based neural networks such as backpropagation or CMAC networks to a control task which only feeds back a scalar error signal ("critic") instead of the required "teacher". The heuristics of the approach are (a) reinforcement, (b) random search, and (c) forgetting. Preliminary simulation results of an inverse kinematics problem suggest the usefulness of the approach. First steps towards a theoretical analysis are made.

1. INTRODUCTION

The Widrow-Hoff rule or delta rule, summarized by Rumelhart et al. [5] as $\Delta w_{ji} = \eta \cdot \delta_j \cdot i_i$, has become one of the central ingredients of neural networks that are capable of learning. Many successful classical [8,1] as well as more recent [3-7] neural network applications are based on the delta rule. Recently, Cybenko [2] proved that a backpropagation-type neural network with a single hidden layer can approximate any continuous function f: $\mathcal{R}^n \longrightarrow \mathcal{R}$ to arbitrary precision, so that there seems to exist a neural net solution for most practical applications. However, not all applications supply a "teacher" signal that delta rule-type learning seems to require.

This study points out that (a) standard delta rule theory (including backpropagation) leaves room for the replacement of the usual "teacher" by a more general feedback signal, and (b) that a simple heuristical way of estimating a teacher signal may serve as an adapter between e.g. a backpropagation network and a control task to be learned that only feeds back scalar error signals.

2. THEORY

2.1. Generalization of error minimization

Typical delta rule-based neural networks are trained with a data set $T := \{(\underline{i}_1, \underline{t}_1), \ldots, (\underline{i}_N, \underline{t}_N)\}$, $\underline{i}_k \in \mathcal{R}^m$, $\underline{t}_k \in \mathcal{R}^n$, $1 \leq k \leq N$. After learning, the network outputs $\underline{o}$ should equal (or approximate) the teacher signals: $\underline{o}(\underline{i}_k) \overset{!}{=} \underline{t}_k \quad \forall k \in \{1, \ldots, N\}$. The fed-back error vectors $\underline{e}_k$ are perceived as partial derivatives of a suitably defined error function:

$$\underline{e}_k := (\underline{t}_k - \underline{o}_k) = -\frac{\partial E_k}{\partial \underline{o}}\Big|_{\underline{o}=\underline{o}_k}, \tag{1}$$

$$E_k := \frac{1}{2} \cdot [\underline{t}_k - \underline{o}(\underline{i}_k)]^2 . \tag{2}$$

Delta rule learning is interpreted as gradient descent on E_k.

Please note that the particular quadratic structure is nowhere needed as a prerequisite for gradient descent. Originally it guaranteed convergence to a unique global minimum, but backpropagation always had to cope without that assurance [5].

It therefore appears to be possible to replace E_k by any conveniently defined differentiable error function $\tilde{E}_k$, and to generalize the teachers t_k according to

$$t_k \longrightarrow \tilde{t}_k := o_k - \left.\frac{\partial \tilde{E}_k}{\partial o}\right|_{o=o_k}. \tag{3}$$

Depending on the definition of $\tilde{E}_k$, there might be an additional risk of reaching a local minimum, but the usual delta rule theory (including standard backpropagation [5]) is still applicable.

2.2. Learning inverse kinematics by error minimization via a backpropagation network and a special "adapter"

The following discussion introduces a "critic"-to-"teacher" adapter for learning the differential inverse kinematics of a non-redundant manipulator according to the information flow indicated in figure 1. Although the formalism given here specifies only the 2-degrees-of-freedom case, the generalization for n degrees of freedom is straightforward.

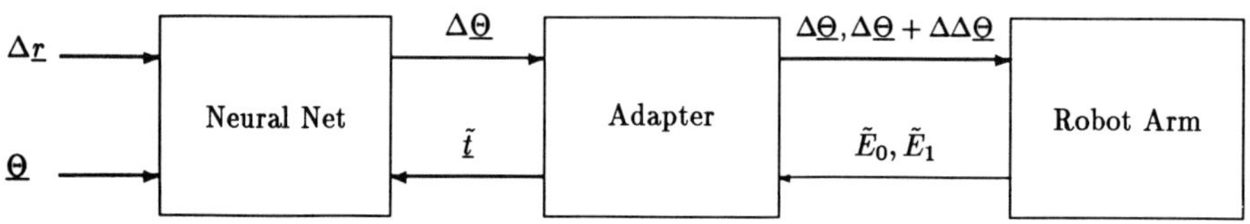

Figure 1: Adaptation of a neural net that needs a teacher to the control task that feeds back only error signals

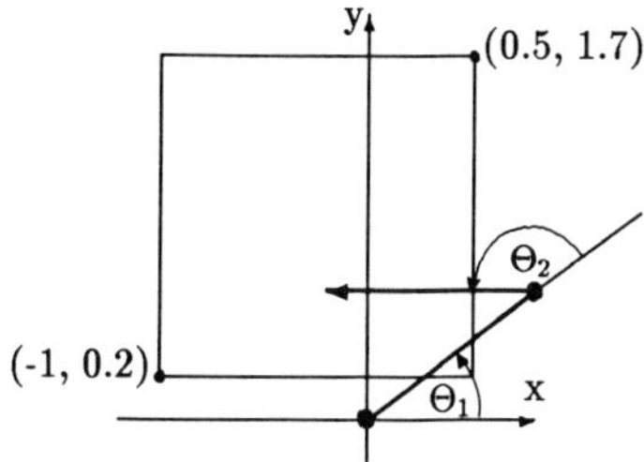

Figure 2: The 2-joint manipulator and the workspace used during training

The differential inverse kinematics function maps a command "move the fingertip by some small distance Δr in user space", and the current state of the arm, here the vector of joint angles Θ, onto the vector $\Delta\Theta$ of joint angle changes that actually move the fingertip by Δr. For a planar 2-joint manipulator (figure 2) with forward kinematics $F : \mathcal{R}^2 \to \mathcal{R}^2$, $(\Theta \mapsto r(\Theta))$, the differential inverse kinematics function

$$D : \mathcal{R}^4 \longrightarrow \mathcal{R}^2, \quad (\Delta r, \Theta) \longmapsto \Delta\Theta \tag{4}$$

satisfying

$$r(\Theta + \Delta\Theta) \overset{!}{=} r(\Theta) + \Delta r \tag{5}$$

with analytical solution

$$\Delta\Theta = \left(\frac{\partial r}{\partial \Theta}\right)^{-1} \cdot \Delta r \tag{6}$$

is to be learned by a backpropagation network of Cybenko type [2] with 4 inputs, 1 hidden layer of h sigmoidal units, and 2 linear output units. Here, input vector $\underline{i}$ corresponds to $(\Delta\underline{r}, \underline{\Theta})$, and output vector $\underline{o}$ to $\Delta\underline{\Theta}$. An error function is easily defined in user space by e.g. the square of distance between desired and actual position of the fingertip:

$$\tilde{E}_k := \frac{1}{2} \cdot [\underline{r}(\underline{\Theta}_k) + \Delta\underline{r}_k - \underline{r}(\underline{\Theta}_k + \Delta\underline{\Theta}_k)]^2 =: \tilde{E}(\Delta\underline{r}_k, \underline{\Theta}_k, \Delta\underline{\Theta}_k) \tag{7}$$

Let's drop the subscript k and the dependence of $\tilde{E}$ on $\Delta\underline{r}_k$ and $\underline{\Theta}_k$ for simplicity of notation.

There are two ways to proceed. A straightforward test of the idea of 2.1 is to assume that $\partial\tilde{E}_k/\partial\underline{o}$ is available, and to use (3) as a teacher. More realistic for a practical application, and more challenging is to assume that only scalar errors $\tilde{E}$ are available. In that case, consider the following heuristical approach for a teacher: Let

$$\underline{\tilde{t}} \;\; := \;\; \Delta\underline{\Theta} + \Delta\Delta\underline{\Theta}, \tag{8}$$

$$\Delta\Delta\underline{\Theta} \;\; := \;\; \begin{cases} \rho \cdot \Delta\hat{\underline{\Theta}} + \nu \cdot \underline{N} & if \quad \tilde{E}(\Delta\underline{\Theta} + \Delta\Delta\underline{\Theta}) < \tilde{E}(\Delta\underline{\Theta}), \\ (1-\epsilon) \cdot \Delta\underline{\Theta} & else \end{cases} \tag{9}$$

with $\underline{N} \in [-1,1]^n$ randomly (e.g., evenly) distributed, $\Delta\hat{\underline{\Theta}} := \Delta\underline{\Theta}/\|\Delta\underline{\Theta}\|$.

Interpretation (figure 1): The robot arm receives the command $\Delta\underline{\Theta}$. After the corresponding move, $\tilde{E}_0 := \tilde{E}(\Delta\underline{\Theta})$ is fed back. Due to inertia, and the influence of noise, it moves on a little further, so $\tilde{E}_1 := \tilde{E}(\Delta\underline{\Theta} + \Delta\Delta\underline{\Theta})$ becomes available. If $\tilde{E}_1 < \tilde{E}_0$, $\Delta\underline{\Theta} + \Delta\Delta\underline{\Theta}$ is accepted as teacher signal. The term $\Delta\Delta\underline{\Theta}$ is made up of a "reinforcement" component with parameter ρ, and a "random search" or noise term with parameter ν. If $\tilde{E}_1 \geq \tilde{E}_0$, i.e. the error does not get smaller, we better try to "forget" the current $\Delta\underline{\Theta}$, because it seems to point to a wrong direction. The factor $(1-\epsilon)$ exponentially reduces $\Delta\underline{\Theta}$ towards 0. Forgetting works approximately in the opposite direction than adding $\Delta\Delta\underline{\Theta}$ does, and it quickly eliminates a wrong direction. Parameters $\rho, \nu,$ and ϵ must be small enough to allow for the implicite linearization of $\partial\tilde{E}/\partial(\Delta\underline{\Theta})$ to be valid. Because $\tilde{E}_1 < \tilde{E}_0$, a descent on the error surface is still in effect, but for lack of more information, the fast gradient descent is no longer possible. Therefore, convergence of the learning network may still be expected, but at a slower rate.

Full or partial reconstruction of the whole gradient $\partial\tilde{E}/\partial(\Delta\underline{\Theta})$ on the basis of more than two error signals is conceivable, depending on the accessibility of values $\tilde{E}_i$. Because the above inverse kinematics problem does not seem to allow for more samples of $\tilde{E}$, if a real robot arm is considered, these variations were not studied.

3. SIMULATION RESULTS

A 2-joint robot arm with unit-length limbs (figure 2) has been trained inside a square workspace window with lower-left corner at $(-1, 0.2)$, and upper-right corner at $(0.5, 1.7)$. For each learning step, the arm was positioned at a randomly chosen point of an equally-spaced 16x16-grid that filled the workspace. The task was to move the fingertip by some little distance $\Delta\underline{r}$. Here, $\|\Delta\underline{r}\| = 0.025$ was fixed, and the direction of $\Delta\underline{r}$ was chosen randomly from 256 equally-spaced directions around the full circle. The backpropagation network conformed to Cybenko [2], with h=34 hidden units. The $\Delta\underline{r}$-inputs had been rescaled by a factor of 40 to roughly match the range of the $\underline{\Theta}$-inputs.

Training of the net with the teacher (6) using the standard momentum method [5] verified that the net was capable of solving the task (4,5) approximately. Learning took a long time, typically $\sim 10^6$ learning steps, so the problem considered is not at all trivial for such a network.

The generalized teacher signal, obtained from combining (3) and (7), appeared to be suitable for training: The network converged at about the same rate as if trained with the original teacher (6).

Training on all grid points $\underline{\Theta}$ by using the adapter (8) & (9) did not converge towards the correct solution, but towards a function that can be approximated by setting $\Delta\underline{\Theta} \propto -\Delta\underline{r}$ for all $\underline{\Theta}$. This occurred for many learning and adapter parameter combinations.

Local training, i.e. keeping Θ constant while varying $\Delta\underline{r}$, did converge towards the correct solution for all Θ that had been tested, especially the Θ's corresponding to the workspace corners. Parameters used were: $\rho = 0.001, \nu = 0.0001, \epsilon = 0.1$. Typically, $\sim 10^5 \ldots 10^6$ learning steps had been performed.

4. DISCUSSION

Preliminary simulation data verified for at least one problem class the presented theory: The error-minimizing properties of delta rule learning can be utilized for a generalization of teaching signals. A control task that feeds back only a scalar error signal has been learned in the absence of a true teacher, though at the expense of additional learning cycles. The observed convergence towards a false function during global training may be due to a "local minimum"-effect, though that remains to be verified. It may also occur, because learning takes place after any single task, not after the presentation of all possible training tasks, as rigorous theory requires.

Extensions of this method towards learning the inverse kinematics of redundant robot arms is possible by adding cost terms to $\tilde{E}$ which e.g. emphasize minimal changes in joint angles, thus eliminating the redundancy. Similarly, obstacle avoidance may be implemented. Studies addressing these extensions are currently prepared.

The observation that a single learning step for some task $(\Delta\underline{r}, \Theta)$ leads to changes in the network function which do not peak around Θ seems to suggest that in order to cure the global convergence problem, the backpropagation topology should be replaced. Topologies like Albus's CMAC [1], Ritter & Schulten's augmented Kohonen map [4], and Daunicht's DEFAnet [3] learn mappings on a more local basis (ref. to the input space). The use of the proposed adapter in combination with these nets is currently studied.

ACKNOWLEDGEMENT

This study was supported by the Ministry of Research and Development (BMFT) of the Federal Republic of Germany (grant No. ITR 8800F6)

REFERENCES

[1] Albus, J.S., A New Approach to Manipulator Control: The Cerebellar Model Articulation Controller (CMAC), J. Dyn. Sys., Meas., Contr. 97 (1975) pp. 220-227.

[2] Cybenko, G., Approximation by Superpositions of a Sigmoidal Function, Math. Control Signals Systems 2 (1989) pp. 303-314.

[3] Daunicht, W.J., DEFAnet - A Deterministic Approach to Function Approximation. IJCNN 1990, in print.

[4] Ritter, H.J., Martinetz, T.M., Schulten, K.J., Topology-Conserving Maps for Learning Visuo-Motor-Coordination, Neural Networks 2 (1989) 159.

[5] Rumelhart, D.E., Hinton, G.E., Williams, R.J., Learning Internal Representations by Error Propagation, in: Rumelhart, D.E., McClelland, J.L., (eds.), Parallel Distributed Processing, Volume 1: Foundations (MIT Press, Cambridge, 1986) pp. 318-362.

[6] Sejnowski, T.J., Rosenberg, C.R., NETtalk: A Parallel Network that Learns to Read Aloud, Technical Report JHU/EECS-86/01, Johns Hopkins University, Baltimore, 1986.

[7] Tesauro, G., Sejnowski, T.J., A Parallel Network that Learns to Play Backgammon, Artificial Intelligence 39 (1989) pp. 357-390.

[8] Widrow, G., Hoff, M.E., Adaptive Switching Circuits, WESCON convention record, Part 4 (1960) pp. 96-104.

Section 11
Selected Applications
for Neural Networks

Parallel Processing in Neural Systems and Computers
R. Eckmiller, G. Hartmann and G. Hauske (Editors)
© Elsevier Science Publishers B.V. (North-Holland), 1990

Parallel Algorithms for Channel Assignment in Cellular Mobile Radio Systems: The Neural Network Approach

M.Duque Antón, D.Kunz

Philips Forschungslaboratorium Hamburg
Postfach 54 08 40, D-2000 Hamburg 54, West Germany

The channel assignment problem (CAP), i.e. the task to assign the channels to the radio base stations such that a minimum number of channels is used is an NP-complete optimization problem occurring during design of cellular radio systems. So far, this problem has been treated by sequential graph colouring algorithms. This paper investigates how far a parallel neural network algorithm, already succesfully applied to other discrete optimization problems [1], may be used for a real-world CAP. It turns out that this algorithm is well suited to this problem. Its flexibility concerning the modelling of technical constraints and the introduction of "soft" optimization criteria makes it very useful, especially under the practical conditions of an actual cellular radio planning task.

1   Introduction

A cellular radio network consists of a number of radio base stations. Depending on his actual position, each mobile subscriber is assigned to one of these base stations. Calls to and from the mobile stations are serviced by a radio link to the actually corresponding base station, using one of the available radio channels. The network operator has to take care that the following two constraints must be satisfied:

- The expected traffic must be serviced. This implies that at base station j there are $traf(j)$ radio channels available to service calls. The value $traf(j)$ is chosen such that the probability is sufficiently high that for an incoming call request an idle channel can be found.

- The service quality must be sufficiently high with respect to interference. This implies that the base stations j, j' may only use the channels i and i', respectively, if the channels (their carrier frequencies) are far enough away from each other, i.e. $| i - i' | \geq c_{jj'}$, in order to insure that the received radio signal is not deteriorated by signals from different calls.

To satisfy the first constraint the operator has to equip each base station with $traf(j)$ transmit and receive units. The channel assignment problem (CAP) now consists of the task to tune these units to radio channels in agreement with the second (interference) constraint using as few as possible, or at least not more than a predefined number of channels. The simplest version of CAP (only co-channel interference is considered, i.e. all $c_{jj'}$ are either 0 or 1) reduces to a graph colouring problem, which is known to be NP-complete. So far this problem has been treated by sequential graph colouring algorithms [2]. These algorithms try to find a sequence in which channels are to be assigned to base stations assigning first channels in regions in which the problem is assumed to be most difficult. By iterative reordering of the assignment sequence one tries to find a sequence with a minimum number of channels satisfying all constraints. This method does only allow to specify hard constraints indicating whether a solution is admissible or not.

On the other hand, Hopfield and Tank [1] showed that artificial neural networks can be applied successfully to other NP-complete optimization problems like that of the travelling salesman. The present study investigates how far these algorithms can be practically applied to the channel assignment problem, too. We will translate the CAP into a neural network setup, show the resulting solutions, and discuss advantages and drawbacks of this novel parallel approach to CAP under the conditions of real-world cellular radio planning tasks.

2   Method

The neural network solution of the CAP essentially follows the ideas of Hopfield and Tank for the travelling salesman problem [1]. First, a suitable network of artificial neurons was mathematically constructed. The neurons were assumed to obey the following differential equation:

$$du_i/dt = \sum_j T_{ij} f(u_j) + I_i - u_i$$

where u_i reveals the state of each neuron, f determines the output of each neuron, T_{ij} represents the coupling weights, and I_i is an additional offset to the network. For details see [1]. The constraints and optimization criteria were translated from terms of mobile radio communication into an energy function which is to be globally minimized. From that function, the external input and the mutual coupling weights of the neurons were derived. The resulting set of differential equations was simulated on a transputer network using a Runge-Kutta algorithm, starting from an arbitrary initial state. The steepness (λ) of the nonlinear function f with

$$V_i = f(u_i), \ f(x) := 1/2(1 + tanh(\lambda x))$$

was stepwise increased until the system reached a stable state in which each neuron showed either maximum or minimum response. This constant λ controls the "temperature" of the system. Thus increasing this constant is equivalent to a stepwise "freezing" of the system, an idea closely related to "simulated annealing", however using a completely different algorithm. The resulting discrete answers were suitably interpreted in terms of the mobile radio system.

3   Channel Assignment Problem with Minimized Spectrum Demand

We first address the question how the response of a neural net can be interpreted in terms of a channel assignment. Assume to have a cellular radio network of n base stations. An upper bound m for the number of required channels could be determined without difficulty using a local criterion. In case of only co-channel interference (interference between the same channel at different base stations) being present, this is simply the maximal degree of the expanded interference graph. Now, assume that each base station is capable of carrying any of these m channels. Take one neuron (i, j) for each channel i in each base station j and additionally one neuron l for each channel l. A high output, i.e. an output close to its maximum value, of a neuron (i, j) is interpreted as the permission to use channel i in base station j. A high output of a neuron l reveals the need to use channel l, i.e. there exists at least one neuron (i, j), $i = l$, which is high.

Now, we turn to the problem to define CAP in terms of an energy function. A possible energy function looks like:

$$E = \frac{1}{2}A \sum_{j=1}^{n} \sum_{j'=1}^{n} \sum_{i=1}^{m} \sum_{|c|<c_{j,j'}} (1 - \delta_{jj'}\delta_{0c}) V_{i,j}V_{i+c,j'} + \frac{1}{2}B \sum_{j=1}^{n} (\sum_{i=1}^{m} V_{i,j} - traf(j))^2$$

$$+C \sum_{l=1}^{m} V_l + D \sum_{j=1}^{n} \sum_{i=1}^{m} V_{i,j}(1 - V_i) \tag{1}$$

The first term favours states, which are compatible with the interference constraints. In the case that radio links to base stations j and j' can interfere with each other (i.e. $c_{j,j'} \geq 1$), $c_{j,j'}$ represents the corresponding minimum allowable channel separation between base stations j and j'. The second term handles the channel demand for each base station. The third term favours states, which use a minimum number of (assigned) channels. The fourth term couples the (l, j) neurons with the channel neuron l.

From the energy function (1), the connection matrix and the external input can be read off:

$$
\begin{aligned}
T_{ij,i'j'} &= -A \sum_{|c| < c_{j,j'}} \left(1 - \delta_{jj'} \delta_{0c}\right) \delta_{i,i'+c} - B \, \delta_{jj'} \\
T_{ij,l} &= D \, \delta_{il} \\
T_{l,l} &= 0 \\
I_{ij} &= B \, traf(j) - D \\
I_l &= -C
\end{aligned}
$$

4 Channel Assignment Problem with Soft Constraints

Under practical conditions, it does not always make sense to minimize the number of required channels ($C = D = 0$, channel neurons l omitted). Usually, the operator is permitted to use a fixed frequency band. If there are several solutions satisfying both channel demand and interference constraints and there is an existing network to which an extension in planned, he may for example prefer a solution that preserves as many of the old channel assignments as possible thus minimizing the retuning effort. If there is no such solution he may look for a solution violating those interference conditions that are least severe. Moreover, the operator has to take into account a variety of further technical constraints as channel restrictions in boundary zones to other operator areas or foreign countries, minimal spacing between channels combined to the same antenna, or other hardware restrictions.

Formulating the optimization problem in terms of an energy function turned out to be very flexible making it easy to include all these technical constraints. Since there is no distinction between constraints and optimization objectives it depends on the choice of parameters whether a condition is a "hard" constraint or a "soft" optimization criterion.

For example, in case of a nework extension, the preference to keep as many as possible from the old channel assignments was easily implemented by adding another term $\left(F \sum_{j=1}^{n} \sum_{i \in PA_j} (1 - V_{ij})\right)$ to the energy function E, where PA_j is the set of preassigned channels for base station j. The importance of this criterion with respect to others is controlled by the relative strength of the energy terms.

5 Results

The network simulation was implemented in C both on a VAX, and on an INMOS T800 transputer network of 9 nodes using the Logical Systems C compiler.

The algorithm was succesfully applied to several artificial examples as well as real-world examples. Since our goal was to test the practical usefulness of an analogous Hopfield network for CAP, we present results made on real-world data. From the area of 24×21 km around Helsinki, topography and morphostructure data (e.g. urban area, forest, open land, ...) were taken. The morphostructure was used to define a traffic density. 25 base station locations were distributed inhomogenously over the area. Using these data the rf field strength distribution was calculated. The result was used to calculate, how much traffic was to be assigned to which base station. A signal-to-interference-ratio of 8 db was permitted to be violated with a probability of 10%. From this specification, it was calculated which two base stations must not use the same channels. Thus the interference matrix (c_{ij}) and the function $traf(j)$ were given. These calculations were carried out using the programm package GRAND [4]. Furthermore, an upper bound for the number of channels, which are required at most was evaluated, computing the maximal degree of the expanded interference graph, which yielded a value of 121. The resulting network for the area of Helsinki consists of 25×121 neurons together with a row of 121 neurons.

To avoid inner local mimima of the energy function, i.e. to ensure that its second derivative is not nonnegative definite (cf. [3]), a small self-excitation of each neuron is implemented, leading to an extension of the energy function by: $-G \sum_{j=1}^{n} \sum_{i=1}^{m} V_{ij}^2$.

A neuron was regarded to be high if the output was greater than 0.9. The corresponding differential equation was evaluated in time steps of 0.1. To avoid the system to be stuck in an instable equilibrium due to the discrete computation, after each time step a small pseudo-random noise value was added to the internal state of each neuron. Each time the number of assigned channels showed no significant change for a fixed simulated time period, the constant λ controling the steepness of the nonlinear function f was increased until all neurons answered either with 0 or 1.

For a parameter choice of

$$\lambda = 0.2, ..., 2.17 \; ; \; A = 5.0 \; ; \; B = 10.0 \; ; \; C = 10.25 \; ; \; D = 10.0 \; ; \; G = 5.0$$

the system converged to an nearly optimum solution within a simulated time of 10.7. The number of channels which was needed by this solution was 78. Analysis of the local clique numbers of the interference graph showed that at least 73 channels are required to solve CAP.

Starting the calculation with only 73 channels and omitting the minimization of number of channels lead to a CAP solution satisfying both the channel demand and all interference constraints. So, with a resticted number of channels an optimum solution was found using the same algorithm.

6 Conclusion

The above neural network algorithm is well suited for the channel assignment problem arising in cellular mobile radio systems. Its parallelity allowed to find a channel assignment without heuristic rules generating the sequence of channel assignments. This is in contrast to the situation for the sequential graph colouring algorithms used so far for that purpose. Formulating the optimization problems in terms of an energy function turned out to be very flexible making it easy to include a variety of additional technical constraints and optimization criteria. Since a distinction between "hard" constraints and "soft" optimization criteria is only expressed by the relative strength of the terms in the energy function, the user is free to choose this relative strength according to his actual planning situation. This point is of great importance in practical situations.

In conclusion, this neural network algorithm is a useful tool for planning cellular mobile radio networks, a field that has rarely been brought into connection to neural networks so far.

References

[1] Hopfield,J.J. and Tank,D.W. *'Neural' computation of decisions in optimization problems*. Biol. Cybern. **52**, 141–152, (1985).

[2] Gamst,A. *A resource allocation technique for FDMA systems*. Alta Frequenza, **LVII**, 89–96, (1988).

[3] Hopfield,J.J. *Neurons with graded response have collective computational properties like those of two-state neurons*. Proc. Natl. Acad. Sci. USA, **81**, 3088–3092, (1984).

[4] Gamst,A., Beck,R., Simon,R. and Zinn,E.G. *An integrated approach to cellular radio network planning*. Proc. 35th Veh. Techn. Conf., pp 21-25, Boulder, CO 21-23 May 1985.

Parallel Processing in Neural Systems and Computers
R. Eckmiller, G. Hartmann and G. Hauske (Editors)
© Elsevier Science Publishers B.V. (North-Holland), 1990

Storing cycles in analog neural networks

T. Gencic, M. Lappe, G. Dangelmayr and W. Güttinger

Institute for Information Sciences, University of Tübingen

Köstlinstr. 6, 7400 Tübingen, FR Germany

1 Introduction

In recent years formal neural networks have attracted increasing interest as models which can perform usefull cognitive tasks and parallel computation. In particular the networks discussed by Hopfield [1] [2] possesses the ability of high level cognitive functions such as content addressable memory and pattern recognition. The key role is played here by the fixed points of the dynamics. They correspond to the stored memories one is retrieving from the initial states which are identified with partial information used to evoke the full memory. More recent work has been devoted to the storage of connections between patterns, i. e., instead of relaxing to a fixed memory the network evolves into a whole sequence of states. We are specifically interested in storing and retrieving cycles in which sequences of states are traversed cyclically in prescibed orders. Various learning rules which produce this behaviour have been developed for networks composed by two–state neurons [3] [4] [5] [6] [7] [8] [9] [10]. These rules commonly rely on time–delayed [3] [4] [5] [6] or time–dependent [9] [10] synapses. In contrast to this work we use the biologically more reasonable continuous Hopfield model [2] which we have implemented by means of simple electronic devices. The cycles are stored by setting up equations for the synaptic weights which produce both fixed points and transitions between the states to be traversed. The resulting equations can either be solved by means of the pseudo–inverse method or by minimizing an appropriate cost or energy function constructed from the cycle's states. The optimization itself can be performed by another network, the so–called master net.

In section 2 we introduce the model and the learning rule and present the results of a computer simulation. In section 3 we describe experiments performed with an electronic network in which the weights can be adjusted continuously.

2 A learning rule for storing cycles in analog neural networks

Our model neuron is described by a continuous sigmoid transfer characteristic $g : \mathbf{R} \to (-1, 1)$ which relates the input voltage u to the output voltage v according to $v = g(u)$. We assume that g is symmetric, i. e., $g(u) = -g(-u)$. This condition is, for example, satisfied for the standard choice $g(u) = \tanh(\lambda u)$ where the parameter λ controls the gain. If N neurons are coupled through synapses J_{ij}, implemented by resistors, the dynamics of the resulting network is governed by a system of o.d.e.'s,

$$\frac{du_i}{dt} = -u_i + \sum_{j=1}^{N} J_{ij} g(u_j), \qquad 1 \leq i \leq N, \tag{1}$$

where the relaxation time of a single neuron has been rescaled to unity. By introducing the inverse function of g, $u = g^{-1}(v)$, (1) can be reformulated as a system of o.d.e.'s for the output voltages v_i,

$$\frac{dg^{-1}(v_i)}{dv_i} \frac{dv_i}{dt} = -g^{-1}(v_i) + \sum_j J_{ij} v_j. \tag{2}$$

If p patterns $\xi^{(\nu)} = (\xi_1^{(\nu)}, \ldots, \xi_N^{(\nu)})$ $(1 \leq \nu \leq p)$ are to be stored as fixed points of the dynamics, i. e., $v_i = \xi_i^{(\nu)}$ is a stationary solution of (2), the weights J_{ij} must satisfy the conditions

$$-g^{-1}(\xi_i^{(\nu)}) + \sum_j J_{ij}\xi_j^{(\nu)} = 0. \tag{3a}$$

The symmetry of g implies that, when (3a) is satisfied, the complementary vectors $-\xi^{(\nu)}$ are also stationary states of (2).

$$
\begin{array}{c}
\xi^{(1)} \rightarrow \xi^{(2)} \rightarrow \xi^{(3)} \rightarrow \xi^{(4)} \rightarrow \xi^{(5)} \rightarrow \xi^{(6)} \\
-\xi^{(6)} \leftarrow -\xi^{(5)} \leftarrow -\xi^{(4)} \leftarrow -\xi^{(3)} \leftarrow -\xi^{(2)} \leftarrow -\xi^{(1)}
\end{array}
$$

Figure 1: A cycle of $p = 6$ states

Assume now the network is to be designed to traverse a cycle of states as is shown in Figure 1. Starting from $\xi^{(1)}$ the system is driven through all patterns up to $\xi^{(p)}$ and then it traverses the sequence of complementary vectors in reversed order. In discrete–time (but still continuous–state) networks this amounts to imposing the conditions

$$-g^{-1}(\xi_i^{(\nu+1)}) + \sum_{j=1}^{N} J_{ij}\xi_j^{(\nu)} = 0 \qquad (1 \leq \nu \leq p), \tag{3b}$$

where $\xi^{(p+1)} = -\xi^{(1)}$. Since, however, the network evolves continuously in time, the condition (3b) does not lead to a cycle in which the system switches abruptly from state to state. Instead, it turns out in numerical simulations that a periodic orbit occurs, but this orbit does not come close to the desired states, even for steep transfer functions. We therefore impose both, the condition (3a) for $\xi^{(\nu)}$ being a fixed point and the transition condition (3b). In the following we refer to the combined equations (3a,b) for the synaptic weights as the cycle equations. We have tested these in a number of numerical simulations. In most cases the dynamics is as desired, i. e., the system remains close to the state $\xi^{(\nu)}$ during a finite amount of time and then switches in a short time step to a state which is close to $\xi^{(\nu+1)}$, although the memory states themselves are not reached exactly. From the viewpoint of dynamical systems theory this behaviour can be explained as follows. The condition (3a) means that the states $\xi^{(\nu)}$ are equilibria, whereas the condition (3b) enforces a periodic orbit which passes through a neighborhood of each of the $\xi^{(\nu)}$. The equilibria are strictly hyperbolic, i. e., they possess stable and unstable manifolds. This means that the state evolves first towards $\xi^{(\nu)}$ near the stable manifold and then is repelled in virtue of the unstable manifold. Since the temporal evolution near fixed points is slow the system remains near an equilibrium for a finite amount of time whereas the transitions from state to state occur on a much shorter time scale. Thus we enforce a kind of 'multi–soliton' into the dynamics of (2). A more detailed analysis, based on a multiple time scale expansion, will be presented elsewhere.

The conditions (3a,b) work well, provided the state of an individual neuron is not forced to switch between positive and negative values at each step of the cycle. If several cycles are to be stored, (3a,b) have to be supplemented by corresponding equations for the other states and cycles. Of course, the total number of conditions must be less or equal to N^2, the number of weights. Figure 2 shows the result of a numerical simulation. We have stored one large cycle traversing ten randomly chosen states $\xi^{(\nu)}$ and their complements $-\xi^{(\nu)}$ in a network of twenty neurons. In the figure the overlaps of the actual state with all memory states are plotted as functions of time.

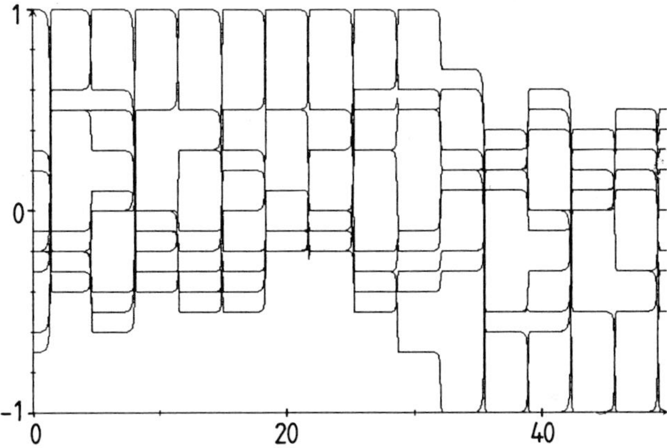

Figure 2: A cycle of ten randomly chosen states (see text).

The cycle equations can be solved either directly, by utilizing the Moore–Penrose pseudo inverse, or by using a generalisation of the master–slave algorithm of Lapedes and Farber [11]. The latter method is based on the energy function

$$E = \sum_{\gamma=0}^{1} \sum_{i=1}^{N} \sum_{\nu=1}^{p} C_{\gamma}^{(\nu)} \left[g^{-1}(\xi_i^{(\nu+\gamma)}) - \sum_j g(U_{ij})\xi_i^{(\nu)} \right]^2, \tag{4}$$

where the $C_{\gamma}^{(\nu)}$ are positive weight factors. Obviously E posseses an absolute minimum if $g(U_{ij}) = J_{ij}$ where the J_{ij} satisfy (3a,b). Following [11] we consider the U_{ij} and $V_{ij} = g(U_{ij})$ as input and output voltages of N^2 neurons. If these are coupled in a network (the 'master net') with synaptic weights J_{ijkl} and input currents I_{ij} given by

$$J_{ijkl} = -2\sum_{\gamma\nu} C_{\gamma}^{(\nu)}\delta_{ik}\xi_j^{(\nu)}\xi_l^{(\nu)}, \qquad I_{ij} = 2\sum_{\gamma\nu} C_{\gamma}^{(\nu)}g^{-1}(\xi_i^{(\nu+\gamma)})\xi_j^{(\nu)}, \tag{5}$$

where δ_{ik} is the Kronecker–symbol, we obtain a network with symmetric couplings whose energy is just given by (4). Thus the dynamics of the master net drives the V_{ij} into a state of minimal energy and therefore leads to the desired synaptic weights, $J_{ij} = V_{ij,\mathrm{min}}$, of our original network (the 'slave net') in which the cycles have to be stored.

3   Experimental results

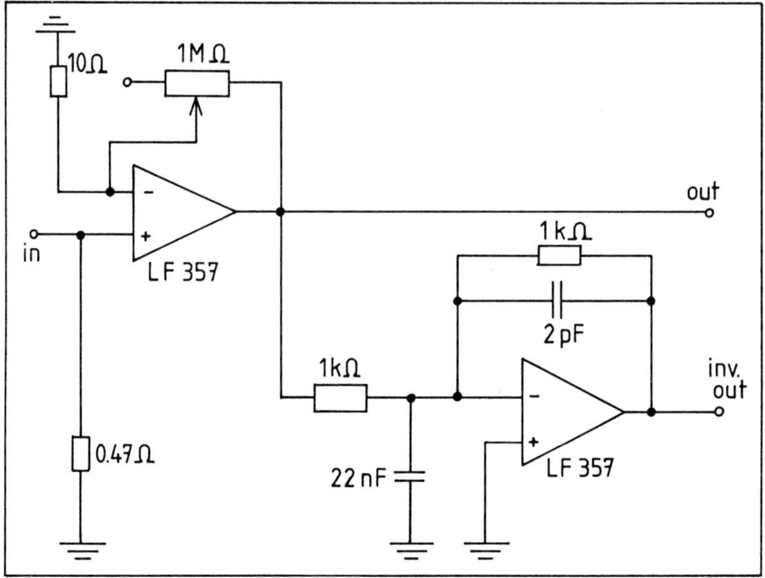

Figure 3: Circuit of the model neuron.

In our experiments a single neuron is modelled by the electronic circuit shown in Figure 3. The operational amplifiers are of type LF357. Its gain is adjustable via the 1MΩ potentiometer; usually we work in a range between 10^3 and 10^4. The 2pF and 22nF condensators are mainly for the stability of the circuit, but the latter condensator also serves as an output capacitance, resulting in a time constant of $22\mu\mathrm{s} \leq \mathrm{RC} \leq$ 1ms. We have built a network of eight of these neurons with coupling resistors in the range between 0 and 50kΩ. In contrast to other electronic implementations [12] the coupling resistors can be adjusted continuously. Inhibitory synapses are modelled by the inverting amplifiers. For a given set of states in a cycle the corresponding weights are determined numerically on the basis of the equations (3a,b) and then the resistors are adjusted according to these values. In this way we were able to store several cycles in the analog network. As an example we present a cycle with two patterns in a network of four neurons. The two memory states and the resulting synaptic weight matrix are given by

$$\begin{aligned}\xi^{(1)} &= (+1,+1,+1,-1)\\ \xi^{(2)} &= (-1,+1,-1,+1)\end{aligned} \qquad \text{and} \qquad \frac{1}{2}\begin{pmatrix} 1 & -3 & 1 & -1\\ 2 & 3 & 2 & -2\\ 1 & -3 & 1 & -1\\ -1 & 3 & -1 & 1\end{pmatrix}$$

In Figure 4 we present the neurons' activities on oscilloscope screens. Figure 4(a) shows the temporal behaviour of the states of neuron 1 (upper scan) and 3 (lower scan). Analogous photographs for neurons 2 and 4 are presented in Figure 4(b). In order to visualize the phase shift between neurons 1 and 2 we show the temporal evolution of their states in Figure 4(c) with a small vertical offset. Finally, in Figure 4(d) these states are plotted versus each other. The quadratic shape indicates that the desired dynamical state is well performed.

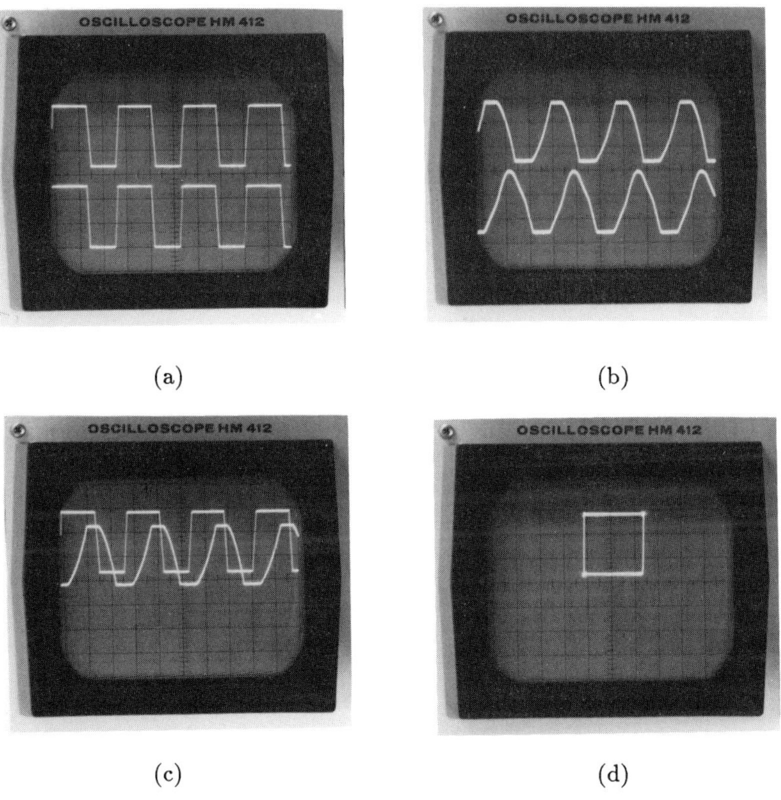

(a)

(b)

(c)

(d)

Figure 4: Performance of the analog network

References

[1] J. J. Hopfield. Neural networks and physical systems with emergent collective computational abilities. *Proc.Nat.Acad.Sci. USA*, 79:2554 – 2558, April 1982.

[2] J. J. Hopfield. Neurons with graded response have collective computational properties like those of two–state neurons. *Proc.Nat.Acad.Sci. USA*, 81:3088 – 3092, Mai 1984.

[3] H. Sompolinsky and I. Kanter. Temporal association in asymmetric neural networks. *Phys.Rev. Letters*, 57(22):2861 – 2864, December 1986.

[4] H. Gutfreund and M. Mezard. Processing of temporal sequences in neural networks. *Phys.Rev. Letters*, 61(2):235 – 238, July 1988.

[5] D. J. Amit. Neural networks counting chimes. *Proc.Nat.Acad.Sci. USA*, 85:2141 – 2145, April 1988.

[6] U. Riedel, R. Kühn, and J. L. van Hemmen. Temporal sequences and chaos in neural nets. *Phys.Rev. A*, 38(2):1105 – 1108, July 1988.

[7] I. Guyon, L. Personnaz, J. P. Nadal, and G. Dreyfus. Storage and retrieval of complex sequences in neural networks. *Phys.Rev. A*, 38(12):6365 – 6372, December 1988.

[8] I. Guyon, L. Personnaz, and G. Dreyfus. Of points and loops. In R. Eckmiller and Ch. v. d. Malsburg, editors, *Neural Computers*, p. 261 – 269, Springer Verlag, 1988.

[9] S. Dehaene, J. P. Changeux, and J. P. Nadal. Neural networks that learn temporal sequences by selection. *Proc.Nat.Acad.Sci. USA*, 84:2727 – 2731, May 1987.

[10] P. Peretto and J. P. Niez. Collective properties of neural networks. In E. Bienenstock et al., editor, *Disordered Systems and Biological Organization*, p. 170 – 185, Springer Verlag, 1987.

[11] A. Lapedes and R. Farber. A self–optimizing, nonsymmetrical neural net for content addressable memory and pattern recognition. *Physica*, 22D:247 – 259, 1986.

[12] C. M. Marcus and R. M. Westervelt. Basins of attraction for electronic neural networks. In D. Z. Anderson, editor, *Neural Information Processing Systems, Denver CO 1987*, p. 524 – 533, AIP, 1988.

Parallel Processing in Neural Systems and Computers
R. Eckmiller, G. Hartmann and G. Hauske (Editors)
© Elsevier Science Publishers B.V. (North-Holland), 1990

RECOGNITION OF PATTERNS AND MOVEMENT PATTERNS BY A SYNERGETIC COMPUTER

H. Haken

Institut für Theoretische Physik und Synergetik
Pfaffenwaldring 57/IV
D 7000 Stuttgart 80 (Vaihingen)

ABSTRACT

Synergetic computers utilize an analogy between pattern formation by systems far from thermal equilibrium and pattern recognition by means of the concept of associative memory. We present an algorithm which can be realized by parallel computers with a new type of "neurones". Our procedure is invariant against translation, rotation, and scaling; it allows the decomposition of scenes and the oscillatory perception of ambiguous patterns. It can also be applied to movement patterns, e.g. recognition of gaits.

1. WHAT IS SYNERGETICS ABOUT?

Physics, chemistry, and biology provide us with numerous examples for the spontaneous (self-organized) formation of patterns. When fluids are heated from below, they may form various patterns, e.g. in form of rolls or hexagons. When specific chemicals such as those of the Belousov-Zhabotinski reaction are poured together, they may spontaneously form stripes or moving concentric rings or rotating spirals. In biology we observe the phenomena of morphogenesis. Synergetics [1], [2] has unearthed a common root of pattern formation by self-organization. When a system is subjected to internal or external control parameters, at specific values of these parameters the macroscopic structure may change spontaneously. At these instability points the dynamics of structure formation is governed by very few variables, the order parameters. The behavior of the subsystems of the system is uniquely governed by these order parameters by means of the slaving principle. In quite a few cases the order parameter equations can be grouped into universality classes indicating the same macroscopic behavior of the systems.

The mathematical approach can be sketched as follows: The system's behavior is described by a state vector q whose components may be space- and time-dependent. The state vector q obeys evolution equations of the form

$$\dot{q} = N(q, \alpha) + F(t) \tag{1.1}$$

where N is a nonlinear function which may contain spatial derivatives,

α is a control parameter, $F(t)$ are fluctuating forces describing internal or external noises. Close to an instability point where the linear stability of the solution q gets lost, q may be expressed by the few order parameters ξ

$$q = f(\xi) \ . \tag{1.2}$$

The order parameters in turn obey equations of the form

$$\dot{\xi} = \tilde{N}(\xi, \alpha) + \tilde{F}(t) \ . \tag{1.3}$$

In a number of cases pattern formation shows the properties of associative memory [3]. For instance when a liquid in a circular vessel is heated from below and one roll in a specific direction is initially created, the system will complement a roll system into that direction. If the initial direction is chosen differently, the total roll system is created with the new direction of the initially generated roll [4]. When the initial state consists of a superposition of two rolls in two different directions but with different strengths, the liquid will eventually form a roll system in the direction of the predominant initial roll.

2. SYNERGETIC COMPUTERS FOR ASSOCIATIVE MEMORY

While synergetics had been concerned with the spontaneous formation of patterns, we now turn the argument around, namely we consider pattern recognition as a specific kind of pattern formation. Since our approach has been described at various occasions [5], here we sketch it only briefly. The prototype patterns are encoded by vectors v_k where k differentiates between the different prototypes or classes of prototypes. We introduce adjoint vectors v_k^+ with the property

$$(v_k^+ \ v_k) = \delta_{kk'} \ . \tag{2.1}$$

We then construct a dynamics which pulls any initially given test pattern vector q into one of the prototype patterns v_k, namely the one to which q has been closest initially. One may show that such a dynamics is provided by the equations

$$\dot{q} = \Sigma \ \lambda_k (v_k^+ \ q) \ v_k - B \sum_{k \neq k'} (v_k^+ \ q)^2 \ (v_{k'}^+ \ q) \ v_k$$

$$- \ C \ |q|^2 \ q + F(t) \tag{2.2}$$

where B and C are positive quantities, $F(t)$ are fluctuating forces which may or may not be included in the approach. The order parameters may be introduced by means of the relation

$$\xi_k = (v_k^+ \ q) \ . \tag{2.3}$$

The brackets in (2.2) and (2.3) are meant as scalar products. One may show that the order parameter equations resulting from (2.2) are given by

$$\dot{\xi}_k = \xi_k(\lambda_k - \sum_{k'} B_{kk'} \, \xi^2_{kk'}) + F_k(t). \tag{2.4}$$

We have made our procedure invariant against simultaneous translation, rotation, and scaling by a combination of Fourier transformations and logarithmic maps [6]. Our procedure was shown to be capable of recognizing for instance faces independently of their positions and orientations. The procedure is robust against noise. It allows the recognition of scenes, for instance composed of several faces which partly hide each other [6]. When ambiguous patterns are offered, the system is capable of showing oscillations as is known from human perception [7]. This is achieved by allowing saturation effects of the attention parameters λ_k by subjecting them to equations of the form

$$\dot{\lambda}_k = \gamma(1-\lambda_k - \xi^2_k) . \tag{2.5}$$

A fast formalism has been developed for learning in which prototype patterns to be learned are presented in a random sequence to the computer [8]. It was shown [5] that the algorithm described by equations (2.2) can be realized on a network of individual neurones either in one layer or with three layers with one hidden layer. But in contrast to conventional neurones, the nonlinearity of the neurones is rather different; they have to multiply and sum up over the inputs from different neurones. The realization by means of a parallel network based on the order parameter equations (2.4) is particularly simple. It was shown that the network can perform logical operations including the XOR operation provided the right-hand side of (2.2) is altered, though only terms up to the fourth order in q need to be included [9]. In the remainder of this contribution I want to sketch more recent results of ours on the recognition of movement patterns.

3. RECOGNITION OF MOVEMENT PATTERNS

It is well-known that movements of man and animals occur in specific patterns, for instance horses may walk, trot or gallop or show still other gaits. Humans have a remarkable capability of distinguishing between different movement patterns. For instance experiments done by Johannson [10] have become quite famous. He fixed light bulbs on the joints of people who moved in the dark. When observers saw the corresponding light spots, they could not identify the object when the light spots were at rest. However, in the very moment the light spots were moving, the observer could immediately say what kind of movement was going on.

In the following we wish to describe a general approach how a machine can recognize movement patterns.

We study the following problem : The movement of a person or animal is characterized by the movement of points at the individual joints. We shall allow that the person or animal moves in a plane which has an angle α between it and the plane of the screen onto which the coordinates are projected. Later on we shall allow that the plane of movement rotates so that α becomes a time-dependent but for the observer unknown function. The coordinates of the screen of the observer are called x and y, those of the moving object in its plane of movement x', y'. We assume that the plane of movement is rotated around the horizontal axis. The relation between x,y and x', y' is given by

$$x = x' \cos \alpha \tag{3.1}$$

$$y = y' . \tag{3.2}$$

Our first task will it be to identify those points which are connected by limbs, the labels of the corresponding points being i and j. We have to search for such combinations i,j for which

$$|\underset{\sim}{x}_i(t) - \underset{\sim}{x}_j(t)|^2 = \text{const.} \tag{3.3}$$

holds where the constant is time-independent. When we write (3.3) in components, we obtain

$$(x_i'(t) - x_j'(t))^2 + (y_i'(t) - y_j'(t'))^2 . \tag{3.4}$$

We introduce the abbreviation

$$\Delta x_{ij} = x_i - x_j \tag{3.5}$$

and use the relation (3.1) so that we obtain

$$\frac{1}{\cos^2 \alpha} (\Delta x_{ij})^2 + (\Delta y_{ij})^2 . \tag{3.6}$$

We first assume that α is time-independent so that we may multiply (3.6) by $\cos^2 \alpha$ to obtain

$$(\Delta x_{ij})^2 + \cos^2 \alpha \, \Delta y_{ij}^{\,2} . \tag{3.7}$$

The condition that (3.3) or (3.7) are time-independent, can be cast into the form

$$\Delta x_{ij} \cdot \Delta \dot{x}_{ij} + \cos^2 \alpha \, \Delta y_{ij} \, \Delta \dot{y}_{ij} = 0 \tag{3.8}$$

where the dot above x_{ij} and y_{ij} indicates the time derivative. Solving (3.8) for $\cos^2 \alpha$, we obtain

$$\frac{\langle \Delta x_{ij} \cdot \Delta \dot{x}_{ij} \rangle}{\langle \Delta y_{ij} \, \Delta \dot{y}_{ij} \rangle} = \cos^2 \alpha_{ij} \equiv A_{ij} . \tag{3.9}$$

We have now to insert on the left-hand side the time sequences of the observed data in order to check whether

$$\cos \alpha_{ij}(t) \tag{3.10}$$

is time-independent or equivalently whether

$$\frac{dA_{ij}}{dt} = 0 \tag{3.11}$$

holds. The relationship (3.11) can be checked by computer calculations. The computer has then to sort out those combinations of i and j where (3.11) is fulfilled. In order to verify that the individually identified pairs ij, kℓ, ... describe motion in the same plane, the computer has to verify

$$A_{ij} = A_{k\ell} \; . \tag{3.12}$$

In the next step of our analysis we introduce the angles of the individual limbs at each joint. We then proceed as in a previous paper [11]; we may now calculate the time-dependence of the angles, Fourier analyse them and use the leading Fourier coefficients as features describing the movement pattern. In the last step of the analysis, the vector $\underline{q}$ of equation (2.2) is made up of the features just mentioned and an index coding for the different movement patterns. In this way our computer program was able to identify in-phase and out-of-phase movements of hands, or different kinds of movements of humans.

We have generalized our approach in two ways. In a first generalization of our approach we assume that there is a finite distance ℓ between the plane in which the center of the man or animal moves and the plane in which the limb movement occurs. The relation (3.1) then generalizes to

$$x = \ell \sin \alpha + x' \cos \alpha \; . \tag{3.13}$$

Again we have to study the time-dependence or time-independence of (3.4). We also treated a second generalization, namely when α is time-dependent. Then we have to check whether the relation

$$\frac{(\Delta x_{ij})^2}{\cos^2 \alpha} + (\Delta y_{ij})^2 = \text{const. ?} \tag{3.14}$$

holds where the right-hand side is time-independent. Differentiation of (3.14) with respect to time yields

$$\frac{\Delta x_{ij} \; \Delta \dot{x}_{ij}}{\cos^2 \alpha} + 2 \frac{(\Delta x_{ij})^2}{\cos^3 \alpha} \sin \alpha \; \dot{\alpha} + \Delta y_{ij} \; \Delta \dot{y}_{ij} = 0 \tag{3.15}$$

or when we introduce suitable abbreviations

$$K(t) + L(t) \frac{\sin \alpha}{\cos \alpha} \; \dot{\alpha} + \cos^2 \alpha \; M(t) = 0 \; . \tag{3.16}$$

We consider (3.16) as a differential equation for α. Making the substitution

$$\eta = - \ln \cos \alpha \tag{3.17}$$

from which

$$\dot{\eta} = \frac{\sin \alpha}{\cos \alpha} \; \dot{\alpha} \tag{3.18}$$

and

$$\ln \cos \alpha = - \eta$$

$$\cos \alpha = e^{-\eta} \qquad (3.19)$$

result, (3.16) can be cast in the form

$$\dot{\eta} + e^{-2\eta} \, \tilde{M}_{ij}(t) = \tilde{K}_{ij}(t) . \qquad (3.20)$$

We consider in addition a different pair $k\ell$ which is assumed to rotate in the same plane as the pair ij again with the fixed distance $k\ell$. We then obtain

$$\dot{\eta} + e^{-2\eta} \, \tilde{M}_{k\ell} = \tilde{K}_{k\ell}(t) . \qquad (3.21)$$

We have used the abbreviations

$$\tilde{K} = - \frac{K}{L} \qquad (3.21a)$$

and

$$\tilde{M} = \frac{M}{L} . \qquad (3.21b)$$

In order to get rid of the time derivative $\dot{\eta}$, we take the difference between (3.20) and (3.21) and obtain

$$e^{-2\eta} \, (\tilde{M}_{ij}(t) - \tilde{M}_{k\ell}) = \tilde{K}_{ij} - \tilde{K}_{k\ell} . \qquad (3.22)$$

We may resolve (3.22) with respect to $e^{-2\eta}$

$$e^{-2\eta} = \frac{\Delta \tilde{K}}{\Delta \tilde{M}} . \qquad (3.23)$$

This possesses the solution

$$- 2\eta = \ln \frac{\Delta \tilde{K}}{\Delta \tilde{M}} . \qquad (3.24)$$

The right-hand side of (3.24) must obey equation (3.20) and (3.21). A different self-consistency check is obtained by establishing (3.23) for a further pair i'j' so that

$$e^{-2\eta} = \frac{\Delta \tilde{K}'}{\Delta \tilde{M}'} \qquad (3.25)$$

holds. Dividing (3.23) by (3.25) we find the relation

$$\frac{\Delta \tilde{K}'}{\Delta \tilde{M}'} = \frac{\Delta \tilde{K}}{\Delta M} \qquad (3.26)$$

which is the self-consistency check. Making these different self-consistency checks for the different pairs, the computer may identify those pairs for which the relation (3.3) holds. We now know the motion of the limbs and may proceed as described above.

A detailed account of this work will be published elsewhere by Fuchs, Haas, Haken and Kelso.

REFERENCES

[1] Haken, H., Synergetics, An Introduction, Springer, Berlin, Heidelberg, New York (1983)

[2] Haken, H., Advanced Synergetics, 2nd ed., Springer, Berlin, Heidelberg, New York (1987)

[3] Kohonen, T., Self-Organization and Associative Memory, 2nd. ed., Springer, Berlin, Heidelberg, New York (1987)

[4] Bestehorn, M., Haken, H., Phys. Review A in print

[5] Haken, H., in H. Haken, ed., Neural and Synergetic Computers, Springer, Berlin, Heidelberg, New York (1988)

[6] Fuchs, A., Haken, H., in H. Haken, ed., Neural and Synergetic Computers, Springer, Berlin, Heidelberg, New York (1988)

[7] Ditzinger, T., Haken, H., Oscillations in the Perception of Ambiguous Patterns, Biol. Cyb. 61, 279-287, Springer, Berlin, Heidelberg, New York (1989)

[8] Haken, H., Haas, R., Banzhaf, W., Biol. Cyb. (1989) in print

[9] Haken, H., to be published in Suppl. to Progr. Theor. Physics

[10] Johansson, G., Visual Perception of Biological Motion and a Model for its Analysis, Perception and Psychophysics (1973)

[11] Haken, H., Kelso, J.A.S., Fuchs, A., Pandya, A.S., Neural Networks, in print

Parallel Processing in Neural Systems and Computers
R. Eckmiller, G. Hartmann and G. Hauske (Editors)
© Elsevier Science Publishers B.V. (North-Holland), 1990

PATTERN RECOGNITION IN DAMAGED HOPFIELD NETWORKS

Eva Koscielny-Bunde

Fachbereich Informatik
Universität Hamburg
D-2000 Hamburg 50, W. Germany

Damaged Hopfield networks are studied where the synaptic connections between pairs of neurons are cut with probability p. We investigate by numerical simulations, how the retrieval of noisy patterns and the maximum storage capacity α_c of the net depends on the concentration p of cut bonds. We find that α_c depends linearly on p, $\alpha_c(p) \cong \alpha_c(0)(1-p)$. In the retrieval phase, the mean overlap $\langle q \rangle$ between input and output patterns decreases exponentially with p, $1-\langle q \rangle \sim \exp[-\gamma(1-p)N/M]$, where M/N is the number of learnt patterns per neuron, and γ seems to approach $1/2$ for large networks.

A prototype for a large number of neural network models is the Hopfield model [1](for recent reviews see e.g. [2, 3]). In the Hopfield network, N neurons are considered, each of them can be in two states $S_i = \pm 1$. Pairs of neurons are coupled by bonds $J_{i,j}$, which represent the synaptic efficacies between them. The couplings $J_{i,j}$ are determined by the M random patterns $\{\xi^\mu\} \equiv (\xi_1^\mu, ..., \xi_N^\mu)$, $\mu = 1, 2, ..., M$, which are to be stored in the network. According to Hebb [4] one chooses

$$J_{i,j} \equiv J_{j,i} = \frac{1}{N} \sum_{\mu=1}^{M} \xi_i^\mu \xi_j^\mu. \tag{1}$$

The state of neuron i at time $t+1$ is obtained from the states of all other neurons at time t by the deterministic dynamic rule

$$S_i(t+1) = sign[\sum_{j=1}^{N} J_{i,j} S_j(t)]. \tag{2}$$

Starting from an initial input pattern $(S_1(0), ..., S_N(0))$ the system evolves towards the local minima of the energy function (cost function) $E = -\frac{1}{2} \sum_{i,j=1}^{N} J_{i,j} S_i S_j$ where each neuron S_i is aligned in parallel with its local field $h_i = \sum_{j=1}^{N} J_{i,j} S_j$.

A particular pattern $\{\xi^\nu\}$ is called stored by the network, if an initial configuration $\{S(0)\}$ near $\{\xi^\nu\}$ develops under the dynamic rule (1) to a state whose overlap

$$q^\nu = \frac{1}{N} \sum_{i=1}^{N} S_i \xi_i^\nu. \tag{3}$$

with $\{\xi^\nu\}$ is large. According to Amit et al. [5], only a certain percentage $\alpha = M/N$ of patterns per neuron can be stored in the Hopfield network. For pure input patterns, as long as α is below a critical value $\alpha_c \cong 0.14$, the system approaches "retrieval states" which have at least 97 percent overlap with the stored patterns.

460   *E. Koscielny-Bunde*

As α decreases to zero, the mean overlap increases exponentially fast towards 1 [5],

$$1 - \langle q \rangle \cong \exp(-1/(2\alpha)). \tag{4}$$

Above α_c, the system flows into a locally stable state, where the remanent overlap is considerably smaller, $q \cong 0.35$. Eq. (4) is rigorous in the limit of large networks, $N \to \infty$.

In this work, we study the effect of damage in the Hopfield network. We assume that in a damaged network, the synaptic efficacies $J_{i,j} = J_{j,i}$ between pairs of neurons follow the Hebb rule (1) only with probability $(1 - p)$, with probability p a bond is broken and $J_{i,j} = J_{j,i} = 0$. We discuss in particular, how the storage capacity and the mean overlap in the retrieval phase, Eq. (4), are modified in damaged networks.

In the numerical study we have considered networks of variable size, from $N = 200$ until $N = 800$ neurons. The numerical simulations consist of two parts [6]. In the first part, M random patterns $\{\xi^\mu\}$ are generated. The symmetric synaptic efficacies $J_{i,j}$ are calculated according to the Hebb rule, and are cutted with probability p. In the second part, one of the nominated patterns is chosen, and a noisy input pattern is generated, where the state of each neuron is changed with probability p_n. Then the dynamic rule (2) is applied and the time dependent overlap $q^\mu(t)$ is calculated. The run stops when $q^\mu(t)$ stays constant for six time steps or oscillations of period 2 with constant amplitude occur. This procedure is repeated for typically 10^3 initial configurations, which are chosen from different nominated patterns.

By averaging the results we obtained, for a given number M of patterns, for fixed initial noise p_n and for fixed damage parameter p, the mean retrieval overlap $\langle q \rangle$.

Fig. 1a shows the maximum number of patterns which can be stored with less than 0.5 percent error, as a function of p, for three values of the initial noise p_n.

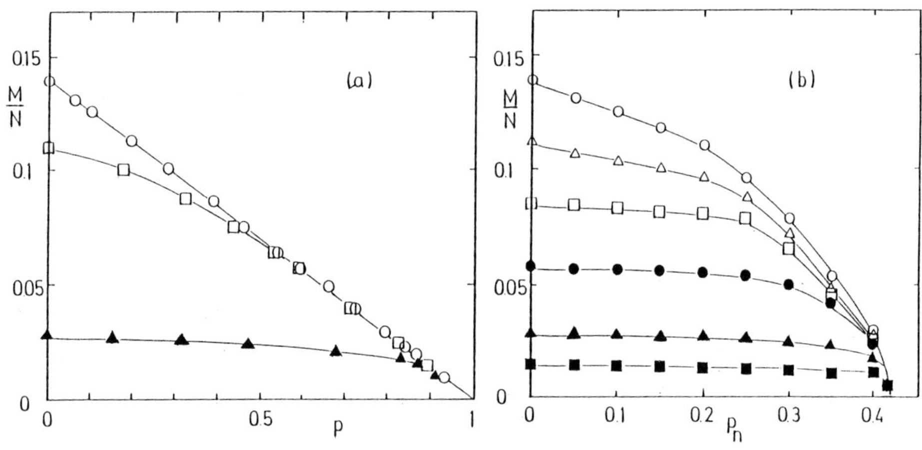

FIGURE 1

Maximum fraction $M/N = \alpha$ of learnt random patterns that are recognized with less than 0.5 percent error (a) as a function of the damage parameter p for the initial noise $p_n = 0(\bigcirc)$, $0.2(\square)$, and $0.4(\blacktriangle)$. and (b) as a function of initial noise p_n, for several values of the damage parameter p: $p = 0(\bigcirc)$, $0.2(\triangle)$, $0.4(\square)$, $0.6(\bullet)$, $0.8(\blacktriangle)$ and $0.9(\blacksquare)$.

The curves separate a retrieval regime from a regime where patterns are retrieved with less than 99.5 percent accuracy ($\langle q \rangle < 0.99$). The maximum storage capacity $\alpha_c(p)$ (which is identical to the curve for $p_n = 0$) is approximately a simple straight line,

$$\alpha_c(p) \cong \alpha_c(0)(1 - p). \tag{5}$$

As the initial noise increases, the curve bends down for small p-values. As can be seen clearly in the Figure, there exists a cross over value $p^x(p_n)$, which increases with increasing noise level p_n. Above p^x, $\alpha \equiv M/N$ is a straight line and follows (5), below p^x α bends down. For *large* noise, the curve becomes rather flat, showing only a *weak* dependence on the concentration p of damaged synaptic efficacies: An increase in the concentration of damaged bonds has comparatively little effect on the number of patterns that can be stored. Fig. 1b shows the maximum number of patterns which can be stored with less than 0.5 percent error, as a function of the initial noise p_n, for various values of the concentration p of damaged synaptic efficacies. The effect of the damage is twofold: The number of random patterns which can be stored is reduced and, with an increasing amount of damage, becomes *less sensitive* to initial noise.

Next we have studied how the final overlap changes, in the retrieval regime, as a function of M/N and p. For $p = 0$, $1 - \langle q \rangle$ is described by the exponential, Eq. (4). It is natural to assume that also in the damaged network $1 - \langle q \rangle$ is described by an exponential. An Ansatz which is consistent with both the exact result Eq. (4) for undamaged networks and the linear dependance of the maximum storage capacity on the damage parameter p, Eq. (5), for small initial noise, is

$$1 - \langle q \rangle = A(M/N, N) \exp(-\gamma(1 - p)N/M), \tag{6}$$

where $A(M/N, N)$ is a slowly varying amplitude. For $N \to \infty$ and $p = 0$ Eq. (6) must be identical to Eq. (4), yielding $\gamma = 1/2$ in this case. We have calculated $\langle q \rangle$ for several fractions $\alpha = M/N$ of stored patterns and various sizes N of the network. Since we are interested in the maximum overlap, we have chosen $p_n = 0$.

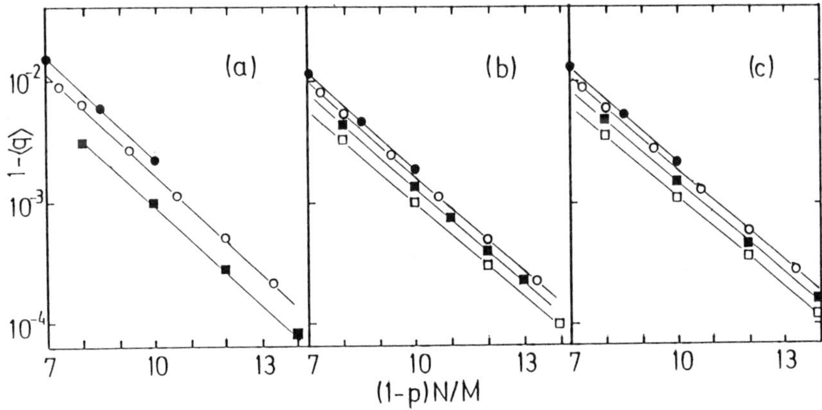

FIGURE 2

Average retrieval overlap $\langle q \rangle$ of the pure patterns as a function function of $(1 - p)N/M$ for several fractions $\alpha = M/N$ of M random patterns and several sizes N of the network: (a) $N = 200$, (b) $N = 400$, and (c) $N = 800$. Different symbols represent different values of α: $\alpha = 0.1(\bullet), 0.075(\bigcirc), 0.05(\blacksquare)$ and $0.025(\square)$. The data are based on averages over (a) 3200, (b) 1600, and (c) 800 configurations.

To test Ansatz (6) we have plotted the logarithm of $1 - \langle q \rangle$ as a function of the combined variable $(1 - p)N/M$, for various fractions α of nominated patterns and various sizes N of the network, N=200, 400, and 800. The results are shown in Fig. 2a,b,c. All curves are straight lines, their slopes γ do not depend on α and decrease continously when N is increased. We find $\gamma \cong 0.61$ for $N = 200$, $\gamma \cong 0.59$ for $N = 400$, and $\gamma \cong 0.56$ for $N = 800$. It seems likely that γ approaches $1/2$ for $N \rightarrow \infty$. In all cases, the curves are located close to each other, but do not yet collapse. This shows that the amplitude A depends, for the values of N considered here, weakly on M/N. For $N = 800$, e.g., we have $A \cong 0.3, 0.4$, and 0.6 for $\alpha = 0.1, 0.05$, and 0.025, respectively. Our results suggest that Eq. (6), with $\gamma = 1/2$, generalizes the result of Amit et al., Eq.(4), to damaged Hopfield networks in the limit of $N \rightarrow \infty$.

I like to thank Profs. F. Schwenkel and S. Havlin for valuable discussions on associative memory and neural networks, and Prof. A. Bunde for critical reading of the manuscript. I also like to thank Prof. H.E. Stanley for the kind hospitality at Boston University, where part of this work has been performed.

REFERENCES

[1] Hopfield, J. J., Proc. Natl. Acad. Sci. USA 79 (1982) 2554.
[2] Eckmiller, R., and v. Malsburg, Ch., eds., Neural Computers,
    (Springer Verlag, Berlin, Heidelberg, 1988).
[3] Sompolinsky, H., Physics Today 41 (Dec.1988) 70 ;
    Domany, E., J. Stat. Phys. 51 (1988) 743.
[4] Hebb, D. O., The Organization of Behavior (Wiley, New York, 1949).
[5] Amit, D. J., Gutfreund, H., and Sompolinsky, H., Ann. of Physics 173 (1987) 30.
[6] Koscielny-Bunde, E., J. Stat. Phys. 58 (1990) issue 5/6.

Section 12
Parallel Processing
in Artificial Intelligence

Parallel Processing in Neural Systems and Computers
R. Eckmiller, G. Hartmann and G. Hauske (Editors)
© Elsevier Science Publishers B.V. (North-Holland), 1990

Parallel Process Interfaces to Knowledge Systems

K.H. Becks[†], W. Burgard[*], A.B. Cremers[*], A. Hemker[†] and A. Ultsch[*]

[*]Department of Computer Science   [†]Department of Physics
University of Dortmund   University of Wuppertal
D-4600 Dortmund 50   D-5600 Wuppertal 1

This paper deals with the integration and combination of connectionist approaches and symbolic knowledge based systems. Our aim is the development of expert systems which are tightly coupled to real world processes. We discuss architectures for such systems and present two possible outlines of an integrated system for the analysis of data produced by a high energy physics experiment.

1. Introduction

The wide field of domains in which expert systems solve complex problems such as diagnosis, construction, planning, scheduling etc. shows the success of the symbolic knowledge representation paradigm. One expert system which performs fault diagnosis in a high energy physics experiment is presented in [Beck89]. Symbolic knowledge representation has a clear advantage: the knowledge chunks are communicable i.e. they can be explained, taught, learned and easily modified. On the other hand, the symbolic approach lacks of inherent methods to refine raw incoming data into structured symbolic information and to tackle with inconsistencies, such as erroneous data, missing values, and unexpected events. Both properties, for example, are required if expert systems have to be connected to complex external processes in order to solve tasks such as data analysis or control. In this case a further important task which has to be carried out is the gathering of sensory information, the influence of the sensors, the processing and filtering of the data in an appropriate and possibly adaptive manner.

Connectionist models, however, claim to capture more precisely more complex and imprecise properties of the real world. They are inherently parallel and furthermore adaptive in the sense that knowledge can be learned directly from experience.

Hence, the combination of symbolic and connectionist approaches or the use of integrated systems seems to be sensible if expert systems have to solve tasks which require a tight coupling to the environment they are embedded in. Different approaches to achieve an integration of expert systems into given environments will be discussed in the following section. For illustration we concentrate on the requirements for such systems in the framework of a high energy physics experiment.

2. Combination of Connectionist and Symbolic Systems

First, connectionist models can be an integral part of an expert system. They can be used to represent the rules of a system (see [Gall88]) or to implement a fuzzy reasoning calculus (see [Beck87]). Furthermore they can be integrated into the reasoning process itself. In the PANDA system (see [Ults89a, Ults89b]) connectionist models are used to learn symbolic rules from given data and to observe the reasoning process. In this system a neural network "learns" structural properties of proofs to the effect that the same proof or similar proofs can be performed more efficiently the next time.

Second, a connectionist system can operate as a sub-system which condensates a high volume of data describing complex states to information which is needed as input for an expert system.

Third, both paradigms can be integrated into one. This was carried out with the development of classifier systems which originally were suggested by John Holland (see [Holl76]) as an outgrowth of his work on genetic algorithms. So far, mainly the machine learning society was interested in classifier systems because they perform a kind of data intensive learning by example, with sparse reinforcement. Although classifier systems are rule based, they work on a subsymbolic level (see [Gold89, Davi87]). The basic idea of classifier systems should enable them to unify the distinct cognitive abilities, which are required to solve a complex problem. On the one hand classifier systems are capable of communicating directly with the environment which enables them to work on low-level tasks, on the other hand symbol-level activities can also be implemented (see Forr85]). Rule-learning also covers the whole range from signal processing rules to problem solving rules, although the internal representation scheme is always subsymbolic.

Because of their ability to interact directly with the environment and their adaptive behaviour, classifier systems are capable of dealing with dynamic environments. All parts of a classifier system are inherently parallel. There are applications of classifier systems in quite different domains ranging from poker playing, gas pipeline control to simulation of an "animal-like automaton". So far classifier systems did not succeed in larger applications, although no principal limitations have yet been encountered.

3.   The DELPHI Experiment

The new electron positron collider LEP at the European High Energy Physics Research Center CERN near Geneva has started operating in July '89. In the collider ring electrons and their anti-particles, the positrons, are accelerated in opposite directions. Each time an electron collides with a positron, in a so called event, many new and partially short-living particles arise. Some particles can be identified with the help of large detector systems, one of which is the DELPHI detector. In order to show the difficulties in analyzing such an event we subsequently will give a short and simplified description of the involved physical processes (for more details we refer to [Davi89]).

The electron-positron annihilation process produces an initial quark-antiquark pair and possibly several gluons. Quarks are elementary particles and are the fundamental constituents of matter. There are six known varieties of quarks distinguished by their flavours: up, down, charm, strange, bottom, top. The top quark is of major interest in the DELPHI experiment.

After the creation of the initial quark-antiquark pair new quark-antiquark pairs with less energy are generated in a fragmentation phase. Fragmentation is a cascading process in which each quark-antiquark pair can produce new pairs with less energy. Quark and anti-quark always have the same flavour e.g. up and anti-up. Quarks of different flavours cluster and form new particles (hadrons). Some produced short-living hadrons decay into long-living particles according to known rules. Most of these stable particles are observed with the help of particle detector systems (see figure 1).

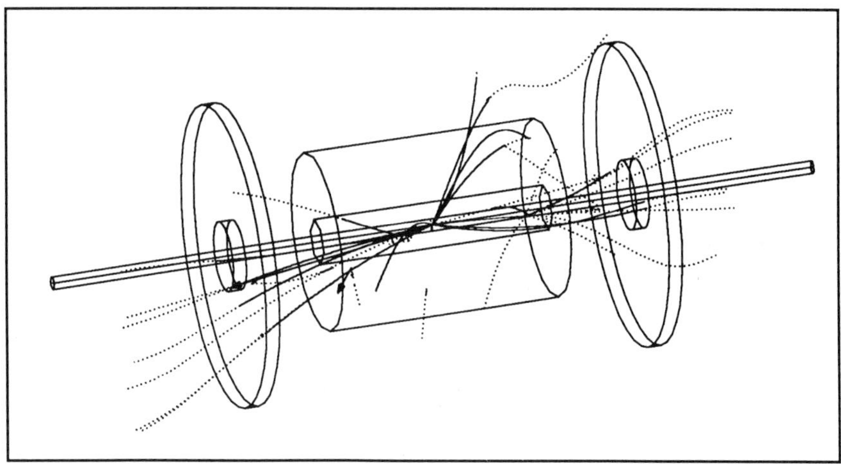

Figure 1: Typical event observed by parts of the DELPHI detector

Typical questions for an event concerning the intermediate states of the physical process are:

- What was the flavour of the initial quarks?
- Were there any exotic particles e.g. Higgs or rare decays?
- How many gluons were emitted resp. how many jets occurred?

It is very difficult to answer these questions because the whole physical process is extremely non-deterministic and the detectors can observe the physical processes only with a limited efficiency.

The fragmentation process itself is not directly observable as well as many decay processes of hadrons. It is very difficult to detect some of the neutral particles and furthermore noise and blind sections in the detector can inhibit the observation and identification of stable particles.

The non-determinism of the physical process has several reasons. The flavours of the produced quark-antiquark pairs are randomly distributed and the depth of the fragmentation cascade is unknown. Although the quark constituents of a hadron are known, it is impossible to observe the quarks which cluster to a given hadron. A further source of uncertainty is that different decay rules are possible for an unstable hadron.

4.   Analysis Strategies

The average number of observed charged particles per event is about forty. Inferring back from these observed particles to the preceding physical process expands an extraordinary complex search space. Nowadays most of the physics analysis programs avoid this problem by looking only at the signature of the final annihilation state to answer questions about an event. The huge amount of data from each event (the number of events is 1-2 per second where each event has an average data size of 150 Kbyte) is reduced to only a few physical, statistical and geometrical parameters which are sensitive to a certain question. It is left to the intuition of the physicist to find out the relevant parameters.

Simulation programs can prove whether a value is sensitive to a certain question. These so called Monte Carlo programs contain the actual knowledge of the elementary physical processes, which could have happened after the collision of electron and positron. With an additional simulation of the detector system it is possible to produce data before starting the real experiment. The simulated data can be used to test and tune the physics analysis programs. Moreover, it is possible to check the underlying physical model by comparing simulated data with real data from the experiment.

Data reduction and flat, statistical based reasoning imply a loss of information. Unusual effects which did not manifest in the selected parameters of the analysis programs cannot be recognized. Differences between simulated and real data cannot be directly interpreted, little deviations are not conspicuous. An ideal system would be able to reconstruct each event using the underlying physical model or would locate the point where an event, that contains new physics, does not match the model. If one attempts to develop an intelligent assistant for the analysis of the event data, one has to build tools which allow a fast condensation of the incoming data. Based on possible architectures for an integration of different paradigms provided in section 2, we now will present an outline of two possible systems which should be able to solve this task. Since this project is still in the starting phase, we only can give proposals for both solutions.

5.   Outlines of an Integrated System for Data Analysis

Fragmentation and particle decay suggests the representation by production rules. With these rules contained in the production memory it is possible to reconstruct the event history by abduction. The initial state is equal to the final physical state and the goal state corresponds to the original electron-positron annihilation state. State transitions are defined by applying fragmentation and decay rules backwards. In this case not the goal itself is interesting but the path leading from the initial state to the goal state. An expert system, which is able to reconstruct the physical processes, i.e. to find this solution path in the problem space, could answer all the questions outlined in the previous section.

The non-determinism of the physical processes requires probabilistic reasoning. Since we will have more than one possible solutions in most cases, there has to be a mechanism to judge the quality of solutions. Heuristics to simplify the search are hardly known or not explicitly formulated. Hence the system must be able to learn such heuristics. Monte Carlo simulation can provide learning examples and the protocol of the simulation can be used to control the learning process and to judge solutions.

As pointed out in section 2 a connectionist system could be used to preprocess the incoming data from the detector. This sub-system would benefit from the advantages of the connectionist approach with regard to suppression of detector noise, parallel processing and condensation of quantitative data to qualitative data. The preprocessed data would be the input of a symbolic knowledge based system, whose task is to reconstruct the physical process. An advantage of this possible solution is that the learned heuristics could easily be interpreted by the physicists. Furthermore, the rules to guide the search could be implemented as connectionist systems, and neural networks could be used to improve the search. In this case, however, an interpretation of the learned rules will be very difficult.

As second possible solution of the reconstruction problem would be the application of a classifier system. If this solution is preferred, the question arises how the different types of information can be represented. Messages should be able to encode signals from the particle detector, as well as it should be possible to represent an intermediate states of the physical process by one or more messages. Classifiers could be used for signal processing, but also to simulate a fragmentation step or a particle decay. There should be room left in the representation of a classifier to encode a situation, in which the processing rule should be preferably applied. In the training phase classifiers should learn those situations. Starting with messages encoding signals from the particle detector, the classifier system should explore effectively and in parallel the problem space to find a solution. A backtracking strategy is not necessary because of the parallel rule activation in classifier systems. The environment can check the validity of physical laws after each cycle, i.e. for each reconstruction step. In the training phase the environment can evaluate possible reconstruction steps by comparing them with the Monte Carlo protocol.

The inherent learning capability of a classifier system makes it possible to integrate subsequent changes of the physical model into the system only by changing the environment. It still remains a problem to interpret the heuristics learned by the system. An additional analysis is necessary to investigate the characteristics of those situations, in which one processing rule should be preferably applied.

6. Conclusion

In this paper we have presented different strategies to couple knowledge based systems with connectionist systems. We have pointed out how such combinations can be used to improve the current expert system technology. Connectionist models can be used to couple the expert system to its environment. To achieve a fast refinement from data to information they can build parallel interfaces between real world processes and conventional knowledge based systems. They can also be used as an integral part of an expert system. We discussed classifier systems which can be seen as homogeneous integrated systems. We have proposed two possible outlines of an integrated system to be used for data analysis in a high energy physics experiment. As this example shows the development of tools which allow a tight coupling of expert systems with connectionist systems as parallel process interfaces seems to be a very important task.

References

[Beck87] Becker, L.A. and Peng, J., Using Activation Networks for Analogical Ordering of Consideration, IEEE Conf. on Neural Networks, San Diego, CA, (1987) pp. 367-371.

[Beck89] Becks, K.-H. and Cremers, A.B. and Hemker, A., Design of an Expert-System-Shell for Fault-Diagnosis in the DELPHI Experiments, KI 1/89, (Oldenburg Verlag, 1989), in German.

[Davi87] Davis, L., Genetic Algorithms and Simulated Annealing, (Morgan Kaufman, Los Altos 1987).

[Davi89] Davies, P.C.W., The New Physics, (Cambridge University Press 1989).

[Forr85] Forrest, S., Implementing Semantic Network Structures using Classifier Systems, in: in: Grevenstette (ed.), Proceedings of an International Conference on Genetic Algorithms and their Applications, (Carnegie-Mellon University, Pittsburgh, 1985).

[Gall88] Gallant S.I., Connectionist Expert Systems, Communications of the ACM, Vol. 31, (1987), pp. 52-169.

[Gold89] Goldberg, D.E., Genetic Algorithms in Search, Optimization and Machine Learning, (Addison-Wesley 1989).

[Holl76] Holland, J.H., Adaptation, Progress in Theoretical Biology, Vol. 4, (1976) pp. 263-294.

[Robe87] Robertson, G.G. and Riolo, R.L., A Tale of Two Classifier Systems (1987).

[Ults89a] Ultsch, A., PANDA: Prolog and Neural Distributed Architectures, Technical Report, Department of Computer Science, (University of Dortmund, 1989), in German.

[Ults89b] Ultsch, A. and Hannuschka,R. and Hartmann, U. and Weber, V., Learning of Control Knowledge for Symbolic Proofs with Backpropagation Networks, this volume.

Parallel Processing in Neural Systems and Computers
R. Eckmiller, G. Hartmann and G. Hauske (Editors)
Elsevier Science Publishers B.V. (North-Holland), 1990

Motion Detection by Correlation and Voting[*]

*Stefan Bohrer[1], Heinrich H. Bülthoff[2], and
Hanspeter A. Mallot[1]*

1. Institut für Neuroinformatik, Ruhr–Universität, D–4630 Bochum, FRG
2. Brown University, Dept. of Brain and Cognitive Sciences, Providence, RI, USA

We discuss the properties of a motion detection algorithm published recently
by Bülthoff *et al.* [1]. Analytical results show that the algorithm can correctly
recover the projected motion in most interesting cases and thus solves the socalled
weak aperture problem. An implementation on a serial computer is presented
that processes image frames within resonable time. Numerical results from this
application are included to illustrate the algorithm's performance.

1 INTRODUCTION

Optical flow is generated on the retina of an observer by objects moving relative to the
observer. The true motion field W is a 3D vector field whose projection on the image
plane is denoted by $W'(\mathbf{x})$ where $\mathbf{x}$ is a 2D vector in the image plane. Unfortunately, the
measurement of this 2D field of image velocities from changes of intensity in subsequent
images, $E_t(\mathbf{x})$, $E_{t+\Delta t}(\mathbf{x})$, is not possible in general [5]. It is, however, possible to compute
suitable optical flows $V(\mathbf{x})$ that are qualitatively similar to the true velocity field in most
cases. We describe a simple, parallel algorithm that computes optical flow from sequences of
real images. The algorithm is consistent with human psychophysics and electrophysiological
data from cortical areas V1 and MT [1, 3].

2 PARALLEL MOTION ALGORITHM

The algorithm can be described by the following three steps:

Shift and Compare: The expected motion displacements are characterized by an 2D interval $D_\delta := [-\delta, \delta] \times [-\delta, \delta]$. For each node $\mathbf{x}$ and permissible displacement $\mathbf{d} \in D_\delta$, a
comparison function $\phi(a, b)$ is evaluated. (Here, a, b denote either greylevels or intensities
of preprocessed images.) The output of this step is a matching strength for each node and
displacement, $m(\mathbf{x}, \mathbf{d}) = \phi(E_t(\mathbf{x}), E_{t+\Delta t}(\mathbf{x} + \mathbf{d}))$.

Local Summation: At each pixel $\mathbf{x}$ the matching strength for corresponding displacements
from the pixels in a neighborhood $P_\nu(\mathbf{x})$ are accumulated. The output of this step is a
combined matching strength which, again, is a point wise function:

$$M_\mathbf{x}(\mathbf{d}) \;:=\; \sum_{\mathbf{y} \in P_\nu(\mathbf{x})} m(\mathbf{y}, \mathbf{d}). \tag{1}$$

---

[*]Supported by the German Federal Department of Research and Technology (BMFT),
Grant No. ITR8800K4

Winner-Take-All: To each pixel $\mathbf{x}$, the displacement that received the highest matching strength M is assigned as its velocity value $V(\mathbf{x})$ by a winner-take-all scheme. That is, V is selected to satisfy the condition

$$M_{\mathbf{x}}(V(\mathbf{x})\Delta t) \;=\; \max_{\mathbf{d}\in D_\delta} M_{\mathbf{x}}(\mathbf{d}). \tag{2}$$

A large vote for one particular displacement is expected if the motion field is locally constant.

3   TRUE AND RECOVERED MOTION

In this Section, we prove an important Theorem for the continuous approximation of the algorithm:

Theorem: Let $E_1(\mathbf{x})$ and $E_2(\mathbf{x})$ be two image frames with $E_2(\mathbf{x}) = E_1(\mathbf{x}+\mathbf{v})$ for some $\mathbf{v} \in D_\delta$. Let further $\phi : \mathbf{R}^2 \mapsto \mathbf{R}$ be a comparison function in the above sense. Then, if $\phi(a,b)$ is of the form $\psi(a-b)$ with $\psi'(0) = 0$, the true displacement vector (u,v) maximizes the combined matching strength M.

Proof: In the continuous approximation, we have:

$$M_{\mathbf{x}}(\mathbf{d}) = \int_{P_\nu(\mathbf{x})} \phi(E_1(\mathbf{y}), E_2(\mathbf{y}+\mathbf{d}))d\mathbf{y}, \tag{3}$$

Where the integral is taken over a 2D range. In the maximum, the gradient of M, $(\partial M/\partial d_1, \partial M/\partial d_2)$, vanishes. That is, we have to show that $\mathrm{grad}\,M(\mathbf{v}) = 0$.

$$\mathrm{grad}\,M_{\mathbf{x}}(\mathbf{v}) \;=\; \int_{P_\nu(\mathbf{x})} \frac{\partial \phi}{\partial b}(E_1(\mathbf{y}), E_2(\mathbf{y}+\mathbf{v}))\,\mathrm{grad}\,E_2(\mathbf{y}+\mathbf{v})d\mathbf{y} \tag{4}$$

$$\;=\; \int_{P_\nu(\mathbf{x})} \frac{\partial \phi}{\partial b}(E_1(\mathbf{y}), E_1(\mathbf{y}))\,\mathrm{grad}\,E_1(\mathbf{y})d\mathbf{y}. \tag{5}$$

We now use the relation $\phi(a,b) = \psi(a-b)$ and obtain:

$$\mathrm{grad}\,M_{\mathbf{x}}(\mathbf{v}) \;=\; \int_{P_\nu(\mathbf{x})} -\psi'(0)\,\mathrm{grad}\,E_1(\mathbf{y})d\mathbf{y} = 0. \qquad \Box \tag{6}$$

Suitable choices of the comparison function that satisfy the condition $\psi'(0) = 0$ are of the type:

$$\phi(a,b) \;=\; |a-b|^n. \tag{7}$$

It is easy to show that functions of the type $\phi(a,b) = ab$ lead to systematic errors in the algorithm.

The above theorem shows that the voting algorithm is not subject to the weak aperture problem. As was argued earlier by Reichardt *et al.* [4], the aperture problem seems to be much less severe than was previously thought (for review see [2]). An illustration of how the algorithm deals with the aperture problem is given in Fig. 1.

A circle moves horizontally within the image plane and its motion is correctly recovered by the motion scheme. Fig. 1b shows the region in which motion was detected and Fig. 1c illustrates the calculated velocity field.

4   IMPLEMENTATION AND EXPERIMENTAL RESULTS

The algorithm is serially implemented on a SUN 4/110 workstation. In contrast to the parallel implementation [1], the computational cost is of the order $\delta^2 \times \nu^2$. In general, there is a trade–off between computation time and memory requirements. We have developed two different kinds of implementation with differing advantages:

a.   b.   c.

Figure 1: Performance of the algorithm for a synthetic image pair, i.e., a shifted circle. **a.** Input image (only one frame shown). **b.** Regions of detected motion. **c.** Needle–plot. In algorithms subjected to the aperture problem, only the motion components orthogonal to the contour can be recovered. The voting scheme yields correct results, even if only small voting domains ($\delta = 2$) are used.

Minimal Storage: For each pixel x_{ij}, only those matching strength $m(\mathbf{x}, \mathbf{d})$ are computed and stored that have not been computed previously. Note that $m(\mathbf{x_1}, \mathbf{x_2} - \mathbf{x_1}) = m(\mathbf{x_2}, \mathbf{x_2} - \mathbf{x_1})$. *(Overlap–Scheme)*

Maximal Speed: For each pixel x_{ij}, all required values of m are stored. *(Nested Loops–Scheme)*

A number of experiments with different kinds of circular and rectangular displacement- and voting-windows was made and various voting rules with a considerable level of complexity were tested against each other. Nevertheless, the simple rule of choosing the maximal excitation by a fixed, space–invariant threshold yielded the best results. This is consistent with the results of Sect. 3, where the winner–take–all mechanism was shown to give correct motion estimates.

A surprising result is the fact that very small voting regions (no more than 7×7 pixel) are sufficient to recover the projected velocity field correctly. Due to this fact, natural image sequences can be processed in a reasonable time. A typical example of a natural scene is shown in Fig. 2.

Fig. 2 shows a sequence of two frames. The resolution is 128×256 pixels in noninterlaced mode. A rectangular displacement area of 27×21 pixels and a relatively small rectangular voting area of 7×7 pixels was chosen. Computation time was roughly 25 minutes with the nested–loop–version and some 90 minutes for the overlap-scheme on a SUN 4/110 workstation.

5 FUTURE WORK

- Parallel implementation on a MIMD architecture (Transputer).

- Adaptive adjustment of voting– and displacement–domains. Expected space-variance (e.g. from optical flow, cf. [3]) can thus be represented by a map of local magnification factors given by the relative sizes of the displacement window.

- Combination of a 1D version of the algorithm with inverse perspective mapping for obstacle avoidance [3].

- Development of a looming- (i.e., "blasting-bomb-") detector based on the local histograms of the $m(\mathbf{x}, \mathbf{d})$.

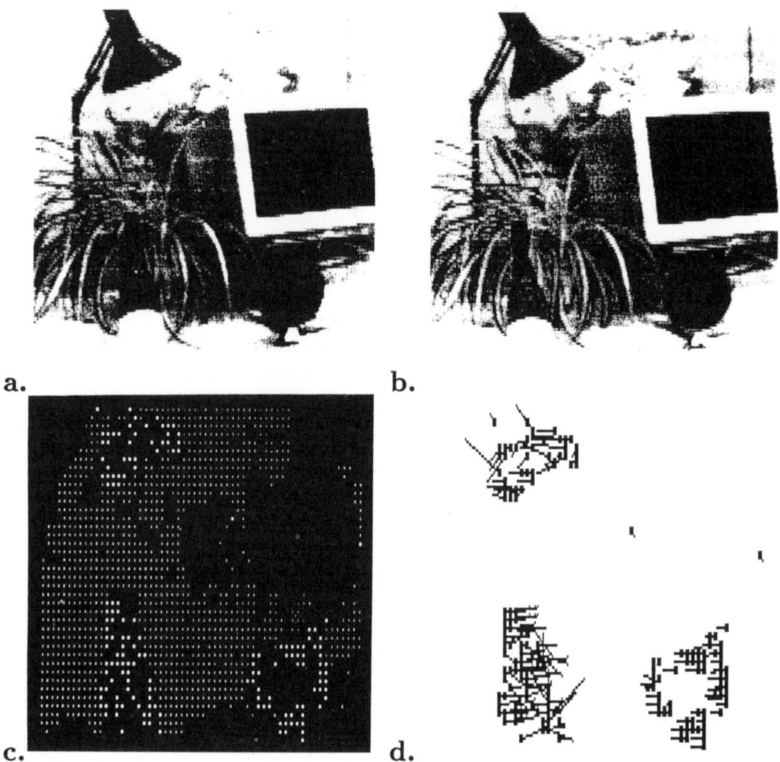

Figure 2: Performance for a natural scene. **a.**, **b.** Motion sequence: the lamp is rotated upwards while the bottle and the bomb are shifted towards each other. **c.** Intensity-coded length of motion displacement. The image is segmented into regions of coherent motion. **d.** Needle-plot.

References

[1] H. H. Bülthoff, J. J. Little, and T. Poggio. A parallel algorithm for real–time computation of optical flow. *Nature*, 337:549, 1989.

[2] E. C. Hildreth and C. Koch. The analysis of visual motion: From computational theory to neuronal mechanisms. *Ann. Rev. Neurosci.*, 10:477 –533, 1987.

[3] H. A. Mallot, H. H. Bülthoff, and J. J. Little. Neural architecture for optical flow computation. Artif. Intell. Lab. Memo. 1067, Massachusetts Institute of Technology, 1989.

[4] W. Reichardt and R. W. Schlögl. A two dimensional field theory for motion computation. First order approximation; translatory motion of rigid patterns. *Biol. Cybern.*, 60:23 – 35, 1988.

[5] A. Verri and T. Poggio. Motion field and optical flow: Quantitative properties. *IEEE Trans. Pattern Analysis and Machine Intell.*, 11:490 – 498, 1989.

Parallel Processing in Neural Systems and Computers
R. Eckmiller, G. Hartmann and G. Hauske (Editors)
© Elsevier Science Publishers B.V. (North-Holland), 1990

The development of "symbolic behaviour" in natural and artificial neural networks[*]

Gordon D.A. Brown & Mike Oaksford

Department of Psychology
University College of North Wales
Bangor, Gwynedd, LL57 2DG
United Kingdom

Earn/BITNET: PSSØØ1@uk.ac.bangor.vaxa

Introduction

The ability of connectionist architectures to produce *symbolic behaviour* is currently controversial (e.g. Chater & Oaksford, 1989; Clark, 1987; Fodor & Pylyshyn, 1988; Pinker & Prince, 1988). Questions therefore arise concerning the criteria against which to evaluate the ability of connectionist models to provide an empirically realistic account of the ontogenesis of symbolic behaviour. In this paper we evaluate the performance of a connectionist model in an experimental paradigm developed in the behaviour analytic tradition, and use the modelling process to enhance our understanding of what it means for a system to exhibit symbolic behaviour.

Behavior analysts have claimed to be investigating "symbolic behaviour" for many years, albeit in a more limited sense than is generally employed within cognitive approaches. This work resulted in a set of widely adopted criteria for symbolic behaviour, based on the ability to treat two perceptually dissimilar stimuli as "equivalent". It was suggested that the presence of equivalence may be assessed by three experimental tests (e.g. Sidman & Tailby, 1982). In a "matching to sample" task a subject responds first to a stimulus **A**, and then learns that in the context of **A** a subsequent response to **B** (rather than to **C** or **D**, say) will elicit a reward. A typical **A** stimulus might be a colour, while a typical **B** stimulus might be a shape. The training of the **AB** relation is the first step in the assessment of equivalence. Tests are then conducted to see if the subject *spontaneously* produces certain generalisations: (We omit discussion of the "reflexivity" test.)

Transitivity: **Train AB and BC; test for AC.**   Example: Subjects are first trained to associate **RED**-MIDDLE (a red colour patch in the middle of a display) to **Y**-OUTSIDE (a Y-shape on the outside of the display). They are then trained to associate **Y**-MIDDLE to **GREEN**-OUTSIDE.   Finally, subjects are presented with **RED**-MIDDLE, and given the opportunity to choose **GREEN**-OUTSIDE. If subjects spontaneously produce this transitive generalisation, they have passed the "transitivity test".

Symmetry: **Train AB; test for BA.** Example: Subjects are first trained to associate **RED**-MIDDLE to **Y**-OUTSIDE. They are then presented with **Y**-MIDDLE, and given the opportunity to choose **RED**-OUTSIDE. If subjects spontaneously choose **RED**-OUTSIDE over other alternatives, they have passed the "symmetry test".

These tests have been interpreted as distinguishing behaviour which relies on simple association from that requiring the ability to allow stimuli to stand in for one another in a putatively

---

[*] This work was supported by a grant from the Medical Research Council (U.K.), reference no. G8809938N, to the first-named author, and by a grant from the Economic and Social Research Council (U.K.), reference no. R000231282, to the second-named author. We are grateful to many of our colleagues at UCNW for our introduction to the relevant learning literature.

"symbolic" and context-independent manner. This is because, regardless of position, the stimuli can be interpreted as standing in for each other if the behaviour emerges spontaneously. The ability to pass equivalence tests emerges contemporaneously with the ability to name items and stimulus relations (e.g. Dugdale & Lowe, 1989). Animals and pre-linguistic children are unable to pass equivalence tests.

Brown & Oaksford (1989) showed that a connectionist architecture can pass the above symmetry and transitivity tests. However, that model did not show how this ability could emerge ontogenetically. The model used a *code-generation* architecture (e.g. Ackley et al., 1985). Unsupervised code generation takes place via an *input* learning how to reproduce itself as *output* using symmetrical input and output units. If there are N input units, and each input vector involves just one input unit being turned on, $Log_2 N$ hidden units will allow a binary encoding over the smaller number of hidden units. With eight input and eight output units, and an appropriate learning rate parameter, just three hidden units may develop a simple binary encoding of the inputs.

To now account for the developmental data, we varied the number of hidden units available to investigate the relation between the number of hidden units, the number of associations the network must remember, and the development of "symbolic" behaviour on equivalence tasks. Thus, in the model, development = hidden unit "pruning". Three factors justify this identification. First, the massive *cell-death* observed at certain developmental stages (e.g. Hamburger, 1949) may relate to the pruning of multi-layer networks to achieve economical storage and efficient generalisation. Secondly, the principled reduction of the number of hidden units, eliminating those that contribute least to the performance of the model (then relearning) can lead to fast learning and efficient encoding (e.g. Mozer & Smolensky, 1989). Finally, requiring a code generation net with a fixed number of hidden units to encode an ever-increasing number of items will force it to use as few hidden units as possible to represent as much of the variance as it can. This is equivalent to giving it fewer hidden units to encode the same number of associations. In the simulations we only examine the effect of varying the number of hidden units directly, but we assume that the developmental process may also be driven by the ever-increasing storage demands on a given network.

The model

The simulation has two distinct phases. First, the code-generation architecture is used to generate a hidden-unit vector for each stimulus input in an equivalence task. Second, the hidden unit vectors are associated with one another by associative learning. The learned associations encode the experimentally-presented contingencies.

(i) Formation of internal representations

Each possible stimulus was represented as a six-bit vector. Two bits encode *position* in the array where the stimulus appears ("MIDDLE" or "OUTSIDE"); one of the remaining four bits is turned on for each of the possible stimuli (**RED** or **GREEN**; **Y** or **|**). With two positions, and four objects, there are eight vectors to encode. To introduce extra variation into the input, and to avoid making the task too "easy", a seventh input unit's activation level was varied throughout each epoch. Note that *all* the stimulus/position combinations are made available to the code generation process; this assumes that final code generation will not take place until presentation of the test stimuli in a real experiment. Alternatively, more information about invariant parts of the stimulus array could have been made available during the training phase. In our current work we are exploring these more realistic alternatives.

As would be expected, three hidden units suffice to encode the eight "core" input vectors and a perspicuous binary encoding resulted in which each hidden unit took on an activation close to 0 or 1: an RMS error of 0.05 was achieved in 2,716 epochs of training with a learning rate of 0.05 and a momentum term of 0.9 (standard backpropagation was used throughout). The codes for the four stimulus items were spread over two of the three hidden units such that the activation levels over those units were the same for a given stimulus in whichever position it appeared. The remaining hidden unit was always "ON" for "MIDDLE" positions and "OFF" for "OUTSIDE" positions. In other words, *"position-independent" internal representations had been formed for each of the stimulus items.*

Five hidden units, in contrast, could encode 32 input vectors, so such a network has considerable surplus capacity. The same learning parameters as in the three-hidden-unit version were used. The network was not in this case forced to choose common representations for the related stimuli, because the network could solve the task without so doing. The units' activations approached the extreme value of 0 and 1 more closely in the three-unit case than in the five-unit case: in the former case the average smallest deviation from an extreme value was 0.05, while in the latter case it was 0.14. So when the hidden units are not overloaded, they will not be forced to choose maximally efficient binary encodings to achieve a given level of encoding accuracy relative to the case where they are forced to encode more economically.

(ii) Association of internal representations

We now consider how the internally-generated codes may be used in the performance of "equivalence tasks". Consider the simple **AB** training paradigm, with **A** and **B** represented by their hidden unit vectors obtained above. We simplify by assuming simultaneous presentation of **A** and **B**. Hence, these representations can associate themselves with bidirectional associative learning. To effect the learning, we assume that **B** becomes co-active with **A** when choice of **B** is reinforced. Given suitable internal codes, as in the three hidden unit case, it is obvious that simple associative learning gives the basic equivalence results. The net will exhibit **symmetry** if it has a context-independent internal code for **A** that is bidirectionally associated with an internal code for **B**, and consequently one can readily evoke the other. Testing for symmetry in the network involves training the hidden unit vectors produced in the initial training phase for the association **RED**-MIDDLE -> **Y**-OUTSIDE, and for the association **GREEN**-MIDDLE -> | -OUTSIDE, and then in a test phase presenting the hidden unit vector that represents **RED**-MIDDLE. If an equivalence relation between **RED** and **Y** has been formed during initial training, then the network should choose the **Y** stimulus when faced with a choice between **Y**-OUTSIDE and | -OUTSIDE. Initial training used the standard delta-rule with a learning rate of 0.05 and a linear transformation on the output units. To simulate the formation of *bidirectional* associations between hidden unit vectors, the association **Y**-OUTSIDE -> **RED**-MIDDLE is taught as well as the association **RED**-MIDDLE -> **Y**-OUTSIDE and so on.

In the three-unit case an error criterion of 0.05 was met after 349 passes through the training set. The novel input corresponding to **Y**-MIDDLE in the above example was then presented. The resulting output vector predictably matched most closely the **RED** rather than the **GREEN** choice of the two OUTSIDE stimuli, with a total mean absolute output error of 1.16 in the former case, and 1.96 in the latter. This will always be the case when the stimuli to be associated are the binary context-independent hidden unit vectors generated in the three-hidden-unit code-generation net. So the network can pass the symmetry test. When the same associations are made but using the codes generated in the code-generation network with **five** hidden units, the equivalent error scores are 1.15 for the symmetrical response, and 2.20 for the alternative response (**GREEN**-OUTSIDE in the present example). It is obvious that the network will pass this test more than the 50% that would be expected by chance alone just to the extent that it has formed context-independent representations.

The three-hidden-unit net shows **transitivity** because it has a context-independent internal code for **A** that is bidirectionally associated with an internal code for **B** which is in turn bidirectionally associated with an internal code for **C**, so **C** will clearly be a preferred response when **A** is presented. The associations were learned in 23 epochs of training, and resulted in an error of 1.05 for the "correct", transitive response, and an error of 1.74 for the "incorrect" response. When the same associations were tested for the five-unit case, a mean error score of 1.67 was obtained for both possible responses. Thus this network could not be guaranteed to pass the transitivity test, since the representations were not context-independent enough to survive the two-stage association.

Both networks were able to pass the "symmetry" test, but only the net that had been forced to develop economical, context-independent internal representations was able to pass the more demanding transitivity test. A larger network, faced with noisier input and more hidden units, would be unlikely to pass the symmetry test reliably. The simulation results accord well with the developmental data, which show that children go through a developmental stage at which they are able to pass on symmetry but not on transitivity (Beasty, 1987).

Conclusions

We have shown that a simple neural network architecture can account for the observed empirical data and provide a computationally explicit account of how developmentally increasing informational demands on a code-generation network can lead to successful equivalence-task performance. Performance on the equivalence tasks separates animals and pre-linguistic humans from language-using humans; on these tasks, the code generation architecture performs on the verbal side of the verbal / non-verbal divide. But although it is of interest to have shown how the behaviour observed by behaviour analysts can emerge developmentally in simple neural net architectures, we note that the network achieves this level of performance while remaining isolated from a high-level inferential economy of the type that would be necessary to demonstrate that the representations function equivalently in a full logical sense (Brown & Oaksford, 1989). This suggests to us that psychologists' equivalence tasks, despite having given rise to some interesting developmental observations, do not capture all that is necessary to demonstrate symbol use. Such a demonstration requires that symbols combine productively and systematically with one another to form a dynamic inferential economy.

Returning to the potential of the neural network architecture that we have discussed: we speculate that storage-driven code generation may account for a number of general developmental phenomena. Brown and Watson (1987) have argued that age-of-acquisition effects in single word reading reflect the requirement of information about word pronunciations to be stored more economically as vocabulary size increases, and it is also possible that code-generation could provide an alternative explanation of the data that motivated the controversial verb-tense-learning model (cf. Pinker & Prince, 1988).

Bibliography

Ackley, D.H., Hinton, G.E., & Sejnowski, T.J. (1985). A learning algorithm for Boltzmann machines. *Cognitive Science,* 9, 147-169.

Beasty, A. (1987). *The role of language in the development of equivalence relations: A developmental study.* Unpublished PhD thesis: University of Wales.

Brown, G.D.A., & Watson, F.L. (1987). First in, first out: Word learning age and spoken word frequency as predictors of word familiarity and word naming latency. *Memory & Cognition,* 15(3), 208-216.

Brown, G.D.A., & Oaksford, M. (1990). Symbolic behaviour and code generation: The emergence of "equivalence relations" in neural networks. In *Proceedings of the Tenth European Meeting on Cybernetics and systems Research.* World Scientific Publishing Corp. (in press).

Clark, A. (1987). Being there: Why implementation matters to cognitive science. *Artificial Intelligence Review,* 1, 231-244.

Chater, N., & Oaksford, M. (1989). Autonomy, implementation and cognitive architecture: A reply to Fodor and Pylyshyn. *Cognition,* 33 (in press).

Dugdale, N., & Lowe, C.F. (1989). Naming and stimulus equivalence. In D.E. Blackman & H. Lejeune (Eds.), *Behaviour analysis in theory and practice: Contributions and controversies.* Brighton: Lawrence Erlbaum Associates (in press).

Fodor, J.A., & Pylyshyn, Z.W. (1988). Connectionism and cognitive architecture: A critical analysis. *Cognition,* 28, 3-71.

Mozer, M.C., & Smolensky, P. (1989). Skeletonization: A technique for trimming the fat from a network via relevance assessment. In D. Touretzky (Ed.), *Advances in Neural Information Processing Systems 1.* Morgan Kaufmann, 107-115.

Pinker, S., & Prince, A. (1988). On language and connectionism: Analysis of a parallel distributed processing model of language acquisition. *Cognition,* 28, 73-193.

Sidman, M., & Tailby, W. (1982). Conditional discrimination vs. matching to sample: An expansion of the testing paradigm. *Journal of the Experimental Analysis of Behaviour,* 37, 5-22.

Parallel Processing in Neural Systems and Computers
R. Eckmiller, G. Hartmann and G. Hauske (Editors)
© Elsevier Science Publishers B.V. (North-Holland), 1990   479

APPARENT MOTION AND OTHER MYSTERIES

Jerome A. FELDMAN

International Computer Science Institute and UC Berkeley

The exquisite spatial capabilities of the primate visual system are even more remarkable when the adverse temporal conditions are taken into account. Changes of ecological importance can be significantly faster than the integration times of photoreceptors or the signaling ability of optic nerve fibers. The remarkable dynamic range ($\sim 10^{10}$) of the system is achieved by fairly rapid adaptation and a variety of large and small eye movements must also be taken into account. This paper is an attempt to outline a connectionist model of temporal change in the understanding of visual scenes. It is a direct continuation of an effort begun about a decade ago, the early results of which were presented and criticized in [Feldman 1985], and is an abbreviated version of [Feldman 1988].

The normal functioning of the visual system is so robust that it is often necessary to use abnormal conditions to study it. One major source of constraints for our work comes from studies of apparent motion. Example 1 will help illustrate some of the issues of concern:

Example 1:

If a subject is first shown the two small circles labeled 1 and, after a delay of 50-500 milliseconds, the two circles labeled 2, a strong perception arises that two dots simultaneously moved at an appropriate speed. The distance between the dots can be several degrees, far beyond the range of receptive fields of neurons in the retina or primary visual cortex. The continuous spatio-temporal change (slip) which is typically the basic motion cue is totally absent. Also, it is interesting that the motion path (trajectory) must be retroactively determined by the second flash. Most importantly, the display of Example 1 is ambiguous – the dots can be seen as moving clockwise or counter-clockwise. A remarkable fact is that, under these and most other conditions, the system always chooses consistent interpretations for all the moving dots [Ramachandran 1985].

When the dots in Example 1 are far apart, there are several influences on which of the competing motions is perceived. Slight biases in the spatial and temporal differences are effective, as are matching the shapes, colors or contrasts of two dots. A subject can almost always choose which motion to see. Adding even one (17 m.sec) frame of slip will bias the choice as will priming from previous examples, neighboring displays, etc. Adding a static, low-contrast (blur) path between two dots is compelling [Shepard & Zare 1982]. All of these effects are representative of general properties of normal motion and change processing.

We have been trying for some time to model these and related phenomena in a way that makes biological and computational sense. Example 1 is obviously highly artificial, but such ambiguities arise in any situation with similar-looking moving objects. Our starting assumption is that the visual system evolved to construct plausible real world scenarios that make sense out of the ongoing spatio-temporal flux in the context of the animal's current internal state. Our starting point is the Four-Frames model of [Feldman 1985].

The basic idea was that vision is carried out by a collection of interacting networks grouped into four distinct representational frames of reference (cf. Figure 1). The units in all these networks should be thought of as abstract neurons, computing activity levels and communicating by simple codes. The representation of information in the first frame is intended to model the view of the world that changes with each eye movement. The second frame must deal with the phenomena surrounding what has been called the illusion of a stable visual world. The model assumes that the first (retinotopic) frame computes proximal stimulus features and the second captures distal (constancy, intrinsic) features in addition to being stable. The latter is therefore called the Stable Feature Frame. An important aspect of the model is the assumption that phenomenal perception as well as recognition is based on constancy, not retinotopic features.

The fourth, or environmental frame, is intended to model an animal's representation of the space around it at a given moment. It captures the information that enables one to locate quickly the source of a stimulus from sound, wind, smell, or verbal cue, as well as maintaining the relative location of visual phenomena not currently in view. For a variety of reasons, the model proposes a single allocentric environmental frame that gets mapped, by situation links, to the current situation and the observer's place in it.

The third representational frame is the observer's general knowledge of the world, including items not dealing with either vision or space. We follow the conventional wisdom in assuming that this knowledge is captured in propositional (relational) form, modeled in this case by a kind of semantic network. One class of knowledge encoded will be the visual appearance of objects encoded as a collection of relationships among primitive parts. Since the other three representations are geometrically organized, the collection of semantic knowledge is referred to as the world knowledge formulary, to emphasize its nature as a collection of conceptual relations.

The central problem of vision is linking visual-feature information to the knowledge of how objects in the world can appear. The problem of going from a set of visual features to the description of a situation is called the *indexing* problem, by analogy to looking up something in an index. Obviously enough, it is more effective to index with invariant, real-world features than with their retinal manifestations and facilitating this is the primary function of the Stable Feature Frame. Recognition of an object or situation is modeled as a mutually reinforcing coalition of active nodes in the world knowledge frame. The mutual excitation of feature and model networks also involves top-down, context, links from visual elements to the feature units that are appropriate (cf. Figure 1).

When temporal change is included, things get much more complex. One reason that it has been difficult to isolate the temporal properties of the visual system is that the internal time scale of the computing elements is of the same order as the events they are trying to compute. Nature has been constrained to elements with millisecond operation times and every aspect of the system exhibits signs of this constraint. The rise time of photo-receptor potentials, the transmission times of axons and the firing rate of neurons all lie in the range of a few to a few tens of milliseconds. Pulfrich illusions caused by reducing the light to one eye show that even the notion of relative time of visual events can be easily confounded. An additional conceptual problem arises because much of the information in the system is transmitted by a temporal code (spike rate) which requires more time than the events it is describing. It does not seem likely that we will be able to treat temporal change as just another visual property like color or spatial scale. The fundamental complexity of temporal issues in form vision extends through all conceptual and anatomical levels. A large body of evidence suggests that there are two parallel pathways extending through many anatomical levels and characterized, at least in part, by different temporal characteristics [Maunsell 1988].

Another observation is that temporal change in visual information has many possible real-world causes and is detected by several mechanisms of the visual system. All of these interact in complex ways and it is difficult to isolate one mechanism experimentally or theoretically. Figure 2 presents the basic organization. The inputs (top) represent various ways that the visual system obtains information pertaining to temporal change. The lower half depicts the kinds of real-world events to which the system can attribute the changes. These include a perception of self-motion, the movement of articulated objects within a scene and non-rigid shape changes. We are also able to detect other sources of image change such as changes in illumination. One important point is that there is no simple relation between the kind of real-world change and its manifestations. The situation is exactly analogous to the problems in spatial vision where the brightness at a point is a joint function of illumination, distance, albedo and orientation. The visual system apparently solves these inverse problems by best-fit in a parameter space embodied in neural networks. Since the solution is normally rapid and robust, the networks must embody much of the solution in their structure.

The main inputs to the change processing network come from three sources: pursuit eye movements, local (short-range) slip detection, and the matching (correspondence) of features displaced in space and time. In addition, estimates of depth (from stereopsis, etc.) and top-down contextual expectations play a central role in the interpretation of visual change. Much of our effort is concerned with what is known about the inputs and outputs of Figure 2 and their interactions. But, as in [Feldman 1985], constancy feature calculations form just the base of our concerns; the critical step remains the indexing into world knowledge. The indexing process that links the feature frame with world knowledge retains its central role. A specific motion primitive, the trajectory, is hypothesized as the key link between features and objects.

One should think of the visual system (and the change subsystem) as constantly trying to compute the most plausible explanation of the spatio-temporal flux and proprioceptive feedback it is confronting. We see apparent motion when that is the best fit to a visual happening. We see it differently depending on fixation, suggestion and a variety of other factors. The basic operation underlying apparent motion is the matching of two objects that appear in successive frames. There is a large literature in this area and enough data to provide a serious challenge for any model; in fact, none has been proposed. But a central aspect of our story is that the match mechanism revealed by apparent motion experiments is a fundamental component of normal change perception and must be accounted for. I will outline a proposed model of the match mechanism and its relation to other change cues. The design and implementation of a closely related model has been carried out by Tom Olson as part of his dissertation [Olson 1989].

One basic question about matching concerns whether local or more global features are used to match two objects. An intuitive solution can be derived from considering the apparent motion sequence:

Example: 2

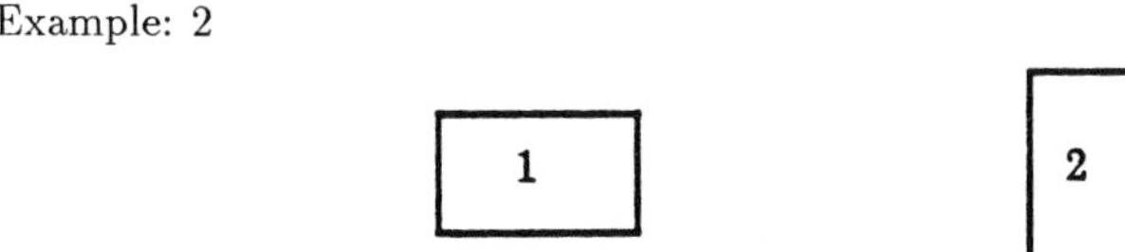

All of the local features are the same for both objects, but we always see movement from 1) to 2) as including a rotation – the overall shape is critical in matching. Experimental confirmation for this intuition can be found, e.g., [Ramachandran 1985] who found good apparent motion between squares defined by texture differences in images with no shared texture elements. A wide variety of experimental results fit the same general pattern [Feldman 1988]. This also seems to be a good way to look at the vector analysis proposed by Johanssen [1973] to account for our ability to recognize people from just of the motion of lights at their joints. The overall body motion is recognized and other motions are seen as relative to the main one. Notice, by the way, that the need for relative motion calculations preclude any pure signal processing model of motion analysis.

The first step in building any model is to specify its inputs and outputs. In this case, the output representation is critical because it is the link between visual change detection and the recognition of occurrences in the world. Recall that the entire change perception mechanism is part of the Stable Feature Frame and that its output will be used as part of the indexing process that accesses world knowledge (cf. Figure 1).

The output of the perceived change network will be in terms of trajectories. A trajectory, more precisely a trajectory segment, will be a pair consisting of a circular arc (with a straight line as a special case) and a constant speed.

$$t = (a,s).$$

This particular representation was chosen because it is simple and tractable and because a wide range of data suggests that primates deal with motion in a way that is consistent with this form. Notice that constant speed along a curved trajectory is not constant velocity. Our representation is such that any change in speed or any deviation from the current path requires a new trajectory segment. These second order changes turn out to be important indexing cues. There are also undoubtedly higher order motion primitives such as rotation and expansion patterns used in indexing, but these are not considered here.

The output space of trajectories must be explicitly represented, as always in connectionist models. We will suppress efficiency questions for now and assume (counter-factually) that there are enough units to represent every trajectory of interest. For concreteness, let us suppose that each circular trajectory segment is represented by its center point, radius, start point, arc-distance and speed. In two-dimensional space this would require six parameters if we assume the start point is coded as a position on the circle, or five if it is represented directly. Straight line trajectories could be represented similarly as a line, start point, distance and speed. Of course, one really needs three-dimensional trajectories and the parameterization gets out of hand quickly. This is a standard problem in connectionist models and Olson [Olson 1989] has some specific suggestions for efficient encodings for this domain. Any such encoding will entail an inability to make certain simultaneous judgments and thus model (more or less successfully) various experimental findings.

The important points here are that constant-speed trajectory segments are the output of change perception and that they are represented explicitly. This allows us to employ the standard connectionist device of having the various trajectory-units compete to explain the different pieces of evidence presented as input to the match subnetwork. The computational idea is the standard one of competing trajectory units, each receiving activation from a variety of sources. Obviously enough, priming, blur paths and slip cues could be directly wired to all trajectory units that are consistent with that input source. An obstacle could inhibit paths that go through it. A connectionist rendition of the match process is more complex and more interesting.

First, consider the case where all of the image elements are identical dots, a case that has been extensively explored experimentally [Kolers 1972]. The first and most serious issue is that any match process would seem to call for a buffer of the information in the first (or nth) frame while the next frame was being processed. This is not, of course, restricted to apparent motion – any feature matching implies relatively low-level memory and there is no physiological evidence for multiple buffer-like storage in the visual system. Olson [1989] has developed an ingenious scheme for matching dot images, based on the quantitative decay of activation at image points and comparator circuits sensitive to time and distance separations. Some such mechanism will be needed and no local the slip detection will suffice – matching can take place with separation of many degrees and hundreds of milliseconds.

But matching is not restricted to just dots and our story must be elaborated. Consider the earlier example of the sequence of horizontal and vertical rectangles. With appropriate timing, this is always seen as the continuous rotation of a rectangle while moving along a curved path. The current model treats this, and similar phenomena, as a separate but related set of parameter fitting relaxations. Our assumption is that the objects 1 and 2 are represented in the Stable Feature Frame as a vector of properties, as in the original model. If objects 1 and 2 are matched (by a complex relaxation), the discrepancy in orientation activates the unit representing say, a clockwise rotation of 90. This unit is consistent with an upward circular arc and sends activation to it so that we see the combined motion. According to the literature [Kolers 1972], some discrepancies (size, orientation, affine change) yield continuous perceptions where others (color, topology) do not.

We are now in a position to consider change in more complex situations like a trotting horse or a crowd scene. Recognition of complex motion is presumed to be heavily dependent on having available models to integrate the individual trajectories. Subjects who easily perceive moving light display presentations of moving people will totally fail if the displays are inverted. The current design has parallel indexing through both a form and a motion hierarchy based on shape features and on trajectories. The match system being modeled here can only compute trajectories for a few points of interest at a time. It does need the ability to compute trajectories relative to frames that are themselves moving and mechanisms for this are being worked out by Nigel Goddard as part of his forthcoming thesis [Goddard 1990]. Goddard is using mechanisms like those described here to recognize Johanssen [1973] style moving light displays. This is one test of the expanded model. As Figure 1 suggests, the change processing network of Figure 2 has important interactions with all of the four frames of the earlier model. A more detailed discussion of how temporal change affects the Four-Frames model can be found in [Feldman 1988].

ACKNOWLEDGEMENTS

This work is part of a continuing collaborative effort involving Nigel Goddard, Brian Madden and Thomas Olson. It has been supported in part by the Office of Naval Research under grants N00014–89–J–1323 and N00014–84–K–0655.

REFERENCES

Feldman, J.A., (1985). Four frames suffice: a provisional model of vision and space, Behavioral and Brain Sciences 8, pp. 265-289.

Feldman, J.A., (1988). Time, Space and Form in Vision, TR–88–011, International Computer Science Institute, Berkeley, California.

Goddard, W., (1990). Recognition of moving light displays, Ph.D. dissertation, Computer Science Department, University of Rochester, forthcoming.

Johanssen, G., (1973). Visual perception of biological motion and a model for its analysis, Perception and Psychophysics 14, pp. 201-211.

Kolers, P.A., (1972). Aspects of Motion Perception, Pergamon Press.

Maunsell, J.H.R., (1987). Physiological evidence for two visual subsystems, in: L.M. Vaina (ed.), Matters of Intelligence (D. Reidel Publishing Company), pp. 59-87.

Olson, T.J., (1989). A two-stage model of motion understanding, TR305 and Ph.D. dissertation, Computer Science Department, University of Rochester.

Ramachandran, V.S., (1985). Apparent motion of subjective surfaces, Perception 14, pp. 127-134.

Shepard, R.N. & Zare, S.L., (1982). Path-guided apparent motion, Science 220, pp. 632-634.

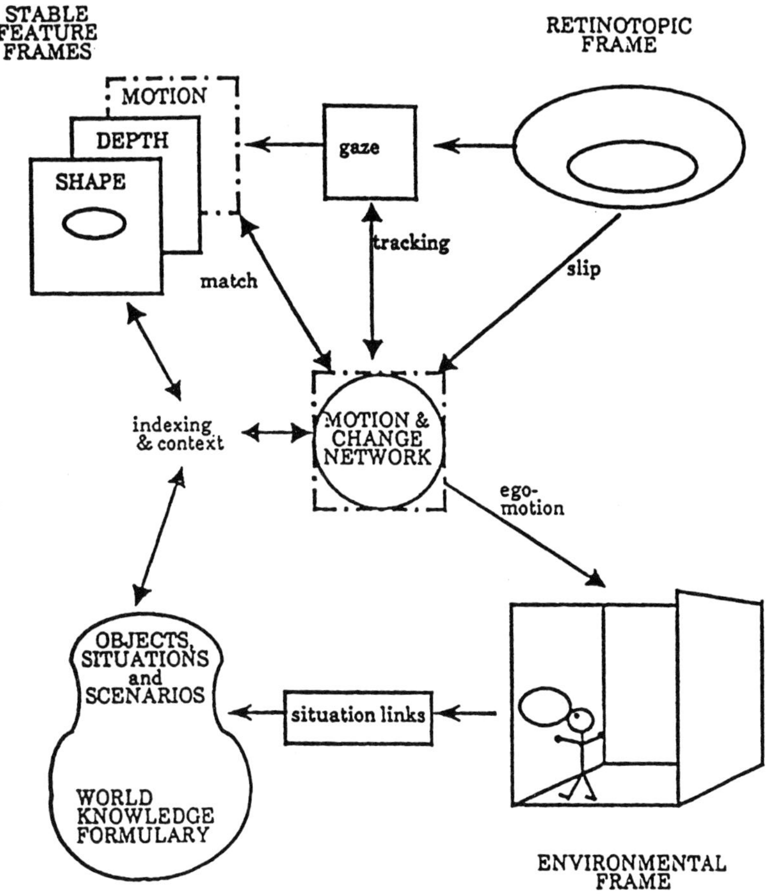

Figure 1: Four Frames and Temporal Change

J.A. Feldman

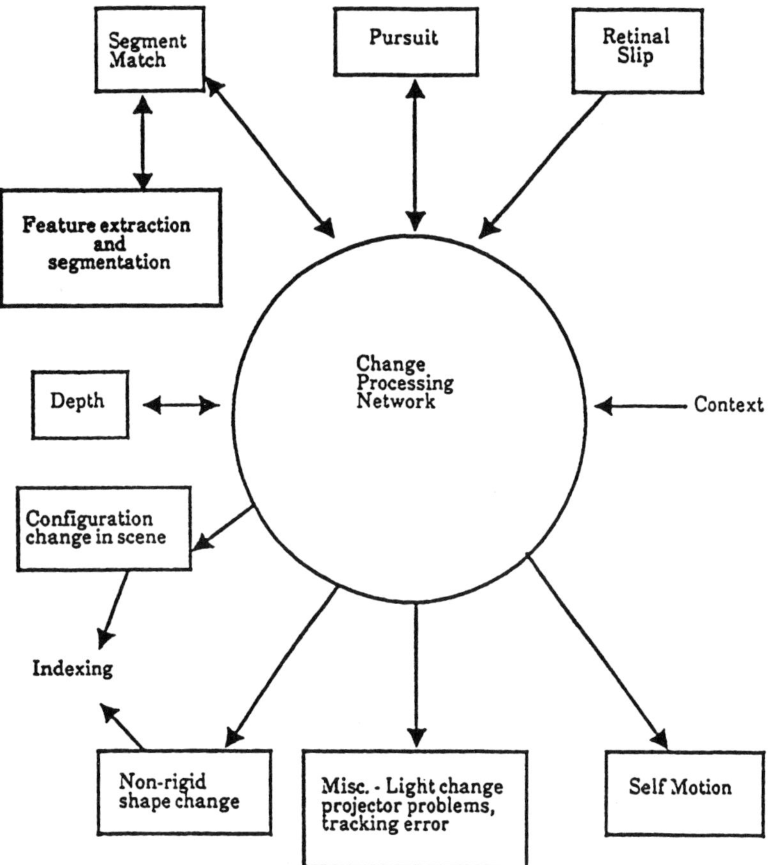

Figure 2

Parallel Processing in Neural Systems and Computers
R. Eckmiller, G. Hartmann and G. Hauske (Editors)
© Elsevier Science Publishers B.V. (North-Holland), 1990

LEARNING TO PRODUCE DISCRIMINATIVE OBJECT DE-
SCRIPTIONS: ON THE REPRESENTATION OF RULES IN A
PDP-NET

Roland MANGOLD-ALLWINN

Department of Psychology
University of Mannheim
D 6800 Mannheim, West Germany

A major issue of PDP-research is to study the
representation of structures and rules in
connectionist models. For the production of
context-adequate object descriptions it can be
demonstrated, that rule-based associations are
learned faster than random combinations, and
that there are ways to identify the rules in
the final set of weights.

1. RULE-LEARNING IN CONNECTIONIST MODELS

The perspective of *associationism* on human information
processing is that stimuli (= input patterns) as they
are presented to the system elicit responses (= output
patterns) that have been connected with them previously. For example, in reading aloud a written word
triggers a speech execution program, that results in
the acoustic encoding of the word. In principle, associations between input and output patterns can be at
random, or they can follow certain rules. For example,
forming the past tense of verbs is governed by rules
for regular verbs, but it must be learned by rote for
irregular ones.

In *symbolic models* rules are coded explicitly, mostly
in the form of if-then productions. This allows an easy
interpretation of the system's knowledge and its behavior. For example, an explanation module can be implemented easily in rule-based expert systems by simply
referring to the productions, that are currently executed. However, to build such a system the rules governing the relations between input and output patterns
must be known beforehand.

In a *PDP-model*, on the other hand, statistical correlations between input and output patterns are extracted
automatically, and associations which do not follow any

rules can be learned as well. As learning rules is an
effective way for reducing the diversity of pattern
combinations (and makes generative behavior possible),
one would assume, that a PDP-net can be trained faster
with rule-based as opposed to random combinations of
input-output patterns. Another indicator for an effective learning of rules is when a net can produce correct output patterns as responses to input patterns it
never has seen before [1].

2. INTERPRETATION OF RULES CODED IN A CONNECTIONIST MODEL

But even if improved learning performance or generalizing behavior suggests that rule-like relationships
have been discovered within a set of patterns, extracting those rules is by no means an easy task. (This is
the reason, why it is so difficult to establish an explanation module for neural expert systems.) How it is
possible to make the weight matrix of a connectionist
model interpretable is demonstrated by McMillan and
Smolensky [2]. According to their results the weights
in McClelland and Rumelhart's [1] pattern associator
for learning to form the past tense of English verbs,
can be conceptualized as a combination of 11 "simpler"
weight matrices, each of which is specialized for one
single class of verbs. This result is linguistically
meaningful insofar, as Bybee and Slobin [3] differentiate 11 ways of forming the past tense of verbs (3 irregular and 8 regular ones).

3. A PDP-NETWORK FOR THE PRODUCTION OF CONTEXT-DISCRIMINATIVE OBJECT DESCRIPTIONS

With the connectionist model described in this paper we
tried to simulate the *context-adequate description of
objects*, which is conceived of as a rule-based speech
production process. If a target object - a small white
circle - must be referred to in the context of two
similar objects in a way, that a listener unambiguously
can identify it, attributes (or attribute combinations)
of the target can be specified, which are not attributes of the context objects. (If the context objects can be small or large, white or black, a circle
or a square, context-adequate object descriptions consist of various combinations of size, color, and shape
specifications.) However, in the case that more than
one discriminative target description is possible,
those with the least number of attributes will be preferred. And among descriptions with the same number of
attributes a preference order is determined by visual
attribute salience (color > size > shape; cf. Mangold &
Pobel [4]). Following these rules a total set of 49
pairs of context object constellations and correct target descriptions was constructed.

A PDP-network with one layer of six hidden units was
trained with these 49 input-output-patterns. Six units
were used to code the attributes of both context objects; for each object the activation of the size unit
was -1 for small and +1 for large, of the color unit -1
for white and +1 for black, of the shape unit -1 for
circle and +1 for square. There were three output
units, one for each of the three possible attributes of
the target object. An activation of greater .50 of the
output units meant, that the corresponding attribute
was part of the target description. Below that threshold the attribute was counted as not being specified.

4. LEARNING PERFORMANCE FOR RULE-BASED AND RANDOM PAT-
TERN ASSOCIATIONS

The network was trained with *backpropagation of error*
with these 49 rule-based combinations of input-output
patterns. A maximum of 47 correct reproductions could
be obtained. In order to analyze performance as related
to the kind of description to be produced this procedure was performed 100 times, and each time after 50
training epochs the number of hits was recorded. For
all patterns the mean number of correct descriptions in
100 trials was 79.70. As can be seen from figure 1
descriptions with one attribute could learned better
than descriptions with two. Within one-attribute
descriptions the ranking of learning performance was
analogous to attribute salience. No principle of oder-
ing could be discovered for two-attribute descriptions.

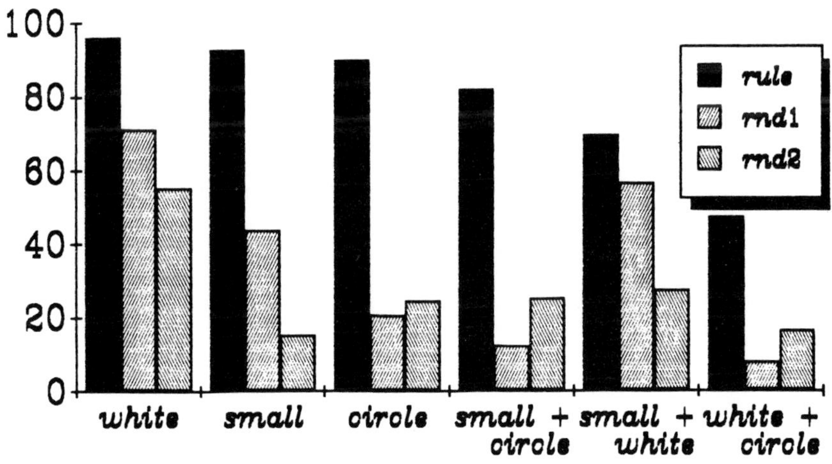

FIGURE 1
Learning performance for rule-based and random object
descriptions

Two more sets of the same input and output patterns
were constructed, but this time the associations were
at random. The network was trained again under the same
conditions as reported before; this time for 100 trials
performance as averaged accross all patterns was as low
as 35.54 and 27.62, respectively. It is evident from
this, that rule-based object descriptions can be learned more efficiently than random combinations. Although
in the two random-assignment cases there was an advantage of the single attribute description "white", unlike the rule-based case no systematic correlations
between outputs to be produced and learning performance
could be detected.

ACKNOWLEDGEMENTS

This research was made possible by grant MA 1121/1-1
from the Deutsche Forschungsgemeinschaft to the author.

REFERENCES

[1] Rumelhart, D.E. and McClelland, J.L., On learning
    the past tenses of English verbs, in: McClelland,
    J.L. and Rumelhart, D.E., (eds.), Parallel distri-
    buted processing. Vol. 2: Psychological and bio-
    logical models (MIT Press, Cambridge, 1986) pp.
    216-271.
[2] McMillan, C. and Smolensky, P., Analyzing a
    connectionist model as a system of soft rules
    (Technical Report CU-CS-393-88, Department of Com-
    puter Science, University of Colorado at Boulder,
    1988).
[3] Bybee, J. and Slobin, D., Rules and schemata in the
    development and use of the English past tense.
    Language 58 (1982) 265-289.
[4] Mangold, R. and Pobel, P., Informativeness and
    instrumentality in referential communication.
    Journal of Language and Social Psychology 7 (1989)
    181-191.

Parallel Processing in Neural Systems and Computers
R. Eckmiller, G. Hartmann and G. Hauske (Editors)
© Elsevier Science Publishers B.V. (North-Holland), 1990

TOWARD A COMPUTATIONAL ARCHITECTURE FOR MONOCULAR PREATTENTIVE SEGMENTATION

Heiko NEUMANN, H.Siegfried STIEHL

Universität Hamburg, Fachbereich Informatik
Bodenstedtstr.16, D-2000 Hamburg 50

We report on ongoing research toward a computational architecture for preattentive segmentation of static monocular images of 3-D scenes. The components of the underlying feed-back network – which is based upon basic research by Grossberg and Mingolla – will be briefly described and exemplary results from the experimental validation of the network capabilities will be presented.

1. INTRODUCTION AND MOTIVATION

Psychological results of vision research gave rise to the hypothesis that in mammalians two principal visual subsystems exist: *i)* early preattentive parallel processing (e.g. cortical detection/coding of visual primitives and perceptual organization), and *ii)* cognitive serial processing (e.g. attention-driven visual search and model-based recognition).

Only recently [1] and [2], based upon results from experimental psychology and neurophysiology, reported on evidence for the existence of different specific visual pathways which' outputs are hypothesized to be fed into a complex pooling mechanism (e.g. monocular shading/occluding boundary pooling). Such findings, along with the proof of existence of e.g. orientation sensitive cortical cells for edge/bar/slit detection as well as end-stopped cells, spurred on the design of computational architectures for mammalian visual information processing. By now at least three competing academic schools – working on biologically influenced computational vision – have been established which draw upon either *i)* Marr's paradigm, focusing on e.g. token-oriented early vision, local ('reduction tube like') shape-from-x processes, and model-driven recognition ([3]), *ii)* the Gestaltist's paradigm, focusing on e.g. perceptual organization, token-based monocular 3-space inference, and verification of image structure driven object hypothesis via 3-D model projections ([4]), or *iii)* Grossberg's paradigm, favoring a neural network architecture along with a holistic form-and-color-in-depth representation ([5]).

However, even in the case of early computational vision a number of basic questions have not been fully answered up to now, e.g. to name a few problems open to further discussion: *i)* generalizable definition of spatiotemporal structure in discrete images, *ii)* generally applicable processes, e.g. receptive fields of different specificity and subsequent cell interaction in visual pathways and subsystems, *iii)* general mechanisms for cooperative pooling of the variety of image domain evidence, and *iv)* general (non-linear) feed-back network structures (e.g. number of layers, complexity of artificial cortical cells, layer interaction types, subsymbolic/symbolic transition level, dynamic behavior, etc.) and network processes (hierarchical feed-forward/feed-back potential propagation, pooling of results from different visual pathways, etc.).

2. COMPUTATIONAL ARCHITECTURE AND SOME RESULTS

In our research we are concerned with the design and validation of a computational architecture for monocular preattentive segmentation (related to Cavanagh's luminance pathway or, respectively, Livingstone's parvo-interblob channel) of visual primitives (static edges, bars, slits, both repetitive lines and dots as well as line segment junctions and isophote fields) which is invariant against the particular scale of the local image structure. The feed-back network, being the core of the computational architecture, is an extension of the Grossberg/Mingolla scheme ([6], [7]) (which again has its roots in Grossberg's theory for preattentive visual computation ([5])). An initial filtering cascade (incorporating scaled directionally sensitive filters of different specificity), parallelizable competitive/cooperative processes for detection of local features, and a distributed potential field representation of visual image structure are the main features of the computational architecture. The current domain of application are discrete noisy, monocular and static retinotopic intensity arrays e.g. CCD-camera images of 3-D laboratory scenes.

The *first* component of the current computational architecture (designed for experiments on contrast detection) is a hierarchical, viz feed-forward, filtering cascade in which *i)* scaled isotropic center-surround operators are convolved with the intensity array and subsequently *ii)* scaled directionally-sensitive anisotropic operators are convolved with the operator output from *i)*. Operators are here meant to be weighted computational receptive fields the operationalization of which is a convolution. The impulse response functions are the Laplacian and first order partial derivatives of a scaled support-limited Gaussian regularization filter, respectively. Output of the cascade is a potential field $J_{xy\varepsilon}$ coding local contrast evidence at loci (x, y) with respect to a discrete orientation ε (with discrete equiangular spacing; $\Delta\varepsilon = \pi/8$). Since a positional as well as a directional uncertainty inheres in $J_{xy\varepsilon}$, the potential field is fed forward as net input into a feed-back network processing to allow for, e.g., evidence accumulation at true contrast loci as well as for pooling of spatially aligned and compatible potentials to close gaps. Hence in the *second* component an intra-orientational positional competition between potentials $J_{xy\varepsilon}$ in a local circular neighborhood $\mathcal{N}$ takes place. The competition scheme basically draws upon mutual inhibition in an on-center off-surround interaction, where the weighting function over $\mathcal{N}$ is inversely proportional to the distance from a local reference cell. The competition generates a potential $W_{xy\varepsilon}$ by using a specific form of a membrane equation (see e.g. [6])

$$\dot{W}_{xy\varepsilon} = -A \cdot W_{xy\varepsilon} + (B - W_{xy\varepsilon}) \cdot \text{net}^+ - (C + W_{xy\varepsilon}) \cdot \text{net}^-$$

The particular forms of the equations used for describing the different competition processes can be found in [8] and [9]. Starting with the $W_{xy\varepsilon}$ potential field, the *third* component carries out an inter-orientational directional inhibition which is defined as a pooling of potentials utilizing on/off-cell dipoles for the purpose of inhibiting potentials of nearly orthogonal local orientations. This processing step thus reduces directional uncertainty of the initial operator output at loci of e.g. corners and line terminations. The resulting potential $X_{xy\varepsilon}$ is again input to a *fourth* component which realizes an inter-orientational/inter-positional competition of potentials for all directions within the local neighborhood $\mathcal{N}$ from above. The ultimate goal of the competition within this component is the computation of a normalized potential $Y_{xy\varepsilon}$ which is input to the *fifth* component of the computational architecture. In this last component, a directional cooperation of spatially aligned potentials $Y_{xy\varepsilon}$ at distant loci takes place to excite a Z potential if distant loci of evidenced local contrast are associated with compatible orientations. For this long-range cooperation task, a spatially circular weighting function has been defined as a set of orientationally sensitive fan-shaped receptive fields with equiangular spacing (viz, a rosette). The computation of the final cooperation potentials $Z_{xy\varepsilon}$ is based upon both sufficient excitation based upon compatible Y potentials within single fans and pooling of positive excitations, $S_{xy\varepsilon}$, from at least two different fans of the entire rosette ([9]). Prior to feed-back to the *second* component, the potential field Z is additionally undergoing a positional competition process from which the final feed-back potentials V are derived.

A number of computational experiments have been carried out on both synthetic and CCD-camera images. For differently scaled edges (step and ramp), the operator responses in scale-space uniquely signal the different types of scaled intensity variation. For the bar primitive fine scaled local contrast detectors respond to each individual flank of inverse polarity, while the appropriate bar detectors maximally respond on a coarser scale. The application to synthetic standard stimuli for investigating proximity grouping capability (see e.g. [10]) resulted in the expected network behavior on the basis of scale-space processing with receptive fields of different specificity (Fig.1). The net behavior of the iterative loop was assessed by using synthetic T-junction configurations, line ends and real camera images. Fig.2 shows both J- and V-potentials for a synthetic T-junction. Our experiments on the stability of the feed-back loop demonstrate convergence after 2 iterations for data without artifacts. In order to demonstrate the capability of emergent segmentation even in the case of artifacts, synthesized operator potentials in a synthetically generated potential field for a T-junction were deleted at locations in a neighborhood of the junction point. After 3 iterations the gap was closed with filled-in activities. Fig.3 shows the processing results (viz, potential fields) for a windowed part of a CCD-camera image containing a contrast junction configuration at the top of an illuminated pyramid.

3. RÉSUMÉ

We have briefly described ongoing research towards a computational architecture for monocular preattentive segmentation of static line-like image structure related to contrast discontinuities. Exemplary results have been reported and discussed in short. Currently we are investigating basic pooling principles for different cortical filters, inclusion of approaches to discrete scale-space processing, and noise stability. Future research will focus on a multi-scale boundary/feature contour system including more sophisticated interaction of cells of different specificity, grouping mechanisms for computational perceptual organization which also take curvature (with different radii) into account (see e.g. [11]), and the role of brightness diffusion ([12], [13]) interacting with processes of the boundary contour system on different spatial scales.

REFERENCES

[1] Cavanagh, P., Pathways in Early Vision, in: Pylyshyn, Z.W. (ed.), Computational Processes in Human Vision: An Interdisciplinary Perspective (Ablex Publ. Corp., Norwood, 1988), 239-261.

[2] Livingstone, M., Art, Illusion and the Visual System, Scientific American, 1 (1988), 68-75.

[3] Marr, D., Vision (W.H.Freeman & Co, San Francisco, 1982).

[4] Lowe, D.G., Perceptual Organization and Visual Recognition (Kluwer, Boston, 1985).

[5] Grossberg, S., Cortical Dynamics of Three-Dimensional Form, Color, and Brightness Perception: I. Monocular Theory, II. Binocular Theory, Perception & Psychophysics 41 (1987), 87-158.

[6] Grossberg, S. and Mingolla, E., Neural Dynamics of Perceptual Grouping: Textures, Boundaries, and Emergent Segmentations, Perception & Psychophysics 38 (1985), 141-171.

[7] Grossberg, S. and Mingolla, E., Neural Dynamics of Surface Perception: Boundary Webs, Illuminants, and Shape-from-Shading, CVGIP 37:1 (1987), 116-165.

[8] Neumann, H., Extraction of Image Domain Primitives with a Network of Competitive/Cooperative Processes, in: Hoeppner, W. (ed.), Informatik Fachberichte 181, Proc. GWAI-88 (1988), 265-274.

[9] Neumann, H. and Stiehl, H.S., A Competitive/Cooperative (Artificial Neural) Network Approach to the Extraction of N-th Order Edge Junctions, in: Burkhardt, H., Höhne, K.H. and Neumann, B. (eds.), Informatik Fachberichte 219, Proc. 11. DAGM Symp. (1989), 256-263.

[10] Rock, I., An Introduction to Perception (Collier MacMillan, London, 1975).

[11] Parent, P. and Zucker, S.W., Trace Inference, Curvature Consistency, and Curve Detection, IEEE Trans. on PAMI 11:8 (1989), 823-839.

[12] Perona, P. and Malik, J., Scale-Space and Edge Detection Using Anisotropic Diffusion, Univ. of California, Computer Science Division, Report No. UCB/CSD 88/483 (1988).

[13] Grossberg, S. and Todorović, D., Neural Dynamics of 1D and 2D Brightness Perception: A Unified Model of Classical and Recent Phenomena, Perception & Psychophysics 43 (1988), 241-277.

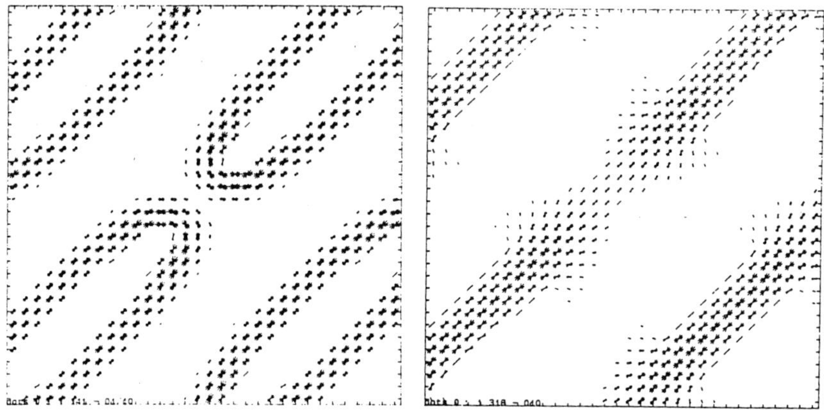

Figure 1: Dominant responses for differently scaled edge (left) and bar (right) detectors

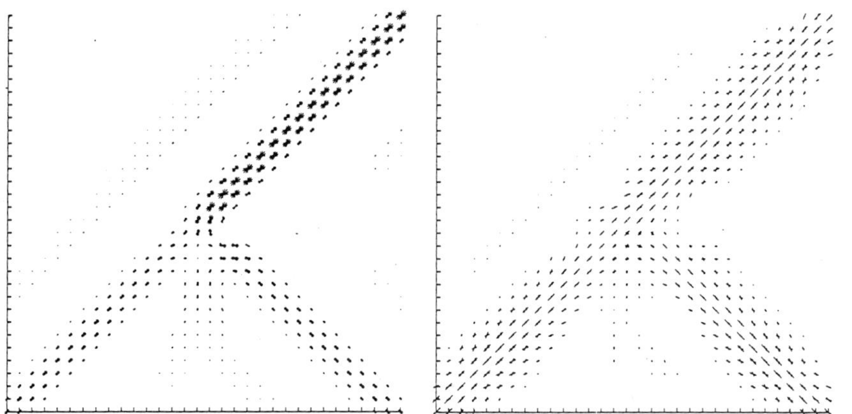

Figure 2: Potentials for a synthetic contrast T-junction (left: J-potentials, right: V-potentials)

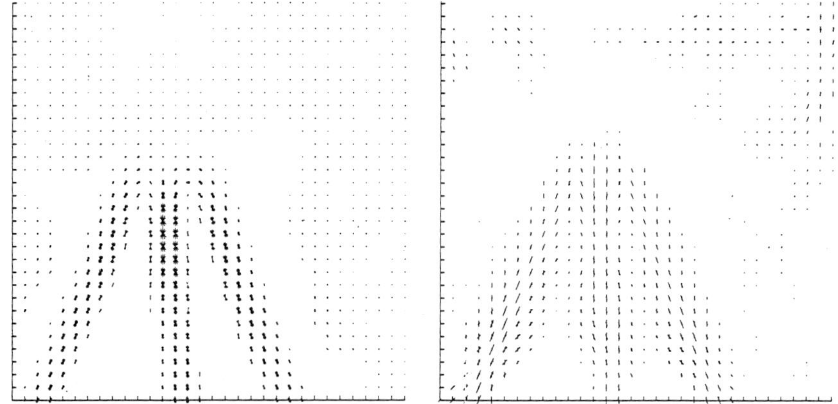

Figure 3: Processing result for a part of a CCD-camera image window (left: J-potentials, right: V-potentials)

Parallel Processing in Neural Systems and Computers
R. Eckmiller, G. Hartmann and G. Hauske (Editors)
Elsevier Science Publishers B.V. (North-Holland), 1990

MODELLING ATTENTION
IN A CONNECTIONIST SPEECH PRODUCTION MODEL[*]

Ulrich SCHADE and Hans-Jürgen EIKMEYER

Faculty of Linguistics and Literary Science
Bielefeld University
Bielefeld, Federal Republic of Germany

1. INTRODUCTION

Cognitive modelling is an important branch of AI. It collects empirical data about cognitive processes and tries to explain them by models. These models give rise to new experiments which in turn may lead to new empirical data. Natural Language Processing has always been of special interest to the AI community. Therefore, modelling speech perception and speech production is one of the prominent areas of cognitive modelling.

In the last few years several connectionist speech production models (e.g. Berg [1,2], Dell [3,4,5], MacKay [6], Schade [7,8], Stemberger [9,10]) have been proposed. Since connectionist models are parallel by definition, they can explain empirical effects conventional models (cf. e.g. Fromkin [11], Garrett [12,13]) can only explain by doubtful assumptions (cf. Stemberger [10], Berg [2]).

The cognitive processes of speech production are not directly observable. However, there is indirect evidence: These processes sometimes run into problems resulting in speech errors. They provide evidence for the processes of speech production since they are regular and confirm to rules. Therefore, a model of speech production has to offer explanations for occurrences of errors and their regularities.

2. THE SHATTUCK-HUFNAGEL EXPERIMENT

Some surprising data on speech production arose from an experiment by Shattuck-Hufnagel [14], replicated by Willshire [15]. This data had not been discussed before 1987 when Shattuck-Hufnagel herself tried to explain them (cf. Levelt [16]). The experiment concerns the *initialness effect*. Due to this effect, consonants preceding the vowel of a syllable (the onset) are involved in errors more often than those following the vowel (the coda, cf. Shattuck-Hufnagel [17]). Thus, errors as in (1) are more common than those exemplified by (2). Both

[*] Research on this subject has been supported by the German Research Foundation (DFG), Forschergruppe "Kohärenz", project "Gesprochenes Deutsch".

examples are quoted from Fromkin [18].

(1) Katz and Fodor   → fats and kodor
(2) gone to seed   → god to seen

Connectionist models allow for the initialness effect automatically (cf. Schade [7]). The Shattuck-Hufnagel experiment, however, seems to run counter to this effect: Some subjects involved in this experiment were asked to utter tongue twisters like (3a). Others had to utter the critical words of the tongue twister in list form (3b).

(3a) sentence: „From the leap of the note to the nap of the lute."
(3b) list: [leap, note, nap, lute]

The results show more onset errors than coda errors (64:19) if a sentence is uttered. This is in line with the initialness effect. However, error rates change with respect to the list form (76:96). The results were replicated and completed by Willshire [15], the rates being (184:101) for the sentence form and (151:215) for the list form.

A simulation of the experimental setting using the connectionist model described by Schade [7] led to somewhat different results. The simulation correctly predicted an increasing error rate for list form utterances. But it also predicted more onset errors than coda errors, though these rates should be nearly identical.

A deeper examination of Willshire's results offers an explanation for the difference between experiment and simulation. Willshire [15, p.22f] reports that her subjects suffered from stress throughout the experiment, since they had difficulties to memorize the items and their sequence. Thus, 22% of all errors had been word substitutions and word omissions. Stress and the rate of word errors were reduced (to 4%) in one of Willshire's control experiments [15, p.51f]. This control experiment differed from the original one in only one respect: The subjects could see a written presentation of the items to be produced during the experiment. Similar to the prediction of the simulation, the error rate of the list form utterance was (72:57) in the control experiment.

It is evident that subjects find it more difficult to repeat the correct sequence in list form. The syntactic structure which helps to solve the task in the sentence form setting is missing. Therefore, the subjects had to concentrate on the sequence in the list form setting. Thus, the initial parts of the words had to be given special attention as the "tip of the tongue" experiments (Brown & McNeill [19]) and the work of Meyer [20] demonstrate. Focussing the word initial parts prevents some onset errors and, consequently, the error rate for the coda position can exceed the onset position's rate.

3. TOWARDS MODELLING ATTENTION

In order to simulate the original Shattuck-Hufnagel experiment it is thus necessary to find a connectionist model for attention. We developed the following topology for a connectionist

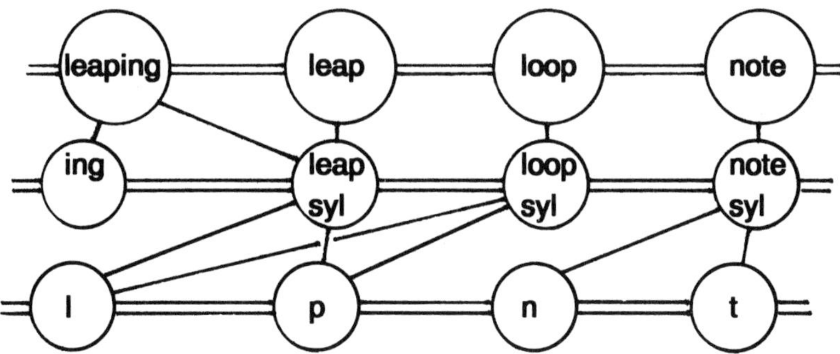

FIGURE 1

speech production model which is also designed to handle attention. Linguistic units like words, syllables, and segments of syllables are represented by one node each. Excitatory connections between these nodes result from the syntagmatic relations between units. A syllable is connected to its segments, and a word to its syllables. Inhibitory connections result from the paradigmatic relations of the units, describing possible choices at certain structural positions. There is lateral inhibition among all segments of syllables, among all syllables, and among all words (cf. figure 1, where circles represent nodes, normal lines excitatory connections, and double lines inhibitory connections).

In its simplest form, the varying degree of attention is modelled by modifying the strength of lateral inhibition. Given that the position of a certain unit is getting special attention (e.g. the onset of a word), the lateral inhibition among the possible choices for this position is strengthened. Simulations using this strategy increase the contrast in activation values for the possible choices. The target node (i.e. the node which receives highest activation on the normal course of events) will normally win the race since it is activated by higher level nodes. Due to the increase in contrast, random disturbations are less effective and the error rate goes down. The results of the Shattuck-Hufnagel experiment have been simulated in this manner.

4. CONCLUSION

Exerting control on the degree of attention according to the described simulation method is presently performed manually. However, it should be regarded as a first step towards a general connectionist model of attention. In order to deal with the whole problem of attention (cf. e.g. Neumann [21]) one has to define and simulate the control of attention as well. This is a challenging task to be undertaken in the future.

5. REFERENCES

[1]   Berg, T., The Problems of Language Control: Editing, Monitoring, and Feedback. *Phychological Research* (1986) 48, pp. 133–144.

[2]   Berg, T., *Die Abbildung des Sprachproduktionsprozesses in einem Aktivierungsflußmo-dell* (Tübingen: Niemeyer, 1988).

[3]   Dell, G.S., Positive Feedback in Hierarchical Connectionist Models: Applications to Language Production. *Cognitive Science* (1985) 9, pp. 3–23.

[4]   Dell, G.S., A Spreading-Activation Theory of Retrieval in Sentence Production. *Psychological Review* (1986) 93, pp. 283–321.

[5]   Dell, G.S., The Retrieval of Phonological Forms in Production: Tests of Prediction from a Connectionist Model. *Journal of Memory and Language* (1988) 27, pp. 124–142.

[6]   MacKay, D.G., *The Organisation of Perception and Action: A Theory for Language and Other Cognitive Skills* (New York, NY: Springer, 1987).

[7]   Schade, U., ‚Fischers Fritz fischt fische Fische' – Konnektionistische Modelle der Satzproduktion. *Kolibri 6* (Bielefeld University, 1987).

[8]   Schade, U., A Note on K.Bock's "Syntactic Adjustment Effect" Problem, in: Retti, J. and Leidlmair, K., (eds.), *5. Österreichische Artificial-Intelligence-Tagung* (Berlin, Heidelberg: Springer, 1989) pp. 218–223.

[9]   Stemberger, J.P., Lexical Access and Serial Order in Sentence Production. *Proceedings of the 5th Conference of the Cognitive Science Society* (Rochester, NY, 1983).

[10]   Stemberger, J.P., An Interactive Activation Model of Language Production, in: Ellis, A.W. (ed.), *Progress in the Psychology of Language: Vol. 1* (London: Erlbaum, 1985) pp. 143–168.

[11]   Fromkin, V.A., The Non-Anomalous Nature of Anomalous Utterances. *Language* (1971) 47, pp. 27–52.

[12]   Garrett, M.F., The Analysis of Sentence Production, in: Bower, G. (ed.), *Pychology of Learning and Motivation* (New York, NY: Academic Press, 1975) 9, pp. 133–177.

[13]   Garrett, M.F., Levels of Processing in Speech Production, in: Butterworth, B. (ed.), *Language Production: Vol. 1* (London: Academic Press, 1980) pp. 177–220.

[14]   Shattuck-Hufnagel, S., Position of Errors in Tongue Twisters and Spontaneous Speech: Evidence for two Processing Mechanisms? *MIT Research Laboratory of Electronics, Speech Comunication, Working Papers* (1982) 1, pp. 1–8.

[15]   Willshire, C., *Speech Error Distributions in two Kinds of Tongue Twisters*, unpublished bachelor thesis (Monash University, Clayton, Victoria, 1985).

[16]   Levelt, W.J.M., *Speaking: From Intention to Articulation* (Cambridge: MIT, 1989).

[17]   Shattuck-Hufnagel, S., The Role of the Word-onset Consonants in Speech Planing: New Evidence from Speech Error Patterns, in: Keller, E. and Gopnik, M. (eds.), *Motor and Sensory Processes of Language* (Hillsdale, NJ: Erlbaum, 1987) pp. 17–51.

[18]   Fromkin, V.A., Appendix, in: Fromkin, V.A. (ed.), *Speech Errors as Linguistic Evidence* (Paris: Mouton, 1973) pp. 241–269.

[19]   Brown, R. & McNeill, D., The "Tip of the Tongue" Phenomenon. *Journal of Verbal Learning and Verbal Behaviour* (1966) 5, pp. 325–337.

[20]   Meyer, A.S., *Phonological Encoding in Language Production: A Priming Study*, unpublished doctoral dissertation (Nijmegen University, 1988).

[21]   Neumann, O., Beyond Capacity: A Functional View of Attention, in: Heuer, H. and Sanders, A.F. (eds.), *Perspectives on Perception and Action* (Hillsdale, NJ: Erlbaum, 1987) pp. 361–394.

Parallel Processing in Neural Systems and Computers
R. Eckmiller, G. Hartmann and G. Hauske (Editors)
Elsevier Science Publishers B.V. (North-Holland), 1990

LEARNING OF CONTROL KNOWLEDGE FOR SYMBOLIC PROOFS WITH BACKPROPAGATION NETWORKS

A.Ultsch, R.Hannuschka, U.Hartmann, V.Weber

Institute of Informatics
University of Dortmund
D-4600 Dortmund, Germany

1. INTRODUCTION

This paper presents the application of a connectionist network to optimize symbolic proofs. By symbolic proofs we understand proofs in first order logic. Prolog interpreters are an implementation of theorem provers for special first order formulas, called horn clauses, based on the resolution principle [Robinson 65]. The usage of Prolog interpreters for symbolic proofs implies a certain proof strategy: first, selection of the leftmost partial goal in the resolvent as goal to be solved and second, selection of the clauses in written order for the resolution of the selected partial goal. In case of failure of a partial goal, the interpreter backtracks systematically to the last choice made without analyzing the cause of failure. Even for simple programs, this implicit control strategy is not sufficient to obtain efficient computations [Naish 84]. Explicit control by using control constructs like cut is, however, contradictory to Prolog as a declarative programming language. The need for the distinction of logic and control has therefore long been recognized [Kowalski 75].

We describe an approach to learn and store control knowledge in a connectionist network. The system utilizes a three-layer backpropagation network. A meta-interpreter generates training patterns encoding successful Prolog proofs. Trained with these examples of proofs the network generalizes a control strategy to select clauses. Another meta-interpreter uses the network to compute optimized proofs.

2. TRAINING THE SYSTEM

Fig.1a shows the generation of the training data. A generating meta-interpreter (GMI) containing the Prolog program is asked to prove a goal.

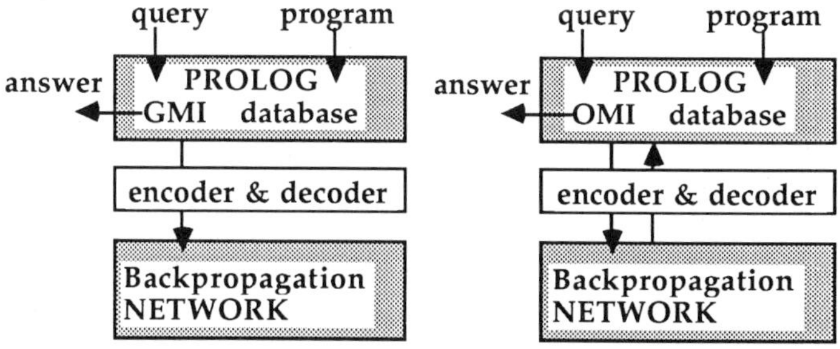

Fig. 1a: generation of training data   b: using the network in proofs

The GMI constructs the optimal proof for the given goal, i.e. the proof with the minimal number of resolutions. The optimal proof is found by generating all possible proofs and comparing them with reference to the number of resolutions. For an optimal proof each clause-selection-situation is recorded. A clause-selection-situation is described by the features of the partial goal to be proved and the clause which is selected to solve that particular goal. The clause is described by a unique identification and two different sorts of information concerning the structure of arguments are used: the types of arguments and their possible identity.

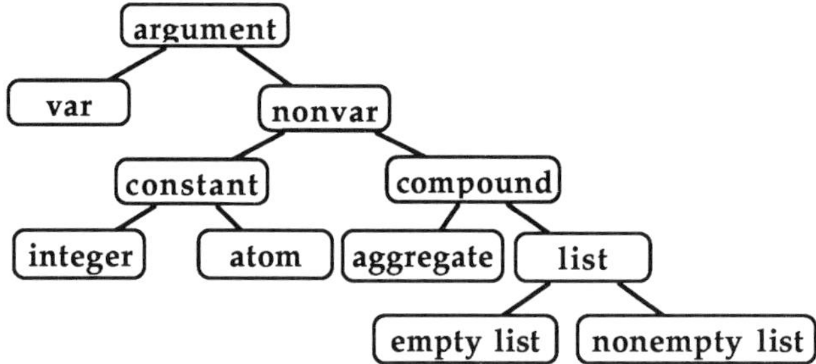

Fig. 2: type tree of Prolog types

For the types of arguments a hierarchical ordering of the possible argument types, as shown in Fig.2, is used. In the tree a parent-node represents a generalization of its son nodes. For example an argument may be a 'var' (variable) or a 'nonvar' (instantiated argument). The encoder (see Fig. 1a) takes the clause-selection-situation and produces a binary training pattern for the network. The encoding preserves similarities among the types, e.g. the code for 'integer' is more similar to the code for 'atom' than to the code for 'list'. The used network is a backpropagation network with an input layer, one hidden layer and an output layer. The activation function is the standard sigmoid function [Rumelhart/McClelland 86]. The number of hidden units is equal to the number of input units. The network is trained with the encoded training patterns until it is able to reproduce the choice of a clause for a partial goal. The usage of the trained net in a proof is shown in Fig. 1b. A query is passed to the optimizing meta-interpreter (OMI). For each partial goal the OMI presents the description of the partial goal as input to the net and obtains a candidate-clause for resolution. With this candidate the resolution is attempted. If resolution fails, the OMI uses Prolog search strategy as default.

3. FIRST RESULTS

To test our approach we use a Prolog program consisting of a predicate with 64 different clauses with 4 arguments. To produce training data three different classes of queries are used: class A, requiring two identical arguments, class B, requiring three identical arguments, and class C, requiring all arguments to be identical. The training set consists only of queries having atoms as argument type (see Fig.2) in order to test for generalization (see below).

The left part of Fig. 3a shows the mean number of unifications for Prolog to prove the goals. The network immediately finds the right clause for examples having been trained before, as illustrated in the right part of Fig. 3a. In a second experiment, goals are used with different types of arguments. This experiment tests the ability to generalize to proofs of similar structure. The results are given in Fig. 3b. While Prolog needs an average number of unifications between 33 and 64, our system demands a range of 15 to 38 unifications. Column D in Fig. 3b gives the results of a test involving only queries with type integer as arguments. This shows in particular that the number of unifications is drastically reduced, if the type in the tested query is very similar to the type the network is trained with.

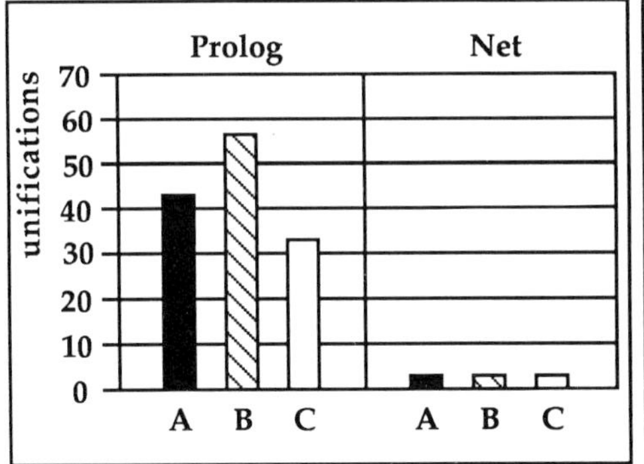 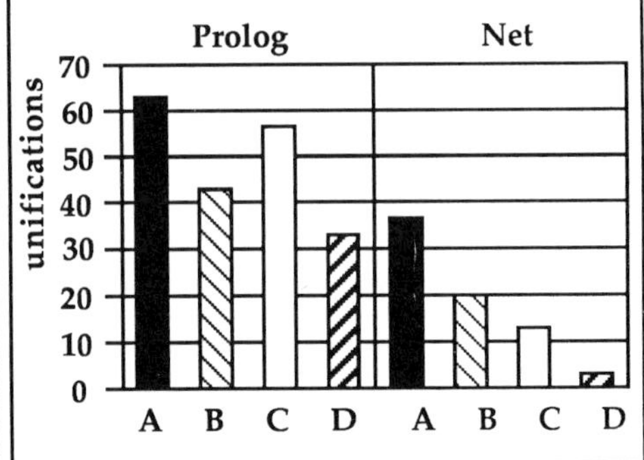

Fig. 3a: clause selection for trained queries   b: clause selection for new queries

It has to be noted that all experiments are made with Prolog strategy as default if no clause is found by the network. This means in particular that there may be an overhead of one additional resolution compared with Prolog for some queries. The results in Fig. 3a and 3b, however, show that averaged over a number of queries the network heuristic for clause selection is superior to Prolog strategy.

4. DISCUSSION

Up to now our system improves simple Prolog programs. Transferring the results to arbitrary Prolog programs should be done with great care. Nevertheless a significant speedup for proofs can be observed. It is clear that in a symbolic system storing all proofs of the past would result in the same speedup. The latter method, however, would exhaust every storage media sooner or later. In contrast to such a method a network of limited size is used to store the proof information. We argue, however, that the ability of the network to extract to structure of the proof captures the essential features without having to remember all the details. In the network a distributed representation of the learned concepts is used. The interference of different concepts makes networks capable to learn new concepts without the need for new storage. The interference is also the reason of the ability to generalize. As the first results exemplify, the learned strategy is also applicable to other proofs of the same structure. Since we are learning proof heuristics, it is clear that there will be always examples where a learned strategy is not applicable at all.

5. OTHER APPROACHES

The representation of control knowledge for Prolog programs has been proposed before. For example Kasif suggests to represent control knowledge as expression in a procedural language [Kasif et al. 83]. Others propose a declarative representation of control knowledge [Gallaire/Lasserre 82, Genesereth/Ginsberg 85, Ultsch 87]. This means conditions are formulated on which certain clauses may be selectable to prove a partial goal. All these approaches require a human expert to formulate the control knowledge. Unlike the declarative approaches mentioned before, we don't limit the description of the arguments of partial goals to the instantiation status. Our approach uses a much more detailed description of instantiated arguments. To improve efficiency the selection of clauses often depends on instantiations of the arguments to certain structures.

Suttner introduces control to the selection of axiom in theorem provers [Suttner 89]. A proof is observed and certain features of the proof are counted, e.g. the number of variables in the theorem. Disadvantageous to this approach is the loss of information about the positions of the variables. Our approach considers the positions of arguments in encoding the features of a goal.

6. CONCLUSION

In this paper a new approach to learn and memorize control knowledge for Prolog programs is presented. Our system utilizes a three layer backpropagation network. Trained with examples of successful proofs, the network learns a control strategy. Proof strategies are declaratively described by certain features like the types and the possible identity of the arguments of a clause. Similarities of these features are used to train the network in order to generalize to similar clause-selection-situations. First experimental results show the ability of the network to learn the presented description of proofs. Our system is able to reproduce information without using too much memory, because the knowledge is stored in a connectionist network of limited size. The system is able to generalize a strategy for a clause selection to new, but similar proofs. An important advantage of the system is its ability to learn proof heuristics from examples. No programmer is needed to define explicit control knowledge for the program.

Because of the encouraging results described in this paper our further research will deal with the representation of other kinds of control knowledge for Prolog programs. We will extend the system to control the selection of literals and backtrack points. An important extension of our model should be a system which observes a Prolog program and recognizes situations in which it is likely that the proof will not terminate.

ACKNOWLEDGEMENTS

We would like to thank all members of the student research group PANDA (Prolog And Neural Distributed Architectures, Institute of Informatics, University of Dortmund) for their encouragement. Special thanks to M. Mandischer for his support concerning the usage of programming tools.

REFERENCES

[Gallaire/Lasserre 82]   H. Gallaire, C. Lasserre: *Metalevel Control for Logic Programs,* in: Clark/Tärlund (Eds.): Logic Programs, Academic Press, London, 1982, pp.173-188

[Genesereth/Ginsberg 85]   M.R. Genesereth, M.L. Ginsberg: *Logic Programming* in: Communications of the ACM, Vol. 28 N° 9, 1985, pp.933-941

[Kasif et al. 83] S. Kasif, M. Kohli, J. Minker: *Prism: A Parallel Inference System for Problem Solving,* in: Proceedings of the 8th IJAI, Karlsruhe 1983, pp.544-546

[Kowalski 75]   R. Kowalski: *Algorithm = Logic + Control,* in: Communications of the ACM, Vol. 33 N° 7, July 1979, pp.424-436

[Naish 84] L. Naish: *Prolog Control Rules,* Technical Report, No.13, University of Melbourne, Australia, 1984

[Robinson 65]   J.A. Robinson: *A machine-oriented Logic based on the Resolution Principle,* in: Journal of the ACM, Vol. 12 N° 1, Jan. 1965, pp.23-44

[Rumelhart/McClelland 86] D.E. Rumelhart, J.L. McClelland: *Parallel Distributed Processing : Explorations in the Microstructure of Cognition,* MIT Press, 1986

[Suttner 89] C.B. Suttner: *A feature-based heuristic module for the SETTHEO theorem prover,* Forschungsgruppe Künstliche Intelligenz, TU München, 1989

[Ultsch 87]   A. Ultsch: *Control for Knowledge-based Information Retrieval,* Dissertation, Swiss Federal Institute of Technology, Zurich, 1987

Parallel Processing in Neural Systems and Computers
R. Eckmiller, G. Hartmann and G. Hauske (Editors)
© Elsevier Science Publishers B.V. (North-Holland), 1990

Intelligent Dimensional Data-reduction by a Topological Map

The interpretation and use of an insurance database

L. VERCAUTEREN, R.A. VINGERHOEDS, L. BOULLART

Ghent State University, Automatic Control Laboratory

Grotesteenweg Noord 2, 9710 GENT - Zwijnaarde, Belgium.

Topological Mapping can be applied succesfully to detect patterns in a database. It can also be used to make distinct classifications in this database. This paper gives an example of such industrial applications, after the treatment of some theoretical and practical aspects.

1. INTRODUCTION

The topic of this paper is the use of Topological Mapping for making classifications in databases. First, some theoretical aspects and features are treated, and some possibilities are suggested. Second, an application is showed, in which the system learns to make a classification of insurance requests, based on the database of the insurance company. The reader is assumed to be familiar with Vector Quantisation in a Topological Map as proposed by Kohonen (see ref. 2 and 3).

2. SPATIAL, SUB-SYMBOLIC AND SYMBOLIC DATA

Most of the data used by men and machine can be placed in one of the following three groups: spatial data, sub-symbolic data and symbolic data.

Spatial data is the data used in Image Processing and/or Signal Processing. The place in or sequention of the presentation of this data-structure is important. In an image eg., one pixel has normally no meaning. But it has an important meaning in its relations to its neighbours in the input space. Higher level processing is based on the detection of local features in the input. It is practical spoken impossible to use an input space of an eye or an ear without making use of the coherence of this space. This coherence is visible in the data structure and is two dimensional in a static image (purely spatial), two dimensional in the sensory input of the skin (also purely spatial), and two dimensional in the auditorial input (one dimension is frequence, and one dimension is time). The use of the coherence of the input space for data reduction is demonstrated by Fukushima (see ref. 1) with the so called Neocognitron. A clear example where the importance of position of the variable in the input is important, is with figures (9000 is not the same as 0009). The larger the data structure and the smaller the variability of each variable or figure, the greater becomes the importance of the availability of the coherence structure : 9 (decimal) or 1001 (binary). The coherence (in this example) is our convention about the meaning of a particular figure in the data structure. One is often using another coherence, without noticing it: the coherence within one figure (1 comes before 2, 2 before 3 , and so on). The data structure of one digital figure is a convention, indicating a graduality, related to the object where it stands for. An AI engineer has to be aware of this convention.

Sub-symbolic data is information without a relevant meaning on its own. The coherence giving relevance to this data is given by the simultaneously or sequential appearance of certain input values. Sub-symbolic data can also possess spatial coherence, but this is not strictly needed for its processing, or it is not known. In some cases, it is possible to give a meaning to a certain variable of the input data. This variable is then called a feature. The profit of these variables must be given by a self-organising and self-learning system. The lack of meaning is a freedom given to the system working with this data.

Symbolic data is data of which each variable has a relevant meaning on its own. Processing data of this kind is eg. done by a formule or by an expert system. The way of processing is dependent on the presentation form of the variables. Some of them are continuous, or piecewise continuous, while others are clearly discrete (with two or more possible values).

Topological Mapping ignores all kinds of spatial information, and can therefore only be applied to sub-symbolic or symbolic data. Most of the inputs do also have to be continuous (not in the analytic, but in the intelligent sense). There exist better technics to attack problems with only discrete or binary variables.

3. INTELLIGENT ASSOCIATION OF DATA

Two variables x and y can be related in a lot of different ways. In fact, there exist as many solutions as there are different functions z=f(x,y). The choice of the possible functions f and the possible dynamical behaviour during the learning stage(s), are very important. It is therefore crucial to make certain hypotheses concerning the presentation form of the data and about the possible functions f. One can stay general by postulating that each variable of the data is piecewise continuous between two strict borders (eg. 0 and 100), and that the variables have to be scaled in respect to there significance (see further). Binary data (0/1) can become 'continuous' (0/100). By doing this, some of the input information will be lost, namely the discrete appearance of the variable. This will be catched by the processing of the data during learning.

The processing function f will reduce the n dimensions of the input to one dimension, the output of the neuron. The key of association used in Topological Mapping is the fact that the simultaneous appearance of certain values of the variables has a significant meaning. The n-dimensional input vector will be compared with the n-dimensional weight-vector of each neuron. This comparison associates the n values in the weight vector. This process has an equivalent in the Boolean algebra, namely the AND-function.

After learning, each neuron will possess a weight vector. In case of a clear matching of the input with one of the neurons, there is no discussion of the winner of the Network. When there is no clear winner, an Euclidian distance calculation in the n-dimensional input space can be taken between the inputs and the weights (see ref. 3). The neuron with the smallest distance is the winner. To get a nice map, and a good interpretation of the 'distance', it is necessary to scale the input variables equally. Hence, the scaling must be done in respect to the significance of the variables in the classification problem.

One is free to choose the output of the network after learning. Because of the summation in the distance measurement, the input vector I can have less then n inputs during working, eg. n-1. The weight(s) not used for recognition can be used as output of the neuron, in case of winning the n-1 dimensional distance competition. This associative recall can be used for a linkage between the input and a wanted output, for the completion of a partial input pattern, or for a selective attention application when some of the 'input'-variables are given by another part of the system (feedback).

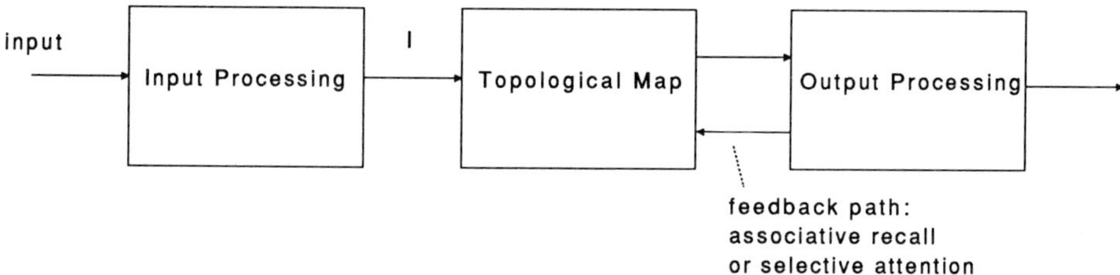

In case of a 'spatio'-temporal pattern, this feedback can consist of the output results of the network at one or more earlier time steps. In this case, the learning and convergence of the network becomes very complex. This is not the topic of this paper. Evenso, partial mapping or adaption of the influence of each input variable to the Euclidean distance during learning, can be suggested.

It is important to see that there is no equivalent for the Boolean OR-function in the Topological Map. There occurs only a straightforward classification based on the AND-association.

4. REDUCTION OF THE DIMENSION OF A DATA SPACE

Suppose one has to deal with an n-dimensional input as described above. These inputs have to be classified in an m-dimensional space (with m<=n). This transformation of an n dimensional (data) space to an m dimensional (intelligence) space is a conversion from point to point. The continuous character of the input variables must be found back as much as possible in the weights of the neighbouring neurons in the m dimensional network.

Concerning the choice of m, there exists four possibilities. First, m dimensional inputs are transformed by some piecewise point to point transformations to the n dimensional inputs. Hereby, the network transforms them back, as good as possible to an m dimensional space. Second, the programmer defines in advance an m dimensional space, well knowing what they mean, and then tries to map the n dimensional input space into this space by guiding the learning process. Third, m is choosen randomly (mostly m=2 for easier interpretation) to explore the input space for possible classifications and patterns. Fourth, m can be equal to n. After carefull learning, the weights will map the input space in such a way that the transformation of the network does not affect the generality of the input space, but only performs a simplification and a scaling in the sense of the appearance of the input variables. The resolution of this transformation is dependent on the number of neurons in the network, and the number of dimensions. Sometimes, it can be usefull to put an n to n Topological map in the preprocessing part of the system, to help and eventually speed up the convergence of the n to m Topological map by presenting a better scaled, and more simple input.

Two neighbouring neurons in the network will normally have almost similar weight vectors. However, sometimes, these vectors may differ drastically. These discontinuaties may refer to a bad mapping or to 'jumps' in the learning space. A bad mapping can be caused by a bad convergence or by an unfortunate projection of the n dimensional space on the m dimensional space. These effects can be overcome by a better choice of m or by a better guidance of the learning.

5. EXAMPLE : DIRECTED CLASSIFICATION OF AN INSURANCE DATABASE

This example is a simulation of the decision activities of a car insurance. The company receives a request for an insurance, with information on the client and on the car. An expert has to decide whether the company takes this risk or not. This decision is based on experience which is an extraction of the cases handled in the past (stored in a database:) and general company rules. In the database information of the request, the decision of the company, and in case of a positive decision a validation good/bad is stored. The validation is based on the profit of the company in this particular case.

The construction of the system has four stages : the construction of the input processing unit, the construction and learning of the Topological Map, the construction and learning of the output unit that can assign to each case a percentage good/bad, and the building of an interface to make the system applicable for non-experts. In this paper, we will not discuss the interface part.

5.1. The input processing unit

Of the data available some data has to be scaled and can be used directly in the system. Other parts have to be processed further by (for example) an expert system. Finally data may be present for administrative reasons only. In this example an 11-dimensional input space was used: client-adress, age, genus, fysical condition, civil state, profession, professional use, accident history, type car, age car, price car. Some of the input variables can be obtained by direct scaling while others need to be transformed from a binary input to an analoge input and may then scaled. The other inputs are proceessed by an expert system, based on the policy and experience of the insurance experts. The 'City' can be translated to a figure related to the 'wealth' in that city. The 'Fysical Condition' will be a figure that expresses the expected risk for the company (0 : blind, 10 : chance for hart attack, epilepsy,..., 100 : in good health). The 'Profession' translation is also based on experience. Some of them are giving less trust in the client then other one's. The same remarks about 'Accident History' and 'Type car'.

The learning examples from the database of the insurance company will have an extra special binary input, namely the success of the insurance: good means profit, bad means loss. In this example, we did not use or touch the insurance-premiums.

5.2. The Topological Map

The intelligence space is two dimensional, namely one dimension 'client', and one dimension 'car'. This statement is purely intuitive and can only be proved or verified by experiments. However, in some cases it is possible to define clearly m. The intelligence space of $(x,y,x*y)$ is two dimensional (surface), while the intelligence space of $(x,y,x+y)$ is three dimensional (volume). In case of sub-symbolic or symbolic input from a database, may not be so clear.

The learning of the Topological Map (10X10 neurons) is performed in three stages: a one dimenional orientation of 'car' and 'client', a two dimensional mapping and orientation, and a final two dimensional fine tuning of the map. The learning rate in these three stages must decrease but must also be attended carefully to come to a nice mapping.

a. One dimensional orientation

The weights of the Topological Map will be initialised with the results of this mapping. Firstly, only the inputs 9, 10 and 11 are considered.

A one dimenional map is taken whereby the examples considered as 'good' can only win in the first five neurons, while the examples considered as 'bad' can only win in the last five neurons. The learning radius however, is ignoring the good/bad border. The same process can be followed for the 'client'-dimension.

The respective weights obtained of this learning are then used as initial values for the further learning.

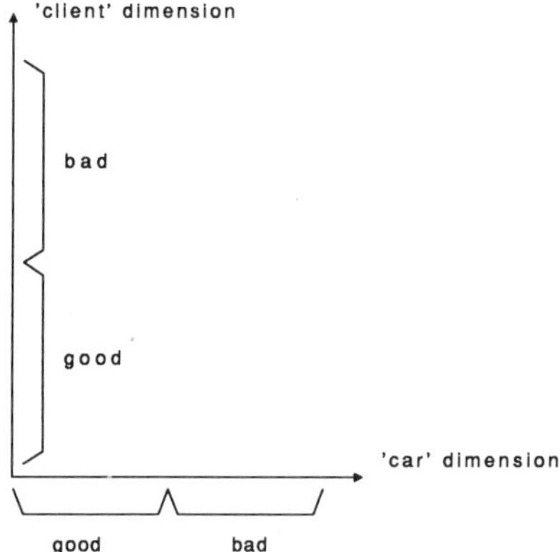

b. Two dimensional mapping and orientation

The two dimensional map is divided into two areas, namely a good and a bad area.

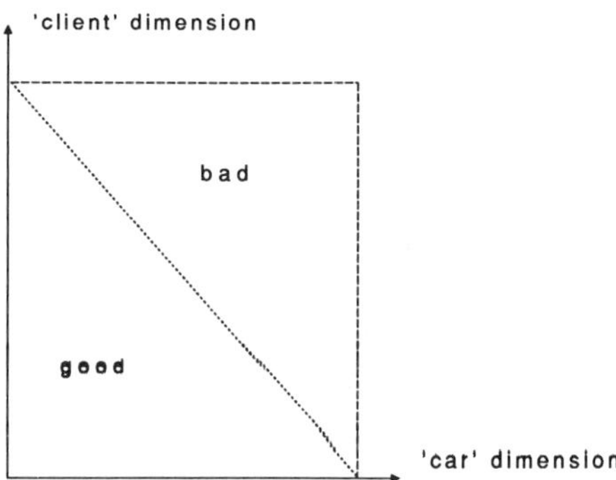

Learning inputs that are labeled 'good' can only win in the good area, while 'bad' one's are only able to win in the bad area of the network.

By this organisation, the mapping is forced to respect the good/bad distinction, while the client/car orientation is more free to develop during the learning process. Experiments showed that this initial orientation remained respected provided that the learning rates where not to big.

When the learning seemes to converge, the proces is stopped and the network is fine-tuned in the third stage.

c. Fine tuning of the network

During this stage, the Learning examples are no longer forced to win in a certain area. The learning rate and radius are very small (eg. 0.008 and 1.1). The initialised good/bad border remains, but is not that straight after learning. After this learning, the network is ready for interpretation and application.

5.3. The output processing unit

The whole set of Learning examples is send through the network. Each neuron receives two variables, indicating the number of resp. good and bad winnings during this proces. After this run, these figures are then transferred to a percentage, indicating the risk for the company when a arbitrary request is activating a certain neuron. This gives two areas in the network, one with a percentage lower then 50%, indicating good risks, and one with a percentage higher then 50% indicating bad risks.

It depends on the policy of the company using this system, to decide how to use these figures. One can for instance implement the interface in such a way that the winning neurons with a percentage between 40% and 60%, will be handled by a human expert. This can be achieved by eg. a parameter setting.

A great advantage of the system is that it runs very fast on every computer onces the weights are fixed. The learning stage, on the other hand, is consuming a lot of computer time. One can also provide a real time updating of the percentages during working. The weights however must remain fixed.

6. CONCLUSIONS

Topological Mapping is an engineering tool that is very usefull in the interpretation and use of existing databases. It is able to find intelligent correlations and to make accurate classifications of real world data. Hence, it is important to understand that these relations need to be detectable; they have to be present in the learning examples.

The industrial application of this tool must also be well-considered and adaptive. Mostly, it is embedded in a system that is also using other (eg. classification) techniques, and even Expert Systems. In doing so , Topological Mapping is becoming a very powerfull tool.

7. REFERENCES

1. Fukushima, K. (1987), 'Neural Network Model for selective attention in visual pattern recognition and associative recall', Applied Optics, Vol. 26, No. 23, pp. 4985-4992.

2. Kohonen, T. (1984), 'Self Organisation and Associative Memory', Springer Verlag, New York.

3. Kohonen, T. (1988), 'The "Neural" Phonetic Typewriter', IEEE Computer, March, pp. 11-22.

4. Vercauteren, L., et al (1989), 'Pattern Directed Real-World Interface', in: 'Connectionism in Perspective', ed. R. Pfeiffer, et al., North Holland, Amsterdam, pp. 455-461.

8. ACKNOWLEDGEMENT

This paper present research results of the Belgian National inventive program for fundamental research in Artificial Intelligence, initiated by the Belgian State - Prime Minister's Office - Science Policy Programming. The scientific responsibility is assumed by the authors.

Parallel Processing in Neural Systems and Computers
R. Eckmiller, G. Hartmann and G. Hauske (Editors)
Elsevier Science Publishers B.V. (North-Holland), 1990

PARALLEL ASSOCIATIVE PROCESSES IN INFORMATION RETRIEVAL*

Manfred WETTLER and Reinhard RAPP

Department of Psychology
University of Paderborn, FRG

We present an autoassociative net which describes the associative connections between 269 words in a scientific reference database. This net is used to predict which words professional searchers will use when generating queries on the basis of written problem descriptions. These predictions are compared with online searches conducted in natural settings.

1. THE TASK OF A DATABASE SEARCHER

Scientific bibliographic databases are collections of up to several millions references to documents. Typically they include the titles, abstracts, keywords and additional bibliographic data of these publications. They can be retrieved by queries in so-called query-languages. The use of such languages presupposes a special training. Therefore most database searches are not performed by the end-users themselves but by professional intermediaries who generate the queries on the basis of written problem descriptions given to them by the end-users.

Example 1. Problem description: Studies on the *effects* and efficiency of *therapies* with *alcoholics*, *treatment* of alcoholism, treatment results, catamnestic studies, follow up studies, *evaluation* of treatment *methods*.

Query:

```
1   503 FIND alcoholi$
2 1001 FIND (psychotherapeutic outcome$ OR treatment effectiveness)
3   418 FIND follow up stud$
4    73 FIND 1 AND (2 OR 3)
```

Example 1 shows a problem description and the corresponding query generated by a professional intermediary. The searcher has omitted several words of the problem description (results, effects, evaluation, methods etc.), introduced a new term (psychotherapy), used several truncations, and has combined the different terms by logical connectors.

The generation of such queries has been described as a cognitive process directed by the goal to satisfy the information-need of the end-user. This kind of description, however, does not allow to predict which specific strategies the searcher will use, which words of the problem description he will discard, and which additional terms he will use in the query.

In the model presented here such lexical decisions are described as the result of associative processes. These are induced when the searcher directs his attention to a problem description. The activation of these words leads to a propagation of activities to all associatively related words. The associative strength between two words depends on the probability of their joint occurrence. The words which become most active during this propagation of activities will be used in the query and the other words will be omitted.

*This research was supported by the Deutsche Forschungsgemeinschaft (project No. 524/88) and by the Heinz-Nixdorf-Foundation

2. THE MODEL AND ITS IMPLEMENTATION

The calculation of the associative strengths between the words is based on the assumption that each co-occurrence of two words raises their associative connection, and that each occurrence of one word without the other reduces it. Moreover, we assume that professional searchers have experienced a representative sample of the database. Thus, their estimates of the frequencies and co-occurrences of the words in the database should be more or less correct. With these assumptions the associative strength between two words may be calculated with the following formula:

$$w_{ij} = \frac{p(x_i = 1 \, \& \, x_j = 1)}{p(x_i = 1)\,p(x_j = 1)} - 1 \tag{1}$$

w_{ij} is the connective strength between two words x_i and x_j, and $p(x_i = 1 \, \& \, x_j = 1)$ the probability that these words will occur together in an arbitrary document of the database. This formula gives symmetrical associations with positive and negative strengths. Using this formula we calculated the associative strengths between 269 words. They have been chosen at random from all the words in the database PSYCINFO with an absolute frequency higher than 2000.

The propagation of activities proceeds in cycles according to a linear summation model. This means that the activity of word k after cycle t is the sum of the activities of all other words in the previous cycle $(t-1)$ multiplied by the respective associative strengths. Additionally, all words of the problem description receive a constant amount of extra activation during each cycle. This process is described by the following formula:

$$a_k(t) = \sum_i a_i(t-1)w_{ik} + input_k(t) \tag{2}$$

Figure 1 shows the development of activities when a subset of the words of example 1 is stimulated during 10 cycles. The reason for the selection of these words will be given in section 3.

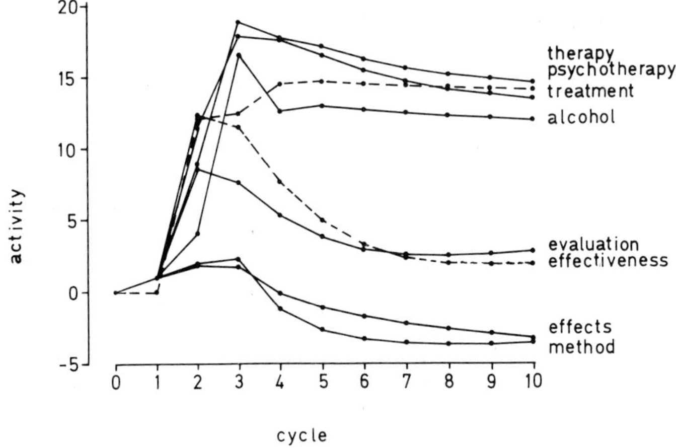

FIGURE 1

Activity curves of words from the problem description (solid lines) and new words (dotted lines) in example 1.

In this, and in a large number of other examples, two effects can be observed which can also be found in the protocols of the searchers: the activation of synonyms of stimulus-words (psychotherapy) and the deactivation of irrelevant words (e.g. results, effects, methods).

3. QUANTITATIVE COMPARISON OF THE ACTIVITIES OF WORDS AND THEIR USE IN THE QUERIES.

To test the validity of the associative model we analyzed how well the model predicts the lexical decisions of professional intermediaries conducting online searches. This analysis was based on 27 problem descriptions and the corresponding search protocols. In all protocols at least five words of our vocabulary or narrow synonymes thereof are used. In the examples presented here these words are written in italics.

For each of these 27 examples we ran 10 propagation cycles with all those words of the problem description as stimuli that are part of our vocabulary. We then ranked these words according to their activities after the 10th cycle. If the model gave a perfect prediction of the behaviour of the searchers then the most active words should be included in the queries. In the example given above the problem description includes six words which are part of our vocabulary (treatment, effects, therapy, alcohol, evaluation, methods). Out of these six words two have been used in the query (alcohol and treatment). Only one of these two words, therapy, belongs to the two most active words of the problem description (see figure 1). The associative model thus predicted correctly that the word "treatment" became part of the query, and that the words "methods", "effects", and "evaluation" were left out. It made the wrong prediction that the word "therapy" would be included in the query. The model predicted correctly that the word "psychotherapy", which is not mentioned in the problem description, was included in the query.

	predicted in query	predicted not in query
observed in query	46	36
observed not in query	36	96

TABLE 1

Predicted and observed use of the words from the problem descriptions in the queries

Table 1 shows the predictions of the model and the actual lexical decisions of the searchers for the 214 words which appear in the 27 examples and are part of our lexicon. The predictions of the model are confirmed in 67% of the cases, and they are disconfirmed in 33%. At first view, this result might not seem overwhelming. However, in evaluating these data one must take into account two facts which prevent a perfect or nearly perfect fit between the model and the observations: First, the size of our vocabulary does not permit an adequate representation of the problem descriptions and of the queries. Second, we do not know which words would be chosen by other searchers. The agreement between the predictions of the model and the lexical choices of one searcher certainly will not be higher than the agreement between different searchers.

4. QUALITATIVE ANALYSIS OF ADDITIONAL EXAMPLES

In this section we present and analyze two more examples which illustrate the performance of the associative model.

Example 2. Problem description: *Changes* in the *self-concept* of partners following client-centered partner-*therapy.*

```
1     26 FIND partner therapy
2    539 FIND CT=self concept
3      3 FIND 1 AND 2
4    581 FIND marriage$/PQ
5     22 FIND 3 OR (2 AND 4)
```

Example 2 shows that the model not only predicts the acceptance and rejection of the words of the problem description but also the introduction of a new word (Fig. 2). The activity curves indicate that the activation of this new word continues after the second cycle. Therefore this process can not be explained by the direct associations between the new words and the stimuli. The connectionist model grasps higher order depencies between words. This is not possible with the traditional statistical models in Information Retrieval.

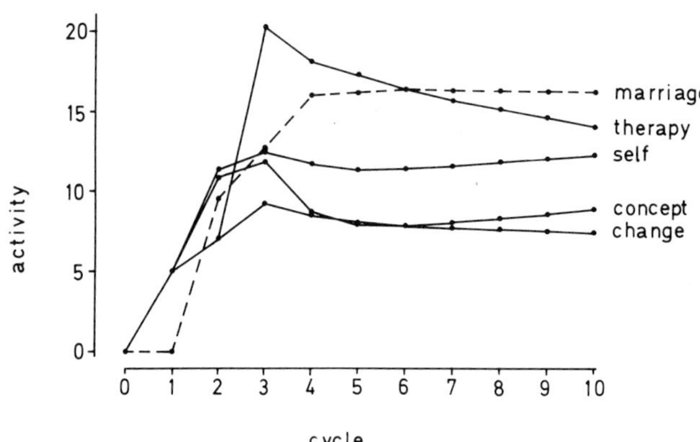

FIGURE 2

Activity curves of words from the problem description (solid lines) and new words (dotted lines) in example 2.

Example 3. Problem description: Castration *anxiety.* Its *correlation* to pre- and perinatal *development* of *male* and *female* genitalia.

```
1       51 FIND CT=castration anxiety
2      413 FIND CT D (female genitalia OR male genitalia)
3 135273 FIND (pidgeon$ OR monkey$ OR rat$ OR mouse OR mice OR rabbit$)
4      175 FIND 2 NOT 3
5       61 FIND 1 OR 2 AND PY>1982
```

In this example our vocabulary misses two central terms: castration and genitalia. Nevertheless, the activities of the vocabulary words after the propagation correspond to the lexical choices of the searcher: she dropped "correlation" and "development", which got low activities, and she introduced the word "rats" which has the highest activity after the propagation. This correspondence demonstrates that the introduction of new words, which are not part of the search problem, and which are integrated in the query by the NOT-operator, is also the result of associative processes: a large part of psychological literature has a behaviouristic orientation, where the words "male" and "female" are used to indicate attributes of experimental animals. This explains the activation of "rats". The decision of the searcher to exclude publications about rats by the NOT-operator is the result of a later interpretation of the role of the associatively evoked word.

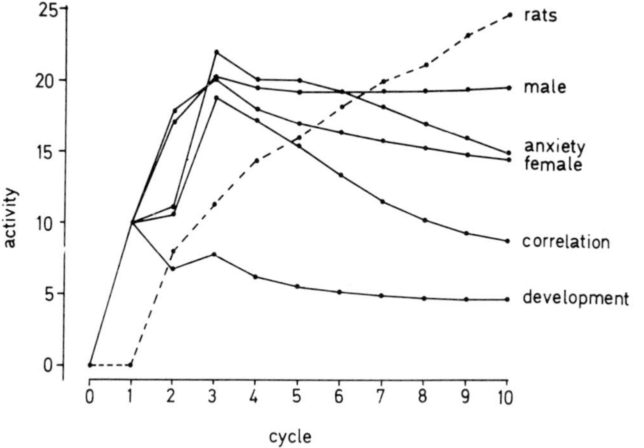

FIGURE 3

Activity curves of words from the problem description (solid lines) and new words (dotted lines) in example 3.

Parallel Processing in Neural Systems and Computers
R. Eckmiller, G. Hartmann and G. Hauske (Editors)
© Elsevier Science Publishers B.V. (North-Holland), 1990

Distributed Parallel Cooperative Problem-Solving with Voting and Election System of Neural Learning Networks

Byoung-Tak ZHANG and Gerd VEENKER

Division I – Artificial Intelligence
Institute for Computer Science
University of Bonn
D-5300 Bonn, FRG

In contrast to symbolic problem-solving systems, neural network systems have many salient properties such as parallelism, learnability, and fault-tolerance [1,2,5]. But it seems difficult, if not impossible, to construct an expert system in a single gigantic neural network [3,6]. In this paper we introduce a class of structured networks called VEN's (*Voting and Election Networks*) in which several different types of neural learning networks are organized into a voting and election system. The distributed parallel cooperative inference mechanism of the generic VEN system is discussed, an election scheme based on a network of sigma-pi units is presented, and the experimental results and observations on a two-person board game are reported. The VEN architecture appears to provide a framework for constructing knowledge-based problem-solving systems using neural networks as building blocks, while still maintaining the strengths of the neural computing.

1 The Voting and Election Model

The *voting and election network* (VEN) is a network whose components are neural learning networks that are organized into a voting and election system. The voting and election system consists of a *distributor*, one or more *voters*, and an *elector* (see Fig.1). The distributor distributes the input data among the voters. The voters cast their votes (hypotheses or constraints) for all candidates in parallel. It is the responsibility of the elector to determine the final solution that best satisfies the multiple hypotheses.

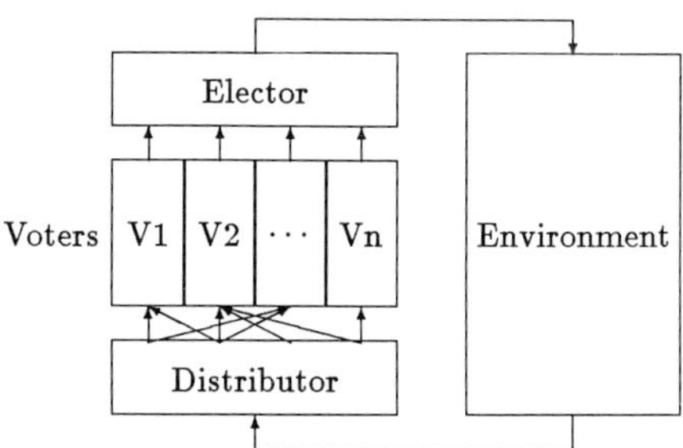

Fig.1. The organization of a VEN system with its environment

A *VEN system* is a VEN or a VEN of VEN systems. In VEN systems, the input data is partially shared among the voters, but the knowledge for solving a problem is divided into knowledge modules which are not shared among the voters in parallel voting. In general, a voter alone can not solve a problem, rather all the voters must cooperate and/or compete through the arbitration of the elector. It is noticeable that there are no constraints on the architecture and learning mechanism of the voter itself.

2   Distributed Parallel Cooperative Inference

Inferences in a voting and election system of neural networks (i.e. VEN) are performed at two different levels: at the level of neurons (*micro-inferences*) and at the level of problem states (*macro-inferences*). The problem-solving process of the system is a sequence of *conclusive* macro-inferences, each step of which corresponds to a state transition (or move) in the problem space.

At the beginning of each problem-solving step, current input data, $\mathbf{s}(t)$, describing the current situation of the problem is distributed among voting networks by the distributor $\mathcal{D}$, i.e.

$$\mathcal{D}(\mathbf{s}(t)) = \{\mathbf{x}_i\}.$$

In every conclusive inference step, a large number of *hypothetical* macro-inferences

$$\mathcal{V}_i(\mathbf{x}_i) = \mathbf{v}_i = (v_{i1}v_{i2}\ldots v_{im})^T$$

are made simultaneously in each voting network where massively parallel micro-inferences are performed. The v_{ij} with $0 \le v_{ij} \le 1$ is the vote of the voter i that represents the probability of the candidate j being elected. The hypothetical macro-inference is duplicated as many times as the number of voters.

In general, election is a cooperative process that draws the conclusive macro-inferences by satisfying maximally the hypotheses of the voters. An election scheme will be described in the following section.

Continuous sequential macro-inference (i.e. problem-solving) is carried out by chaining (or feeding back) the output to the input of the next macro-inference step—possibly by means of an interaction with the environment.

3   The Strategic Election Scheme

Formally, election is a constraint-satisfaction (or optimization) process that produces a vector $\mathbf{y}$

$$\mathbf{y} = \mathcal{E}(\mathbf{V}, \mathcal{S})$$

as a function of the votes $\mathbf{V}$ (*short-term constraints*) and the strategy $\mathcal{S}$ (*long-term constraints*), where $y_j = 1$ for the best candidate and $y_j = 0$ for the others. The function $\mathcal{E}$ and the strategy $\mathcal{S}$ can be implemented in various ways.

We present here an election network which consists of *sigma-pi* units (see Fig.2). The election involves two stages. In the first stage, the election units act as pi-units (p_j) and the votes v_{ij} for the candidate j, each v_{ij} weighted by the corresponding connection weights s_{ij}, are multiplied to a product

$$(1) \qquad y_j(0) = \prod_{i=1}^{n} s_{ij}v_{ij} = \prod_{i=1}^{n} a_i b_j v_{ij} = b_j \prod_{i=1}^{n} a_i v_{ij} = b_j p_j$$

which builds the input to the next stage. In (1) the matrix

$$\mathbf{S} = (s_{ij}) = \mathbf{a}\mathbf{b}^T = (a_i b_j)$$

represents the *privileges* that reflect the relative importance of the connections between the voters i and the candidates j. The vectors **a** and **b** will be called *privilege vectors* and the matrix **S** *privilege matrix* or *strategy matrix*. The separation of **a** and **b** vectors and the use of the intermediate units p_j make the strategy modification easier.

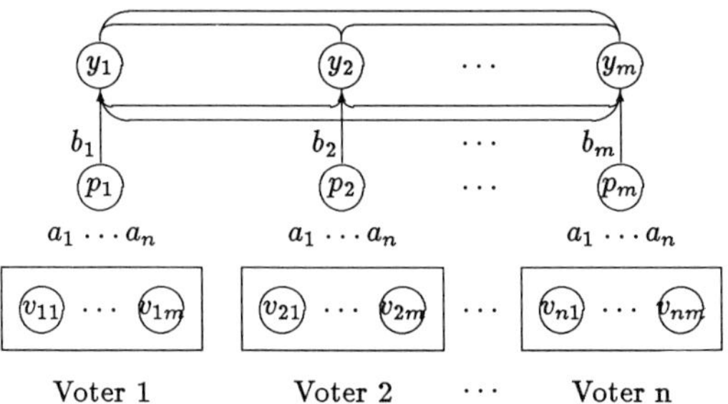

Fig.2. The election network

In the second stage, the election units act as sigma-units (y_j) and the winner-take-all [1] competition process

$$(2) \qquad y_j(t) = \epsilon \sum_{k=1}^{m} c_{jk} y_j(t-1)$$

is repeated until a candidate (or l candidates in case of l-winner-take-all) takes all the votes. It is worth noting from Equations (1) and (2) that a candidate has a very low probability of winning if a voter casts him a strictly negative vote (near 0). This is the effect of the characteristic of the pi-units.

4 An Example: Playing a Two-Person Board Game

To test the voting and election model and the strategic election scheme, we have constructed a VEN system which learns to play a two-person board game. In the game, the two players move their figures alternately on a 3 by 3 game board. The figure can move only one square horizontally or vertically at a time. A player wins the game when he conquers the home of the opponent or captures the opponent's figure. The home of the player is the position from which the player started the game. A player can capture the opponent's figure if it is in the position to which the player can move legally.

Attempting to move a step, in the mind of the player, at least the five *little minds* are cooperating and/or fighting: reaching the opponent's home, capturing the opponent's figure, not being captured, avoiding loops, and moving legally. We have implemented these little minds as the voters in the VEN system. Each voter was implemented as a three-layer (or two-connection-layer) feedforward network. The election network described above was used in election. The five voting networks were trained with examples from typical situations by a back-propagation learning algorithm [5]. A result of voting and election for board positions to move to is shown in Fig.3 (see [7] for details).

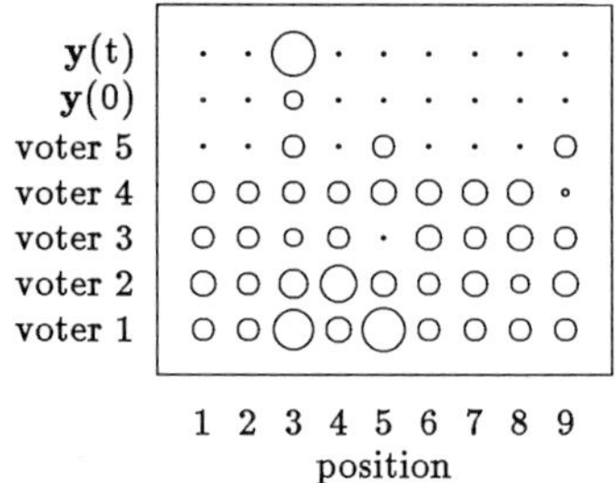

Fig.3. The voting and election for board positions.
The diameter of circles reflects the strength of hypotheses.

5   Concluding Remarks

The game-playing VEN system has predicted optimal moves for the situations, for which it is not trained explicitly as a whole. From the experiments with the system we have made the following observations. The voting and election formulation of problem-solving makes it possible to realize a synchronic distribution of the entire burden among voting networks. This renders parallel processing feasible and the macro-level distribution of knowledge possible. The knowledge distribution helps further in acquiring knowledge easily in terms of a single neural network, which also results in the enhancement of reusability of the knowledge. Besides the cooperative election, the indirect interaction and partial duplication of knowledge among voting networks through the sharing of data and training examples also contribute to the robustness of the inference. Although these observations should be proved more rigorously in different problem domains, the VEN architecture appears to provide a framework for constructing distributed parallel cooperative problem-solving systems for Artificial Intelligence. In the future, attempts will be made on the self-organization of the VEN systems using the (co-)evolutionary learning concepts, e.g. [3,4].

Acknowledgements

We would like to thank Heinz Mühlenbein, Jörg Kindermann, Knut Möller, and the members of the Arbeitsgemeinschaft Neuronale Netze for useful discussions and comments. Thanks are also extended to the POSCO Scholarship Society for the financial support.

References

[1] Feldman, J.A. and Ballard, D.H., Connectionist models and their properties, *Cognitive Science 6*, pp. 205-254 1982.

[2] Hinton, G.E., Connectionist learning procedures, *Artificial Intelligence 40*, pp. 185-234, 1989.

[3] Mühlenbein, H. and Kindermann, J., *Distributed problem solving by coevolution*, Internal Report, German National Research Center for Computer Science (GMD), 1989.

[4] Mühlenbein, H. and Kindermann, J., The dynamics of evolution and learning—Towards genetic neural networks, In: Pfeifer, R. et al.(eds.), *Connectionism in Perspective*, Elsevier, 1989.

[5] Rumelhart, D.E and McClelland, J.L. (eds.), *Parallel Distributed Processing*, MIT Press, 1986.

[6] Smolensky, P., Connectionist AI, symbolic AI, and the brain, *Artificial Intelligence Review 1*, pp. 95-109, 1987.

[7] Zhang, B.T., *Playing a two-person board game with a voting and election system of neural learning networks*, AGNN-Memo, Inst. for Comp. Sci., Div.I-AI, Univ. of Bonn, 1989.

Section 13
Optical and Molecular Computing

Parallel Processing in Neural Systems and Computers
R. Eckmiller, G. Hartmann and G. Hauske (Editors)
© Elsevier Science Publishers B.V. (North-Holland), 1990

NOVEL LOGIC AND ARCHITECTURES FOR MOLECULAR COMPUTING

John R. BARKER

Department of Electronics and Electrical Engineering
University of Glasgow
Glasgow G12 8QQ, United Kingdom

Fabrication and design criteria plus the possibility of one bit per carrier transport lead to the consideration of small neighbourhood, asynchronous cellular automata supported by molecular scale unary or binary conservative logic as good targets for molecular electronics. A preliminary experimental route is described based on attaching pre-fabricated molecules to a nano-lithographically prepared solid substrate.

1.   INTRODUCTION

Recently the first practical steps have been taken towards performing electronic operations on individual molecules whether they be alone or embedded in some large assembly [1,2]. These might eventually lead to molecular-scale electronic systems for computing or other applications. Progress has been possible due to advances in synthetic organic chemistry[1-3], the availability of nanometre scale inorganic technology[4] and the molecular scale instrumentation afforded by scanning tunnelling microscopy[5] and atomic force microscopy[6]. Whereas previous excursions into molecular computing have been largely theoretical with an over-emphasis on massively complex systems[3,7] it is now clear that the way forward must involve less ambitious goals which are forced on us by the practical problems of fabrication and verification. For example, various groups are now considering : (i) the synthesis of a molecule or molecular complex with a <u>designed</u> switching/transport function; (ii) the confirmed <u>emplacement</u> of such a "designer molecule" into a controlled environment; (iii) a demonstrated electrical or optical <u>interface</u> to the emplaced molecule; (iv) <u>characterisation</u> of the emplaced molecule, including excite and probe measurements and/or transport measurement; (v) <u>demonstration</u> of a controlled molecular scale switch/ gate and/or wire; (vi) demonstration of <u>self-organisation</u> in either assembly or the dynamics of a molecular assembly which is of potential electronic or optoelectronic exploitation.

Although theorists have argued that there are no physical limits in principle to building computing circuits on a molecular or indeed atomic scale[2-3,7-13] there are special problems over and above the difficulty of finding the analogues of memories, gates, switches, amplifiers, wires and transmission lines. There have been many proposals for suitable molecules to fulfill these functions but it is not known whether or not such systems could be "wired-up" and interconnected or to what extent thermal and quantum fluctuations would overwhelm any device(s) as is the case for many semiconductor nanostructures. The problems are particulary acute for granular

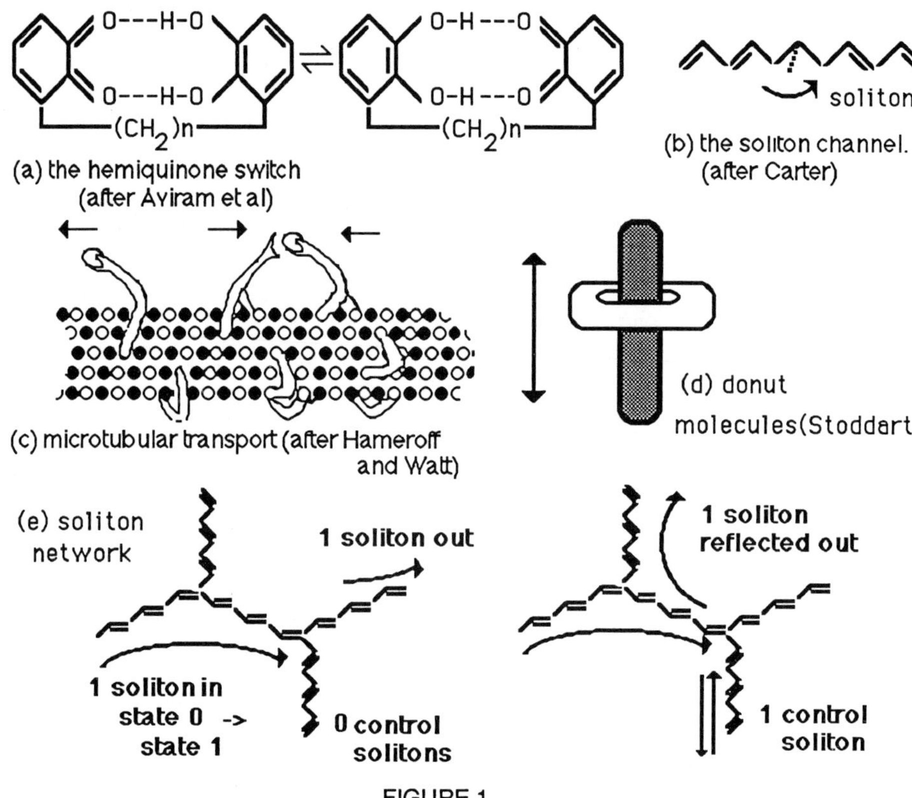

FIGURE 1

Examples of granular molecular transport/switching models

electronic systems which we define as systems in which information is carried by
discrete carriers. Examples, sketched in fig.1, include the experimentally feasible,
hemiquinone proton-tunnelling structure[2,15] , the donut-cylinder molecules[Stoddart,
private communication), the recently proposed molecular NAND gate and molecular
shift registers (Lindsey [2]; Hopfield et al [2]), natural systems such as the matter-
transporting microtubule[14], and hypothesised schemes such as soliton channels and
valves. Each proposed structure deals with <u>individual</u> information-carrying entities (let
us call these *carriers*). This may be compared with today's silicon microelectronics
where, each bit is encoded as a very large number (not necessarily fixed) of electrons
within a pulse. The associated reservoirs and sinks of charge carriers may be tapped
and manipulated to provide macro-currents which may be readily amplified or
curtailed. Restoring logic is deployed to offset signal leakage and decay . There is a
ubiquitous and continuous creation, destruction and restoration of the physical signals.
Even with systolic and so-called steering logic structures the information flow is being
passed from one underlying group of charge carriers to another, much as the
underlying local motion of water molecules supports the passage of an ocean wave.
Molecular scale logic cannot afford such luxury it will require a logic scheme that
closely matches the physical representation of information; <u>conservative logic</u> in either
its unary or binary form seems the most profitable route. There is an associated cost in

complexity because garbage lines are required to recover many of the usual logic functions. These play an analogous role to power rails and power distribution lines in conventional circuits.

The ocean wave analogy reminds us that a short-range (local) to-and-fro movement of particles (carriers) may yet support a long-range wave propagation and if it is modulated, a globally broadcast information flow. This property is exploited in the local connectivity of VLSI systolic arrays, and in the abstract architectures of local neighbourhood <u>cellular automata</u> [20,23] which have been advocated for ultra-dense molecular scale circuits[8-12]. Cellular automata structures have many of the properties required to overcome the complexity problems implied by molecular electronics, namely simple regular architectures, simple local connectivity, fault-tolerance, good self-organising properties, a potential for floating architecture schemes [12]. Three problems however, arise which alter one's view of suitable practicable candidates: support for conservative logic, synchronisation, the problem of fluctuations.

2. MODELS FOR MOLECULAR SCALE LOGIC

Neural network structures for molecular electronics have been considered{eg Neuschi and Menhart in [2]} but are currently out of favour because of the wiring complexity (design and fabrication) plus the problems of using single bit-single carrier molecular logic to represent fan-in, fan-out, amplification and long-distance transport. Fortunately, it is easy to prove that the equivalent long-range cellular automata may be mapped into a short range (eg nearest neighbour) cellular automaton with a larger state space[12].

<u>One-bit one-carrier representations</u> are familiar from the days of magnetic bubble technology which was one of the first practical applications for conservative logic [16-17] which steers data through a system such that the total number of bit zeros and bit ones in the input data is equal to the total numbers present in the output. The basic building block for conservative logic is usually taken as the computation-universal, invertible Fredkin gate illustrated in figure 2(a). There is no bit destruction/creation or signal amplification involved, but functions such as fan-out can only be emulated at the expense of introducing external constant input data streams and output garbage (don't care) data streams. This is inevitable if standard logical functions are emulated because replication of a signal as in fan-out requires us to physically bring in extra carriers: the power supply problem remains with us and indeed becomes more complex. From a computational point of view there are recognised difficulties with conservative logic implemented by Fredkin gates because in general the logic computation depends dynamically on the control signal (c in fig.2a). For example in figure 2b we illustrate the exclusive-or XOR operation in a form which keeps separate the control and data lines. This operation may be cascaded to compute any number of additional XORs as for example in the 2-D replicator cellular automaton discussed by Barker[11,12]. In the latter case this clean separation is actually lost in the basic cell of the replicator, where the fan-out stage leads to mixing of control and data lines and a large number of constants and garbage lines impinge on the cell. In principle the problem may be overcome by building the Fredkin gate from sub-gates which make no

distinction between the input lines. One method, discussed by Fredkin and Toffoli[14] builds the Fredkin gate from the cascading of six interaction gates which are invertible gates which use no control-specific signals. The composite Fredkin gates may be cascaded into arbitrary reversible sequential circuits. The interaction gate, a reversible computation-universal gate[12,18-20] , is shown in figure2(d). It is related to the Priese gate, figure 2(c), a computation - universal gate which is a fundamental route switcher using a control to route an input signal between two alternative output paths. The similarity to the soliton net in figure 1d is discussed in [12].

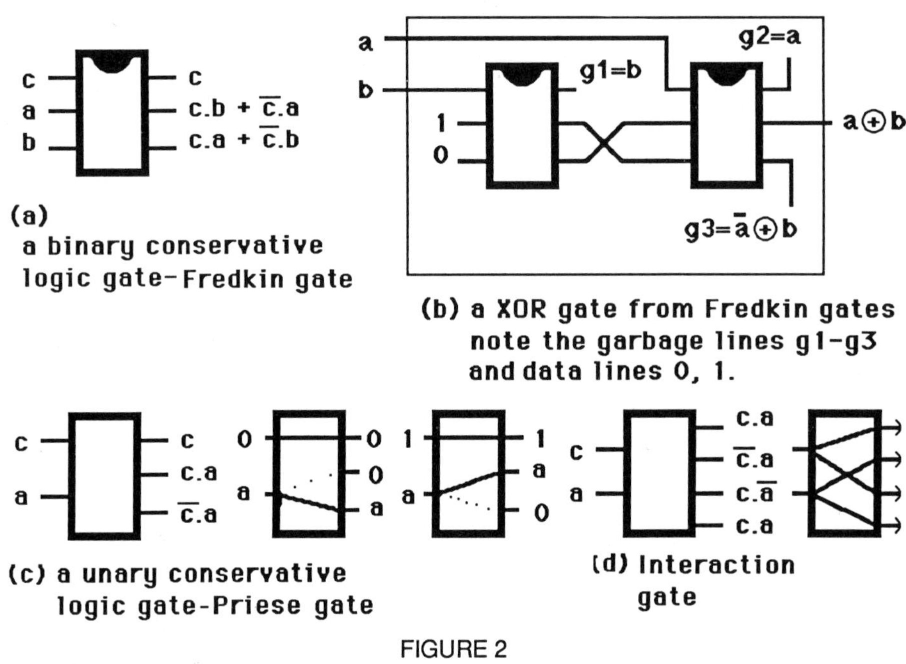

FIGURE 2
Conservative logic gates

Priese[18] has shown that these gates provide a powerful means to implement abstract <u>asynchronous cellular networks</u> thus breaking the rigid clocking scheme inherent in many cellular automata. This may prove vital for any future molecular logic. Earlier studies of asynchronous switching networks may be found in ref.[21].

3. IMPLICATIONS FOR PRACTICAL MOLECULAR DEVICES

We have seen that cellular automata provide a useful architectural target both from the standpoint of fabrication and from logical issues. If each cell corresponds to a molecular assembly, the state of each cell must be a well-defined property such as charge, excitation energy, conformational setting, spin *et cetera*. Inter-molecular interactions (eg soliton valving, resonant electron transfer, interfacial charge transfer) are necessary for cells to communicate. There are two distinct ways in which information processing might be represented. In the first, analogously to VLSI circuits, information is represented by a transported quantity such as charge which is routed throughout the system network using switches. In the second approach, analogous to phonons or spin waves, there is local exchange of energy and momentum but not material. A hypothetical example would have a molecule switch between different conformational states, the effect of which would be to induce a similar transition in its neighbours. Such a process, in analogy to a set of coupled non-linear oscillators, needs a mechanism for amplifying the perturbation: an energy source, not necessarily localised, is required. Thermodynamic considerations require that a molecular cellular automaton must be kept out of thermal equilibrium. If the internal state of a cell corresponds to an occupied physical state of the molecule that state must be sufficiently long-lived for processing events to take place. A super-block cellular automaton architecture which might achieve this is discussed in [12]

4. FABRICATION SCHEMES

Routes to the fabrication of molecular computing components are reviewed in [1-3,12]. An interim hybrid approach is being developed by us (Barker, Beaumont,Britland, Connolly, Coggins, Moores,Wilkinson, unpublished) in which the problem of the electronic interface is being tackled first. Nanolithographic techniques borrowed from semiconductor electronics are routinely used (for quantum dot fabrication) to fabricate polymer dot arrays (typically polystyrene) on a solid substrate (the Glasgow group has achieved 20 nm diameter dots on 20nm spaced arrays). We are investigating the use of such patterned substrates as sites for the growth of small numbers of organic molecules using the Merrifield-Shepherd techniques (Birr[22]), and for the attachment of pre-synthesised molecules. There are good possibilities for progressing to conducting polymer sites which are connected through the substrate to a solid state nanoelectronic circuit. The latter will provide a means for directly electrically addressing the polymer sites and hence the molecular array. The overall aim is to study the direct positioning of small numbers of significant organic molecules (ultimately single molecules) on precisely defined sites and to study the electrical interface. The proposed fabrication scheme for molecular arrays offers the possibility of directly testing hypotheses of molecular wires and switches for single molecules.There are additional advantages in the proposed approach to setting up molecular arrays which derive from the ability to generate signal patterns from the nanoelectronic substrate on a block and super-block basis.

Recent low temperature experiments on linear arrays of very small metal-insulator-metal tunnel junctions[23] have demonstrated the occurrence of stable soliton propagation and switching due to the correlated single electron tunnelling arising from Coulomb blockade effects. The temperature dependence is controlled by the junction capacitance C, according to $kT < e^2/2C$, where e is the electronic charge, and at nanometre dimensions the transition temperature may in principle exceed 300K. The normal influence of fluctuations is strongly suppressed. A generalisation of the inorganic capacitor array is possibly the use of arrays of molecules as the basic capacitative tunneling elements.

REFERENCES

1.   Lehn J-M ., Angew. Chem. Int. Ed. Engl. 27 (1988) 89.
2.   Aviram, A. (editor) Molecular Electronics-Science and Technology (Academic Press - to be published)
3.   Carter, F. (editor) Molecular Electronic Devices II (Marcel Dekker, New York,1987).
4.   Read, M., and Kirk, W.P., Physics and fabrication of nanostructures (Academic Press ,1989).
5.   Binnig, G., Rohrer, H.,Gerber, Ch., and Weibel., E., Phys.Rev.Letter., 49 (1982) 57.
6.   Binnig, G., Quate, C.F., Gerber, Ch., Phys.Rev.Lett., 56 (1986) 930.
7.   Langton, C.G. (editor),Artificial Life (Addison-Wesley, 1989).
8.   Barker, J.R., Hybrid Circuits 14, 19 (1987).
9.   Barker, J.R., Molecular Electronic Devices II, (Marcel Dekker, Inc: New York , 1987) 639 .
10.   Barker, J.R., Molecular Electronic Devices II, (Marcel Dekker, Inc: New York , 1987) 674 .
11.   Barker, J.R., Alta Frequenza, vol LVIII (1989) 249.
12.   Barker, J. R., in Molecular Electronics - Science and Technology, edited by A. Aviram, to be published.
13.   Feynman, R.P., Foundations of Physics, 16 (1986) 507.
14.   Hameroff, S., Rasmussen, S., and Mansson, B., Artificial Life (Academic Press, 1989) 521-554.
15.   Aviram A., P. E. Seiden and M. Ratner , IBM internal publication RC no.5919, (1976).
16.   Fredkin, E., and T. Toffoli, Int. J. Theor. Phys. 21, 219 (1982).
17.   Toffoli, T.,Lecture Notes in Comp.Sci. 85, 632 (1980).
18.   Priese, L.,J.Comp.Syst.Sci. 17, 237 (1978).
19.   Keller, R.M., IEEE Trans. Computers, C-23, 21 (1974).
20.   Priese, L., J. Cybernetics 6, 101, (1976).
21.   Miller, R.E., Switching Theory, vols 1,2, Wiley-New York, (1965).
22.   Birr , C., Aspects of the Merrifield Peptide Synthesis Springer- Verlag, Berlin, Heidelberg, NY (1978).
23.   Likhaerev, k., Bakhvalov, B., Kazachan, G.S., and Serdyukova, S.I., IEEE Trans. Magnet. 25 (1989) 1436.

Parallel Processing in Neural Systems and Computers
R. Eckmiller, G. Hartmann and G. Hauske (Editors)
© Elsevier Science Publishers B.V. (North-Holland), 1990

NEW LASER TECHNIQUES FOR QUASI-MOLECULAR STORAGE

D. Haarer

Physikalisches Institut and BIMF, University of Bayreuth,
Postfach 101251, D-8580 Bayreuth, F.R.G.

An optical scheme is described which allows to address small groups of molecules within a large molecular ensemble. It uses the frequency domain and the narrow bandwidth properties of optical lasers to achieve the molecular selection processes. Realistic schemes can adress as few as 10^4 molecules. This is an improvement of about 10^4 as compared to present optical schemes.

I.   INTRODUCTION

With the 'intrinsic' limitations of the present semiconductor-based computing schemes becoming more and more visible, our human mind is challenged to think about new schemes to handle data. What are, from the physics viewpoint, the limitations of present computers? Here we only point out a few limitations, without claiming to provide a complete list.

a) Size-limitations
For structuring the presently used semiconductors devices, the diffraction limit of light will limit the gate dimensions for the foreseeable future to dimensions of a micron or, at the best, a fraction of a micron. From a 'molecular' viewpoint one can, however, accomodate about 10^{16}-10^{18} molecules within a cube of one micron length and, hence, one is far away from processing and storing information on a molecular level /1/.

b) 'Particle-effects'
Even if one could limit the size of switching elements to sub-micron dimensions by using x-ray and e-beam techniques, the presently used metal-oxide-semiconductor field-effect transitors (MOSFET) would show fluctuations which are due to single electronic interface states /2,3/. With decreasing device size, there will be a limit of miniaturization which is given by the discrete nature of electrons and by quantum effects.

The above difficulties can be put into perspective, if one compares the human brain with its roughly 10^{15} bits with presently known storage devices. If one would make up a store of the size of the human brain, using 60 MByte magnetic discs (50 Watt power consumption) one would have to wire up about 2 Million discs with a total power consumption of about 0.1 Gigawatt. If one would use the present 1 M bit chip ($ 10 a piece) one would have to spend about 10 Billion Dollars to assemble a semiconductor store of the size of the human brain. The above calculations should not be taken as accurate technical estimates; they should, however, point out that biological systems have more power- and space-efficient schemes than our present technical schemes.

II.   SCHEMES FOR PROCESSING MOLECULAR INFORMATION

II.1. Biological Schemes

It seems that one of the main secrets of biological systems is the selectivity

by which molecules can induce certain activities. From the large variety of
examples, a specific example shows the way in which regulatory proteins
(called actuators or repressors) can bind to the DNA-helix and thus increase
or decrease its transcription /4/. Fig. 1 (taken from Ref. 4) shows how large
molecules with 'recognition helices' can find patterns of base pairs along the
DNA-helix. This example has been chosen with a certain arbitariness; yet it
shwos a typical molecular mechanisms which nature has developed during its
evolution to use single macromolecules to perform certain functions.

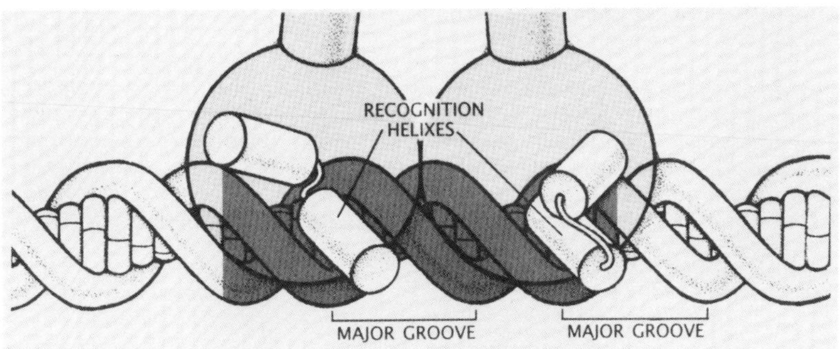

FIGURE 1
Two repressor molecules with helical chains (two barrel-like structures) which
can run along the DNA-helix and recognize base pair patterns (this figure is
taken from Ref. 4, which should also be consulted for details)

So far we have no technical schemes which resemble the above biological
schemes. The main reasons are the following:

a)   Synthetic molecules of the size of biological molecules cannot be re-
       produced with the accuracy which is needed to perform biological pro-
       cesses with the above described precision: The mutation rate per base
       pair per replication of drosophila-DNA is on the order of 10^{-10} /5/; the
       accuracy of the translation of the genetic information into amino acids is
       on the order of 10^{-5} to 10^{-4} erros per amino acid /6/.

b)   Technical schemes do not use so much the conformational degrees of
       freedom (compared to macromolecules) but use electrical signals instead.
       For 'electrical' systems the above quoted accuracies are very remarka-
       ble!

Since it is very unlikely that technical schemes will evolve along patterns
which are similar to our evolutionary scheme, one has to think about other
mechanisms to process information by using molecular schemes. One scheme is
based on the usage of high resolution lasers and will be described in more
detail.

II.2. The Energy Selective Scheme of Information Processing,
       Using Laser Photons

If one incorporates identical molecules (for instance molecules of a given dye
species) into an amorphous matrix, there is a high likelyhood that each
molecule will have a different 'local environment'. This local environment will,
in general, lead to a broadening of the molecular electronic transitions and,
hence, is called 'inhomogeneous broadening'. Fig. 2 shows such a broadening
mechanisms for three molecules in an amorphous material /7/.

Inspite of this matrix broadening mechanism, the molecular transitions can, at low temperatures, be rather narrow. This fact can be used to laser-select individual molecular groups at low temperatures. If we assume that the molecules considered are photochemically active, then one can select and transform a very small fraction of the dopant molecules by using a narrow-band laser beam (narrow compared to the molecular linewidth) and by working at low temperatures to keep the molecular transitions sufficiently narrow. If one performs such a 'narrow-band photochromic reaction' one burns a hole in the bell-shaped absorption spectrum of the dye molecules (typically inhomogeneous absorption bands have a Gaussians envelope). This hole is quite different

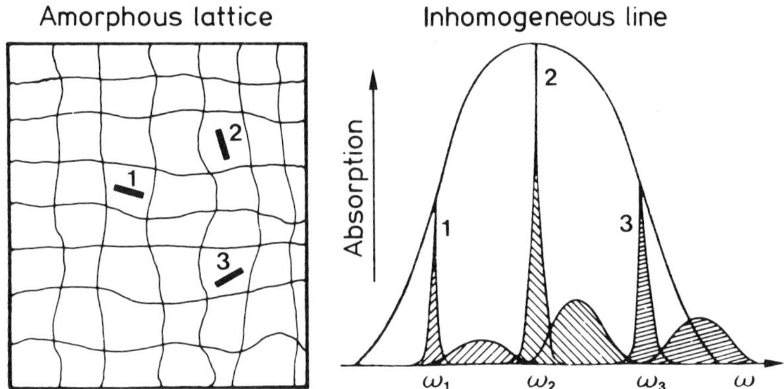

FIGURE 2

Schematic view of the electronic absorption lines of three identical molecules, embedded into an amorphous matrix. The molecules are labelled 1,2,3; they absorb at the frequencies ω_1, ω_2 and ω_3 respectively. The broad sidebands at the high frequency side of the transitions are caused by lattice vibrations. They can be ignored for the purpose of this article (see also Ref. 7).

from a hole in the common sense; it is a hole in the <u>frequency space</u>, as defined by the absorption spectrum of dye molecules.

After irradiation at a given frequency within the inhomogeneous band, a hole appears in the spectrum (Fig. 3b). Its width is strongly temperature dependent and is given by molecular relaxation processes (see for instance Ref. 8).

It is the narrow width of a photochemical hole, compared to the width of the total absorption line, which allows the definition of an attractive data storage scheme: A typical inhomogeneous bandwidth $\Delta\omega_i$ is on the order of several 100 cm^{-1}. The homogenenous bandwidth, on the other hand, can be on the order of 10^{-3} cm^{-1} at low temperatures. This allows, in principle, the independent discrimination of 10^3 to 10^4 holes within a single absorption band and thus distinguish 10^3 to 10^4 subensembles of a large ensemble of molecules. The multiplexing factor f_m is defined as

$$f_m = \frac{\Delta\omega_i}{2\Delta\omega_h}$$

f_m gives the number of holes which can be burnt into one spectrum. Fig. 4 shows the absorption spectrum of the dye molecule free base phthalocyanine (H_2Pc) in PMMA at 2 K. (PMMA = polymethylmethacrylate = plexiglas) /7,9/.

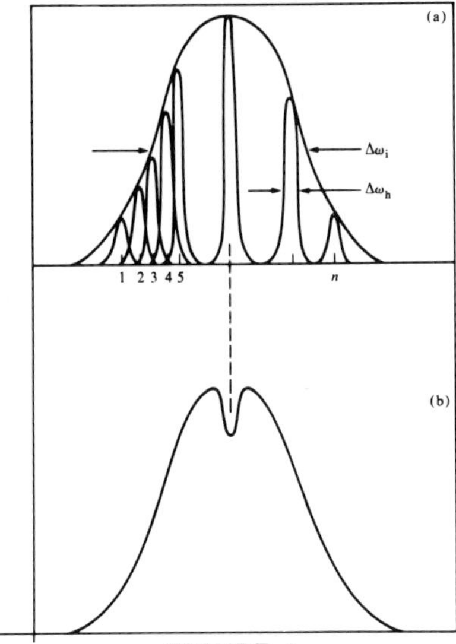

FIGURE 3a

Electronic absorption lines of n molecular 'sites'. A 'site' is a molecule–matrix configuration with a certain energy. $\Delta\omega_i$ is the inhomogeneous bandwidth; $\Delta\omega_h$ the homogeneous bandwidth of the molecular absorbes (see text)

FIGURE 3b

Hole in the absorption band after irradiation into the center frequency.

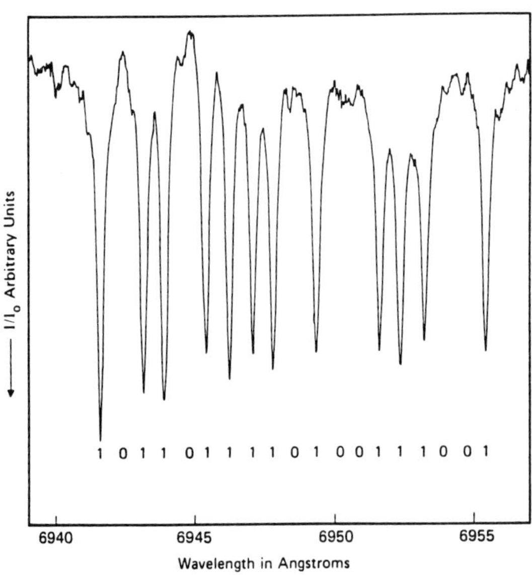

FIGURE 4

Hole pattern of the dye molecule H_2pc in PMMA (see text)

The hole width is less than 10^{-2} cm^{-1}; the total width of the absorption line (which appears flat in Fig. 4 due to the scale of the axis) is about 10^{2} cm^{-1}. Therefore the multiplexing factor f_m is on the order of 10^{4}.

III. TECHNICAL APPLICATIONS OF THE LASER SELECTION MECHANISM

The technical applications of the above laser scheme are quite obvious: Whereas a normal optical data storage scheme is limited to 10^{8} bits/cm^2 due to the diffraction limit of visible light, the hole burning storage scheme can enhance this figure by four orders of magnitude to 10^{12} bits/cm /10/. This is made obvious by Fig. 5. In each x,y-position of a two dimensional scheme one can, in principle, store 10^{4} frequency bits.

There is, however, a price which has to be paid for this enhancement of optical data storage schemes. The holes are only narrow at low temperatures, because relaxation processes broaden the individual molecular transitions. Therefore the storage density is reduced by about one order of magnitude, if going from 2 K to 10 K.

The present state of the technical development of this interesting scheme is summarized in Ref. 9 and 11. The main advantages and limitations will be discussed in the lecture. They can be summarized as follows:

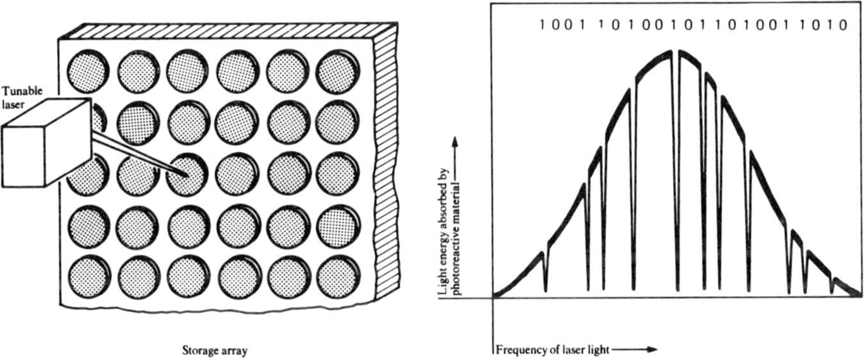

FIGURE 5

Schematic description of a three dimensional x,y-ν-storage scheme. (Fig. 5a) shows the typical x,y-configuration with the circles representing the spatial bits. Fig. 5b shows a schematic representation of a 'bit pattern' in the frequency domain. A different frequency bit pattern can be stored in each x,y-spot.

Advantages

a)   Optical hole burning is, at the time being, the only feasible scheme for achieving data densities up to 10^{12} bits/cm^2.

b)   The hole burning scheme can address as little as 10^{4} molecules with visible light and with high data rates.

c)   Optical hole burning can be used for <u>Multiplexing-Schemes</u> i.e. schemes in which information can be handled in various parallel frequency channels.

d)   The scheme would be rather useful for optical computers and will, in principle, work with semiconductor laser diodes.

e)   If implemented at low temperatures, an optical hole burning disk of 20 cm radius could carry up to 10^{14} bits i.e. represent a store of the size of the human brain.

<u>Disadvantages</u>

a)   Low temperatures are needed for high multiplexing factors ($f_m > 10$), making the involved technologies difficult

b)   Gating schemes are needed for reading and processing data without deterioration

c)   The architecture for memories of the size of hole burning memories has not been developed yet.

IV.   CONCLUDING REMARKS

Hole burning spectroscopy has two aspects. One aspect of data storage which we have highlighted in this article. There is, however, a second aspect, namely the spectroscopy of amorphous materials /8/. Here recent experimental efforts have yielded new insights into the structure of amorphous materials, a class of materials which is still not well understood.

ACKNOWLEDGEMENTS

We acknowledge support by the 'Stiftung Volkswagenwerk', by the 'Deutsche Forschungsgemeinschaft' and by the 'Fonds der Chemischen Industrie'. We also acknowledge the stimulating contributions of many of my PhD and Diploma Students in Bayreuth.

REFERECES

/1/   D. Haarer, Advanced Materials I, 361 (1989) or Angew.Chem., Adv.Mater. Bd. 101, 1576 (1989)

/2/   M. Schulz, Surf.Sci. <u>132</u>, 422 (1983)

/3/   K.S. Ralls, W.J. Skocpol, L.D. Jackel, K.E. Howard, L.H. Fetter, R.W. Epworth and D.M. Tennant, Phys.Rev.Lett. <u>52</u>, 228 (1984)

/4/   M.Ptashne, 'How Gene Activators Work', Scientific American, Jan. 1989

/5/   J.W. Drake, Nature <u>221</u>, 1132 (1969)

/6/   M. Yarus, Progr.Nucleic Acid Res. <u>23</u>, 195 (1980)

/7/   J. Friedrich and D. Haarer, Angew.Chem.Int.Ed.Engl. <u>23</u>, 113 (1984)

/8/   D. Haarer in 'Persistent Spectral Hole Burning: Science and Applications'. Topics in Current Physics Vol. 44 Ed. W.E. Moerner, Springer Verlag 1988 pp. 79 ff

/9/　　D. Haarer, Jap.Journ.Appl.Physics $\underline{26}$, Suppl. $\underline{26}$, 227 (1987)

/10/　　G. Castro, D. Haarer, R.M. Macfarlane, H.P. Trommsdorff; US Patent 4.101.976 (1978)

/11/　　W. E. Moerner in 'Persistent Spectral Hole Burning: Science and Applications'. Topics in Current Physics Vol. 44 Ed. W.E. Moerner, Springer Verlag 1988 pp. 251 ff.

Parallel Processing in Neural Systems and Computers
R. Eckmiller, G. Hartmann and G. Hauske (Editors)
© Elsevier Science Publishers B.V. (North-Holland), 1990

CHAOS, COOPERATION AND ASSOCIATIVE MEMORY
IN NONLINEAR PICTORIAL FEEDBACK SYSTEMS

Gerd Häusler

Department of Optics, Physics Institute
University of Erlangen
Erwin-Rommel-Str. 1

D 8520 Erlangen

We investigate pictures circulating through a sequence of
local nonlinearity and convolution. Such pictures can be
considered as specific examples of high dimensional nonlinear dynamical systems or of 'neural convolution networks'. Dependent on the parameters they exhibit spatial
and temporal chaos, or evolution of stable structures, or
they show autoassociativity. The latter example is specifically interesting, because 'neural convolution networks'
can autoassociatively restore pictures independent of
there position and they can be implemented by Fast Fourier
Transformation.

1. INTRODUCTION

Nonlinear Dynamical Systems may exhibit effects from a rich class
of behavior, such as deterministic chaos, morphogenesis or
associative recall of memory contents. Specifically interesting
are systems with many variables, i. e. many dimensions, such as
pictures. Unfortunately, the implementation of high dimensional
systems on a serial computer is somewhat frustrating, because of
time cost and difficult interpretation of the results: for
example, a neural net for an input vector with N elements
requires matrix-vector multiplications with an N^2 matrix. For a
TV-picture with $N=512^2$ pixels as an input signal, the coupling
matrix would have already about $6 \cdot 10^{10}$ elements. It is nearly
impossible today, to handle such a system.

What can we do, if we do not want to reduce the number of pixels
to, say, $N=32 \cdot 32$ (which makes a bad picture but requires already
a 10^6 element matrix)? We have to restrict the generality of
operations performed on the picture. Such a very 'natural'
operation for pictures is convolution. This operation is usually
applied to pictures by any reasonable imaging system, such as our
eye, and by many systems for optical imformation processing.

For our purposes, two features of convolution are important: The
first is spatial invariance: Each pixel interacts with its
neighbour pixels in the same manner, independent of its position
within the picture. One could say, the 'natures of law' in our
'picture world' are space independent. Hence, we can expect to
have a good model for effects like the formation of crystals
or the recognition of objects independent of their spatial
position. The second feature of convolution is concerning its
implementation: the convolution matrix is cyclic, the same

information is contained within each row. The operation can
advantageously be performed in Fourier space, instead of in the
space domain, requiring only the order of N·ldN operations,
instead of N^2 operations. A 512^2 Fourier-transformation can
already be done within less than 1 second, with commercially
available hardware. A further aspect of the Fourier algorithm is
its potential to be realized by optical means.

In the following we will briefly sketch the mathematics of the
considerations above. A reasonably general nonlinear dynamical
system is the following:

(1) $u_k(\mathbf{x};t+1) = NL_k[(u_1(\mathbf{x};t),\ldots u_N(\mathbf{x};t)]$;   with k=1,...,N

There are N coupled variables u_k dependent on space $\mathbf{x}$ and time t.
NL_k are nonlinear functions. The u_k may be, for example,
concentrations of interacting chemical reagents. In order to make
our system suitable for the 'optical implementation', we introduce some simplifications: We drop the space dependence in
Eq. (1) but identify the the u_k by the intensity of a pixel k, in
our picture. Furthermore we drop the index k of the nonlinear
function NL, i. e. the function is the same, for each pixel. Now
we introduce a further specialization: NL be a sequence of a
local nonlinearity and a general linear transformation:

(2) $u_k(t+1) = NL[\sum_{l=1}^{N} T_{kl}u_l(t)\ ]$

This is the basic equation of a processing step in a neural
network, with an N^2 coupling matrix T_{kl}. There NL is usually a
saturation type nonlinearity, such as the sigmoid function. We
use mainly the logistic parabola NL(u)=4au(1-u)+b, with free
parameters a,b. Now we introduce space invariance by a cyclic
matrix with elements $T_{kl}=h_{(k-1)modN}$:

(3) $u_k(t+1) = NL[\sum_{l=1}^{N} h_{(k-1)modN}\ u_l(t)\ ]$

With vectorial notation we get an equivalent equation:

(4) $\mathbf{u}(t+1) = \mathbf{NL}[\ \mathbf{u}(t)*\ \mathbf{h}\ ]$,

where "*" denotes the convolution operator and $\mathbf{h}$ is the impulse
response or 'point spread function' of the system. The iterations
of Eqs. (3) and (4) are visualized in Fig. 1.

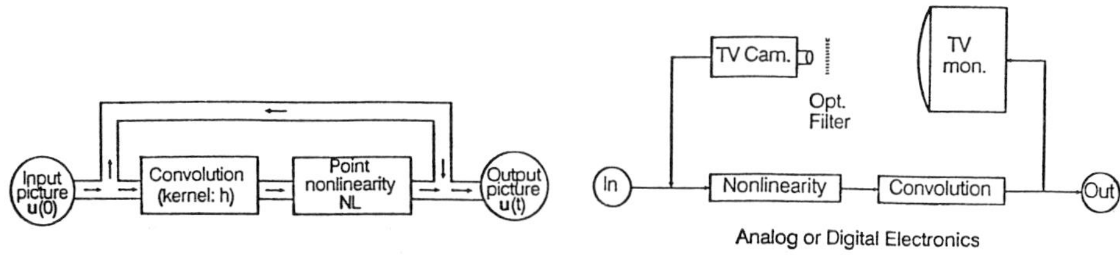

FIGURE 1: Shift invariant nonlinear feedback system and its
optical-electronical implementation

2. SOME EXPERIMENTS LEADING TO CHAOS AND SELF ORGANISATION

Dependant on the parameters of the nonlinearity and the kind of convolution kernel, the pictorial feeback system of Fig. 1 displays different kind of behavior [1], which is shown in Fig. 2. With a purely positive point spread function (low pass filter), only chaotic behavior was exhibited (Fig. 2a). With a bipolar point spread function ('lateral inhibition') the evolution of stable pictures ('fixed points') could be observed (Fig. 2b).

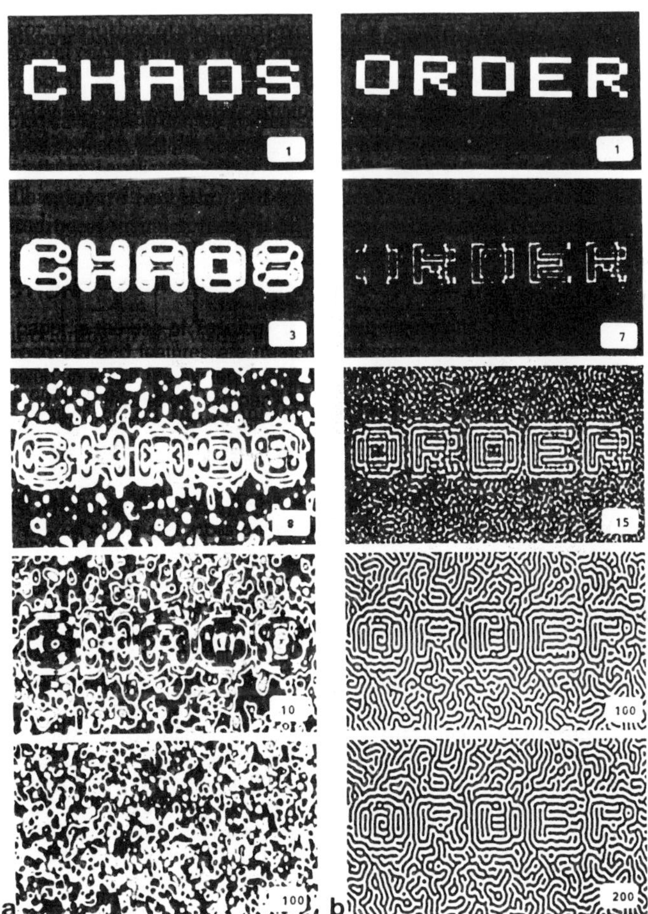

FIGURE 2: Examples of deterministic chaos and cooperative behavior in nonlinear pictorial feedback ssystems.

3. AUTOASSOCIATIVE RESTAURATION OF STRONGLY PERTURBED PICTURES

It should be noted that the fixed point of Fig. 2b is stable in spite of a considerable amount of noise added in each cycle. Principally, this property can be used for the autoassociative

restauration of perturbed pictures. One first example was   shown
already in [2]: After about 2000 iterations the system displays a
stable fixed point (Fig. 3a). We perturb the stable iterate (Fig.
3b). After about 30 more iterations the original piture is
restored (Fig. 3c). Such an autoassociative recall works   for
perturbations with a size of up to 20*20 pixels. The experiments
prove that   our nonlinear pictorial feedback system displays
autoassociativity as neural networks do.

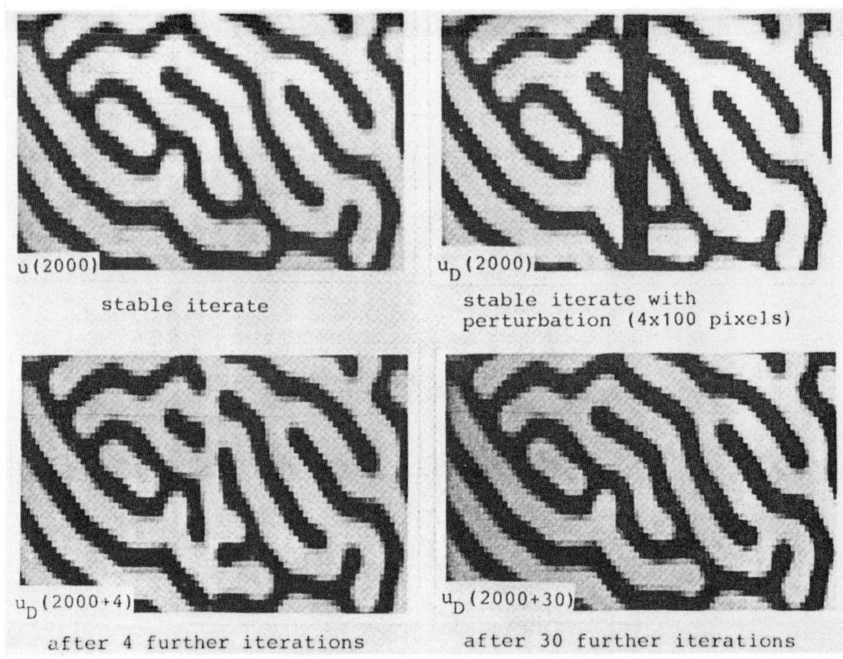

FIGURE 3: Associative restauration of an arbitrary fixed point

However, the question arises, how to create a fixed point that
looks like a desired picture? So far the fixed points were
arbitrary, dependent on the starting conditions and the point
spread function. Is it possible to 'design' a desired fixed
point? Neural networks are trained for certain fixed points by
'learning rules' to find a proper coupling matrix. This is very
similar to our systems, with the only difference that our
coupling matrix is cyclic. Hence we have only N free parameters
to be trained instead of N^2 free parameters in conventional
neural networks.

Now we briefly explain how it is possible to associatively
restore binary pictures (with intensities 1/2 and 1).

The basic idea is the following: The perturbed picture will be
repeatedly sent through a sequence of space invariant linear
filtering and local nonlinearity. The filter is designed in a way
that the important frequencies f_i (frequencies with high contrast) **of the desired fixed point** are transmitted without loss of
contrast. The contrast of other frequencies f_u will be slightly
decreased by the filter.

In this first step the frequencies f_u from potential perturbations of the picture will be slightly reduced. However, the
object frequencies f_u are also decreased and the picture will
suffer from some errors introduced by the filter. This error can
be partially compensated for by the local nonlinearity in the
space domain. The nonlinearity is a logistic parabola which is
designed in a way that it has two stable fixed points at the
intensity levels 1/2 and 1. Intensities that are falsified by the
filter, are drawn in the direction to the correct levels by the
nonlinearity. However, the nonlinearity causes small errors in
the spectrum. These errors are again partially repaired by one
more filter step. And so forth.

The system can be considered as a specific neural feedback system
with space invariant coupling, or as an iteration algorithm that
switches back and forth between the space domain and the Fourier
domain. The usefulness of such 'switching algorithms' has been
demonstrated already by Fienup [3], Gerchberg and Saxton [4] and,
recently by Dainty et al [5] for phase retrieval, superresolution
and blind deconvolution.

In Figs. 4 and 5 some results of associative restauration are
displayed.

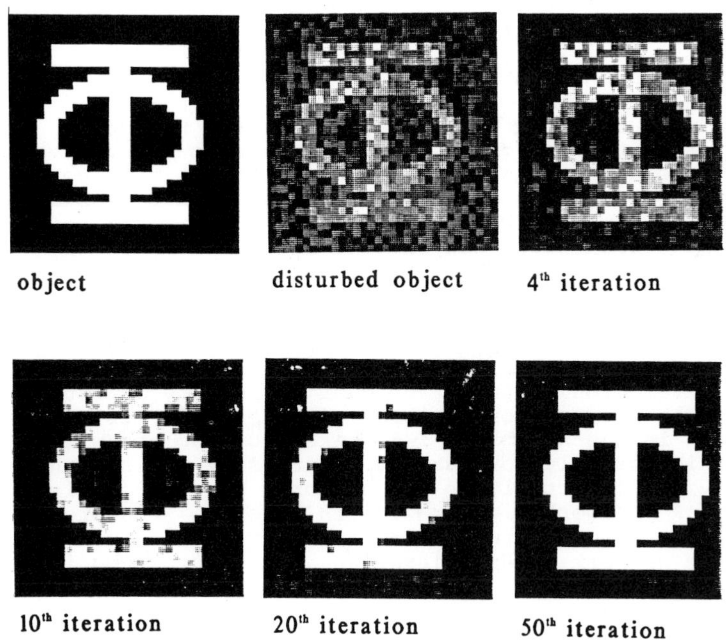

object   disturbed object   4th iteration

10th iteration   20th iteration   50th iteration

FIGURE 4: Restauration of a letter perturbed by additive noise.
          (Signal-to-noise ratio = 1:1)

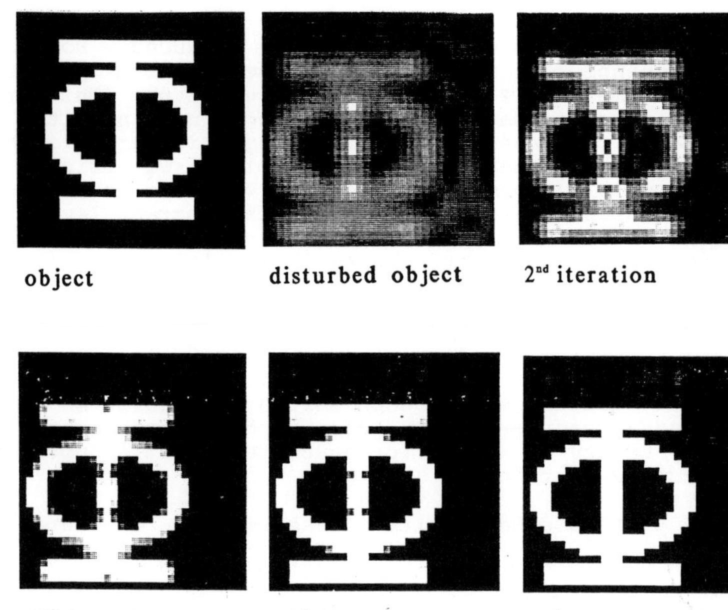

object   disturbed object   2nd iteration

10th iteration   30th iteration   50th iteration

FIGURE 5: Associative restoration of a picture perturbed by low pass filtering **and shift.**

4. SOME CONCLUDING REMARKS

We introduced a strong constraint to neural networks: a space invariant structure of the coupling between neurons. Nevertheless such networks display associative behavior and it is possible to train such a network for desired fixed points (objects with binary intensity distribution). Space invariance offers two advantages: the network can be very fast implemented by Fourier transformation, and it restores even shifted objects. – What about data efficiency? The network contains only N degrees of freedom, hence we can only train one object vector with N components. In conventional (space variant) networks we can train about $.15*N^2$ data within an N^2 coupling matrix [6,7]. In our 'Neural Convolution Network' the data efficiency is 100%. What is the price that we pay for the high efficiency? First, we loose redundancy in the network. This makes no problems with digital systems which usually work safely. More serious is, that we can store only one object within one network. However, it is possible to build K networks in parallel in order to restore K input objects. We demonstrate as an application of such a neural convolution network, shift invariant character recognition, in a further paper.

[1] G. Häusler, G. Seckmeyer, T. Weiß, Appl. Opt. 25 (1986) 4656
[2] M. Fang, G. Häusler, Proc. of SPIE Conf. 667 (1986) 214
[3] J. R. Fienup, Opt. Lett. 3 (1978) 27
[4] R. W. Gerchberg, W. G. Saxton, Optik 35 (1972) 237
[5] G. R. Ayers, J. C. Dainty, Opt. Lett. 13 (1988) 547
[6] G. Palm, Biol. Cybernetics 36 (1980) 19
[7] J. J. Hopfield, Proc. Natl. Acad. Sci. USA 79 (1982) 2554

Parallel Processing in Neural Systems and Computers
R. Eckmiller, G. Hartmann and G. Hauske (Editors)
© Elsevier Science Publishers B.V. (North-Holland), 1990

OPTICALLY CONTROLLED MULTISTABLE QUANTUM SYSTEMS: POSSIBLE REALIZATION OF A MOLECULAR COMPUTER? [*]

G. MAHLER and W.G. TEICH

Institut für theoretische Physik, Universität Stuttgart
Pfaffenwaldring 57, 7000 Stuttgart 80, FRG

1. INTRODUCTION

The search for ever faster and more powerful computing devices has led to a steady decrease of the size of the components of an information processing system. This success has been achieved within the framework of "conventional" microelectronics, based on the transistor. Alternative technologies, such as computers based on Josephson-junctions, have been proposed, but none of them could compete either technically or economically with the enormous progress in microelectronics [1]. This was so, despite the fact that specific properties like the switching time have been shown to be much better than in conventional devices; but this could not compensate disadvantages of the system as a whole. For new concepts, it is therefore necessary not only to consider isolated switching devices, but to judge the complete system design. It is foreseeable that the size of the components will reach the nanometer regime in the near future. For such small dimensions quantum effects will dominate the system, and the concept of the conventional transistor will break down. The quest is for new concepts which allow information processing on the length scale of individual atoms and molecules. An example is "molecular electronics" [2] which is supposed to use characteristic properties of individual atoms or molecules, such as discrete energy eigenstates, to incorporate information processing. On the other hand, the limitations of the classical von Neumann architecture have been recognized recently and, independently of the physical realization, new computer architectures such as hypercubes, neural networks, and synergetic computers are discussed. All of them have in common that information is processed in parallel. In this context cellular automata (CA) [3] can be considered as an idealized mathematical model for parallel processing, similar to the Turing machine, which is an idealized model for a serial architecture.

A computer stores and manipulates information [4]. In order to accomplish this task, any physical system representing a computer must possess some fundamental properties [5]. *Multistability* is required in order to represent information. The system must possess several stable attractors on the time scale of computation. A computer must communicate with its surroundings (input and output), i.e. a reliable *preparation* and *measurement* of the attractors must be possible. On a molecular scale, this raises questions regarding quantum fluctuations and the nature of the quantum mechanical measurement process (i.e., enhancement of quantum signals). These properties are sufficient for the storage of information. In order to process information, a minimal amount of *controllability* is required. Control means to switch selectivly from one stable attractor to another. This assures an independent preparation and measurement of various subunits (which is necessary for "fair coding") and the realization of simple logical operations or transition rules.

2. THE MODEL

The specific model we consider is a multistable quantum system, based on a semiconductor

---

[*] Work supported by the Deutsche Forschungsgemeinschaft (SFB 329).

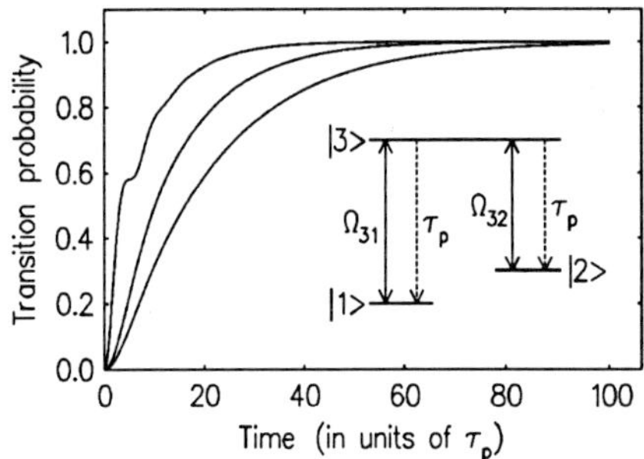

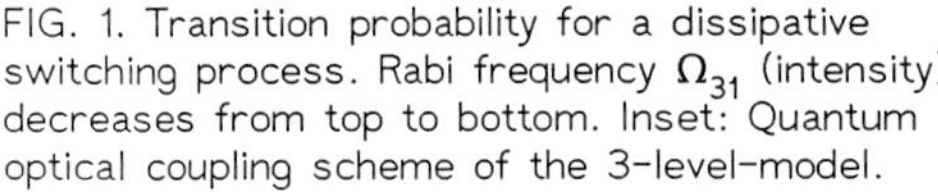

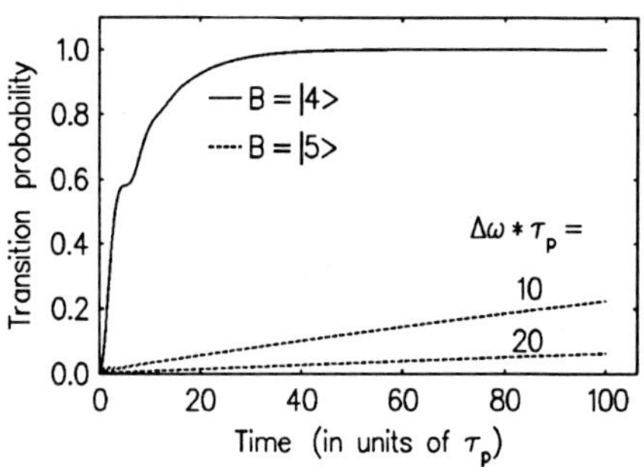

FIG. 1. Transition probability for a dissipative switching process. Rabi frequency Ω_{31} (intensity) decreases from top to bottom. Inset: Quantum optical coupling scheme of the 3-level-model.

FIG. 2. Conditional transition probability for two coupled cells. A laser pulse of frequency $\omega_{32}(4)$ is applied.

heterostructure or on complex macromolecules, which is coupled optically to a macroscopic control system (laser, photodetector) [5–9]. The problem of addressing individual subunits of molecular size is solved within frequency space by persistent hole burning. Multistability is realized in form of localized charge-transfer excitations. Localization selection rules assure the stability of the charge-transfer state on the time scale of computation. The hierarchical structure of the system leads to vastly different interaction energies; this allows to decompose the system into independent subunits (modularity), coupled via dipole-dipole interaction.

2.1. Preparation: Dissipative switching dynamics

The minimal model for a dissipative switching dynamics is given by a 3-level-system (cell) (Fig. 1, inset). The ground state $|1\rangle$ and the metastable charge-transfer state $|2\rangle$, which decays into the ground state on a time scale τ_d, form the two stable attractors of the system. The transient state $|3\rangle$, on the other hand, decays into either state on a time scale $\tau_p \ll \tau_d$. This time scale spreading can be due to symmetry selection rules as, e.g., in atomic systems, or it can be tailored by the localization of the respective wave functions (localization selection rules) [7]. It is the basis for multistability and a reliable preparation of the attractors on the time scale of computation.

A switching process proceeds, e.g., by applying a laser pulse of frequency $\omega_{31} = (E_3 - E_1)/\hbar$ which drives the transition $|1\rangle \leftrightarrow |3\rangle$ (E_i is the energy of state $|i\rangle$). The cell is excited into the transient state $|3\rangle$, from which it can decay into the metastable state $|2\rangle$ or back into the ground state $|1\rangle$, from which it will be reexcited again by the laser pulse. State $|2\rangle$ is the only stable attractor of the system as long as the laser is on, and for long enough pulse durations the cell will be in state $|2\rangle$ after the laser has been switched off. Since the switching dynamics is a stochastic process it must be described in terms of a transition probability [5], which is determined by the intensity and duration of the applied laser pulse (Fig. 1). The dissipative character of the switching dynamics assures that the switching process is not sensitive to initial conditions and incorrect light pulse parameters (frequency, intensity, and pulse length). The switching process is reversible: By applying a laser pulse of frequency $\omega_{32} \neq \omega_{31}$ the cell can be switched back to state $|1\rangle$.

2.2. Measurement: Resonance fluorescence

To measure the state of a cell, a fourth state $|4\rangle$, strongly coupled to the ground state $|1\rangle$ (time scale τ_p), but only weakly coupled to the metastable state $|2\rangle$ (time scale τ_d), is added to the cell [9]. Applying a laser pulse of frequency ω_{41}, the cell will scatter photons of the same frequency, if it is in state $|1\rangle$. Being in state $|2\rangle$ no photons will be detected. The basis for the amplification of the quantum signal is the time scale spreading $\tau_p \ll \tau_d$. It allows to

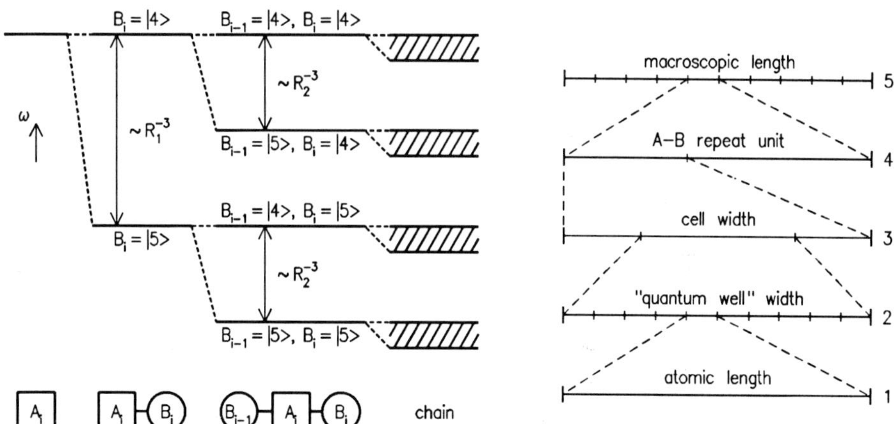

FIG. 3. Renormalization of the transition frequency due to the dipole-dipole interaction.

FIG. 4. Hierarchical levels defined by various length scales of the system.

repeat the scattering experiment over and over again during the lifetime of the metastable state $|2\rangle$ to arrive at a reliable prediction about the state of the system.

2.3. Control: Conditional switching dynamics

The minimal model for a controlled quantum system are two coupled cells A and B. Each of the two cells consists of a 3-level-model as discussed in section 2.1, but with different transition frequencies ω_{31}, ω_{32} and ω_{64}, ω_{65}. Any transition within each cell can be addressed selectivly within frequency space. The two cells are separated by a distance R large enough to assure localized excitations within each cell, but small enough in order for the two cells to "communicate" via the Coulomb interaction. It leads to an energy renormalization which, to lowest order, is given by the dipole-dipole interaction. Due to different localizations of the wave functions of the states $|1\rangle$, $|2\rangle$, and $|3\rangle$, each state has a different dipole moment and the transition frequencies of each cell depend on the state of the other cell. For the frequency selectivity $\Delta\omega$ we find [8]

$$\Delta\omega = \hbar\omega_{31}(4) - \hbar\omega_{31}(5) = \frac{p_A\,p_B}{4\pi\varepsilon\varepsilon_0 R^3}\,F(\Theta_A, \Theta_B) \tag{1}$$

$\omega_{31}(i)$ is the conditional transition frequency between states $|1\rangle$ and $|3\rangle$ of cell A, if cell B is in state $|i\rangle$. p_A and p_B measure the magnitude of the change of the static dipole moment between states $|1\rangle$ and $|2\rangle$ (cell A) or $|4\rangle$ and $|5\rangle$ (cell B), respectively, and the factor $F(\Theta_A, \Theta_B)$ accounts for the direction of the charge transfer in cell A and B. $\varepsilon\varepsilon_0$ is the static dielectric constant of the embedding material.

By applying, e.g., a laser pulse of frequency $\omega_{32}(4)$ and small enough bandwidth, a conditional switching dynamics can be induced in cell A: Cell A is switched from state $|2\rangle$ to state $|1\rangle$ only if cell B is in state $|4\rangle$ (Fig. 2). The control which can be achieved in this way is sufficient for an independent preparation of the two cells A and B and to realize all elementary logical operations [8]. The single laser pulse of frequency $\omega_{32}(4)$, e.g., represents a logical "OR" for suitable coding.

2.4. Architecture: Adaptive cellular structure

As a minimal model for a system design, we finally consider a linear arrangement of alternating cells A and B [8]. All cells A and all cells B are switched alternatingly, depending on the state of its respective left and right nearest neighbors (which must be stationary during the conditional switching process). The influence of all other cells on the respective transition frequencies is compensated by a large enough bandwidth $\delta\omega$ of the laser pulse, i.e. $\delta\omega$ accounts for the shift of the transition frequencies of all other cells (Fig. 3). For $\delta\omega < \Delta\omega$ local transition rules can be realized. An asymmetric arrangement of the cells guarantees that all eight possible conditional transition frequencies for each cell are distinct.

Since individual cells cannot be addressed selectively neither in real space (the wavelength of the applied laser pulse is much larger than the extension of a cell) nor in frequency space, the preparation of the cellular structure must be performed with the help of a shift operation: Starting from a cell C, physically distinct from cells A and B, any inhomogeneous state of the cellular structure can be prepared or measured.

Once the state of the cellular structure is prepared, the dynamics of the system evolves in parallel and is controlled by the external light field. The system adapts to different light pulse sequences, defined by various frequencies, intensities, and pulse length. The behavior includes the dynamics of a deterministic 1D CA [8] as well as the stochastic simulation of the 1D kinetic Ising model [5].

3. DISCUSSION

Optically controlled multistable quantum systems provide model systems where not only reversible switching processes or the representation of elementary logical functions can be demonstrated, but which can be extended to a complete system design. The problem of addressing individual quantum systems within an array of identical modules is solved within frequency space and by using a sequential input and output procedure. Once prepared, the dynamics evolves fully in parallel. The basis for the complex and controllable dynamics is a hierarchical structure of the system (Fig. 4). Various interaction energy scales allow to decompose the quantum system into separated subsystems which are coupled via the dipole–dipole interaction.

Fundamental limitations of the concept are given by the limited selectivity in frequency space. For finite detuning any transition will be driven. This leads to the unwanted back–reaction in the case of an isolated cell [5] or to undesired switching processes in the case of coupled cells. Both effects can be minimized by respective light pulse parameters, but not eliminated completely. Further error correction can only be achieved by redundancy.

To allow for the possibility of malfunctioning cells, it would be desirable to extend the system to a 2D network of coupled cells. Due to the constriction $\delta\omega < \Delta\omega$ and the nonlocal nature of the Coulomb interaction, this is difficult to achieve. One way out would be to consider a hybrid system of a 2D dielectric cellular structure and a metal film which modifies the Coulomb interaction in such a way, that the influence of neighboring cells dominates the renormalization of the transition frequency.

(FOOT)NOTES AND REFERENCES

[1]   R.W. Keyes, Rev. Mod. Phys. **61**, 279 (1989).
[2]   F.L. Carter (Ed.), *Molecular Electronic Devices* (Marcel Dekker, New York, 1982).
[3]   S. Wolfram (Ed.), *Theory and Applications of Cellular Automata*
       (World Scientific, Singapure, 1986).
[4]   R. Landauer, Berichte der Bunsen-Gesellschaft für Physikalische Chemie **80**, 1041 (1976).
[5]   W.G. Teich and G. Mahler, Physica Scripta **40**, 688 (1989).
[6]   K. Obermayer, G. Mahler, and H. Haken, Phys. Rev. Lett. **58**, 1792 (1987).
[7]   K. Obermayer, W.G. Teich, and G. Mahler, Phys. Rev. B **37**, 8096 (1988).
[8]   W.G. Teich, K. Obermayer, and G. Mahler, Phys. Rev. B **37**, 8111 (1988).
[9]   W.G. Teich, G. Anders, and G. Mahler, Phys. Rev. Lett. **62**, 1 (1989).

Parallel Processing in Neural Systems and Computers
R. Eckmiller, G. Hartmann and G. Hauske (Editors)
© Elsevier Science Publishers B.V. (North-Holland), 1990

LEARNING IN OPTICAL NEURAL NETWORKS

Demetri Psaltis, David Brady, and Ken Hsu

California Institute of Technology, Pasadena, California 91125

I. Introduction.

The basic mechanism used in almost all neural network learning algorithms is a form of Hebbian learning in which the strength of the connection between two neurons is modified by the product of the activations of the neurons. This simple mechanism can be simulated optically by the modification of the strength of a hologram which interconnects two points in space. In general, the strength of a holographic interconnection is modified in proportion to the product of the amplitudes of the two optical beams that are used to record the hologram. Thus, holography is well matched to neural network learning algorithms and it is rather straightforward to design optical system architectures that holographically implement the various algorithms [1-5]. Photorefractive crystals are commonly used as the holographic media in these systems. When a photorefractive crystal is exposed to light, the charge distribution within the crystal is rearranged [6]. Typically, electrons are optically excited from the bright regions of the light pattern and are transferred to the dark regions. The rearrangement of the charge distribution within the crystal causes a corresponding modification in the index of refraction of the medium which in turn is sensed by a read-out optical beam. In this manner a hologram is recorded. However, the redistribution of the existing charge pattern also causes the previously recorded holograms to be erased. This gradual erasure is similar to the "forgeting" that is often part of learning algorithms and can be a useful mechanism. For instance, forgetting prevents saturation of the interconnection weights and it causes experiences that are not frequently reinforced to decay. The main focus of this chapter is the examination of this erasure or forgetting mechanism in photorefractive holography and several methods for adapting it so that it conforms with the requirements of neural network learning algorithms.

II. An optical neural network architecture.

The basic module for a volume holographic neural system is sketched in Fig. 1. The activity of the ith neuron in this system is represented by an optical signal on the ith pixel at the input plane. A connection is made from the ith neuron to the jth neuron via a holographic grating coupling the mode excited by the ith input with the mode incident on the jth output pixel. The signals diffracted from all the input modes onto the jth output mode are summed at the jth output pixel. The summed signal is then processed (e.g. thresholded) by a nonlinear optoelectronic device that is placed at each node. The strength of the connection is proportional to the the product of the activity of the ith pixel on the input plane and the activity of the jth pixel at the training plane. Light from the jth pixel on the training plane excites the mode which couples into the jth pixel on the output plane. The strength of the connection between the ith and j^{th} neurons is

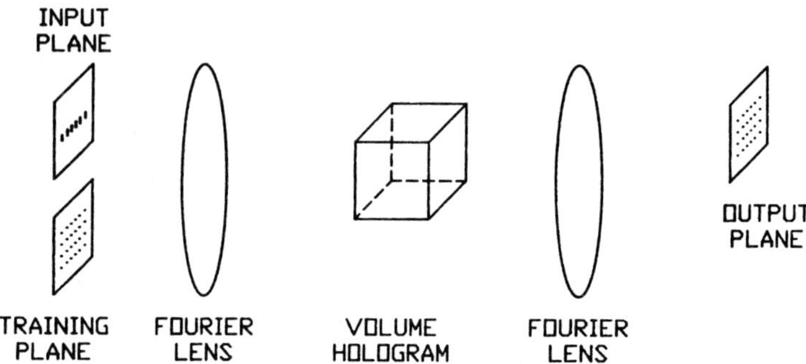

Figure 1. Volume holographic neural system.

increased if the interference pattern between the corresponding modes is in phase with the hologram which stores the connection and is decreased if the interference pattern and the hologram are out of phase.

In addition to connecting the modes used to record it, a grating coupling a pair of modes results in undesired connections between other pairs of modes. To prevent this degeneracy from constraining the nature of the transformations which can be implemented using a hologram, the modes which are coupled by the hologram must be restricted. Since each mode corresponds to a unique pixel on the input or output plane, only a subset of the available pixels can be used in the interconnection system. Suppose that the input and output planes each consist of N^2 pixels. If the same number of pixels are to be used on the input and output plane, only $N^{\frac{3}{2}}$ of the pixels on each plane may be used in an unconstrained interconnection system. A grid wich samples pixels on the input and output planes such that each pair of input-output pixels may be independently interconnected is shown in Fig. 2. The input pixels are sampled densely as shown at the top of the figure. The output pixels are arranged in lines separated by the width of the input grid. In this figure, $N = 100$.

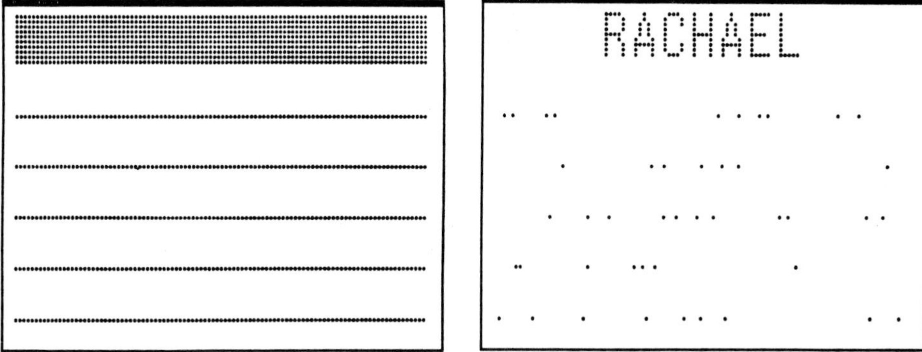

Figure 2. Sampling grids.

Figure 3. Code-name pair.

A hologram recorded in a photorefractive crystal remains stable in the dark, but the mobile charge created by writing a new hologram increases the conductivity of the material and causes the charge patterns corresponding to previous holograms to decay exponentially in time. The time constant of the decay is inversely proportional to the intensity of the writing beams. The decay of previously recorded holograms limits the number of exposures which can be usefully recorded in a photorefractive crystal. We have described previously, [7], an exposure schedule by which an arbitrary number of holograms, M, each with identical diffraction efficiency, may be recorded in a photorefractive material. Unfortunately, recording each hologram with the same strength necessitates a fall off in the efficiency with each successive exposure. The end result is that the diffraction efficiency of each hologram falls off in proportion to $\frac{1}{M^2}$.

A single layer associative memory in which the strengths of the interconnections are formed as the sum of the outer products of the stored vectors can be simulated using photorefractive holograms in the system of Fig. 1 by exposing the hologram with a sequence of appropriately sampled patterns on the input and training planes. As an example of a memory of this type, we have stored up to twenty associations between random image-name pairs of the sort shown in Fig. 3 using this approach. The sampling grids of Fig. 2 where used in these experiments. The reconstruction fidelity when only a few patterns are recorded was good and we were able to verify the fall off in diffraction efficiency with the square of the number of associations recorded. Mok et. al. have recently recorded up to 500 holograms in a similar experiment [8].

III. Short and long term memory.

An adaptive system responds in real time to some function of its input and control signals. In general, the input and control signals will be generated externally and will not drive the system to its "optimal" state in a minimum number of steps. For a system implemented using photorefractive crystals, it is desirable to minimize the number of learning steps in order to preserve the diffraction efficiency, and thus the dynamic range, of the recorded hologram. One method by which the effective number of recording steps can be minimized is to record the hologram in a two step process using a "short term memory" to condense the information stored in each short series of exposures and a "long term memory" to store the connections which emerge over the duration of the learning process. This approach is particularly applicable to systems which adapt continuously, in which case an indefinitely long sequence of exposures controls a finite number of independent connections in the hologram.

In this chapter we describe and present experimental results from a preliminary implementation of short term-long term storage. The basic idea is to use two holographic media to periodically refresh each other. The architecture of the system is shown in Fig. 4. A series of holograms between a reference plane wave and a set of signal beams is recorded in a cesium doped strontium barium nio-

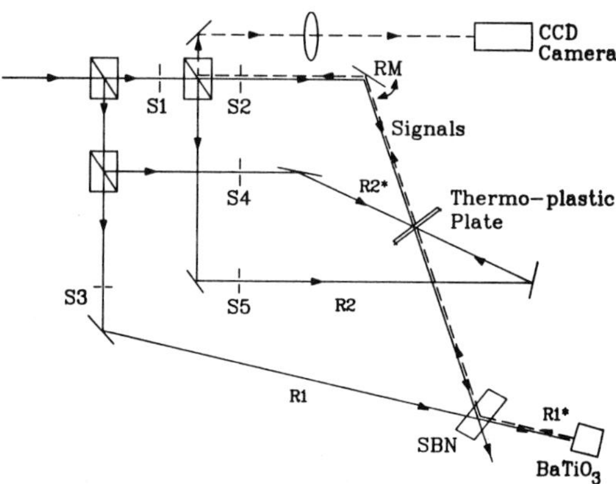

Figure 4. Experimental system layout.

bate crystal (SBN:Ce). Shutters $S4$ and $S5$ are closed during this operation. In our experiments we used plane waves generated by rotation of the mirror RM as signal beams. The diffraction efficiency of the recorded holograms is monitored continuously using the phase conjugate of the reference wave. The path of the diffracted conjugate to an output CCD is shown as a dashed line in the figure. A self-pumped $BaTiO_3$ phase conjugate mirror is used to generate the conjugate wave. When the diffraction efficiency of the photorefractive holograms reaches unacceptably low levels, the recorded holograms are copied from the SBN to a second holographic medium, which in our experiments was a thermoplastic plate. The thermoplastic hologram is formed using the diffracted phase conjugate reference and a back-traveling reference wave. Shutters $S2$ and $S4$ are closed. The hologram written on the plate is copied back to the SBN with the original intensities in the signal and reference beams. The original reference beam and the conjugate to the thermoplastic reference are used to create this hologram. Shutters $S1$ and $S5$ are closed during this step. The result is a rejuvenated hologram of each of the signal beams in the SBN. The diffraction efficiency of each hologram is now proportional to $\frac{1}{M}$ as opposed to the previous $\frac{1}{M^2}$. Since the diffraction efficiency per hologram for M superposed patterns is at best $\frac{1}{M}$, this copying scheme allows us to implement adaptation under multiple exposures with no cost compared to recording the same information in a single exposure.

The temporal behavior of photorefractive holograms may be described by growth in the amplitude of the space charge density proportional to $(1 - e^{-\alpha I t})$ during recording and decay proportional to $e^{-\alpha I t}$ during the recording of successive holograms. I is the recording intensity. The amplitude of the space charge corresponding to the mth hologram when M holograms are recorded is

$$A_m = A_o(1 - e^{-\alpha I t_m})\exp(-\sum_{m'=m+1}^{M}\alpha I t_{m'}), \tag{1}$$

where A_o is the saturation diffraction efficiency and t_m is the recording time of the mth hologram. A_m is a constant for all m if

$$t_m = (\alpha I)^{-1} \log\left(\frac{1 + (m-1)\chi}{1 + (m-2)\chi}\right), \qquad (2)$$

where $\chi = \frac{A_1}{A_o}$.

To begin recording a series of holograms in the system of Fig. 4, we record m_1 holograms on the SBN following the schedule of eqn. (2) for $\chi = 1$. The amplitude of the space charge for each hologram is $\frac{A_o}{m_1}$ at this point. When only a few holograms are recorded, the diffraction efficiency of the optical field may be nonlinear in the space charge amplitude due to pump depletion. We assume, however, that m_1 is large enough that the diffraction has fallen to the linear regime. In this case the diffraction efficiency in intensity for each of the stored holograms is $\frac{\eta_o}{m_1{}^2}$, where η_o is the saturation diffraction efficiency for a single hologram. We now copy the summed holograms in the SBN onto the thermoplastic plate by using the phase conjugate of the reference beam to read out the crystal. Copying the hologram on the thermoplastic back onto the SBN with the original total intensities in the signal and reference beams results in a restoration of the photorefractive hologram with $\sqrt{m_1}$ times greater amplitude. The reduction by a factor of $(\sqrt{m_1})^{-1}$ in the amplitude of each hologram results from a reduction in the modulation depth with which each hologram is recorded due to the sharing by all m_1 signals of the intensity available in the signal beam. This factor is inherent in the simultaneous recording of m_1 signals. However, the total diffraction efficiency, summed over all m_1 holograms, is restored to its saturation value.

At this point we begin recording another series of new holograms on the SBN using the schedule of eqn. (2) with $\chi = \frac{1}{\sqrt{m_1}}$. We make m_2 exposures in this cycle. In order to maintain a constant diffraction efficiency from the thermoplastic, m_2 is selected such that the total diffraction efficiency of the summed hologram on the SBN falls back to its value after the first m_1 exposures, i. e. $\frac{\eta_o}{m_1}$. After m_2 exposures we copy back to the thermoplastic, back to the SBN and again make holograms until the total diffraction efficiency falls again to $\frac{\eta_o}{m_1}$. From here the process may proceed indefinitely. Each time $M = \sum m_i$ holograms are copied back and forth, the diffraction efficiency for each hologram is restored to $\frac{1}{M}$.

Fig. 5 is a log-log plot of experimental results for recording holograms in using the exposure schedule of Eq. 2 and using periodic copying. The diffraction efficiency of the recorded holograms was monitored in each case using the diffracted phase conjugate reference and the CCD shown in Fig. 4. The solid line in Fig. 5 corresponds to the theoretical M^{-2} decay in the diffraction efficiency per hologram with no copying between short and long term storage. The $*$'s are experimental data points for the mean diffracted power of the stored holograms. The dashed line shows the theoretical decay in diffraction per hologram when periodic copying is

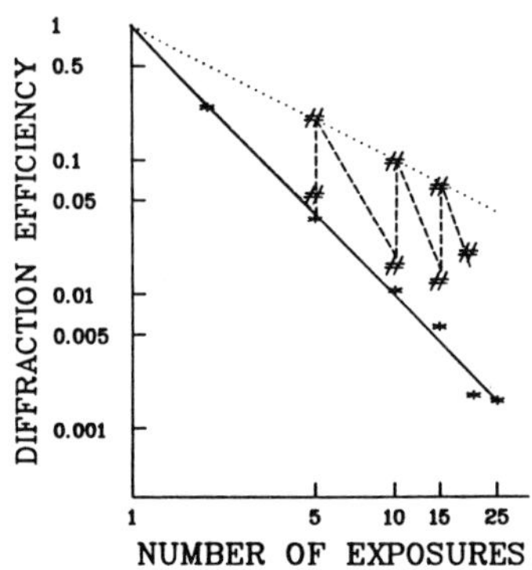

Figure 5. Mean diffraction efficiency vs. number of exposures.

used. The first 5 holograms are recorded exactly as in Fig. 2. These five are then copied to the thermoplastic and back to the SBN, restoring the diffraction efficiency per hologram to the M^{-1} line, which is the dotted line in the figure. Thermoplastic holograms are also made after 10 and 15 exposures. The #'s represent experimental data points.

Acknowledgement

This work is supported by the Defense Advanced Research Projects Agency and the Air Force Office of Scientific Research. The authors thank Rockwell International and Ratnakar Neurgaonkar for the SBN crystal used in these experiments. Portions of this work were presented at the International Joint Conference on Neural Networks, Washington, DC, in January 1990.

References

[1] D. Psaltis and N. H. Farhat, Opt. Let. **10**,(2), 98(1985).

[2] Y. S. Abu-Mostafa and D. Psaltis, Scientific American, **256(3)**,88(1987).

[3] D. Z. Anderson, Opt. Lett., **11**,45 (1986).

[4] A. Yariv and S. K. Kwong, Opt. Lett., **11**,482 (1986).

[5] Special issue on Neural Networks, Appl. Opt., **26**, no. 23, 1987.

[6] See, for example, P. Gunter and J.-P. Huignard, eds., *Photorefractive materials and their applications I and II*, Springer-Verlag, (Berlin)1989.

[7] Demetri Psaltis, David Brady, and Kelvin Wagner, Appl. Opt., **27**, 1752-1759, 1 May 1988.

[8] F. Mok, M. Tackitt, H. M. Stoll, OSA Annual Meeting, *1989 Technical Digest Series*, **12**, Optical Society of America, Washington, DC, 74(1989).

REFERENCES FROM ALL CONTRIBUTIONS

Aarts, E.H.L. and Korst, J.H.M., Simulated Annealing and Boltzmann Machines, (Wiley, Chichester, 1989).

Abeles, M., Local Cortical Circuits, An Electrophysiological Study, (Springer, Berlin, New York, 1982).

Abelson, H. and diSessa, A., Turtle geometry, (MIT Press, Cambridge, 1980, paperback 1986).

Abu–Mostafa, Y.S. and Psaltis, D., optical neural computers, Scientific American 256, (1987) p.88.

Ackley, D.H., Hinton, G.E. and Sejnowski T.J., A learning algorithm for Boltzmann machines, Cognitive Science 9, (1985) pp.147–169.

Adorf, H.-M., Connectionism and neural networks, in: Knowledge–Based Systems in Astronomy, F. Murtagh and A. Heck (eds.), (Springer, Berlin, 1989) pp.215–245.

Adrian, E.D. and Matthews, R., The action of light on the eye I, The discharge of impulses in the optic nerve and its relation to the electric changes in the retina, J. Physiol. 63, (1927) pp.378–414.

Aertsen, A.M.H.J. and Gerstein, G.L., Evaluation of neuronal connectivity, Sensitivity of cross–correlation, Brain Research 340, (1985) pp.341–354.

Aertsen, A.M.H.J., Gerstein, G.L., Habib, M.K., and Palm, G., Dynamics of neuronal firing correlation, Modulation of 'Effective Connectivity', J. of Neurophysiology 61, (1989) pp.900–917.

Ahlsen, G., Lindström, S., and Fu–Sun, L., Exp. Brain Research 58, (1985) pp.134–143.

Ahmed, H.M., Delosme, J.M., and Morf, M., Highly concurrent computing structures for matrix arithmetic and signal processing, IEEE Computer 15, (1982) pp.65–82.

Aiple, F. and Krüger, J., Neuronal synchrony in monkey striate cortex: Interocular signal flow and dependency on spike rates, Exp. Brain Research 72, (1988) pp.141–149.

Ajioka, Y., Anzai, Y., and Aiso, H., Adaptive Junction, A spatio–temporal neuron model, IJCNN–90 Washington D.C., in press.

Ajioka, Y., Anzai, Y., and Aiso, H., The discrimination of rotating patterns and the local representation on adaptive junction, IEICE Tech. Report NC 89, (JAPAN, 1989) in press.

Ajioka, Y., Ishikawa, M., Akamatsu, M., and Anzai, Y., The mechanisms of learning on adaptive junction, Symposium on Paradigms of Learning and Their Applications, Proc. IPS Symposium 89, (JAPAN, 1989) pp.11–20.

Albano, A.M., Abraham, N.B., de Guzmam, G.C., Tarroja, M.F.H., Brandy, D.K., Gioggia, R.S., Rapp, P.E., Zimmermann, I.D., Greenbaum, N.N., and Bashore, T.R., Laser and brains, complex systems with low–dimensional attractors, in: Dimensions and entropies in chaotic systems, G. Mayer–Kress (ed.), (Springer, Berlin, 1986), pp.231–240.

Albert, A., Regression and the Moore–Penrose Pseudoinverse, (Academic Press, New York, 1972).

Albus, J.S., A new approach to manipulator control, The cerebellar model articulation controller (CMAC), J. Dynamical Systems, Meas., Contr.97, (1975) pp.220–227.

Aleksander, I. and Morton, H.B., An Introduction to Neural Computing, Ch.5, (Chapman and Hall, UK, and MIT Press, USA, 1989).

Aleksander, I. and Morton, H.B., An overview of weightless neural nets, Proc. IJCNN 90, Washington, (1990).

Aleksander, I. and Stonham, T.J., A guide to pattern recognition using random access memories, IEE J. Computer and Digital Technology 2, (1979) pp.29–40.

Aleksander, I., Neural Computing Architectures, (North Oxford Academic Publ., London, 1989).

Aleksander, I., The logic of connectionist systems, in: Eckmiller, R., Malsburg v.d., C. (eds.), Neural Computers, (Springer, Berlin, 1988) pp.189–197.

Aleksander, I., The logic of connectionist systems, in: Aleksander, I. (ed.), Neural Computing Architectures, (North Oxford Academic Publ. London, 1989) pp.133–155.

Aleksander, I., Thomas, W.V., and Bowden, P.A., WISARD, A radical step forward in image recognition, Sensor Review 4, (1984) pp.120–124.

Almeida, L.B., Backpropagation in nonfeedforward networks, Neural computing architectures, I. Aleksander (ed.), (North Oxford Academic Publ., London, 1989) pp.74-91.

Amari, S. and Maginu, K., Statistical neurodynamics of associative memory, Neural Networks 1, (1988) pp.63–73.

Amari, S., Learning patterns and pattern sequences by selforganizing nets of threshold elements, IEEE Trans. Computer C–21, (1972) pp.1197–1206.

Amari, S.–I., Neural theory of association and concept formation, Biol. Cybern. 26, (1977) pp.175–185.

Amit, D.J., Gutfreund, H, and Sompolinsky, H., Storing infinite numbers of patterns in a spin–glass model of neural networks, Phys. Rev. Letters 55, (1985) pp.1530–1533.

Amit, D.J., Gutfreund, H. and Sompolinsky, H., Statistical mechanics of neural networks near saturation, Annual Phys. (N.Y.) 173, (1987) pp.30–67.

Amit, D.J., Neural networks counting chimes, Proc. Natl. Acad. Sci. USA 85, (1988) pp.2141–2145.

Anderson, D.Z., Coherent optical eigenstate memory, Opt. Lett. 11, (1986) p.45.

Anderson, J.R., Skill acquisition: Compilation of weak–method problem solutions, Psychol. Rev. 94, (1987) pp.192–210.

Anderson, P., Gillow, M., and Rudjord, T., J. of Physiol. 185, (1966) p.418.

Anderson, S., Merrill, J., and Port, R., Dynamic speech categorization with recurrent networks, in: D. Touretzky, G. Hinton, and T. Sejnowski (eds.), Proc. 1988 Connectionist Models Summer School, (Morgan Kaufmann, Palo Alto, 1988) pp.398–406.

Anninos, P.A. and Anogianakis, G., Computer simulation studies to deduce the structure and function of the human brain, in: Cotterill, R.M.J. (ed.), Computer Simulation in Brain Science, (Cambridge University Press, 1988).

Anninos, P.A. and Cyrulnic, R., J. Theoret. Biol. 66, (1977) p.695.

Anninos, P.A. and Kokkinidis, M., J. Theoret. Biol. 109, (1984) p.95.

Anninos, P.A., A neural model with multiple memory domains, in: Eckmiller, R. and v.d.Malsburg, C. (eds.), Neural Computers, (Springer, Berlin, 1988).

Anninos, P.A., Beek, B., Csermely, T.J., Harth, E.M., and Pertile, G., J. Theoret. Biol. 26, (1970) pp.121.

Anninos, P.A., Zenone, S., and Elul, R., J. Theoret. Biol. 103, (1983) p.339.

Aoki, K., Ikezawa, O., Migubayashi, N., and Yamamoto, K., Chaotic noise in the firing wave instability of the current filaments in GaAs, Physica 134 B, (1985) pp.288–292.

Arbib, M.A., The Metaphorical Brain 2, (Wiley, New York, 1989).

Asanuma, H., Fernandez, J., Scheibel, M.E., and Scheibel, A.B., Exp. Brain Res. 20, (1974) p.315.

Askew, C.R., Carpenter, D.B., Chalker, J.T., Hey, A.J.G., Moore, M., Nicole, D.A., and Pritchard, D.J., MC simulation on Transputer Arrays, Par. Computer 6, (1988) pp.247–258.

Ayers, G.R. and Dainty, J.C., Iterative blind deconvolution method and its applications, Opt. Lett. 13, (1988) pp.547–549.

Babloyantz, A. and Destexhe, A., Is the normal heart a periodic oscillator, Biol. Cybern. 58, (1988) pp.203–211.

Babloyantz, A. and Destexhe, A., Low dimensional chaos in the instance of epilepsy, Proc. Natl. Acad. Sci. USA 83, (1986) pp.3513–3517.

Babloyantz, A., Evidence of chaotic dynamics of brain activity during the sleep cycle, in: Dimensions and entropies in chaotic systems, G. Mayer–Kress (ed.), (Springer, Berlin, 1986) pp.241–245.

Bach, M. and Krüger, J., Correlated neuronal variability in monkey visual cortex revealed by a multi–microelectrode, Exp. Brain Research 61, (1981) pp.451–456.

Baird, B., Nonlinear Dynamics of Pattern Formation and Pattern Recognition in the Rabbit Olfactory Bulb, Physica 22 D, (1986) pp.150–175.

Baldi, P. and Hornik, K., Neural Networks and Principal Component Analysis, Learning from Examples Without Local Minima, Neural Networks 2, (1989) pp.53–58.

Baldi, P. and Venkatesh, S.S., Number of stable points for Spinglasses and neural networks of higher order, Phys. Rev. Lett. 58, (1987) pp.913–916.

Ballard, D.H., Eye movements and spatial cognition, Technical Report TR 218, Computer Sci. Department, Univ. of Rochester, (1987).

Ballard, D.H., Hinton, G.E., and Sejnowski, T.J., Parallel visual computation, Nature 306, (1983) pp.21–26.

Bannister, J.A. and Trivedi, K.S., Task allocation in fault tolerant distributed systems, in: Hard Real–Time Systems (Tutorial), J.A. Stankovic and K. Ramamritham (eds.), IEEE Computer Society Press, (1988) pp.256–272.

Baron, R.J., The Cerebral Computer, An Introduction to the Computational Structure of the Human Brain, (Lawrence Erlbaum Ass., Hillsdale, New Jersey, 1987).

Barto, A.G. and Jordan, M.I., Gradient following without backpropagation in layered networks, in: IEEE First Int. Conf. on Neural Networks, San Diego, (1987) pp.629–636.

Barto, A.G., Sutton, R.S., and Anderson, C.W., Neuronlike adaptive elements that can solve difficult learning control problems, IEEE Transactions on Systems, Man, and Cybern. SMC–13, (1983) pp.834–846.

Bauer, H.–U. and Geisel, T., Feedback multilayer perceptrons for time–warping invariant signal processing, Phys. Rev. A, (1989) in press.

Bauer, H.–U. and Geisel, T., Motion and direction detection in local neural nets, Int. J. Neural Systems, (1989) in press.

Baum, E.B., Moody, J., and Wilczek, F., Internal representations for associative memory, Biol. Cybern. 59, (1988) pp.217–228.

Becker, L.A. and Peng, J., Using activation networks for analogical ordering of consideration, IEEE Conf. on Neural Networks, San Diego, (1987) pp.367–371.

Becks, K.–H., Cremers, A.B., and Hemker, A., Design of an expert–system–shell for fault–diagnosis in the DELPHI Experiments, KI 1/89, (Oldenburg Verlag, 1989) in German.

Behme, H. and Gramss, T., Speicherung und Reproduktion gesprochener Worte mittels neuronaler Netzwerke, in: Fortschritte der Akustik – DAGA '89, (DPG–GmbH, Bad Honnef, 1989) pp.319–322.

Bell, E.T., Men of mathematics – The lives and achievements of the great Mathematicians from Zeno to Poincaré, (Simon & Schuster, New York, 1937, 1965, and 1986).

Benati, M., Gaglia, S., Morasso, P., Tagliasco, V., and Zaccaria, R., Anthropomorphic robotics, Vol.I, Representing mechanical complexity, Biol. Cybern. 38, (1980) pp.125–140.

Berg, T., Die Abbildung des Sprachproduktionsprozesses in einem Aktivierungs-flußmodell, (Tübingen: Niemeyer, 1988).

Berg, T., The Problems of Language Control: Editing, Monitoring, and Feedback, Psychol. Res. 48, (1986) pp.133–144.

Best, J., Reuss, S., and Dinse, H.R.O., Lamina–specific differences of visual latencies following photic stimulation in the cat striate cortex, Brain Research 385, (1986) pp.356–360.

Beunder, M., Kernhof, J., and Hoefflinger, B., The CMOS GATE FOREST: An efficient and flexible high performance ASIC design environment, IEEE J. Solid State Circuits 23, pp.387–399.

Bisset, D.L., Filho, E.C.D.B.C., and Fairhurst, M.C., A comparative study of neural network structures for practical application in a pattern recognition environment, in: First IEE Int. Conf. on Artificial Neural Networks, (London, UK, Oct.1989) pp.378–382.

Boch, R., Behavioral modulation of neuronal activity in monkey striate cortex, excitation in the absence of active central fixation, Exp. Brain Research 64, (1986) pp.610–614.

Böhme, J.F. and Yang, B., Highly parallel and pipelined computing using CORDIC processors and applications to computer graphics, signal processing and linear algebra, in: Proc. German–Chinese Electronics Week, (VDE Verlag, Düsseldorf, March 1990) in press.

Boer de, E., On the residue and auditory pitch perception, Handb. Sens. Physiol. V, Part 3, (1976) pp.479–583.

Bonhoeffer, T., Staiger, V., and Aertsen, A.M.H.J., Synaptic plasticity in rat hippocampal slice cultures: Local ”Hebbian” conjunction of pre– and postsynaptic stimulation leads to distributed synaptic enhancement, Proc. Natl. Acad. Sci. USA 86, (1989) pp.8113–8117.

Booker, L.B., Classifier systems that learn internal world models, in: Machine Learning 3, (Oct. 1988) pp.161–192.

Bottini, S., An After–Shannon measure of the storage capacity of an associative noise–like coding memory, Biol. Cybern. 59, (1988) pp.151–159.

Bourlard, H. and Wellekens, C.J., Multilayer perceptrons and automatic speech recognition, in: Proc. of IEEE First Int. Conf. on Neural Networks, San Diego, (1987) pp.407–416.

Boven, K.–H. and Aertsen, A.M.H.J., Functional connectivity in neuronal systems, fast modulation of synaptic efficacy reflects dynamics of network activity, in: Dynamics and Plasticity in Neuronal Systems, N. Elsner, W. Singer (eds.), (Thieme, Stuttgart, 1989), p.444.

Bradburn, D.S, Reducing transmission error effects using a self–organizing network, Proc. IJCNN 89, Int. Joint Conf. on Neural Networks, Washington D.C., (1989) pp.II/531–537.

Braitenberg, V., Cell assemblies in the cerebral cortex, in: R. Hein and G. Palm (eds.), Theoret. Approaches to Complex Systems, Lecture Notes in Biomathematics 21, (Springer, Berlin, 1978) pp.171–188.

Braitenberg, V., in: Palm, G., Aertsen, A.M.H.J. (eds.), Brain Theory, (Springer, New York, 1986) pp.81–96.

Briggs, J., Fire in the crucible, (St. Martin's Press, New York, 1988).

Brown, G.D.A. and Oaksford, M., Symbolic behaviour and code generation, The emergence of "equivalence relations" in neural networks, in: Proc. Tenth European Meeting on Cybern. and Systems Research, (World Scientific Publishing Corp., 1990) in press.

Brown, G.D.A. and Watson, F.L., First in, first out: Word learning age and spoken word frequency as predictors of word familiarity and word naming latency, Memory & Cognition 15, (1987) pp.208–216.

Brown, R. and McNeill, D., The 'Tip of the Tongue' phenomenon, J. Verbal Learning and Verbal Behaviour 5, (1966) pp.325–337.

Brown, R., Topology – A geometric account of general topology, homotopy types and the fundamental groupoid, (Horwood, Chichester, 1988).

Bruge, F., Martorana, V., And Fornili, S.L., Concurrent statistical simulation on a Transputer–based systems, in: Proc. CONPAR'88, C.U. Jesshope and K.D. Reinartz (eds.), (Cambridge University Press, 1989).

Bruge, F., San Biagio, P.L., and Fornili, S.L., Transputer–based upgrading of a laser photon correlator, Rev. Sci. Instrum. 60, (1989) pp.222–225.

Bülthoff, H.H., Little, J.J., and Poggio, T., A parallel algorithm for real–time computation of optical flow, Nature 337, (1989) pp.549–553.

Büttner, W., Grundlagenforschung und Softwareentwicklung im Siemensproject 'Neurodemonstrato', in: W. Brauer (ed.) Wissensbasierte Systeme, (Springer, Heidelberg, 1989).

Buhmann, J. and Schulten, K., Noise–driven temporal association in neural networks, Europhys., Lett. 4, (1987) pp.1205–1209.

Buhmann, J., Lange, J., v.d.Malsburg, C., Vorbrüggen, J.C., and Würtz, R.P., Object recognition in the dynamic link architecture – parallel implementation on a transputer network, in: Kosko, B. (ed.), Neural Networks: A Dynamical Systems Approach to Machine Intelligence, (Prentice–Hall, New York, 1990).

Burr, D.C., Morrone, M.C., and Spinelli, D., Evidence for edge and bar detectors in human vision, Vision Research 29, (1989) pp.419–431.

Bybee, J. and Slobin, D., Rules and schemata in the development and use of the English past tense, Language 58, (1982) pp.265–289.

Caianiello, E., Outline of a theory of thought–processes and thinking machines, J. Theoret. Biology 1, (1961) pp.204–235.

Campenhausen v., C., Die Sinne des Menschen, (Georg Thieme, Stuttgart, New York, 1981).

Caputo, J.G. and Atten, P., Metric entropy, An experimental means for characterizing and quantifying chaos, Phys. Rev. A 35, (1987) pp.1311–1316.

Caramazza A., Some aspects of language processing revealed through the analysis of acquired aphasia, The Lexical System, Ann. Rev. Neurosci. 11, (1988) pp.395–421.

Carpenter, G.A. and Grossberg, S., A massively parallel architecture for a self–organizing neural pattern recognition machine, Computer Vision, Graphics, and Image Processing 37, (1987) pp.54–115.

Carpenter, G.A. and Grossberg, S., ART 2: self–organization of stable category recognition codes for analog input patterns, Applied Optics 26, (1987) pp.4919–4930.

Carpenter, G.A. and Grossberg, S., ART 3: Hierarchical search using chemical transmitters in self–organizing pattern recognition architectures, Neural Networks, (1990) in press.

Carpenter, G.A. and Grossberg, S., The ART of adaptive pattern recognition by a self–organizing neural network, IEEE Computer: Special issue on Artificial Neural Systems 21, (1988) pp.77–88.

Carpenter, G.A., Neural network models for pattern recognition and associative memory, Neural Networks 2, (1989) pp.243–257.

Carter, F.L. (ed.), Molecular Electronic Devices, (Marcel Dekker, New York, 1982).

Cavanagh, P., Pathways in Early Vision, in: Pylyshyn, Z.W. (ed.), Computational Processes in Human Vision, An Interdisciplinary Perspective, (Ablex Publ. Corp., Norwood, 1988) pp.239–261.

Chate, H. and Manneville, P., Transition to turbulence via spatiotemporal intermittency, Phys. Rev. Lett. 58, (1987) pp.112–115.

Chater, N. and Oaksford, M., Autonomy, implementation and cognitive architecture, A reply to Fodor and Pylyhsyn. Cognition 33, (1989) in press.

Chauvin, Y., A back–propagation algorithm with optimal use of hidden units, in: D.S. Touretzky (ed.), Advances in Neural Information Processing Systems I, (Morgan Kaufmann, San Mateo, 1989) pp.519–526.

Chen, H.H., Lee, Y.C., Sun, G.Z., Lee, H.Y., Maxwell, T., and Giles, C.L., High order correlation model for associative memory, J.S. Denker (ed.), Neural Networks for Computing, (Amer. Inst. Physics, 1986) pp.86–92.

Churchland, P., Neurophilosophy, Toward a Unified Sci. of the Mind–Brain, (MIT Press, Cambridge, MA, 1986).

Clark, A., Being there: Why implementation matters to cognitive science, Artificial Intelligence Review 1, (1987) pp.231–244.

Clark, J.W., Rafelski, J. and Winston, J.V., Brain without mind: Computer simulation of neural networks with modifiable neuronal interactions, Physics Reports 123, (1985) pp.215–273.

Clarkson, T.G., Gorse, D., and Taylor, J.G., Hardware Realisable models of Neural Networks, in: Proc. of the First IEE Int. Conf. on Neural Networks, pp.242–246.

Cohen, M.A. and Grossberg, S., Absolute stability of global pattern formation and memory storage in neural networks, IEEE Trans. Systems Man and Cybern. 13, (1983) pp.815–825.

Cohen, M.A., Grossberg, S., and Storck, D., Recent Developments in a Neural Model of Real–Time Speech Analysis and Synthesis, Proc. IEEE First Int. Conf. on Neural Networks, San Diego, (1987) pp.IV/443–453.

Cohen, M.A., Grossberg, S., and Stork, D., Speech perception and production by a self–organizing neural network, in: Lee, Y.C. (ed.), Evolution, Learning, Cognition, and advanced Architectures, (World Scientific Publishers, Hong Kong, 1988) pp.217–231.

Combes, J.M., Grossmann, A., and Tchamitchian, P. (eds.), Wavelets, Time–Frequency Methods and Phase Space, (Springer, Berlin, 1989).

Cotter, N.E., Smith, K., and Gasper, M., A pulse width modulation design approach and programmable logic for artificial neural networks, in: Proc. of 5th MIT Conf. Advanced Research in VLSI, (MIT Press, Cambridge, 1988) pp.1–15.

Cottrell, G.W., Munro, P.W., and Zipser, D., Image compression by back propagation, A demonstration of extensional programming, in: Advances in Cognitive Sci. 2, N.E. Sharkey (ed.), (Abbex, Norwood, NJ, 1989).

Courant, R. and Robbins, H., What is Mathematics?, (Oxford University Press, New York, 1941 and 1969).

Crick, F., Function of the thalamic reticular complex: The searchlight hypothesis, Proc. Natl. Acad. Sci. USA 81, (1984) pp.4586–4590.

Crisanti, A. and Sompolinsky, H., Dynamics of spin systems with randomly asymmetric bonds, Ising spins and Glauber dynamics, Physical Review A 37, (1988) pp.4865–4874.

Cruse, H. and Brüwer, M., The human arm as a redundant manipulator, the control of path and joint angles, Biol. Cybern., 57 (1987) pp.137–144.

Cruse, H., Constraints for joint angle control of the human arm, Biol. Cybern. 54, (1986) pp.125–132.

Cruse, H., Dean, J., and Suilman, M., The contributions of diverse sense organs to the control of leg movement by a walking insect, J. Computer Physiology A 154, (1984) pp.695–705.

Cruse, H., The control of path and joint angles in a human arm. in: L. Personnaz, G. Dreyfus (eds.), Neural Networks from Models to Applications, (IDSET, Paris, 1989).

Cruse, H., The control of the anterior extrem position of the hindleg of a walking insect, Carausius morosus, Physiol. Entomol 4, (1979) pp.121–124.

Cruse, H., The function of the legs in the free walking stick insect, Carausius morosus, J. Computer Physiology 112, (1976) pp.235–262.

Crutchfield, J.P. and Kaneko, K., Are attractors relevant to turbulence, Phys. Rev. Lett. 26, (1988) pp.2715–2718.

Crutchfield, J.P., Farmer, J.D., Packard, N.H., and Shaw, R.S., Chaos, Sci. Am. 225, (1986) pp.38–49.

Cun le, Y., Generalization and network design strategies, in: R. Pfeiffer, Z. Schreter, F. Fogelman–Solié, L. Steels (eds.), Proc. – Connectionism in Perspective, SGAICO, (Elsevier Sci. Publishers B.V., 1989).

Cybenko, G., Approximation by superpositions of a sigmoidal function, Math. Control Signals Systems 2, (1989) pp.303–314.

Damasio, A.R., The brain binds entities and events by multi–regional activation from convergence zones, Neural Computation 1, (1989) pp.123–132.

Dammasch, I.E., Structural realization of a Hebb–type learning rule, in: Cotterill, R.M.J. (ed.), Models of Brain Function, (Cambridge University Press, 1989) in press.

Dammasch, I.E., Wagner, G.P., and Wolff, J.R., Self–stabilization of neuronal networks, The compensation algorithm for synaptogenesis, Biol. Cybern. 54, (1986) pp.211–222.

Daugman, J.G., Complete discrete 2–D Gabor transforms by neural networks for image analysis and compression, IEEE Trans. ASSP–36, (1988) pp.1169–1179.

Daunicht, W.J., A biophysical approach to the spatial function of eye movements, extraocular proprioception and vestibulo–ocular reflex, Biol. Cybern. 58, (1988) pp.225–233.

Daunicht, W.J., Control of manipulators by neural networks, IEE Proc. 136, (1989) pp.395–399.

Daunicht, W.J., DEFAnet – a deterministic approach to function approximation, IJCNN 1990, Washington, (1990) in press.

Davies, P.C.W., The new Physics, (Cambridge University Press, 1989).

Davis, L., Genetic Algorithms and Simulated Annealing, (Morgan Kaufman, Los Altos 1987).

Davis, P.J. and Hersh, R., Descartes Dream – The World according to Mathematics, (Harcourt Brace Javanovich, 1986, and Penguin Books, London, 1988).

Dean, J. and Wendler, G., Stick insect locomotion on a walking wheel: Interleg coordination of leg position, J. Exp. Biology 103, (1983) pp.75–94.

deCallatay, A., Biological aspects of neural networks, in: Artificial Neural Networks, (Presses Polytechniques Romandes, 1989) pp.1–15.

deCallatay, A., Brain model with periodic processing, Curr. Mod. Biol. 2, (1969) pp.307–319.

deCallatay, A., Can artificial intelligence help in finding how brains may work?, in: Brain Dynamics: Progress and Perspectives, E. Basar, T. Bulloch (eds.), (Springer, Heidelberg, 1989) pp.214–232.

deCallatay, A., Logic programs directly processed in a network of content addressable memories, Future Generations Computer Systems 4, (1988) pp.117–131.

deCallatay, A., Natural and Artificial Intelligence: Processor Systems Compared to the Human Brain, (Elsevier, North Holland, Amsterdam, 1986).

Dehaene, S., Changeux, J.P., and Nadal, J.P., Neural networks that learn temporal sequences by selection, Proc. Natl. Acad. Sci. USA 84, (1987) pp.2727–2731.

DeJong, K.A., Learning with genetic algorithms: An overview, in: Machine Learning 3, No.2/3, (1988).

Dell, G.S., A spreading–activation theory of retrieval in sentence production, Psych. Rev. 93, (1986) pp.283–321.

Dell, G.S., Positive feedback in hierarchical connectionist models, Applications to language production, Cognitive Sci. 9, (1985) pp.3–23.

Dell, G.S., The retrieval of phonological forms in production: Tests of prediction from a connectionist model, J. Memory and Language 27, (1988) pp.124–142.

Deneubourt, J.L., Pasteels, J.M., and Verhaeghe, J.C., Probabilistic behaviour in ants: A strategy of errors?, in: J. Theoret. Biol. 105, (1983).

Derrida, B., Gardner, E., and Zippelius, A., An exactly sovable asymmetric neural network model, Europhys. Lett. 4, (1987) pp.167-173.

Descartes, R., Discours DE LA METHODE pour bien conduire sa raison & chercher – La verité dans les sciences, plus LA DIOPTRIQUE. LES METEORES. Et LA GEOMETRIE. Qui sont des essais de cete METHODE, (Ian Maire, Leyden, 1637).

DeYoe, E.A. and Van Essen, D.C., Concurrent processing streams in monkey visual cortex, Trends in Neurosciences 11, (1988) p.219.

Diederich, S. and Opper, M., Learning of correlated patterns in spin–glass networks by local learning rules, Phys. Rev. Lett. 58, (1987) pp.949–952.

Dingledine, R., N–methyl–aspartate activates voltage–dependent calcium conductance in rat hippocampal pyramidal cells, J. Physiol. 343, (1983) pp.385–405.

Dinse, H.R.O., Informationsverarbeitung im visuellen System der Katze, Neuronale Grundlagen, (Thieme Verlag, Stuttart, 1989).

Dinse, H.R.O., Krüger, K., and Best, J., Temporal aspects of cortical information processing: Cortical architecture, oscillations and non–separability of spatio–temporal receptive field organization, in: Krüger, J. (ed.), Neural Cooperativity: Models and Experiments, (Springer, Berlin, Heidelberg, 1989).

Ditzinger, T. and Haken, H., Oscillations in the perception of ambiguous patterns, Biol. Cybern. 61, (Springer, Berlin, 1989) pp.279–287.

Divenyi, P.L., Is pitch a learned attribute of sounds? Two points in support of Terhardt's theory, J. Acoust. Soc. Am. 66, (1979) pp.1210–1213.

Divko, R. and Schulten, K., Stochastic spin models for pattern recognition, in: Neural Networks for Computing, J.S. Denker (ed.), American Institute of Physics Publication, (1986) pp.129–134.

Domany, E., Neural Networks: A biased overview, J. Stat. Physics 51, (1988) pp.743–775.

Doyle, P.G. and Snell, J.L., Random Walks and Electric Networks, (The Mathematical Association of America, 1984).

Dugdale, N. and Lowe, C.F., Naming and stimulus equivalence, in: D.E. Blackman and H.Lejeune (eds.), Behaviour Analysis in Theory and Practice, Contributions and Controversies, (Lawrence Erlbaum Associates, Brighton, 1989) in press.

Dvorak, I. and Siska, J., On some problems encountered in the estimation of the correlation dimension of the EEG, Phys. Lett. 118 A, (1986) p.63.

Dwoyer, D.L., Hussaini, M.Y., and Voigt, R.G. (eds.), Finite elements – Theory and application, (Springer, New York, 1988).

Eckhorn, R., Bauer, R., Jordan, W., Brosch, M., Kruse, W., Munk, M., and Reitboeck, H.J., Coherent oscillations: A mechanism of feature linking in the visual cortex? Multiple electrode and correlation analysis in the cat, Biol. Cybern. 60, (1988) pp.121–130.

Eckhorn, R., Reitboeck, H.J., Arndt, M., and Dicke, P., A neural network for feature linking via synchronous activity, in: Cotterill, R.M.J. (ed.), Models of Brain Function, (Cambridge University Press, 1989) pp.1–18.

Eckhorn, R., Reitboeck, H.J., Arndt, M., and Dicke, P., Feature linking via stimulus–evoked oscillations: Experimental results from cat visual cortex and functional implications from a network model, Proc. IJCNN–89, IEEE, (1989) pp.I/723–730.

Eckmann, J.P. and Ruelle, D., Ergodic theory of chaos and strange attractors, Rev. Modern Physics 57, (1985) p.617.

Eckmiller, R. and v.d.Malsburg, C. (eds.), Neural Computers, (Springer Verlag, Berlin, 1988, reprinted 1989 and 1990).

Eckmiller, R., Beckmann, J., Werntges, H., and Lades, M., Neural kinematics net for a redundant robot arm, IJCNN–89, (1989) pp.333–340.

Eckmiller, R., Concept of a 4–joint machine with neural net control for the generation of 2–dimensional trajectories, in: Neural Networks 1, Suppl. 334 (1988).

Eckmiller, R., Electronic simulation of the vertebrate retina, IEEE Trans. BME–22, (1975) pp.305–311.

Edelman, G.M. and Finkel, L., Neuronal group selection in the cerebral corbex, in: Edelman, G.M., Cowan, W.M. and Gall, W. (eds.), Dynamic Aspects of Neocortical Function, (Wiley, New York, 1984) pp.653–695.

Edelman, G.M. and Mountcastle, V.B., The Mindful Brain, (MIT Press, Cambridge, 1978).

Edmondson, A.C., A Fuller explanation – The synergetic geometry of R. Buckminster Fuller, (Birkhäuser, Boston, 1987).

Eigen, M. and Winkler, R., Das Spiel, (Piper, München, 1975).

Elman, J.L., Finding structure in time, CRL Tech. Report 8801, University of California, San Diego, (1988).

Erb, M., Aertsen, A.M.H.J., and Palm, G., Functional connectivity in neuronal systems: Context–dependance of effective network organization does not require synaptic plasticity, in: Dynamics and Plasticity in Neuronal Systems, N. Elsner, W. Singer (eds.), (Thieme, Suttgart, New York, 1989) p.445.

Ermentrout, G.B. and Cowan, J.D., Temporal oscillations in neuronal nets, J. Math. Biol. 7, (1979) pp.265–280.

Ewert, J.-P. and v.Seelen, W., Neurobiologie und System–Theorie eines visuellen Muster–Erkennungsmechanismus bei Kröten, Kybernetik 14, (1974) pp.167–183.

Ewert, J.-P. and Wietersheim v., A., Musterauswertung durch tectale und thalamus / praetectale Nervennetze im visuellen System der Kröte (Bufo bufo L), J. Comp. Physiol. 92, (1974) pp.131–148.

Ewert, J.-P., Neuroethology of releasing mechanisms: Prey–catching in toads, Behavioral Brain Sci. 10, (1987) pp.337–405.

Ewert, J.-P., Quantitative Analyse von Reiz–Reaktions–Beziehungen bei visuellem Auslösen der Beutefang–Wendereaktion der Erdkröte (Bufo bufo L), Pflügers Archiv 308, (1969) pp.225–243.

Fahlman, S.E., Faster learning variations of back–propagation: An empirical study, in: Touretzky, D., Hinton, G. and Sejnowski, T. (eds.), Proc. Connectionist Models Summer School, (1988) pp.38–51.

Fang, M., Häusler, G., and Weiss, T., Chaos and cooperation in nonlinear pictorial feedback systems, Proc. of SPIE, Vol.667, (1986) pp.214–219.

Farhat, N.H., Psaltis, D., Prata, A., and Paek, E., Optical implementation of the Hopfield model, Applied Opt. 24, (1985) pp.1469–1475.

Faure, B. and Mazaré, G., A VLSI asynchronous cellular architecture dedicated to multilayered neural networks, in: Neural Networks – from Models to Applications, Personnaz, L. and Dreyfus, G. (eds.), (IDSET, Paris, 1989) pp.710-719.

Feldman, A.G., Functional tuning of the nervous system during control of movement or maintenance of a steady posture: Vol.III, Mechanographic analysis of the execution by man of the simplest motor tasks, Biofizika 11, (1967) pp.667–675.

Feldman, J.A. and Ballard, D.H., Connectionist models and their properties, Cognitive Science 6, (1982) pp.205–254.

Feldman, J.A., Fanty, M.A., and Goddard. N.H., Computing with structured neural networks, Computer 21, (1988) pp.91–102.

Feldman, J.A., Fanty, M.A., Goddard, N.H., and Lynne, K.J., Computing with structured connectionist networks, Comm. of the AMC 31, (1988) pp.170– 187.

Feldman, J.A., Four frames suffice: a provisional model of vision and space, Behavioral and Brain Sci.s, Vol.8, (1985) pp.265–289.

Feldman, J.A., Time, Space and Form in Vision, TR–88–011, Internal Computer Sci. Institute, Berkeley, California, (1988).

Fender, K. and Julez, B., Extension of Panum's fusional area in binoculary stabilized vision, J. Opt. Soc. Am. 87, (1967) pp.819–830.

Field, D.J., Relations between the statistics of natural images and the response properties of cortical cells, J. Opt. Soc. Am. A 4(12), (1987) pp.2379–2394.

Fienup, J.R., Reconstruction of an object from the modulus of its Fourier transform, Opt. Lett. 3, (1978) pp.27–29.

Fincham, D., Parallel computers and molecular simulation, Molec. Simulation 1, (1987) pp.1–45.

Fischer, B., Overlap of receptive field centers and representation of the visual field in the cat's optic tract, Vision Res. 13, (1973) pp.2113–2120.

Fitts, P.M., The information capacity of the human motor system in controlling the amplitude of movement, J. Exp. Psychol. 47, (1954) pp.381–391.

Fodor, J.A. and Pylyshyn, Z.W., Connectionism and cognitive architecture: A critical analysis, Cognition 28, (1988) pp.3–71.

Földiak, P., Adaptive network for optimal linear feature extraction, in: Proc. IJCNN, Washington, (1989) p.I/401.

Fogelman-Soulie, F., Weisbuch, G., Random iterations of threshold networks and associative memory, SIAM J. Computing 16, (1987), pp.203-219.

Fohlmeister, J., Electrical processes involved in the encoding of nerve impulses, Biol. Cybern. 36, (1980) pp.103–108.

Forgy, C.L., RETE: a fast algorithm for the many pattern / many object pattern match problem, Artificial Intelligence J. 19, (1982) pp.17–37.

Forrest, B.M., Roweth, D., Stroud, N., Wallace, D.J., Wilson, G.V., Implementing neural network models on parallel computers, Comp. J. 30, (1987) pp.413-419.

Forrest, S., Implementing semantic network structures using classifier systems, in: Grevenstette (ed.), Proc. Int. Conf. on Genetic Algorithms and their Applications, (Carnegie–Mellon University, Pittsburgh, 1985).

Fox, G.C., Johnson, M., Lyzenga, G., Otto, S., Salmon, J., and Walder, D., Solving Problems on Concurrent Processors, (Prentice–Hall, Englewood Cliffs, NJ, 1988).

Freeman, W.J. and van Dijk, B.W., Spatial patterns of visual cortical fast EEG during conditioned reflex in a Rhesus monkey, Brain Res. 422, (1987) pp.267–276.

Freeman, W.J., Mass Action in the Nervous System, (Academic Press, New York, 1975).

Freeman, W.J., Nonlinear neural dynamics in olfaction as a model for cognition, in: Basar, E., Dynamics of Sensory and Cognitive Processing by the Brain, (Springer, Berlin, 1988) pp.19–29.

Freeman, W.J., Simulation of chaotic EEG patterns with a dynamic model of the olfactory system, Biol. Cybern. 56, (1987) p.139.

French, A.S. and Stein, R.B., A flexible neuronal analog using integrated circuits, IEEE Trans. Biomed. Eng. 17, (1970), pp.248–253.

Friedlander, B., Lattice filters for adaptive filtering, Proc. IEEE 70, (1982) pp. 829–867.

Friedrich, J. and Haarer, D., Angewandte Chemie, Int. Ed. Eng. 23, (1984) p.113.

Fromkin, V.A., Appendix, in: Fromkin, V.A. (ed.), Speech Errors as Linguistic Evidence, (Mouton, Paris, 1973) pp.241–269.

Fromkin, V.A., The non–anomalous nature of anomalous utterances, **Language 47**, (1971) pp.27–52.

Fuchs, A. and Haken, H., Computer simulations of pattern recognition as a dynamical process of a synergetic system, in: H. Haken (ed.), Neural and Synergetic Computers, (Springer, Berlin, Heidelberg, 1988) pp.16–28.

Fujii, M. and Fukushima, K., A neural network model for binocular parallax extraction, Tech. Report, IEICE Japan, MBE–88–148, (March 1989) in Japanese.

Fukushima, K., A feature extractor for curvilinear patterns, a design suggested by the mammalian visual system, Kybernetik 7, (1970) pp.153–160.

Fukushima, K., A neural network model for selective attention in visual pattern recognition, Biol. Cybern. 55, (1986) pp.5–15.

Fukushima, K., Cognitron: A self–organizing multilayered neural network, Biol. Cybern. 20, (1975) pp.121–136.

Fukushima, K., Neocognitron: A hierarchical neural network capable of visual pattern recognition, Neural Networks 1, (1988) pp.119–130.

Fukushima, K., Neocognitron: A self–organizing neural network model for a mechanism of pattern recognition unaffected by shift in position, Biol. Cybern. 36, (1980) pp.193–202.

Fukushima, K., Neural network model for selective attention in visual pattern recognition and associative recall, Applied Optics 26, (1987) pp.4985–4992.

Funahashi, K., On the approximate realization of continuous mappings by neural networks, Neural Networks 2, (1989) pp.183–192.

Gabor, D., Theory of communication, J. Inst. Elec. Eng. 93, (London), (1946) pp.429–457.

Gallaire, H. and Lasserre, C., Metalevel control for logic programs, in: Clark / Tärlund (eds.), Logic Programs, (Academic Press, London, 1982) pp.173–188.

Gallant, S.I., Connectionist expert systems, Comm. of the ACM 31, (1987) pp.52–169.

Gamst, A., A resource allocation technique for FDMA systems, Alta Frequenza 57, (1988) pp.89–96.

Gamst, A., Beck, R., Simon, R., and Zinn, E.G., An integrated approach to cellular radio network planning, Proc. 35th Veh. Techn. Conf., Boulder, CO, May 1985, pp.21–25.

Gardiner, C.W., Handbook of Stochastic Methods, (Springer, Berlin 1983).

Gardner, E. and Derrida, B., Optimal storage properties of neural network models, J. Physics A 21, (1988) pp.271–284.

Gardner, E., Optimal basins of attraction in randomly sparse neural network models, J. Physics A 22, (1989) pp.1969–1974.

Gardner, E., The space of interactions in neural network models, J. Physics A 21, (1988) pp.257–270.

Garrett, M.F., Levels of processing in speech production, in: Butterworth, B. (ed.), Language Production 1, (Academic Press, London, 1980) pp.177–220.

Garrett, M.F., The analysis of sentence production, in: Bower, G. (ed.), Psychology of Learning and Motivation 9, (Academic Press, New York, 1975) pp.133–177.

Geman, S., Stochastic relaxation, Gibbs distribution, and the Bayesian restoration of images, IEEE PAMI–6, (1984) pp.721–741.

Genesereth, M.R. and Ginsberg, M.L., Logic programming, in: Comm. of the ACM 28, (1985) pp.933–941.

Gerchberg, W. and Saxton, O., A practical algorithm for the determination of phase from image and diffraction plane pictures, Optik 35, (1972) pp.237–246.

Gerstein, G.L., Bedenbaugh, P., and Aertsen, A.M.H.J., Neuronal assemblies, IEEE Trans. Biomedical Engineering 36, (1989) pp.4–14.

Ghosh, J. and Hwang, D., Mapping neural networks onto message–passing multicomputers, J. Parallel and Distributed Computing 6, (1989) pp.291–330.

Gilbert, C.D. and Wiesel, T.N., Morphology and intracortical projections of functionally characterized neurons in the cat visual cortex, Nature 280, (1979) pp.120–125.

Glauber, R.J., Time–dependent statistics of the Ising model, J. Math. Physics 4, (1963) pp.294–307.

Glünder, H., A dualistic view of motion and invariant shape analysis, in: Simon, J.C. (ed.), From Pixels to Features, (Elsevier, Amsterdam, 1989) pp.323–332.

Glünder, H., Correlative velocity estimation: visual motion analysis, independent of object form, in arrays of velocity–tuned bilocal detectors, J. Opt. Soc. Am. A 7, (1990) pp.317–329.

Glünder, H., Invariant description of pictorial patterns via generalized autocorrelation functions, in: Meyer–Ebrecht, D. (ed.), ASST'87, (Springer, Berlin, 1987) pp.84–87.

Gold, B., Lippmann, R.P., and Malpass, M.L., Some neural net recognition results on isolated words, in: Proc. First Int. Conf. on Neural Networks, IEEE, (1987) pp.IV/427–434.

Goldberg, D.E., Genetic Algorithms in Search, Optimization, and Machine Learning, (Addison–Wesley, Reading, 1989).

Goldberg, D.E., Optimal initial population size for Bibary–coded genetic algorithms, Tech. Report TCGA, No.85001, Univ. of Alabama, Tuscaloosa, (1985).

Gonzalez, D.L., Piro, O., Global bifurcations and phase portrait of an analytically solvable nonlinear oscillator: Relaxation oscillations and saddle–node collisions, Physical Rev. A 36, (1987) pp.4402–4410.

Goodglass, H., Wingfield, A., and Hyde, M.R., Category specific dissociations in Namint and recognition by aphasic patients, Cortex 22, (1986) pp.87–102.

Gorse, D. and Taylor, J.G., Physical Lett. A 131, (1989) pp.326–332.

Gorse, D. and Taylor, J.G., Physica D 34, (1989) p.90.

Graf, D.H. and LaLonde, W.R., Neuroplanners for hand / eye coordination, Proc. IJCNN 89, Washington D.C., (1989) pp.II/543–548.

Grassberger, P. and Procaccia, I., Characterisation of strange attractors, Physical Rev. Lett. 50, (1983) pp.346–349.

Grassberger, P. and Procaccia, I., Measuring the strangeness of strange attractors, Physica 9 D, (1983) pp.189–208.

Gray, C.M. and Singer, W., Stimulus–specific neuronal oscillations in orientation columns of cat visual cortex, Proc. Natl. Acad. Sci. USA 86, (1989) pp.1698–1702.

Gray, C.M. and Singer, W., Stimulus–specific neuronal oscillations in the cat visual cortex: A cortical functional unit, Soc. Neurosci. Abstracts 13, (1987) pp.404.3.

Gray, C.M., König, P., Engel, A.K., and Singer, W., Oscillatory responses in cat visual cortex exhibit inter–columnar synchronization which reflects global stimulus proberties, Nature 338, (1989) pp.334–337.

Grefenstette, J.J., Optimization of control parameters for genetic algorithms, IEEE Trans. Systems, Man, and Cybernetics SMC–16, (1986).

Grossberg, S. (ed.), Neural Networks and Natural Intelligence, (MIT Press, Cambridge, MA, 1988).

Grossberg, S. and Ellias, S.A., Biol. Cybern. 20, (1975) pp.69–98.

Grossberg, S. and Mingolla, E., Neural dynamics of perceptual grouping: Textures, boundaries, and emergent segmentations, Perception and Psychophysics 38, (1985) pp.141–171.

Grossberg, S. and Mingolla, E., Neural dynamics of surface perception: Boundary webs, illuminants, and shape–from–shading CVGIP 37, (1987) pp.116–165.

Grossberg, S. and Todorović, D., Neural dynamics of 1D and 2D brightness perception: A unified model of classical and recent phenomena, Perception and Psychophysics 43, (1988) pp.241–277.

Grossberg, S., Adaptive pattern classification and universal recording, Vol.II: Feedback, expectation, olfaction, illusions, Biol. Cybern. 23, (1976) pp.187–202 and pp.121–134.

Grossberg, S., Cortical dynamics of three–dimensional form, color and brightness perception, Vol.I: Monocular theory and Vol.II: Binocular theory, Perception and Psychophysics 41, (1987) pp.87–158.

Grossberg, S., Nonlinear neural networks: Principles, mechanisms and architectures, Neural Networks 1, (1988) pp.17–61.

Grossberg, S., Studies of Mind and Brain: Neural principles of learning, perception, development, cognition, and motor control, (Reidel Press, Boston, 1982).

Grossmann, T., Lerning by choice of internal representations, Complex systems 2, (1989) p.555.

Grünbaum, B. and Shepard, G.C., Tilings and Patterns, (Freeman, New York, 1987).

Guckenheimer, J. and Holmes, P., Nonlinear Oscillations, Dynamical Systems, and Bifurcations of Vector Fields, (Springer, Berlin, 1983).

Gumm, H.P. and Hergert, F.B., Neural Pascal (NP), Siemens Tech. Report INF2–ANN–5–89, München, (1989).

Gunter, P. and Huignard, J.–P. (eds.), Photorefractive Materials and Their Applications, Vol.I and Vol.II, (Springer, Berlin, 1989).

Gutfreund, H. and Mézard, M., Processing of temporal sequences in neural networks, Physical Rev. Lett. 61, (1988) pp.235–238.

Guyon, I., Personnaz, L., and Dreyfus, G., Of points and loops, in: R. Eckmiller and C. v.d.Malsburg (eds.), Neural Computers, (Springer, Heidelberg, 1988) pp.261–269.

Guyon, I., Personnaz, L., Nadal, J.P., and Dreyfus, G., Storage and retrieval of complex sequences in neural networks, Physical Rev. A 38, (1988) pp.6365–6372.

Haarer, D., Advanced Materials I, (1989) p.361; and in: Angewandte Chemie, Adv. Mater., Bd.101, (1989) p.1576.

Haarer, D., in: Persistent Spectral Hole Burning: Sci. and Applications, Topics in Current Physics, Moerner, W.E. (ed.), Vol.44 (Springer, 1988) pp.79ff.

Haarer, D., Jap. J. Appl. Physics 26, Suppl.26, (1987) pp.227.

Häusler, G., Seckmeyer, G., and Weiss, T., Chaos and cooperation in nonlinear pictorial feedback systems, Vol.I–III, Appl. Opt. 25, (1986) pp.4656–4672.

Haken, H. (ed.), Neural and Synergetic Computers, (Springer, Berlin, Heidelberg, New York, 1988).

Haken, H., Haas, R., and Banzhaf, W., Biol. Cybern., (1989) in press.

Haken, H., Kelso J.A.S., and Bunz, H., A theoretical model of phase transitions in human hand movements, Biol. Cybern. 51, (1985) pp.347–356.

Haken, H., Synergetics – An Introduction, (Springer, Berlin, Heidelberg, New York, 1983) 3rd edition.

Hamilton, A., Murray, A.F., and Tarassenko, L., Programmable analog pulse–firing neural networks, in: D.S. Touretzky (ed.), Advances in Neural Inf. Proc. Systems I, (Morgan Kaufmann, San Mateo, CA, 1989) pp.671–677.

Hampshire, J.B. and Waibel, A.H., A novel objective function for improved phoneme recognition using time–delay neural networks, Tech. Report CMU–CS–89–118, Computer Sci. Department, Carnegie Mellon Univ., Pittsburgh, PA, (March 1989).

Hancock, P.J.B., Data representation in neural nets: An empirical study, in: D. Touretzky, G. Hinton, T. Sejnowski (eds.), Proc. 1988 Connectionist Models Summer School, (Morgan Kaufmann, San Mateo, CA, 1989) pp.11–20.

Hartline, H.K., The response of single optic nerve fibres of the vertebrate eye to illumination of the retina, Am. J. Physiol. 121, (1938) pp.400–415.

Hartmann, G., Processing of continuous lines and edges by the visual system, Biol. Cybern. 47, (1983) pp.43–50.

Hartstein, A. and Koch, R.H., A neural network capable of forming associations by example, Neural Networks 2, (1989) pp.395–403.

Hassenstein, B., Reichardt, W., Systemtheoretische Analyse der Zeit–, Reihenfolgen– und Vorzeichenauswertung bei der Bewegungsrezeption des Rüsselkäfers 'Chlorophanus', Z. Naturforschung 11b, (1956) pp.513–524.

Hauser, R., Horner, H., Makhaniok, M., Männer, R., Architectural considerations for NERV – a general purpose neural network simulation system, Proc. WOPPLOT 89, Wildbad Kreuth, FRG, (1989) in print.

Haykin, S., Adaptive Filter Theory, (Prentice–Hall, Englewood Cliffs, NJ, 1986).

Hebb, D.O., The Organization of Behavior, (Wiley, New York, 1949).

Hecht–Nielsen, R., Counterpropagation networks, in: Proc. IEEE First Int. Conf. Neural Networks, San Diego, Vol.II, (1987) pp.19–31.

Heistermann, J., Parallel algorithms for learning in neural networks with evolution strategy, Parallel Computing, (1989).

Hildreth, E.C. and Koch, C., The analysis of visual motion: From computational theory to neuronal mechanism, Annual Rev. Neuroscience 10, (1987) pp.477–533.

Hillis, W.D., The Connection Machine, (MIT Press, Cambridge, 1985).

Hillis, W.D. and Steele, G.L., Data parallel algorithms, in: Comm. of the ACM 29, (1986) pp.1170–1183.

Hinton, G.E. and Anderson J.A. (eds.), Parallel Models of Associative Memory, (Lawrence Erlbaum Ass., Hillsdale, New Jersey, 1989) updated edition.

Hinton, G.E., Connectionist learning procedures, Artificial Intelligence 40, (1989) pp.185–234.

Hinton, G.E., McClelland, J.L. and Rumelhart, D.E., Distributed representations, in: Parallel Distributed Processing: Explorations in the Microstructure of Cognition, Ch.3, (MIT Press, Bradford Books, 1986).

Hinton, G.E., Sejnowski, T.J., and Ackley, D.H., Boltzmann machines: Contraint satisfaction networks that learn, Carnegie–Mellon University, Tech. Report CMU–CS–84–119, (1984).

Hirai, Y. and Fukushima, K., A model of neural network extracting binocular parallax, Biol. Cybern. 18, (1975) pp.19–29.

Hirai, Y., Kamada, K., Yamada, M., and Ooyama, M., A digital neuro–chip with unlimited connectability for large scale neural networks, Proc. IJCNN 1989, Washington D.C., (1989) pp.II/163–169.

Ho, A.W. and Furmanski, W., Pattern recognition using neural networks in hypercubes, in: Proc. Third Conf. on Hypercube Concurrent Computers and Applications, Vol.2, Fox, G.C. (ed.), (ACM Press, New York, 1988) pp.1011–1021.

Ho, A.W., A Back–propagation navigation controller for land and space vehicles, Tech. Report C^3P–735, Calif. Inst. Technology, (1989).

Hofbauer, J. and Sigmund, K., Evolutionstheorie und dynamische Systeme, (Paul Parey, Berlin, Hamburg, 1984).

Holland, J.H., Adaptation in Natural and Artificial Systems, (University of Michigan Press, Ann Arbor, 1975).

Holland, J.H., Adaptation, Progress in Theoretical Biol. 4, (1976) pp.263–294.

Holland, J.H., Properties of the bucket brigade, in: Proc. Int. Conf. on Genetic Algorithms, (Lawrence Erlbaum Associates, Hilldale, NJ, 1985) pp.1–7.

Hopfield, J.J. and Tank, D.W., 'Neural' computation of decisions in optimization problems, Biol. Cybern. 52, (1985) pp.141–152.

Hopfield, J.J. and Tank, D.W., Computing with neural circuits: A model, Science 233, (1986) pp.625–633.

Hopfield, J.J., Collective computation, content–addressably memory, and optimization problems, in: Complexity in Information Theory, Y.S. Abu–Mosafa (ed.), (Springer, Berlin, 1989) pp.99–114.

Hopfield, J.J., Neural networks and physical systems with emergent collective computational abilities, Proc. Natl. Acad. Sci. USA 79, (1982) pp.2445–2558.

Hopfield, J.J., Neurons with graded responses have collective computational properties like those of two–state neurons, Proc. Natl. Acad. Sci. USA 81, (1984) pp.3088–3092.

Hornik, K., Stinchcombe, M., and White, H., Multilayer feedforward networks are universal approximators, Neural Networks 2, (1989) pp.359–366.

Horowitz, E. and Sahni, S., Fundamentals of Computer Algorithms, (Computer Sci. Press, 1978).

Houtsma, A.J.M. and Goldstein, J.L., The central origin of the pitch of complex tones: evidence from musical interval recognition, J. Acoust. Soc. Am. 51, (1972) pp.520–529.

Hubel, D.H., Eye, Brain and Vision, The Scientific American Lib., New York, (1988).

Hubel, D.H. and Wiesel, T.N., Receptive fields and functional architecture in two nonstriate visual areas (18 and 19) of the cat, J. Neurophysiology 28(2), (1965) pp.229–289.

Hubel, D.H. and Wiesel, T.N., Receptive Fields, binocular interaction and functional architecture in the cat's visual cortex, J. Physiol. Lond. 160, (1962) pp.106–154.

Hutchinson, J.M. and Koch, C., Simple analog and hybrid networks for surface interpolation, in: Denker, J.S. (ed.), Neural Networks for Computing, American Inst. of Physics, Snowbird UT 151, (1986) pp.235–240.

Jenkin, M. and Jepson, A.D., The measurement of binocular disparity, in: Computational Prosesses in Human Vision: An interdisciplinatory perspective, Z. Pylyshin (ed.), (Ablex Press, Norwood, NJ, 1988) pp.69–98.

Jockusch, S., Panagos, G. and Strube, H.W., Anwendung neuronaler Netzwerde auf die Schätzung von Sprachparametern, in: Fortschritte der Akustik – DAGA '89, (DPG–GmbH, Bad Honnef, 1989) pp.315–318.

Johansson, G., Visual perception of biological motion and a model for its analysis, Perception and Psychophysics 14, (1973) pp.201–211.

Jones, J.P. and Palmer L.A., An evaluation of the two–dimensional Gabor filter model of simple receptive fields in cat striate cortex, J. Neurophysiol. 58, (1987) pp.1233–1258.

Jordan, M.I., Attractor dynamics and parallelism in a connectionist sequential machine, Proc. Eighth Ann. Conf. of the Cognitive Sci. Soc., (1986) pp.531–546.

Jordan, M.I., Generic constraints on underspecified target trajectories, Proc. of IJCNN–89, IEEE, Vol.1, (1989) pp.217–225.

Josin, G., Neural–space generalization of a topological transformation, Biol. Cybern. 59, (1988) pp.283–290.

Julesz, B., Textons, the elements of texture perception, and their interactions, Nature 290, (1981) pp.91–97.

Kaas, J.G., Merzenich, M.M., and Killakey, H.P., The reorganization of somatosensory cortex following peripheral nerve damage in adult and developing mammals, Ann. Rev. Neurosci. 6, (1983) pp.325–356.

Kämmerer, B. and Küpper, W., Perceptrons and multilayer perceptrons in speech recognition: Improvements from temporal warping of the training material, Proc. of nEuro'88, L. Personnaz and G. Dreyfus (eds.), IDSET Paris, (1989) pp.531–540.

Kammen, D.M. and Yuille, A.L., Spontaneous symmetry–breaking energy functions and the emergence of orientation selective cortical cells, Biol. Cybern. 59, (1988) pp.23–31.

Kammen, D.M., Holmes, P.J., and Koch, C., Cortical architecture and oscillations in neuronal networks: Feedback versus local coupling, in: Models of Brain Function, R.M.J. Cotterill (ed.), (Cambridge University Press, Cambridge, 1989).

Kanter, I. and Sampolinsky, H., Associative recall of memory without errors, Physical Rev. A 35, (1987) pp.380–392.

Kapur, D. and Mundy, J.L. (eds.), Geometric reasoning, (MIT Press Cambridge, 1989).

Kasif, S., Kohli, M., and Minker, J., Prism: A Parallel Inference System for Problem Solving, in: Proc. 8th IJAI, Karlsruhe, (1983) pp.544–546.

Kauffman, L.H., On knots, (Princeton University Press, Princeton, 1987).

Kawasaki, M., Rose, G., and Heiligenberg, W., Temporal hyperacuity in single neurons of electric fish, Nature 336, (1988) p.173.

Kelso, J.A.S., Schöner, G., Scholz, J.P., and Haken, H., Phase–locked modes, phase transitions, and component oscillators in biological motion, Physica Scripta 35, (1987) pp.79–87.

Kelso, J.A.S., Southard, D.L., and Goodman, D., On the nature of human interlimb coordiation, Sci. 203, (1979) pp.1029–1031.

Kelso, J.A.S., Tuller, B., Vatikiotis–Bateson, E., and Fowler, C.A., Functionally specific articulatory cooperation following jaw perturbations during speech: Evidence for coordinative structures, J. Exp. Psychol., Hum. Perc. Perf. 10, (1984) pp.812–832.

Kelso, S.R., Ganong, A.H., and Brown, T.H., Hebbian synapses in hippocampus, Proc. Natl. Acad. Sci. USA 83, (1986) pp.5326–5330.

Kertesz, A. (ed.), Localization in Neuropsychology, (Academic Press, New York, 1983).

Keyes, R.W., Physics of digital devices, Rev. Modern Physics 61, (1989) pp.279–287.

Kickert, W.J.M. and Lemke, H.R.v.N., Application of a fuzzy controller in a warm water plant, Automatica 12, (1976) pp.301–308.

Kinzel, W., Learning and pattern recognition in spin glas models, Z. für Physik B, Condensed Matter 60, (1985) pp.205–213.

Kirkpatrick, S., Gelatt, C.P.Jr., and Vecci, M.P., Optimization by simulated annealing, Science 220, (1983) pp.671–680.

Klar, H. and Ramacher, U., Microelectronics for Artificial Neural Networks, (VDI–Verlag GmbH, Düsseldorf, 1989).

Knudsen, E.I., Lac du, S., and Esterley, S.D., Computational maps in the brain, Ann. Rev. of Neurosci. 10, (1987) pp.41–66.

Koch, C. and Poggio, T., Biophysics of computation: neurons, synapses and membranes, in: Synaptic Function, G. Edelman, W.E. Gall, W.M. Cowan (eds.), (Wiley, New York, 1987) pp.637–698.

Koenderinck, J., The concept of local sign, in: A.J. van Doorn et al (eds.), Limits in Perception, (VNU Sci. Press., 1984) pp.495–549.

Kohonen, T., A new model for randomly organized associative memory, Int. J. Neurosci. 5, (1973) pp. 27–29.

Kohonen, T., An introduction to neural computing, Neural Networks 1, (1988) pp.3–16.

Kohonen, T., Analysis of a simple self–organizing process, Biol. Cybern. 44, (1982) pp.135–140.

Kohonen, T., Clustering, taxonomy, and topological maps of patterns, Proc. Sixth Int. Conf. on Pattern Recognition, München, (1982) pp.114–128; and: IEEE Computer Soc. Press, Silver Springs (1982) pp.114-128.

Kohonen, T., Correlation matrix memories, IEEE Trans. Computer C–21, (1972) pp.353–359.

Kohonen, T., Introduction of the principle of virtual images in associative memory, Acta Polytechn. Scandinavica 29, (1971).

Kohonen, T., Mäkisara, K., and Saramäki, T., Phonotopic maps – Insightful representation of phonological features for speech recognition, Proc. Seventh Int. Conf. Pattern Recognition, Montreal, Canada, (IEEE Computer Soc., Silver Spring, 1984) pp.182–185.

Kohonen, T., Reuhkala, E., Makisara, K., and Vainio, L., Associative recall of images, Biol. Cybern. 22, (1976) pp.159–168.

Kohonen, T., Self–Organization and Associative Memory, (Springer, Berlin, 1984), (Springer, New York, 1987), (2nd Ed. Springer, Berlin, Heidelberg, 1988), (3rd Ed. Springer, Berlin, Heidelberg, 1989).

Kohonen, T., Self–organized formation of topologically correct feature maps, Biol. Cybern. 43, (1982) pp.59–69.

Kohonen, T., The 'neural' phonetic typewriter, IEEE Computer 21, (1988) pp.11–22.

Kohonen, T., Torkkola, K., Shozakai, M., Kangas, J., and Ventä, O., Microprocessor implementation of a large vocabulary speech recognizer and phonetic typewriter for Finnish and Japanese, Proc. European Conf. on Speech Technology, CEP Consultants Ltd., Edinburgh, (1987) pp.377–380.

Koikkalainen, P. and Oja, E., Specification and implementation environment for neural networks using communicating sequential processes, Proc. IEEE Int. Conf. on Neural Networks San Diego, July 24.–27. 1988, pp.I/533–540.

Kolers, P.A., Aspects of Motion Perception, (Pergamon Press, 1972).

Koscielny–Bunde, E., Effect of damage in neural networks, J. Statistical Physics 58, (1990) in press.

Kosko, B., Adaptive bidirectional associative memories, Appl. Opt. 26, (1987) pp.4947–4959.

Kowalski, R., Algorithm = Logic + Control, in: Comm. of the ACM 33, (1979) pp.424–436.

Krauth, W. and Mezard, M., Learning algorithms with optimal stability in neural networks, J. Physics A 20 L, (1987) pp.745–752.

Krone, G., Mallot, H.A., Palm, G., and Schüz, A., Spatio–temporal receptive fields: A dynamical model derived from cortical architectonics, Proc. Royal Soc. Lond. B 226, (1986) pp.421–444.

Krüger, J. and Aiple, F., Multi–microelectrode investigation of monkey striate cortex: spike train correlations in the infragranular layers, J. Neurophysiol. 60, (1988) pp.798–828.

Kürten, K.E., Critical phenomena in model neural networks, Phys. Lett. A 129(3), (1988) pp.157–160.

Kürten, K.E., Dynamical phase transitions in short–ranged and long–ranged neural network models, J. Phys. France 50, (1989) pp.2313–2323.

Kuczewski, R.M., Myers, M.H., Crawford, W.J., Neurocomputer Workstations and Processors: Approaches and Application, Proc. IEEE 1st Int'l Conf. on Neural Networks III, (1987) pp.487–426.

Kung, H.T., Why systolic architectures?, IEEE Computer 15, (1982).

Kung, S.Y., On Supercomputing with Systolique / Wavefront Array Processors, Proceeding of the IEEE 72, (1984).

Kupferstein, M. and Rubinstein, J., Implementation of an Adaptive Neural Controller for Sensory–Motor Coordination, IEEE Control Systems Magazine 9, (1989) pp.25–30.

Kuramoto, Y., Chemical Oscillations, Waves, and Turbulence, (Springer, Berlin, New York, Tokyo, 1984).

Lal, R. and Friedlander, M.J., Gating of retinal transmission by afferent position and movement signals, Sci. 243, (1989) p.93–96.

Landauer, R., Fundamental limitations in the computational process, Berichte der Bunsen–Gesellschaft für Physikalische Chemie 80, (1976) pp.1048–1059.

Lang, K.J. and Hinton, G.E., The development of the time–delay neural network architecture for speech recognition, Tech. Report CMU–CS–88–152, Department of Computer Science, Carnegie Mellon University (1988).

Lapedes, A.S. and Farber, R., A self–optimizing, nonsymmetrical neural net for content addressable memory and pattern recognition, Physica 22 D, (1986) pp.247–259.

Lapedes, A.S. and Farber, R., Nonlinear signal processing using neural networks: Prediction and system modeling, Tech. Report LA–UR–87, Los Alamos National Laboratory, (1987) submitted to Proc. IEEE.

Lattard, D. and Mazaré, G., Image reconstruction using an original asynchronous cellular array, IEEE Int. Symp. on Circuits and Systems, (1989) pp.13–17.

Lawley, D.N. and Maxwell, A.E., Factor Analysis as a Statistical method, (Butterworths, London, 1963).

Layne, S.P, Mayer–Kress, G., and Holzfuss, J., Problems associated with dimensional analysis of electroencephalogram data, in: Dimensions and Entropies in Chaotic Systems, G. Mayer–Kress (ed.), (Springer, Berlin, 1986), pp.246–256.

Lee, D.N., The optic flow field: The foundation of vision, Phil. Trans. Royal Soc. Lond. B 290, (1980) pp.169–179.

Lemon, R., The output map of the primate motor cortex, Trends in Neur. Science 11, (1988) pp.501–506.

Levelt, W.J.M., Speaking: From Intention to Articulation, (MIT Press, Cambridge, 1989).

Lewis, P.S., Multichannel adaptive least squares – relating the 'Kalman' recursive least squares (RLS) and least squares lattice (LSL) adaptive algorithms, in: Proc. IEEE ICASSP, (April 1988) pp.1926–1929.

Linsker, R., An application of the principle of maximum information preservation to linear systems, in: Touretzky, D.S. (ed.), Advances in Neural Information Processing Systems 1, (Morgan–Kaufmann, San Mateo, CA, 1989).

Linsker, R., From basic network principles to neural architecture (series), Proc. Natl. Acad. Sci. USA 83, (1986) pp. 7508–7512, pp.8390–8394, and pp.8779–8783.

Linsker, R., How to generate ordered maps by maximizing the mutual information between input and output signals, IBM Res. Report RC 14624, No.65530, (22. May 1989).

Linsker, R., Self–organization in a perceptual network, IEEE Computer 21, (1988) p.105–117.

Lippmann, R.P., An introduction to computing with neural nets, IEEE ASSP Magazine, (April 1987) pp.4–22.

Lippmann, R.P., Review of neural networks for speech recognition, Neural Computation 1, (1989) pp.1–38.

Livingstone, M., Art, illusion and the visual system, Scientific American 1, (1988) pp.68–75.

Lohmann, A.W. and Wirnitzer, B., Triple correlations, Proc. IEEE 72, (1984) pp.889–901.

Lorenz, E.N., Deterministic nonperiodic flow, J. Atmospheric Sci. 20, (1963) pp.130–141.

Lowe, D.G., Perceptual Organization and Visual Recognition, (Kluwer, Boston, 1985).

Ludermir, T.B., A feedback RAM–network for temporal pattern recognition, Neural Systems Eng. Report, Imperial College, Dept. of Electrical Eng., (1989).

Ludermir, T.B., Stability and temporal pattern recognition, in: Proc. IJCNN–90, IEEE, Washington D.C., (1990).

MacGregor, R.J. and Lewis, E.R., Neural Modeling, Electrical Signal Processing in the Nervous System, (Plenum, New York, 1977).

MacGregor, R.J., Neural and Brain Modeling, (Academic Press, London, New York, 1987).

MacKay, D.G., The Organisation of Perception and Action: A Theory for Language and Other Cognitive Skills, (Springer, New York, 1987).

Männer, R., Horner, H. Hauser, R., Genthner, A., multiprocessor simulation of neural networks with NERV, Proc. Supercomputing 89, Reno, NV, 1989, pp.457-465.

Mainzer, K., Die Evolution intelligenter Systeme, Zeichen im Gehirn?, Semiotik und Künstliche Intelligenz, Z. für Semiotik 12, (1990).

Mainzer, K., Symmetrien der Natur, Ein Handbuch zur Natur- und Wissenschaftsphilosophie, (DeGruyter Verlag, Berlin, 1988).

Mallot, H.A. and Brittinger, R., Towards a network theory of cortical areas, in: R. Cotterill (ed.), Models of Brain Function, (Copenhagen, 1989) in press.

Mallot, H.A., An overall description of retinotopic mapping in the cat's visual cortex areas 17, 18 and 19, Biol. Cybern. 52, (1985) pp.45–51.

Mallot, H.A., Point images, receptive fields, and retinotopic mapping, Trends in Neurosciences 10, (1987) pp.310–311.

Mallot, H.A., Schulze, E., and Storjohann, K., Neural network strategies for robot navigation, in: G. Dreyfus and L. Personnaz (eds.), Neural Networks from Models to Applications, (IDSET, Paris, 1989) pp.560–569.

Malsburg v.d., C. and Bienenstock, E., Statistical coding and short–term synaptic plasicity, A scheme for knowledge representation in the brain, in: Bienenstock, E., Fogelman Soulié and Weisbuch, G. (eds.), Disordered Systems and Biological Organization, (Springer, Berlin, 1986).

Malsburg v.d., C. and Schneider, W., A neural cocktail–party processor, Biol. Cybern. 54, (1986) pp.29–40.

Malsburg v.d., C. and Singer, E., Principles of cortical network organization, in: Rakic, P. and Singer, W., Neurobiology of Neocortex, (Wiley, New York, 1988) pp.69–99.

Malsburg v.d., C. and Willshaw, D.J., How to lable nerve cells so that they can interconnect in an ordered fashion, PNAS USA 74, (1977) pp.5176–5178.

Malsburg v.d., C., Nervous structures with dynamical links, Bericht der Bunsengesellschaft, Physikalische Chemie 89, (1985) pp.703–710.

Malsburg v.d., C., Pattern recognition by labeled graph matching, Neural Networks 1, (1988) pp.141–148.

Mandelbrot, B., The fractal geometry of nature, (Freeman, New York, 1982).

Mangold, R. and Pobel, P., Informativeness and instrumentality in referential communication, J. Language and Social Psychology 7, (1989) pp.181–191.

Marcelja, S., Mathematical description of the responses of simple cortical cells, J. Opt. Soc. of Am. A 70, (1980) pp.1297–1300.

Marcus, C.M. and Westervelt, R.M., Basins of attraction for electronic neural networks, in: D.Z. Anderson (ed.), Neural Information Processing Systems, Denver CO 1987, (AIP, 1988) pp.524–533.

Marko, H., A biological approach to pattern recognition, IEEE Trans. SMC–4, (1974) pp.34–39.

Marks, K.M. and Goser, K.F., Analysis of VLSI process data based on self–organizing feature maps, Proc. Neuro–Nimes '88, Nimes, France, (1988) pp. 337–347.

Marr, D. and Poggio, T., A computational theory of human stereo vision, Proc. Royal Soc. Lond. B 204, pp.301–328.

Marr, D., Vision, (Freeman & Co., San Francisco, 1982).

Mars, P. and Poppelbaum, W.J., Stochastic and deterministic averaging processors, IEE Digital Electronics and Computings Series Peregrinus Ltd., (1981).

Martinetz, T., Ritter, H., and Schulten, K., 3–D–neural net for learning visuomotor–coordination of a robot arm, Proc. IJCNN–89 II, Washington D.C, (1989) pp.351–356.

Martinetz, T., Ritter, H., and Schulten, K., Kohonen's self–organizing map for modelling the formation of the auditory cortex of a bat, in: SGAICO–Proc. 'Connectionism in Perspective', Zürich, (1988).

Massen, R., Stochastische Rechentechnik, (Hanser Verlag, 1977).

Matsumoto, N., Schwippert, W.W., and Ewert, J.–P., Intracellular activity of morphologically identified neurons of the grass frog's optic tectum in response to moving configurational visual stimuli, J. Comp. Physiol. 159, (1986) pp.721–739.

Maunsell, J.H.R., Physiological evidence for two visual subsystems, in: L.M. Vaina (ed.), Matters of Intelligence (D. Reidel Publishing Company, 1987) pp.59–87.

Mayer–Kress, G. and Kaneko, K., Spatiotemporal chaos and noise, J. Statistical Physics 54, (1989) pp.1489–1508.

Mazaré, G. and Payan, E., A programmable highly parallel architecture for digital signal processing, IEEE Int. Symp. on Circuits and Systems, (1989) pp.1332–1336.

McClelland, J.L. and Rumelhart, D.E., Explorations in parallel destributed processing, A Handbook of Models, Programs, and Exercises, (MIT Press, Cambridge, Mass, 1988).

McCulloch, W.S. and Pitts, W.H., A logical calculus of the ideas immanent in neuron activity, Bull. Math. Biophysics 5, (1943) pp.115–133.

McEliece, R.J., Posner, E.C., Rodemich, E.R., and Venkatesh, S.S., The capacity of the Hopfield associative memory, IEEE Trans. Inf. Th. IT–33, (1987) p.461.

McLaughlin, J.A. and Raviv, J., Nth–order autocorrelations in pattern recognition, Information and Control 12, (1968) pp.121–142.

McMillan, C. and Smolensky, P., Analyzing a connectionist model as a system of soft rules, Tech. Report CU–CS–393–88, Dept. of Computer Sci., Univ. of Colorado at Boulder, (1988).

Mead, C., Analog VLSI and Neural Systems, (Addison–Wesley, Reading, MA, 1989).

Merzenich, M.M., Jenkins, W.M., and Middlebrooks, J.C., Observations and hypotheses on special organizational features of the central auditory nervous system, in: Edelman, G.M., Cowan, W.M. and Gall, W. (eds.), Dynamic Aspects of Neocortical Function, (Wiley, New York, 1984) p.397–424.

Merzenich, M.M., Kaas, J.H., Wall, J.T., Sur, M., Nelson, R.J., and Felleman, D.J., Progression of change following median nerve section in the cortical representation of the hand in areas 3b and 1 in adult owl and squirrel monkeys, Neuroscience 10, (1983) pp.639–665.

Merzenich, M.M., Nelson, R.J., Kaas, J.H., Stryker, M.P., Jenkins, W.M., Zook, J.M., Cynader, M.S., and Schoppmann, A., Comp. Neurol. 258, (1987) p.281.

Merzenich, M.M., Nelson, R.J., Stryker, M.P., Cynader, M.S., Schoppmann, A., and Zook, J.M., Somatosensory cortical map changes following digit amputation in adult monkeys, J. Comp. Neurol. 224, (1984) pp.591–605.

Merzenich, M.M., Recanzone, G.H., Jenkins, W.M., Allard, T., and Nudo, R.J., in: Rakic, P., Singer, W. (eds.), Neurobiology of Neocortex, Dahlem Konferenzen 1988, (Wiley, New York, 1988) p.41.

Mézard, M. and Nadal, J.P., Learning in feedforward layered networks: the tiling algorithm, J. Physics A 22, (1989) pp.2191–2203.

Mézard, M., Nadal, J.P., and Toulouse, G., Solvable models of working memories, J. Physique 47, (1986) pp.1457–1462.

Migliore, M., Martorana, V., and Sciortino, F., An algorithm to find all paths between two nodes in a graph, J. Comp. Phys., (1989) in press.

Miller, W.T., real–time application of neural networks for sensor–based control of robots with vision, IEEE Trans. on Systems, Man, and Cybernetics 19, (1989) pp.825–831.

Minsky, M., A framework for representing knowledge, in: The Psychology of Computer Vision, Winston P.H. (ed.), (McGraw Hill, New York, 1975) pp.211–277.

Minsky, M.L., The Soc. of Mind, (Simon and Schuster, New York, 1986).

Minsky, M.L. and Papert, S., Perceptrons, (MIT Press, Cambridge, MA, 1969 and 1988).

Mitzdorf, U. and Singer, W., Prominent excitatory pathways in the cat visual cortex (A17 and A18), A current source density analysis of electrically evoked potentials, Exp. Brain Res. 33, (1978) pp.371–394.

Mitzdorf, U., Current source density method and application in cat cerebral cortex, Investigation of evoked potentials and EEG phenomena, Physiol. Rev. 65, (1985) pp.37–100.

Miyakawa, H. and Kato, H., Active properties of dendritic membrane examined by current source density analysis in hippocampal CA1 pyramidal neurons, Brain Res. 399, (1986) pp.303–309.

Moerner, W.E., Persistent spectral hole burning, science and applications, in: Topics in Current Physics 44, Moerner, W.E. (ed.), (Springer, 1988) pp.251ff.

Moore, B.J.C. and Glasberg, B.R., Formulae describing frequency selectivity as a function of frequency and level, and their use in calculating excitation patterns, Hearing Res. 28, (1987) pp.209–225.

Morasso, P., Neural models of cursive script handwriting, Proc. IJCNN 89, Int. Joint Conf. on Neural Networks, Washington D.C., (1989) pp.II/539–542.

Morgan, F., Geometric measure theory, (Academic Press, Boston, 1988).

Mozer, M.C. and Smolensky, P., Skeletonization, A technique for trimming the fat from a network via relevance assessment, in: D. Touretzky (ed.), Advances in Neural Information Processing Systems 1, (Morgan Kaufmann, 1989) pp.107–115.

Mühlenbein, H. and Kindermann, J., The dynamics of evolution and learning – Towards genetic neural networks, in: R. Pfeifer, et al (eds.), Connectionism in Perspective, (Elsevier, Amsterdam, 1989) pp.173–197.

Mugnaini, E. and Oertel, W.H., An atlas of the distribution of GABAergic neurons and terminals in the rat CNS as revealed by GAD immunohistochemistry, in: Björklund, A. and Hökfelt, T. (eds.), Handbook of Chemical Neuroanatomy, Vol.4, Part 1, Ch.10, (Elsevier, Amsterdam, 1985) pp.436–608.

Myers, C.E., Output functions for probabilistic logic nodes, in: Proc. First IEE Int. Conf. of Neural Networks, London, (1989) pp.310–314.

Myerson, J., Manis, P.B., Mieyin, F.M., and Allman, J.M., Magnification in striate cortex and retinal ganglion cell layer of owl monkey: A quantitative comparison, Sci. 198, (1977) pp.855–857.

Naish, L., Prolog control rules, Tech. Report No.13, Univ. of Melbourne, Australia, (1984).

Nakano, K., Associatron, a model of associative memory, IEEE Trans. on Systems, Man and Cybernetics SMC–2, (1972) pp.380–388.

Nedler, J.A. and Mead, C., A SIMPLEX method for function minimization, Computer J. 7, (1965) p.308.

Nelson, M.E., Furmanski, W., and Bower, J.M., Simulating neurons and networks on parallel computers, in: Methods in Neuronal Modeling, From Synapses to Networks, T.J. Sejnowski and T.A. Poggio (eds.), (MIT Press, Cambridge, MA, 1989).

Nelson, P.P., General input–output relations for the electrical behaviour of neurons, Kybernetes 9, (1980) pp.123–131.

Neumann v., J., Probabilistic Logics, in: Automata Studies, C.E. Shannon (ed.), (Princeton University Press, 1956).

Neumann, O., Beyond capacity: A functional view of attention, in: Heuer, H. and Sanders, A.F. (eds.), Perspectives on Perception and Action, (Erlbaum, Hillsdale, NJ, 1987) pp.361–394.

Newman, C.M., Memory capacity in neural network models, rigorous lower bounds, Neural Networks 1, (1988) p.223.

Nicholson, C. and Freeman, J.A., Theory of current source density analysis and determination of conductivity tensor for anuran cerebellum, J. Neurophysiol. 38, (1975) pp.356–368.

Nicolis, G. and Prigogine, I., Exploring Complexity, (Piper Verlag, München, 1987).

Nijhuis, J.A.G. and Spaanenburg, L., Fault tolerance of neural associative memories, IEE Proc. E 136, (1989) pp.389–394.

Nijhuis, J.A.G., Spaanenburg, L., and Warkowski, F., Structure and application of NNSIM: A general–purpose neural network simulator, Microprocessing and Microprogramming 27, (Aug. 1989) pp.189–194.

Nijhuis, J.A.G., Spaanenburg, L., Warkowski, F., Höfflinger, B., Darianian, M., Martini, C., Waldschmidt, K., and Kollbach, D., DUITS: Driver–User Interface for Tablet Symbols, Proc. 2nd PROMETHEUS workshop, Stockholm, Sweden, 1 (Oct. 1989) pp.350–358.

Noble, B. and Daniel, J.W., Applied Linear Algebra, (Prentice–Hall, 1977).

Novak, J.L. and Wheeler, B.C., Two dimensional current source density analysis of propagation delays for components of epileptiform bursts in rat hippocampal slices, Brain Res. 497, (1989) pp.223–230.

Obermayer, K., Mahler, G., and Haken, H., Multistable quantum systems: Information processing at microscopic levels, Phys. Rev. Lett. 58, (1987) pp.1792–1795.

Obermayer, K., Teich, W.G., and Mahler, G., Structural basis of multistationary quantum systems, Vol.I, Effective single–particle dynamics, Phys. Rev. B 37, (1988) pp.8096–8110.

Oja, E., A simplified neuron model as a principal component analyzer, J. Math. Biol. 15, (1982) pp.267–273.

Oja, E., Neural networks: Principal components, and subspaces, Int. J. Neural Systems 1, (1989) pp.61–68.

Olson, T.J., A two–stage model of motion understanding, TR305 and Ph.D. dissertation, Computer Sci. Department, University of Rochester, (1989).

Opper, M., Learning in neural networks: Solvable dynamics, Europhys. Lett. 8, (1989) pp.389–392.

Opper, M., Learning times of neural networks, Exact solution for a PERCEP-
TRON algorithm, Phys. Rev. A 38, (1988) pp.3824–3826.

Packard, N.H., Crutchfield, J.P., Farmer, J.D., and Shaw, R.S., Geometry from
a time series, Phys. Rev. Lett. 45, (1980) pp.712–716.

Paetsch, W. and Kaneko, M., A three fingered, multijointed gripper as a working tool for experimental use, Proc. of the 2nd EIOOT Symposium, Toulouse,
France, (1989).

Paik, E., Gungner, D., and Skrzypek, J., UCLA SFINX – A neural network simulation environment, IEEE First Int. Conf. on Neural Networks, San Diego,
CA, (1987) pp.367–375.

Palm, G., Aertsen, A.M.H.J., and Gerstein, G.L., On the significance of correlation amoung neuronal spike trains, Biol. Cybern. 59, (1988) pp.1–11.

Palm, G., On associative memory, Biol. Cybern. 36, (1980) pp.19–31.

Palumbo, D.L. and Butler, R.W., A performance evaluation of the software–
implemented fault–tolerance computer, J. Guidance 9, (1986) pp.175–180.

Parent, P. and Zucker, S.W., Trace inference, curvature consistency, and curve
detection, IEEE Trans. PAMI 11, (1989) pp.823–839.

Parga, N. and Visasoro, M.A., The ultrametric organization of memories in a
neural network, J. Physique 47, (1986) pp.1857–1864.

Parker, T.S. and Chua. L.O., Chaos: a tutorial for engineers, Proc. IEEE 75,
(1987) pp.982–1010.

Paton, J.A. and Nottebohm, F., neurons generated in the adult brain are recruited into functional circuits, Science 225, (1984) pp.1046–1048.

Pearlmutter, B.A., Learning state space trajectories in recurrent neural networks,
Tech. Report CMU–CS–88–191, Carnegie Mellon University, Pittsburgh, PA,
(1988).

Pearson, J.C., Finkel, L.H., and Edelman, G.M., Plasticity in the organization of
adult cerebral maps: A computer simulation based on neuronal group selection,
J. Neuroscience 12, (1987) pp.4209–4223.

Peeling, S.M. and Moore, R.K., Isolated digit recognition experiments using the
multi–layer perceptron, Speech Communication 7, (1988) pp.403–409.

Penrose, A., On best approximation solutions of linear matrix equations, Proc. Cambridge Phil. Soc. 52, (1956) pp.17–19.

Peretto, P. and Niez, J.P., Collective properties of neural netorks, in: E. Bienestock et al. (ed.), Disordered Systems and Biological Organization, (Springer, 1987) pp.170–185.

Perkel, D.H., Gerstein, G.L., and Moore, G.P., Neuronal spike trains and stochastic point processes II, Simultaneous spike trains, Biophys. J. 7, (1967) pp.419–440.

Perona, P. and Malik, J., Scale–space and edge detection using anisotropic diffusion, Univ. of California, Berkeley, Computer Sci. Division, Report No. UCB–CSD–88–483, (1988).

Perrett, D.I., Rolls, E.T., and Caan, W.C., Visual neurons responsive to faces in the monkey temporal cortex, Exp. Brain Res. 47, (1982) pp.329–342.

Personnaz, L., Guyon, I., and Dreyfus, G., Information storage and retrival in spin–glass like neural networks, J. Physique Lett. 46, (1985) pp.L/359–365.

Personnaz, L., Guyon, I., and Dreyfus, G., Neural network design for efficient information retrieval, in: Disordered Systems and Biological Organisation, E. Bienestock, F. Fogelman, G. Weisbuch (eds.), (Springer, Berlin, New York, 1986) p.4217.

Peterson, C. and Anderson, J.R., Neural networks and NP–complete optimization problems, A performance study on the graph bisection problem, Complex Systems 2, (1988) pp.59–89.

Peterson, C. and Soderberg, B., A new method for mapping optimization problems onto neural networks, Int. J. Neural Systems 1, (1989) pp.3–22.

Peterson, I., The mathematical tourist – Snapshots of modern mathematics, (Freeman, New York, 1988).

Petkov, N., Systolische Algorithmen und Arrays, (Akademie–Verlag, Berlin, 1989).

Pettey, C.P., Leuze, M.R. and Grefenstette, J.J., A parallel genetic algorithm, in: Grefenstette, J.H. (ed.), Proc. Second Int. Conf. on Genetic Algorithms (Lawrence Erlbaum, Hillsdale, 1987).

Pineda, F.J., Generalization of back–propagation to recurrent neural networks, Phys. Rev. Lett. 59, (1987) pp.2229–2232.

Pinker, S. and Prince, A., On language and connectionism, Analysis of a parallel distributed processing model of language acquisition, Cognition, 28, (1988) pp.73–193.

Poggio, T. and Koch, C., Synapses that compute motion, Scientific American 256, (1986) pp.46–71.

Poggio, T., Torre, V., and Koch, C., Computational vision and regularization theory, Nature 317, (1985) pp.314–319.

Poggio, T., Woodward, Y., and Torre, V., Optical flow: computational properties and networks, biological and analog, in: Durbin, R., Miall, C., and Mitchison, G. (eds.), The Computing Neuron, (Addison–Wesley, Wokingham, 1989) pp.355–370.

Pollack, J.B., Recursive auto–associative memory, devising compositional distributed representations, Proc. Tenth Ann. Conf. of the Cognitive Sci. Soc., (1988) pp.33–39.

Pomerleau, D.A., ALVINN: and autonomous land vehicle in a neural network, in: Touretzky, D.S. (ed.), Advances in Neural Information Processing Systems 1, (Morgan Kaufmann Publishers, San Mateo, CA, 1989) pp.305–313.

Pool, R., Science 243, (1989): Is it chaos, or is it just noise, pp.25–28; Ecologists flirt with chaos, pp.310–313; Is it healthy to be chaotic, pp.604–607; Quantum chaos: Enigma wrapped in a mystery, pp.893–895; Is something strange about the weather?, pp.1290–1293.

Protzel, P.W. and Arras, M.K., Fault–tolerance of optimization networks: Treating faults as additional contraints, in: Proc. IJCNN–90, Washington D.C., (1990).

Protzel, P.W., Comparative performance measure for neural networks solving optimization problems, in: Proc. IJCNN–90, Washington D.C., (1990).

Psaltis, D. and Farhat, N.H., Optical information processing based on an associative–memory model of neural nets with threshold and feedback, Opt. Lett. 10, (1985) p.98–100.

Psaltis, D. and Park, C.H., Nonlinear discriminant functions and associative memories, in: Neural Networks for Computing, (AIP, 1986) pp.370–375.

Psaltis, D., Brady, D., and Wagner, K., Applied Opt. 27, (1988) pp.1752–1759.

Qian, N. and Sejnowski, T.J., Predicting the secondary structure or globular proteins using neural network models, J. Molecular Biol. 202, (1988) pp.865–884.

Ramachandran, V.S., Apparent motion of subjective surfaces, Perception 14, (1985) pp.127–134.

Rapp, P.E., Zimmermann, I.D., Albano, A.M., de Guzman, G.C. and Greenbaun, N.N., Dynamics of spontaneous neural activity in the simian motor cortex, the dimension of chaotic neurons, Physique Lett. 110 A, (1985) pp.335–338.

Rechenberg, I., Evolutionsstrategie: Optimierung technischer Systeme nach Prinzipien der biologischen Evolution, (Friedrich Frommann–Holzboog, Stuttgart, Bad Cannstatt, 1973) pp.1–143.

Reggia, J., Methods for deriving competitive activation mechanisms, Proc. IJCNN 1989, Washington D.C., (1989) pp.357–363.

Reichardt, W. and Schlögl, R.W., A two dimensional field theory for optical flow computation, First order approximation, translatory motion of rigid patters, Biol. Cybern. 60, (1988) pp.23–35.

Reitboeck, H.J. and Brody, T.P., A transformation with invariance under cyclic permutations for applications in pattern recognition, Information and Control 15, (1969) pp.130–154.

Reitboeck, H.J., A 19 – Channel matrix drive with individually controllable fiber microelectrodes for neurophysiological applications, IEEE–SMC 13, (1983) pp. 676–682.

Reitboeck, H.J., Eckhorn, R., Arndt, M., and Dicke, P., A model of feature linking via correlated neural activity, in: Haken, H., Synergetics of Cognition, (Springer, 1989).

Renyi, A., Wahrscheinlichkeitsrechnung, (VEB Verlag, Berlin, 1962).

Richmond, B.J., Optican, L.M., Podell, M., and Spitzer, H., Temporal encoding of two–dimensional patterns by single units in primate inferior temporal cortex I, Response characteristics, J. Neurophysiology 57, (1987) pp.132–146.

Riedel, U., Kühn, R., and van Hemmen, J.L., Temporal sequences and chaos in neural nets, Phys. Rev. A 38, (1988) pp.1105–1108.

Ritter, H.J. and Kohonen, T., Self–organizing semantic maps, Biol. Cybern. 61, (1989) pp.241–254.

Ritter, H.J. and Schulten, K.J., Convergence properties of Kohonen's topology conserving maps, Biol. Cybern. 71, (1988) pp.59–71.

Ritter, H.J. and Schulten, K.J., Extending Kohonen's self–organizing mapping algorithm to learn ballistic movements, in: R. Eckmiller, C. v.d.Malsburg (eds.), Neural Computers, (Springer, Heidelberg, 1987) pp.393–406.

Ritter, H.J. and Schulten, K.J., On the stationary state of Kohonen's self–organizing sensory mapping, Biol. Cybern. 54, (1986) pp.99–106.

Ritter, H.J., Martinetz, T.M., and Schulten, K.J., Topology conserving maps for learning visuo–motor–coordination, Neural Networks 2, (1989) p.159–168.

Robertson, G., Population size in classifier systems, in: Proc. Fifth Int. Conf. on Machine Learning, (Morgan Kaufmann, San Mateo, 1988).

Robinson, A.J.R. and Fallside, F., A dynamic connectionist model for phoneme recognition, Proc. of nEuro 88, L. Personnaz and G. Dreyfus (eds.), IDSET Paris, (1989) pp.541–550.

Robinson, A.J.R. and Fallside, F., Static and dynamic error propagation networks with application to speech coding, in: Neural Information Processing Systems, D.Z. Andersen (ed.), American Institute of Physics, (1987) pp.632–641.

Robinson, J.A.R., A machine–oriented logic based on the resolution principle, in: Comm. the ACM–12, (1965) pp.23–44.

Rock, I., An Introduction to Perception, (Collier MacMillan, London, 1975).

Rohwer, R., Renals, S., and Terry, M., Unstable connectionist networks, in: Speech Recognition, ICASSP 88, (1988) pp.426–428.

Rosenberg, C.R. and Sejnowski, T.J., NETTALK: a parallel network that learns to read aloud, Complex Systems 1, (1987) p.145.

Rosenblatt, R., Principles of Neurodynamics, (Spartan Books, Washington D.C., 1962).

Rujan, P., A geometric approach to learning in neural networks, IJCNN, Washington, 1989, pp.II/105–109.

Rumelhart, D.E. and McClelland, J.L., On learning the past tenses of English verbs, in: McClelland, J.L. and Rumelhart, D.E. (eds.), Parallel Distributed Processing 2, Psychological and Biological Models, (MIT Press, Cambridge, 1986) pp.216–271.

Rumelhart, D.E. and McClelland, J.L. and the PDP Group, Parallel Distributed Processing, Explorations in the Microstructure of Cognition, Vol.1 and Vol.2, (MIT Press, Cambridge, MA, 1986).

Rumelhart, D.E., Hinton, G.E., and McClelland, J.L., A general framework for parallel distributed processing, in: Rumelhart, D.E. and McClelland, J.L. (eds.), Parallel Distributed Processing 1, (MIT Press, 1986).

Rumelhart, D.E., Hinton, G.E., and Williams, R.J., Learning internal representations by error propagation, in: Parallel Distributed Processing, Explorations in the Microstructure of Cognition 1, (MIT Press, Cambridge, M.A., 1986) pp.318–362.

Saarinen, J. and Kohonen, T., Self–organized formation of colour maps in a model cortex, Perception 14, (1985) pp.711–719.

Salomon, R., Adaptiv geregelte Lernrate bei Back–propagation, Tech. Report 89–24, Technische Universität Berlin, Forschungsberichte Fachbereich Informatik, (1989).

Samuel, A.L., Some studies in machine learning using the game of checkers, IBM J. Research and Development 3, (1959) pp.210–229.

Sanger, T.D., Stereo disparity computation using gabor filters, Biol. Cybern. 59, (1988) pp.405–418.

Satou, M. and Ewert, J.–P., The antidromic activation of tectal neurons by electrical stimuli applied to the caudal medulla oblongata in the toad (Bufo bufo L), J. Comp. Physiol. 157, (1985) pp.739–748.

Saund, E., Dimensionality–reduction using connectionist networks, IEEE Trans. PAMI–13, (1989) pp.304–314.

Schade, U., A Note on K. Bock's 'Syntactic adjustment effect' problem, in: Retti, J. and Leidlmair, K. (eds.), Vol.5, Österreichische Artificial–Intelligence–Tagung, (Springer, Berlin, Heidelberg, 1989) pp.218–223.

Schiffmann, W.H., Selbstorganization neuronaler Netze nach den Prinzipien der Evolution, Fachbericht Physik Nr.7, Universität Koblenz, (1989) pp.15–21.

Schmidhuber, J.H., Recurrent networks adjusted by adaptive critics, in: IJCNN, Washington D.C., (1990).

Schmidhuber, J.H., The neural bucket brigade, in: R. Pfeifer, Z. Schreter, Z. Fogelman, and L. Steels (eds.), Connectionism in Perspective, (Elsevier, North Holland, Amsterdam, 1988) pp.429–437.

Schöner, G. and Kelso, J.A.S., A dynamic theory of behavioral change, J. Theoret. Biol. 135, (1988) pp.501–524.

Schöner, G. and Kelso, J.A.S., Dynamic pattern generation in behavioral and neural systems, Sci. 239, (1988) pp.1513–1520.

Schuerg–Pfeiffer, E., Behavior–correlated properties of tectal neurons in freely moving toads, in: Ewert, J.–P. and Arbib, M.A. (eds.), Visuomotor Coordination: Amphibians, Comparisons, Models, and Robots, (Plenum Press, New York, 1989) pp.451–480.

Schürmann, B., Sampling learning, recall and filtering in stable, adaptive neural systems with graded response, Proc. IJCNN, Washington D.C., (1990) in press.

Schürmann, B., Stability and adaptation in artificial neural systems, Phys. Rev. A 40, (1989) pp.2681–2688.

Schwartz, E.L., Columnar architecture and computational anatomy in primate visual cortex, Segmentation and feature extraction via spatial frequency coded difference mapping, Biol. Cybern. 42, (1982) pp.157–168.

Schwartz, E.L., Spatial mapping in primate visual sensory projection and relevance to perception, Biol. Cybern. 25, (1977) pp.181–195.

Schwartz, J.T., Sahir, M., and Hopcroft, J., Planning, Geometry, and Complexity of Robot Motion, (Ablex Publ. Corp., Norwood, NJ, 1984).

Schwippert, W.W., Beneke, T.W., and Ewert, J.–P., Responses of medullary neurons to moving visual stimuli in the common toad II, Intracellular recording and cobalt–lysine labeling studies, J. Comp. Physiol., (1990) in press.

Seelen v., W., Informationsverarbeitung in homogenen Netzen von Neuronenmodellen I, Biol. Cybern. 5, (1968) pp.133–148.

Seelen v., W., Mallot, H.A., and Giannakopoulos, F., Characteristics of neuronal systems in the visual cortex, Biol. Cybern. 56, (1987) pp.37–49.

Sejnowski, T., Skeleton filters in the brain, in: Parallel Models of Associative Memory, Hinton, G.E. and Anderson, J.A. (eds.), (Lawrence Erlbaum Associates, Hillsdale, 1981), pp.189–212.

Sejnowski, T.J., Rosenberg, C.R., NETtalk: A parallel network that learns to read aloud, Tech. Report JHU–EECS–86–01, Johns Hopkins University, Baltimore, (1986).

Serra, R., Andretta, M., Compiani, M., and Zanarini, G., Introduction to the Physics of Complex Systems, The mesoscopic approach to fluctuation, non linearity and self–organization, (Pergamon Press, Oxford, 1986).

Shamma, S., Neural networks for speech processing and recognition, Proc. First Int. Conf. on Neural Networks, IEEE, (1987) pp.IV/397–405.

Shannon, C.E. and Weaver, W., The Mathematical Theory of Communication, (University of Illinois Press, Urbana, 1949).

Shastri, L, Semantic Nets, An Evidential Formalization and Its Connectionist Realization, (Morgen Kaufmann, Los Altos, CA, 1988).

Shastri, L., A connectionist approach to knowledge representation and limited inference, Cognitive Science 12, (1988) pp.331–392.

Shattuck–Hufnagel, S., The role of the word–onset consonants in speech planing, New evidence from speech error patterns, in: Keller, E. and Gopnik, M. (eds.), Motor and Sensory Processes of Language, (Erlbaum, Hillsdale, NJ, 1987) pp.17–51.

Shaw, G.L., Silverman, J.C., and Pearson, J.C., in: Palm, G., Aertsen, A.M.H.J. (eds.), Brain Theory, (Springer, New York, 1986) pp.177.

Shaw, R.S., The Dripping Faucet as a Model Chaotic System, (Ariel Press, Santa Cruz, 1984).

Shackleford, J.B., Neural data structures, programming with Neurons, Hewlett–Packard J., (1989) pp.69–78.

Shepard, R.N. and Zare, S.L., Path–guided apparent motion, Sci. 220, (1982) pp.632–634.

Sherman, S.M. and Koch, C., The control of retinogeniculate transmission in the mammalian lateral geniculate nucleus, Exp. Brain Res. 63, (1986) pp.1–20.

Sidman, M. and Tailby, W., Conditional discrimination versus matching to sample, An expansion of the testing paradigm, J. the Exp. Anal. Behaviour 37, (1982) pp.5–22.

Silverman, D.J., Shaw, G.L., and Pearson, J.C., Associative recall properties of the trion model of cortical organization, Biol. Cybern. 53, (1986) pp.259–271.

Simic, P., Statistical mechanics as the underlying theory of 'elastic' and 'neural' potimizers, NETWORK: Computation in Neural Systems 1, (1990) pp.1–15.

Singer, W. et al., Restriction of visual experience to a single orientation affects the oganization of orientation columns in cat visual cortex, a study with Deoxyglucose, Exp. Brain Res. 41, (1981) pp.199–215.

Singer, W., Control of thalamic transmission by corticofugal and ascending reticular pathways in the visual system, Physiol. Rev. 57, (1977) pp.386–420.

Skarda, A. and Freeman, W.J., How brains make chaos in order to make sense of the world, Behavioral and Brain Sci.s 10, (1987) pp.161–195.

Smith, A.W. and Zipser, D., Encoding sequential structure, experience with real–time recurrent learning algorithm, Proc. IJCNN–89 I, (1989), pp.645–648.

Smolensky, P., Connectionist AI, symbolic AI, and the brain, Artificial Intelligence Rev. 1, (1987) pp.95–109.

Smolensky, P., On the proper treatment of connectionism, Behavioral and Brain Science 11, (1988) pp.1–74.

Sompolinsky, H. and Kanter, I., Temporal association in asymmetric neural networks, Phys. Rev. Lett. 57(22), (1986) pp.2861–2864, Phys. Rev. Lett, 61(2), (December 1986) pp.2861-2864.

Sompolinsky, H., Statistical mechanics of neural networks, Physics Today 41, (1988) pp.70–80.

Sparks, D.L. and Nelson, J.S., Sensory and motor maps in the mammalian superior colliculus, TINS 10, (1987) pp.312–317.

Stanfill, C. and Waltz, D., Toward memory–based reasoning, CACM 29, Vol.12, (1987) pp.1213–1228.

Stein, R.B., Leung, K.V., Mangeron, D., and Oguztöreli, M.N., Improved neuronal models for studying neural networks, Kybernetik 15, (1974) pp.1–9.

Stemberger, J.P., An interactive activation model of language production, in: Ellis, A.W. (ed.), Progress in the Psychology of Language 1, (Erlbaum, London, 1985) pp.143–168.

Stemens, K.H., Implicit versus explicit computation, A commentary, Behavioral Brain Sci. 10, (1987) pp.387–388.

Steriade, M. and Deschenes, M., The thalamus as a neuronal oscillator, Brain Res. Rev. 8, (1984) pp.1–63.

Stevens, J.K. and Gerstein, G.L., Spatiotemporal organization of cat lateral geniculate receptive fields, J. Neurophysiology 39, (1976) pp.213–238.

Stevens, K.H., Implicit versus explicit computation. A commentary. Behav. Brain Sci., Vol.10, (1987) pp.387-388.

Stornetta, W.S., Hogg, T., and Huberman, B.A., A dynamical approach to temporal pattern processing, in: Neural Inf. Proc. Systems, D.Z. Anderson (ed.), (AIP, New York, 1988) pp.750–759.

Stotzka, R., Männer, R., Self–organization of a multilayered feedforward neural network, Proc. Int. Joint Conf. on Neural Networks, Washington D.C., (1990) in press.

Suga, N. and O'Neill, W.E., Neural axis representing target range in the auditory cortex of mustache bat, Science 206, (1979) pp.351–353.

Sutton, R.S., Learning to predict by the methods of temporal differences, Machine Learning 3, (1988) pp.9–44.

Takens, F., Detecting strange attractors in turbulence, in: Dynamical systems and turbulence, Rand and Young (eds.), (Springer, Berlin, 1981) pp.366–381.

Tanese, R., A parallel genetic algorithm for a hypercube, in: Grefenstette, J.J. (ed.), Proc. Second Int. Conf. on Genetic Algorithms, (Lawrence Erlbaum, Hillsdale, 1987).

Tank, D.W. and Hopfield, J.J., Neural computation by concentrating information in time, Proc. Natl. Acad. Sci. USA 84, (1987) pp.1896–1900.

Tank, D.W.and Hopfield, J.J., Simple 'neural' optimization networks, An A/D converter, signal decision circuit, and a linear programming circuit, IEEE Trans. Circuit Systems CAS–33, (1986) pp.533–541.

Taylor, J.G., J. Theoret. Biol. 36, (1972) p.513.

Teich, W.G. and Mahler, G., Optically controlled multistabiliby in nanostructured semiconductors, Physica Scripta 40, (1989) pp.688–693.

Teich, W.G., Anders, G., and Mahler, G., Transition between Incompatible Properties, A dynamical model for quantum measurement, Phys. Rev. Lett. 62, (1989) pp.1–4.

Teich, W.G., Obermayer, K., and Mahler, G., Structural basis of multistationary quantum systems II, Effective few–particle dynamics, Phys. Rev. B 37, (1988) pp.8111–8121.

Tesauro, G. and Sejnowski, T.J., A parallel network that learns to play backgammon, Artificial Intelligence 39, (1989) pp.357–390.

Thompson, J.M.T. and Stewart, H.B., Nonlinear Dynamics and Chaos – Geometrical methods for engineers and scientists, (Wiley, New York, 1986, reprinted 1988).

Thorpe, S.J. and Imbert, M., Biological constraints on connectionist models, in: Pfeiffer R. et al (ed.), Connectionism in Perspective, (Elsevier, Amsterdam, North–Holland, 1989) pp.63–92.

Thorpe, S.J., Traitement d'images chez l'homme, Techniques et Sci.s Informatiques 7, (1988) pp.517–525.

Toffoli, T. and Margolus, N., Cellular Automata Machines, (MIT Press, 1987).

Tolat, V.V. and Peterson, A.M., A self–organizing neural network for classifying sequences, in: Proc. IJCNN, Washington D.C., IEEE, (1989) pp.II/561–568.

Tomberg, J., Ritoniemi, T., Kaski, K., and Tenhunen, H., Fully digital neural network implementation based on pulse density modulation, Proc. CICC '89, (San Diego, 1989) pp.12./7.1–7.4.

Tomlinson, R.W.W. and Schwarz, D.W.F., Perception of the missing fundamental in non–human primates, J. Acoust. Soc. America 84, (1988) pp.560–565.

Touretzky, D.S. and Hinton, G.E., A distributed connectionist production system, Cognitive Sci. 12(3), (1988) pp.423–466.

Toyama, K., Functional connections of the visual cortex studied by cross–correlation techniques, in: Rakic, P. and Singer, W., Neurobiology of Neocortex, (Wiley, New York, 1988) pp.203–217.

Traunmüller, H. and Lacerda, F., Perceptual relativity in identification of two–formant vowels, Speech Communication 6, (1987) pp.143–157.

Treisman, A.M. and Gelade, G., A feature–integration theory of attention, Cognitive Psychology 12, (1980) pp.97–136.

Tryba, V., Marks, K.M., Rückert, U., and Goser, K., Selbstorganisierende Karten als lernende klassifizierende Speicher, ITG Fachbericht 102, (1988) pp.407–419.

Ts'o, D.Y., Gilbert, C.D., and Wiesel, T.N., Relationships between horizontal interactions and functional architecture in cat striate cortex as revealed by cross–correlation analysis, J. Neuroscience 6, (1986) pp.1160–1170.

Tunturi, A.R., Physiological determination of the arrangement of the afferent connections to the middle ectosylvian auditory area in the dog, Amer. J. Physiology 162, (1950) pp.489–502.

Ultsch, A., PANDA: Prolog and Neural Distributed Architectures, Tech. Report, Department of Computer Sci., University of Dortmund, (1989) in German.

Van Essen, D.C., Functional organization of primate visual cortex, in: A. Peters and E.G. Jones (eds.), Cerebral Cortex 3, Visual Cortex, (Plenum Press, New York, London, 1985) pp.259–329.

Van Hemmen, J.L. and Zagrebnov, V.A., Storing extensively many weighted patterns in a saturated neural network, J. Phys. A 20, (1987) pp.3989–3999.

Van Hemmen, J.L., Grensing, D., Huber, A., and Kühn, R., nonlinear neural networks I, General theory, J. Statistical Phys. 50, (1988) pp.231–257.

Vercauteren, L. et al., Pattern directed real–world interface, in: Connectionism in Perspective, R. Pfeiffer, et al (ed.), (Elsevier, North Holland, Amsterdam, 1989) pp.455–461.

Verri, A. and Poggio, T., Motion field and optical flow, Quantitative properties, IEEE Trans. Pattern Analysis and Machine Intelligence 11, (1989) pp.490–498.

Volder, J.E., The CORDIC trigonometric computing technique, in: IRE Trans. on Electronic Computers, Vol.8, pp.330–334, (1959).

Waerden v.d., B.L., Geometry and algebra in ancient civilizations, (Springer, Heidelberg, 1983).

Wagh, M.D. and Kanetkar, S.V., A class of translation invariant transforms, IEEE Trans. on Acoustics, Speech, and Signal Processing 25, (1977) pp.203–205.

Waibel, A., Hanazawa, T., Hinton, G., Shikano, K., and Lang, K., Phoneme recognition, Neural networks versus hidden markov models, Proc. ICASSP '88, (1988) pp.107–110.

Waibel, A.H., Modular construction of time–delay neural networks for speech recognition, Neural Computation 1, (1989) pp.39–46.

Wang, H.T., Mathur, B., and Koch, C., Computing optical flow in the primate visual system, Neural Computation 1, (1989) pp.92–103.

Warden, M., Floeder, W., and Waldner, F., Quantitative description of chaotic periods between relaxation oscillations in an antiferromagnet, in: Proc. of 23rd Congress Ampere on Magnetic Resonance, B. Maravigilia, F. DeLuca, R. Campanella (eds.), (Rome, 1986) pp.272–273.

Warkowski, F., Leenstra, J., Nijhius, J.A.G., and Spaanenburg, L., Issues in the test of artificial neural networks, Proc. ICCD '89, Boston, (1989) pp.487–490.

Watrous, R.L. and Shastri, L., Learning phonetic features using connectionist networks, An experiment in speech recognition, Proc. First Int. Conf. on Neural Networks, IEEE, (1987) pp.IV/381–388.

Watson, A.B., Efficiency of a model human image code, J. Opt. Soc. Am. A 4, (1987) pp.2401–2417.

Werbos, P.J., Generalization of backpropagation with application to a recurrent gas market model, Neural Networks 1, (1988) pp.339–356.

Widrow, B. and Hoff, M.E., Adaptive Switching Circuits, Western electronic Show and Convention, Convention Record IV, (1960) pp.96–104.

Widrow, B., Winter, R.G., and Baxter, R.A., Layered neural nets for pattern recognition, IEEE Trans. on Acoustics, Speech, and Signal Processing 36, (1988) pp.1109–1118.

Williams, R.J. and Zipser, D., A learning algorithm for continually running fully recurrent networks, Tech. Report ICS, Report 8805, Univ. California, San Diego, (1988).

Williams, R.J. and Zipser, D., Experimental analysis of the realtime recurrent learning algorithm, Connection Sci. 1, (1989) pp.87–111.

Williams, R.J., Toward a theory of reinforcement–learning connectionist systems, Tech. Report NU–CCS–88–3, College of Comp. Sci., Northeastern University, Boston, MA, 1988.

Willshaw, D.J., Holography, associative memory, and inductive generalization, in: Hinton, G.E. and Anderson, J.A. (eds.), Parallel models of associative memory, (Lawrence Erlbaum Assoc., Hillsdale, New Jersey, 1981) pp.83–104.

Wilson, H.R. and Cowan, J.D., A mathematical theory of the functional dynamics of the cortical and thalamic nervous tissue, Kybernetik 13, (1973) pp.55–80.

Wilson, H.R. and Cowan, J.D., Excitatory and inhibitory interactions in localized populations of model neurons, Biophys. J. 12, (1972) pp.1–24.

Wilson, M.A. and Bower, J.M., The simulation of large–scale neural networks, in: Koch, C. and Segev, I. (eds.), Methods in Neuronal Modeling, (MIT Press, Cambridge, 1989) pp.291–334.

Wolff, J.R., Some morphogenetic aspects of the development of the central nervous system, in: Immelmann, K. et al (ed.), Behavioral Development, (Cambridge University Press, 1981) pp.164–190.

Wolfram, S. (ed.), Theory and Applications of Cellular Automata, (World Scientific, Singapure, 1986).

Yaglom, I.M., Felix Klein and Sophus Lie – Evolution of the Idea of Symmetry in the Nineteenth Century, (Birkhäuser, Basel, 1988).

Yariv, A. and Kwong, S.K., Opt. Lett. 11, (1986) pp.482.

Yuille, A.L. and Grzywacz, N.M., A computational theory for the perception of coherent visual motion, Nature 333, (1988) pp.71–74.

Zadeh, L.A., Outline of a new approach to the analysis of complex systems and decision processes, IEEE Trans. Systems, Man and Cybernetics 1, (1973) pp.28–44.

Zak, U., Terminal attractors in neural networks, Neural Networks 2, (1989) pp. 259–274.

Zeki, S. and Shipp, S., The functional role of cortical connections, Nature 335, (1988) pp.311–316.

Zetzsche, C. and Barth, E., Fundamental limits of linear filters in the visual processing of two–dimensional signals, Vision Research, (1990) in press.

Zetzsche, C. and Schönecker, W., Orientation selective filters lead to entropy reductions in the processing of natural images, Perception 16, (1987) p.229.

LIST OF CONTRIBUTORS

Aarts, E.H.J. (Netherlands)   NL-5600 JA Eindhoven
Philips Research Labs., WB3   277

Acker, R. (FRG)   Dept. of Control Theory and Robotics
Techn. Universität Darmstadt   D-6100 Darmstadt   229

Adamopoulos, A. (Greece)   Dept. of Medicine
University of Thraki   Alexandroupolis   49

Aertsen, A. (FRG)   MPI for Biol. Cybernetics
Max-Planck-Institut   D-7400 Tübingen   53,75,83

Aiple, F. (FRG)   Neurolog. Univ. Clinic
Universität Freiburg   D-7400 Freiburg   97

Aiso, H. (Japan)   Department of Electrical Engineering
Keio University   Yokohama 223   331

Ajioka, Y. (Japan)   Department of Computer Science
Keio University   Yokohama 223   331

Aleksander, I. (UK)   Dept. of Electrical Engineering
Imperial College   London SW7 2BZ   225

Anlauf, J.K. (FRG)   Dept. of Theoret. Physics
Justus-Liebig-Universität   D-6300 Giessen   153

Anninos, P. (Greece)   Dept. of Medicine, Neurol. & Med.
University of Thraki   G-68100 Alexandroupolis   49

Anzai, Y. (Japan)   Department of Electrical Engineering
Keio University   Yokohama 223   331

Arndt, M. (FRG)   Dept. of Biophysics
Philipps Universität   D-3550 Marburg   101

Arnoldi, M. (FRG)   D-8044 Unterschleißheim/München
Kratzer Automatisierungs GmbH   323

Ayala, G.F. (Italy)   Dept. of Psychophysiology
University of Palermo   I-90143 Palermo   291

Baran, R.H. (USA)   Dept. of Mathematics
Towson State University   Towson, MD 21204   19

Barker, J.R. (UK)   Dept. Electronics & Electrical Engineering
University of Glasgow   Glasgow, G12 8QQ   519

Bauer, H.-U. (FRG)   Dept. of Theoret. Physics
Universität Frankfurt   D-6000 Frankfurt/Main   375

Becker, J.D. (FRG)   Neurolog. Univ. Clinic
Universität Freiburg   D-7800 Freiburg   121

Beckmann, J. (FRG)   Dept. of Biophysics
Heinrich-Heine-Universität   D-4000 Düsseldorf   405

Becks, K.H. (FRG)   Dept. of Physics
Universität Wuppertal   D-5600 Wuppertal 1   465

Behme, H. (FRG)   Dept. of Physics
Universität Göttingen   D-3400 Göttingen   379

Beneke, T.W. (FRG)   Div. of Neurobiology
Universität Kassel   D-3500 Kassel   109

Best, J. (FRG)   Dept. of Neuroinformatics
Ruhr-Universität Bochum   D-4630 Bochum   61,113

Biehl, M. (FRG)   Dept. of Theoret. Physics
Justus-Liebig-Universität   D-6300 Giessen   153

Bisset, D.L. (UK)   Electronic Engineering Laboratories
University of Kent   Canterbury, Kent, CT2 7NT   343

Böhme, J.F. (FRG)   Dept. of Electrical Engineering
Ruhr-Universität Bochum   D-4630 Bochum   43

Böller, D. (FRG)   D-8044 Unterschleißheim/München
Kratzer Automatisierungs GmbH   323

Bohrer, S. (FRG)   Dept. of Neuroinformatics
Ruhr-Universität Bochum   D-4630 Bochum   471

Boullart, L. (Belgium)   Automatic Control Lab.
Ghent State University   B-9710 Gent-Zwijnaarde   503

Boven, K.-H. (FRG)   MPI for Biol. Cybernetics
Max-Planck-Institut   D-7400 Tübingen   53

Brady, D. (USA)   Dept. of Electrical Engineering
Calif. Inst. of Technology   Pasadena, CA 91125   543

Brandt, B. (USA)   Computation and Neural Systems Program
Calif. Inst. of Technology   Pasadena, CA 91125   147

Brauer, W. (FRG)   Dept. of Computer Science
Technische Univ. München   D-8000 München   323

Brown, G.D.A. (UK)   Dept. of Psychology
Univ. College of North Wales   Bangor, Gwynedd, LL57 2DG   475

Brüwer, M. (FRG)   Div. of Biol. Cybernetics
Universität Bielefeld   D-4800 Bielefeld   409

Bülthoff, H.H. (USA)   Dept. of Brain and Cognitive Science
Brown University   Providence, RI 02912   471

Burgard, W. (FRG)
Universität Dortmund
Dept. of Computer Science
D-4600 Dortmund 50 — 465

de Callatay, A. (Belgium)
IBM
European Systems Research Inst.
B-1310 La Hulpe — 233

Carpenter, G.A. (USA)
Boston University
Center for Adaptive Systems
Boston, MA 02215 — 383

Cecconi, F. (Italy)
National Research Council
Institute of Psychology
I-00137 Roma — 237

Chen, C.-C. (Belgium)
Free University of Brussels
Artificial Intelligence Laboratory
B-1050 Bruxelles — 15

Coughlin, J.P. (USA)
Towson State University
Dept. of Mathematics
Towson, MD 21204 — 19

Cremers, A.B. (FRG)
Universität Dortmund
Dept. of Computer Science
D-4600 Dortmund 50 — 465

Cruse, H. (FRG)
Universität Bielefeld
Div. of Biol. Cybernetics
D-4800 Bielefeld — 409

Dammasch, I.E. (FRG)
Universität Göttingen
Department of Anatomy
D-3400 Göttingen — 241

Dangelmayr, G. (FRG)
Universität Tübingen
Inst. for Information Science
D-7400 Tübingen — 445

Daunicht, W.J. (FRG)
Heinrich-Heine-Universität
Dept. of Biophysics
D-4000 Düsseldorf — 413,417

Dean, J. (FRG)
Universität Bielefeld
Div. of Biocybernetics
D-4800 Bielefeld — 57

Dicke, P. (FRG)
Philipps Universität
Dept. of Biophysics
D-3550 Marburg — 101

Dinse, H.R. (FRG)
Ruhr-Universität Bochum
Dept. of Neuroinformatics
D-4630 Bochum — 61,65

Divko, R. (FRG)
Technische Universität München
Physics Department
D-8046 Garching — 335

Drüe, S. (FRG)
Universität Paderborn
Electrical Engineering Dept.
D-4790 Paderborn — 361

Duque Anton, M. (FRG)
Deutsche Philips GmbH
Research Laboratory
D-2000 Hamburg 54 — 441

Eckhorn, R. (FRG)
Philipps Universität
Dept. of Biophysics
D-3550 Marburg — 101

Eckmiller, R. (FRG)
Heinrich-Heine-Universität
Dept. of Biophysics
D-4000 Düsseldorf — 5,417

<table>
<tr><td>Eikmeyer, H.J. (FRG)
Universität Bielefeld</td><td>Linguistics and Literary Science
D-4800 Bielefeld</td><td>495</td></tr>
<tr><td>Engel, A.K. (FRG)
Max-Planck-Institut</td><td>Div. of Neurophysiology
D-6000 Frankfurt 71</td><td>105</td></tr>
<tr><td>Englisch, H. (GDR)
Karl-Marx-Universität</td><td>Section Informatics
DDR-7010 Leipzig</td><td>245</td></tr>
<tr><td>Eppler, W. (FRG)
Universität Karlsruhe</td><td>Dept. of Computer Design
D-7500 Karlsruhe</td><td>249</td></tr>
<tr><td>Ernst, H.P. (Switzerland)
University of Zürich</td><td>AI-Lab, Institute for Informatics
CH-8057 Zürich</td><td>283</td></tr>
<tr><td>Ewert, J.-P. (FRG)
Universität Kassel</td><td>Div. of Neurobiology
D-3500 Kassel</td><td>109</td></tr>
<tr><td>Fang, L. (Australia)
Univ. of South Australia</td><td>Dept. of Computer Science
Bedford Park, S.A. 5042</td><td>253</td></tr>
<tr><td>Fang, M. (FRG)
Universität Erlangen</td><td>Dept. of Optics
D-8520 Erlangen</td><td>339</td></tr>
<tr><td>Feldman, J.A. (USA)
Int. Comp. Sci. Inst. (ICSI)</td><td>Berkeley, CA 94704</td><td>479</td></tr>
<tr><td>Filho, E.C. de B.C. (UK)
University of Kent</td><td>Electronic Engineering Laboratories
Canterbury, Kent, CT2 7NT</td><td>343</td></tr>
<tr><td>Fornili, S.L. (Italy)
University of Palermo</td><td>Physics Department
I-90143 Palermo</td><td>291</td></tr>
<tr><td>Fukushima, K. (Japan)
Osaka University</td><td>Dept. of Biophysical Eng., Fac.Eng.Sci.
Toyonaka, Osaka 560</td><td>351</td></tr>
<tr><td>Geiger, H. (FRG)
Kratzer Automatisierungs GmbH</td><td>D-8044 Unterschleißheim/München</td><td>323</td></tr>
<tr><td>Geisel, T. (FRG)
Universität Frankfurt</td><td>Dept. of Theoret. Physics
D-6000 Frankfurt/Main</td><td>375</td></tr>
<tr><td>Gencic, T. (FRG)
Universität Tübingen</td><td>Inst. for Information Science
D-7400 Tübingen</td><td>445</td></tr>
<tr><td>Geva, S. (Australia)
Queensland Univ. of Technology</td><td>Faculty of Information Technology
Brisbane, Q 4001</td><td>257</td></tr>
<tr><td>Giannakopoulos, F. (FRG)
Ruhr-Universität Bochum</td><td>Dept. of Neuroinformatics
D-4630 Bochum</td><td>113</td></tr>
<tr><td>Giefing, G.-J. (FRG)
Ruhr-Universität Bochum</td><td>Dept. of Neuroinformatics
D-4630 Bochum</td><td>125</td></tr>
<tr><td>Glünder, H. (FRG)
Ludwig-Maximilians-Univ.</td><td>Dept. of Medical Psychology
D-8000 München 2</td><td>357</td></tr>
</table>

Goebel, R. (FRG)
Universität Marburg
Dept. of Psychology
D-3550 Marburg
157

Gorse, D. (UK)
University College
Dept. of Computer Science
London WC1E 6BT
161

Gray, C.M. (FRG)
Max-Planck-Institut
MPI for Brain Research
D-6000 Frankfurt 71
105

Grossberg, S. (USA)
Boston University
Center for Adaptive Systems
Boston, MA 02215
383

Güttinger, W. (FRG)
Universität Tübingen
Inst. for Information Science
D-7400 Tübingen
445

Gumm, H.P. (USA)
SUNY at New Paltz
Dept.of Computer Science
New Paltz, N.Y. 12561
287

de Haan, P.E. (FRG)
Inst. Microelectronics (IMS)
D-7000 Stuttgart 80
299

Haarer, D. (FRG)
Universität Bayreuth
Physics Department
D-8580 Bayreuth
525

Häusler, G. (FRG)
Universität Erlangen
Dept. of Optics
D-8520 Erlangen
339,533

Haken, H. (FRG)
Universität Stuttgart
Dept. of Theoret. Physics & Synergetics
D-7000 Stuttgart 80
451

Hannuschka, R. (FRG)
Universität Dortmund
Dept. of Computer Science
D-4600 Dortmund
499

Hartmann, G. (FRG)
Universität Paderborn
Electrical Engineering Dept.
D-4790 Paderborn
361

Hartmann, U. (FRG)
Universität Dortmund
Dept. of Computer Science
D-4600 Dortmund
499

Hauser, R. (FRG)
Universität Heidelberg
Physics Institute
D-6900 Heidelberg
315

Hauske, G. (FRG)
Technische Universität München
Dept. of Communication Science
D-8000 München 2
3

Heinke, D. (FRG)
Ruhr-Universität Bochum
Dept. of Neuroinformatics
D-4630 Bochum
113

Heistermann, J. (FRG)
Siemens AG
Corporate Research and Development
D-8000 München 83
165

Hemker, A. (FRG)
Universität Wuppertal
Department of Physics
D-5600 Wuppertal 1
465

Hergert, F.B. (FRG)
Siemens AG
Corporate Research and Development
D-8000 München 83
287

Hertz, J.A. (Denmark)   DK-2100 Kobenhavn
Niels Bohr Inst. and Nordita   183

Hess, G. (FRG)   Microelectronic Circuits and Systems
Fraunhofer Institut   D-4100 Duisburg   311

Ho, A.W. (USA)   Dept. of Physics
Calif. Inst. of Technology   Pasadena, CA 91125   421

Hollatz, J. (FRG)   Corporate Research and Development
Siemens AG   D-8000 München 83   213

Hormel, M. (FRG)   Dept. of Control Theory and Robotics
Techn. Universität Darmstadt   D-6100 Darmstadt   427

Hosticka, B. (FRG)   Dept. of Electronic Elements and Circuits
Universität Duisburg   D-4100 Duisburg   311

Hsu, K. (USA)   Dept. of Electrical Engineering
Calif. Inst. of Technology   Pasadena, CA 91125   543

Jockusch, S. (FRG)   Physics Department
Universität Göttingen   D-3400 Göttingen   169

Kammen, D.M. (USA)   Computation and Neural Systems Program
Calif. Inst. of Technology   Pasadena, CA 91125   147

Karayiannis, N.B. (Canada)   Dept. of Electrical Engineering
University of Toronto   Toronto, M5S 1A4   173

Kesper, M. (FRG)   Microelectronic Circuits and Systems
Fraunhofer Institut   D-4100 Duisburg   311

Kindermann, J. (FRG)   D-5205 Sankt Augustin
Ges. Math. & Datenverarbtg.(GMD)   217

König, P. (FRG)   MPI for Brain Research
Max-Planck-Institut   D-6000 Frankfurt 71   105,117,139

Kohonen, T. (Finland)   Lab. of Computer and Information Science
Helsinki Univ. of Technology   SF-02150 Espoo   177

Koscielny-Bunde, E. (FRG)   Dept. of Computer Science
Universität Hamburg   D-2000 Hamburg 50   459

Kowalewski, F. (FRG)   Physics Department
Universität Göttingen   D-3400 Göttingen   391

Krogh, A. (Denmark)   DK-2100 Kobenhavn
Niels Bohr Inst. and Nordita   183

Krüger, J. (FRG)   Neurolog. Univ. Clinic
Universität Freiburg   D-7800 Freiburg   97,121

Krüger, K. (FRG)   Dept. of Neuroinformatics
Ruhr-Universität Bochum   D-4630 Bochum   61

Kruse, W. (FRG)   Dept. of Biophysics
Philipps Universität   D-3550 Marburg   101

Kühnel, H. (FRG)   Physics Department
Techn. Universität München   D-8046 Garching   187

Kürten, K.E. (FRG)   Dept. of Neuroinformatics
Ruhr-Universität Bochum   D-4630 Bochum   191

Kunz, D. (FRG)   Research Laboratory
Deutsche Philips GmbH   D-2000 Hamburg 54   441

Kurz, A. (FRG)   Dept. of Control Theory and Robotics
Techn. Universität Darmstadt   D-6100 Darmstadt   229

Lades, M. (FRG)   Dept. of Biophysics
Heinrich-Heine-Universität   D-4000 Düsseldorf   417

Lappe, M. (FRG)   Inst. for Information Science
Universität Tübingen   D-7400 Tübingen   445

Leuthäusser, I. (FRG)   Corporate Research and Development
Siemens AG   D-8000 München 83   213

Li, T. (Australia)   Department of Computer Science
Monash University   Clayton, VIC 3168   253

Ludermir, T.B. (UK)   Neural Systems Engineering Group
Imperial College   London, SW7 2BT   395

Männer, R. (FRG)   Physics Institute
Universität Heidelberg   D-6900 Heidelberg   315

Mahler, G. (FRG)   Dept. of Theoret. Physics
Universität Stuttgart   D-7000 Stuttgart 80   539

Mainzer, K. (FRG)   Dept. of Philosophy
Universität Augsburg   D-8900 Augsburg   9

Mallot, H.-A. (FRG)   Dept. of Neuroinformatics
Ruhr-Universität Bochum   D-4630 Bochum   125,129,471

v.d. Malsburg, C. (USA)   Dept. of Computer Science
Univ. of Southern California   Los Angeles, CA 90089-0782   37

Manderick, B. (Belgium)   Artificial Intelligence Laboratory
Free University of Brussels   B-1050 Bruxelles   31,195

Mangold-Allwinn, R. (FRG)   Dept. of Psychology
Universität Mannheim   D-6800 Mannheim   487

Martinetz, T. (USA)   Beckman Inst., Dept. of Physics
University of Illinois   Urbana-Champaign, Il 61801   431

Massen, R. (FRG)   D-7750 Konstanz
Transferzentrum f. Bilddaten   23

Mazare, G. (France)	Laboratoire de Genie Informatique	
INPG	F-38031 Grenoble Cedex	27
Mecklenburg, K. (FRG)	Physics Department	
Universität Koblenz	D-5400 Koblenz	205
Merzenich, M.M. (USA)	Dept. Otolaryng., Coleman Laboratories	
University of California	San Francisco, CA 94143	65
Migliore, M. (Italy)	Inst. for Interdisciplinary Appl. of Physics	
National Research Council	I-90123 Palermo	291
Miksa, M. ((FRG)	Dept. of Mathematics/Computer Science	
Universität Marburg	D-3500 Marburg	295
Mokry, B. (Switzerland)	AI-Lab, Institute for Informatics	
Universität Zürich	CH-8057 Zürich	283
Moyson, F. (Belgium)	Artificial Intelligence Laboratory	
Free Univ. of Brussels	B-1050 Bruxelles	195
Neumann, H. (FRG)	Dept. of Computer Science	
Universität Hamburg	D-2000 Hamburg 50	491
Nijhuis, J.A.G. (FRG)	D-7000 Stuttgart 80	
Inst. Microelectronics (IMS)		299,319
Nischwitz, A. (FRG)	D-8044 Unterschleißheim/München	
Kratzer Automatisierungs GmbH		323
Oaksford, M. (UK)	Dept. of Psychology	
Univ. College of North Wales	Bangor, Gwynedd, LL 57 2 DG	475
Obermayer, K. (USA)	Beckman-Inst./ Dept. of Physics	
University of Illinois	Urbana-Champaign, IL 61801	71
Ossen, A. (FRG)	Dept. of Applied Computer Science	
Techn. Universität Berlin	D-1000 Berlin 10	201
Palm, G. (FRG)	Vogt-Inst. for Brain Research	
Heinrich-Heine-Universität	4000 Düsseldorf	83
Parisi, D. (Italy)	Dept. of Psychology	
National Research Council	I-00137 Roma	237
Payan, E. (France)	Laboratoire de Genie Informatique	
INPG	F-38031 Grenoble Cedex	27
Petkov, N. (FRG)	Dept. of Informatics	
Universität Erlangen-Nürnberg	D-8520 Erlangen	303
Plenz, D. (FRG)	MPI for Biol. Cybernetics	
Max-Planck-Institut	D-7400 Tübingen	75
Prange, S.J. (FRG)	Dept. of Microelectronics	
Techn. Universität Berlin	D-1000 Berlin 12	79

Preißl, H. (FRG)   MPI for Biol. Cybernetics
Max-Planck-Institut   D-7400 Tübingen   83

Protzel, P.W. (USA)   Inst.f. Computer Appl. Science & Eng.
NASA Research Center   Hampton, Virginia 23665   307

Psaltis, D. (USA)   Dept. of Electrical Engineering
Calif. Inst. of Technology   Pasadena, CA 91125   543

Radons, G. (FRG)   Dept. of Theoret. Physics
Universität Kiel   D-2300 Kiel   261

Rapp, R. (FRG)   Dept. of Psychology
Universität Paderborn   D-4790 Paderborn   509

Recanzone, G.H. (USA)   Dept. Otolaryng., Coleman Laboratories
University of California   San Francisco, CA 94143   65

Reichardt, W. (FRG)   MPI for Biol. Cybernetics
Max-Planck-Institut   D-7400 Tübingen   133

Reitboeck, H.J. (FRG)   Dept. of Biophysics
Philipps Universität   D-3550 Marburg   101

Richert, P. (FRG)   Microelectronic Circuits and Systems
Fraunhofer Institut   D-4100 Duisburg   311

Ritter, H. (USA)   Beckman-Inst., Dept. of Physics
University of Illinois   Urbana-Champaign, IL 61801   71,431

Rubner, J. (FRG)   Physics Department
Techn. Universität München   D-8046 Garching   365

Schade, U. (FRG)   Fac. of Linguistics and Literary Science
Universität Bielefeld   D-4800 Bielefeld   495

Schaller, H.N. (FRG)   Dept. of Data Processing
Techn. Universität München   D-8000 München 2   265

Schiffmann, W. (FRG)   Physics Department
Universität Koblenz   D-5400 Koblenz   205

Schillen, T.B. (FRG)   MPI for Brain Research
Max-Planck-Institut   D-6000 Frankfurt 71   117,135,139

Schmidhuber, J. (FRG)   Dept. of Informatics
Techn. Universität München   D-8000 München   209

Schöner, G.F. (FRG)   Dept. of Neuroinformatics
Ruhr-Universität Bochum   D-4630 Bochum   87

Schreter, Z. (Switzerland)   AI-Lab, Institute for Informatics
Universität Zürich   CH-8057 Zürich   283

Schürmann, B. (FRG)   Corporate Research and Development
Siemens AG   D-8000 München 83   213

Schulten, K. (USA)   Beckman-Inst., Dept. of Physics
Universität of Illinois   Urbana-Champaign, IL 61801   71,335,365,431

Schuster, H.G. (FRG)   Dept. of Theoret. Physics
Universität Kiel   D-2300 Kiel   143,261

Schwarz, M. (FRG)   Microelectronic Circuits and Systems
Fraunhofer Institut   D-4100 Duisburg   311

Schwippert, W.W. (FRG)   Div. of Neurobiology
Universität Kassel   D-3500 Kassel   109

v. Seelen, W. (FRG)   Dept. of Neuroinformatics
Ruhr-Universität Bochum   D-4630 Bochum   129

Seytter, F. (UK)   Dept. of Artifical Intelligence
University of Edinburgh   Edinburgh, EHI 2QL   369

Singer, W. (FRG)   Div. of Neurophysiology
Max-Planck-Institut   D-6000 Frankfurt 71   105

Sitte, J. (Australia)   Faculty of Information Technology
Queensland Univ. of Technology   Brisbane, Q 4001   257

Spaanenburg, L. (FRG)   D-7000 Stuttgart 80
Inst. Microelectronics (IMS)   299,319

Spiessens, P. (Belgium)   Artificial Intelligence Laboratory
Free University of Brussels   B-1050 Bruxelles   31

Stiehl, H.S. (FRG)   Dept. of Computer Science
Universität Hamburg   D-2000 Hamburg 50   491

Stotzka, R (FRG)   Physics Institute
Universität Heidelberg   D-6900 Heidelberg   315

Strube, H.W. (FRG)   Physics Department
Universität Göttingen   D-3400 Göttingen   391

Tavan, P. (FRG)   Physics Department
Techn. Universität München   D-8046 Garching   187,365

Taylor, J.G. (UK)   Dept. of Mathematics
King's College   London, WC2R 2LS   161

Teich, W.G. (FRG)   Dept. of Theoret. Physics
Universität Stuttgart   D-7000 Stuttgart 80   539

Thorpe, S.J. (France)   Dept. of Neurosciences
Univ. Pierre et Marie Curie   F-57006 Paris   91

Tomlinson, R.W.W. (FRG)   Dept. of Zoology
Tech. Hochschule Darmstadt   D-6100 Darmstadt   399

Treurniet, W. (Canada)   Dept. of Communications
Communications Res. Center   Ottawa, Ontario K2H 8S2   399

Ultsch, A. (FRG) Universität Dortmund	Dept. of Computer Science D-4600 Dortmund 50	465,499
Van Hemmen, J.L. (FRG) Techn. Universität München	Physics Department D-8046 Garching	245
Veenker, G. (FRG) Universität Bonn	Inst. for Computer Science D-5300 Bonn	513
Venetsanopoulos, A.N. (Canada) University of Toronto	Dept. of Electrical Engineering Toronto M5S 1A4	173
Vercauteren, L. (Belgium) Ghent State University	Automatic Control Lab. B-9710 Gent-Zwijnaarde	503
Vingerhoeds, R.A. (Belgium) Ghent State University	Automatic Control Lab. B-9710 Gent-Zwijnaarde	503
Vorbrüggen, J.C. (FRG) Max-Planck-Institut	MPI for Brain Research D-6000 Frankfurt 71	37
Wagner, P. (FRG) Universität Kiel	Dept. of Theoret. Physics D-2300 Kiel	143
Warkowski, F. (FRG) Inst. Microelectronics (IMS)	D-7000 Stuttgart 80	299,319
Waschulzik, T. (FRG) Kratzer Automatisierungs GmbH	D-8044 Unterschleißheim/München	323
Weber, V. (FRG) Universität Dortmund	Dept. of Computer Science D-4600 Dortmund	499
Werner, D. (FRG) Universität Kiel	Dept. of Theoret. Physics D-2300 Kiel	261
Werntges, H. (FRG) Heinrich-Heine-Universität	Dept. of Biophysics D-4000 Düsseldorf	413,417,435
Wettler, M. (FRG) Universität Paderborn	Dept. of Psychology D-4790 Paderborn	509
Whittle, P. (UK) University of Cambridge	Statistical Laboratory Cambridge, CB2 1SB	269
Wilson, W.H. (Australia) Univ. of South Australia	Dept. of Computer Science Bedford Park, S.A. 5042	253
Windheuser, C. (FRG) Ges. Math. & Datenverarbtg.(GMD)	D-5205 Sankt Augustin	217
Wörgötter, F. (USA) Calif. Inst. of Technology	Computation and Neural Systems Program Pasadena, CA 91125	147
Wolff, J.R. (FRG) Universität Göttingen	Department of Anatomy D-3400 Göttingen	241

Würtz, R.P. (FRG)   MPI for Brain Research
Max-Planck-Institut   D-6000 Frankfurt 71   37

Yang, B. (FRG)   Dept. of Electrical Engineering
Ruhr-Universität Bochum   D-4630 Bochum   43

Zetsche, C. (FRG)   Dept of Communication Engineering
Techn. Universität München   D-8000 München 2   273

Zhang, B.-T. (FRG)   Inst. for Computer Science
Universität Bonn   D-5300 Bonn 3   513

Zwietering, P.J. (Netherlands)   Dept. of Mathematics and Computing Science
Eindhoven Univ. of Technology   NL-5600 MB Eindhoven   277

AUTHOR INDEX

A
Aarts, E.H.L.   277, 280
Abbott, L.F.   156
Abeles, M.   56, 108
Abelson, H.   8
Abraham, N.B.   86
Abu–Mostafa, Y.S.   18, 548
Acker, R.   229
Ackley, D.H.   280, 328, 478
Adamopoulos, A.   49
Adorf, H.–M.   268
Adrian, E.D.   100
Aertsen, A.M.H.J.   53, 56, 70, 75, 78, 83, 86
Ahlsen, G.   116
Ahmed, H.M.   46
Aiple, F.   97, 100, 123
Aiso, H.   331, 334
Ajioka, Y.   331, 334
Akamatsu, M.   334
Albano, A.M.   86
Albert, A.   416
Albus, J.S.   430, 438
Aleksander, I.   164, 225, 228, 398
Allard, T.   69, 70
Allman, J.M.   128, 132
Almeida, L.B.   378
Amari, S.–I.   216, 260, 342
Amit, D.J.   156, 216, 272, 450, 462
Anders, G.   542
Anderson, C.W.   212, 416
Anderson, D.Z.   548
Anderson, J.A.   182
Anderson, J.R.   236, 256
Anderson, P.   52
Anderson, S.   221
Andretta, M.   199
Anlauf, J.K.   153, 156
Anninos, P.   49, 52
Anogianakis, G.   52
Anzai, Y.   331, 334
Aoki, K.   86
Arbib, M.A.   112
Arndt, M.   82, 101, 104
Arnoldi, M.   323, 328
Arras, M.K.   310

Asanuma, H.   70
Askew, C.R.   294
Atten, P.   86
Ayala, G.F.   291
Ayers, G.R.   538

B
Babloyantz, A.   86
Bach, M.   123
Baird, B.   146
Baldi, P.   186, 190, 260
Ballard, D.H.   94, 108, 128, 138, 252, 516
Bandy, D.K.   86
Bannister, J.A.   310
Banzaf, W.   457
Baran, R.H.   19
Barker, J.R.   519
Baron, R.J.   12
Barth, E.   276
Barto, A.G.   164, 212, 416
Bashore, T.R.   86
Bauer, H.–U.   86, 375, 378
Bauer, R.   100, 104, 108, 138, 142, 150
Baum, E.B.   276
Baxter, R.A.   176
Beasty, A.   478
Beaulieu, C.   420
Beck, R.   444
Becker, D.   121
Becker, L.A.   470
Beckmann, J.   405, 434
Becks, K.H.   465, 470
Bedenbaugh, P.   56, 70, 86
Beek, B.   52
Behme, H.   379, 382
Bell, E.T.   8
Benati, M.   412
Beneke, T.W.   109, 112
Berg, T.   497, 498
Best, J.   61, 64, 70, 113, 150
Bestehorn, M.   457
Bethke, A.D.   36

612   *Author Index*

Beunder, M.   322
Biehl, M.   153, 156
Bienenstock, E.   328
Bisset, D.L.   343, 349
Boch, R.   123
Böhlau, P.   248
Böhme, J.F.   43, 46
Böller, D.   323, 328
Boer de, E.   402
Bös, S.   248
Bohrer, S.   471
Bonhoeffer, T.   78
Booker, L.B.   240
Bottini, S.   276
Boullart, L.   503
Bourlard, H.   221
Boven, K.-H.   53, 56, 86
Bowden, P.A.   164, 398
Bower, J.M.   150, 294
Bradburn, D.S.   182
Brady, D.   543, 548
Braitenberg, V.   70, 132, 194
Brandt, B.   147
Brauer, W.   323
Briggs, J.   8
Brittinger, R.   132
Brody, T.P.   342
Brosch, M.   86, 100, 104, 108, 138, 142
Brown, G.D.A.   475, 478
Brown, R.   8, 498
Brown, T.H.   360
Brüwer, M.   409, 412
Bruge, F.   294
Bülthoff, H.H.   360, 471, 474
Bueno–Lopez, J.L.   120
Büttner, W.   298
Buhmann, J.   41, 378
Bunz, H.   90
Burgard, W.   465
Burr, D.C.   41
Butler, R.W.   310
Bybee, J.   490

C
Caan, W.C.   94
Caianiello, E.   194
Callatay de, A.   233, 236
Campenhausen v., C.   116

Caputo, J.G.   86
Caramazza, A.   182
Carnevali, P.   240
Carpenter, D.B.   294
Carpenter, G.A.   172, 244, 260, 349, 383, 389
Carter, F.L.   542
Castani , F.   26
Castro, G.   531
Cavanagh, P.   493
Cecconi, F.   237, 240
Chalker, J.T.   294
Changeux, J.P.   450
Chate, H.   86
Chater, N.   478
Chauvin, Y.   204
Chen, C.-C.   15, 18
Chen, H.H.   260
Chua, L.O.   86
Churchland, P.   12
Clark, A.   478
Clark, J.W.   194
Clarkson, T.G.   164
Cohen, M.A.   22, 378, 389
Combes, J.M.   41
Compiani, M.   199
Cornu–Emieux, R.   30
Cotter, N.E.   314
Cottrell, G.W.   186
Coughlin, J.P.   19
Courant, R.   8
Cowan, J.D.   116, 146, 222
Crawford, W.J.   318
Cremers, A.B.   465, 470
Crick, F.   108, 120, 138, 142
Crisanti, A.   22
Cruse, H.   60, 409, 412
Crutchfield, J.P.   86
Csermely, T.J.   52
Cybenko, G.   438
Cynader, M.S.   70, 74
Cyrulnic, R.   52

D
Dainty, J.C.   538
Damasio, A.R.   104, 120, 138, 142
Dammasch, I.E.   241, 244
Dangelmayr, G.   445
Daniel, J.W.   176

DAP 36
Darianian, M. 302
DARPA 302
Daughman, J.G. 276
Daunicht, W.J. 413, 416, 417, 420, 438
Davies, P.C.W. 470
Davis, L. 470
Davis, P.J. 8
Dean, J. 57, 60
Dehaene, S. 450
DeJong, K.A. 36, 199
Dell, G.S. 498
Delosme, J.M. 46
Deneubourg, J.L. 199
Derrida, B. 156, 194, 244, 248
Descartes, R. 8
Deschenes, M. 100
Destexhe, A. 86
DeYoe, E.A. 94
Dicke, P. 82, 101, 104
Diederich, S. 156
Dingledine, R. 360
Dinse, H.R.O. 61, 64, 65, 70, 150
diSessa, A. 8
Ditzinger, T. 475
Divenyi, P.L. 402
Divko, R. 335, 338
Domany, E. 462
Doyle, P.G. 408
Drake, M. 530
Dreyfus, G. 156, 260, 450
Drüe, S. 361
Dugdale, N. 478
duLac, S. 94
Duque Antón, M. 441
Dvorak, I. 86
Dwoyer, D.L. 8

E

Eckhorn, R. 82, 86, 100, 101, 104, 108, 138, 142, 150, 364
Eckmann, J.P. 86
Eckmiller, R. 5, 8, 12, 314, 417, 420, 434, 462
Edelman, G.M. 70, 74
Edmondson, A.C. 8
Eickmeyer, H.-J. 495
Eigen, M. 168

Ellias, S.A. 328
Elman, J.L. 160, 240, 378
Elul, R. 52
Engel, A.K. 86, 100, 104, 105, 108, 120, 138, 142, 146, 150, 248
Englisch, H. 245, 248
Eppler, W. 249
Epworth, R.W. 530
Ermentrout, G.B. 146
Ernst, H.P. 283
Ersue, E. 430
Esterley, S.D. 94
Ewert, J.-P. 109, 112

F

Fahlman, S.E. 160
Fairhurst, M.C. 349
Fallside, F. 190, 212, 378
Fang, L. 253
Fang, M. 339, 538
Fanty, M.A. 302, 349
Farber, R. 264, 450
Farhat, N.H. 342, 548
Farmer, J.D. 86
Faure, B. 30
Feldman, A.G. 90
Feldman, J.A. 94, 252, 302, 349, 479, 484, 516
Felleman, D.J. 74
Fender, D. 116
Ferandez, J. 70
Fernandes, C.G. 398
Fetter, L.H. 530
Field, D.J. 276
Fienup, J.R. 538
Filho, E.C.D.B.C. 343, 349
Fincham, D. 294
Finkel, L. 70, 74
Fischer, B. 128
Fitts, P.M. 90
Floeder, W. 86
Fodor, J.A. 478
Földiak, P. 190
Fogelman-Soulie, F. 318
Fohlmeister, J. 82
Forgy, C.L. 236
Fornili, S.L. 291, 294
Forrest, B.M. 318
Forrest, S. 470

Fowler, C.A.   90
Fox, G.C.   294, 423, 424
Freeman, J.A.   78, 252
Freeman, W.J.   86, 104, 108, 120, 138, 142, 364
French, A.S.   364
Friedlander, B.   45
Friedlander, M.J.   94
Friedrich, J.   530
Fromkin, V.A.   498
Fu–Sun, L.   116
Fuchs, A.   457
Fujii, M.   356
Fukushima, K.   351, 356, 507
Funahashi, K.   394
Furmanski, W.   294, 423

G
Gabor, D.   41, 338
Gaglia, S.   412
Gaines, B.R.   26
Gallaire, H.   502
Gallant, S.I.   470
Gamst, A.   444
Ganong, A.H.   360
Gardiner, C.W.   264
Gardner, E.   156, 194, 244, 248
Garrett, M.F.   498
Gasper, M.   314
Geiger, H.   323
Geisel, T.   375, 378
Gelade, G.   108
Gelatt, C.D.   168, 338
Gelatt, C.P.Jr.   264
Gelfand   434
Geman, S.   338
Gencic, T.   445
Genesereth, M.R.   502
Genthner, A.   318
Gerchberg, W.   538
Gerstein, G.L.   56, 64, 70, 86, 100
Geva, S.   257
Ghosh, J.   298
Giannakopoulos, F.   64, 113, 116, 128, 132
Giefing, G.-J.   125
Gilbert, C.D.   64, 108
Giles, C.L.   260
Gillow, M.   52

Ginsberg, M.L.   502
Gioggia, R.S.   86
Glasberg, B.R.   402
Glauber, R.J.   22
Glünder, H.   357, 360, 372
Goddard, N.H.   302, 349
Goddard, W.   484
Goebel, R.   157
Gold, B.   378
Goldberg, D.E.   36, 168, 470
Goldstein, J.L.   402
Gonzalez, D.L.   90
Goodglass, H.   181
Goodman, D.   90
Gorse, D.   161, 164, 228
Goser, K.F.   182
Graf, D.H.   182
Grajski, K.A.   70
Gramss, T.   382
Grassberger, P.   86
Gray, C.M.   86, 94, 100, 104, 105, 108, 120, 138, 142, 146, 150, 364
Greenbaum, N.N.   86
Grefenstette, J.J.   36
Grensing, D.   248
Grossberg, S.   22, 172, 216, 260, 328, 349, 378, 383, 389, 493
Grossmann, A.   41, 156
Grünbaum, B.   8
Grzywacz, N.M.   360
Guckenheimer, J.   12
Güttinger, W.   445
Guic–Robles, E.   70
Gumm, H.P.   287, 290
Gungner, D.   302
Gunter, P.   548
Gurewitz, E.   423
Gutfreund, H.   156, 216, 272, 450, 462
Guyon, I.   156, 260, 450
Guzman de, G.C.   86

H
Haan de, P.E.   299
Haarer, D.   525, 530, 531
Haas, R.   475
Habib, M.K.   56, 70, 86
Häusler, G.   339, 533, 538
Haken, H.   12, 90, 199, 451, 457, 542

Hamilton, A.   314, 322
Hampshire, J.B.   222
Hanazawa, T.   394
Hancock, P.J.B.   204
Handelman   434
Hannuschka, R.   470, 499
Harth, E.M.   52
Hartline, H.K.   64
Hartmann, G.   361, 364
Hartmann, U.   470, 499
Hartstein, A.   244
Hassenstein, B.   372
Hauser, R.   315, 318
Hauske, G.   3
Haykin, S.   45
Hebb, D.O.   56, 108, 222, 244, 368, 372, 462
Hecht–Nielsen, R.   172
Heiligenberg, W.   94
Heinke, D.   113
Heistermann, J.   165, 168
Hemker, A.   465, 470
Hepp, D.O.   108, 372
Hergert, F.B.   287, 290
Hersh, R.   8
Hertz, J.A.   183
Hess, G.   311
Hey, A.J.G.   294
Hildreth, E.C.   474
Hillis, W.D.   15, 18, 36, 199, 318
Hinton, G.E.   18, 108, 138, 160, 182, 204, 216, 252, 280, 328, 356, 394, 424, 438, 478, 516
Hinton, T.J.   328
Hirai, Y.   322, 356
Ho, A.W.   421, 423, 424
Höfflinger, B.   302, 322
Hofbauer, J.   168
Hoff, M.E.   156, 420, 434, 438
Hogg, T.   378
Holland, J.H.   36, 168, 199, 212, 240, 470
Hollatz, J.   213
Holmes, P.J.   12, 146, 150
Holzfuss, J.   86
Hopcroft, J.   408
Hopfield, J.J.   22, 156, 194, 216, 228, 244, 256, 260, 268, 310, 318, 342, 378, 424, 444, 450, 462, 538
Hormel, M.   427
Horner, H.   318

Hornik, K.   186, 190, 394
Horowitz, E.   256
Hosticka, B.   311
Houtsma, A.J.M.   402
Howard, K.E.   530
Hradek, G.T.   70
Hsu, K.   543
Hubel, D.H.   64, 112, 318, 356, 368, 372
Huber, A.   248
Hubermann, B.A.   378
Huignard, J.P.   548
Hussaini, M.Y.   8
Hutchinson, J.M.   302
Hwang, K.   298
Hyde, M.R.   181

I
Ikezawa, O.   86
Imbert, M.   94
IMMOS LTD   30, 294
Ishikawa, M.   334
Ito, T.   356

J
Jackel, L.D.   530
Jaris, P.X.   64
Jenkin, M.   338
Jenkins, W.M.   69, 70
Jepson, A.D.   338
Jockusch, S.   169, 172
Johansson, G.   457, 484
Johnson, M.   294
Jones, J.P.   41, 368
Jordan, M.I.   160, 164, 240, 334, 378
Jordan, W.   86, 100, 104, 108, 138, 142, 150
Josin, G.   412
Julesz, B.   108, 138, 276
Julez, B.   116

K
Kaas, J.H.   70, 74
Kämmerer, B.   378
Kamada, K.   322

Kammen, D.M.   146, 147, 150, 186
Kaneko, K.   86
Kaneko, M.   430
Kanetkar, S.V.   342
Kangas, J.   182
Kanter, I.   260, 378, 450
Kapur, D.   8
Karayiannis, N.B.   173
Kasif, S.   502
Kaski, K.   322
Kato, H.   78
Kauffman, S.A.   164
Kauffman, L.H.   8
Kawasaki, M.   94
Kelso, J.A.S.   90, 457
Kelso, S.R.   360
Kepler, T.B.   156
Kernhof, J.   322
Kertesz, A.   182
Kesper, M.   311
Keyes, R.W.   542
Kickert, W.J.M.   252
Killakey, H.P.   74
Kindermann, J.   217, 516
Kinzel, W.   156, 244, 342
Kirkpatrick, J.   338
Kirkpatrick, S.   168, 264, 424
Klar, H.   82
Knudsen, E.I.   94
Koch, C.   94, 116, 146, 150, 236, 302, 338, 378, 474
Koch, R.H.   244
Koenderinck, J.   364
König, P.   86, 100, 104, 105, 108, 117, 120, 138, 139, 142, 146, 150
Kohli, M.   502
Kohonen, T.   52, 74, 82, 172, 177, 181, 182, 190, 222, 232, 276, 342, 382, 402, 412, 416, 434, 457, 507
Koikkalainen, P.   286
Kokkinidis, M.   52
Kolers, P.A.   484
Kollbach, D.   302
Koller, J.   423
Korst, J.H.M.   280
Koscielny–Bunde, E.   459, 462
Kosko, B.   216
Kouno, T.   356
Kowalewski, F.   391
Kowalski, R.   502
Krauth, W.   156

Krogh, A.   183
Krone, G.   64, 100, 116, 132
Krüger, J.   97, 100, 121, 123
Krüger, K.   61, 70, 150
Kruse, W.   86, 100, 101, 104, 108, 138, 142, 150
Kuczewski, R.M.   318
Kühn, R.   248, 378, 450
Kühnel, H.   187
Küpper, W.   378
Kürten, K.E.   191, 194
Kung, H.T.   30
Kung, S.Y.   30
Kunz, D.   441
Kuperstein, M.   434
Kuramoto, Y.   146
Kurz, A.   229
Kwong, S.K.   548

L
Lacerda, F.   382, 394
Lades, M.   417, 434
Lal, R.   94
LaLonde, W.R.   182
Landauer, R.   542
Lane   434
Lang, K.   394
Lang, K.J.   18
Lange, J.   41
Lapedes, A.S.   264, 450
Lappe, M.   445
Lasserre, C.   502
Lattard, D.   30
Lawley, D.N.   190, 368
Layne, S.P.   86
le Cun, Y.   204
Lee, D.N.   372
Lee, H.Y.   260
Lee, Y.C.   260
Leenstra, J.   322
Lemke, H.R.v.N.   252
Lemon, R.   74
Leung, V.   82
Leuthäusser, I.   213
Leuze, M.R.   36
Levelt, W.J.M.   498
Lewis, E.R.   52
Lewis, P.S.   45
Li, T.   253

Lindström, S.   116
Linsker, R.   186, 190, 318, 368
Lippmann, R.P.   276, 378
Little, J.J.   360, 474
Livingstone, M.   493
Lohmann, A.W.   360
Lorenz, E.N.   86
Lowe, C.F.   478
Lowe, D.G.   493
Ludermir, T.B.   395, 398
Lynne, K.J.   302
Lyzenga, G.   294

M
Macfarlane, R.M.   531
MacGregor, R.J.   52, 56, 294
MacKay, D.G.   498
Mäkisara, K.   182, 402
Männer, R.   315, 318
Maginu, K.   260
Mahler, G.   539, 542
Mainzer, K.   9, 12
Makhaniok, M.   318
Malik, J.   493
Mallot, H.A.   64, 100, 116, 125, 128, 129, 132, 471, 474
Malpass, M.L.   378
Malsburg v.d., C.   8, 12, 37, 41, 56, 70, 74, 108, 120, 138, 142, 150, 328, 364, 462
Mandelbrot, B.   8
Manderick, B.   31, 195, 199
Mangeron, D.   82
Mangold–Allwin, R.   487, 490
Manis, P.B.   128, 132
Manneville, P.   86
Marcelja, S.   41
Marcinowski, M.   276
Marcus, C.M.   450
Margolus, N.   8
Marko, H.   276
Marks, K.M.   182
Marr, D.   108, 138, 338, 493
Mars, P.   26
Martinetz, T.M.   74, 182, 431, 434, 438
Martini, C.   302
Martorana, V.   294
Massen, R.   23, 26

Mathur, B.   378
Matsumoto, N.   112
Matthews, R.   100
Maunsell, J.H.R.   484
Maxwell, A.E.   190, 368
Maxwell, T.   260
Mayer–Kress, G.   86
Mazaré, G.   27, 30
McClelland, J.L.   60, 168, 176, 204, 208, 236, 252, 264, 298, 302, 305, 328, 412, 490, 502, 516
McCulloch, W.S.   52, 244, 314
McEliece, R.J.   272
McLaughlin, J.A.   360
McMillan, C.   490
McNeill, D.   498
Mead, C.   82, 302, 314, 408, 420
Mecklenburg, K.   205
Mel, B.W.   18, 434
Merrill, J.   221
Merzenich, M.M.   65, 69, 70, 74
Meyer, A.S.   498
Mézard, M.   156, 216, 450
Middlebrooks, J.C.   69
Mieyin, F.M.   128, 132
Migliore, M.   291, 294
Migubayashi, N.   86
Miksa, M.   295
Militzer, J.   430
Miller, W.T.   434
Mingolla, E.   493
Minker, J.   502
Minsky, M.L.   194, 236, 260, 360
Mitzdorf, U.   78
Miyakawa, H.   78
Miyake, S.   356
Moerner, W.E.   531
Mok, F.   548
Mokry, B.   283
Moody, J.   276
Moore, B.J.C.   402
Moore, G.P.   100
Moore, M.   294
Moore, R.K.   394
Morasso, P.   182, 412
Morf, M.   46
Morgan, F.   8
Morrone, M.C.   41
Morton, H.B.   228
Mountcastle, V.B.   70
Moyson, F.   195, 199

Mozer, M.C.   204, 478
Mühlenbein, H.   516
Mugnaini, E.   244
Mundy, J.L.   8
Munk, M.   86, 100, 104, 108, 138, 142,   150
Munro, P.W.   186
Murray, A.F.   314, 322
Myers, C.E.   164, 228
Myers, M.H.   318
Myerson, J.   128, 132

N
N.N.   26
Nadal, J.P.   156, 216, 450
Naish, L.   502
Nakano, K.   342
Nedler, J.A.   420
Nelson, J.S.   74
Nelson, M.E.   294
Nelson, P.P.   82
Nelson, R.J.   70, 74
Neumann von,   26
Neumann, H.   491, 493
Neumann, K.   168
Neumann, O.   498
Newman, C.M.   272
Nicholson, C.   78
Nicole, D.A.   294
Nicolis, G.   199
Niez, J.P.   450
Nijhuis, J.A.G.   299, 302, 319, 322
Nilsson, N.J.   424
Nischwitz, A.   323, 328
Noble, B.   176
Nottebohm, F.   172
Novak, J.L.   78
Nudo, R.J.   69, 70

O
O'Neill, W.E.   74
Oaksford, M.   475, 478
Obermayer, K.   71, 542
Objois, P.   30
Ochs, M.T.   70
Oertel, W.H.   244
Oguztöreli, M.N.   82

Oja, E.   186, 190, 286, 368
Olson, T.J.   484
Ooyama, M.   322
Opper, M.   156
Optican, L.M.   150
Ossen, A.   201
Otto, S.   294

P
Packard, N.H.   86
Paek, E.   342
Paetsch, W.   430
Paik, E.   302
Palm, G.   56, 64, 70, 83, 86, 100, 116, 132, 276, 538
Palmer, L.A.   41, 368
Palumbo, D.L.   310
Panagos, G.   172
Pandya, A.S.   457
Papert, S.   194, 236, 260, 360
Parent, P.   493
Parga, N.   248
Parisi, D.   237, 240
Park, C.H.   260
Parker, T.S.   86
Pasteels, J.M.   199
Pastur, L.A.   248
Patarnello, S.   240
Paton, J.A.   172
Payan, E.   27, 30
Pearlmutter, B.A.   212, 222
Pearson, J.C.   70, 74, 100
Peeling, S.M.   394
Peng, J.   470
Penrose, A.   416
Peretto, P.   450
Perkel, D.H.   100
Perona, P.   493
Perrett, D.I.   94
Personnaz, L.   156, 260, 450
Pertile, G.   52
Peterson, A.M.   222
Peterson, C.   256
Peterson, I.   8
Petkov, N.   303, 305
Pettey, C.P.   36
Pineda, F.J.   378
Pinker, S.   478
Piro, O.   90

Pitts, W.H. 52, 244, 314
Plaice, J.A. 30
Plenz, D. 75
Plumbley, M.D. 190
Pobel, P. 490
Podell, M. 150
Poggio, T. 94, 236, 338, 360, 474
Pollack, J.B. 160
Pomerleau, D.A. 264
Pool, R. 86
Poppelbaum, W.J. 26
Port, R. 221
Posner, E.C. 272
Prange, S.J. 79, 82
Prata, A. 342
Preißl, H. 83
Prigogine, I. 199
Prince, A. 478
Pritchard, D.J. 294
Procaccia, I. 86
Protzel, P.W. 307, 310
Psaltis, D. 18, 260, 342, 543, 548
Ptashne, M. 530
Pylyshyn, Z.W. 478

Q
Qian, N. 264

R
Radons, G. 261, 264
Rafelski, J. 194
Ralls, K.S. 530
Ramachandran, V.S. 484
Ramacher, U. 82
Rapp, P.E. 86
Rapp, R. 509
Rasmusson, W.J. 70
Raviv, J. 360
Recanzone, G.H. 65, 69, 70
Rechenberg, I. 82, 168, 208
Reggia, J. 256
Reichardt, W. 133, 372, 474
Reitboeck, H.J. 82, 86, 100, 101,
  104, 108, 138, 142, 150, 342
Renals, S. 394
Renyi, A. 86
Reuhkala, E. 402

Reuss, S. 64
Richert, P. 311
Richmond, B.J. 150
Riedel, U. 378, 450
Riolo, R.L. 470
Ritoniemi, T. 322
Ritter, H. 71, 74, 182, 372, 382,
  408, 412, 431, 434, 438
Robbins, H. 8
Robertson, G.G. 36, 470
Robinson, A.J.R. 212, 378, 502
Rock, I. 493
Rodemich, E.R. 272
Rohwer, R. 394
Rolls, E.T. 94
Rose, G. 94
Rosenberg, C.R. 264, 438
Rosenblatt, R. 356
Roweth, D. 318
Rubinstein, J. 434
Rubner, J. 190, 365, 368
Rudjord, T. 52
Rückert, U. 182
Ruelle, D. 86
Rujan, P. 156
Rumelhart, D.E. 60, 160, 168, 176,
  204, 208, 216, 236, 252, 264, 298,
  302, 305, 328, 356, 394, 412, 438,
  424, 490, 502, 516

S
Saarinen, J. 182
Sahir, M. 408
Sahni, S. 256
Salem, J.B. 199
Salmon, J. 294
Salomon, R. 204
Sampolinsky, H. 260
Samuel, A.L. 212
San Biagio, P.L. 294
Sanger, T.D. 186, 338
Saramäki, T. 182
Satou, M. 112
Saund, E. 204
Saxton, O. 538
Schade, U. 495, 498
Schaller, H.N. 265
Scheibel, A.B. 70
Scheibel, M.E. 70

Schiffmann, W.H.   205, 208
Schillen, T.B.   117, 120, 135, 138, 139, 142
Schlögl, R.W.   474
Schmidhuber, J.H.   209, 212
Schneider, W.   108, 120, 138, 142, 150
Schönecker, W.   276
Schöner, G.   87, 90, 93
Scholz, J.P.   90
Schoppmann, A.   70, 74
Schreiner, C.E.   64, 70
Schreter, Z.   283
Schürg–Pfeiffer, E.   112
Schürmann, B.   213, 216
Schütte, A.   248
Schüz, A.   64, 100, 116, 132
Schulten, K.J.   71, 74, 182, 190, 335, 338, 365, 368, 378, 382, 412, 431, 434, 438
Schulz, M.   530
Schulze, E.   128, 132
Schuster, H.G.   143, 261, 264
Schuz, A.   64
Schwartz, E.L.   128, 132
Schwartz, J.T.   408
Schwarz, D.W.F.   402
Schwarz, M.   311
Schwippert, W.W.   109, 112
Sciortino, F.   294
Seckmeyer, G.   538
Seelen v., W.   64, 112, 128, 129, 132
Sejnowski, T.J.   56, 138, 264, 280, 328, 438, 478
Serra, R.   199
Seytter, F.   369, 372
Shackleford, J.B.   268
Shamma, S.   378
Shannon, C.E.   190
Shastri, L.   160, 252, 378
Shattuck–Hufnagel, S.   498
Shaw, G.L.   70, 100
Shaw, R.S.   86
Shepard, G.C.   8
Shepard, R.N.   484
Sherman, S.M.   116
Shikano, K.   394
Shipp, S.   94
Shozakai, M.   182
Sidman, M.   478
Sigmund, K.   168

Silverman, D.J.   100
Silverman, J.C.   70, 100
Simic, P.   423, 424
Simon, R.   444
Singer, W.   78, 86, 94, 100, 104, 105, 108, 116, 120, 138, 142, 146, 150, 364, 372
Siska, J.   86
Sitte, J.   257
Skarda, A.   86
Skocpol, W.J.   530
Skrzypek, J.   302
Slobin, D.   490
Smith, A.W.   334
Smith, K.   314
Smolensky, P.   204 236, 252, 478, 490, 516
Snell, J.L.   408
Soderberg, B.   256
Sompgy, P.   120
Sompolinsky, H.   22, 156, 216, 236, 272, 378, 450, 462
Southard, D.L.   90
Spaanenburg, L.   299, 302, 319, 322
Sparks, D.L.   74
Spiessens, P.   31
Spinelli, D.   41
Spitzer, H.   150
Staiger, V.   78
Stanfill, C.   236
Stcherbina, M.V.   248
Steele, G.L.Jr.   18, 199
Steels, L.   199
Stein, R.B.   82, 364
Stemberger, J.P.   498
Steriade, M.   100
Stevens, J.K.   64
Stevens, K.H.   112
Stewart, H.B.   8, 86
Stiehl, H.S.   491, 493
Stinchcombe, M.   394
Stoll, H.M.   548
Stonham, T.J.   398
Storck, D.   378
Storjohann, K.   128, 132
Stork, D.   378, 389
Stornetta, W.S.   378
Stotzka, R.   315, 318
Stroud, N.   318
Strube, H.W.   172, 391
Stryker, M.P.   70, 74

Suga, N. 74
Suilmann, M. 60
Sun, G.Z. 260
Sur, M. 74
Suttner, C.B. 502
Sutton, R.S. 212, 416

T
Tackitt, M. 548
Tagliasco, V. 412
Tailby, W. 478
Takens, F. 86
Tanese, R. 36
Tank, D.W. 256, 268, 310, 378, 424, 444
Tarassenko, L. 314, 322
Tavan, P. 187, 190, 365, 368
Taylor, J.G. 161, 164, 228
Tchamitchian, P. 41
Teich, W.G. 539, 542
Tenhunen, H. 322
Tennant, D.M. 530
Terry, M. 394
Tesauro, G. 438
Thomas, I.V. 164
Thomas, W.V. 398
Thompson, J.M.T. 8, 86
Thorpe, S.J. 91, 94
Todorvic, D. 493
Toffoli, T. 8
Tolat, V.V. 222
Tomberg, J. 322
Tomlinson, R.W.W. 399, 402
Torkkola, K. 182
Torre, V. 338, 360
Toulouse, G. 216
Touretzky, D.S. 160, 252
Toyama, K. 108
Traunmüller, H. 382, 394
Treisman, A.M. 108
Treurniet, W. 399
Trivedi, K.S. 310
Tryba, V. 182
Ts'o, D.Y. 108
Tuller, B. 90
Tunturi, A.R. 74

U
Ultsch, A. 465, 470, 499, 502

V
Vainio, L. 402
Van Dijk, B.W. 104
Van Essen, D.C. 94, 132
Van Hemmen, J.L. 245, 248, 378, 450
Vatikiotis–Bateson, E. 90
Vecchi, M.P. 168, 186, 264, 338, 424
Veenker, G. 513
Venetsanopoulos, A.N. 173
Venkatesh, S.S. 260, 272
Ventä, O. 182
Vercauteren, L. 503, 507
Verhaeghe, J.C. 199
Verri, A. 474
Vingerhoeds, R.A. 503
Virasoro, M.A. 248
Voigt, R.G. 8
Volder, J.E. 46
Vorbrüggen, J.C. 37, 41

W
Waerden v.d., B.L. 8
Wagh, M.D. 342
Wagner, G.P. 244
Wagner, K. 548
Wagner, P. 143
Waibel, A.H. 222, 378, 394
Waldner, F. 86
Waldschmidt, K. 302
Walker, D. 294
Wall, J.T. 74
Wallace, D.J. 318
Waltz, D. 236
Wang, H.T. 378
Warden, M. 86
Warkowski, F. 299, 302, 319, 322
Waschulzik, T. 323, 328
Watrous, R.L. 378
Watson, A.B. 276
Watson, F.L. 478
Weaver, W. 190
Weber, V. 470, 499
Weisbuch, G. 318

Weiss, T.   538
Wellekens, C.J.   221
Wendler, G.   60
Werbos, P.J.   378
Werner, D.   261, 264
Werntges, H.   413, 416, 417, 434, 435
Westervelt, R.M.   450
Wettler, M.   509
Wheeler, B.C.   78
White, H.   394
Whittle, P.   269, 272
Widrow, B.   156, 176, 434
Widrow, G.   420, 438
Wiesel, T.N.   64, 108, 112, 356, 368, 372
Wiethersheim v., A.   112
Wilczek, F.   276
Williams, R.J.   160, 204, 212, 216, 334, 356, 394, 428, 438
Willshaw, D.J.   74, 276
Willshire, C.   498
Wilson, G.V.   318
Wilson, H.R.   116, 146, 222
Wilson, M.A.   150
Wilson, W.H.   253
Windheuser, C.   217
Wingfield, A.   181
Winkler, R.   168
Winston, J.V.   194
Winter, R.G.   176
Wirnitzer, B.   360
Wörgötter, F.   147
Wolff, J.R.   241, 244
Wolfram, S.   199, 542
Wond, Y.-F.   423
Wong, I.   423, 424
Woodward, Y.   360
Würtz, R.P.   37, 41

Y
Yaglom, I.M.   8
Yamada, M.   322
Yamamoto, K.   86
Yang, B.   43, 46
Yariv, A.   548
Yarus, M.   530
Yuille, A.L.   186, 360

Z
Zaccaria, R.   412
Zadeh, L.A.   252
Zagrebnov, V.A.   248
Zak, M.   222
Zanarini, G.   199
Zare, S.L.   484
Zeki, S.   94
Zenone, S.   52
Zetzsche, C.   273, 276
Zhang, B.T.   513, 516
Zimmermann, I.D.   86
Zinn, E.G.   444
Zippelius, A.   194, 244, 248
Zipser, D.   160, 186, 212, 334
Zook, J.M.   70, 74
Zucker, S.W.   493
Zwietering, P.J.   277, 280 277, 280

SUBJECT INDEX

A

Adaptive
junction   331
networks   191
search techniques   23, 31
system   539
Amorphous materials   525
Analog
circuits   79
coding   91
neural network   445, 299
Anti-Hebbian rule   187, 365
Assembly formation   105
Associative
memory   177, 191, 213, 233, 241, 257, 273, 331, 427, 451
recall   351
Asynchronous systems   27, 75
Auditory preprocessing   391
Autoassociative networks   257

B

Backpropagation   157, 205, 261, 391, 487, 499
Bark spectra   391
Binding problem   105, 117, 135, 139
Binocular interaction   113
Biological neurons   75, 311, 361
Boltzmann machines   277

C

Cellular automaton   539
Channel assignment   441
Chaos   83, 261, 533
Character recognition   343
Classifier system   421, 465
Coding   121
Combinatorial optimizations   253, 277
Competitive learning   169, 217, 253
Computational neuroscience   5, 491
Connection machine   71
Connectionist models   15, 283, 495, 499, 509
Convergence   173
CORDIC processor   43
Correlation analysis   53, 65, 83, 97, 161, 471

C

Cortex   71, 97, 121, 147
Cortical plasticity   65
Counterpropagation   169
Current source density   75

D

Damaged network   459
Delay differential equations   117, 135, 139
Delta rule   435
Dendritic spikes   75
Dendritic-tree-structures   323
Document retrieval   233
Dynamic
feature linking   101
organization   53, 83, 133
patterns   87
Dynamics
of neural networks   445
of receptive field organization   61

E

Edge, detection   491
Error minimization   435
Event learning   233
Evolution   5, 165

F

Face recognition   37
Fast fourier transform, parallel   37
Fault
-injection   307
-tolerance   307, 319
Feature detectors   169, 177, 187, 229, 233, 335, 361, 365, 379
Feedback
internal   209
Feedback network   129, 265, 323, 331, 375, 395
Feedforward neural networks   201, 303
Forgetting   413, 435
Fovea   125
Fuzzy controller   249

G
Gabor transformation   335
Genetic algorithms   31, 165, 205
Givens rotation   43
Glauber dynamics   19
Gradient descent   161, 231
Grand-mother neuron   233
Graph, labeled   37
Grasping control   427

H
Hebbian learning   183, 187, 365
Hierarchical system   201, 335, 539
Hopfield nets   19, 225, 241, 245, 265,
    269, 459
Hypercube   421

I
Image compression   125
Inhibitory interneurons   241
Inner-product model   245
Intracortical feedback   61
Inverse kinematics   435

K
Knowledge representation   465, 499
Kohonen map   71, 229

L
Laser photochemistry   525
Lateral inhibition   495
Lattice filter   43
Learning
    algorithm   153, 157, 165, 173,
    187, 191, 487, 499
    autonomous   421
    graded   435
    on-line   209
    open loop   375
    short-term   323
    supervised   331, 435
    unsupervised   183, 413
Learning   165, 499
Leg coordination   57
Lexicalization   509
Line detection   299
Load Balancing   307
Local
    computations   209, 241, 539
    minima, Escape from   261
Logic programming   233, 499

Low level vision   273
Low-activity network   245

M
Map coloring   265
Map
    adaptive   71
    topographic   65, 71
    topology conserving   431
Mapped filters   129
Mapping
    acoustic-articulatory   169
    machines   5
    retinotopic   125, 129
Massive parallelism   27, 31
Master - slave algorithm   445
Matched filter   187
Matrix by vector multiplication   303
Maximum storage capacity   459
Missing fundamental   399
Molecular
    electronics   539
    storage   525
Motion
    analysis   357, 471, 479
    perception   479
Motor control   15, 57, 87, 409, 451
Multilayer network   217, 303, 315,
    351, 391
Multiple
    coding   97, 129
    neuron recordings   53, 65, 121
Multiplying synapses   323, 357
Multiprocessor system   315
Multistability   265, 375, 539
Mutation   205

N
Natural language processing   273,
    471, 495
Navigation   421
Network simulation   75, 135, 205,
    287, 295, 315, 323
Neural
    computers   9, 311
    hardware   319
    network applications   295
    networks   19, 23, 53, 75, 101, 117,
    135, 139, 165, 169, 173, 225, 229,
    249, 253, 277, 307, 331, 351, 409,
    421, 427

processor 311
Neuroinformatics 9
Neuronal Oscillations 133
Neurophilosophy 9
Nonlinear
  dynamics 117, 135, 139
  network 113

O
Object
  oriented programming 287, 295
  recognition 37
Obstacle avoidance 405
Occam 37, 283
Optical
  control 539
  flow 471
  multiplexing 525
  storage 525
Optimization 307, 335, 413, 421, 441
Orientation
  selectivity 61
  sensitive cells 315
Oscillation 61, 97, 105, 117, 133, 135, 139, 147

P
Parallel
  algorithms 165
  computers 15, 43, 109, 205, 283, 295, 315, 451, 471, 513, 539
Path planning 421
Pattern,
  Activity 121
Pattern recognition 331, 343, 451, 459
Patterns 173, 241
Performance degradation 307
Periodic orbits 445
Phase Description 133
Phonotopic Map 379
Photo resistors 405
Photochemical hole burning 525
Photochronism 525
Pitch perception 399
Point process 83
Principal component analysis 183, 187, 365
Probabilistic classifier 161, 395
Proprioception 57

Prototyping 299, 319
Pseudo inverse 445
Pulse coding 23, 311

R
RAM network 395
Random
  patterns 153
  search 315, 413, 435
Receptive fields 65, 133, 365
Recurrent networks 157, 209, 375, 391
Redundant degrees of freedom 409, 413, 431
Reinforcement learning 161, 413, 435
Representation
  internal 177, 201, 323
  of time 157
Resistive network 405
Retina 125, 299
Rhythmic firing behavior 147
Robot control 431
Rule discovery 487

S
Sampling, space-variant 125
Scene segmentation 105, 117, 135, 139, 351
Selective attention 351
Selforganization 9, 71, 177, 229, 315, 343, 351, 361, 365
Self-organizing map 177
Sequence
  analysis 375
  learning 379
Sigma-pi units 323, 357, 513
Simulated annealing 261, 277, 335
Somatosensory system 65, 71
Space-variant filters 125
Sparce coding 191, 273
Spatial filters 273
Spatio-temporal
  dynamics 61, 75
  pattern 331
Speech
  errors 495
  production 487, 495
  recognition 15, 379, 391
Spike trains 91, 97, 121
Stability analysis 183

Static scenes   491
Stereo vision   335
Stochastic computers   23, 261
Storage capacity   257
Structured networks   169, 323, 479
Synaptic weights   173, 241
Synchronization   133, 361
Synergetic computer   451
Systolic algorithms   43, 303

T
Temporal
    coding   91, 105, 117, 135, 139, 361
    pattern recognition   395
Threshold logic units   241, 357
Time warping invariance   375
Transputers   19, 37, 283, 295, 421, 471

U
Uncorrelated features   187

V
Velocity invariance   391
Vision   121, 331, 479
Visomotor-coodination   431
Visual
    cortex   61, 101, 133
    feedback   431
    pattern recognition   109, 351
    system   91, 273, 491

W
Wavelets   37
Widrow-Hoff rule   431
Winner-takes-all   91, 323, 431
Word recognition   379, 391